WITHDRAWN
UTSA LIBRARIES

FLUID–STRUCTURE INTERACTIONS

SLENDER STRUCTURES AND AXIAL FLOW

VOLUME 1

FLUID–STRUCTURE INTERACTIONS

SLENDER STRUCTURES AND AXIAL FLOW

VOLUME 1

MICHAEL P. PAÏDOUSSIS
Department of Mechanical Engineering,
McGill University,
Montreal, Québec, Canada

ACADEMIC PRESS
SAN DIEGO   LONDON   NEW YORK   BOSTON
SYDNEY   TOKYO   TORONTO

This book is printed on acid-free paper.

Copyright © 1998 by ACADEMIC PRESS

All Rights Reserved.
No part of this publication may be reproduced or transmitted in any form or by any means, electronic or mechanical, including photocopy, recording, or any information storage and retrieval system, without permission in writing from the publisher.

Academic Press
525 B Street, Suite 1900, San Diego, California 92101-4495, USA
http://www.apnet.com

Academic Press Limited
24–28 Oval Road, London NW1 7DX, UK
http://www.hbuk.co.uk/ap/

ISBN 0-12-544360-9

A catalogue record for this book is available from the British Library

Library of Congress Catalog Card Number: 98-86469

Typeset by Laser Words, Madras, India
Printed in Great Britain by WBC Book Manufacturers, Bridgend, Mid-Glamorgan

98 99 00 01 02 03 WB 9 8 7 6 5 4 3 2 1

Library
University of Texas
at San Antonio

Contents

Preface

A word about *la raison d'être* of this book could be useful, especially since the first question to arise in the prospective reader's mind might be: *why another book on flow-induced vibration*?

Flow-induced vibrations have been with us since time immemorial, certainly in nature, but also in artefacts; an example of the latter is the Aeolian harp, which also makes the point that these vibrations are not always a nuisance. However, in most instances they *are* annoying or damaging to equipment and personnel and hence dangerous, e.g. leading to the collapse of tall chimneys and bridges, the destruction of heat-exchanger and nuclear-reactor internals, pulmonary insufficiency, or the severing of offshore risers. In virtually all such cases, the problem is 'solved', and the repaired system remains trouble-free thereafter — albeit, sometimes, only after a first and even a second iteration of the redesigned and supposedly 'cured' system failed also. This gives a hint of the reasons why a book emphasizing (i) *the fundamentals* and (ii) *the mechanisms giving rise to the flow-induced vibration* might be useful to researchers, designers, operators and, in the broadest sense of the word, students of systems involving fluid–structure interactions. For, in many cases, the aforementioned problems were 'solved' without truly understanding either the cause of the original problem or the reasons why the cure worked, or both. Some of the time-worn battery of 'cures', e.g. making the structure stiffer via stiffeners or additional supports, usually work, but often essentially 'sweep the problem under the carpet', for it to re-emerge under different operating conditions or in a different part of the parameter space; moreover, as we shall see in this book, for a limited class of systems, such measures may actually be counterproductive.

Another answer to the original question 'Why yet another book?' lies in the choice of the material and the style of its presentation. Although the discussion and citation of work in the area is as complete as practicable, the style is not encyclopaedic; it is sparse, aiming to convey the main ideas in a physical and comprehensible manner, and in a way that is *fun to read*. Thus, the objectives of the book are (i) to convey an understanding of the undoubtedly fascinating (even for the layman) phenomena discussed, (ii) to give a complete bibliography of all important work in the field, and (iii) to provide some tools which the reader can use to solve other similar problems.

A second possible question worth discussing is 'Why the relatively narrow focus?' By glancing through the contents, it is immediately obvious that the book deals with axial-flow-related problems, while vortex-induced motions of bluff bodies, fluidelastic instability of cylinder arrays in cross-flow, ovalling oscillations of chimneys, indeed all cross-flow-related topics, are excluded. Reasons for this are that (i) some of these topics are already well covered in other books and review articles; (ii) in at least some cases, the fundamentals are still under development, the mechanisms involved being incompletely understood; (iii) the cross-flow literature is so vast, that any attempt to cover it, as well as axial-flow problems, would by necessity squeeze the latter into one chapter or two, at most.

After extensive consultations with colleagues around the world, it became clear that there was a great need for a monograph dealing exclusively with axial-flow-induced vibrations and instabilities. This specialization translates also into a more cohesive treatment of the material to be covered. The combination of axial flow and slender structures implies, in many cases, the absence or, at most, limited presence of separated flows. This renders analytical modelling and interpretation of experimental observation far easier than in systems involving bluff bodies and cross-flow; it permits a better understanding of the physics and makes a more elegant presentation of the material possible. Furthermore, because the understanding of the basics in this area is now well-founded, this book should remain useful for some time to come.

In a real sense, this book is an anthology of much of the author's research endeavours over the past 35 years, at the University of Cambridge, Atomic Energy of Canada in Chalk River and, mainly, McGill University — with a brief but important interlude at Cornell University. Inevitably and appropriately, however, vastly more than the author's own work is drawn upon.

The book has been written for engineers and applied mechanicians; the physical systems discussed and the manner in which they are treated may also be of interest to applied mathematicians. It should appeal especially to researchers, but it has been written for practising professionals (e.g. designers and operators) and researchers alike. The material presented should be easily comprehensible to those with some graduate-level understanding of dynamics and fluid mechanics. Nevertheless, a real attempt has been made to meet the needs of those with a Bachelor's-level background. In this regard, mathematics is treated as a useful tool, but not as an end in itself.

This book is not an undergraduate text, although it could be one for a graduate-level course. However, it is not written in text-book format, but rather in a style to be enjoyed by a wider readership.

I should like to express my gratitude to my colleagues, Professor. B.G. Newman for his help with Section 2.2.1, Professors S.J. Price and A.K. Misra for their input mainly on Chapters 3 and 6, respectively, Dr H. Alighanbari for input on several chapters and Appendix F, and Professor D.R. Axelrad for his help in translating difficult papers in German.

I am especially grateful and deeply indebted to Dr Christian Semler for some special calculations, many suggestions and long discussions, for checking and rechecking every part of the book, and particularly for his contributions to Chapter 5 and for Appendix F, of which he is the main creator. Also, many thanks go to Bill Mark for his willing help with some superb computer graphics and for input on Appendix D, and to David Sumner for help with an experiment for Section 4.3.

I am also grateful to many colleagues outside McGill for their help: Drs D.J. Maull and A. Dowling of Cambridge, J.M.T. Thompson of University College London, S.S. Chen of Argonne, E.H. Dowell of Duke, C.D. Mote Jr of Berkeley, F.C. Moon of Cornell, J.P. Cusumano of Penn State, A.K. Bajaj of Purdue, N.S. Namachchivaya of the University of Illinois, S. Hayama and S. Kaneko of the University of Tokyo, Y. Sugiyama of Osaka Prefecture, M. Yoshizawa of Keio, the late Y. Nakamura of Kyushu and many others, too numerous to name.

My gratitude to my secretary, Mary Fiorilli, is unbounded, for without her virtuosity and dedication this book would not have materialized.

Finally, the loving support and constant encouragement by my wife Vrisseïs (*Βρισηΐς*) has been a *sine qua non* for the completion of this book, as my mother's exhortations to 'be laconic' has been useful. For what little versatility in the use of English this volume may display, I owe a great deal to my late first wife, Daisy.

Acknowledgements are also due to the Natural Sciences and Engineering Research Council of Canada, FCAR of Québec and McGill University for their support, the Department of Mechanical Engineering for their forbearance, and to Academic Press for their help and encouragement.

Michael P. Païdoussis
McGill University,
Montreal, Québec, Canada.

Artwork Acknowledgements

A number of figures used in this book have been reproduced from papers or books, often with the writing re-typeset, by kind permission of the publisher and the authors. In most cases permission was granted without any requirement for a special statement; the source is nevertheless always cited.

In some cases, however, the publishers required special statements, as follows.

- Figure 3.4 from Done & Simpson (1977),[†] Figure 3.10 from Païdoussis (1975), Figures 3.24 and 3.25 from Naguleswaran & Williams (1968), Figures 3.32, 3.33 and 3.49–3.51 from Païdoussis (1970), and Figures 3.78–3.80 from Païdoussis & Deksnis (1970) are reproduced by permission of the Council of the Institution of Mechanical Engineers, U.K.
- Figures 3.3 and 5.11(b) from Thompson (1982b) by permission of Macmillan Magazines Ltd.
- Figures 3.55, 3.58 and 3.60 from Chen & Jendrzejczyk (1985) by permission of the Acoustical Society of America.
- Figures 3.54 and 3.59 from Jendrzejczyk & Chen (1985), Figure 3.71 from Herrmann & Nemat-Nasser (1967), Figure 5.29 from Li & Païdoussis (1994), Figures 5.43 and 5.45–5.48 from Païdoussis & Semler (1998), and Figure 5.60 from Namachchivaya (1989) and Namachchivaya & Tien (1989a) by permission of Elsevier Science Ltd., The Boulevard, Langford Lane, Kidlington, OX5 1GB, U.K.
- Figure 5.16 from Sethna & Shaw (1987) and Figures 5.57(a,b) and 5.58 from Champneys (1993) by permission of Elsevier Science-NL, Sara Burgerhartstraat 25, 1055 KV Amsterdam, The Netherlands.
- Figure 4.38(a) from Lighthill (1969) by permission of Annual Review of Fluid Mechanics.

[†]See bibliography for the complete reference.

1
Introduction

1.1 GENERAL OVERVIEW

This book deals with the dynamics of slender, mainly cylindrical or quasi-cylindrical, bodies in contact with axial flow — such that the structure either contains the flow or is immersed in it, or both. *Dynamics* is used here in its generic sense, including aspects of *stability*, thus covering both self-excited and free or forced motions associated with fluid–structure interactions in such configurations. Indeed, flow-induced instabilities — instabilities in the linear sense, namely, divergence and flutter — are a major concern of this book. However, what is rather unusual for books on flow-induced vibration, is that considerable attention is devoted to the *nonlinear behaviour* of such systems, e.g. on the existence and stability of limit-cycle motions, and the possible existence of *chaotic oscillations*. This necessitates the introduction and utilization of some of the tools of modern dynamics theory.

Engineering examples of slender systems interacting with axial flow are pipes and other flexible conduits containing flowing fluid, heat-exchanger tubes in axial flow regions of the secondary fluid and containing internal flow of the primary fluid, nuclear reactor fuel elements, monitoring and control tubes, thin-shell structures used as heat shields in aircraft engines and thermal shields in nuclear reactors, jet pumps, certain types of valves and other components in hydraulic machinery, towed slender ships, barges and submarine systems, etc. Physiological examples may be found in the pulmonary and urinary systems and in haemodynamics.

However, much of the work in this area has been, and still is, 'curiosity-driven',[†] rather than applications-oriented. Indeed, although some of the early work on stability of pipes conveying fluid was inspired by application to pipeline vibrations, it soon became obvious that the practical applicability of this work to engineering systems was rather limited. Still, the inherent interest of the extremely varied dynamical behaviour which this system is capable of displaying has propelled researchers to do more and more work — to the point where in a recent review (Païdoussis & Li 1993) over 200 papers were cited in a not-too-exhaustive bibliography.[‡] In the process, this topic has become a new paradigm in dynamics, i.e. a new *model dynamical problem*, thus serving two purposes: (i) to illustrate known dynamical behaviour in a simple and convincing manner;

[†]With the present emphasis on utilitarianism in engineering and even science research, the characterization of a piece of work as 'curiosity-driven' stigmatizes it and, in the minds of some, brands it as being 'useless'. Yet, some of the highest achievements of the human mind in science (including medical and engineering science) have indeed been curiosity-driven; most have ultimately found some direct or indirect, and often very important, practical application.

[‡]See also Becker (1981) and Païdoussis (1986a, 1991).

(ii) to serve as a vehicle in the search for new phenomena or new dynamical features, and in the development of new mathematical techniques. More of this will be discussed in Chapters 3–5. However, the foregoing serves to make the point that the curiosity-driven work on the dynamics of pipes conveying fluid has yielded rich rewards, among them (i) the development of theory for certain classes of dynamical systems, and of new analytical methods for such systems, (ii) the understanding of the dynamics of more complex systems (covered in Chapters 6–11 of this book), and (iii) the direct use of this work in some *a priori* unforeseen practical applications, some 10 or 20 years after the original work was done (Païdoussis 1993). These points also justify why so much attention, and space, is devoted in the book to this topic, indeed Chapters 3–6.

Other topics covered in the book (e.g. shells containing flow, cylindrical structures in axial or annular flow) have more direct application to engineering and physiological systems; one will therefore find sections in Chapters 7–11 entirely devoted to applications. In fact, since 'applications' and 'problems' are often synonymous, it may be of interest to note that, in a survey of flow-induced vibration problems in heat exchangers and nuclear reactors (Païdoussis 1980), out of the 52 cases tracked down and analysed, 36% were associated with axial flow situations. Some of them, notably when related to annular configurations, were very serious indeed — in one case the repairs taking three years, at a total cost, including 'replacement power' costs, in the hundreds of millions of dollars, as described in Chapter 11.

The stress in this book is on the fundamentals as opposed to techniques and on physical understanding whenever possible. Thus, the treatment of each sub-topic proceeds from the very simple, 'stripped down' version of the system, to the more complex or realistic systems. The analysis of the latter invariably benefits from a sound understanding of the behaviour of the simpler system. There are probably two broad classes of readers of a book such as this: those who are interested in the subject matter *per se*, and those who skim through it in the hope of finding here the solution to some specific engineering problem. For the benefit of the latter, but also to enliven the book for the former group, a few 'practical experiences' have been added.

It must be stressed, however, for those with limited practical experience of flow-induced vibrations, that these problems can be very difficult. Some of the reasons for this are: (i) the system as a whole may be very complex, involving a multitude of components, any one of which could be the real culprit; (ii) the source of the problem may be far away from the point of its manifestation; (iii) the information available from the field, where the problem has arisen, may not contain what the engineers would really hope to know in order to determine its cause. These three aspects of practical difficulties will be illustrated briefly by three examples.

The first case involved a certain type of boiling-water nuclear reactor (BWR) in which the so-called 'poison curtains', a type of neutron-absorbing device, vibrated excessively, impacting on the fuel channels and causing damage (Païdoussis 1980; Case 40). It was decided to remove them. However, this did not solve the problem, because it was then found that the in-core instrument tubes, used to monitor reactivity and located behind the curtains, vibrated sufficiently to impact on the fuel channels — 'a problem that was "hidden behind the curtains" for the first two years'! Although this may sound amusing at this point, neither the power-station operator nor the team of engineers engaged in the solution of the problem can have found it so at the time.

The second case also occurred in a nuclear power station, this time a gas-cooled system (Païdoussis 1980; Case 35). It involved excessive vibration of the piping — so excessive that the sound associated with this vibration could be heard 3 km away! The excitation source was not local; it was a vortex-induced vibration within the steam generator, quite some distance away. A similar but less spectacular such case involved the perplexing vibration of control piping in the basement of the Macdonald Engineering Building at McGill University, which occurred intermittently. The source was eventually, and quite by chance, discovered to be a small experiment involving a plunger pump (to study parametric oscillations of piping, Chapter 4) three floors up!

Another case involved a boiler (Païdoussis 1980; Case 23), and the report from the field stated that 'There is severe vibration on this unit. The forced draft duct, gas duct and superheater-economizer sections all vibrate. The frequency I would guess to be 60–100 cps. It feels about like one of those 'ease tired feet' vibration machines'. A very colourful description, but lacking in the kind of detail and quantitative information one would wish for. The difficulty of instrumenting the troublesome operating system *a posteriori* should also be remarked upon.

To be able to deal with practical problems involving flow-induced vibration or instability, one needs first of all a certain breadth of perspective to be able to recognize in what class of phenomena it belongs, or at least in what class it definitely does *not* belong. Here experience is a great asset; reference to books with a broader scope would also be recommended [e.g. Naudascher & Rockwell (1994), Blevins (1990)]. Once the field has been narrowed, however, to be able to solve and to redesign properly the system, a thorough familiarity with the topic is indispensable. If the problem is one of axial flow, then here is where this book becomes useful.

A final point, before embarking on more specific items, should also be made: despite what was said at the beginning of the discussion on practical concerns — that applications and problems are often synonymous — *flow-induced vibrations are not necessarily bad.* First of all, they are omnipresent; a fact of life, one might say. They occur whenever a structure is in contact with flowing fluid, no matter how small the flow velocity. Admittedly, in many cases the amplitudes of vibration are very small and hence the vibration may be quite inconsequential. Secondly, even if the vibration is substantial, it may have desirable features, e.g. in promoting mixing, dispersing of plant seeds, making music by reed-type wind instruments; as well as for wave-generated energy conversion, or for the enhancement of marine propulsion (Chapter 4). Recently, attempts have been made 'to harness' vibration in heat-exchange equipment so as to augment heat transfer, so far without spectacular success, however. Even chaotic oscillation, usually a term with negative connotations, can be useful, e.g. in enhancing mixing (Aref 1995).

1.2 CLASSIFICATION OF FLOW-INDUCED VIBRATIONS

A number of ways of classifying flow-induced vibrations have been proposed. A very systematic and logical classification is due to Naudascher & Rockwell (1980, 1994), in terms of the *sources of excitation* of flow-induced vibration, namely, (i) extraneously induced excitation, (ii) instability-induced excitation, and (iii) movement-induced excitation. Naudascher & Rockwell consider flow-induced excitation of both body and fluid oscillators, which leads to a 3×2 tabular matrix within which any given situation can be accommodated; in this book, however, we are mainly concerned with flow-induced *structural* motions, and hence only half of this matrix is of direct interest. The structure, or 'body oscillator', is any component with a certain inertia, either elastically supported or flexible (e.g. a flexibly supported rigid mass, a beam, or a shell). Thus, in a one-degree-of-freedom system, the equation of which may generally be written as $\ddot{x} + \omega_n^2 x + g(x, \dot{x}, \ddot{x}) = f(t)$, the first two terms must be present, i.e. the structure, if appropriately excited, must be able to oscillate!

Extraneously induced excitation (EIE) is defined as being caused by fluctuations in the flow or pressure, independently of any flow instability and any structural motion. An example is the turbulence buffeting, or turbulence-induced excitation, of a cylinder in flow, due to surface-pressure fluctuations associated with turbulence in the flow. *Instability-induced excitation* (IIE) is associated with a flow instability and involves local flow oscillations. An example is the alternate vortex shedding from a cylindrical structure. In this case it is important to consider the possible existence of a control mechanism governing and perhaps enhancing the strength of the excitation: e.g. a fluid-resonance or a fluidelastic feedback. The classical example is that of lock-in, when the vortex-shedding frequency is captured by the structural frequency near simple, sub- or superharmonic resonance; the vibration here further organizes and reinforces the vortex shedding process. Finally, in *movement-induced excitation* (MIE) the fluctuating forces arise from movements of the body; hence, the vibrations are self-excited. Flutter of an aircraft wing and of a cantilevered pipe conveying fluid are examples of this type of excitation. Clearly, certain elements of IIE with fluidelastic feedback and MIE are shared; however, what distinguishes MIE is that in the absence of motion there is no oscillatory excitation whatsoever.

A similar classification, related more directly to the nature of the vibration in each case, was proposed earlier by Weaver (1976): (a) forced vibrations induced by turbulence; (b) self-controlled vibrations, in which some periodicity exists in the flow, independent of motion, and implying some kind of fluidelastic control via a feedback loop; (c) self-excited vibrations. Other classifications tend to be more phenomenological. For example, Blevins (1990) distinguishes between vibrations induced by (a) steady flow and (b) unsteady flow. The former are then subdivided into 'instabilities' (i.e. self-excited vibrations) and vortex-induced vibrations. The latter are subdivided into: random, e.g. turbulence-related; sinusoidal, e.g. wave-related; and transient oscillations, e.g. water-hammer problems.

All these classifications, and others besides, have their advantages. Because this book is essentially a monograph concerned with a subset of the whole field of flow-induced vibrations, adherence to a single classification scheme is not so crucial; nevertheless, the phenomenological classification will be used more extensively. In this light, an important aim of this section is to sensitize the reader to the various types of phenomena of interest and to some of the physical mechanisms causing them.

1.3 SCOPE AND CONTENTS OF VOLUME 1

Chapter 2 introduces some of the concepts and methods used throughout the book, both from the fluids and the structures side of things. It is more of a refresher than a textbook treatment of the subject matter, and much of it is developed with the aid of examples. At least some of the material is not too widely known; hence, most readers will find something of interest. The last part of the chapter introduces some of the differences in dynamical behaviour as obtained via linear and nonlinear analysis, putting the emphasis on physical understanding.

Chapters 3 and 4 deal with the dynamics, mainly the stability, of straight (as opposed to curved) pipes conveying fluid: both for the inherently conservative system (both ends supported) and for the nonconservative one (e.g. when one end of the pipe is free). The fundamentals of system behaviour are presented in Chapter 3 in terms of linear theory, together with the pertinent experimental research. Chapter 4 treats some 'less usual' systems: pipes sucking fluid, nonuniform pipes, parametric resonances, and so on, and also contains a section on applications. The nonlinear dynamics of the system, as well as chaotic oscillations, are presented in Chapter 5, wherein may also be found an introduction to the methods of modern nonlinear dynamics theory.

The ideas and methods developed and illustrated in Chapters 3–5 are of importance throughout the rest of the book, since the fundamental dynamical behaviour of the systems in the other chapters will be explained by analogy or reference to that presented in these three chapters; hence, even if the reader has no special interest in the dynamics of pipes conveying fluid, reading Chapter 3 is *sine qua non* for the proper understanding of the rest of the book.

Chapter 6 deals with the dynamics of curved pipes conveying fluid, which, surprisingly perhaps, is distinct from and analytically more complex than that of straight pipes.

1.4 CONTENTS OF VOLUME 2†

The pipes considered in Chapters 3–6 are sufficiently thick-walled to suppose that ideally, their cross-section remains circular while in motion, so that the dynamics may be treated via beam theory. In Chapter 7, thin-walled pipes are considered, which must be treated as thin cylindrical shells. Turbulence-induced vibrations, as well as physiological applications are discussed at the end of this chapter.

Chapters 8 and 9 deal with the dynamics of cylinders in axial flow: isolated cylinders in unconfined or confined flow in Chapter 8, and cylinders in clusters in Chapter 9. The stability and turbulence-induced vibrations of such systems are also discussed. Engineering applications are also presented: e.g. submerged towed cylinders, and clustered cylinders such as those used in nuclear reactor fuel bundles and tube-in-shell heat exchangers. Chapter 10 deals with plates in axial flow.

Chapter 11 treats a special, technologically important, case of the material in Chapters 7 and 8: a single cylinder or shell in a rigid or flexible tube, subjected to annular flow in the generally narrow passage in-between. This chapter also closes with discussion of some engineering applications.

Chapter 12 presents in outline some topics involving axial flow not treated in detail in this book, and Chapter 13 contains some general conclusions and remarks.

†Volume 2 is scheduled to appear later, but soon after Volume 1.

2
Concepts, Definitions and Methods

As the title implies, this also is an introductory chapter, where some of the basics of the dynamics of structures, fluids and coupled systems are briefly reviewed with the aid of a number of examples. The treatment is highly selective and it is meant to be a refresher rather than a substitute for a more formal and complete development of either solid or fluid mechanics, or of systems dynamics.

Section 2.1 deals with the basics of discrete and distributed parameter systems, and the classical modal techniques, as well as the Galerkin method for transforming a distributed parameter system into a discrete one. Some of the definitions used throughout the book are given here. A great deal if not all of this material is well known to most readers; yet, some unusual features (e.g. those related to nonconservative systems or systems with frequency-dependent boundary conditions) may interest even the *cognoscenti*.

The structure of Section 2.2, dealing with fluid mechanics, is rather different. Some generalities on the various flow regimes of interest (e.g. potential flow, turbulent flow) are given first, both physical and in terms of the governing equations. This is then followed by two examples, in which the fluid forces exerted on an oscillating structure are calculated, for: (a) two-dimensional vibration of coaxial shells coupled by inviscid fluid in the annulus; (b) two-dimensional vibration of a cylinder in a coaxial tube filled with *viscous* fluid.

Finally, in Section 2.3, a brief discussion is presented on the dynamical behaviour of fluid–structure-interaction systems, in particular the differences when this is obtained via nonlinear as opposed to linear theory.

2.1 DISCRETE AND DISTRIBUTED PARAMETER SYSTEMS

Some systems, for example a mathematical simple pendulum, are *sui generis* discrete; i.e. the elements of inertia and the restoring force are not distributed along the geometric extent of the system. However, what distinguishes a discrete system more precisely is that its configuration and position in space at any time may be determined from knowledge of a numerable set of quantities; i.e. the system has a finite number of degrees of freedom. Thus, the simple pendulum has one degree of freedom, even if its mass is distributed along its length, and a double (compound) pendulum has two.

The quantities (variables) required to completely determine the position of the system in space are the *generalized coordinates*, which are not unique, need not be inertial, but must be equal to the number of degrees of freedom and mutually independent (Bishop &

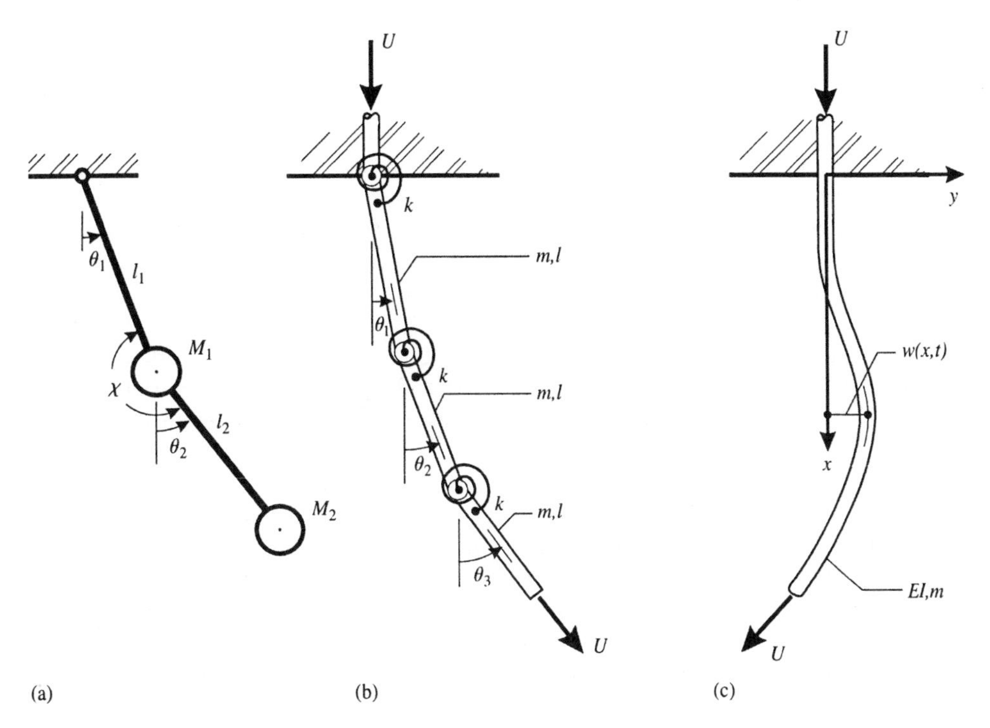

Figure 2.1 (a) A mathematical double pendulum involving massless rigid bars of length l_1 and l_2 and concentrated masses M_1 and M_2; (b) a three-degree-of-freedom ($N = 3$) articulated pipe system conveying fluid, with rigid rods of mass per unit length m and length l, interconnected by rotational springs of stiffness k, and generalized coordinates $\theta_i(t)$, $i = 1, 2, 3$; (c) a continuously flexible cantilevered pipe conveying fluid, the limiting case of the articulated system as $N \to \infty$, with $EI = kl$ (see Chapter 3). In most of this chapter $U = 0$.

Johnson 1960; Meirovitch 1967, 1970). Thus, for a double pendulum [Figure 2.1(a)], the two angles, θ_1 and θ_2 may be chosen as the generalized coordinates, each measured from the vertical; or, as the second coordinate, the angle χ between the first and the second pendulum may be used. Closer to the concerns of this book, a vertically hung articulated system consisting of N rigid pipes interconnected by rotational springs (Chapter 3) has N degrees of freedom; the angles, θ_i, of each of the pipes to the vertical may be utilized as the generalized coordinates [Figure 2.1(b)]. Contrast this to a flexible pipe [Figure 2.1(c)], where the mass *and* flexibility (as well as dissipative forces) are distributed along the length: it is effectively a beam, and this is a *distributed parameter*, or *'continuous', system*; in this case, the number of degrees of freedom is infinite. Discrete systems are described mathematically by ordinary differential equations (ODEs), whereas distributed parameter systems by partial differential equations (PDEs). If a system is linear, or linearized, which is admissible if the motions are small (e.g. small-amplitude vibrations about the equilibrium configuration), the ODEs may generally be written in matrix form. This is very convenient, since computers understand matrices very well! In fact, a number of generic matrix equations describe most systems (Pestel & Leckie 1963; Bishop *et al.* 1965; Barnett & Storey 1970; Collar & Simpson 1987; Golub & Van Loan, 1989) and they may be solved with the aid of a limited number of computer subroutines [see, e.g. Press *et al.* (1992)]. Thus, a damped system subjected to a set of external forces may be described by

$$[M]\{\ddot{q}\} + [C]\{\dot{q}\} + [K]\{q\} = \{Q\}, \tag{2.1}$$

where $[M]$, $[C]$ and $[K]$ are, respectively, the mass, damping and stiffness matrices, $\{q\}$ is the vector of generalized coordinates, and $\{Q\}$ is the vector of the imposed forces; the overdot denotes differentiation with time.

On the other hand, the form of PDEs tends to vary much more widely from one system to another. Although helpful classifications (e.g. into hyperbolic and elliptic types, Sturm–Liouville-type problems, and so on) exist, the fact remains that the equations of motion of distributed parameter sytems are more varied than those of discrete systems, and so are the methods of solution. Also, the solutions are generally considerably more difficult, if the equations are tractable at all by other than numerical means. Furthermore, the addition of some new feature to a known problem (i.e. to a problem the solution of which is known), is not easily accommodated if the system is continuous. Consider, for instance the situation of the articulated pipe system which can be described by an equation such as (2.1), and the ease with which the addition of a supplemental mass at the free end can be accommodated. Then, contrast this to the difficulties associated with the addition of such a mass to a continuously flexible pipe: since the boundary conditions will now be different, this problem has to be solved from scratch, even if the solution of the problem without the mass (i.e. the solution of the simple beam equation) is already known. Hence, it is often advantageous to transform distributed parameter systems into discrete ones by such methods as the Galerkin (or Ritz–Galerkin) or the Rayleigh–Ritz schemes (Meirovitch 1967).

In this section, first the standard methods of analysis of discrete systems will be reviewed. Then, the Galerkin method will be presented via example problems, as well as methods for dealing with the forced response of continuous systems. Along the way, a number of important definitions and classifications of systems, e.g. conservative and nonconservative, self-adjoint, positive definite, etc., will be introduced.

2.1.1 The equations of motion

The equations of motion of discrete systems are generally derived by either Newtonian or Lagrangian methods. In the latter case, for a system of N degrees of freedom and generalized coordinates q_r, the Lagrange equations are

$$\frac{\mathrm{d}}{\mathrm{d}t}\left(\frac{\partial T}{\partial \dot{q}_r}\right) - \frac{\partial T}{\partial q_r} + \frac{\partial V}{\partial q_r} = Q_r, \qquad r = 1, 2, \ldots, N, \tag{2.2}$$

where T is the kinetic energy and V the potential energy of some or all of the conservative forces acting on the system, while Q_r are the generalized forces associated with the rest of the forces (Bishop & Johnson 1960; Meirovitch 1967, 1970).

For continuous (distributed parameter) systems, the equations of motion may be obtained either by Newtonian methods (by taking force and moment balances on an element of the system) or by the use of Hamilton's principle and variational techniques, i.e. by using

$$\delta \int_{t_1}^{t_2} (T - V + W)\,\mathrm{d}t = 0, \tag{2.3}$$

where δ denotes the variational operator and W is the work done by forces not included in V. The use of Hamilton's principle is especially convenient in cases of unusual boundary conditions, because the equation(s) of motion and boundary conditions are determined in a unified procedure [see, for example, Meirovitch (1967)].

Special forms or interpretation of (2.2) and (2.3) may be necessary for 'open systems', where the mass is not conserved, e.g. with in-flow and out-flow of mass and momentum, as is common in fluidelastic systems. These, however, will be discussed in the chapters that follow (e.g. in Section 3.3.3).

2.1.2 Brief review of discrete systems

A system is conservative if all noninertial forces may be derived from a potential function, i.e. if they are all functions of position alone; thus, if the system is displaced from a to b, the work is not path-dependent (or, equivalently, if the system is returned to a by whatever path, the total work done is null). For a conservative system, the equations of motion may be written as

$$[M]\{\ddot{q}\} + [K]\{q\} = \{Q\}, \tag{2.4}$$

a special form of (2.1); the matrices are of the same order as the number of degrees of freedom, N. Provided that (i) the generalized coordinates are measured from the (stable) equilibrium configuration, (ii) the potential energy is zero at equilibrium, and (iii) the constraints are scleronomic — conditions that are not difficult to satisfy in many cases — the $[M]$ and $[K]$ matrices are symmetric.

Constraints are auxiliary kinematical conditions; e.g. in Figure 2.1(a) the mass M_1 cannot move freely in the plane but must remain at a fixed distance l_1 from the point of support. The two constraint equations that must implicitly be satisfied for the system of Figure 2.1(a) are what makes this system have two and not four degrees of freedom. If a constraint equation may be reduced to a form $f(x, y, z, t) = 0$, then the constraint is said to be *holonomic*; a subclass of this is when the constraint equation does not contain time explicitly, in which case the constraint is said to be *scleronomic* (Meirovitch 1970; Neĭmark & Fufaev 1972). Thus, if l_1 were a prescribed function of time, the constraint would be holonomic but not scleronomic.†

The homogeneous form of equation (2.4), representing *free motions* of the system,

$$[M]\{\ddot{q}\} + [K]\{q\} = \{0\}, \tag{2.5}$$

may be re-written as

$$\{\ddot{q}\} + [W]\{q\} = \{0\}, \tag{2.6}$$

in which $[W] = [M]^{-1}[K]$ — provided that $[M]$ can be inverted, i.e. if it is nonsingular. Oscillatory solutions are sought, of the form

$$\{q\} = \{A\}e^{i\Omega t}, \tag{2.7}$$

†These words derive from the Greek: ὅλος = whole or total and νόμος = law, hence holonomic means totally demarcated or defined; the first component of scleronomic is from σκληρός = hard, hence the word denotes a hard and fast rule!

where $\{A\}$ is a column of unknown amplitudes and Ω the circular frequency. Substituting (2.7) into (2.6) and defining $\lambda \equiv \Omega^2$, leads to the standard eigenvalue problem,

$$(\lambda[I] - [W])\{A\} = \{0\}, \tag{2.8}$$

where $[I]$ is the unit matrix. Nontrivial solution of (2.8) requires that

$$\det([W] - \lambda[I]) = 0, \tag{2.9}$$

which is the characteristic equation, from which the eigenvalues, $\lambda_i, i = 1, 2, \ldots, N$, and hence the corresponding eigenvectors, $\{A\}_i$ or $\mathbf{A}_i$, may be found. The free-vibration characteristics of the system are fully determined by the eigenvalues (and hence the eigenfrequencies $\Omega_i = \lambda_i^{1/2}$) and the corresponding eigenvectors. The latter may be viewed as shape functions. Thus, for the double pendulum of Figure 2.1(a), if $M_1 = 2M$, $M_2 = M$ and $l_1 = l_2 = l$, one obtains $\lambda_1 = \frac{1}{2}(g/l)$ and $\lambda_2 = 2(g/l)$. The first- and second-mode eigenvectors are, respectively $\{1, 1\}^{\mathrm{T}}$ and $\{1, -2\}^{\mathrm{T}}$, which means that, for motions purely in the first mode (at Ω_1), the second pendulum oscillates with the same angular amplitude as the first, and in the same direction; while in the second mode (at Ω_2), the second pendulum has twice the amplitude of the first, but in the opposite sense. Pure first-mode motions could be generated via initial conditions $\{q(0)\} = \{1, 1\}^{\mathrm{T}}$, $\{\dot{q}(0)\} = \{0\}$, and similarly for second-mode motions. Other initial conditions generate motions which involve — can be synthesized from — both eigenvectors and both eigenfrequencies.

As a consequence of $[M]$ and $[K]$ being symmetric, the eigenvalues are *real* (as in the foregoing example),† and the following *weighted orthogonality* holds true for the eigenvectors:

$$\{A\}_j^{\mathrm{T}}[K]\{A\}_i = 0, \qquad \{A\}_j^{\mathrm{T}}[M]\{A\}_i = 0 \qquad \text{for} \quad i \neq j; \tag{2.10}$$

if $[W]$ is symmetric too — recall that the product of two symmetric matrices is not necessarily symmetric — then direct orthogonality also applies, i.e. $\{A\}_j^{\mathrm{T}}\{A\}_i = 0$ for $i \neq j$. Relations (2.10) hold true, provided that the eigenvalues are distinct; the case of repeated eigenvalues will be treated later.

Since $[M]$ is, or can be, derived from the kinetic energy, which is a positive definite function, $[M]$ is a *positive definite* matrix (Meirovitch 1967; Pipes 1963).‡ If $[K]$ is also positive definite, then so is the system, and the eigenvalues are all positive. If $[K]$ is only positive, the system is said to be *semidefinite*, and it may have zero eigenvalues — e.g. if the system as a whole is unrestrained.

For the *forced response*, equation (2.4) has to be solved. This may be done in many ways, e.g. by the use of Laplace transforms or by modal analysis. This latter will be reviewed briefly in what follows. First, the *modal matrix* is defined,

$$[A] = [\{A\}_1\{A\}_2 \cdots \{A\}_N]; \tag{2.11}$$

then, the so-called *expansion theorem* is invoked, stating that any vector, including $\{q\}$, in the vector space spanned by $[A]$ may be expressed ('synthesized') in terms of the

†This is physically reasonable — see equation (2.7).

‡If the determinant of successive submatrices, each containing the left-hand corner element are all positive, then the matrix is *positive definite*. That is, for a 3×3 matrix $[M]: m_{11} > 0, m_{11}m_{22} - m_{21}m_{12} > 0$ and $\det[M] > 0$; and similarly for higher order matrices. If any of the determinants is zero, then $[M]$ is said to be only *positive* rather than positive definite.

eigenvectors making up $[A]$. Hence, the coordinate transformation

$$\{q\} = [A]\{y\} \tag{2.12}$$

is introduced, in which y_i, $i = 1, \ldots, N$, are the *normal* or *principal coordinates*. Substituting (2.12) into (2.4), and pre-multiplying by $[A]^T$ leads to

$$[P]\{\ddot{y}\} + [S]\{y\} = [A]^T\{Q\} = \{F\}, \tag{2.13}$$

in which

$$[P] = [A]^T[M][A], \qquad [S] = [A]^T[K][A] \tag{2.14}$$

are diagonal, in view of the relations (2.10).

The system (2.13) has therefore been *decoupled.* Each row reads $p_i\ddot{y}_i + s_i\, y_i = F_i(t)$, which is easily solvable, subject to the initial conditions $\{y(0)\} = [A]^{-1}\{q(0)\}$ and $\{\dot{y}(0)\} = [A]^{-1}\{\dot{q}(0)\}$. The response in terms of the original coordinates may then be obtained by application of (2.12).

In case of repeated eigenvalues, or if $[M]$ or $[K]$ are not symmetric but the eigenvalues are still real, provided that linearly independent eigenvectors may be found,[†] one may proceed as follows: (i) equation (2.4) is pre-multiplied by $[M]^{-1}$, (ii) transformation (2.12) is introduced, and (iii) the equation is decoupled by pre-multiplication by $[A]^{-1}$; this leads to

$$\{\ddot{y}\} + [\lambda]\{y\} = [A]^{-1}[M]^{-1}\{Q\}, \tag{2.15}$$

where $[A]^{-1}[W][A] = [\lambda]$ has been utilized, and $[\lambda]$ is the diagonal matrix of the eigenvalues.

If *damping* is present, then the full form of equation (2.1) applies — provided, of course, that the damping is viscous or that it may be approximated as such. In this case, eigenvalues and eigenvectors are no longer real. The procedure that follows applies to cases where $[M]$, $[K]$ and $[C]$ are symmetric — the latter being so if $[C]$ is derived from a dissipation function, for instance (Bishop & Johnson 1960). The following partitioned matrices and vectors of order $2N$ are defined:

$$[B] = \begin{bmatrix} [0] & [M] \\ [M] & [C] \end{bmatrix}, \quad [E] = \begin{bmatrix} -[M] & [0] \\ [0] & [K] \end{bmatrix}, \quad \{\Phi\} = \begin{Bmatrix} \{0\} \\ \{Q\} \end{Bmatrix}, \quad \{z\} = \begin{Bmatrix} \{\dot{q}\} \\ \{q\} \end{Bmatrix}, \tag{2.16}$$

and equation (2.1) may now be reduced into the first-order form

$$[B]\{\dot{z}\} + [E]\{z\} = \{\Phi\}. \tag{2.17}$$

The procedure henceforth parallels that of the conservative system. Assuming solutions of the form $\{z\} = \{A\} \exp(\lambda t) \equiv \{A\} \exp(\mathrm{i}\Omega t)$, the reduced equation (2.17) eventually leads to the eigenvalue problem

$$(\lambda[I] - [Y])\{A\} = \{0\}, \tag{2.18}$$

where $[Y] = -[B]^{-1}[E]$. The eigenvalues, λ_i, and eigenvectors $\{A\}_i$, $i = 1, 2, \ldots, 2N$, may now be determined. The λ_i occur in complex conjugate pairs,[‡] and the eigenvectors

[†]Hence, in principle and if desired, a set of orthogonal eigenvectors may be determined via the Gram–Schmidt procedure.

[‡]Note that, even for a conservative mass–spring one-degree-of-freedom system, one obtains $\Omega = \pm\sqrt{k/m}$, where the negative value is usually ignored (see Section 2.3); here $\Omega_i \equiv \lambda$, so $\lambda_{1,2} = 0\mathrm{i} \pm (k/m)^{1/2}$.

for $i = 1, \ldots, N$ are those for $i = N+1, \ldots, 2N$, multiplied by λ_i. Since the λ_i are complex, so are the Ω_i — the real part of Ω being associated with the frequency of oscillation and the imaginary part with damping (see Section 2.3); recall that $\lambda_i = \mathrm{i}\Omega_i$.

A modal matrix, $[A]$, is then constructed, and the transformation $\{z\} = [A]\{y\}$ introduced. In view of the weighted orthogonality of the $\{A\}_i$, for a set of distinct eigenvalues, one obtains

$$[P]\{\dot{y}\} + [S]\{y\} = [A]^{\mathrm{T}}\{\Phi\} = \{\Psi\}, \tag{2.19}$$

where $[P] = [A]^{\mathrm{T}}[B][A]$ and $[S] = [A]^{\mathrm{T}}[E][A]$ are diagonal. Hence, each row reads $\dot{y}_i - \lambda_i y_i = \alpha_i \Psi_i$, $i = 1, 2, \ldots, 2N$, which is easily solvable. As before, the solution in terms of $\{q\}$, and redundantly in terms of $\{\dot{q}\}$, is obtained by $\{z\} = [A]\{y\}$.

In fluidelastic systems $[C]$ and $[K]$ are often nonsymmetric, and the foregoing decoupling procedure then needs to be modified (Meirovitch 1967). To that end, the adjoint of eigenvalue problem (2.18) is defined,

$$\left(\lambda[I] - [Y]^{\mathrm{T}}\right)\{\tilde{A}\} = \{0\}, \tag{2.20}$$

the eigenvalues of which are the same as those of (2.18), but the eigenvectors, $\{\tilde{A}\}_i$, are different. Then, the original system may be decoupled by introducing in (2.17) the transformation $\{z\} = [A]\{y\}$, and (ii) making use of the biorthogonality properties

$$\{\tilde{A}\}_i^{\mathrm{T}}\{A\}_j = 0, \quad \{\tilde{A}\}_i^{\mathrm{T}}[Y]\{A\}_j = \{0\}, \quad \text{for} \quad i \neq j, \tag{2.21}$$

which lead to a decoupled equation, similar, in form at least, to (2.19).

2.1.3 The Galerkin method via a simple example

As already mentioned, it is advantageous to analyse distributed parameter (or *continuous*) systems by transforming them into discrete ones by the Galerkin method (or, for that matter, by collocation or finite element techniques), and then utilizing the methods outlined in Section 2.1.2. The Galerkin method will be reviewed here by means of an example.

Consider a uniform cantilevered pipe of length L, mass per unit length m, and flexural rigidity EI. The simplest equation describing its flexural motion is

$$EI\frac{\partial^4 w}{\partial x^4} + m\frac{\partial^2 w}{\partial t^2} = 0, \tag{2.22}$$

where $w(x, t)$ is the lateral deflection — according to the Euler–Bernoulli beam theory, as opposed to the Timoshenko or other higher order theories. The boundary conditions are

$$w\bigg|_{x=0} = 0, \quad \frac{\partial w}{\partial x}\bigg|_{x=0} = 0, \quad EI\frac{\partial^2 w}{\partial x^2}\bigg|_{x=L} = 0, \quad EI\frac{\partial^3 w}{\partial x^3}\bigg|_{x=L} = 0. \tag{2.23}$$

The solution of this problem is well known [e.g. Bishop & Johnson (1960)]. After separation of variables, with separation constant λ_r^4, the spatial equation admits a solution consiting of exponentials of $\pm\lambda_r$ and $\pm\lambda_r\mathrm{i}$. Substitution into (2.23) gives a system of four homogeneous equations, the condition for nontrivial solution of which leads to the characteristic equation,

$$\cos\lambda_r L \cosh\lambda_r L + 1 = 0. \tag{2.24}$$

This transcedental equation yields an infinite set of eigenvalues, the first three of which are

$$\lambda_1 L = 1.875\,10, \qquad \lambda_2 L = 4.694\,09, \qquad \lambda_3 L = 7.854\,76; \tag{2.25}$$

the corresponding natural- or eigenfrequencies are

$$\Omega_r = (\lambda_r L)^2 \left(\frac{EI}{mL^4}\right)^{1/2}. \tag{2.26}$$

The modal shapes or eigenfunctions are

$$\phi_r(x) = \cosh \lambda_r x - \cos \lambda_r x - \sigma_r (\sinh \lambda_r x - \sin \lambda_r x), \tag{2.27}$$

where

$$\sigma_r = \frac{\sinh \lambda_r L - \sin \lambda_r L}{\cosh \lambda_r L + \cos \lambda_r L}. \tag{2.28}$$

Before proceeding further, an *important note* should be made. It is customary in vibration theory and in classical mathematics to define the eigenvalue as being essentially the square or, as in equation (2.26), the square-root of the frequency, except possibly for a dimensional factor as in (2.26); the main point is that *a positive eigenvalue here is associated with a positive eigenfrequency*. In dynamics and stability theory, however, solutions are expressed as being proportional to $\exp(\mathrm{i}\Omega t)$ or $\exp(\lambda t)$, so that Ω and λ are 90° out of phase; a positive eigenvalue in this case would represent *divergent motion*, i.e. an unstable system! This can lead to confusion, no doubt. However, these different meanings and notations are so deeply embedded in these fields [cf. equations (2.26) and (2.36)] that, in the author's opinion, trying to unify the notation and meanings would create even more confusion. Instead, the context and occasional reminders will be preferable, to make the reader aware of which of the two notations for eigenvalue is being used.

When a concentrated mass M_e is added at the free end of the pipe,† the equation of motion is the same, but the boundary conditions are

$$w\bigg|_{x=0} = 0, \qquad \frac{\partial w}{\partial x}\bigg|_{x=0} = 0, \qquad EI\,\frac{\partial^2 w}{\partial x^2}\bigg|_{x=L} = 0, \qquad EI\,\frac{\partial^3 w}{\partial x^3}\bigg|_{x=L} = M_e\,\frac{\partial^2 w}{\partial t^2}\bigg|_{x=L}; \tag{2.29}$$

hence there is a shear force at the free end, associated with the inertia of the supplemental mass. Of course, for a simple problem like this, it is possible to proceed in the normal way and determine the eigenvalues and eigenfunctions of the modified problem. It will nevertheless be found convenient to transform such systems into discrete ones by the Galerkin method. To this end, for the problem at hand, the end-shear is transferred from the boundary conditions into the equation of motion, which may be re-written as

$$EI\,\frac{\partial^4 w}{\partial x^4} + \left[m + M_e\,\delta(x-L)\right]\frac{\partial^2 w}{\partial t^2} = 0, \tag{2.30}$$

†The main purpose here is purely tutorial; nevertheless, the dynamics of a pipe conveying fluid with an added mass at $x = L$ is considered in Chapter 5 (Section 5.8.3), and it is shown to add a lot of zest to the dynamics of the system.

where $\delta(x-L)$ is the Dirac delta function; boundary conditions (2.29) then reduce to (2.23). According to Galerkin's method, the solution of (2.30) may be expressed as

$$w(x,t) \simeq w_N(x,t) = \sum_{j=1}^{N} \psi_j(x) q_j(t), \tag{2.31}$$

where the $\psi_j(x)$ are appropriate *comparison functions*, i.e. functions in the same domain, $\mathscr{D} = [0, L]$, satisfying all the boundary conditions (both geometrical and natural†), and $q_j(t)$ are the generalized coordinates of the discretized system which will eventually emerge by application of this method (Meirovitch 1967). It is now clear why it is advantageous to recast this problem into the form of equations (2.30) and (2.23), for it is then possible to use $\psi_j(x) \equiv \phi_j(x)$, i.e. to use the eigenfunctions given by (2.27) as suitable comparison functions: suitable, since they satisfy the boundary conditions associated with (2.30), and also convenient, since they are already known.

When approximation (2.31) is substituted into the left-hand side of (2.30), the result will generally not be zero, but equal to an error function, which may be denoted by $\mathscr{E}[w_N]$. Galerkin's method requires that

$$\int_{\mathscr{D}} \mathscr{E}[w_N]\psi_r(x)\,\mathrm{d}\mathscr{D} = 0, \qquad r = 1, 2, \ldots, N; \tag{2.32}$$

i.e. over the domain, the integrated error, weighted by $\psi_r(x)$,‡ should be zero (Finlayson & Scriven 1966).

Thus, in this example, substituting approximation (2.31) with $\psi_j(x) = \phi_j(x)$ into equation (2.30), multiplying by $\phi_r(x)$ and integrating over $\mathscr{D} = [0, L]$, leads to

$$\sum_{j=1}^{N} \{EI\lambda_j^4 q_j L\delta_{rj} + [mL\delta_{rj} + M_e\phi_r(L)\,\phi_j(L)]\ddot{q}_j\} = 0, \qquad r = 1, 2, \ldots, N, \tag{2.33}$$

in view of the orthogonality of eigenfunctions (2.27), i.e.

$$\int_0^L \phi_r(x)\phi_j(x)\,\mathrm{d}x = L\delta_{rj}, \tag{2.34}$$

where δ_{rj} is the Kronecker delta (0 for $r \neq j$ and 1 for $r = j$). Clearly the system is now discretized. Thus, if a two-mode approximation ($N = 2$) is utilized, equation (2.33) may be written in the following matrix form:

$$\begin{bmatrix} mL + M_e\phi_1^2(L) & M_e\phi_1(L)\phi_2(L) \\ M_e\phi_1(L)\phi_2(L) & mL + M_e\phi_2^2(L) \end{bmatrix} \begin{Bmatrix} \ddot{q}_1 \\ \ddot{q}_2 \end{Bmatrix} + EI \begin{bmatrix} \lambda_1^4 L & 0 \\ 0 & \lambda_2^4 L \end{bmatrix} \begin{Bmatrix} q_1 \\ q_2 \end{Bmatrix} = \{0\}. \tag{2.35}$$

The eigenvalues and eigenfrequencies of this matrix system are approximations of the lowest two of the continuous system; thus, if $M_e = \frac{1}{2}mL$, then $\Omega_1 = 2.018(EI/mL^4)^{1/2}$, $\Omega_2 = 17.165(EI/mL^4)^{1/2}$. The corresponding eigenvectors give, in a

---

†*Geometrical* boundary conditions are of the type $w\big|_{x=0} = 0$, while *natural* ones involve forces or moments, e.g. $EI(\partial^3 w/\partial x^3)\big|_{x=L} = 0$.

‡The weighting function comes in 'naturally' if Galerkin's method is derived via variational techniques.

sense, the 'mix' of first- and second-mode eigenfunctions of the original system, necessary to approximate the eigenfunctions of the modified one; thus, for this example,

$$\{A\}_1 = \begin{Bmatrix} 1 \\ -0.02 \end{Bmatrix}, \qquad \{A\}_2 = \begin{Bmatrix} 1 \\ 1.48 \end{Bmatrix}.$$

In general, N must be sufficiently large to assure convergence. Table 2.1 shows that convergence can be very rapid. The exact values, by solving (2.22) with boundary conditions (2.29), are $\Omega_r(EI/mL^4)^{-1/2} = 2.0163, 16.901, 51.701$ for $r = 1, 2, 3$.

Galerkin's method will now be expressed formally in a generalized form, useful for further development. The eigenvalue problem associated with equations (2.22) and (2.30) may be expressed as

$$\mathcal{L}[w] = \lambda\mathcal{M}[w], \tag{2.36}$$

subject to the appropriate boundary conditions. Generally, $\mathcal{L}$ and $\mathcal{M}$ are linear differential operators, although $\mathcal{M}$ in many cases is a scalar, and $\lambda(=\Omega^2)$ is the eigenvalue. In the case of equation (2.30), $\mathcal{L} = EI(\partial^4/\partial x^4)$ and $\mathcal{M} = m + M_e\delta(x - L)$. The equivalent to statement (2.31) now is

$$w_N(x) = \sum_{j=1}^{N} a_j\psi_j(x). \tag{2.37}$$

The elements of the mass and stiffness matrices [cf. equation (2.1)], the two matrices in (2.35), may be obtained by

$$m_{rj} = \int_0^L \psi_r(x)\mathcal{M}[\psi_j(x)]\,\mathrm{d}x, \qquad k_{rj} = \int_0^L \psi_r(x)\mathcal{L}[\psi_j(x)]\,\mathrm{d}x. \tag{2.38}$$

In the case where M_e is incorporated in $\mathcal{M}$ and the boundary conditions are (2.23), this is a standard problem. If, however, M_e is left out of the equations of motion, boundary conditions (2.29) may be re-written as

$$w(0) = 0, \qquad w'(0) = 0, \qquad w''(L) = 0, \qquad EIw'''(L) = -\lambda M_e w(L), \tag{2.39}$$

in which $(\ )' \equiv \partial/\partial x$, and the problem is unusual in that the eigenvalue appears in the boundary conditions. Hence, strictly (Friedman 1956), the domain $\mathcal{D}$ depends upon λ. In this example, for the calculations with equation (2.22) and boundary conditions (2.29) leading to the 'exact results' to which those of Table 2.1 were compared, we have proceeded by blithely ignoring this subtlety (by retaining $\mathcal{D} = [0, 1]$), yet still obtained the correct results. However, this is not always true, as will be seen in Section 2.1.4.

Table 2.1 Approximations to the lowest three eigenfrequencies of the modified cantilevered pipe for various N in the case of $M_e = \frac{1}{2}mL$.

N	2	4	6	8	10
$\Omega_1(EI/mL^4)^{-1/2}$	2.0184	2.0166	2.0164	2.0163	2.0163
$\Omega_2(EI/mL^4)^{-1/2}$	17.166	16.936	16.912	16.906	16.904
$\Omega_3(EI/mL^4)^{-1/2}$	-	52.125	51.826	51.754	51.728

2.1.4 Galerkin's method for a nonconservative system

Consider next that a fluid of constant velocity U and mass per unit length M is flowing through the pipe in the example of Section 2.1.3, i.e. the pipe with the extra mass M_e at the free end. As shown in Chapter 3, the equation of motion in this case is

$$EI\,\frac{\partial^4 w}{\partial x^4} + MU^2\,\frac{\partial^2 w}{\partial x^2} + 2MU\,\frac{\partial^2 w}{\partial x \partial t} + (m+M)\,\frac{\partial^2 w}{\partial t^2} = 0, \tag{2.40}$$

with boundary conditions (2.23) or (2.29) for $M_e = 0$ and $M_e \neq 0$, respectively.

For $M_e \neq 0$, the problem is solved by the same two methods as before: (a) with M_e included in the equation of motion, with a Dirac delta function, and boundary conditions (2.23); (b) with equation (2.40) as it stands and boundary conditions (2.29). Table 2.2 gives the results for $\Gamma \equiv M_e/[(m+M)L] = 0.3$ and $\beta \equiv M/(m+M) = 0.1$ for two values of the dimensionless flow velocity $u = (M/EI)^{1/2}LU$. Two interesting observations may be made from the results of Table 2.2. First, for $u = 2$, the eigenfrequencies are no longer real; in fact, for all $u \neq 0$ they need not be real because the system is *nonconservative*. Second, the eigenfrequencies for $u = 2$ (again, for all $u \neq 0$) as obtained by the two methods are not identical as they should have been.

That the system is nonconservative may be assessed by calculating the rate of work done by all the forces acting on the pipe. If it is zero, then there is no net energy flow in and out of the system, which must therefore be conservative; otherwise, the system is nonconservative. In this case,

$$\frac{\mathrm{d}W}{\mathrm{d}t} = -\int_0^L \frac{\partial w}{\partial t}\,\mathcal{L}[w]\,\mathrm{d}x \tag{2.41}$$

is found not to be zero by virtue of the forces represented by the second and third terms in (2.40)† — see Chapter 3. Viewed another way, this means that it is not possible to derive these forces from a potential; like dissipative forces, for instance, they are nonconservative, at least for this set of boundary conditions.

The second observation suggests that, for $u \neq 0$, the results from either method (a) or (b) must be wrong. Indeed, those of method (b), utilizing equations (2.40) and (2.29) as they stand, are wrong because of the remark made at the end of Section 2.1.3. There *is*

Table 2.2 The lowest two eigenfrequencies calculated by two different methods for different u; $\Gamma = 0.3$, $\beta = 0.1$. In method (a) the extra mass, M_e, is included in the equation of motion via a Dirac delta function, while in (b) it is accounted for in the boundary conditions.

	$u = 0$		$u = 2$	
	Method (a)	Method (b)	Method (a)	Method (b)
$\Omega_1[EI/(m+M)L^4]^{-1/2}$	2.36	2.36	2.71 + 0.660i	2.18 + 1.16i
$\Omega_2[EI/(m+M)L^4]^{-1/2}$	17.58	17.58	16.48 + 0.084i	16.34 + 1.56i

†In this problem, the definition of $\mathcal{L}$ is not clear-cut, because of the mixed derivative. However, by taking $\mathcal{L}[w] = [EI(\partial^4/\partial x^4) + MU^2(\partial^2/\partial x^2) + 2MU(\partial^2/\partial x\,\partial t)]w$, one obtains $(\mathrm{d}W/\mathrm{d}t) = -MU[(\partial w/\partial t)^2 + U(\partial w/\partial x)(\partial w/\partial t)]\big|_{x=L} \neq 0$.

a way of solving the problem correctly while utilizing boundary conditions (2.29), but the meaning of the domain $\mathcal{D}$ has to be expanded (Friedman 1956; Meirovitch 1967); an example is given in Chapter 4 (Section 4.6.2).

2.1.5 Self-adjoint and positive definite continuous systems

The eigenvalue problem of equation (2.36), and thereby the system, is said to be *self-adjoint*[†] if for any two comparison functions, u and v,

$$\int_{\mathcal{D}} u\mathcal{L}[v]\,\mathrm{d}\mathcal{D} = \int_{\mathcal{D}} v\mathcal{L}[u]\,\mathrm{d}\mathcal{D}, \qquad \int_{\mathcal{D}} u\mathcal{M}[v]\,\mathrm{d}\mathcal{D} = \int_{\mathcal{D}} v\mathcal{M}[u]\,\mathrm{d}\mathcal{D}, \tag{2.42}$$

are satisfied. A consequence of self-adjointness is that the eigenvalues are *real*. Another consequence is that a generalized or weighted orthogonality of the eigenfunctions then holds true for nonrepeated eigenvalues; thus,

$$\int_{\mathcal{D}} \phi_r\mathcal{M}[\phi_s]\,\mathrm{d}\mathcal{D} = 0, \qquad \int_{\mathcal{D}} \phi_r\mathcal{L}[\phi_s]\,\mathrm{d}\mathcal{D} = 0 \qquad \text{for} \qquad \lambda_r \neq \lambda_s. \tag{2.43}$$

Furthermore, if

$$\int_{\mathcal{D}} u\mathcal{L}[u]\,\mathrm{d}\mathcal{D} > 0 \qquad \text{and} \qquad \int_{\mathcal{D}} u\mathcal{M}[u]\,\mathrm{d}\mathcal{D} > 0 \tag{2.44}$$

for all nonzero u, the operators are *positive definite*, and hence so is the system. The consequence of this is that the eigenvalues of such a system are positive — refer to Section 2.3 for the significance of this and back to Section 2.1.3 for further clarification of the different usage of the word ‘eigenvalue’. In cases where $\mathcal{L}$ is only positive, rather than positive definite, i.e. when the first integral (2.44) can be zero for some nonzero u, while $\mathcal{M}$ remains positive definite, the system is called positive *semidefinite*, and admits solutions with $\lambda = 0$.

Clearly, for the system of equations (2.22) and (2.23), the problem is self-adjoint. To illustrate the case of a non-self-adjoint system in as simple a manner as possible while still keeping in the framework of the examples already discussed, consider the system

$$\mathcal{L}[w] = EI(\mathrm{d}^4/\mathrm{d}x^4) + P(\mathrm{d}^2/\mathrm{d}x^2), \qquad \mathcal{M}[w] = m; \tag{2.45}$$

$$w(0) = 0, \qquad w'(0) = 0, \qquad EIw''(L) = 0, \qquad EIw'''(L) = 0. \tag{2.46}$$

This could represent a cantilevered beam, subjected to a compressive tangential ‘follower’ force P, such that the boundary conditions remain unaffected. A follower force is one retaining the same orientation to the structure in the course of motions of the system, in this case remaining tangential to the free end.[‡] By applying the integrals (2.43) it is found that the integrated out parts do not vanish [since $Pu(1)v'(1)$ and $Pu'(1)v(1)$ are not zero for the boundary conditions given].

[†]If $\mathcal{L}$ and $\mathcal{M}$ are complex operators, the equivalent property is for the eigenvalue problem to be *Hermitian*.
[‡]In fact, such a compressive follower force could be generated by a light rocket engine $[M_e/(mL) \simeq 0]$ mounted on the free end of the cantilever, so that the force of reaction is always tangential to the free end.

2.1.6 Diagonalization, and forced vibrations of continuous systems

The equation of motion associated with the problem defined by (2.45) is

$$EI\,\frac{\partial^4 w}{\partial x^4}+P\,\frac{\partial^2 w}{\partial x^2}+m\,\frac{\partial^2 w}{\partial t^2}=0, \tag{2.47}$$

with the boundary conditions as given in (2.46). This clearly represents free motions of the system; hence, of interest are the eigenfrequencies and the corresponding eigenfunctions and how they vary with P (or its nondimensional counterpart, PL^2/EI). This can be done by direct application of the Galerkin method with $w_N=\sum_j \phi_j(x)q_j(t)$, in which the cantilever-beam eigenfunctions (2.27) are used as comparison functions, since they satisfy boundary conditions (2.46), which are identical to (2.23). In this way, one obtains an equation similar to (2.35), i.e.

$$[M]\{\ddot{q}\}+[K]\{q\}=\{0\}, \tag{2.48}$$

but with only $[M]$ being diagonal, while $[K]$ is nondiagonal. In fact, the elements of $[K]$ are

$$k_{rj}=EI\lambda_r^4 L\delta_{rj}+P\int_0^L \phi_r\phi_j''\,\mathrm{d}x,$$

the prime denoting differentiation with respect to x.

Suppose now that this system is subjected also to a distributed force, $F(x,t)$, so that the equation of motion is

$$EI\,\frac{\partial^4 w}{\partial x^4}+P\,\frac{\partial^2 w}{\partial x^2}+m\,\frac{\partial^2 w}{\partial t^2}=F(x,t); \tag{2.49}$$

see Figure 2.2. After discretization by the Galerkin procedure, we obtain

$$[M]\{\ddot{q}\}+[K]\{q\}=\{Q\}. \tag{2.50}$$

If this had been a self-adjoint conservative system, matrices $[M]$ and $[K]$ in equation (2.50) would both be symmetric. For the problem at hand, however, the system is non-self-adjoint, as remarked earlier, and hence $[K]$ is asymmetric, by virtue of the fact that $\int_0^L \phi_r\phi_j''\,\mathrm{d}x \neq \int_0^L \phi_j\phi_r''\,\mathrm{d}x$. Hence, the decoupling procedure leading to equation (2.15) should be adopted.

Before proceeding further, however, it is useful to transform equation (2.49) into dimensionless form, which serves to introduce the kind of dimensionless terms appearing frequently in the following chapters. Hence, defining

$$\begin{aligned}&\xi=x/L, \qquad \eta=w/L, \qquad \tau=(EI/mL^4)^{1/2}t,\\ &\mathcal{P}=PL^2/EI, \qquad f=FL^3/EI, \qquad \omega=(EI/mL^4)^{-1/2}\Omega\end{aligned} \tag{2.51}$$

and taking, as a concrete example, $f=f_0\,\xi\sin(\omega_f\tau)$ — representing a triangularly distributed load along the beam, as shown in Figure 2.2 — substitution into (2.49) yields

$$\eta''''+\mathcal{P}\eta''+\ddot{\eta}=f_0\,\xi\,\sin(\omega_f\tau), \tag{2.52}$$

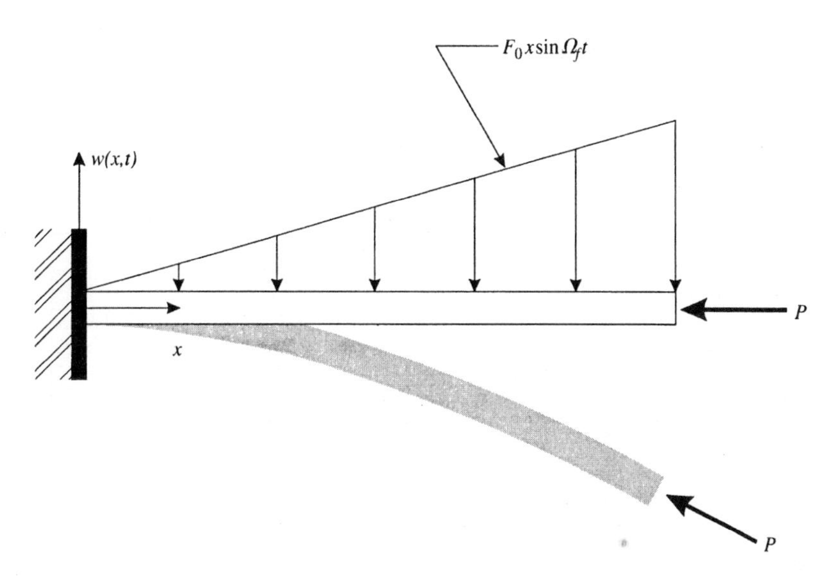

Figure 2.2 A cantilevered beam subjected to a tangential, follower compressive load, P, and to a time-dependent distributed force, $F_0 x \sin \Omega_f t$.

in which primes and overdots denote, respectively, partial differentiation with respect to ξ and τ. The discretized form of (2.52) is

$$[I]\{\ddot{q}\} + [K]\{q\} = \{Q\} \sin(\omega_f \tau), \tag{2.53}$$

and the elements of $[K]$ and $\{Q\}$ are

$$k_{ij} = \lambda_i^4 \delta_{ij} + \mathcal{P} \int_0^1 \phi_i \phi_j'' \, \mathrm{d}\xi, \qquad Q_i = \int_0^1 f_0 \, \xi \phi_i \, \mathrm{d}\xi, \tag{2.54}$$

in which the $\phi_i \equiv \phi_i(\xi)$, the dimensionless version of (2.27). The decoupled equation, corresponding to equation (2.15), is

$$\{\ddot{y}\} + [\Lambda]\{y\} = [A]^{-1}\{Q\} \sin(\omega_f \tau) = \{\Psi\} \sin(\omega_f \tau), \tag{2.55}$$

in which $[\Lambda]$ is the diagonal matrix of the eigenvalues; the solution therefore is

$$y_k = \alpha_k \cos \Lambda_k^{1/2} \tau + \beta_k \sin \Lambda_k^{1/2} \tau + [\Psi_k / (\Lambda_k - \omega_f^2)] \sin(\omega_f \tau), \qquad k = 1, 2, \ldots, N. \tag{2.56}$$

Numerical results for the case of $\mathcal{P} = 1$, $f_0 = 7$, $\omega_f = 0.6$ are shown in Figure 2.3: (a) for $\alpha_k = \beta_k = 0$, i.e. showing only the particular solution, and (b,c) for $\eta(1, 0) = 0.15$, $\dot{\eta}(1, 0) = 1.5$. The dimensionless natural frequencies, obtained with $N = 4$, are found to be $\omega_1 = 3.64$, $\omega_2 = 21.73$, $\omega_3 = 61.32$ and $\omega_4 = 120.5$; ω_f is chosen to be far below all of them.

In Figure 2.3(a), where the homogeneous part of the solution is totally absent, it is seen that the response is a pure sinusoid with period $T = 2\pi/\omega_f = 10.47$. The effect of the homogeneous part of the solution, however, complicates the response, as shown

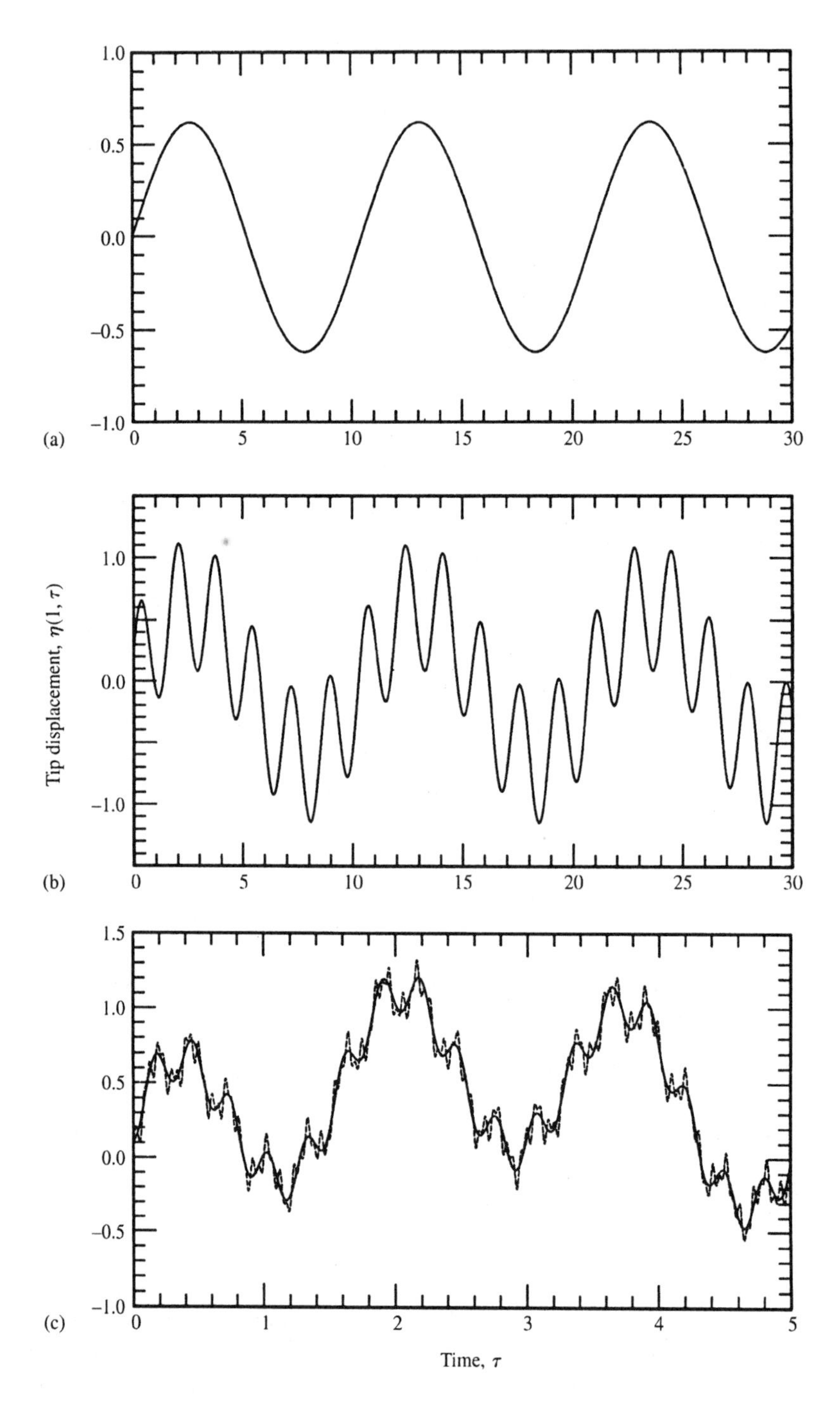

Figure 2.3 Solutions to equation (2.52) showing $\eta(1, \tau)$ versus τ, for $\mathcal{P} = 1$, $f_0 = 7$ and $\omega_f = 0.6$: (a) the particular solution alone [i.e. $\alpha_k = \beta_k = 0$ in equation (2.56)], which would correspond to the steady-state solution if damping were included; (b) full solution for $N = 1$; (c) full solution for $N = 2$ (——) and $N = 4$ ($\cdots$) on an expanded scale of τ.

in Figure 2.3(b), obtained with $N = 1$. A higher frequency component, at ω_1, is now superposed on the solution. Two observations should be made: (i) since, unrealistically, there is no damping in the system, the effect of initial conditions persists in perpetuity, whereas, with even a small amount of damping, the steady-state response would be like that in Figure 2.3(a); (ii) since ω_1/ω_f is not rational, the response is not periodic but quasiperiodic, although the effect of 'unsteadiness' in the response time-trace is just barely visible. This is more pronounced in Figure 2.3(c), plotted on an expanded time-scale, showing calculations with $N = 2$ and $N = 4$; in the latter case, the contribution of all four eigenmodes is visible. On the other hand, the period associated with the forcing frequency is hardly discernible in the time-scale used in Figure 2.3(c).

The fact that the response in Figure 2.3(b,c) is quasiperiodic is most apparent in the phase plane, as shown for example in Figure 2.4. It is seen that the response evolves by winding itself around a torus, the projection of which is shown in the figure, instead of tracing a planar curve, as would be the case for periodic motion.

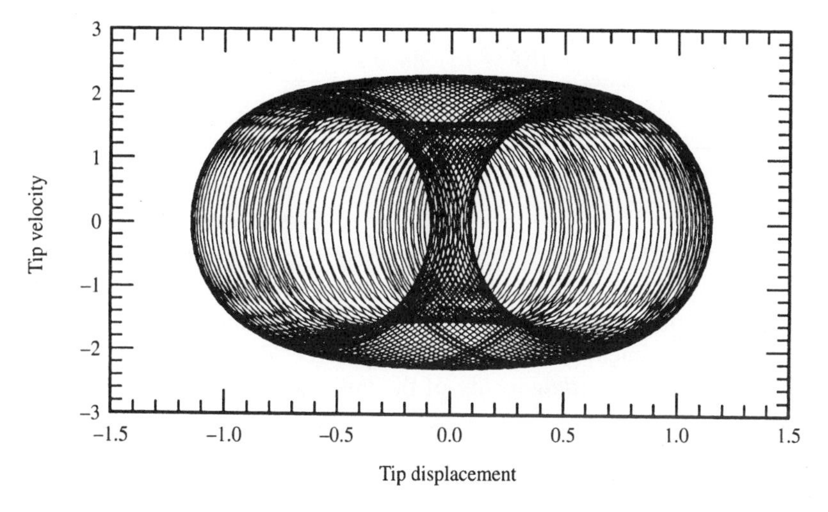

Figure 2.4 The response of Figure 2.3(b) plotted in the phase plane: the dimensionless tip velocity, $\dot{\eta}(1, \tau)$, versus displacement, $\eta(1, \tau)$.

There is another, more general method for obtaining the response of such a system, specific to non-self-adjoint problems (Washizu 1966, 1968; Anderson 1972). This begins with the determination of the *adjoint problem*.[†] If the eigenfunctions of the homogeneous form of equation (2.47), i.e. of the compressively loaded beam, are $\chi_i(\xi)$ and those of the adjoint problem $\psi_i(\xi)$, the adjoint problem is defined through the new operators $\mathscr{L}^*$ and $\mathscr{M}^*$, such that

$$\int_{\mathscr{D}} \psi(\xi)\mathscr{L}[\chi(\xi)]\,\mathrm{d}\mathscr{D} - \int_{\mathscr{D}} \chi(\xi)\mathscr{L}^*[\psi(\xi)]\,\mathrm{d}\mathscr{D} = C[\chi(\xi), \psi(\xi)]_{\mathscr{D}} \tag{2.57}$$

in which it is required that the so-called concomitant, C, vanish. A similar expression for $\mathscr{M}$ should be satisfied, but since for the problem at hand $\mathscr{M}$ is a scalar, we immediately

[†]Sometimes referred to as the *adjugate* problem (Collar & Simpson 1987).

have $\mathcal{M}^* = \mathcal{M}$. In the nondimensional notation used here, $\mathcal{D} = [0, 1]$ and $\xi = x/L$. This problem has in fact been solved by Chen (1987), but it is not difficult to reproduce the results. One finds $\mathcal{L}^* = \mathcal{L}$, but a new set of boundary conditions for the adjoint problem, namely

$$\psi(0) = \psi'(0) = 0, \qquad \psi''(1) + \mathcal{P}\psi(1) = 0, \qquad \psi'''(1) + \mathcal{P}\psi'(1) = 0. \tag{2.58}$$

Solving the two eigenvalue problems, one obtains

$$\begin{aligned}
&\chi(\xi) = A_1 \sin p\xi + A_2 \cos p\xi + A_3 \sinh q\xi + A_4 \cosh q\xi,\\
&\psi(\xi) = B_1 \sin p\xi + B_2 \cos p\xi + B_3 \sinh q\xi + B_4 \cosh q\xi,\\
&\quad p = \left[(\tfrac{1}{4}\mathcal{P}^2 + \lambda)^{1/2} + \tfrac{1}{2}\mathcal{P}\right]^{1/2}, \qquad q = \left[(\tfrac{1}{4}\mathcal{P}^2 + \lambda)^{1/2} - \tfrac{1}{2}\mathcal{P}\right]^{1/2},^{\dagger}\\
&A_1 = 1, \qquad A_2 = -(p^2 \sin p + pq \sinh q)/(p^2 \cos p + q^2 \cosh q),\\
&A_3 = -p/q, \qquad A_4 = -A_2,\\
&B_1 = 1, \qquad B_2 = -\frac{\left[(\mathcal{P} - p^2) \sin p - (p/q)(\mathcal{P} + q^2) \sinh q\right]}{\left[(\mathcal{P} - p^2) \cos p - (\mathcal{P} + q^2) \cosh q\right]},\\
&B_3 = -p/q, \qquad B_4 = -B_2.
\end{aligned} \tag{2.59}$$

The characteristic equation is

$$\mathcal{P}^2 + 2\lambda(1 + \cos p \cosh q) + \mathcal{P}\sqrt{\lambda} \sin p \sinh q = 0,$$

and it is the same for both problems; hence, so are the eigenvalues.

The essence of this method is that it achieves direct decoupling of the equations of motion via the so-called *biorthogonality* of the initial and adjoint eigenfunctions, viz.

$$\int_0^1 \psi_r \mathcal{L}[\chi_j]\, \mathrm{d}\xi = k_{rj}\delta_{rj}, \qquad \int_0^1 \psi_r \mathcal{M}[\chi_j]\, \mathrm{d}\xi = m_{rj}\delta_{rj}. \tag{2.60}$$

By introducing $\eta_N = \sum \chi_j(\xi) q_j(\tau)$ into equation (2.52), then multiplying by $\psi_r(\xi)$ and integrating over $\mathcal{D}$, the system is decoupled in a single operation, by virtue of relations (2.60), yielding

$$m_j \ddot{q}_j + k_j q_j = f_j \sin \omega_f \tau, \qquad j = 1, 2, \ldots, N. \tag{2.61}$$

Calculations with the same set of parameters produce virtually identical results as those shown in Figure 2.3 for $N = 4$.[‡] What is more surprising is that the rate of convergence with N is not better with this method than with the previous one. Clearly, therefore, in this particular case, there is no advantage in utilizing this second, more general but more laborious, procedure rather than the first. Similar conclusions are reached by Anderson (1972), who tested a very similar problem, essentially by the same two methods — although very small differences are found in that case in the results obtained by the two methods.

[†] A typographical error in p and q is noted in Chen (1987, Appendix C).

[‡] The results obtained by integrating the equations numerically are also identical, although in that case it took about one order of magnitude longer in time to obtain them.

2.2 THE FLUID MECHANICS OF FLUID–STRUCTURE INTERACTIONS

2.2.1 General character and equations of fluid flow

Trying to give a selective encapsulation of the 'fluids' side of fluid–structure interactions is more challenging than the equivalent effort on the 'structures' side, as attempted in Section 2.1. Solution of the equations of motion of the fluid is much more difficult. The equations are in most cases inherently nonlinear, for one thing; moreover, unlike the situation in solid mechanics, linearization is not physically justifiable in many cases, and solution of even the linearized equations is not trivial. Thus, complete analytical and, despite the vast advances in computational fluid dynamic (CFD) techniques and computing power, *complete* numerical solutions are confined to only some classes of problems. Consequently, there exists a large set of approximations and specialized techniques for dealing with different types of problems, which is at the root of the difficulty remarked at the outset. The interested reader is referred to the classical texts in fluid dynamics [e.g. Lamb (1957), Milne-Thomson (1949, 1958), Prandtl (1952), Landau & Lifshitz (1959), Schlichting (1960)] and more modern texts [e.g. Batchelor (1967), White (1974), Hinze (1975), Townsend (1976), Telionis (1981)]; a wonderful refresher is Tritton's (1988) book.

Excluding non-Newtonian, stratified, rarefied, multi-phase and other 'unusual' fluid flows,† the basic fluid mechanics is governed by the continuity (i.e. conservation of mass) and the Navier–Stokes (i.e. conservation of momentum) equations. For a homogeneous, isothermal, incompressible fluid flow of constant density and viscosity, with no body forces, these are given by

$$\nabla \cdot \mathbf{V} = 0, \tag{2.62}$$

$$\frac{\partial \mathbf{V}}{\partial t} + (\mathbf{V} \cdot \nabla)\mathbf{V} = -\frac{1}{\rho}\nabla p + \nu \nabla^2 \mathbf{V}, \tag{2.63}$$

where $\mathbf{V}$ is the flow velocity vector, p is the static pressure, ρ the fluid density and ν the kinematic viscosity. The fluid stress tensor (Batchelor 1967),

$$\sigma_{ij} = -p\,\delta_{ij} + 2\mu e_{ij}, \tag{2.64}$$

used in the derivation of (2.63), is also *directly* useful for the purposes of this book: its components on the surface of a body in contact with the fluid determine the forces on the body; μ is the dynamic viscosity coefficient, and e_{ij} are the components of strain in the fluid. In cylindrical coordinates, for example, where $i, j = (r, \theta, x)$ and $\mathbf{V} = \{V_r, V_\theta, V_x\}^{\mathrm{T}}$, the components of $e_{ij}(= e_{ji})$ are

$$e_{xx} = \frac{\partial V_x}{\partial x}, \qquad e_{rr} = \frac{\partial V_r}{\partial r}, \qquad e_{\theta\theta} = \frac{1}{r}\frac{\partial V_\theta}{\partial \theta} + \frac{V_r}{r},$$
$$e_{r\theta} = \frac{1}{2}\left[r\frac{\partial}{\partial r}\left(\frac{V_\theta}{r}\right) + \frac{1}{r}\frac{\partial V_r}{\partial \theta}\right], \quad e_{\theta x} = \frac{1}{2}\left[\frac{1}{r}\frac{\partial V_x}{\partial \theta} + \frac{\partial V_\theta}{\partial x}\right], \quad e_{xr} = \frac{1}{2}\left[\frac{\partial V_r}{\partial x} + \frac{\partial V_x}{\partial r}\right]. \tag{2.65}$$

†Non-Newtonian fluids are nevertheless in the majority, in the process industries and biological systems, for instance. Polymer melts, lubricants, paints, and fluids involved in synthetic-fibre-, plastics- and food-processing are generally non-Newtonian, rheological fluids (Barnes *et al*. 1989).

Equations (2.62) and (2.63) together with appropriate boundary conditions, including equations matching the motion of a moving boundary (which could be part of the structure of interest), should in principle be sufficient to solve problems involving incompressible fluids. Similarly for compressible fluids, but the equations in this case are more complex and will not be presented here. Possible boundary conditions for a body surface moving with velocity $\mathbf{v}_w$ in the fluid are

$$\mathbf{V} \cdot \mathbf{n} = \mathbf{v}_w \cdot \mathbf{n} \qquad \text{and} \qquad \mathbf{V} \times \mathbf{n} = \mathbf{v}_w \times \mathbf{n}, \tag{2.66}$$

the first matching the normal components of fluid and solid-surface velocities, and the second being a form of the no-slip boundary condition, matching fluid and body velocities parallel to the surface; $\mathbf{n}$ is the unit normal to the surface.

By 'solution' of the fluid equations we mean the determination of the velocity and pressure fields, $\mathbf{V}$ and p. For fluid–structure interaction problems in which the forces induced by the fluid on the structure are the only concern, most of the information on $\mathbf{V}$ and p is 'thrown away'. This is because the forces on the structure may be determined by the pressure and viscous stresses *on the body surface*, cf. equations (2.64) and (2.65). This allows for approximate treatment of some classes of problems, which will be discussed in what follows. Indeed, the rest of this preamble will introduce, in general terms, some of the broad classes of admissible simplifications and hopefully guide the reader towards other ones.

The topic of *turbulent flows* [subsection (f)] is treated at considerably greater length than the other classes of flows. The reasons for this anomaly are that turbulence is more complex and generally less well remembered than the rest, at least by those not in constant touch with it. Nevertheless, the concepts and some of the relations to be recalled will be needed later on, e.g. in treating turbulence-induced vibrations of pipes and cylinders in axial flow; see Chapters 8 and 9 in Volume 2.

(a) High Reynolds number flows; ideal flow theory

If U is a characteristic flow velocity (e.g. a mean flow velocity in the system) and D a characteristic dimension, the Reynolds number is $\text{Re} = UD/\nu$. If equation (2.63) is written in dimensionless form, the last term is divided by Re; hence, for sufficiently high Re this term is negligible, and the Navier–Stokes equations reduce to the so-called Euler equations. Thus, away from any solid boundaries, the fluid is considered to be essentially inviscid. Close to a boundary, in the boundary layer, the effects of viscosity are predominant, but they may be treated separately. In such cases, precluding situations of large-scale turbulence and separated flow regions, the pressure field is determined as if the flow were inviscid and then the shear stresses on the body are determined by boundary layer theory or via empirical information.[†] This is the treatment adopted for slender cylindrical structures in axial flow in Chapters 8 and 9. Strictly, this approach constitutes but a first approximation; in general, the boundary-layer and inviscid-flow calculations should be matched iteratively.

For sufficiently high Re, the flow becomes turbulent and, if the effects of turbulence cannot be ignored, this introduces new complexity [see subsection (f)].

[†]The key idea making this possible is that of a constant pressure across the boundary layer.

(b) Potential flow theory

Many interesting inviscid flows (e.g. a uniform flow approaching a body) are initially irrotational, i.e. the vorticity, $\boldsymbol{\omega}$, is everywhere zero: $\boldsymbol{\omega} = \nabla \times \mathbf{V} = \mathbf{0}$. Hence, by Kelvin's theorem, such flows remain irrotational;[†] the flow is then referred to as potential flow and is associated with the velocity potential, ϕ, where $\mathbf{V} = \nabla\phi$. Euler's equations in this case simplify to the well known unsteady Bernoulli, or Bernoulli–Lagrange, equation

$$\frac{\partial \phi}{\partial t} + \tfrac{1}{2}V^2 + \frac{p}{\rho} = 0, \tag{2.67a}$$

where p is measured relative to the stagnation pressure of the free stream.[‡] This form of the equation applies if there are no body forces. If there are, for example due to gravity, the following form may be more useful:

$$\frac{\partial \phi}{\partial t} + \tfrac{1}{2}V^2 + \frac{p}{\rho} + gz = 0, \tag{2.67b}$$

where z is the vertical height. There exists a highly developed mathematical treatment of potential flow — see, e.g. Lamb (1957), Streeter (1948), Milne-Thomson (1949, 1958), Karamcheti (1966), Batchelor (1967).

(c) Very low Reynolds number flows

In this case, when $\mathrm{Re} \to 0$, inertial effects become negligible, and the Navier–Stokes equations reduce to the equations of creeping flow,

$$\nabla p = \mu \nabla^2 \mathbf{V}. \tag{2.68}$$

A number of well known solutions exist, e.g. for the plane Couette and Poiseuille flows, classical lubrication theory (Lamb 1957), Stokes flow past a sphere and constant pressure-gradient laminar flow through pipes; but, surprisingly perhaps, not for low-Re two-dimensional cross-flow over a cylinder (Stokes' paradox).

(d) Linearized flows

In some problems there is one dominant steady flow-velocity component, while all others are perturbations thereof, say induced by structural motion, e.g. $\mathbf{V} = U\mathbf{i} + \mathbf{v}$, where $\|\mathbf{v}\| \ll U$; $\mathbf{i}$ is the unit vector in the x-direction. In such cases, the Navier–Stokes equations may be linearized and simplified considerably. Thus, if U is steady, i.e. not time-dependent, and spatially uniform, the Navier–Stokes equations reduce to

$$\frac{\partial \mathbf{v}}{\partial t} + U\frac{\partial \mathbf{v}}{\partial x} = -\frac{1}{\rho}\nabla p + \nu \nabla^2 \mathbf{v}. \tag{2.69}$$

[†]Interestingly, this is not so if there is a density gradient to the fluid!

[‡]Thus, the integration constant that would otherwise appear on the right-hand side reduces to zero. This constant, $C(t)$, is generally a function of time if, unusually, the hydrostatic pressure varies with time.

In other cases, e.g. when fluid motion is entirely caused by small-amplitude oscillatory motion of a structure, all components of **V** may be small, and (2.69) is further simplified to

$$\frac{\partial \mathbf{v}}{\partial t} = -\frac{1}{\rho}\nabla p + \nu\nabla^2 \mathbf{v}. \tag{2.70}$$

Because there is no mean flow velocity in this case, the Reynolds number as such does not exist. Hence, to decide whether viscous effects are important or not, the *'oscillatory Reynolds number'* is used instead. For a circular cylinder of diameter D, this may be defined as $\beta = |\dot{A}|D/\nu$, where $|\dot{A}|$ is the amplitude of the oscillatory velocity of the body. Further, denoting the amplitude of motion by ϵD, $\epsilon \ll 1$, and the oscillation frequency by Ω, one obtains $|\dot{A}| = \Omega\epsilon D$ and hence $\beta = \Omega\epsilon D^2/\nu$, from which it is obvious that this is a modified Stokes number. Clearly, if β is sufficiently large, then viscous effects become unimportant, and the approximation

$$\frac{\partial \mathbf{v}}{\partial t} = -\frac{1}{\rho}\nabla p \tag{2.71}$$

may be used (see, Section 2.2.2 and Chapter 11). This may be combined with the continuity equation to give

$$\nabla^2 p = 0, \tag{2.72}$$

the Laplace equation. In terms of the velocity potential, ϕ, the continuity equation and equation (2.71) may be written as

$$\nabla^2 \phi = 0 \tag{2.73a}$$

and

$$\frac{\partial \phi}{\partial t} = -\frac{p}{\rho}. \tag{2.73b}$$

(e) Slender-body theory

A particular class of linearized flows pertains to slender bodies, i.e. bodies of small cross-sectional dimensions as compared to their length [e.g. for a body of revolution of radius $R(x)$, if $R(x) \ll L$] and no abrupt changes of cross-section ($\mathrm{d}R/\mathrm{d}x \ll 1$), with the flow being irrotational and along the long axis of the body or at a small angle to that axis [Figure 2.5(a)]. Let the body be defined by

$$F(r, \theta, x) = r - R(x) = 0. \tag{2.74}$$

The flow field may be expressed as

$$\mathbf{V} = \mathbf{V}_\infty + \nabla\phi, \tag{2.75}$$

where ϕ is associated with the perturbations to the flow associated with the presence of the body and satisfies

$$\nabla^2 \phi = 0 \tag{2.76}$$

and the boundary conditions

$$(\mathbf{V}_\infty + \nabla\phi)\cdot\nabla F = 0 \qquad \text{on} \qquad F(r, \theta, x) = 0 \tag{2.77a}$$

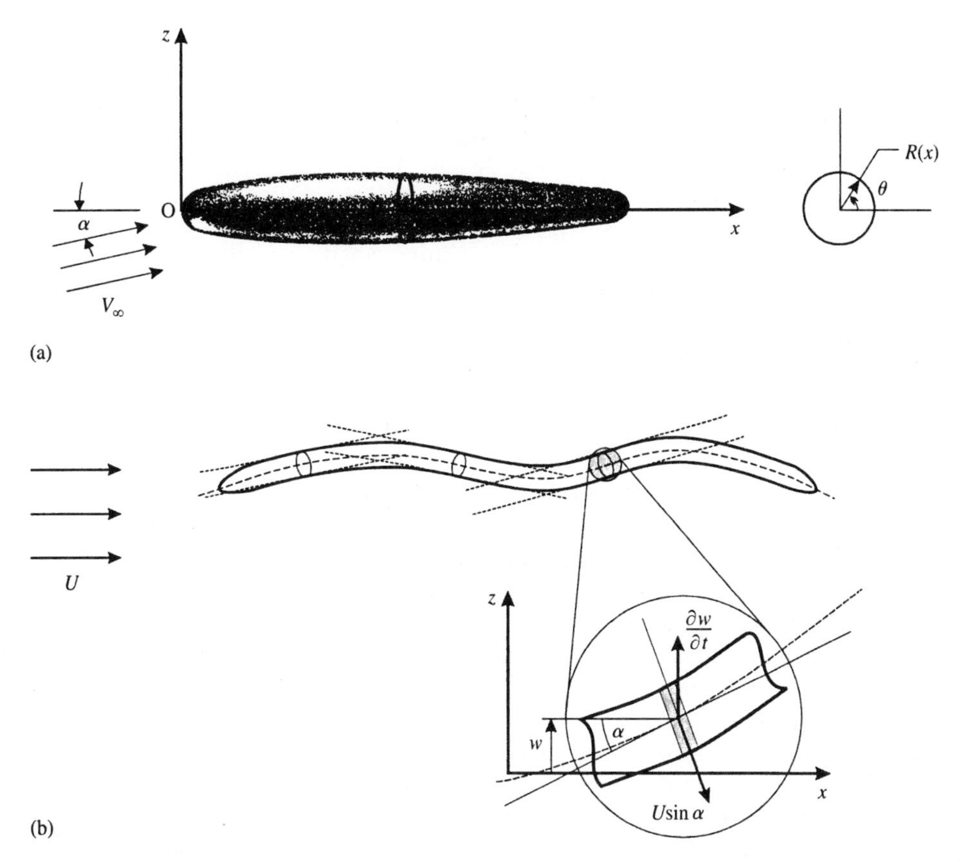

Figure 2.5 (a) A slender body of revolution in uniform flow at a small angle of attack, α. (b) A flexible slender body performing lateral oscillations of long wavelength, such that each segment may be considered to be part of an infinitely long cylinder; $\alpha = \tan^{-1}(\partial w/\partial x)$.

and

$$\nabla\phi = 0 \qquad \text{at infinity.} \tag{2.77b}$$

If the angle of attack is α, then in the (r, θ, x)-frame equations (2.77a,b) lead to (Karamcheti 1966)

$$(V_\infty \sin\alpha \sin\theta + u_r)\frac{\partial F}{\partial r} + (V_\infty \cos\alpha + u_x)\frac{\partial F}{\partial x} = 0 \qquad \text{on} \qquad F(r, \theta, x) = 0 \tag{2.78}$$

and

$$u_r = u_\theta = u_x = 0 \qquad \text{at infinity,} \tag{2.79}$$

in which u_r, u_θ and u_x are the components of $\nabla\phi$, and $\partial F/\partial r = 1$, $\partial F/\partial x = -\mathrm{d}R/\mathrm{d}x$ from (2.74). Hence, the surface condition (2.78) becomes

$$\left.\frac{\partial\phi}{\partial r}\right|_{r=R} = (V_\infty \cos\alpha + u_x)\frac{\mathrm{d}R}{\mathrm{d}x} - V_\infty \sin\alpha \sin\theta, \qquad 0 \le x \le L. \tag{2.80}$$

The essence of slender-body theory is to take advantage of the linearity of the problem and to express it as the superposition of the following two problems: (i) the axisymmetric flow past the body of revolution with flow velocity $V_\infty \cos\alpha$, and (ii) the cross-flow around the body with flow velocity $V_\infty \sin\alpha$ (Ward 1955; Karamcheti 1966). Thus, defining $\phi = \phi_1 + \phi_2$ and $u_x = u_{x1} + u_{x2}$, equation (2.80) may be re-written as

$$\left.\frac{\partial \phi_1}{\partial r}\right|_{r=R} = (U + u_{x1})\frac{\mathrm{d}R}{\mathrm{d}x} \simeq U\frac{\mathrm{d}R}{\mathrm{d}x}, \tag{2.81a}$$

$$\left.\frac{\partial \phi_2}{\partial r}\right|_{r=R} = u_{x2}\frac{\mathrm{d}R}{\mathrm{d}x} - W\sin\theta, \tag{2.81b}$$

where

$$U = V_\infty \cos\alpha, \qquad W = V_\infty \sin\alpha. \tag{2.82}$$

The solution to (2.81a) is usually obtained by representing the body through a distribution of singularities (e.g. sources and sinks) along the centreline, while the solution to (2.81b) may be obtained via standard potential-flow analysis (Streeter 1948; Milne-Thomson 1949; Karamcheti 1966).

Consider next a very slender cylindrical body for which $\mathrm{d}R/\mathrm{d}x \simeq 0$, or exactly 0, except near the extremities [Figure 2.5(b)]. The body is subjected to an oscillatory lateral displacement $w(x, t)$ in the $\theta = \frac{1}{2}\pi$ plane. Then, according to slender-body theory, the flow can be regarded as compounded of (a) the steady flow around the stretched-straight body, which we shall ignore here [and hence (2.81a) also] since $\mathrm{d}R/\mathrm{d}x$ is nearly or exactly zero over most of the length of the body, and (b) the flow due to displacements $w(x, t)$ (Lighthill 1960). Hence, only the velocity component related to (2.81b) remains, namely $(\partial\phi_2/\partial r)|_{r=R} \simeq -W$. The lateral velocity of the fluid relative to the moving body is made up of (i) the component of U normal to the inclined body, equal to $-U\sin\alpha$, where $\alpha = \tan^{-1}(\partial w/\partial x)$, and (ii) the lateral velocity of the body, $\partial w/\partial t$, reversed, if at that instant the body is moving upwards as in the inset of Figure 2.5(b). Therefore, for sufficiently small α, one may write

$$\left.\frac{\partial \phi}{\partial r}\right|_{r=R} \equiv V(x,t) = \frac{\partial w}{\partial t} + U\frac{\partial w}{\partial x}, \tag{2.83}$$

on the implicit assumption that, locally, the body shape differs little from that of a long (infinite) cylinder C_x of the same cross-section all the way along. Thus, according to the slender-body approximation, this lateral flow near any point of the cylinder is identical with the two-dimensional potential flow that would result from the motion of C_x through fluid at rest, with velocity $V(x, t)$. Lighthill (1960) then goes on to obtain the rate of change of lateral momentum of the fluid passing over the flexible body,

$$L(x,t) = -\rho\left(\frac{\partial}{\partial t} + U\frac{\partial}{\partial x}\right)\{A(x)V(x,t)\}, \tag{2.84}$$

where $A(x)$ is the slowly varying (or constant) cross-sectional area along the length of the body. This equation is further discussed in Chapters 8 and 9, where the slender-body approach is used extensively.

(f) Turbulent flows

Due to a three-dimensional instability of laminar flow, the flow field becomes turbulent: the flow velocity and pressure are no longer steady but contain randomly fluctuating components. Two-dimensional disturbances in the laminar flow field eventually become three-dimensional, and this is soon followed by turbulence. The critical Reynolds number for the onset of turbulence is best stated in terms of the width of the flow† and depends on the shape of the laminar velocity profile; it is typically $\mathcal{O}(10^2)$ for profiles with inflection points and $\mathcal{O}(10^3)$ or more for profiles of single curvature. Thus, boundary layers in falling pressures are a good deal more stable than those that are suffering a pressure rise; similarly, jets and wakes are also very unstable.

When turbulence appears, as originally observed and described by Reynolds for pipe flow, the flow field may be expressed as $\mathbf{V} + \mathbf{v}$ and $P + p$, where the lower-case quantities represent the fluctuating components about the mean (with zero average) and $\mathbf{V}$ and P are the mean components; $\mathbf{V} = \{U_1, U_2, U_3\}^{\mathrm{T}}$ and $\mathbf{v} = \{u_1, u_2, u_3\}^{\mathrm{T}}$ in an $\{x_1, x_2, x_3\}$-frame. Substitution into the Navier–Stokes equations and averaging yields

$$\frac{\partial U_i}{\partial t} + U_j \frac{\partial U_i}{\partial x_j} = -\frac{1}{\rho}\frac{\partial P}{\partial x_i} + \frac{\partial}{\partial x_j}\left(\nu_m \frac{\partial U_i}{\partial x_j} - \overline{u_i u_j}\right), \qquad i, j = 1, 2, 3, \tag{2.85}$$

where the indicial notation is utilized, in which repeated indices imply summation; e.g. $U_j(\partial U_i/\partial x_j) = \sum_{j=1}^{3} [U_j(\partial U_i/\partial x_j)]$. The new term $-\overline{u_i u_j}$ is the correlation of u_i and u_j, obtained by multiplying the two, integrating over a long time (appropriate to the flow under investigation), and then dividing by the time interval. The quantity $-\rho\overline{u_i u_j}$ represents additional normal and shear stresses due to additional momentum transfer associated with the velocity fluctuations,‡ the so-called *Reynolds stresses*. Thus, in a simple two-dimensional shear flow predominantly in the x_1-direction, the viscous shearing stress $\mu(\partial U_1/\partial x_2)$ is increased by $-\rho\overline{u_1 u_2}$, which has the same sign as $\partial U_1/\partial x_2$ and is sometimes written as $\mu_t(\partial U_1/\partial x_2)$, where the subscript t is for 'turbulence'; $\nu_t = \mu_t/\rho$ is the so-called kinematic *eddy viscosity*. To differentiate the quantities associated with viscous stresses from those related to turbulence, or equivalently the quantities associated with velocity fluctuations at the molecular (Brownian) scale from the turbulent ones, the subscript m (for 'molecular') is introduced, as in ν_m in equation (2.85); ν_m here is the same as ν in equation (2.63).

The Reynolds stresses are generally much larger than the viscous ones, except near walls, in the viscous sublayer (Hinze 1975); on the wall itself, all turbulent fluctuations vanish. One of the central problems of turbulent flows is the derivation of satisfactory relations for Reynolds stresses in terms of the mean flow field (Townsend 1961).

The spatial structure of a turbulent flow may be described statistically by correlation functions or by spectra. The general space-time correlation function between, say, u_i at point $\mathbf{x}$ and u_j at point $\mathbf{x} + \mathbf{r}$ is defined by

$$R_{ij}(\mathbf{x}, \mathbf{r}, \tau) = \overline{u_i(\mathbf{x}, t) u_j(\mathbf{x} + \mathbf{r}, t + \tau)}, \tag{2.86a}$$

† The width of a jet or a wake, or the thickness of a boundary layer.

‡ This is an essential characteristic of turbulence. As noted by Townsend (1961), 'a sharp increase in friction, or in heat and mass transfer is frequently used to determine the onset of turbulent motion if direct observation of the fluctuations is inconvenient'.

where τ is a time delay in the measurement of u_i and u_j. For homogeneous turbulence, R_{ij} depends only on the separation between the two points $r = \|\mathbf{r}\|$. For a uniform flow field *in a given direction*, e.g. for fully developed turbulent flow in a pipe, R_{ij} depends on the separation r, but also on the direction, hence on $\mathbf{r}$. In this latter case,

$$R_{ij}(\mathbf{r}, \tau) = \overline{u_i(0, t)u_j(\mathbf{r}, t + \tau)}. \tag{2.86b}$$

Keeping with this latter form, one distinguishes *spatial correlations*,

$$R_{ij}(\mathbf{r}, 0) = \overline{u_i(0, t)u_j(\mathbf{r}, t)}, \tag{2.86c}$$

in which u_i and u_j are associated with different points in space, but the same time; and *temporal correlations*, involving the same point in space and a time delay τ,

$$R_{ij}(\mathbf{r}, \tau) = \overline{u_i(\mathbf{r}, t)u_j(\mathbf{r}, t + \tau)}, \tag{2.86d}$$

autocorrelations for $i = j$, and cross-correlations for $i \neq j$.

The spatial correlation, when plotted versus a particular component of $\mathbf{r}$, indicates the distance over which motion at one point significantly affects that at another. It may be used to assign a *length scale* to the turbulence, defined as $L_k = (1/v^2)\int_0^\infty R_{ij}(r_k, 0)\,\mathrm{d}r_k$, where v^2 is a normalizing factor, e.g. $v^2 = \overline{u_i^2}$, and r_k is a particular component of $\mathbf{r} = \{r_1, r_2, r_3\}^{\mathrm{T}}$ in the $\{x_1, x_2, x_3\}$-frame used here.[†] For flow in the x-direction, e.g. for fully developed pipe flow, the integral (or macro-) scale, associated mainly with the largest, most energetic eddies, is defined by

$$L_1 = \frac{\displaystyle\int_0^\infty R_{11}(r_1, 0)\,\mathrm{d}r_1}{\overline{u_1^2}}. \tag{2.87}$$

For points r_2 apart, in the cross-stream direction, L_2 may be defined in a similar way, with r_2 taking the place of r_1; in terms of the normalized form of the correlation function (the coherence), $\overline{R}_{11}$, L_2 is given by

$$L_2 = \int_0^\infty \overline{R}_{11}(r_2, 0)\,\mathrm{d}r_2, \qquad \overline{R}_{11}(r_2, 0) = \frac{\overline{u_1(0)u_1(r_2)}}{[\overline{u_1^2(0)}]^{1/2}[\overline{u_1^2(r_2)}]^{1/2}}. \tag{2.88}$$

The correlation in the streamwise (longitudinal) direction generally decays from 1 at $r = 0$ to zero at sufficiently large r, smoothly and without change in sign (Figure 2.6); whereas the cross-stream (lateral) correlation generally has a negative part for intermediate r, before it too decays to zero for large enough r (Tritton 1988).

The temporal correlations are functions of the time delay τ for measurements at the same point; they give a measure of the *time scale* of turbulence. For small times, or over small enough distances, turbulence may be considered to be advected past the point of observation without change in structure. This is *Taylor's hypothesis*, as a result of which a temporal correlation is equal to the corresponding spatial correlation for $\tau = r_1/U_1$; thus, according to this hypothesis, the eddies of the turbulence are convected without change over a sufficiently short distance, r, as further discussed in Chapter 9.

[†] Alternative definitions, for experimental convenience, are sometimes utilized; e.g. by defining the scale as the distance to where R_{ij} plotted versus r_k becomes negative, or to where it is reduced to 1/e.

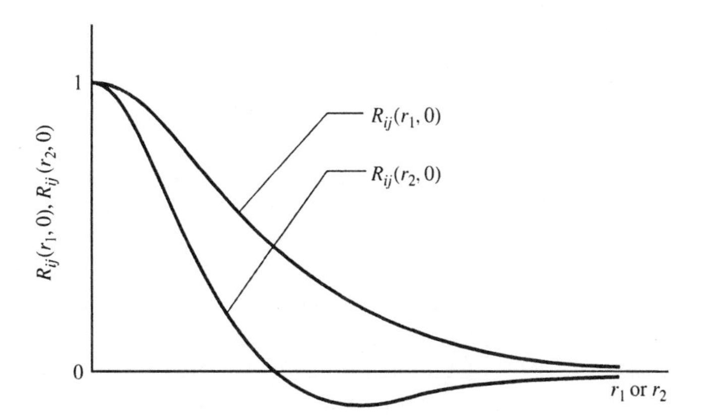

Figure 2.6 Typical form of correlation functions: $R_{ij}(r_1, 0)$ for points i and j separated by a variable r_1 in the streamwise direction; and $R_{ij}(r_2, 0)$ for points separated by r_2 in the cross-stream direction — following Tritton (1988).

The Fourier transform of the autocorrelation function gives the frequency spectrum of the turbulence at a given point,

$$F_{ij}(\omega) = \frac{1}{2\pi} \int_{-\infty}^{\infty} R_{ij}(\mathbf{r}, \tau)\, \mathrm{e}^{-\mathrm{i}\omega\tau}\, \mathrm{d}\tau, \tag{2.89}$$

where ω is the radian frequency. The $F_{ij}(\omega)$ give a measure of the energy spectrum of the turbulence. Hence, a peak in the spectrum denotes a dominant frequency, which could excite an underlying structure, for instance. The energy spectrum is often described in terms of the wavenumber $\mathbf{k}$, generally a 3-D vector, $\mathbf{k} = \{k_1, k_2, k_3\}^{\mathrm{T}}$, with each $k_i = 1/2\pi\lambda_i$, λ_i being the wavelength of turbulent fluctuations associated with a frequency ω_i. Thus, the equivalent of (2.89) in terms of $\mathbf{k}$ is

$$F_{ij}(\mathbf{k}) = \frac{1}{(2\pi)^3} \iiint_{-\infty}^{\infty} R_{ij}(\mathbf{r})\, \mathrm{e}^{-\mathrm{i}\mathbf{k}\cdot\mathbf{r}}\, \mathrm{d}^3\mathbf{r}. \tag{2.90}$$

This may be expressed as a function of a scalar variable by averaging it over all directions of $\mathbf{k}$; thus,

$$\Phi_{ij}(k) = \int F_{ij}(\mathbf{k})\, \mathrm{d}A(k), \tag{2.91}$$

where $k = \|\mathbf{k}\|$, and the integration is over the surface of a sphere of which $\mathrm{d}A$ is an element, so that $\Phi_{ij}(k)$ is the contribution to the energy tensor $\overline{u_i u_j}$ from wavenumbers whose magnitudes lie between k and $k + \mathrm{d}k$ (Batchelor 1960, Chapter III).

Another quantity of interest is the turbulence intensity, which may be defined by

$$\mathrm{Tu} = \left(\tfrac{2}{3}K\right)^{1/2}/U \tag{2.92}$$

for sensibly one-dimensional flow, where

$$K = \tfrac{1}{2}\overline{u_i u_i} \equiv \tfrac{1}{2}\left[\overline{u_1^2} + \overline{u_2^2} + \overline{u_3^2}\right] \tag{2.93}$$

is the turbulence kinetic energy per unit mass. In view of the foregoing, this may also be written as

$$K = \int_0^\infty E(k)\,\mathrm{d}k = \tfrac{1}{2}\Phi_{ii}(k), \tag{2.94}$$

in which $E(k)$ is the *energy spectrum function*, i.e. the density of contributions to the kinetic energy on the wavenumber magnitude axis (Batchelor 1960).

Some progress has been made in understanding the changing scales of turbulence, as measured by its spectra and expressed in terms of the scalar wavenumber k. The spectra at low k (large eddies) often retain something of the original unsteady laminar flow; but, with increasing k, there is a continual stretching of the eddies by the medium scales, which causes a transfer of turbulence energy to large k (small eddies) and also randomizes the orientation of the eddies so that turbulence becomes locally isotropic. If the Reynolds number is very large, the intermediate spectrum is inertial (i.e. it sensibly does not depend on viscosity), and it may be shown by dimensional analysis that the spectrum is proportional to $k^{-5/3}$. For the smallest eddies, where $k > \frac{1}{5}(\epsilon/\nu^3)^{1/2}$, the Kolmogoroff wavenumber, viscosity takes over and causes a decay of the cascading energy with dissipation rate ϵ to heat. This structure, as described in the foregoing, enables a dramatic assumption to be made, namely that away from walls, the Reynolds stresses are independent of ν_m. In this one respect, turbulent flow may often be easier to analyse than laminar flow.

In analysing the boundary layer near walls, the so-called *law of the wall* is often used. In this discussion, 2-D or axisymmetric boundary layers only are considered. Let U_1 be the streamwise flow velocity in the boundary layer and $x_2 = y$ the distance perpendicularly away from the wall. Then, near enough to the wall, $U_1 = U_1(\rho, \mu, U_\tau, y)$, where $U_\tau = (\tau_w/\rho)^{1/2}$ is the skin-friction velocity and τ_w is the shear stress at the wall; thus, U_1 is independent of outer parameters, such as the overall boundary-layer thickness, the free-stream velocity U, and the pressure gradient when not too large. Thus,

$$\frac{U_1}{U_\tau} = \mathcal{F}\left(\frac{yU_\tau}{\nu}\right), \tag{2.95}$$

which is the law of the wall. Rotta (1962) predicts the functional form of $\mathcal{F}$ by noting that changes in U_1 in most of the region outside the viscous sublayer are independent of μ, because the shear stress is almost entirely due to $-\rho\overline{u_1u_2}$ there. Dimensional analysis then leads to $(y/U_\tau)(\partial U_1/\partial y) = 1/\mathrm{K} \simeq 0.41$, a universal constant named after von Kármán. After integration, this gives

$$\frac{U_1}{U_\tau} = \frac{1}{\mathrm{K}} \ln\left(\frac{yU_\tau}{\nu}\right) + B, \tag{2.96}$$

where $B = 5.5$ for a smooth wall. This proof applies to rough walls, 'fully rough walls' (where μ is unimportant even near the wall), and ribletted walls for which there is a drag reduction. The only thing that changes is the value of B, which is lower for rough walls, increasingly with the roughness, and slightly higher for ribletted walls.

The law of the wall has been accepted for the purposes of CFD (Computational Fluid Dynamics), where it often becomes the inner boundary condition, but it must be noted that the corresponding law for turbulence intensity is not exactly true when comparing, say, boundary-layer flow and pipe flow; i.e. $\sqrt{\overline{u^2}}/U_\tau \neq \mathcal{F}(yU_\tau/\nu)$.

In some of the work to be presented later (e.g. in Chapter 7, Volume 2), particular forms of the foregoing for pipe flows — containing considerable empirical input — is utilized. Thus, for pipe flow, a friction factor, f, is sometimes defined via

$$U_\tau = \left(\frac{\tau_w}{\rho}\right)^{1/2} = \left(\frac{1}{8} f U^2\right)^{1/2}, \tag{2.97}$$

where U is the mean flow velocity; f is given empirically, for instance by the Colebrook equation,

$$\frac{1}{\sqrt{f}} = -2\,\log_{10}\left\{\frac{k_s/D}{3.7} + \frac{2.51}{\mathrm{Re}\sqrt{f}}\right\}, \tag{2.98}$$

where Re is the Reynolds number based on the diameter, D, and k_s/D is the relative roughness.

Reverting now to equation (2.85) for a more general analysis of turbulent flow, it is noted that $-\overline{u_i u_j}$ is often not measured, but modelled mathematically. For example, by means of Boussinesq's eddy viscosity concept, one may write

$$-\overline{u_i u_j} = \nu_t \left(\frac{\partial U_i}{\partial x_j} + \frac{\partial U_j}{\partial x_i}\right) - \frac{2}{3} K \delta_{ij}, \tag{2.99}$$

where K is as given by (2.93); ν_t is the *eddy viscosity* which, unlike ν_m (or $\nu = \nu_m$ in laminar flow), it is not a constant but is dependent on the flow field. The chosen form depends on the turbulence model adopted — see, for instance, Launder & Spalding (1972), Jones & Launder (1972), Launder & Sharma (1974), Rodi (1980), Lesieur (1990), So *et al*. (1991), Wilcox (1993).

Perhaps the simplest model is based on Prandtl's mixing-length hypothesis for 2-D or axisymmetric flows, in which

$$\nu_t = l^2 \left|\frac{\mathrm{d}U}{\mathrm{d}y}\right|, \tag{2.100}$$

where l is Prandtl's mixing length, $y = x_2$ is the coordinate measured away from the wall, and $U = U_1$ is the mean flow velocity.[†] In the case of smooth pipes, for instance, Nikuradse's measurements yield the following empirical expression (Schlichting 1960):

$$\frac{l}{R} = 0.14 - 0.08\left(1 - \frac{y}{R}\right)^2 - 0.06\left(1 - \frac{y}{R}\right)^4, \tag{2.101}$$

R being the pipe radius.

There are many other models, including so-called two-equation models, for turbulent flow (Wilcox 1993). One of the first and most popular was pioneered by Launder and Spalding. It is based on two scalar functions, already defined: $K = \frac{1}{2}\overline{u_i u_i}$, the average turbulence (kinetic) energy per unit mass; and ϵ, the rate of decay of turbulence energy per unit mass, which is also the rate of transfer of energy from the large eddies to smaller ones, and hence, in this latter capacity, it is independent of viscosity. In this so-called,

[†]Incidentally, this is the equation, with $l \propto y$, originally used by Prandtl to prove the law of the wall.

K–ε model,† $\nu_t \propto K^2/\epsilon$. Equations may be written for K and ϵ, namely

$$\rho\frac{\partial K}{\partial t} + \rho U_j\frac{\partial K}{\partial x_j} = \tau_{ij}\frac{\partial U_i}{\partial x_j} - \rho\epsilon + \frac{\partial}{\partial x_j}\left[\mu_m\frac{\partial K}{\partial x_j} - \tfrac{1}{2}\rho\overline{u_i u_i u_j} - \overline{pu_j}\right], \tag{2.102}$$

$$\rho\frac{\partial \epsilon}{\partial t} + \rho U_j\frac{\partial \epsilon}{\partial x_j} = 2\mu_m\left\{[\overline{u_{i,k}u_{j,k}} + \overline{u_{k,i}u_{k,j}}]\frac{\partial U_i}{\partial x_j} + \overline{u_k u_{i,j}}\frac{\partial^2 U_i}{\partial x_k \partial x_j}\right.$$
$$\left. + \overline{u_{i,k}u_{i,l}u_{k,l}} + \nu_m\overline{u_{i,kl}u_{i,kl}}\right\} + \frac{\partial}{\partial x_j}[\mu_m\overline{u_j u_{i,l}u_{i,l}} - 2\nu_m\overline{p_{,l}u_{j,l}}], \tag{2.103}$$

in which p is the fluctuating pressure and τ_{ij} the Reynolds stress tensor,

$$\tau_{ij} = 2\mu_t e_{ij} - \tfrac{2}{3}\rho K\delta_{ij}, \tag{2.104}$$

with e_{ij} being the mean strain-rate tensor [cf. relations (2.65)]; $u_{i,k} \equiv \partial u_i/\partial x_k$, $p_{,l} \equiv \partial p/\partial x_l$ and so on. Since the correlations in (2.102) and (2.103) are effectively impossible to measure, these very complex equations have been simplified by various approximations.

The 'standard form' of the K–ϵ model is expressed in terms of the following equations and relationships (Wilcox 1993):

Eddy viscosity

$$\mu_t = \rho C_\mu K^2/\epsilon; \tag{2.105a}$$

Turbulence kinetic energy

$$\rho\frac{\partial K}{\partial t} + \rho U_j\frac{\partial K}{\partial x_j} = \tau_{ij}\frac{\partial U_i}{\partial x_j} - \rho\epsilon + \frac{\partial}{\partial x_j}\left[\left(\mu_m + \frac{\mu_t}{\sigma_K}\right)\frac{\partial K}{\partial x_j}\right]; \tag{2.105b}$$

Dissipation rate

$$\rho\frac{\partial \epsilon}{\partial t} + \rho U_j\frac{\partial \epsilon}{\partial x_j} = C_{\epsilon 1}\frac{\epsilon}{K}\tau_{ij}\frac{\partial U_i}{\partial x_j} - C_{\epsilon 2}\rho\frac{\epsilon^2}{K} + \frac{\partial}{\partial x_j}\left[\left(\mu_m + \frac{\mu_t}{\sigma_\epsilon}\right)\frac{\partial \epsilon}{\partial x_j}\right]; \tag{2.105c}$$

Closure coefficients

$$C_{\epsilon 1} = 1.44, \quad C_{\epsilon 2} = 1.92, \quad C_\mu = 0.09, \quad \sigma_K = 1.0, \quad \sigma_\epsilon = 1.3; \tag{2.105d}$$

Auxiliary relations

$$\omega = \epsilon/(C_\mu K) \quad \text{and} \quad l = C_\mu k^{3/2}/\epsilon, \tag{2.105e}$$

ω being the so-called specific dissipation rate and l the turbulence length scale. Thus, the K and ϵ equations contain five empirical constants which have been inferred from standard measurements.

It has been found necessary to adjust the closure coefficients somewhat to agree with different classes of measurements, but in the hands of a skilled practitioner this approach is usually much better than integral methods. [In integral methods, equations for entrainment, momentum, mechanical energy and so on are written integrated-up across the flow at any

†This is usually written as the k–ϵ model, but an upper case K is used here to avoid confusion with the wavenumber k.

downstream station, and a selected group of these integral equations is solved, often by a relatively simple numerical method (White 1974; Schetz 1993).]

Presuming now that ν_t has been determined, substitution of (2.99) into (2.85) gives

$$\frac{\partial U_i}{\partial t} + U_j \frac{\partial U_i}{\partial x_j} = -\frac{1}{\rho}\frac{\partial P_t}{\partial x_i} + (\nu_m + \nu_t)\frac{\partial^2 U_i}{\partial x_j^2} + \frac{\partial \nu_t}{\partial x_j}\left(\frac{\partial V_i}{\partial x_j} + \frac{\partial V_j}{\partial x_i}\right), \tag{2.106}$$

where $P_t = P + \frac{2}{3}\rho K$ is the turbulent 'total pressure'. Equation (2.106) may be written in the usual, but perhaps less convenient, form

$$\frac{\partial \mathbf{V}}{\partial t} + (\mathbf{V}\cdot\nabla)\mathbf{V} = -\frac{1}{\rho}\nabla P_t + (\nu_m + \nu_t)\nabla^2\mathbf{V} + (\nabla\nu_t\cdot\nabla)\mathbf{V} + (\nabla\mathbf{V})\cdot\nabla\nu_t, \tag{2.107}$$

where $\nabla\mathbf{V}$ is the so-called dyad, a vector (Wills 1958; Tai 1992). Examples of the use of these equations and/or the ideas summarized in this subsection are presented in Chapters 7–10.

(g) Empirical formulations

As intimated in the foregoing, mixed analytical-empirical formulations of the fluid-dynamic forces may be the only convenient way to analyse some fluid–structure interaction problems (e.g. provided that there is no large-scale flow separation, by analysing the flow as if it were inviscid, thereby obtaining the pressure-related forces, and adding empirical expressions for the viscous stresses acting on the body surface). Indeed, in many cases involving complex flows, e.g. cross-flow of heat-exchanger tube arrays, the very foundation of the theoretical model may be empirical or quasi-empirical.

In analysing the empirical (experimental) data, it is convenient to express the unsteady fluid loading, $F(t)$, acting on an oscillating structure in terms of components in phase with acceleration, velocity and displacement of the structure, locally linearized; thus, for a one-degree-of-freedom system,

$$F(t) = -m'\ddot{z} - c'\dot{z} - k'z. \tag{2.108}$$

When this is substituted in the equation of motion of the structure, $m\ddot{z} + c\dot{z} + kz = F(t)$, one obtains

$$(m + m')\ddot{z} + (c + c')\dot{z} + (k + k')z = 0, \tag{2.109}$$

hence the appellation of m', c' and k' as the added mass, added damping and added stiffness [e.g. Naudascher & Rockwell (1994, Chapter 3)].

For example, for a long cylinder of cross-sectional area A and length L, oscillating in unconfined dense fluid of density ρ, the added mass per unit length is $m^* = m'/L = \rho A$, if end effects are negligible. If the cylinder is in a conduit of complex geometry, m' may be determined analytically, numerically or experimentally, and the added mass per unit length expressed by

$$m^* = \frac{m'}{L} = C_m \rho A. \tag{2.110}$$

In general, C_m will be a function of geometry, viscosity and frequency (hence of the oscillatory Reynolds number), amplitude, and other factors as discussed in Sections 2.2.2 and 2.2.3 and by others (Chen 1987; Gibert 1988; Naudascher & Rockwell 1994). In many cases the approximation is made that the added mass in quiescent (stagnant) and flowing fluid is the same, although this is not rigorous. Such an approximation is definitely shaky if the flow is grossly unsteady or accelerating. Thus, in the extreme case of oscillatory flow, $C_m = 2$ instead of 1 [see, e.g. Sarpkaya & Isaacson (1981)], as a result of induced buoyancy — i.e. because of the presence of a pressure gradient.[†]

The added damping may similarly be expressed in terms of a damping coefficient C_d, which may be defined in different ways, e.g.

$$c^* = \frac{c'}{L} = C_d \omega \rho A \qquad \text{or} \qquad C'_d \rho U D, \tag{2.111}$$

for oscillations in quiescent or flowing fluid; other definitions are possible.

Added stiffness may arise due to buoyancy, asymmetry[‡] or proximity to other solid boundaries. For example, if a body lies close to a wall or a free surface and it is subjected to flow, there will be a fluid force acting on it, because the flow field is nonuniform. If the body is displaced towards or away from the aforementioned boundary by Δz, this force will change by ΔF. The quantity $\Delta F/\Delta z$ is the so-called added stiffness, and it depends purely on displacement and not on velocity or acceleration. Hence, one may similarly define a stiffness force coefficient by

$$k^* = \frac{k'}{L} = C_k \omega^2 \rho A \qquad \text{or} \qquad C'_k \tfrac{1}{2} \rho U^2. \tag{2.112}$$

In equation (2.109), m, c and k are devoid of fluid effects; i.e. in an experimental system they should ideally be measured in vacuum. Also, unless there exists a mathematical model the linearization of which yields (2.108), m', c' and k' must be determined experimentally, e.g. by conducting experiments first in vacuum (practically in still air) and then in fluid (say, in water) or fluid flow; it is noted that although the c' coefficient of the fluid force determined thereby is easily separable from the rest, since the velocity-dependent component is in quadrature (90° out of phase) with displacement, more than one experiment would be necessary, and in some cases it is virtually impossible, to separate m' and k' since they are 180° out of phase with each other (hence, they differ only in sign).

The rest of Section 2.2 is devoted to the presentation of two simple but representative analyses — in abridged form — which illustrate the use of the foregoing and also introduce some useful nomenclature for the chapters that follow. In both cases, the mean flow is zero. Problems involving a mean axial flow, the prime concern of this book, are dealt with in the other chapters.

2.2.2 Loading on coaxial shells filled with quiescent fluid

Consider two long, thin coaxial shells, with the annular space between them filled with quiescent, inviscid, dense fluid (e.g. water), while within the inner shell and outside the outer one the fluid is of much smaller density (e.g. air) or a vacuum; see Figure 2.7. The

[†]One way of looking at the difference between a cylinder oscillating in quiescent fluid ($C_m = 1$) and a cylinder in oscillatory flow ($C_m = 2$) is that in the former case the flow velocity at infinity tends to zero, whereas in the latter it has the full amplitude of the oscillation: clearly two very different flow fields.

[‡]For example, in the case of an iced conductor in uniform wind, rotation of the noncircular-section conductor clearly results in a change in the static forces experienced by it; see, e.g. Den Hartog (1956).

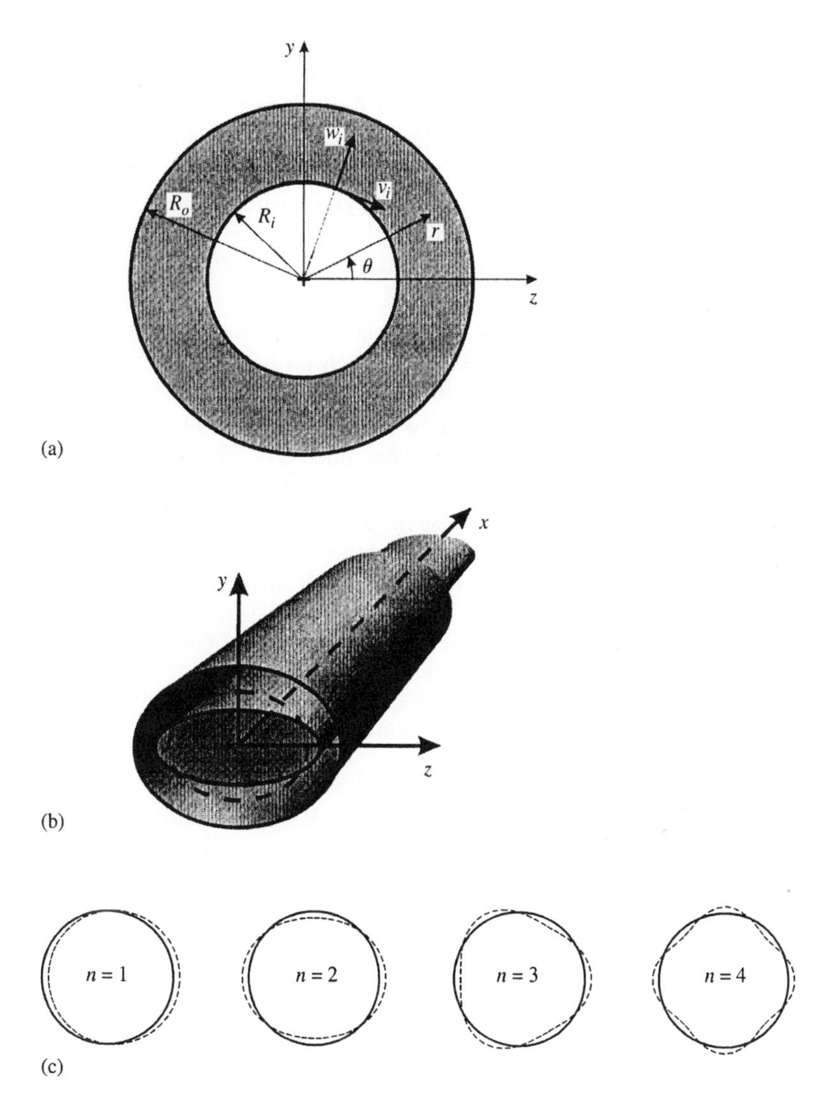

Figure 2.7 (a) Cross-sectional view of two coaxial thin shells at rest, with the annulus filled with a dense fluid; v_i and w_i are the displacements of the inner shell in the circumferential and radial direction, respectively. (b) Three-dimensional view of the shell instantaneously deformed in the $n = 2$ circumferential mode, with little axial variation (either because the shell is long and the mode of axial deformation is small, or because idealized 2-D deformation has been assumed). (c) Definition of the $n = 1$–4 circumferential modes.

shells are free to vibrate in a low axial mode number (e.g. in the first, beam-like mode), so that gradients of displacements in the longitudinal direction are negligible, as compared to the transverse directions [i.e. in the plane of Figure 2.7(a)]. Alternatively, one could assume that the mode of oscillation is purely two-dimensional, as shown in the example of Figure 2.7(b). Hence, the displacement of the mean surface of the shell, generally of the form $\{u, v, w\}^{\mathrm{T}}$, with u, v and w being, respectively, the axial, circumferential and radial components, in this case simplifies to $\{v, w\}^{\mathrm{T}}$.

The vibration of the shells induces oscillatory flow in the fluid, and the task is to determine the unsteady fluid loading on the shells, resulting thereby. By further assuming that the amplitude of shell vibrations is small, it follows that all fluid velocities are also small and hence governed by equations (2.71)–(2.73a,b). Because the shells are very long, end effects are negligible; also, because the mode of deformation is such that motion is essentially two-dimensional as outlined in the previous paragraph, motion-induced flow variations in the direction of the long axis of the shells are negligible. Hence, in this case, $\partial p/\partial x = 0$ and $\phi = \phi(r, \theta, t)$; the analysis is therefore carried out in the plane of Figure 2.7(a).

The solution to the fluid flow resulting from this motion may be obtained via equations (2.73a,b), although (2.72) could be used equally well (Gibert 1988). The eigenmodes of each shell are of the form $\{v, w\}^{\mathrm{T}} = \{v_n \sin n\theta,\ w_n \cos n\theta\}^{\mathrm{T}}$, where the relation between v_n and w_n is dependent on the shell equations used [e.g. Flügge (1960)], which need not concern us here; n is the circumferential wavenumber. The cross-sectional deformation for $n = 1$–4 is shown in Figure 2.7(c).

Consider first the case where the outer shell is replaced by a rigid immobile cylinder of inner radius R_o, and let v_i and w_i be the displacement components of the inner shell. Furthermore, consider oscillation in the nth mode, such that

$$v_i(\theta, t) = v_{ni}(t) \sin n\theta, \qquad w_i(\theta, t) = w_{ni}(t) \cos n\theta, \tag{2.113}$$

in which it is understood that v_{ni} and w_{ni} are harmonic functions, e.g. $w_{ni}(t) = \overline{w}_{ni} \exp(\mathrm{i}\Omega t)$. The corresponding velocity potential is

$$\phi = \overline{\phi}(r, \theta)\, \mathrm{e}^{\mathrm{i}\Omega t}. \tag{2.114}$$

The boundary conditions for the fluid are

$$V_r\Big|_{R_i} = \frac{\partial \phi}{\partial r}\Big|_{R_i} = \frac{\partial w_i}{\partial t} = \frac{\mathrm{d}w_{ni}}{\mathrm{d}t} \cos n\theta, \tag{2.115a}$$

$$V_r\Big|_{R_o} = \frac{\partial \phi}{\partial r}\Big|_{R_o} = 0. \tag{2.115b}$$

The solution of the Laplace equation for $\overline{\phi} = \overline{\phi}(r, \theta)$, after separation of variables, gives

$$\overline{\phi}(r, \theta) = \sum_{n=1}^{\infty} \{r^n [A_n \cos n\theta + B_n \sin n\theta] + r^{-n} [C_n \cos n\theta + D_n \sin n\theta]\}; \tag{2.116}$$

application of the boundary conditions yields

$$A_n = \frac{1}{n R_i^{n-1}} \left[1 - \left(\frac{R_o}{R_i}\right)^{2n}\right]^{-1} \frac{\mathrm{d}w_{ni}}{\mathrm{d}t}, \qquad C_n = R_o^{2n} A_n, \qquad B_n = D_n = 0. \tag{2.117}$$

The pressure on the inner shell and the outer cylinder in the nth circumferential mode may be determined through equation (2.73b), yielding

$$p_{i,ni} \equiv p_{ni}\Big|_{R_i} = -\rho R_i \frac{1}{n} \left[\frac{1 + (R_o/R_i)^{2n}}{1 - (R_o/R_i)^{2n}}\right] \cos n\theta \, \frac{\mathrm{d}^2 w_{ni}}{\mathrm{d}t^2}, \tag{2.118a}$$

$$p_{o,ni} \equiv p_{ni}\bigg|_{R_o} = -\rho R_i \frac{1}{n}\left[\frac{2(R_o/R_i)^n}{1-(R_o/R_i)^{2n}}\right] \cos n\theta \frac{\mathrm{d}^2 w_{ni}}{\mathrm{d}t^2}. \tag{2.118b}$$

The subscript notation i, ni indicates the pressure on cylinder i due to nth mode vibration of cylinder i, whereas o, ni indicates the pressure on cylinder o due to the same vibration of cylinder i.

Next, the loading on the shell and on the outer cylinder may be obtained from the principle of virtual work, i.e. via

$$\begin{aligned} \delta W_{i,ni} &= \int_0^{2\pi} \{(-p_{i,ni} R_i \,\mathrm{d}\theta)(\delta v_{ni} \sin n\theta + \delta w_{ni} \cos n\theta)\}, \\ \delta W_{o,ni} &= \int_0^{2\pi} \{(p_{o,ni} R_o \,\mathrm{d}\theta)(\delta v_{ni} \sin n\theta + \delta w_{ni} \cos n\theta)\}. \end{aligned} \tag{2.119}$$

As seen in equations (2.118a,b), $p_{i,ni}$ and $p_{o,ni}$ are functions of cos $n\theta$; hence, in view of the orthogonality of sin $n\theta$ and cos $n\theta$, only the δw_{ni} component of the virtual displacement contributes to the virtual work. Therefore, the forces on the inner shell and the outer cylinder due to motions of the inner shell in the nth mode, denoted by $F_{i,ni}$ and $F_{o,ni}$, respectively, are given by

$$F_{i,ni} = -\rho\pi R_i^2 \frac{1}{n}\left[\frac{(R_o/R_i)^{2n}+1}{(R_o/R_i)^{2n}-1}\right]\frac{\mathrm{d}^2 w_{ni}}{\mathrm{d}t^2}, \tag{2.120a}$$

$$F_{o,ni} = \rho\pi R_i R_o \frac{1}{n}\left[\frac{2(R_o/R_i)^n}{(R_o/R_i)^{2n}-1}\right]\frac{\mathrm{d}^2 w_{ni}}{\mathrm{d}t^2}. \tag{2.120b}$$

In effect, to obtain these forces, the pressure field was transformed into a surface-force field and projected onto the modal deformation vector in the eigenspace of this system. Further, it is noted that if the shell oscillates in more than one mode, $F_{i,ni}$ and $F_{o,ni}$ will still be the same, because, when projected onto the nth mode eigenvector, the contribution of the additional modes is zero, as a result of orthogonality of the cos $n\theta$ for different n, as per relationships (2.118a,b) and (2.119).

Similarly, if it is the outer shell that is flexible and oscillating while the inner one is rigid, proceeding in the same manner one finds

$$F_{i,no} = F_{o,ni}\frac{\mathrm{d}^2 w_{no}/\mathrm{d}t^2}{\mathrm{d}^2 w_{ni}/\mathrm{d}t^2}, \qquad F_{o,no} = -\rho\pi R_o^2 \frac{1}{n}\left[\frac{(R_o/R_i)^{2n}+1}{(R_o/R_i)^{2n}-1}\right]\frac{\mathrm{d}^2 w_{no}}{\mathrm{d}t^2}. \tag{2.121}$$

There are obvious symmetries in the coefficients of $\mathrm{d}^2 w_{ni}/\mathrm{d}t^2$ and $\mathrm{d}^2 w_{no}/\mathrm{d}t^2$ in (2.120a,b) and (2.121), which will be discussed later in subsection (d).

In the foregoing, *rigid-body transverse motions* of the cylinder were considered as a particular case of shell motions with $n = 1$ [Figure 2.7(c)]. For transverse rigid-body motions, however, the eigenvector or eigenfunction of motion becomes trivial, simplifying to motion along specified *directions*; thus, one can then think of motions in the Cartesian directions y and z, and work out the loading associated with oscillatory displacements v_c and w_c in these directions, as shown in Figure 2.8. In this case the boundary conditions

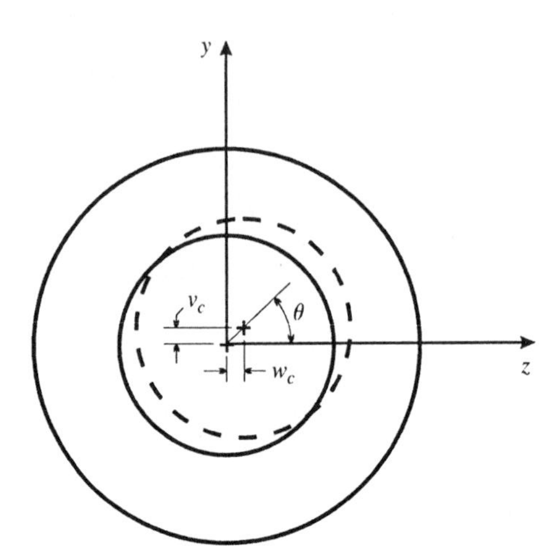

Figure 2.8 Rigid-body motion of the inner cylinder, in the y and z directions within the fluid-filled annulus (see Figure 2.7).

for the fluid become

$$\left.\frac{\partial \phi}{\partial r}\right|_{R_i} = \frac{\mathrm{d}v_c}{\mathrm{d}t}\sin\theta + \frac{\mathrm{d}w_c}{\mathrm{d}t}\cos\theta, \qquad \left.\frac{\partial \phi}{\partial r}\right|_{R_o} = 0. \tag{2.122}$$

Proceeding as before, the pressure on the surface of the inner and outer cylinders is determined and it is a function of both $\mathrm{d}^2v_c/\mathrm{d}t^2$ and $\mathrm{d}^2w_c/\mathrm{d}t^2$; in fact, the coefficients of these accelerations are identical to those in (2.118a,b), but with $n = 1$; e.g.

$$p_{i,1i} \equiv \left. p_{1i}\right|_{R_i} = -\rho R_i \frac{1 + (R_o/R_i)^2}{1 - (R_o/R_i)^2}\left[\sin\theta\frac{\mathrm{d}^2v_c}{\mathrm{d}t^2} + \cos\theta\frac{\mathrm{d}^2w_c}{\mathrm{d}t^2}\right]. \tag{2.123}$$

Then, the forces on the cylinder may be determined (i) either as before, by considering the virtual work associated with virtual displacements $\delta\overline{v}_c \sin\theta + \delta\overline{w}_c\cos\theta$, or (ii) directly, by integrating the pressure on the rigid cylinder via

$$F_{i,1i}^{y} = \int_0^{2\pi} -p_{i,1i}R_i\sin\theta\,\mathrm{d}\theta, \qquad F_{i,1i}^{z} = \int_0^{2\pi} -p_{i,1i}R_i\cos\theta\,\mathrm{d}\theta, \tag{2.124}$$

and similarly for $F_{o,1i}^{y}$ and $F_{o,1i}^{z}$; thus, if following (i), the projection of the force field onto the 'mode' concerned, has an immediate physical meaning in this case! It is obvious that the same results as in equations (2.120a,b) are obtained, as should be, but with $\mathrm{d}^2w_{ni}/\mathrm{d}t^2$ replaced by either $\mathrm{d}^2v_c/\mathrm{d}t^2$ or $\mathrm{d}^2w_c/\mathrm{d}t^2$.

A number of important conclusions are reached and insights gained from these results in the following.

(a) The added mass concept

As is clear from (2.120a,b), the fluid loading is associated entirely with *accelerations* of the structures, and hence accelerations of the fluid. This is physically reasonable: infinitely slowly generated displacement of the shell away from its equilibrium position cannot, in the absence of flow, result in a force; in an inviscid fluid, neither can a velocity of the body.[†]

It is customary to define a *virtual* or *added mass*, by expressing the fluid loading in the form of a d'Alembert (mass) × (acceleration) term. For ease of interpretation, consider first the case of $n = 1$ [see Figure 2.7(c)], so that the shell (only the inner one for simplicity) oscillates transversely as a whole, without deformation of its cross-section — essentially as a beam or a rigid cylinder would. Then considering $w_{1i} \cos\theta\big|_{\theta=0} = w_c$ and $v_c = 0$ (Figure 2.8), the equation of motion of the cylinder in the z-direction may be written as

$$M\ddot{z} + C\dot{z} + Kz = -\left[\rho\pi R_i^2 \frac{(R_o/R_i)^2 + 1}{(R_o/R_i)^2 - 1} L\right]\ddot{z}. \tag{2.125}$$

M, C and K could be the modal mass, damping and stiffness elements in a one axial-mode Galerkin approximation for the structure, or one can think of a long rigid cylinder of mass M, flexibly supported by a spring of stiffness K and a dashpot with damping coefficient C; L is the length of the shell. The quantity in square brackets is defined as the added mass, and may be denoted by M', so that equation (2.125) may be written as

$$(M + M')\ddot{z} + C\dot{z} + Kz = 0, \tag{2.126}$$

thus making obvious the usefulness of this concept and the appellation of 'added' mass. Dividing this added mass by the fluid mass of the volume occupied by ('displaced' by[‡] the presence) of the shell, gives the so-called added mass coefficient,

$$C_m^{i,1i} = \frac{M'}{\rho\pi R_i^2 L} = \frac{(R_o/R_i)^2 + 1}{(R_o/R_i)^2 - 1}. \tag{2.127}$$

For shell-type motions, $n > 1$, one cannot associate added mass or added mass coefficients with motions *in a particular direction* as in (2.125) and (2.127), but rather with motions associated with *particular modes of deformation*, e.g. the nth circumferential mode. In any case, for the analysis of shell motions, forces due to the fluid *per unit surface area* are more pertinent, as is done in Chapter 7. The added mass coefficient, however, is defined in the same way as in the foregoing; thus, corresponding to the forces in (2.120a,b) and (2.121), we have

$$C_m^{i,ni} = C_m^{o,no} = \frac{1}{n}\frac{(R_o/R_i)^{2n} + 1}{(R_o/R_i)^{2n} - 1}, \tag{2.128a}$$

$$C_m^{o,ni} = -C_m^{i,no} = \frac{1}{n}\frac{2(R_o/R_i)^n}{(R_o/R_i)^{2n} - 1}; \tag{2.128b}$$

see also Chen (1987; Chapter 4).[§]

[†]For a body in unbounded fluid this is a consequence of the d'Alembert paradox (stating that an ideal fluid flow exerts zero net force on any body immersed in it). In the presence of solid boundaries this is generally not so, and velocity-dependent forces may arise, but they are proportional to the square of the velocity (Duncan *et al*. 1970), and so, in the present context, they are negligible.

[‡]'Displaced', in the original sense in Archimedes' 'experiment' in Syracuse, when he immersed himself in his bath, thus displacing an equal volume of the fluid — and evoking the famous *eureka*!

[§]Note, however, a typographical error in equation (4.39) therein.

(b) The added mass from the kinetic energy

The classical way of introducing the added (or 'virtual') mass concept is via energy considerations (Milne-Thomson 1949; Duncan *et al*. 1970). As this gives new insights, it is presented here, parenthetically, following the treatment of Duncan *et al*. (1970).

Consider a rigid body moving rectilinearly with velocity U at the instant considered in unconfined fluid, otherwise at rest. The velocity of the fluid thereby generated, at any point, is proportional to U, and hence the velocity components may be written as $u = Uu'$, $v = Uv'$, $w = Uw'$. Hence, the total kinetic energy of the fluid (over the whole region occupied by it) is

$$T = \tfrac{1}{2}\rho U^2 \iiint (u'^2 + v'^2 + w'^2)\,\mathrm{d}x\,\mathrm{d}y\,\mathrm{d}z = \tfrac{1}{2}\rho U^2 \kappa, \tag{2.129}$$

where κ is a constant, for motion in any given direction. Next, suppose that the velocity of the body is variable, and let F be the force exerted by the body on the fluid. Then, by elementary energy considerations, the change in kinetic energy is equal to the work done by F, say in the z-direction, i.e.

$$F\,\mathrm{d}z = \mathrm{d}T = \kappa\rho U \frac{\mathrm{d}U}{\mathrm{d}t}\,\mathrm{d}t,$$

which gives

$$F = \kappa\rho \frac{\mathrm{d}U}{\mathrm{d}t}, \tag{2.130}$$

and the force on the body is the negative of that. In (2.130), $\mathrm{d}U/\mathrm{d}t$ is the body acceleration and, hence, by definition, $\rho\kappa$ is the added mass.

For 2-D oscillations of a circular cylinder in unbounded inviscid fluid, $v = (Ua^2/r^2)\sin 2\theta$, and $w = (Ua^2/r^2)\cos 2\theta$, and $v^2 + w^2 = Ua^2/r^2$; hence, in this case

$$\kappa = \frac{1}{U^2}\int_0^{2\pi}\int_0^{\infty} (v^2 + w^2) r\,\mathrm{d}\theta\,\mathrm{d}r = \pi a^2,$$

per unit length, and the added mass, also per unit length, is

$$m' = \frac{M'}{L} = \rho\pi a^2. \tag{2.131}$$

Thus, the well-known result is obtained that the added mass of a long cylinder oscillating in unconfined fluid is equal to the displaced mass of fluid. This corresponds exactly to the result in equation (2.127) for $R_0 \to \infty$, as it should.

It is worthwhile taking this one step further, to the case where there is an obstacle or boundary in the fluid; κ is then not a constant but a function of position, i.e. $\kappa(z)$. In this case, by following the same procedure one finds

$$F = \kappa\rho \frac{\mathrm{d}U}{\mathrm{d}t} + \frac{1}{2}\frac{\mathrm{d}\kappa}{\mathrm{d}z}\rho U^2; \tag{2.132}$$

i.e. there is now a quadratic velocity-dependent component, which for small-amplitude motion is of second order, as already remarked in the first footnote of subsection 2.2.2(a).

It should also be noted that in cases of symmetric confinement of the fluid, this term may entirely vanish.

(c) Magnitude considerations: wide and narrow annuli

By re-writing expressions (2.120a,b) in terms of R_i/R_o and taking the limit $(R_i/R_o) \to 0$, i.e. as the outer cylinder radius becomes essentially infinite, one obtains

$$F_{i,ni} = -\rho\pi R_i^2 \frac{1}{n}\frac{\mathrm{d}^2 w_{ni}}{\mathrm{d}t^2}, \qquad C_m = \frac{1}{n} \tag{2.133a}$$

and

$$F_{o,ni} = 0, \qquad \text{as} \qquad (R_i/R_o) \to 0. \tag{2.133b}$$

Equation (2.133a) for $n = 1$ yields the result just obtained in (2.131) in another way, that the added mass for $R_o \to \infty$ is equal to the 'displaced' mass of fluid. Also, the physically reasonable result is obtained in equation (2.133b) that, for an infinitely distant outer cylinder, the effect of accelerations of the inner one is infinitely faint.

In the other limit, writing $R_o = R_i + h$ and $R_i \simeq R_o \simeq R$, and taking h to be small,

$$F_{i,ni} = F_{o,ni} \simeq -\rho\pi R^2 \frac{1}{n^2}\frac{R}{h} L. \tag{2.134}$$

This expression shows that for thin shells, and also for light, hollow cylinders in narrow annuli, the added mass can easily exceed and be several times larger than the structural mass; i.e. $M' \gg M$ in (2.126), for instance. Expression (2.134) would suggest that the added mass becomes infinitely large as $h \to 0$. This is not so, however, because the Stokes number becomes small before that limit is reached, signalling that the limit of applicability of inviscid theory has been surpassed; for oscillations of the shell or cylinder of amplitude ϵh and frequency ω, where ϵ is a small number, the Stokes (or oscillatory Reynolds) number is $\beta = \epsilon\omega h R/\nu \equiv \epsilon\omega(h/R)R^2/\nu$. An alternative, and more general, pertinent Stokes number is $\beta = \omega h^2/\nu$. In either case, it is clear that as $h \to 0$, or $h/R \to 0$ and $\epsilon < 1$, β becomes sufficiently small for viscous effects not to be negligible (see Section 2.2.3). Furthermore, in addition to the added damping, the forces associated with shell motions become extremely large, as seen from (2.134), due to the very large accelerations in the narrow fluid annulus; hence, sustained oscillation under the circumstances does not occur.

It is finally noted in (2.120a,b), (2.133a) and (2.134) that the added mass becomes smaller as n is increased, which is reasonable in physical terms: the hills and valleys associated with deformation of the shell are half a circumference apart for $n = 1$, while they are much closer for large n; hence the fluid accelerations are correspondingly smaller for the larger n, and so is the added mass.

(d) Fluid coupling and the added mass matrix

If both shells are flexible, the only thing that changes in the formulation is that boundary condition (2.115b) needs to take a form similar to (2.115a). Recalling the meaning of influence coefficients in solid mechanics, by analogy (and as already done in the foregoing) one can think of a force on the inner shell due to nth mode motion of the inner shell, $F_{i,ni}$, or of a force on the inner shell due to motion of the outer one, $F_{i,no}$, and so on.

It is easy, therefore, to appreciate that in this case there exists an added mass *matrix*, of the form

$$\begin{bmatrix} m_{ii} & m_{io} \\ m_{oi} & m_{oo} \end{bmatrix}, \tag{2.135}$$

which couples hydrodynamically the motions of the two shells; here the subscript n has been suppressed. The corresponding vector is $\{\mathrm{d}^2 w_{ni}/\mathrm{d}t^2, \mathrm{d}^2 w_{no}/\mathrm{d}t^2\}^{\mathrm{T}}$; m_{ii} and m_{oi} are the negatives of the coefficients of $\mathrm{d}^2 w_{ni}/\mathrm{d}t^2$ in (2.120a,b), while m_{io} and m_{oo} are the corresponding quantities from (2.121). It is obvious that the matrix must be symmetric, as a consequence of the reciprocity principle in mechanics.

Consider next the situation of rigid-body motion ($n = 1$) of both the inner and outer cylinders. In this case

$$[M]\ddot{\mathbf{x}} + [C]\dot{\mathbf{x}} + [K]\mathbf{x} = -[M']\ddot{\mathbf{x}}, \tag{2.136}$$

where $\mathbf{x} = \{y_i, y_o, z_i, z_o\}^{\mathrm{T}}$ and

$$[M'] = \begin{bmatrix} m_{ii}^{yy} & m_{io}^{yy} & 0 & 0 \\ m_{oi}^{yy} & m_{oo}^{yy} & 0 & 0 \\ 0 & 0 & m_{ii}^{zz} & m_{io}^{zz} \\ 0 & 0 & m_{oi}^{zz} & m_{oo}^{zz} \end{bmatrix}, \tag{2.137}$$

in which $m_{ii}^{yy} = m_{ii}^{zz} = \rho\pi R_i^2 L[(R_o/R_i)^2 + 1]/[(R_o/R_i)^2 - 1]$ and so on, as given by expressions (2.120a,b) and (2.121) for $n = 1$. Thus, *coupling of the motions* of the two cylinders arises. This means that if, for example, the inner cylinder is given some initial displacement or velocity at $t = 0$, the outer cylinder would also vibrate for $t > 0$.

It is noted in (2.137) that, because of symmetry, there is no fluid coupling between y- and z-motions; i.e. acceleration of one cylinder in one direction generates a symmetric flow field, with no force resultant in the other direction. Generally, however, for asymmetric systems, such cross-coupling does exist, and matrix (2.137) would be fully populated, i.e. m_{ii}^{yz} and similar terms would no longer be null; furthermore, $m_{ii}^{yy} \neq m_{ii}^{zz}$, and so on.

(e) Effects of various parameters on added mass

Tables, figures and lists of results for added mass coefficients in a variety of systems are given by Blevins (1979), Chen (1987), Gibert (1988) and Naudascher & Rockwell (1994). Hence, we shall confine ourselves here to making some general comments on parameters affecting the added mass, of which the reader should be aware.

(i) *General effects of geometry.* In general, proximity to other structures affects the added mass of the vibrating one; e.g. proximity to a rigid wall signifies increased accelerations (for inviscid fluid) and hence larger added mass, as already remarked in the foregoing, especially in connection with the system of two coaxial cylinders or shells (Figure 2.7). Of equal interest is the case of eccentrically located cylinders (see also Chapter 11). A useful result (Gibert 1988) is that the added mass coefficient, C_m^{ecc}, is given by

$$\frac{C_m^{\mathrm{ecc}}}{C_m} = \frac{2(r-1)\left[r - 1 - \sqrt{e(2r-2-e)}\right]}{(r-1-e)^2}, \qquad \text{for} \qquad r < 1.1, \tag{2.138}$$

where C_m is given by equation (2.127), $e =$ (smallest gap between the cylinders)$/R_i$, $r = R_o/R_i$. For larger values of r, the results are given in figure form. Results for a variety of other systems may be found in the compilations of Chen (1987) and Blevins (1979).

(ii) *Aspect ratio effects.* As two-dimensionality of the flow is violated, the validity of the foregoing deteriorates. A particularly simple example illustrating this is a simply-supported cylindrical beam oscillating in a narrow annulus [so that the approximations leading to (2.134) are valid], the ends of which are open to large cavities (Gibert 1988). It is found that

$$\frac{C_m^{\text{beam}}}{C_m} = \frac{1}{2[1 + (\pi R/L)^2]}, \tag{2.139}$$

where C_m is given by expression (2.134) for $n = 1$. Clearly, the shorter the beam, the smaller is C_m^{beam}, as compared to an infinitely long one. The physical reason is that, near the ends, the fluid takes the easy way around the beam, partly in the third (axial) direction; hence, less than the total force that would be obtained by 2-D analysis is realized. A more general analysis (Païdoussis *et al.* 1984; Chen 1987), not making the assumption of a narrow annulus, gives

$$C_m^{\text{beam}} = \frac{\mathrm{I}_1'(2\pi R_o/L)\mathrm{K}_1(2\pi R_i/L) - \mathrm{I}_1(2\pi R_i/L)\mathrm{K}_1'(2\pi R_o/L)}{\mathrm{I}_1'(2\pi R_i/L)\mathrm{K}_1'(2\pi R_o/L) - \mathrm{I}_1'(2\pi R_o/L)\mathrm{K}_1'(2\pi R_i/L)}, \tag{2.140}$$

where I_1 and K_1 are, respectively, the first-order modified Bessel functions of the first and second kind; the primes denote derivatives with respect to the argument. The effect of R_i/L is strong for $1 < R_o/R_i < 2$, but relatively weak for wider annuli.

(iii) *Effects of compressibility and two-phase flow.* If the flow is compressible, the wave equation, $\nabla^2\psi + k^2\psi = 0$, $k^2 = \omega/c$, needs to be solved instead of the Laplace equation, $\nabla^2\phi = 0$. Hence, the results are found to depend also on an oscillatory Mach number, $M_k = \omega R_i/c$, where c is the speed of sound. The effect of compressibility for $M_k \leq 0.2$ is rather weak (Chen 1987).

It has been found (Carlucci 1980; Carlucci & Brown 1983) that in gas–liquid two-phase flows the measured added mass is generally considerably lower than that predicted by homogeneous mixture theory [in which average quantities are assumed for the mixture; e.g. if the void fraction is α and the densities of the liquid and gaseous phases are ρ_l and ρ_g, the mixture density is $\rho = (1 - \alpha)\rho_l + \alpha\rho_g$]. Since the two-phase flow may be considered as a flow with the density of the liquid phase and the compressibility of the gaseous one, it was supposed that the discrepancy may have been due to the neglected effects of compressibility (Païdoussis & Ostoja-Starzewski 1981). Also, the effect of random variations in the surrounding fluid density, inherent in two-phase flows was investigated (Klein 1981). These effects, although qualitatively working in the right direction (Chapter 8), proved incapable of accounting fully for the discrepancy quantitatively, and the search for more elaborate models continues.

(iv) *Amplitude effects.* All of the foregoing apply to cases where the amplitude of oscillation is small enough for separation in the cross-flow not to occur. This brings into play another dimensionless number, the Keulegan–Carpenter number, $\mathrm{KC} = 2\pi V_o/(\omega D)$, where V_o is the amplitude in velocity fluctuations. For a harmonically oscillating cylinder in quiescent flow, this reduces to $\mathrm{KC} = 2\pi(A/D)$, where A is the amplitude of motion of the cylinder. If KC < 4, separation generally does not occur (Sarpkaya & Isaacson 1981; Naudascher & Rockwell 1994). If KC > 8 approximately, the flow field is entirely different, with the cylinder now oscillating in the remnants of vortices shed from previous cycles of oscillation; this type of flow, arising also in wave-induced oscillatory flows, has been studied extensively in conjunction with offshore mechanics applications (Sarpkaya & Isaacson 1981).

(f) Numerical calculations of added mass

Some early attempts to calculate the added mass by numerical (CFD) methods are due to Levy & Wilkinson (1975), Païdoussis *et al.* (1977) and Yang & Moran (1979), for instance. Nowadays, any CFD package capable of heat transfer calculations, hence of solving the Laplace equation, would be suitable — based on finite element, finite difference or other methods. A few examples of finite-element (FEM) based packages are FIDAP from Fluid Dynamics International, U.S.A., and CASTEM 2000 from Commissariat à l'Energie Atomique, France; and finite-volume (FVM) based packages FLOW3D from Harwell Laboratories, U.K., and PHEONICS from Cham Ltd, U.K.

Other numerical methods also exist, e.g. based on spectral methods (Mateescu, Païdoussis & Sim 1994a,b), finite difference methods (Mateescu, Païdoussis & Bélanger 1994a,b), or the boundary integral equation method (BIEM) (Groh 1992).

2.2.3 Loading on coaxial shells filled with quiescent viscous fluid

Consider the same system as in Figure 2.7(a), but with only the inner cylinder free to oscillate, and then only as a beam ($n = 1$) or as a rigid body in the plane of the paper, while the outer one is rigid and immobile. The annular space is filled with a quiescent viscous fluid. Again, the task is to determine the fluid forces generated by harmonic motion of the inner cylinder.

If the cylinders are sufficiently long, the flow is essentially two-dimensional in cross-flow. Writing equation (2.63) in Cartesian coordinates and eliminating the pressure between the two equations, or simply taking the curl of (2.63), one obtains a single equation

$$\frac{\partial \omega}{\partial t} + u_y \frac{\partial \omega}{\partial y} + u_z \frac{\partial \omega}{\partial z} = \nu \left(\frac{\partial^2 \omega}{\partial y^2} + \frac{\partial^2 \omega}{\partial z^2} \right), \tag{2.141}$$

in terms of the vorticity,

$$\omega = \frac{\partial u_y}{\partial z} - \frac{\partial u_z}{\partial y}; \tag{2.142}$$

u_z and u_y are the flow velocity components in the z and y directions, which may be expressed in terms of the stream function: $u_z = \partial\psi/\partial y$, $u_y = -\partial\psi/\partial z$. The continuity equation (2.62), is satisfied automatically. Moreover, since $\omega = -\nabla^2\psi$, equation (2.141) leads to (Schlichting 1960, chapter IV)

$$\frac{\partial}{\partial t}(\nabla^2\psi) + \left[\frac{\partial \psi}{\partial y}\frac{\partial \nabla^2\psi}{\partial z} - \frac{\partial \psi}{\partial z}\frac{\partial \nabla^2\psi}{\partial y} \right] = \nu\, \nabla^4\psi. \tag{2.143}$$

For small motions, this reduces to

$$\frac{\partial}{\partial t}\nabla^2\psi = \nu\nabla^4\psi. \tag{2.144}$$

The boundary conditions match the fluid velocity on the solid surfaces to those of the two cylinders. In polar coordinates, $u_r = -(1/r)(\partial\psi/\partial\theta)$ and $u_\theta = \partial\psi/\partial r$; hence,

$$u_r\Big|_{R_i} = a\cos\theta\,\mathrm{e}^{\mathrm{i}\Omega t}, \qquad u_\theta\Big|_{R_i} = -a\sin\theta\,\mathrm{e}^{\mathrm{i}\Omega t}, \qquad u_r\Big|_{R_o} = u_\theta\Big|_{R_o} = 0, \tag{2.145}$$

where a is the velocity amplitude of the inner cylinder in the $\theta = 0$ plane.

This problem was solved by Wambsganss *et al*. (1974) — see also Chen *et al*. (1976). It may be verified that if

$$\nabla^2\psi = 0 \tag{2.146a}$$

is satisfied, so is equation (2.144); similarly if

$$\nabla^2\psi - \frac{1}{\nu}\frac{\partial\psi}{\partial t} = 0. \tag{2.146b}$$

Hence, a general solution in the form $\psi = \psi_1 + \psi_2$ is sought, with ψ_1 and ψ_2 satisfying (2.146a) and (2.146b), respectively. The form of the boundary conditions suggests

$$\psi_1 = F_1(r)\sin\theta\,\mathrm{e}^{\mathrm{i}\Omega t}, \qquad \psi_2 = F_2(r)\sin\theta\,\mathrm{e}^{\mathrm{i}\Omega t}, \tag{2.147}$$

and hence F_1 and F_2 must satisfy

$$\begin{aligned} &\frac{\mathrm{d}^2F_1}{\mathrm{d}r^2} + \frac{1}{r}\frac{\mathrm{d}F_1}{\mathrm{d}r} - \frac{1}{r^2}F_1 = 0,\\ &\frac{\mathrm{d}^2F_2}{\mathrm{d}r^2} + \frac{1}{r}\frac{\mathrm{d}F_2}{\mathrm{d}r} - \left(\frac{1}{r^2} + \lambda^2\right)F_2 = 0, \qquad \lambda = \left(\frac{\mathrm{i}\Omega}{\nu}\right)^{1/2}. \end{aligned} \tag{2.148}$$

Each of these equations provides two independent solutions, hence four in total, as required and sufficient for the solution of equation (2.144), namely

$$\psi = \psi_1 + \psi_2 = a[A_1 r^{-1} + A_2 r + A_3\mathrm{I}_1(\lambda r) + A_4\mathrm{K}_1(\lambda r)]\sin\theta\,\mathrm{e}^{\mathrm{i}\Omega t}, \tag{2.149}$$

in which the constants A_1 to A_4 are determined via the boundary conditions. Once ψ is determined, the flow field is completely known and hence the stresses on the cylinders may be evaluated through equations (2.64) and (2.65). The force per unit length is given by

$$F = -\rho\pi R_i^2 a\Omega[\mathcal{R}\mathrm{e}(H)\sin\Omega t + \mathcal{I}\mathrm{m}(H)\cos\Omega t] \tag{2.150}$$

(Chen *et al*. 1976), where

$$\begin{aligned} H = \{&2\alpha^2[\mathrm{I}_0(\alpha)\mathrm{K}_0(\beta) - \mathrm{I}_0(\beta)\mathrm{K}_0(\alpha)] - 4\alpha[\mathrm{I}_1(\alpha)\mathrm{K}_0(\beta) + \mathrm{I}_0(\beta)\mathrm{K}_1(\alpha)]\\ &+ 4\alpha\gamma[\mathrm{I}_0(\alpha)\mathrm{K}_1(\beta) + \mathrm{I}_1(\beta)\mathrm{K}_0(\alpha)] - 8\gamma[\mathrm{I}_1(\alpha)\mathrm{K}_1(\beta) - \mathrm{I}_1(\beta)\mathrm{K}_1(\alpha)]\}\\ &\div \{\alpha^2(1-\gamma^2)[\mathrm{I}_0(\alpha)\mathrm{K}_0(\beta) - \mathrm{I}_0(\beta)\mathrm{K}_0(\alpha)] + 2\alpha\gamma[\mathrm{I}_0(\alpha)\mathrm{K}_1(\beta)\\ &- \mathrm{I}_1(\beta)\mathrm{K}_0(\beta) + \mathrm{I}_1(\beta)\mathrm{K}_0(\alpha) - \mathrm{I}_0(\beta)\mathrm{K}_1(\beta)] + 2\alpha\gamma^2[\mathrm{I}_0(\beta)\mathrm{K}_1(\alpha)\\ &- \mathrm{I}_0(\alpha)\mathrm{K}_1(\alpha) + \mathrm{I}_1(\alpha)\mathrm{K}_0(\beta) - \mathrm{I}_1(\alpha)\mathrm{K}_0(\alpha)]\} - 1, \end{aligned} \tag{2.151}$$

in which

$$\alpha = \lambda R_i, \qquad \beta = \lambda R_o, \qquad \gamma = R_i/R_o, \tag{2.152}$$

and I_n and K_n are modified Bessel functions of the first and second kind, respectively. It is noted that, by virtue of the presence of $\sqrt{\mathrm{i}}$ in the argument (in λ), H is complex. To evaluate H, therefore, one can either (i) evaluate $\mathrm{J}_n(\overline{\alpha}\sqrt{\mathrm{i}})$ and $\mathrm{Y}_n(\overline{\alpha}\sqrt{\mathrm{i}})$, the ordinary Bessel functions, utilizing the expressions and tables in Jahnke & Emde (1945), for instance, for the real and imaginary parts of each of them, and then convert to I_n and K_n, or (ii) utilize the ber, bei, ker and kei functions,

$$\mathrm{i}^n \mathrm{I}_n(x\sqrt{\mathrm{i}}) = \mathrm{ber}_n\, x + \mathrm{i}\,\mathrm{bei}_n\, x, \qquad \mathrm{i}^{-n}\mathrm{K}_n(x\sqrt{\mathrm{i}}) = \mathrm{ker}_n\, x + \mathrm{i}\,\mathrm{kei}_n\, x$$

and the expressions given by Dwight (1961).†

Expressing the force F of equation (2.150) in terms of added mass and added damping as in equations (2.110) and (2.111), one can write

$$F = +C_m \rho A \frac{\mathrm{d}^2 z}{\mathrm{d}t^2} - C_d \Omega \rho A \frac{\mathrm{d}z}{\mathrm{d}t}; \tag{2.153}$$

hence

$$C_m = \mathcal{R}\mathrm{e}(H) \qquad \text{and} \qquad C_d = -\mathcal{I}\mathrm{m}(H). \tag{2.154}$$

The results for $\mathcal{R}\mathrm{e}(H)$ and $\mathcal{I}\mathrm{m}(H)$ for various Stokes numbers $\mathrm{S} = \Omega R_i^2/\nu$ are given in Figure 2.9. Several observations may be made, as follows:

(i) both C_m and C_d increase dramatically as R_o/R_i is reduced towards unity, but C_d rises more rapidly;
(ii) for sufficiently high S, the values of C_m approach those obtained by inviscid theory ($\mathrm{S} = \infty$), but increasingly diverge from inviscid theory as S is diminished;
(iii) for sufficiently narrow annuli, the results for C_m sensibly collapse onto a single curve — in the scale of the figure.

Chen *et al*. (1976), Yeh & Chen (1978) and Chen (1981, 1987) give a number of useful approximations for H. These have been rechecked, corrected in some cases, and rewritten into a congruous set, in terms of the parameters

$$\begin{gathered}\gamma = \frac{R_i}{R_o}, \qquad g = \frac{1-\gamma}{\gamma} \equiv \frac{R_o}{R_i} - 1, \qquad \mathrm{S} = \frac{\Omega R_i^2}{\nu}, \\ \alpha = \lambda R_i \equiv \left(\frac{\mathrm{i}\Omega R_i^2}{\nu}\right)^{1/2}, \qquad \beta = \lambda R_o \equiv \left(\frac{\mathrm{i}\Omega R_o^2}{\nu}\right)^{1/2},\end{gathered} \tag{2.155}$$

as follows:

---

†It may be of interest that difficulties are encountered in trying to obtain solutions by standard software packages, including some symbolic manipulation systems. Thus, neither *Maple* nor *Matlab* could do it; *Mathematica* could, but it was painfully slow.

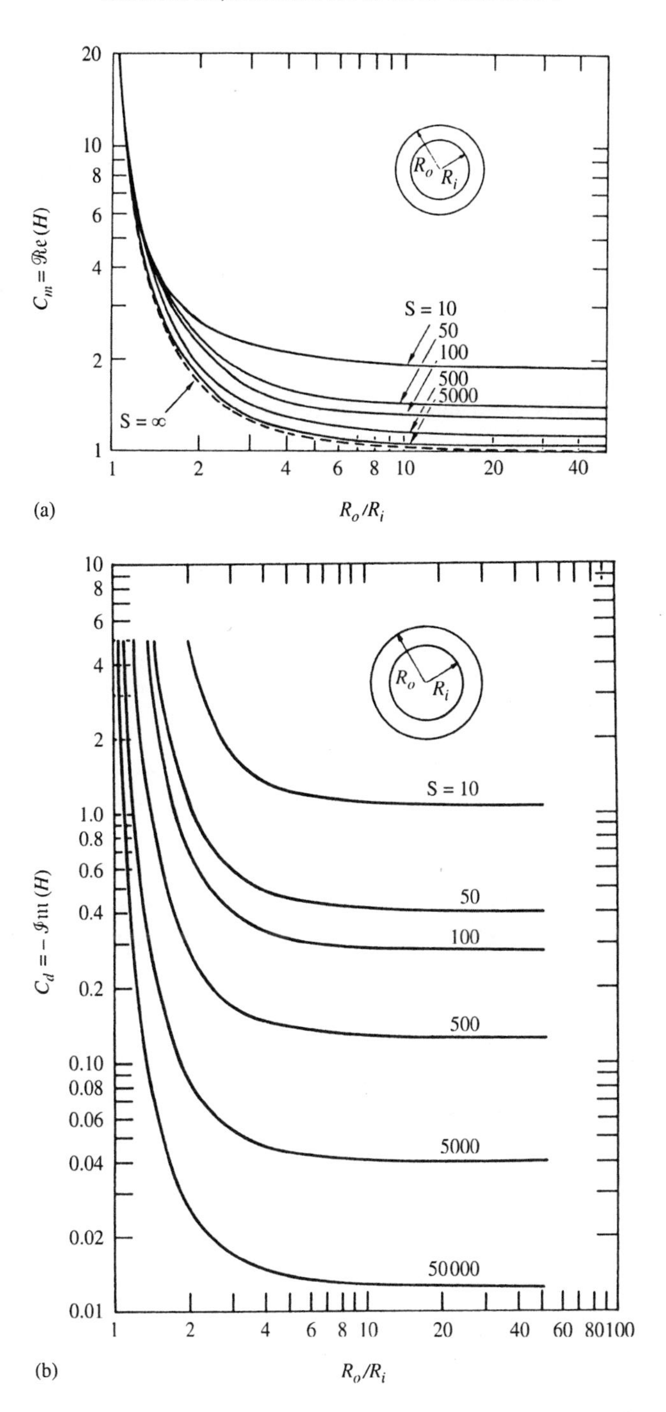

Figure 2.9 (a) The real and (b) negative of the imaginary part of H, given by equation (2.151), equal to the added mass and viscous damping coefficient, respectively, for a cylinder of radius R_i oscillating with frequency Ω in a viscous fluid within a coaxial rigid cylinder of radius R_o, for a number of values of the Stokes number, $S = \Omega R_i^2/\nu$. From Chen *et al.* (1976).

(a) for large α and β, wide or narrow annuli [i.e. $S > 500$ and $g \geq 0.005$ (or $\gamma \leq 0.995$)],

$$H = \frac{[\alpha^2(1+\gamma^2) - 8\gamma]\sinh(\beta-\alpha) + 2\alpha(2-\gamma+\gamma^2)\cosh(\beta-\alpha) - 2\gamma^2\sqrt{\alpha\beta} - 2\alpha\gamma\sqrt{\gamma}}{\alpha^2(1-\gamma^2)\sinh(\beta-\alpha) - 2\alpha\gamma(1+\gamma)\cosh(\beta-\alpha) + 2\gamma^2\sqrt{\alpha\beta} + 2\alpha\gamma\sqrt{\gamma}}; \tag{2.156a}$$

(b) for very wide annuli and large S [$S > 300$ and $g > 40$ (or $\gamma < 0.025$)],

$$H = 1 + \frac{4\mathrm{K}_1(\alpha)}{\alpha \mathrm{K}_0(\alpha)}; \tag{2.156b}$$

(c) for the same range of S and g as in (b), an easier approximation is also valid, namely

$$H = 1 + \frac{4}{\alpha}; \tag{2.156c}$$

(d) for moderately wide annuli and large S ($S > 10^4$ and $g > 0.1$, or $S > 2 \times 10^3$ and $g > 0.2$),

$$H = \frac{[\alpha(1+\gamma^2) + 2(2-\gamma+\gamma^2)]}{\alpha(1-\gamma^2) - 2\gamma(1+\gamma)}; \tag{2.156d}$$

(e) for fairly narrow annuli ($g > 0.05$) and $S > 10^4$,

$$H = \frac{\alpha(1+\gamma^2)\sinh(g\alpha) + 2(2-\gamma+\gamma^2)\cosh(g\alpha) - 4\gamma\sqrt{\gamma}}{\alpha(1-\gamma^2)\sinh(g\alpha) - 2\gamma(1+\gamma)\cosh(g\alpha) + 4\gamma\sqrt{\gamma}}, \tag{2.156e}$$

although approximation (2.156a) is superior and almost as easy to compute;

(f) for very narrow gap and very large S ($g \ll 1$, $S \gg 1$, $g^2 S \gg 1$; e.g. $g < 0.05$, $S > 10^7$, $g^2 S > 10^4$),

$$H = \frac{1+\gamma^2}{1-\gamma^2} + \frac{\sqrt{2}}{g^2\sqrt{S}}(1-\mathrm{i}). \tag{2.156f}$$

In order to utilize these expressions it is recalled that $\sqrt{\mathrm{i}} = \frac{1}{2}\sqrt{2}(1+\mathrm{i})$, a complex quantity, arising because of the form of α and β in equations (2.155); hence, $\sin(A + B\mathrm{i}) = \sin A \cosh B + \mathrm{i} \cos A \sinh B$, etc.

Another set of approximations were derived by Sinyavskii *et al*. (1980), based on the boundary-layer approximation and valid for $S \gg 1$, namely

$$C_m = \frac{1+\gamma^2}{1-\gamma^2} + \frac{2\sqrt{2}}{\sqrt{S}}\frac{1+\gamma^3}{(1-\gamma^2)^2}, \qquad C_d = \frac{2\sqrt{2}}{\sqrt{S}}\frac{1+\gamma^3}{(1-\gamma^2)^2}. \tag{2.157}$$

For zero confinement ($\gamma = 0$), $C_d = 2\sqrt{2}/\sqrt{S}$ corresponds exactly to the expression derived by Batchelor (1967; section 5.13).

2.3 LINEAR AND NONLINEAR DYNAMICS

Consider a one-degree-of-freedom linear system subjected to fluid loading, $F(t)$; the equation of motion is written as

$$m\ddot{x} + c\dot{x} + kx = F(t), \tag{2.158}$$

and $F(t)$ may be expressed as

$$F(t) = -m'\ddot{x} - c'\dot{x} - k'x, \tag{2.159}$$

in which m' is the added or virtual mass of the fluid associated with acceleration of the body, c' is the fluid damping term associated with the velocity of the body, and k' is the fluid added stiffness, as discussed in Section 2.2.1(g). Hence, the equation of motion may be written as

$$(m + m')\ddot{x} + (c + c')\dot{x} + (k + k')x = 0. \tag{2.160}$$

It is noted that the form of equation (2.159) implies that there is no external forcing of the system: all fluid loading is associated with motion. In general, the coefficients associated with the linearized forces in (2.159) are not constant, but depend on flow velocity, amplitude and frequency of motion, fluid viscosity, and so on. For the purpose of this introduction, however, let us neglect most of these effects and take $m' = \text{const.}$, $c' = c'(U)$, $k' = k'(U)$, where U is a characteristic flow velocity in the system. Hence, equation (2.160) may be written as

$$\ddot{x} + 2\zeta(U)\Omega_n(U)\dot{x} + \Omega_n^2(U)x = 0, \tag{2.161}$$

where, as denoted, the damping factor, ζ, and the natural frequency, Ω_n, are functions of U, which is the only variable parameter of this system.

If $c'(U) > 0$ and $k'(U) > 0$ for all U, then the response of the fluid-loaded system is qualitatively the same as that of the mechanical system: only damped oscillations would be observed, with higher or lower frequency, depending on whether added mass or fluid stiffness effects predominate [i.e. whether $(k + k')/(m + m') >$ or $< k/m$], and with higher or lower damping (ζ), depending on whether $(c + c')/(m + m') >$ or $< c/m$.

If, however, $k'(U)$ can become negative, and $|k'(U)| = k$ for some critical value of U, U_c, then the overall stiffness of the system vanishes — and for $U > U_c$ may become negative — which signifies that the system is then statically unstable. The premier example of this (albeit for a system with more than one degree of freedom) is the static instability, or *divergence*, of an articulated or continuously flexible pipe with supported ends conveying fluid (see Chapter 3); it is similar to the divergence, or *buckling*, of a column subjected to an end load. At that point, i.e. when $|k'(U)| = k$, x becomes indeterminate: i.e. the static equilibrium position $x_{st} = 0$ is replaced by a condition where an infinite set of static equilibria are possible (Ziegler 1968) according to linear theory.

Similarly, if $\zeta(U_c) < 0$ [i.e. if $c'(U_c) < 0$ and sufficiently large], this implies a negative damping: instead of the oscillations dying out with time, they are amplified exponentially. A good example of this is the oscillatory instability (in the linear sense), or *flutter*, of a cantilevered pipe conveying fluid (see Chapter 3).

Mathematically, the evolution of a system towards divergence or flutter may be tracked by plotting the complex eigenvalues or, equivalently, the eigenfrequencies in the complex

Argand plane, as U is varied. Figure 2.10 shows the development of (a) divergence and (b) flutter in these two representations. The solution to (2.161) may be expressed as

$$x = A\mathrm{e}^{-\zeta\Omega_n t}\sin(\Omega_n\sqrt{1-\zeta^2}t+\phi), \tag{2.162}$$

or, in terms of the eigenvalues λ and eigenfrequencies Ω, by

$$x = A\mathrm{e}^{\Re\mathrm{e}(\lambda)t}\sin[\Im\mathrm{m}(\lambda)+\phi] = A\mathrm{e}^{-\Im\mathrm{m}(\Omega)t}\sin[\Re\mathrm{e}(\Omega)+\phi], \tag{2.163}$$

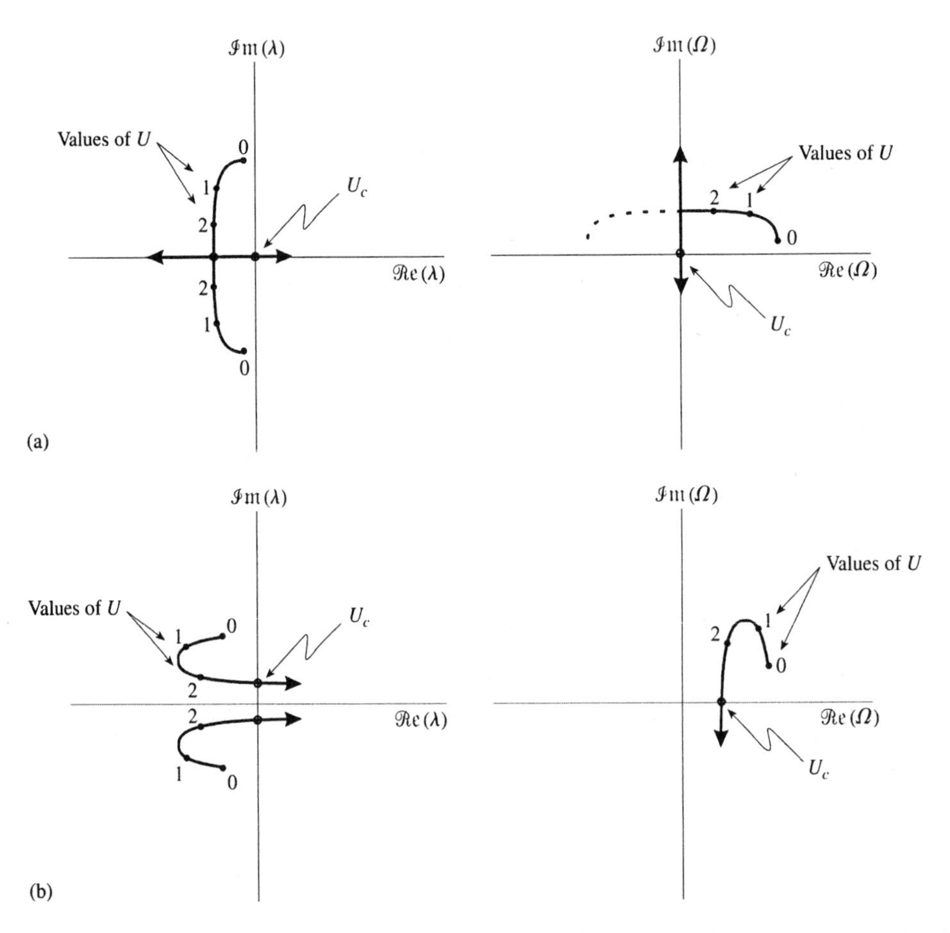

Figure 2.10 Typical Argand diagrams showing the evolution of a system with increasing U, from stability ($U < U_c$), to instability (in the linear sense; $U \geq U_c$): (a) divergence; (b) flutter. The diagrams on the left show this evolution in the eigenvalue- or λ-plane, and those on the right in the frequency- or Ω-plane.

where $\Omega\mathrm{i} = \lambda$ and $\mathrm{i} = \sqrt{-1}$, and $\Re\mathrm{e}$ and $\Im\mathrm{m}$ denote the real and imaginary components. Clearly, $\Re\mathrm{e}(\Omega) = \Im\mathrm{m}(\lambda)$ is proportional to the frequency of oscillation, while $\Im\mathrm{m}(\Omega) = -\Re\mathrm{e}(\lambda)$ is proportional to damping; in fact, for sufficiently small ζ, $\Re\mathrm{e}(\Omega) \simeq \Omega_n$ and $\Im\mathrm{m}(\Omega)/\Re\mathrm{e}(\Omega) \simeq \zeta$. In the λ-plane (left-hand panels of Figure 2.10), it is common to show both of the complex conjugate eigenvalue loci [note that even for a conservative system with zero damping, the solutions are $\lambda_{1,2} = \pm(\Omega_n^2)^{1/2}\mathrm{i}$]. In the frequency-plane, however, the branch associated with negative frequency, being more mathematical than physical, is often suppressed; nevertheless, the bifurcation of the eigenfrequency locus in the upper-right panel of Figure 2.10 is more easily comprehensible if both branches are shown.

Linear theory can only predict *the onset* of divergence or flutter; solution (2.163) with $\mathcal{R}e(\lambda) > 0$ or $\mathcal{I}m(\Omega) < 0$ would suggest that the motion is amplified indefinitely. This is normally not so, but it is only through nonlinear theory that we can discover what happens. We know physically that a column subjected to a compressive load or a pipe with supported ends conveying fluid will diverge to one side or the other and then display a new, buckled equilibrium form. In nonlinear theory the linear instability is simply referred to as a bifurcation. In the case of divergence, where the bifurcation is characterized by one zero eigenvalue, it is referred to as a *pitchfork bifurcation*, whereby the original equilibrium, $x = 0$, becomes unstable and two new stable equilibria, $x = \pm|x_{st}|$, are generated — which may evolve with increasing U in the manner shown in Figure 2.11(a).

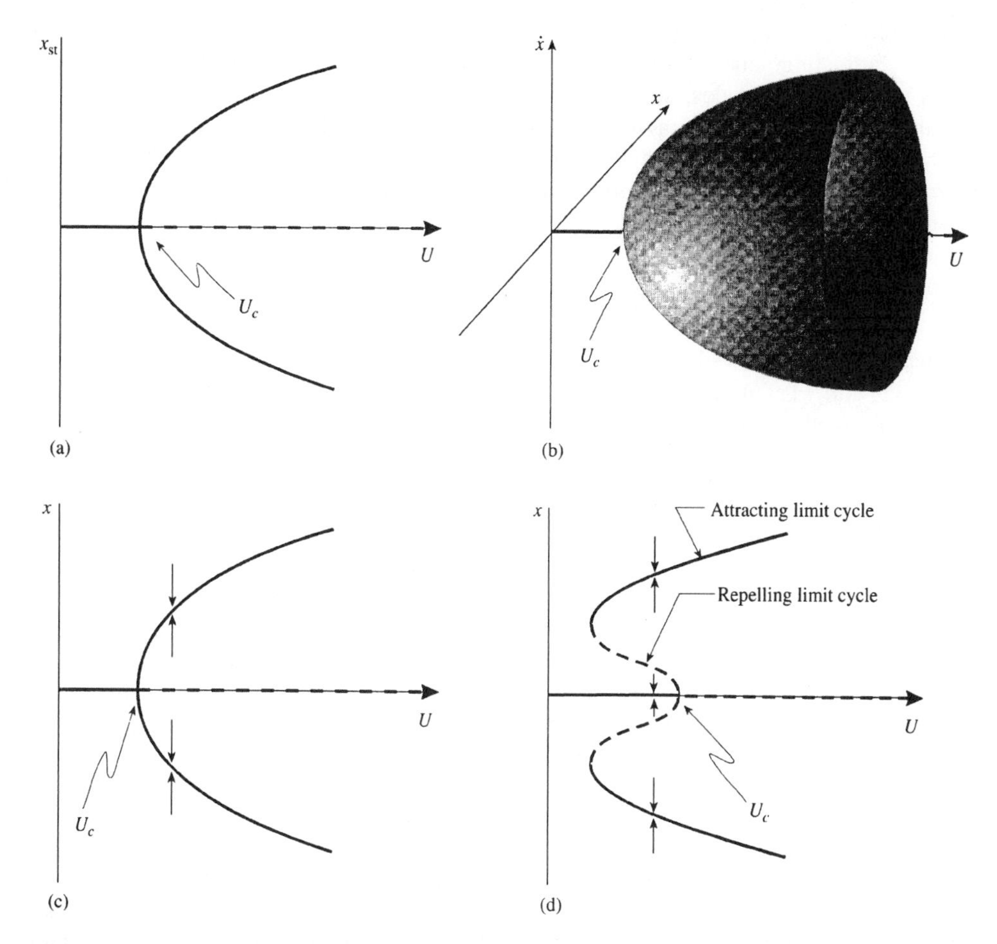

Figure 2.11 Bifurcation diagrams for (a) a supercritical *pitchfork bifurcation* (static loss of stability, or static divergence); (b) a supercritical *Hopf bifurcation* (flutter), shown in 3-D; (c) a supercritical Hopf bifurcation in the $\{x, U\}$-plane; (d) a subcritical Hopf bifurcation. ——, Stable, attracting fixed points or limit cycles; – – –, unstable ones. The small arrows in (c) and (d) reinforce the ideas of attraction/repulsion of solution trajectories towards or away from the pertinent attractors.

Physically, flutter is a self-excited oscillation, which grows from sensibly zero to a steady oscillation of finite amplitude and constant frequency, thus to a closed curve in the phase plane $(x, \dot{x})$, i.e. to a *limit cycle*. Mathematically, the onset of flutter is characterized by a pair of eigenvalues crossing from $\mathcal{R}e(\lambda) < 0$ to $\mathcal{R}e(\lambda) > 0$ as U is increased, such that at $U = U_c$ (i) the pair is purely imaginary, i.e. $\mathcal{R}e(\lambda) = 0$, and (ii) $\mathcal{I}m(\lambda) \neq 0$ [Figure 2.10(b)]. This is defined as a *Hopf bifurcation*. In many cases, the evolution in the phase plane as U is increased is as shown in three-dimensional form in Figure 2.11(b), in which case the Hopf bifurcation is *supercritical*. As shown in Figure 2.11(c), the origin has become unstable and oscillatory solutions of a certain amplitude are possible for $U > U_c$. If the system is perturbed, it will eventually settle down on the limit cycle; hence this is a case of a stable limit cycle.

A *subcritical* Hopf bifurcation is illustrated in Figure 2.11(d), where the limit cycle generated is unstable or 'repelling'; as shown by the small arrows, oscillatory solutions either die out to the stable equilibrium (stable fixed point) or diverge to larger amplitudes. In real physical systems, the existence of this unstable limit cycle usually implies that a stable 'attracting' one [as shown in Figure 2.11(d)] or another kind of stable solution exists at larger amplitudes; so that, the trajectories in the phase plane, repelled by the unstable limit cycle, will gravitate towards the stable fixed (equilibrium) point or the limit cycle beyond. Thus, the system is then said to be *unstable in the small*, but *stable in the large*. A more formal definition of stability is given in Appendix F.1.1.

The behaviour described in the foregoing may be illustrated by a fictitious nonlinear one-degree-of-freedom system, the equation of motion of which is

$$m\ddot{x} + cg(\dot{x}) + kf(x) = 0, \tag{2.164}$$

and which may be viewed as a nonlinear version of equation (2.160) for a specific value of U; $g(\dot{x})$ and $f(x)$ are nonlinear functions. As it is not uncommon for these functions to be odd, let us illustrate the behaviour of such a system by the following particular case:

$$\ddot{x} + 0.02(1 - \dot{x}^2)\dot{x} + (1 - 0.02x^2)x = 0. \tag{2.165}$$

Trajectories in the phase plane are shown in Figure 2.12. Two main features are visible. First, there exists a repelling, unstable limit cycle of amplitude ~ 1.1 around the origin, in the clear white oval between the darker patches near the centre of the figure. One trajectory is shown, slowly spiralling inwards towards the origin (in the dark doughnut-shaped region, although it is noted that the spiralling motion is difficult to see in the scale of the figure); the calculation was discontinued before the trajectory could reach the origin (which would strictly take infinite time). Trajectories with $|x| > 1.1$ spiral outwards. Physically, one can see, by referring to equation (2.165), that if the mean value of $|\dot{x}| \sim \mathcal{O}(1)$ over a cycle, the mean amount of damping would be zero — i.e. the net dissipation, over a cycle of oscillation, vanishes — which is one way of interpreting the existence of a limit cycle; in the 'absence' of damping, the system becomes *effectively* conservative, and a closed curve would be expected in the phase plane, in this case the unstable limit cycle. The second notable feature is the *saddle point* at $|x| = (1/0.02)^{1/2} \simeq 7.1$, which is an *unstable fixed point* (or point of equilibrium),† corresponding to points of static instability (divergence), when the stiffness term vanishes.

†The classical paradigm of a stable fixed point (stable equilibrium) is the point $\{\theta, \dot{\theta}\} = \{0, 0\}$ for a simple pendulum, while $\{\pi, 0\}$ represents an unstable fixed point, a saddle. A characteristic of the saddle is that there are

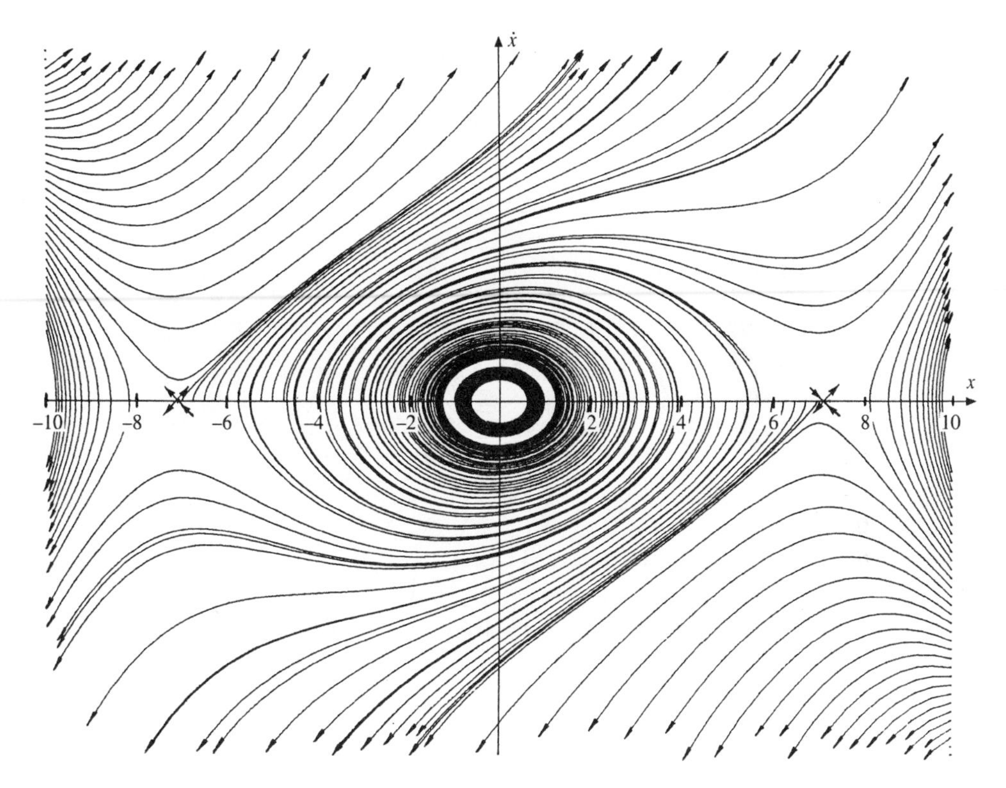

Figure 2.12 Phase-plane trajectories for the system of equation (2.165). The blank oval around the origin shows an unstable limit cycle of amplitude ~ 1.1; the dark oval farther in is a trajectory slowly winding its way towards the stable fixed point at the origin.

For the system of equation (2.165), all solutions, except those within the limit cycle, spiral outwards, as shown. For a physical system, however, one would expect the trajectories to diverge not to infinity, but to another finite state. Viewing the second and third terms of equation (2.165) as particular polynomial approximations to $cg(\dot{x})$ and $kf(x)$ in (2.164), correct to $\mathcal{O}(\epsilon^3)$ for $x, \dot{x} \sim \mathcal{O}(\epsilon)$, one can easily envisage 'more precise' approximations, correct to $\mathcal{O}(\epsilon^5)$, e.g.

$$\ddot{x} + 0.02(1 - 1.1\dot{x}^2 + 0.1\dot{x}^4)\dot{x} + (1 - 2.0069 \times 10^{-2}x^2 + 6.9444 \times 10^{-5}x^4)x = 0. \tag{2.166}$$

Some results in the phase plane are shown in Figure 2.13. It is seen that an attracting limit cycle now exists (dark oval) beyond the repelling one (dashed line) around the origin. In this case, setting $f(x) = 0$ yields five equilibria: the origin, $x = \pm 8.00$ and $x = \pm 15.00$. The origin and $x = x_{st} = \pm 15$ are stable fixed points; whereas $x = \pm 8$ are saddle points, similarly to Figure 2.12. Around the stable fixed points $x_{st} = \pm 15$, 'the flow'[†] is similar

two trajectories in the phase plane leading to it and stopping there; thus, for the pendulum, one can envisage just the right initial conditions which would result in a final state $\{\pi, 0\}$, i.e. with the pendulum inverted. However, there are two more trajectories leading away from the saddle point; the slightest disturbance will cause the pendulum to fall towards the right or the left.

[†]In nonlinear dynamics jargon, looking at trajectories as streamlines, one talks about flow, sources, sinks, etc.

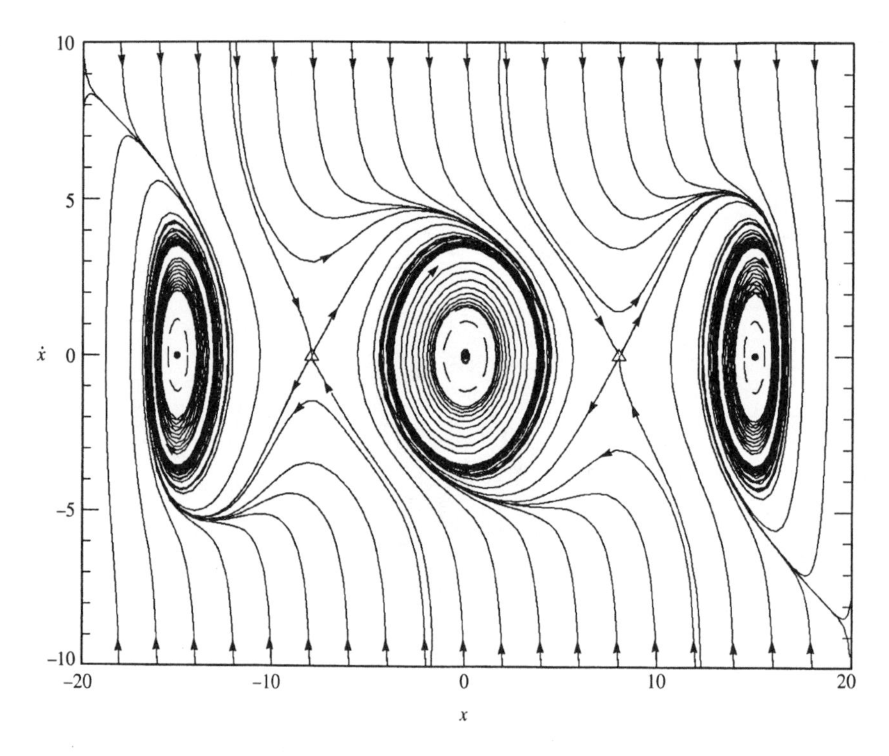

Figure 2.13 Phase-plane trajectories for the system of equation (2.166). Here three stable fixed points are shown, and two unstable ones (saddle points, denoted by $\triangle$). Each stable fixed point is encircled by an unstable limit cycle (– – –), and farther out by a stable limit cycle (dark oval patch).

to that about the origin: an unstable limit cycle (dashed line) and a stable one farther out (dark oval). The main difference is that trajectories beyond, e.g. for $|x| > 20$, cannot escape to another saddle, as there is none. It could be argued that the structure around $\{15, 0\}$ is qualitatively similar to that about $\{0, 0\}$ because (i) both fixed points are stable (hence the two points are statically similar) and (ii) $g(\dot{x})$ is invariant to the transformation $y = x - 15$, $\dot{y} = \dot{x}$; similarly for the dynamics about $\{-15, 0\}$. However, such arguments constitute but *prima facie* evidence and are not always reliable, as will be demonstrated for the system of equation (2.167).

For a physical system, the following dynamical behaviour is implied by the results of Figure 2.13: (i) very small perturbations about the static equilibrium die out, and the system returns to the origin; (ii) perturbations of amplitude larger than that of the unstable limit cycle lead the system away from equilibrium and into limit-cycle oscillations (i.e. to the larger, stable limit cycle); (iii) for still larger perturbations, the system is attracted by either this same limit cycle or beyond, to the other limit cycles, around $x_{st} = \pm 15$.

Usually, all the features described in the foregoing do not occur for the same parameter; as the parameter (U in this case) is varied, some arise, while others disappear. The apparition of any new feature in the system defines a new bifurcation. Thus, for a certain U, perhaps the only notable feature may be the stable fixed point along with the saddle points, which could have arisen earlier via a pitchfork bifurcation. This feature could remain, or disappear via a merging of these two points. At a higher U, the limit cycle(s) may emerge via a Hopf bifurcation.

Figure 2.13, as it stands, serves also to introduce the concept of coexisting *attractors* (the stable fixed points and limit cycles), each with its own *basin of attraction*: i.e. the part of phase space within which trajectories are attracted, as if by a magnet, to this or that state or attractor. The trajectories leading to and emanating from the saddle point (thus tracing an ×-intersection) are referred to as *separatrices*. In this case they separate the basins of attraction of the stable limit cycle around $\{0, 0\}$ from those about $\{\pm 15, 0\}$.

A final point in this regard is the evolution and mutual interference of attractors. Let us say that, as U is varied, the coefficients in (2.166) are altered accordingly, and the equation of motion for another U becomes

$$\begin{aligned}\ddot{x} &+ 0.02\left(1 - 1.06\dot{x}^2 + 0.0625\dot{x}^4\right)\dot{x} \\ &+ \left(1 - 4.444 \times 10^{-2}x^2 + 1.778 \times 10^{-4}x^4\right)x = 0. \qquad (2.167)\end{aligned}$$

As shown in Figure 2.14, the stable limit cycle around the origin no longer exists. Its disappearance, as a result of proximity to the saddle points on either side, constitutes another bifurcation for this system as U is varied. However, the dynamics around the outer fixed points, $|x_{st}| \simeq 15$, remain unaltered.

This case also illustrates the unreliablility of the condition $g(\dot{x}) = 0$ for determining the existence of limit cycles. In the case of the system of equation (2.166) this gives $|\dot{x}_1| = 1$ and $|\dot{x}_2| = 3.16$, which are close to the velocity-amplitudes of the limit cycles around the

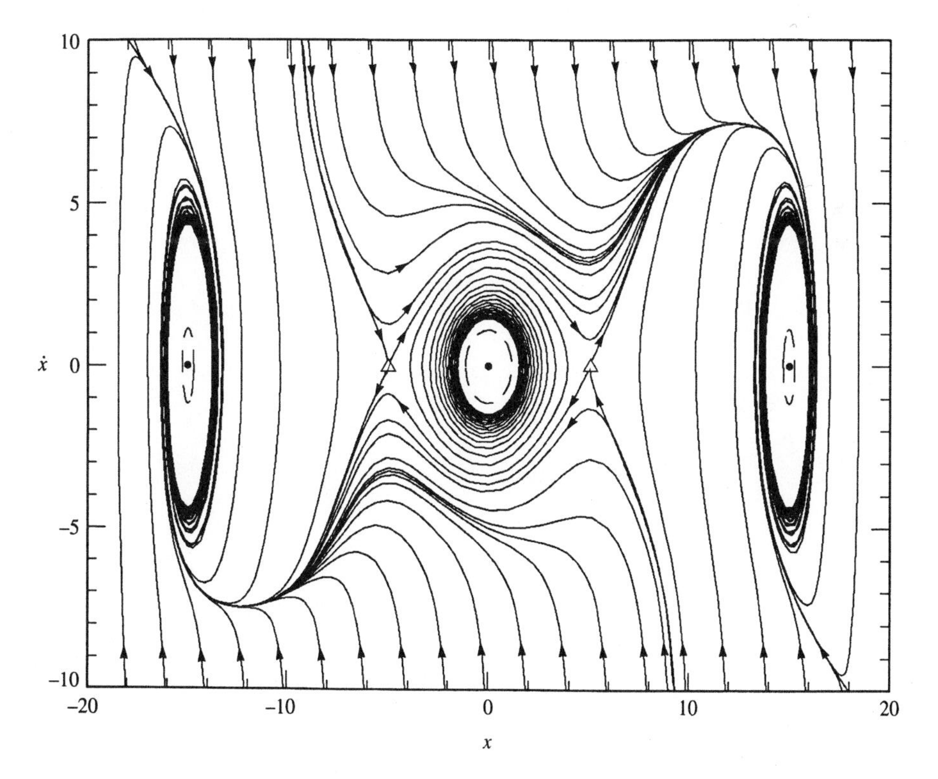

Figure 2.14 Phase-plane trajectories for the system of equation (2.167), showing the disappearance of the stable limit cycle around the origin (cf. Figure 2.13), through proximity to the two saddle points.

3
Pipes Conveying Fluid: Linear Dynamics I

3.1 INTRODUCTION

The study of dynamics of pipes conveying fluid has a fine pedigree. A series of experiments by Aitken (1878) on travelling chains and elastic cords, illustrating the balance between motion-induced tensile and centrifugal forces in this momentum transport system, is perhaps among the earliest work pertinent to the topic at hand. Self-excited oscillations of a cantilevered pipe conveying fluid had been observed by Brillouin as far back as 1885 (Bourrières 1939), but remained unpublished *"dans une Note de laboratoire"*.

The first serious study of the dynamics of pipes conveying fluid is due to Bourrières (1939), who derived the correct equations of motion and carried its analysis remarkably far, reaching admirably accurate conclusions regarding stability, in particular concerning the cantilevered system. This study, published in the year of the outbreak of the Second World War, was effectively 'lost', and researchers rederived everything in ignorance of its existence in the 1950s and 1960s. Bourrières' work was rediscovered by the author in 1973 in the course of delivering a seminar in France, thanks to a comment by Professor A. Fortier of the University of Paris who was in attendance (Païdoussis & Issid 1974).

Certainly, some aspects of the problem have been known for a long time and are in almost everyone's common experience. Thus, the *buckling* (divergence) of a pipe with both ends supported, manifested by the large restraining force that must be exerted by those holding a fire-hose at high discharge rates, is also experienced, albeit highly diminished, by one watering the lawn. The *flutter* of a cantilevered pipe, manifested by the thrashing, snaking motions of a fire-hose accidentally released or by a garden-hose when dropped on the wet grass, is well known to firemen and amateur gardeners alike. In fact, these two phenomena are often, irreverently but graphically, referred to as the *fire-hose* and *garden-hose instability*, respectively.

Nevertheless, the subject is far from being of the 'garden variety' sort. Indeed this has become a new *model problem* in the study of dynamics and stability of structures, on a par with the classical problems of a column subjected to compressive loading and the rotating shaft (Païdoussis & Li 1993). Some reasons why this is so are the following: (i) it is a physically simple system, easily modelled by simple equations, yet capable of displaying a kaleidoscope of interesting dynamical behaviour, both linear and nonlinear; (ii) it is a fairly easily realizable system, thus affording the possibility of theoretical and experimental investigation in concert; (iii) in its many variants, it is a more general problem, with richer dynamical behaviour, than that of the column and in some ways of the rotating shaft, and thus complements them both as a tool for the development of new dynamical theory and methods of analysis (Païdoussis 1987; Païdoussis & Li 1993); (iv) it belongs to a broader class of dynamical systems involving momentum transport: that of axially moving continua, such as high speed magnetic and paper tapes, band-saw blades, transmission chains and belts (Mote 1968, 1972; Wickert & Mote 1990), in paper, fibre and plastic film winding, as well as in extrusion processes.

In terms of the topics covered in this book, all of which deal with axial flow along slender structures, the pipe conveying fluid constitutes the main paradigm, on the basis of which the qualitative dynamics of other systems are explained. This is one of the reasons why so much emphasis is placed on this topic.

This chapter together with Chapter 4 deal with the *linear dynamics of initially straight pipes conveying fluid.* The nonlinear dynamics of the same physical system is the subject of Chapter 5. The dynamics of curved pipes conveying fluid is presented in Chapter 6, and that of shells containing flow in Chapter 7 (Volume 2).

The dynamics of pipes with *steady mean axial flow* is presented first, starting with a discussion of the fundamentals and the derivation of the equations of motion, in Sections 3.2 and 3.3. The dynamics of pipes with supported ends, which is an inherently conservative system (i.e. a conservative system in the absence of dissipative forces), is treated next (Section 3.4), followed by cantilevered pipes, an inherently nonconservative system (Section 3.5), and then hybrid and articulated pipe systems. Other, more complex systems and applications are the subject of Chapter 4.

3.2 THE FUNDAMENTALS

3.2.1 Pipes with supported ends

After Bourrières (1939), the study of pipes conveying fluid was re-initiated by Ashley & Haviland (1950) in an attempt to explain the vibrations observed in the Trans-Arabian Pipeline. Feodos'ev (1951), Housner (1952) and Niordson (1953) were the first to study the dynamics of pipes *supported at both ends*, obtaining the correct linear equations of motion in different ways, and reaching the correct conclusions regarding stability.

If gravity, internal damping, externally imposed tension and pressurization effects are either absent or neglected, the equation of motion of the pipe in Figure 3.1(a–c) takes the particularly simple form

$$EI\,\frac{\partial^4 w}{\partial x^4} + MU^2\,\frac{\partial^2 w}{\partial x^2} + 2MU\,\frac{\partial^2 w}{\partial x \partial t} + (M+m)\,\frac{\partial^2 w}{\partial t^2} = 0, \tag{3.1}$$

where EI is the flexural rigidity of the pipe, M is the mass of fluid per unit length, flowing with a steady flow velocity U, m is the mass of the pipe per unit length, and w is the lateral deflection of the pipe; x and t are the axial coordinate and time, respectively. The fluid forces are modelled in terms of a plug flow model, which is the simplest possible form of the slender body approximation for the problem at hand. This equation will be derived in various ways and forms in Section 3.3. Suffice it to point out here, however, that if one uses the slender body approximation (2.83), together with (2.69) and $\nu = 0$,[†]

[†]As will be seen later, the equation of motion is independent of fluid frictional effects, and equation (3.1) holds true if pressure drop in the pipe is taken into account, i.e. for $\nu \neq 0$.

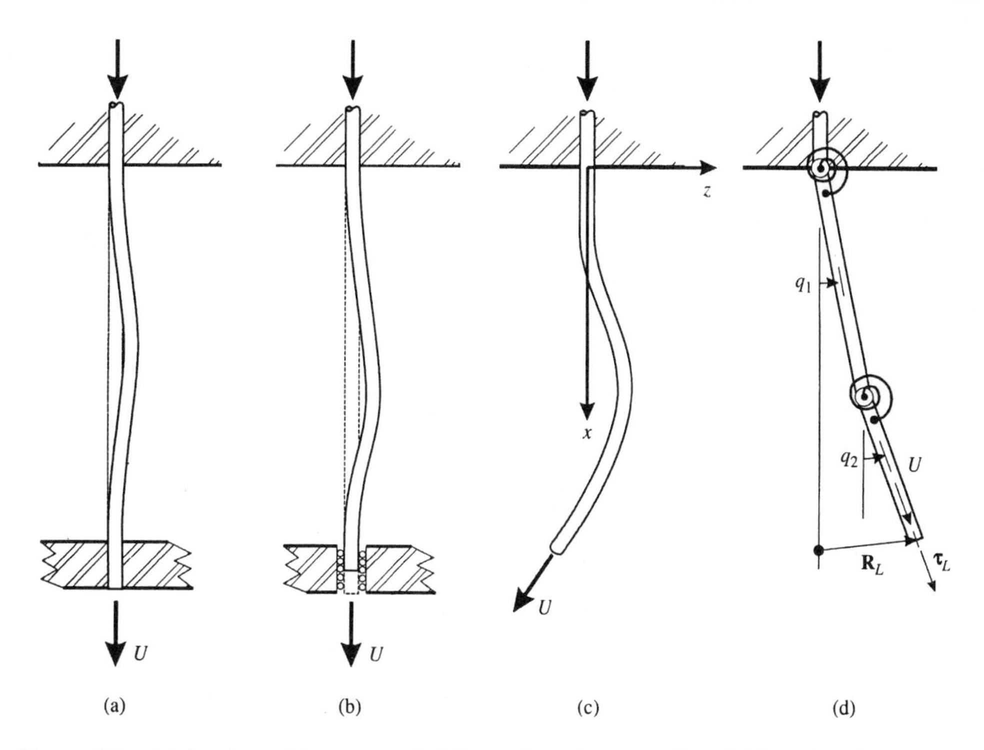

Figure 3.1 (a) A pipe with supported (clamped) ends conveying fluid, where longitudinal movement at the supports is prevented; (b) the same system, but with axial sliding permitted; (c) a cantilevered, continuously flexible pipe conveying fluid; (d) a two-degree-of-freedom articulated version of the cantilevered system, in which $\mathbf{R}_L$ is the position vector of the free end, measured from its position of equilibrium, and $\boldsymbol{\tau}_L$ is the unit vector tangent to the free end.

it is clear how the terms related to fluid acceleration,

$$\left\{\frac{\partial}{\partial t} + U\frac{\partial}{\partial x}\right\}\left[\frac{\partial w}{\partial t} + U\frac{\partial w}{\partial x}\right] = \left[U^2\frac{\partial^2 w}{\partial x^2} + 2U\frac{\partial^2 w}{\partial x \partial t} + \frac{\partial^2 w}{\partial t^2}\right], \tag{3.2}$$

arise in equation (3.1). Here, however, the equation of motion will be considered in purely physical terms.

The first term in equation (3.1) is the flexural restoring force. Upon recalling that $\partial^2 w/\partial x^2 \sim 1/\mathcal{R}$, where $\mathcal{R}$ is the local radius of curvature, it is obvious that the second term is associated with centrifugal forces as the fluid flows in curved portions of the pipe — see Figure 3.1(a–c). Similarly, re-writing $\partial^2 w/\partial x\partial t = \partial\theta/\partial t = \Omega$, the local angular velocity, it is clear that the third term is associated with Coriolis effects: the fluid flows longitudinally with velocity $U\mathbf{i}$, while sections of the pipe rotate with $-\Omega\mathbf{j}$, where $\mathbf{j}$ is normal to (into) the plane of the paper; hence $-2\Omega\mathbf{j} \times U\mathbf{i}$ terms arise. The last term represents the inertial force of the fluid-filled pipe.

Equation (3.1) may be compared to the equation of motion of a beam subjected to a compressive load, P,

$$EI\frac{\partial^4 w}{\partial x^4} + P\frac{\partial^2 w}{\partial x^2} + m\frac{\partial^2 w}{\partial t^2} = 0, \tag{3.3}$$

i.e. equation (2.47). It is clear that the centrifugal force in (3.1) acts in the same manner as a compressive load. In this way, it is easy to see and to understand physically that, with increasing U, the effective stiffness of the pipe is diminished; for sufficiently large U, the destabilizing centrifugal force may overcome the restoring flexural force, resulting in *divergence*, vulgarly known as *buckling* and, in the nonlinear dynamics *milieu*, as a *pitchfork bifurcation*.

In the foregoing argument, it was implicitly assumed that the Coriolis forces do no work in the course of free motions of the pipe, which is true. The rate of work done on the pipe by the fluid-dynamic forces, the only possible source of energy input, in the course of periodic motions is

$$\frac{\mathrm{d}W}{\mathrm{d}t} = -\int_0^L \frac{\partial w}{\partial t} M \left\{ \frac{\partial}{\partial t} + U \frac{\partial}{\partial x} \right\} \left[\frac{\partial w}{\partial t} + U \frac{\partial w}{\partial x} \right] \mathrm{d}x, \tag{3.4}$$

and hence the work done by the fluid forces over a cycle of periodic oscillation of period T is

$$\Delta W = -MU \int_0^T \left[\left(\frac{\partial w}{\partial t} \right)^2 + U \left(\frac{\partial w}{\partial t} \right) \left(\frac{\partial w}{\partial x} \right) \right] \Bigg|_0^L \mathrm{d}t. \tag{3.5}$$

Clearly if the ends of the pipe are positively supported, then $(\partial w/\partial t) = 0$ at both ends, and

$$\Delta W = 0. \tag{3.6}$$

Nonworking velocity-dependent loads are called gyroscopic by Ziegler (1968) and hence this system is classified as a *gyroscopic conservative* system. In Galerkin discretizations of this system, the Coriolis-related velocity-dependent matrix is purely skew-symmetric (antisymmetric) [see, e.g. Done & Simpson (1977) and Section 3.4.1 here].

Because divergence is a static rather than dynamic form of instability, the dynamics of the system may be examined by considering only the time-independent terms in equation (3.1), so effectively equation (3.3) with the inertia term put to zero; whereby, for a simply-supported pipe, the particularly simple result is obtained (Section 3.4.1) for the critical flow velocity U_c, namely that the *dimensionless* critical flow velocity is

$$u_c = \pi, \tag{3.7}$$

where u is defined as

$$u = (M/EI)^{1/2} UL, \tag{3.8}$$

in which L is the length of the pipe. Similarly, for a simply-supported column (Ziegler 1968),

$$\mathcal{P}_c = \pi^2, \qquad \mathcal{P} = PL^2/EI; \tag{3.9}$$

it is clear from equations (3.1) and (3.3) that the equivalent of $\mathcal{P}$ is u^2, rather than u. As expected, the dynamical behaviour of pipes with one or both ends clamped, rather than simply supported, is similar.

The analogy between equations (3.1) and (3.3) and the discussion just made show also how the natural frequencies of the system should develop with increasing U. It is physically obvious in the column problem that, as the compressive load is increased, the effective rigidity (or stiffness) of the system is eroded, to the point where it vanishes;

similarly for the pipe problem, as U is increased. Hence, it is obvious that the frequencies of the system must decrease with increasing U. At u_c, the lowest (fundamental) frequency vanishes as the stiffness in that mode vanishes. In the linear sense, the original straight configuration becomes unstable, and all adjacent deformed states in that mode become possible equilibria. In the nonlinear sense, a pitchfork bifurcation takes place, the original equilibrium is unstable and *two* stable equilibrium states, one on either side, emerge — defined by the nonlinear forces acting on the system, as will be demonstrated in Chapter 5.

However, the analogy of the pipe with supported ends to the column with the same boundary conditions should not be carried too far, because the latter problem is purely conservative, while the former is gyroscopic conservative. As will be shown later, despite the fact that the gyroscopic (Coriolis) forces do no work in the course of free oscillations, they do exert important influence on the overall dynamical behaviour.[†]

Finally, it should be mentioned that, according to linear theory, there should be no difference in the dynamics of systems (a) and (b) of Figure 3.1. In physical terms, however, it is obvious that buckling implies lateral deflection of the pipe. In system (b), once $u \geq u_c$, the pipe may develop large static deflection since it is axially unrestrained. In system (a), on the other hand, where axial sliding of the lower end is prevented, lateral deflection is associated with axial *extension* of the pipe; this implies stretching and hence the generation of a deflection-related axial tension, a nonlinear effect. In practice, this means that the zero-frequency state is never achieved, as will be discussed further in Section 3.4.

3.2.2 Cantilevered pipes

As will be shown, a cantilevered pipe conveying fluid is a nonconservative system, which, for sufficiently high flow velocity, loses stability by *flutter* of the single-mode type, i.e. via a *Hopf bifurcation* — see also Section 3.2.3.

The stability of cantilevered pipes conveying fluid [see Figure 3.1(c)] was first studied by Bourrières (1939), who examined the problem of general motions of an infinitely flexible and inextensible string, and the special case where the string is circulating (travelling) between two fixed supports; he then tackled the problem of one such string within another, which could have flexural rigidity — this of course being equivalent to the case of a pipe conveying fluid. He obtained the general nonlinear equations of motion, but did not develop them fully. Then, he linearized them and proceeded to study such diverse aspects as the difference between spontaneous and perturbation-induced instabilities (cf. Gregory & Païdoussis 1966b), and the wave propagation characteristics; he also attempted to predict the period of self-excited motions, and studied several other aspects of the problem, as well as conducting experiments. On the other hand, he could not calculate the critical flow velocity, which, unlike the case of a pipe with supported ends, requires the use of computers[‡] — of course, then unavailable. Bourrières' was a truly admirable effort, and it is a pity that it was lost to posterity, until recently (Section 3.1). His work did not have

[†]In this respect, as civil servants the world over discovered long ago (and as viewers of BBC's *Yes Minister* have witnessed to their delight), it is not necessary to do actual work in order to exert influence; see also Lynn & Jay (1989).

[‡]Although Païdoussis (1963), in order to check computer calculations — computers then being a relatively new device — did do a hand calculation, thereby demonstrating its feasibility.

any influence on subsequent research, except in an important way on a set of nonlinear studies to be discussed in Chapter 5.

The next study, some 20 years later, was Benjamin's (1961a,b), mainly on the dynamics of articulated cantilevers conveying fluid [Figures 2.1(b) and 3.1(d)], but with an authoritative discussion of the continuous system [Figure 3.1(c)].[†] One of the principal accomplishment, among many, of this work was the establishment of the appropriate form of the Lagrangian equations for this 'open' system (open, in the sense that momentum constantly flows in one end and out the other), namely

$$\frac{\mathrm{d}}{\mathrm{d}t}\left(\frac{\partial T}{\partial \dot{q}_k}\right) - \frac{\partial T}{\partial q_k} + \frac{\partial V}{\partial q_k} = -MU(\dot{\mathbf{R}}_L + U\boldsymbol{\tau}_L)\cdot\frac{\partial \mathbf{R}_L}{\partial q_k}, \tag{3.10}$$

in which T and V are the total kinetic and potential energies of the system, $\mathbf{R}_L$ is the position vector of the free end and $\boldsymbol{\tau}_L$ the unit vector tangent to the free end [Figure 3.1(d)]; q_k are the generalized coordinates, typically the angles made by each of the rigid pipes of the system with the undeformed line of equilibrium. The corresponding statement of Hamilton's principle was also obtained, from which the equations of motion of the continuous system (and the articulated one, if so desired) may be derived.

The equation of motion of the continuous cantilevered system is the same as that of a pipe with supported ends, equation (3.1); this will be derived in Section 3.3, and there are subtle differences in the derivation for these two cases (Section 3.3.3). However, physically, it seems reasonable that the same equation should hold. Similarly, the same expression, equation (3.5), holds true for the work done by the fluid on the pipe over a period T of periodic oscillation, but in this case it is equal to

$$\Delta W = -MU\int_0^T \left[\left(\frac{\partial w}{\partial t}\right)_L^2 + U\left(\frac{\partial w}{\partial t}\right)_L\left(\frac{\partial w}{\partial x}\right)_L\right]\mathrm{d}t \neq 0, \tag{3.11}$$

where $(\partial w/\partial t)_L$ and $(\partial w/\partial x)_L$ are, respectively, the lateral velocity and slope of the free end. In Ziegler's (1968) classification, since some of the forces associated with $\Delta W \neq 0$ are not velocity-dependent [the $MU^2(\partial^2 w/\partial x^2)$ follower load leading to the second term in (3.11)], this is a *circulatory* system. The dynamics of this system was elucidated by means of this expression by Benjamin (1961a) and elaborated by Païdoussis (1970).

For $U > 0$ and sufficiently small for the second term within the square brackets to be much smaller than the first, it is clear that $\Delta W < 0$, and free motions of the pipe are damped — an effect due to the Coriolis forces, which, unlike the case of supported ends, in this case *do* do work. If, however, U is sufficiently large, while over most of the cycle $(\partial w/\partial x)_L$ and $(\partial w/\partial t)_L$ have opposite signs, then $\Delta W > 0$; i.e. the pipe will gain energy from the flow, and free motions will be amplified. The requirement that $(\partial w/\partial x)_L\,(\partial w/\partial t)_L < 0$ suggests that, in the course of flutter, the pipe must execute a sort of 'dragging', lagging motion that one would obtain when laterally oscillating a long flexible blade or baton in dense fluid. This, indeed, is what is observed, as remarked by Bourrières (1939), Benjamin (1961b) and Gregory & Païdoussis (1966b).

[†] 'A continuous system' will henceforth denote the distributed parameter system involving a continuously flexible pipe.

The energy transfer mechanism was also demonstrated in terms of rudimentary representations of the operation of a pump and a radial-flow turbine by Benjamin (1961a), as follows.

Suppose first that in the course of some free motion the pipe rotates about A without bending elsewhere, as shown in Figure 3.2(a). This motion requires transfer of energy from the pipe to the fluid, since the Coriolis forces on the fluid have reactions on the pipe in a direction always opposing motion. [For the motion to continue (with the pipe remaining straight between A and C), work from an external source would have to be done on the pipe, over and above that for bending it at A.] Thus, this energy transfer mechanism causes the fluid to gain kinetic energy in passing through the pipe, and the centripetal acceleration of the fluid results in a suction developing at the inlet, A; on reflection, this is essentially the action of a *centrifugal pump.*

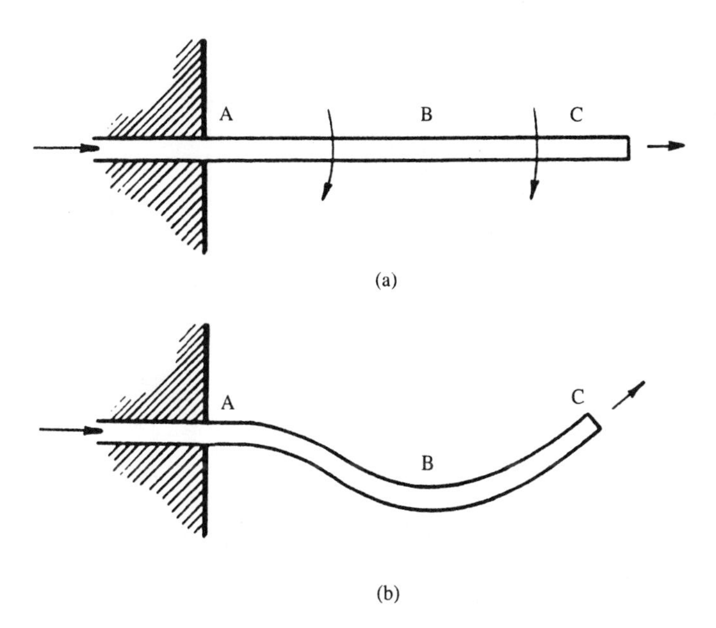

Figure 3.2 Rudimentary representation of (a) a pump and (b) a radial-flow turbine, illustrating the mechanisms of energy transfer in a cantilevered pipe conveying fluid, as proposed by Benjamin (1961a). From Païdoussis (1973a).

Consider next the pipe momentarily 'frozen' in the shape shown in Figure 3.2(b); the change in direction of the momentum of the fluid stream about B gives rise to a reaction on the pipe, resulting in a clockwise couple. In this case, energy is transferred from the fluid to the pipe, causing it to accelerate to a speed at which the rate of energy gain just balances the work done in bending the pipe at B. The energy-transfer mechanism in this case corresponds to that of a *radial-flow turbine.* (It is noted, however, that if the rotation about A becomes sufficiently rapid, pumping action will again prevail.)

In general, in the course of free motions of the system both mechanisms will be operative. If the first predominates, oscillatory motions will be damped; but if the second prevails, they will be amplified continuously, i.e. an oscillatory instability will develop.

A strange characteristic of this system is that, at high flow velocities but before the onset of flutter, supporting the downstream end of the cantilever by one's finger or a pencil causes it to become unstable by divergence (Benjamin 1961b; Gregory & Païdoussis 1966b). So, here is a case where *added support causes instability!* If one tries to remove the finger or pencil slowly, the pipe follows! This shows clearly and physically that the divergence is a negative stiffness instability. This also gives rise to an interesting paradox, discovered by Thompson (1982b) and elucidated in terms of the strange black box of Figure 3.3(a,b). As more weight is placed on the scale, the scale goes *up*.† What could be in the box is shown in Figure 3.3(c). The phenomenon is nonlinear and its discussion properly belongs to Chapter 5 (Section 5.6.1); it has nevertheless been outlined here to whet the appetite, so to speak, for the many interesting aspects of the nonlinear behaviour of this system.

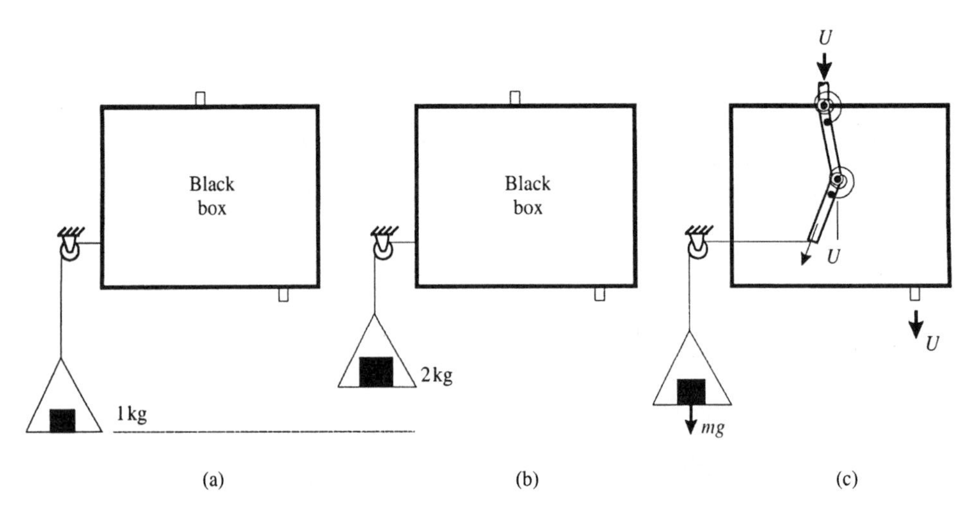

Figure 3.3 Illustration of the negative stiffness mechanism of a buckled pipe conveying fluid, analysed by Thompson (1982b).

The stability of this system was linked to the classical nonconservative problem of a column subjected to a tangential follower-type load at the free end,‡ known as Beck's problem or Nicolai's paradox, by Nemat-Nasser *et al*. (1966), Herrmann (1967) and Herrmann & Nemat-Nasser (1967). Beck's problem may be summarized as follows (Bolotin 1963; Ziegler 1968). As already suggested in Section 3.2.1, the stability of a conservative system may be assessed statically, i.e. by ignoring the time-dependent forces; e.g. in the case of a column with supported ends or of a cantilevered one with a compressive load of fixed orientation. The same may be attempted — as first done by Nicolai in 1928 — for a cantilevered column with a follower load, i.e. a compressive load with fixed orientation *relative to the column*, notably a load always tangential to the free end (as in Figure 2.2). The paradoxical result is then obtained that the system

†A second, but dynamically trivial paradox is that the black box of Figure 3.3 is in fact white!

‡By the analogy between equations (3.1) and (3.3) it may easily be shown that the equivalent of (3.5) is $\Delta W = -P\int_0^T \left[(\partial w/\partial t)(\partial w/\partial x)\right]_0^L \mathrm{d}t \neq 0$, since neither $(\partial w/\partial t)_L$ nor $(\partial w/\partial x)_L$ are zero for all $t \in [0, T]$.

apparently *never loses stability*! The resolution of the paradox is that the system never loses stability *statically*. The critical compressive load was determined by Beck in 1952 (Bolotin 1963) by solving the full equation of motion, equation (3.3). It is given by $\mathcal{P}_c = P_c L^2/EI = 20.05$,[†] at which point *coupled-mode flutter* arises, otherwise known as a *Hamiltonian Hopf bifurcation*, in contrast to the cantilevered pipe, which loses stability by single-mode flutter via an ordinary Hopf bifurcation — see Section 3.2.3.

The fact that the cantilever conveying fluid is not only a nonconservative problem similar to Beck's (a circulatory dynamical system in Ziegler's classification), but is also subject to gyroscopic forces[‡] helps explain the fascination it has exerted, and does so still, on applied mechanicians and mathematicians for the last 30 years. An additional *point fort* of this system is that it can readily be realized and studied experimentally, unlike the original Beck's problem which requires a rocket-engine mounted to the free end of a beam column, or something similar — not an easy task! Indeed, it was implied in a lecture (Païdoussis 1986a) that such a task was much *too* hard to contemplate, which a team of Japanese researchers promptly disproved (Sugiyama *et al*. 1990), by doing the difficult experiment with a solid-fuel rocket, demonstrating the occurrence of flutter and obtaining good agreement with theory — see also Section 3.6.5.

Finally, a few words on the case when the flow is from the free end towards the clamped one: by reinterpreting (3.11) for $U < 0$ it would appear that the system is unstable by flutter for small U (indeed for infinitesimally small U if dissipation is ignored!) and is then stabilized for larger $|U|$, as first pointed out by Païdoussis & Luu (1985) — the inverse behaviour to that described heretofore. More will be said about this in Chapter 4 (Section 4.3), but in what follows we return to the system with $U > 0$.

3.2.3 On the various bifurcations

A general discussion of the evolution of the eigenvalues and the corresponding eigenfrequencies leading to some of the standard bifurcations or linear instabilities was given in Section 2.3. This is reinforced and expanded here for the phenomena of interest in this chapter.

The Argand diagrams for *divergence* via a *pitchfork bifurcation*[§] are shown in Figure 2.10(a). If the system is conservative (zero dissipation), the diagram for the eigenfrequencies is modified. The eigenfrequencies are wholly real for $u < u_c$, and then become wholly imaginary (a conjugate pair), as shown in Figure 3.4(a); hence $\omega = 0$ for $u = u_c$. The corresponding *eigenvalues* are wholly imaginary for $u < u_c$, and then for $u > u_c$ become wholly real; the eigenvalue Argand diagram for each of the cases in Figure 3.4 is obtained via a 90° counterclockwise rotation of the corresponding eigenfrequency diagram. For the pipe and the column with simply-supported ends, $u_c = \pi$ and $\mathcal{P}_c = \pi^2$, respectively — see equations (3.7) and (3.9).

[†]This value of $\mathcal{P}_c$ is about eight times higher than the Euler buckling load for fixed-orientation compression of the cantilevered column, $\mathcal{P}_c = \frac{1}{4}\pi^2$ (Ziegler 1968).

[‡]Unlike the system with supported ends, if this system is discretized, the Coriolis-related matrix is not skew-symmetric; it can of course be decomposed into symmetric and skew-symmetric parts.

[§]Strictly speaking, the type of bifurcation involved is defined by the nonlinear terms in the equation of motion. In this case, the flow-related nonlinearities in the stiffness term are cubic and similar to those in a softening cubic spring. This is what gives rise to two stable static equilibria for $u > u_c$ — cf. equation (2.165) and the discussion following it in Section 2.3.

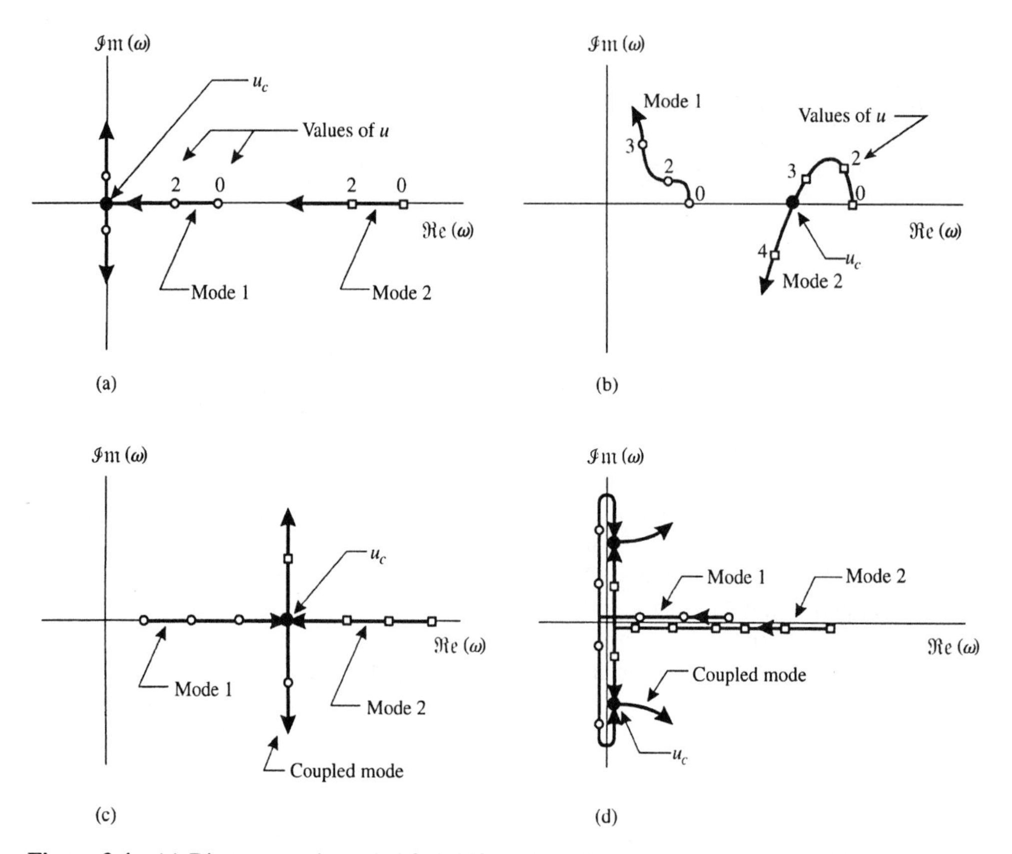

Figure 3.4 (a) Divergence via a pitchfork bifurcation of a conservative system; (b) single-mode flutter of a circulatory system via a Hopf bifurcation; (c) coupled-mode flutter via a Hamiltonian Hopf bifurcation; (d) the 'Païdoussis coupled-mode flutter' [see Done & Simpson (1977)]; ω is the dimensionless form of Ω — see equation (3.73).

The case of the ordinary *Hopf bifurcation* is shown in Figure 3.4(b) for a system with zero structural damping [$\mathfrak{Im}(\omega) = 0$ for $u = 0$]; it is characterized by the crossing of the eigenfrequency locus from the positive to the negative half-plane in the Argand diagram. For $u < u_c$ the system is damped, while for $u > u_c$ it is negatively damped in the second mode, which signifies *single-mode amplified oscillations* or *flutter*.

The Argand diagram for *coupled-mode flutter* of an undamped system via a so-called *Hamiltonian Hopf bifurcation* is shown in Figure 3.4(c). It is called Hamiltonian because (i) for $u < u_c$ there is no damping in the system and (ii) for $u > u_c$ the coalescence of the two modes has resulted in two eigenfrequencies, respectively positively and negatively damped — both characteristics resembling those in a pitchfork bifurcation, generally associated with conservative (Hamiltonian) systems. In this case, however, $\mathfrak{Re}(\omega) \neq 0$ for $u > u_c$, and hence the negative $\mathfrak{Im}(\omega)$-branch leads to flutter, similarly to the ordinary Hopf bifurcation, except that here more than one mode is involved. As discussed by Ziegler (1968), conservative systems lose stability by divergence. If they are gyroscopic, however, they may regain stability, according to linear theory at least, and then be subjected to further linear instabilities as the loading parameter is increased. As will be seen in Section 3.4, bifurcations such as that of Figure 3.4(c) do occur for $u > u_c$ for pipes with supported ends. An example of a system that loses stability by a Hamiltonian Hopf bifurcation is the column subjected to a tangential follower load, a nonconservative circulatory system, for which $u_c^2 \equiv \mathcal{P}_c = 20.05$.

Finally, Figure 3.4(d) shows another form of coupled-mode flutter, for which Done & Simpson's (1977) nomenclature of *Païdoussis' (coupled-mode) flutter* will be retained, to distinguish it from the Hamiltonian Hopf bifurcation of Figure 3.4(c). The distinguishing feature is that in this case the bifurcation originates directly form a divergent state; hence, at the onset of flutter ($u = u_c$), the frequency of oscillation is zero [$\mathcal{R}e(\omega) = 0$], and then $\mathcal{R}e(\omega) \neq 0$ for $u > u_c$. This kind of bifurcation will be found to arise for pipes with supported ends (Section 3.4), as well as for other systems (e.g.in Chapter 8).

3.3 THE EQUATIONS OF MOTION

3.3.1 Preamble

The linear equation of motion for a pipe conveying fluid will be derived in the next two sections by the Newtonian and the Hamiltonian approaches. Before embarking on these derivations, however, it is useful to introduce some basic concepts.

The first is related to the description of the system via either *Eulerian* or *Lagrangian coordinates*, differentiated by the concepts of *spatial position* and *particle individuality*, respectively. In the Eulerian description the coordinates are fixed in space and may not be populated by the same material particles as time varies; these are the coordinates commonly used in fluid mechanics (e.g. in Section 2.2). In the Lagrangian description, coordinates are identified with individual particles (or elemental volumes surrounding marked points in the continuum).

To fix ideas, let us consider the longitudinal vibration of a bar, i.e. a one-dimensional continuum. In the Eulerian description, the position x, fixed in space, may be used as the independent space variable, and the deflection field described as $u(x, t)$; as the bar vibrates, different particles or material points at different times will be located at x. In the Lagrangian description, a given particle may be identified by its position at a given time (say, $t = 0$) or, more usefully, by its position when the bar is undeformed, $x = x_0$. This particle will be at a different x as time varies, but will be identified with x_0 always (Hodge 1970). Clearly, the deflection field may equally be described in terms of $u(x_0, t)$. This is the more 'mechanical' description and it is the foundation of Lagrangian dynamics, for instance.

Similarly, in the case of flexural oscillations of the pipe, treated as a beam, two coordinate systems may be utilized: the Eulerian (x, z) or the Lagrangian (x_0, z_0) — see Figure 3.5(a). The equilibrium configuration is along the x-axis, and hence $(x_0, z_0) \equiv (x_0, 0)$ in this case. The lateral deflection of the pipe may be described as $w(x, t)$ in Eulerian coordinates or $w(x_0, t)$ in the Lagrangian ones; however, as we can see, there is also change in the axial or x-position of each point, i.e. $u(x, t)$ or $u(x_0, t)$. If we consider a point P, which in the undeformed state is at P_0, then its deflection is

$$u = x - x_0 \qquad \text{and} \qquad w = z - z_0 = z. \tag{3.12}$$

In what follows we shall use both sets of coordinates, but the usefulness of this discussion will become most evident when the nonlinear equations of motion are derived in Chapter 5.

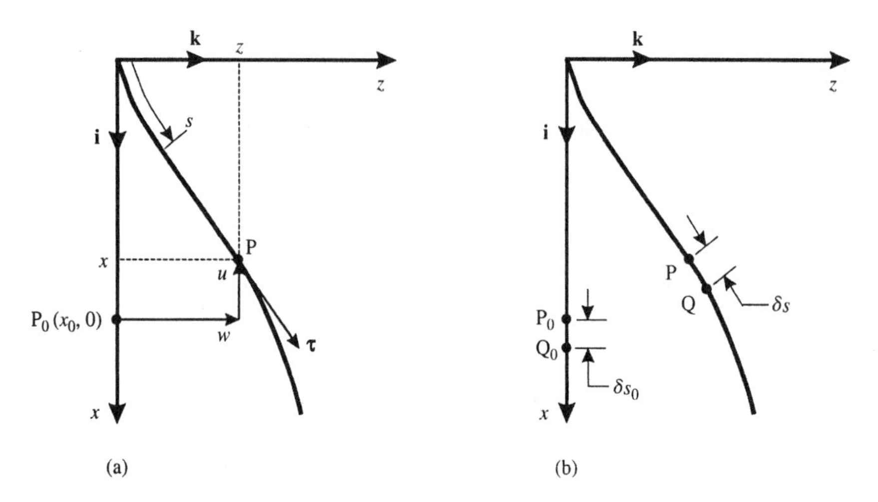

Figure 3.5 (a) The Eulerian coordinate system (x, z) and the Lagrangian one $(x_0, z_0) \equiv (x_0, 0)$ in which the x_0-axis is superposed on the x-axis, showing the deflection of a point $P_0 = P_0(x_0, 0)$ to $P(x, z)$ and the definition of u and w; (b) diagram used for the derivation of the inextensibility condition.

Two further points should be made: (i) whenever Lagrangian coordinates are used, they are used for pipe motions only, not for the fluid; (ii) it is customary to use a curvilinear coordinate s, along the length of the pipe, as shown in Figure 3.5(a) — especially useful if the pipe is considered to be inextensible.

The second concept of importance to be discussed in this section is that of *inextensibility*. For pipes supported as in Figure 3.1(b,c) for instance, where no deflection-dependent axial forces come into play, one may clearly consider the pipe to be inextensible, i.e. the length of its centreline to remain constant during oscillation. However, in the case of a pipe with positively supported ends [Figure 3.1(a)], i.e. with no axial sliding permitted, lateral deflection may occur *only if* the pipe is extensible.

Consider contiguous points P and Q of the deflected pipe, originally (in the undeflected state) at P_0 and Q_0, as in Figure 3.5(b). Then,

$$(\delta s)^2 = (\delta x)^2 + (\delta z)^2, \qquad (\delta s_0)^2 = (\delta x_0)^2 + (\delta z_0)^2 = (\delta x_0)^2,$$

from which one may write

$$(\delta s)^2 - (\delta s_0)^2 = \left[\left(\frac{\partial x}{\partial x_0}\right)^2 + \left(\frac{\partial z}{\partial x_0}\right)^2 - 1\right](\delta x_0)^2. \tag{3.13}$$

If the pipe is inextensible, $\delta s = \delta s_0$ by definition, and the condition of inextensibility may be expressed as

$$\left(\frac{\partial x}{\partial x_0}\right)^2 + \left(\frac{\partial z}{\partial x_0}\right)^2 = 1. \tag{3.14}$$

The inextensibility condition may also be expressed in terms of the displacements (u, w); by invoking (3.12),

$$\left(1+\frac{\partial u}{\partial x_0}\right)^2+\left(\frac{\partial w}{\partial x_0}\right)^2=1. \tag{3.15}$$

In both (3.14) and (3.15), x_0 may be replaced by s.

If the pipe cannot be considered to be inextensible, e.g. in Figure 3.1(a), δx_0 and δs are no longer equal; they must be related through (3.13) which, with the aid of (3.12), leads to

$$\frac{\partial x_0}{\partial s}=\left[\left(1+\frac{\partial u}{\partial x_0}\right)^2+\left(\frac{\partial w}{\partial x_0}\right)^2\right]^{-1/2} \tag{3.16}$$

The final preliminary point that needs be examined is related to the orders of magnitude of the displacements, which define the degree of approximation and simplification that is admissible in the derivations to follow. First, it is reasonable to assume, particularly in linear analysis, that the lateral displacement w is small compared to the pipe length, i.e.

$$w/L \sim \mathcal{O}(\epsilon), \tag{3.17a}$$

where $\epsilon \ll 1$. By expanding (3.15) and neglecting $(\partial u/\partial x_0)^2$ as compared to $2(\partial u/\partial x_0)$, and also replacing x_0 by s, it is clear that

$$u \simeq -\int_0^s \frac{1}{2}\left(\frac{\partial w}{\partial s}\right)^2 \mathrm{d}s, \qquad u/L \sim \mathcal{O}(\epsilon^2); \tag{3.17b}$$

i.e. longitudinal displacements are one order smaller than the lateral ones. It is also well known that, in the Newtonian approach, if all terms are correct to order ϵ, so is the equation of motion. In the Hamiltonian approach, however, since the energies are generally quadratic expressions of displacements and velocities, the various terms should be correct to order ϵ^2. Hence, in the Newtonian derivation of Section 3.3.2 one may take $x = x_0 = s$ and consider only the lateral deflection of the pipe, $w = w(x, t)$. In the Hamiltonian derivation of Section 3.3.3, however, one has to take account of $u(x, t)$ as well, and to take care to differentiate x_0 or s from x, since then generally $x \not\simeq s$ for inextensible pipes and also $x_0 \neq s$ for extensible ones.

3.3.2 Newtonian derivation

Consider the system of Figure 3.1(a–c), a uniform pipe of length L, internal perimeter S, flow-area A, mass per unit length m, and flexural rigidity EI, conveying fluid of mass per unit length M, with mean axial flow velocity U. The flow in the pipe is fully developed turbulent. Consider the undisturbed axis of the pipe to be vertical, along the x-axis, and the effect of gravity to be generally non-negligible. The flow velocity may be subject to small perturbations, imposed externally, so that $\mathrm{d}U/\mathrm{d}t \neq 0$ generally.

The pipe is considered to be slender, and its lateral motions, $w(x, t)$, to be small and of long wavelength compared to the diameter; thus, in accordance with the discussion in Section 3.3.1, the curvilinear coordinate s along the centreline of the pipe and the coordinate x may be used interchangeably. Consider then elements δs of the fluid and the pipe, as shown in Figure 3.6.

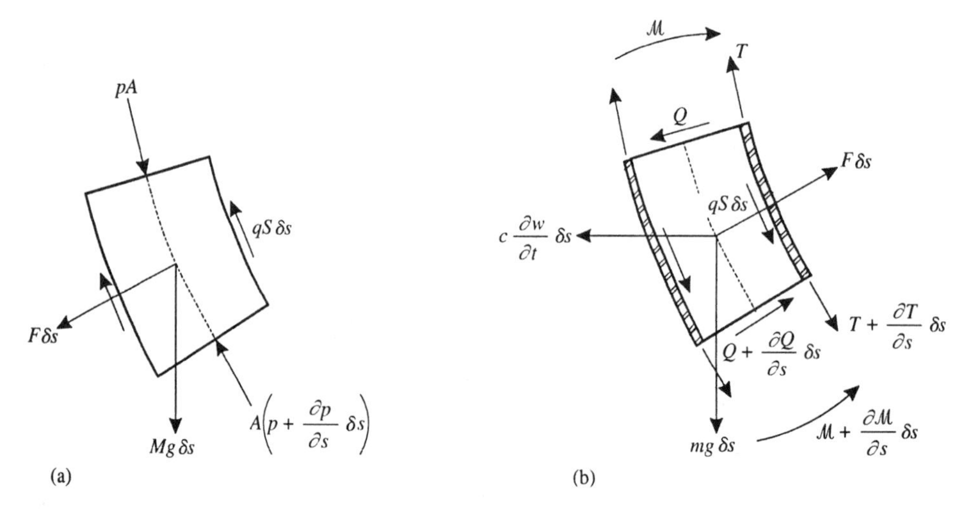

Figure 3.6 (a) Forces acting on an element δs of the fluid; (b) forces and moments on the corresponding element of the pipe.

The fluid element of Figure 3.6(a) is subjected to: (i) pressure forces, where the pressure $p = p(s, t)$ because of frictional losses, and p is measured above the ambient pressure; (ii) reaction forces of the pipe on the fluid normal to the fluid element, $F\,\delta s$, and tangential to it, $qS\,\delta s$, associated with the wall-shear stress q; (iii) gravity forces $Mg\,\delta s$ in the x-direction. Applying Newton's second law in the x- and z-directions, while keeping in mind the small-deflection approximation, yields

$$-A\,\frac{\partial p}{\partial x} - qS + Mg + F\,\frac{\partial w}{\partial x} = Ma_{fx}, \tag{3.18}$$

$$-F - A\,\frac{\partial}{\partial x}\left(p\frac{\partial w}{\partial x}\right) - qS\,\frac{\partial w}{\partial x} = Ma_{fz}, \tag{3.19}$$

where a_{fx} and a_{fz} are the accelerations of the fluid element in the x- and z-direction, respectively. Similarly, for the pipe element of Figure 3.6(b) one obtains

$$\frac{\partial T}{\partial x} + qS + mg - F\,\frac{\partial w}{\partial x} = 0, \tag{3.20}$$

$$\frac{\partial Q}{\partial x} + F + \frac{\partial}{\partial x}\left(T\,\frac{\partial w}{\partial x}\right) + qS\,\frac{\partial w}{\partial x} - c\,\frac{\partial w}{\partial t} = ma_{pz}, \tag{3.21}$$

$$Q = \frac{\partial \mathcal{M}}{\partial x} = -\left(E + E^*\,\frac{\partial}{\partial t}\right) I\,\frac{\partial^3 w}{\partial x^3}, \tag{3.22}$$

where T is the longitudinal tension, Q the transverse shear force, and $\mathcal{M}$ the bending moment; moreover, the pipe is subjected to internal dissipation of the Kelvin–Voigt type (e.g. Shames 1964; Meirovitch 1967; Snowdon 1968), thus following a stress–strain (σ, ε) relationship of the form $\sigma = E\varepsilon + E^*(\mathrm{d}\varepsilon/\mathrm{d}t)$, and also to damping due to friction with the surrounding fluid, expressed in linear form as $c(\partial w/\partial t)$.[†] The subscript f in equations (3.18) and (3.19) identifies the acceleration of the *fluid* and subscript p in (3.21) that of the *pipe*. Terms of second order of magnitude, for example the pipe acceleration in the x-direction, have been neglected, as well as transverse shear deformation and rotatory inertia in accordance with the Euler–Bernoulli beam approximation.

The acceleration of the fluid may be determined in several ways. The simplest is utilized here, while other derivations will be employed when considering variants of the basic system. The basic assumption is that the fluid flow may be approximated as a plug flow, i.e. as if it were an infinitely flexible rod travelling through the pipe, all points of the fluid having a velocity U relative to the pipe; this is a reasonable approximation for a fully developed turbulent flow profile. As it has been assumed that pipe deflections are of long wavelength compared to the diameter, D, and that the pipe is slender, i.e. L/D is large, unsteady secondary-flow effects may be neglected. Hence, the equivalent of a slender-body approximation to the flow is being made. The velocity of the pipe is

$$\mathbf{V}_p = \frac{\partial \mathbf{r}}{\partial t} = \dot{x}\mathbf{i} + \dot{z}\mathbf{k} \tag{3.23}$$

in terms of the unit vectors in the x- and z-directions, defined in Figure 3.5(a), where $\mathbf{r}$ is the position vector to a point measured from the origin; and the velocity of the centre of the fluid element of Figure 3.6(a) is

$$\mathbf{V}_f = \mathbf{V}_p + U\boldsymbol{\tau}, \tag{3.24}$$

where $\boldsymbol{\tau}$ is the unit vector tangential to the pipe,

$$\boldsymbol{\tau} = \frac{\partial x}{\partial s}\mathbf{i} + \frac{\partial z}{\partial s}\mathbf{k}. \tag{3.25}$$

Consequently,

$$\mathbf{V}_f = \left(\frac{\partial}{\partial t} + U\frac{\partial}{\partial s}\right)(x\mathbf{i} + z\mathbf{k}) \equiv \frac{\mathrm{D}\mathbf{r}}{\mathrm{D}t}, \tag{3.26}$$

where $\mathrm{D}(\;)/\mathrm{D}t$ is the material derivative for the fluid element. Recalling that $z = w$ and that $\partial x/\partial s \simeq 1$ and $\partial x/\partial t \sim \mathcal{O}(\epsilon^2) \simeq 0$ in accordance with the assumptions made, this gives

$$\mathbf{V}_f = U\mathbf{i} + \left[\frac{\partial w}{\partial t} + U\frac{\partial w}{\partial s}\right]\mathbf{k}. \tag{3.27}$$

In a similar manner, the acceleration is found to be

$$\mathbf{a}_f = \frac{\mathrm{D}^2\mathbf{r}}{\mathrm{D}t^2} = \frac{\mathrm{d}U}{\mathrm{d}t}\mathbf{i} + \left[\frac{\partial}{\partial t} + U\frac{\partial}{\partial s}\right]^2 w\mathbf{k}, \tag{3.28}$$

in which the bracketed quantity squared represents the successive, double application of the differential operator, and hence

$$\left[\frac{\partial}{\partial t} + U\frac{\partial}{\partial s}\right]^2 w = \frac{\partial^2 w}{\partial t^2} + 2U\frac{\partial^2 w}{\partial s\,\partial t} + U^2\frac{\partial^2 w}{\partial s^2} + \frac{\mathrm{d}U}{\mathrm{d}t}\frac{\partial w}{\partial s}. \tag{3.29}$$

[†]The surrounding fluid is supposed to be sufficiently light (e.g. air) for added-mass effects to be negligible.

[Parenthetically, a more 'fluid mechanical' derivation given by Païdoussis & Issid (1974) will be outlined here, in which an element of the pipe δs is considered containing fluid of volume $\delta\mathcal{V}$. The rate of change of momentum over $\delta\mathcal{V}$ may be written as

$$\frac{\mathrm{d}\mathbf{M}}{\mathrm{d}t} = \iiint_{\delta\mathcal{V}} \left[\frac{\partial \mathbf{V}_f}{\partial t} + (\mathbf{V}_f \cdot \nabla)\mathbf{V}_f\right] \rho\, \mathrm{d}\mathcal{V}, \tag{3.30}$$

where $\mathrm{d}\mathcal{V}$ is a small element within $\delta\mathcal{V}$. Then, by making the plug flow approximation, the velocity $\mathbf{V}_f$ may be approximated by (3.27). Therefore,

$$\frac{\partial \mathbf{V}_f}{\partial t} = \frac{\mathrm{d}U}{\mathrm{d}t}\mathbf{i} + \left(\frac{\partial^2 w}{\partial t^2} + U\frac{\partial^2 w}{\partial s \partial t} + \frac{\mathrm{d}U}{\mathrm{d}t}\frac{\partial w}{\partial s}\right)\mathbf{k},$$

$$(\mathbf{V}_f \cdot \nabla)\,\mathbf{V}_f \simeq U\frac{\partial}{\partial s}\left[U\mathbf{i} + \left(\frac{\partial w}{\partial t} + U\frac{\partial w}{\partial s}\right)\mathbf{k}\right] \simeq \left(U\frac{\partial^2 w}{\partial x \partial t} + U^2\frac{\partial^2 w}{\partial x^2}\right)\mathbf{k}. \tag{3.31}$$

Hence, equation (3.30) yields

$$\frac{\mathrm{d}\mathbf{M}}{\mathrm{d}t} = M\frac{\mathrm{d}U}{\mathrm{d}t}\delta s\,\mathbf{i} + M\left[\frac{\partial}{\partial t} + U\frac{\partial}{\partial x}\right]^2 w\,\delta s\,\mathbf{k}, \tag{3.32}$$

which corresponds to the acceleration as given by (3.28).]

A derivation in which the radial dimensions of the pipe are not ignored is given in Section 4.2, but leads to the same form as above. Therefore, recalling that $s \simeq x$, by using (3.28) or (3.32) one obtains the first two of the following equations:

$$a_{fx} = \frac{\mathrm{d}U}{\mathrm{d}t}, \qquad a_{fz} = \left[\frac{\partial}{\partial t} + U\frac{\partial}{\partial x}\right]^2 w, \qquad a_{pz} = \frac{\partial^2 w}{\partial t^2}; \tag{3.33}$$

the last equation above is the lateral acceleration of the pipe and requires no explanation. Hence, combining (3.19), (3.21), (3.22) and (3.33) one obtains

$$\left(E^*\frac{\partial}{\partial t} + E\right)I\frac{\partial^4 w}{\partial x^4} - \frac{\partial}{\partial x}\left[(T - pA)\frac{\partial w}{\partial x}\right] + M\left[\frac{\partial}{\partial t} + U\frac{\partial}{\partial x}\right]^2 w$$
$$+ c\frac{\partial w}{\partial t} + m\frac{\partial^2 w}{\partial t^2} = 0. \tag{3.34}$$

Also, adding equations (3.18) and (3.20) and using (3.33) yields

$$\frac{\partial}{\partial x}(T - pA) = M\frac{\mathrm{d}U}{\mathrm{d}t} - (M + m)g, \tag{3.35}$$

which integrated from x to L becomes

$$(T - pA)\Big|_{x=L} - (T - pA) = \left[M\left(\frac{\mathrm{d}U}{\mathrm{d}t}\right) - (M + m)g\right](L - x). \tag{3.36}$$

If the flexible pipe discharges the fluid to atmosphere at $x = L$ — the situation shown in Figure 3.1(b,c) — T, which is then entirely due to fluid friction, is zero at $x = L$; unless there is an externally applied tension, denoted by $\overline{T}$ — as could be the case for the system of Figure 3.1(a). The pressure, p, at $x = L$ will also be zero, unless the pipe does not discharge to atmosphere, in which case there may be a mean pressure $\overline{p}$ at $x = L$, over and above that expended to overcome friction (see also Section 3.4.2). Thus, $\overline{T}$ and $\overline{p}$ would act uniformly over the total length of the pipe. Now, if the downstream end is completely fixed, i.e. the system of Figure 3.1(a) rather than (b,c), internal pressurization induces an additional tensile force, which for a thin pipe is equal to $2\nu\overline{p}A$, where ν is the Poisson ratio, as first introduced by Naguleswaran & Williams (1968); i.e. the tendency of the pipe to expand radially and hence to become shorter, induces this tensile force. One may derive this in terms of (i) an axial stress distribution $\sigma_{xx} = T/A_p$ and (ii) the stress distribution due to $\overline{p}$, $\sigma_{rr} + \sigma_{\theta\theta} = 2\overline{p}A/A_p$, where A_p is the cross-sectional area of the pipe material (Sechler 1952); these two are then superposed to give the axial strain $\varepsilon_x = [\sigma_{xx} - \nu(\sigma_{rr} + \sigma_{\theta\theta})]/E$. Now, since no axial movement is allowed at the ends, $\int_0^L \varepsilon_x \, dx = 0$, which yields $T = 2\nu\overline{p}A$. Hence, in general, equation (3.36) may be written as

$$T - pA = \overline{T} - \overline{p}A(1 - 2\nu\delta) + \left[(M + m)g - M\left(\frac{dU}{dt}\right)\right](L - x), \tag{3.37}$$

where $\delta = 0$ signifies that there is no constraint to axial motion at $x = L$, and $\delta = 1$ if there is. Of course, it could be argued that, *in practice*, $\overline{T}$ and $\overline{p}$ can only be imposed if $\delta = 1$, so that one should really write $\delta[\overline{T} - \overline{p}A(1 - 2\nu)]$; still, one can conceive of ingenious theoretical ways in which $\overline{T}$ and $\overline{p}$ may be applied, even for the system of Figure 3.1(b) — e.g. by strings and pulleys and bellows — and hence the form of equation (3.37) will be retained. Now, substitution of (3.37) into (3.34) gives the equation of small lateral motions:

$$\left(E^* \frac{\partial}{\partial t} + E\right) I \frac{\partial^4 w}{\partial x^4} + \left\{MU^2 - \overline{T} + \overline{p}A(1 - 2\nu\delta) - \left[(M + m)g - M\frac{dU}{dt}\right](L - x)\right\} \frac{\partial^2 w}{\partial x^2}$$
$$+ 2MU\frac{\partial^2 w}{\partial x \partial t} + (M + m)g\frac{\partial w}{\partial x} + c\frac{\partial w}{\partial t} + (M + m)\frac{\partial^2 w}{\partial t^2} = 0. \tag{3.38}$$

If gravity, dissipation, tensioning and pressurization effects are either absent or neglected and U is constant, this simplifies to equation (3.1). The derivation given here follows Païdoussis & Issid's (1974). Earlier derivations of the simpler form, equation (3.1), for pipes with supported ends, were made by Feodos'ev (1951), Housner (1952) and Niordson (1953), and for cantilevered pipes by Benjamin (1961a) and Gregory & Païdoussis (1966a). The equation derived by Ashley & Haviland (1950) is wrong, missing the all-important $MU^2(\partial^2 w/\partial x^2)$ term. Similarly, an equation derived by Chen (1971b) for the case of harmonically perturbed flow is partly wrong, in that the first term of equation (3.28) or (3.32), i.e. the axial acceleration effect, is missing, although the last term in (3.29) is present; as a result, instead of the $M(dU/dt)(L - x)(\partial^2 w/\partial x^2)$, a term $M(dU/dt)(\partial w/\partial x)$ is found in Chen's equation of motion.

There are some subtleties in this derivation that are not quite obvious. This is partly the reason for the derivation of Appendix A.

In several calculations in the following, dissipation in the material of the pipe will be modelled not by the Kelvin–Voigt viscoelastic model as in equation (3.38), but by the so-called *hysteretic* or *structural* damping model. As shown by Bishop & Johnson (1960) for instance,[†] for metals and certain types of rubber-like materials, and over frequency ranges of practical interest, energy dissipation can adequately be accounted for by hysteresis; then, when a specimen of such a material is subjected to harmonic loading with a (real) circular frequency Ω, the energy dissipation per cycle can be calculated by taking the Young's modulus to be complex, in the form $E(1 + \mu \mathrm{i})$, where E and μ are constants independent of Ω, and $\mu \ll 1$. This implies that the small stresses related to hysteresis are in quadrature with the principal, linear-elastic stresses. This representation remains a reasonable approximation for lightly damped oscillation — i.e. provided that $\mathcal{I}m(\Omega) \ll \mathcal{R}e(\Omega)$ when $\Omega = \mathcal{R}e(\Omega) + \mathrm{i}\mathcal{I}m(\Omega)$; however, if there is another source of damping (e.g. flow-induced damping in cantilevered pipes conveying fluid) such that the overall damping is large, misleading results may be obtained. Nevertheless, within the limits of its applicability [e.g. close to a flutter boundary or for lightly damped conservative systems where $\mathcal{I}m(\Omega) \ll \mathcal{R}e(\Omega)$], the hysteretic model is very convenient. In that case, the first term of equation (3.38) may be replaced by

$$E(1 + \mu \mathrm{i})I\left(\frac{\partial^4 w}{\partial x^4}\right). \tag{3.39}$$

Finally, a variant of the equation of motion, first introduced by Gregory & Païdoussis (1966a) for experimental convenience (Section 3.5.6) will be discussed. For simplicity, consider the horizontal system with $\mathrm{d}U/\mathrm{d}t = 0$ and neglect dissipation. Then suppose that the downstream end of the pipe is fitted with a convergent nozzle, assumed to be weightless and very short compared to the total length of the pipe. The discharge velocity U_j is given by $U_j = U(A/A_j)$, where A_j is the terminal cross-sectional area of the nozzle flow passage. Equation (3.36) in this case simplifies to

$$(T - pA)\Big|_{x=L} - (T - pA) = 0; \tag{3.40}$$

consideration of momentum at $x = L$ — cf. the second and third terms of equation (2.63) — gives

$$(pA - T)\Big|_{x=L} = MU(U_j - U), \tag{3.41}$$

which, in view of (3.40), applies for all x. Hence, substituting into (3.34), simplified according to the assumptions made here, yields the modified equation of motion

$$EI\,\frac{\partial^4 w}{\partial x^4} + MUU_j\,\frac{\partial^2 w}{\partial x^2} + 2MU\,\frac{\partial^2 w}{\partial x \partial t} + (M + m)\frac{\partial^2 w}{\partial t^2} = 0. \tag{3.42}$$

3.3.3 Hamiltonian derivation

The difficulty in deriving an expression of Hamilton's principle for this problem lies in the fact that the system is *open*, with in-flow and out-flow of mass and momentum. Housner's (1952) derivation of the equation of motion for pipes with supported ends by means of

[†]See also Payne & Scott (1960), Snowdon (1968) and the workshop proceedings edited by Snowdon (1975) and Rogers (1984).

the kinetic and potential energies of the system entirely ignored this aspect, proceeding as if the system were closed, yet fortuitously ended up with the correct equation of motion. Benjamin (1961a,b) was the first to derive a proper statement for Hamilton's principle, in his work related to articulated and continuously flexible cantilevered pipes. Benjamin rightly maintained that Housner's derivation was erroneous, since the proper statement of Hamilton's principle was not used; thus, although the correct equation of motion was spuriously obtained for pipes with supported ends through a fortuitous error in the kinetic energy expression (Benjamin 1961a), there is no question that Housner's derivation would fail if applied to cantilevered pipes. The controversy was resolved by McIver (1973) with the aid of a more general form of Hamilton's principle for open systems, concluding that Benjamin's argument was correct, but Housner's derivation was also 'correct', in a sense, though for unexpected reasons. Hence, in this section Hamilton's principle will be reproduced as per McIver's work, and then the form obtained by Benjamin and the equations of motion will be derived therefrom; finally, Housner's derivation for pipes with supported ends will be considered.

Let us first rewrite the principle of virtual work for a system of N particles, each of mass m_i and subjected to a force $\mathbf{F}_i$. By d'Alembert's principle,

$$\sum_{i=1}^{N} (m_i\ddot{\mathbf{r}}_i - \mathbf{F}_i) \cdot \delta\mathbf{r}_i = 0, \tag{3.43}$$

where $\mathbf{r}_i$ is the position vector of each particle and $\delta\mathbf{r}_i$ the associated virtual displacement compatible with the system constraints. It is first noted that

$$\sum_{i=1}^{N} \mathbf{F}_i \cdot \delta\mathbf{r}_i = \delta W - \delta V, \tag{3.44}$$

is the virtual work by the applied forces, part of which has been expressed in terms of the potential energy V. Then, by re-writing

$$\sum_{i=1}^{N} m_i\ddot{\mathbf{r}}_i \cdot \delta\mathbf{r}_i = \sum_{i=1}^{N} m_i \frac{\mathrm{d}}{\mathrm{d}t}(\dot{\mathbf{r}}_i \cdot \delta\mathbf{r}_i) - \sum_{i=1}^{N} \tfrac{1}{2}m_i\delta(\dot{\mathbf{r}}_i \cdot \dot{\mathbf{r}}_i) = \sum_{i=1}^{N} m_i \frac{\mathrm{d}}{\mathrm{d}t}(\dot{\mathbf{r}}_i \cdot \delta\mathbf{r}_i) - \delta T, \tag{3.45}$$

where T is the kinetic energy of the system, equations (3.43)–(3.45) lead to

$$\delta(T - V) + \delta W - \sum_{i=1}^{N} m_i \frac{\mathrm{d}}{\mathrm{d}t}(\dot{\mathbf{r}}_i \cdot \delta\mathbf{r}_i) = 0. \tag{3.46}$$

Consider next the closed system of Figure 3.7(a) associated with the closed control volume $\mathscr{V}_c(t)$, bounded by the surface $\mathscr{S}_c(t)$, containing a collection of particles of density ρ, each with position vector $\mathbf{r}$ and velocity $\mathbf{u}$. The principle of virtual work in the form just derived may be written as

$$\delta\mathscr{L}_c + \delta W - \frac{\mathrm{D}}{\mathrm{D}t}\iiint_{\mathscr{V}_c(t)} \rho(\mathbf{u} \cdot \delta\mathbf{r})\,\mathrm{d}\mathscr{V} = 0, \tag{3.47}$$

where $\mathscr{L}_c = T_c - V_c$ is the Lagrangian of the closed system, δW is the virtual work by the generalized forces, and $\mathrm{D}/\mathrm{D}t$ is the material derivative following a particle; hence,

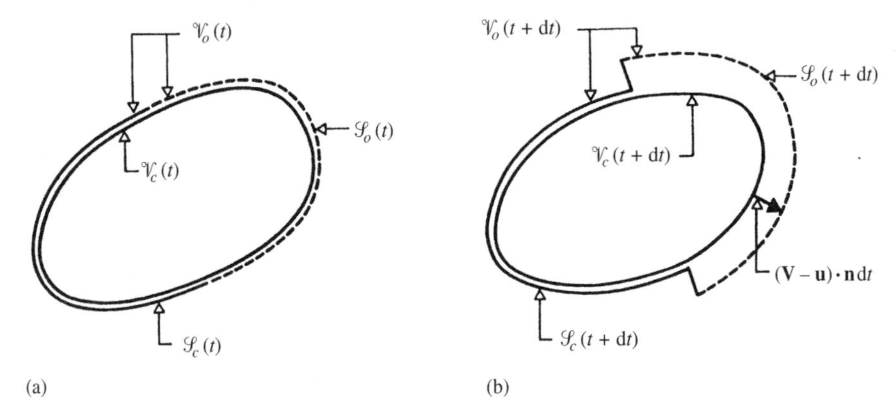

Figure 3.7 Definition of the control volume of the open system under consideration, $\mathscr{V}_o$, and of a fictitious closed system, $\mathscr{V}_c$ coincident with $\mathscr{V}_o$ at time t. The control surfaces $\mathscr{S}_o$ and $\mathscr{S}_c$ are associated with the open and closed parts of the open system. (a) The system at time t, and (b) at time $t + \mathrm{d}t$.

$\mathbf{u} = \mathrm{D}\mathbf{r}/\mathrm{D}t$. Then, Hamilton's principle may be obtained from (3.47) by integrating it between two instants, t_1 and t_2; in accordance with normal variational procedure, the system configuration is prescribed at t_1 and t_2, i.e. $\delta\mathbf{r} = \mathbf{0}$ so that the last term vanishes, and this leads to the familiar form (cf. Section 2.1)

$$\delta \int_{t_1}^{t_2} \mathscr{L}_c \, \mathrm{d}t + \int_{t_1}^{t_2} \delta W \, \mathrm{d}t = 0. \tag{3.48}$$

The extension to open systems is effected by considering a portion $\mathscr{S}_o(t)$ of the surface of the control volume $\mathscr{V}_o(t)$ (Figure 3.7) to be capable of movement with a velocity $\mathbf{V} \cdot \mathbf{n}$ normal to the surface, across which mass may be transported; $\mathbf{n}$ is the outward normal. Thus, $\mathscr{S}_c(t)$ is associated with the closed part of the system and $\mathscr{S}_o(t)$ with the open part. Figure 3.7(a) shows the system at time t, and Figure 3.7(b) at time $t + \mathrm{d}t$. This open system does not necessarily have a constant mass or, if it does, the mass does not necessarily comprise the same particles. On the closed part of the control volume, bounded by $\mathscr{S}_c(t)$, $\mathbf{V} \cdot \mathbf{n} = \mathbf{u} \cdot \mathbf{n}$.

If, at time $t, \mathscr{V}_o(t)$ coincides with $\mathscr{V}_c(t)$ as shown in Figure 3.7(a), Reynolds' general transport equation [e.g. Shames (1992; Chapter 4)]† reads

$$\frac{\mathrm{d}}{\mathrm{d}t} \iiint_{\mathscr{V}_o(t)} \{\quad\} \, \mathrm{d}\mathscr{V} = \frac{\mathrm{D}}{\mathrm{D}t} \iiint_{\mathscr{V}_c(t)} \{\quad\} \, \mathrm{d}\mathscr{V} + \iint_{\mathscr{S}_o} \{\quad\} (\mathbf{V} - \mathbf{u}) \cdot \mathbf{n} \, \mathrm{d}\mathscr{S}, \tag{3.49}$$

in which

$$\frac{\mathrm{D}}{\mathrm{D}t} \iiint_{\mathscr{V}_c(t)} \{\quad\} \, \mathrm{d}\mathscr{V} = \frac{\mathrm{D}}{\mathrm{D}t} \iiint_{\mathscr{V}_o(t)} \{\quad\} \, \mathrm{d}\mathscr{V} \tag{3.50}$$

may be used since $\mathrm{D}\{\quad\}/\mathrm{D}t$ makes it clear that a closed control volume is to be employed.

†Equation (3.49) simply states that the total rate of change in { } is equal to the rate of change in the volume plus that due to influx/efflux through the boundaries.

Hence, utilizing (3.47), (3.49) and (3.50) leads to the following form for the virtual work equation:

$$\delta \mathscr{L}_o + \delta W + \iint_{\mathscr{S}_o(t)} \rho(\mathbf{u} \cdot \delta \mathbf{r})(\mathbf{V} - \mathbf{u}) \cdot \mathbf{n} \, \mathrm{d}\mathscr{S} - \frac{\mathrm{d}}{\mathrm{d}t} \iiint_{\mathscr{V}_o(t)} \rho(\mathbf{u} \cdot \delta \mathbf{r}) \, \mathrm{d}\mathscr{V} = 0. \qquad (3.51)$$

This, integrated over time from t_1 to t_2, at which limits $\delta \mathbf{r} = \mathbf{0}$ again, gives Hamilton's principle for the open system,

$$\delta \int_{t_1}^{t_2} \mathscr{L}_o \, \mathrm{d}t + \int_{t_1}^{t_2} \delta H \, \mathrm{d}t = 0, \qquad (3.52)$$

$$\delta H = \delta W + \iint_{\mathscr{S}_o(t)} \rho(\mathbf{u} \cdot \delta \mathbf{r})(\mathbf{V} - \mathbf{u}) \cdot \mathbf{n} \, \mathrm{d}\mathscr{S}, \qquad (3.53)$$

with $\mathscr{L}_o = T_o - V_o$ being the Lagrangian of the open system.

This is next applied to the case of a cantilevered pipe conveying fluid. For simplicity, the case of no dissipation and a constant flow velocity U is considered. Moreover, it is presumed that the only forces involved in δW are associated with the pressure p, measured above the ambient of the surrounding medium; hence,

$$\delta H = -\iint_{\mathscr{S}_c(t) + \mathscr{S}_i + \mathscr{S}_e(t)} p(\delta \mathbf{r} \cdot \mathbf{n}) \, \mathrm{d}\mathscr{S} + \iint_{\mathscr{S}_i + \mathscr{S}_e(t)} \rho(\mathbf{u} \cdot \delta \mathbf{r})(\mathbf{V} - \mathbf{u}) \cdot \mathbf{n} \, \mathrm{d}\mathscr{S}, \qquad (3.54)$$

where $\mathscr{S}_c(t)$ is the surface covered by the pipe wall, and $\mathscr{S}_i$ and $\mathscr{S}_e(t)$ are the inlet and exit open surfaces for the fluid. Next, it is presumed that any virtual displacement of the pipe does not induce a virtual displacement of the fluid relative to the pipe. Thus, virtual displacements of the fluid relative to the pipe are independent of those of the pipe. Hence, since the fluid is incompressible, there can be no virtual change in the volume of the system, and expression (3.54) simplifies to

$$\delta H = -\iint_{\mathscr{S}_i + \mathscr{S}_e(t)} p(\delta \mathbf{r} \cdot \mathbf{n}) \, \mathrm{d}\mathscr{S} + \iint_{\mathscr{S}_i + \mathscr{S}_e(t)} \rho(\mathbf{u} \cdot \delta \mathbf{r})(\mathbf{V} - \mathbf{u}) \cdot \mathbf{n} \, \mathrm{d}\mathscr{S}. \qquad (3.55)$$

Now, if the fluid entrance conditions are prescribed and constant, the integrals over $\mathscr{S}_i$ are zero. Furthermore, the first integral over $\mathscr{S}_e(t)$ is zero since at the outlet $p = 0$. Hence, the only part remaining is

$$\delta H = \iint_{\mathscr{S}_e(t)} \rho(\mathbf{u} \cdot \delta \mathbf{r})(\mathbf{V} - \mathbf{u}) \cdot \mathbf{n} \, \mathrm{d}\mathscr{S} = -MU(\dot{\mathbf{r}}_L + U\boldsymbol{\tau}_L) \cdot \delta \mathbf{r}_L, \qquad (3.56)$$

in obtaining which $\mathbf{u} = \dot{\mathbf{r}} + U\boldsymbol{\tau}$ [Figure 3.8(a)], $(\mathbf{u} - \mathbf{V}) \cdot \mathbf{n} = U$ at $\mathscr{S}_e(t)$ and $M = \rho A$ have been utilized, A being the open (flow) area. Hence, Hamilton's principle for this system becomes

$$\delta \int_{t_1}^{t_2} \mathscr{L}_o \, \mathrm{d}t - \int_{t_1}^{t_2} MU(\dot{\mathbf{r}}_L + U\boldsymbol{\tau}_L) \cdot \delta \mathbf{r}_L \, \mathrm{d}t = 0, \qquad (3.57)$$

which is identical to that obtained by Benjamin (1961a).[†]

[†]In Benjamin's derivation, as in Figure 3.1(d), $\mathbf{R}_L$ is measured from the $(x, z) = (L, 0)$ position, whereas here $\mathbf{r}_L = L\mathbf{i} + \mathbf{R}_L$ is measured from the origin; however, as $\dot{\mathbf{r}}_L = \dot{\mathbf{R}}_L$ and $\delta \mathbf{r}_L = \delta \mathbf{R}_L$, the two expressions are fully equivalent.

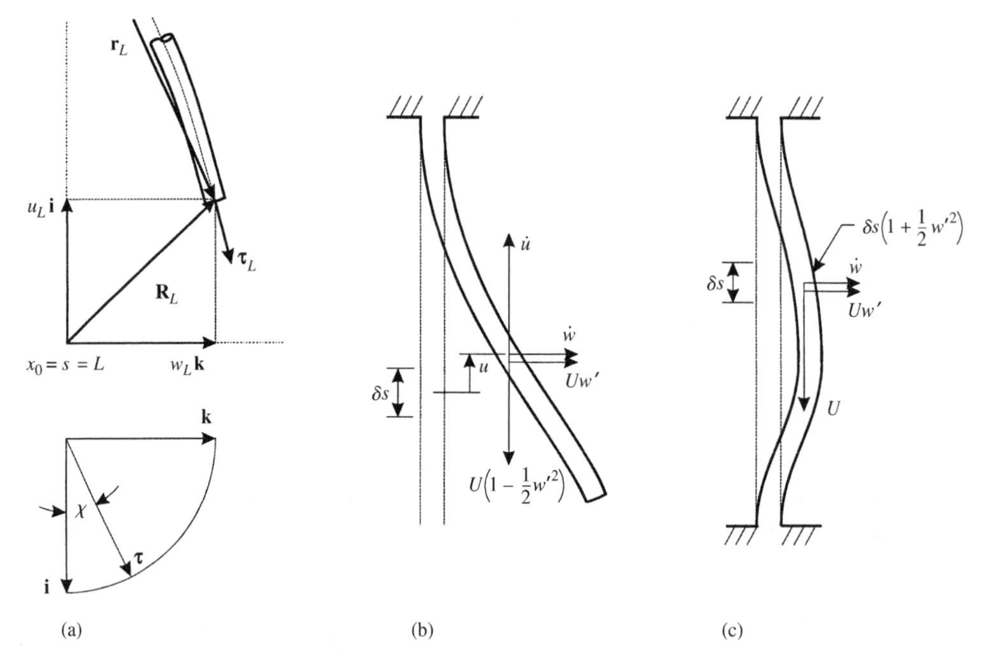

Figure 3.8 (a) Definition of the coordinates and unit vectors associated with movements of the free end of a cantilevered pipe (top), and the relationship between **i**, **k**, $\boldsymbol{\tau}$ and χ for any point along the cantilever (bottom); (b) velocity components for an element of the fluid in a cantilevered pipe; (c) the same for an element of the fluid in a pipe with clamped ends.

The equation of motion is derived next, for a vertical cantilevered pipe, taking into account gravity effects. The pipe is assumed to be inextensible, and use is made of the curvilinear coordinate s. The derivation involves the evaluation of the various terms in the Hamiltonian statement (3.57), following along similar lines to Benjamin's (1961a) and Païdoussis' (1973a), but making use of the notation and relationships developed in Section 3.3.1.

Some useful relationships will be obtained first, as follows: (i) recalling from (3.12) that $u = x - x_0$ with $x_0 = s$ here, then $\dot{x} = \dot{u}$; (ii) from (3.14), $\partial x/\partial s = [1 - (\partial z/\partial s)^2]^{1/2}$ with $z = w$, and hence $\partial x/\partial s \simeq 1 - \frac{1}{2}w'^2$, where $(\ )' = \partial(\ )/\partial s$; (iii) from (3.17b), $u_L = -\int_0^L \frac{1}{2}w'^2\,\mathrm{d}s$. Also, one may write $\dot{\mathbf{r}}_L = \dot{x}_L\mathbf{i} + \dot{z}_L\mathbf{k} = \dot{u}_L\mathbf{i} + \dot{w}_L\mathbf{k}$; from (3.25), $\boldsymbol{\tau}_L = x_L'\mathbf{i} + z_L'\mathbf{k} \simeq [1 - \frac{1}{2}w_L'^2]\mathbf{i} + w_L'\mathbf{k}$; $\delta\mathbf{r}_L = \delta u_L\mathbf{i} + \delta w_L\mathbf{k}$; and the second term of (3.57) may be re-written as

$$\int_{t_1}^{t_2} [MU^2\,\delta u_L + MU(\dot{w}_L + Uw_L')\,\delta w_L]\,\mathrm{d}t, \tag{3.58}$$

correct to $\mathcal{O}(\epsilon^2)$, having made use of the order considerations expressed by (3.17a,b). Hence, by grouping the terms implicitly involving a double integral into the first term, Hamilton's principle is rewritten as

$$\delta\int_{t_1}^{t_2} (\mathcal{L}_o - MU^2u_L)\,\mathrm{d}t - \int_{t_1}^{t_2} MU(\dot{w}_L + Uw_L')\,\delta w_L\,\mathrm{d}t = 0, \tag{3.59}$$

correct to $\mathcal{O}(\epsilon^2)$.

The kinetic energy of the pipe and the fluid may be evaluated by making use of (3.23) and (3.26),

$$T_p = \tfrac{1}{2}m \int_0^L (\dot{x}^2 + \dot{z}^2)\,\mathrm{d}s, \qquad T_f = \tfrac{1}{2}M \int_0^L [(\dot{x} + Ux')^2 + (\dot{z} + Uz')^2]\,\mathrm{d}s, \tag{3.60}$$

in which m and M have been defined in Section 3.3.2; again, the subscripts p and f stand for the pipe and fluid, respectively. The integrands in T_p and T_f may be simplified by noting that $\dot{x} \sim \mathbb{O}(\epsilon^2)$, $x' \simeq 1 - \frac{1}{2}w'^2$, and $x'^2 + z'^2 = 1$ from inextensibility condition (3.14). Hence, recalling also that $\dot{x} = \dot{u}$ and $z = w$, the expressions for T_p and T_f become

$$T_p = \tfrac{1}{2}m \int_0^L \dot{w}^2\,\mathrm{d}s, \qquad T_f = \tfrac{1}{2}M \int_0^L [U^2 + \dot{w}^2 + 2U\dot{w}w' + 2U\dot{u}]\,\mathrm{d}s. \tag{3.61}$$

It is noted that (3.61) could have been obtained directly with the aid of Figure 3.8(b); the various terms are obtained from Cartesian components of (3.24), which may be expressed as $(\dot{w} + U \sin\chi)$ and $(U \cos\chi + \dot{u})$ with $\sin\chi \simeq w'$ and $\cos\chi \simeq 1 - \frac{1}{2}w'^2$, neglecting terms smaller than $\mathbb{O}(\epsilon^2)$.

The potential energy is given by

$$V = V_p + V_f = \tfrac{1}{2}EI \int_0^L w''^2\,\mathrm{d}s + \tfrac{1}{2}(m + M)g \int_0^L \int_0^s w'^2\,\mathrm{d}s\,\mathrm{d}s. \tag{3.62}$$

The component of V associated with gravity may be simplified via integration by parts, as follows:

$$\begin{aligned}\tfrac{1}{2}(m + M)g \int_0^L \int_0^s w'^2\,\mathrm{d}s\,\mathrm{d}s &= \tfrac{1}{2}(m + M)g \left\{ \left[s \int_0^s w'^2\,\mathrm{d}s \right]\Bigg|_0^L - \int_0^L sw'^2\,\mathrm{d}s \right\} \\ &= \tfrac{1}{2}(m + M)g \int_0^L (L - s)w'^2\,\mathrm{d}s. \end{aligned} \tag{3.63}$$

Finally, substituting (3.61)–(3.63) into (3.59) and making use of the standard variational techniques and of the boundary conditions for a cantilever, after considerable manipulation, this reduces to

$$\begin{aligned} -\int_{t_1}^{t_2} \int_0^L \{EIw'''' + MU^2w'' - (M + m)g[(L - s)w']' + 2MU\dot{w}' \\ + (M + m)\ddot{w}\}\,\delta w\,\mathrm{d}s\,\mathrm{d}t = 0. \end{aligned} \tag{3.64}$$

Two items should be remarked upon in the derivation of (3.64). Firstly, the terms in the second integral of (3.59) cancelled out with identical ones originating from the first integral after integration by parts. For instance,

$$\begin{aligned} \delta \int_{t_1}^{t_2} MU^2 u_L\,\mathrm{d}t &= MU^2 \delta \int_{t_1}^{t_2} \int_0^L \tfrac{1}{2}w'^2\,\mathrm{d}s\,\mathrm{d}t = MU^2 \int_{t_1}^{t_2} \int_0^L w'(\delta w)'\,\mathrm{d}s\,\mathrm{d}t \\ &= MU^2 \int_{t_1}^{t_2} w'\,\delta w\Big|_0^L\,\mathrm{d}t - MU^2 \int_{t_1}^{t_2} \int_0^L w''\,\delta w\,\mathrm{d}s\,\mathrm{d}t, \end{aligned}$$

the first part of which becomes $MU^2 \int_{t_1}^{t_2} w'_L \, \delta w_L$, because of the boundary conditions, and cancels the second term of the second integral of (3.59). The expression above also makes it clear that the centrifugal term $MU^2 w''$ does not arise from the kinetic energy, as might have been supposed, but from the second term in the statement of Hamilton's principle, equation (3.57). The second item concerns the term $2U\dot{u}$ in T_f, in equations (3.61). Once the variation is taken, this leads to $\int_0^L 2U \, \delta u \, |_{t_1}^{t_2} = 0$.

For arbitrary variations δw and with $s \simeq x$, the term within the curly brackets in equation (3.64) is the desired equation of motion. It is the same as (3.38), but with $E^* = 0$, $c = 0$ and $\mathrm{d}U/\mathrm{d}t = 0$, in accordance with the assumptions made here.

Consider next a pipe with clamped ends, but allowing sliding at the downstream one [Figure 3.1(b)]. In this case the second integral of (3.59) is zero, but u_L in the first integral is not, and it is again this term rather than the kinetic energy that is responsible for the centrifugal force term in the equation of motion. Everything else remains the same, including the inextensibility condition. After considerable manipulation, the same equation of motion is obtained — but only if u_L is not ignored, whereas it was in Housner's derivation.

Consider finally the case of fully clamped ends — not allowing any sliding at $x = L$. As pointed out by McIver, in this case there is no motion possible at $x = L$, i.e. $\delta x_L = \delta z_L = 0$; that is, the 'contraction' in the sense used by Benjamin and defined for inextensible pipes by $u_L = -\int_0^L \frac{1}{2} w'^2 \, \mathrm{d}s$ is zero in equation (3.59), and hence so is u at any location s along the deformed pipe. In fact, for lateral deformation to occur, there will be some stretching of the pipe as shown in Figure 3.8(c), which results in its cross-sectional shrinking. Thus, the element of the pipe δs is stretched to $\delta s(1 + \frac{1}{2}w'^2)$ and the flow velocity relative to the pipe through the narrower flow passage, $A(1 - \frac{1}{2}w'^2)$, is increased to $U(1 + \frac{1}{2}w'^2)$ for continuity at each location s; hence, the x-component of the flow velocity is $[U(1 + \frac{1}{2}w'^2)]\,(1 - \frac{1}{2}w'^2) \simeq U$. Therefore, in this case, at least approximately to $\mathbb{O}(\epsilon^2)$,

$$T_f = \tfrac{1}{2} M \int_0^L [(\dot{w} + Uw')^2 + U^2] \, \mathrm{d}x,$$

as utilized by Housner — correct, but without the benefit of the refined arguments leading to it. With this expression, i.e. with $\dot{u} = 0$, and with $u_L = 0$, Hamilton's principle (3.59) yields the very same equation of motion as for the sliding end and the cantilevered case — at least to the linear limit. In contrast to the previous two cases, here the centrifugal force term in the equation of motion arises from the kinetic energy.

3.3.4 A comment on frictional forces

A remarkable feature of equations (3.38) and (3.1) is the total absence of fluid-frictional effects, which at first sight might appear to be an idealization. However, within the context of the other approximations implicit in this linearized equation, it may rigorously be demonstrated that fluid-frictional effects play no role in the dynamics of the system, a fact first shown by Benjamin (1961a,b). Consider once more the balance of forces in the axial direction of elements of the fluid and the pipe, i.e. equations (3.18) and (3.20) for the case where $\mathrm{d}U/\mathrm{d}t = 0$ and gravity is inoperative (i.e. for motions in a horizontal plane) to make the argument simplest:

$$-A\,\frac{\partial p}{\partial x} - qS + F\,\frac{\partial w}{\partial x} = 0, \qquad \frac{\partial T}{\partial x} + qS - F\,\frac{\partial w}{\partial x} = 0, \tag{3.65}$$

which, when added give

$$\frac{\partial}{\partial x}(T - pA) = 0. \tag{3.66}$$

Thus, the frictional force qS is replaced by its twin effects: (i) as a tension on the pipe and (ii) as a pressure drop in the fluid. Equation (3.66), when integrated from x to L gives $(T - pA)_x = (T - pA)_L$, the equivalent of equation (3.36). Ignoring externally imposed tensioning and pressurization, which do not enter the argument (and which are discussed in Section 3.4.2), and thus considering for simplicity the fluid to discharge to atmosphere, both p and T vanish at $x = L$, and hence

$$T - pA = 0 \qquad \text{for} \qquad x \in [0, L]. \tag{3.67}$$

It follows that the term related to T and p in equation (3.34), the precursor to the final equation of motion, vanishes, i.e.

$$\frac{\partial}{\partial x}\left[(T - pA)\frac{\partial w}{\partial x}\right] = 0, \tag{3.68}$$

because of (3.66) and (3.67). Therefore, the two effects of friction — tensioning and pressure drop — cancel each other entirely and vanish from the equation of motion, to the order of the linear approximation (Benjamin 1961a; Gregory & Païdoussis 1966a).

This has been verified experimentally (see Sections 3.4.4 and 3.5.6), and also numerically in calculations with shell theory for beam-mode vibrations ($n = 1$) in Chapter 7.

3.3.5 Nondimensional equation of motion

Consider the most general form of the equation of motion derived so far, equation (3.38). It will help further discussion if this equation is generalized a little by considering the possibility that the pipe may be supported all along its length by a Winkler-type elastic foundation, which involves distributed springs of stiffness K per unit length; thus, a term Kw is added to the equation of motion.

The resultant equation may be rendered dimensionless through the use of

$$\xi = \frac{x}{L}, \qquad \eta = \frac{w}{L}, \qquad \tau = \left[\frac{EI}{M+m}\right]^{1/2} \frac{t}{L^2}. \tag{3.69}$$

The dimensionless equation is

$$\begin{aligned} &\alpha\dot{\eta}'''' + \eta'''' + \{u^2 - \Gamma + \Pi(1 - 2\nu\delta) + (\beta^{1/2}\dot{u} - \gamma)(1 - \xi)\}\eta'' \\ &\quad + 2\beta^{1/2}u\dot{\eta}' + \gamma\eta' + k\eta + \sigma\dot{\eta} + \ddot{\eta} = 0, \end{aligned} \tag{3.70}$$

where $(\dot{\ }) = \partial(\ \ )/\partial\tau$ and $(\ \ )' = \partial(\ \ )/\partial\xi$, in which the following dimensionless system parameters have arisen:

$$u = \left(\frac{M}{EI}\right)^{1/2} LU, \qquad \beta = \frac{M}{M+m}, \qquad \gamma = \frac{(M+m)L^3}{EI} g, \qquad \Gamma = \frac{\overline{T}L^2}{EI},$$
$$\Pi = \frac{\overline{p}AL^2}{EI}, \qquad k = \frac{KL^4}{EI}, \qquad \alpha = \left[\frac{I}{E(M+m)}\right]^{1/2} \frac{E^*}{L^2}, \qquad \sigma = \frac{cL^2}{[EI(M+m)]^{1/2}}. \tag{3.71}$$

In general, the system dynamics will depend on all of these parameters.

If the hysteretic damping model is used, it is clear from expression (3.39) that the first two terms of (3.70) should be replaced by

$$(1 + \mu\mathrm{i})\eta''''. \tag{3.72}$$

This corresponds to solutions of (3.70) of the type $\eta(\xi, \tau) = Y(\xi)\exp(\mathrm{i}\omega\tau)$, in which ω is either wholly real or, if complex, such that $\mathfrak{Re}(\omega) \gg \mathfrak{Im}(\omega)$; the hysteretic model may thus be considered as a particular case of the viscoelastic one for which $\alpha\omega = \mu$ or $\alpha\mathfrak{Re}(\omega) = \mu$, respectively. The dimensionless frequency ω is related to the dimensional circular (radian) one, Ω, by

$$\omega = \left(\frac{M+m}{EI}\right)^{1/2} \Omega L^2. \tag{3.73}$$

In the case of an end-nozzle, as discussed at the end of Section 3.3.2, the definitions of u and β in (3.71) need to be modified to

$$u = \left(\frac{M}{EI} UU_j\right)^{1/2} L, \qquad \beta = \frac{M}{M+m}\frac{U}{U_j}. \tag{3.74}$$

With these, the dimensionless form of equation (3.42) is identical to the appropriately simplified equation (3.70), namely

$$\eta'''' + u^2\eta'' + 2\beta^{1/2}u\dot{\eta}' + \ddot{\eta} = 0. \tag{3.75}$$

The usefulness of the end-nozzle emerges from the second of equations (3.74): instead of changing pipes, one may change nozzles to alter β, at least over a range relatively close to the initial β for the pipe without a nozzle.

3.3.6 Methods of solution

Two methods of solution will be given: the first, due to Gregory & Païdoussis (1966a), for the simpler, homogeneous equation of motion; the second, used by Païdoussis (1966) and Païdoussis & Issid (1974), applies to the fuller, nonhomogeneous equation of motion.

(a) First method

The simplest form of the equation of motion, equation (3.1), will be considered first, which in dimensionless form becomes

$$\frac{\partial^4 \eta}{\partial \xi^4} + u^2 \frac{\partial^2 \eta}{\partial \xi^2} + 2\beta^{1/2} u \frac{\partial^2 \eta}{\partial \xi \partial \tau} + \frac{\partial^2 \eta}{\partial \tau^2} = 0, \tag{3.76}$$

subject to the appropriate boundary conditions; e.g. for a pipe with simply-supported ('pinned') ends,

$$\eta = \frac{\partial^2 \eta}{\partial \xi^2} = 0 \quad \text{at} \quad \xi = 0 \quad \text{and} \quad \xi = 1, \tag{3.77}$$

while for a cantilevered pipe,

$$\begin{aligned} \eta = \frac{\partial \eta}{\partial \xi} &= 0 \quad \text{at} \quad \xi = 0, \\ \frac{\partial^2 \eta}{\partial \xi^2} = \frac{\partial^3 \eta}{\partial \xi^3} &= 0 \quad \text{at} \quad \xi = 1. \end{aligned} \tag{3.78}$$

Consider now solutions of the form

$$\eta(\xi, \tau) = \mathcal{Re}[Y(\xi)\, \mathrm{e}^{\mathrm{i}\omega\tau}], \tag{3.79}$$

where ω is the dimensionless circular frequency defined by (3.73). In general, ω will be complex, and the system will be stable or unstable accordingly as the imaginary component of ω, $\mathcal{I}\mathrm{m}(\omega)$, is positive or negative; in the case of neutral stability ω is wholly real. Substituting (3.79) into (3.76) leads to

$$\frac{\mathrm{d}^4 Y}{\mathrm{d}\xi^4} + u^2 \frac{\mathrm{d}^2 Y}{\mathrm{d}\xi^2} + 2\beta^{1/2} u\omega \mathrm{i} \frac{\mathrm{d}Y}{\mathrm{d}\xi} - \omega^2 Y = 0. \tag{3.80}$$

Next, we take a trial solution

$$Y(\xi) = A\mathrm{e}^{\mathrm{i}\alpha\xi}, \tag{3.81}$$

where A is a constant. When this is substituted into equation (3.80), the equation determining the permissible values of the exponent α is obtained, namely

$$\alpha^4 - u^2\alpha^2 - 2\beta^{1/2} u\omega\alpha - \omega^2 = 0, \tag{3.82}$$

and since this equation is of fourth degree, the complete solution of (3.76) is given in general by

$$\eta(\xi, \tau) = \mathcal{Re}\left[\sum_{j=1}^{4} A_j \mathrm{e}^{\mathrm{i}\alpha_j \xi} \mathrm{e}^{\mathrm{i}\omega\tau}\right], \tag{3.83}$$

in which the four A_j must be determined from the boundary conditions. This is illustrated here for the cantilevered system. Making use of (3.78), we find

$$\sum_{j=1}^{4} A_j = 0, \qquad \sum_{j=1}^{4} \alpha_j A_j = 0, \qquad \sum_{j=1}^{4} \alpha_j^2 A_j \mathrm{e}^{\mathrm{i}\alpha_j} = 0, \qquad \sum_{j=1}^{4} \alpha_j^3 A_j \mathrm{e}^{\mathrm{i}\alpha_j} = 0.$$

For nontrivial solution, the determinant of the A_j must vanish, yielding

$$\Delta \equiv \begin{vmatrix} 1 & 1 & 1 & 1 \\ \alpha_1 & \alpha_2 & \alpha_3 & \alpha_4 \\ \alpha_1^2 e^{i\alpha_1} & \alpha_2^2 e^{i\alpha_2} & \alpha_3^2 e^{i\alpha_3} & \alpha_4^2 e^{i\alpha_4} \\ \alpha_1^3 e^{i\alpha_1} & \alpha_2^3 e^{i\alpha_2} & \alpha_3^3 e^{i\alpha_3} & \alpha_4^3 e^{i\alpha_4} \end{vmatrix} = 0. \tag{3.84}$$

Since the roots of (3.82) cannot be expressed in simple explicit form in terms of u, ω and β, and in view of the complexity of (3.84), it is not possible to obtain solutions by direct methods. Three methods of solution were given by Gregory & Païdoussis (1966a): (i) a rather ingenious method of transforming the original problem into one easier to solve numerically in 1966;† (ii) a straightforward numerical method; and (iii) a Galerkin solution. Of these, only (ii) will be outlined here, as follows: (a) starting with a small value of u, say $u = 0.1$, and trial values of $\mathcal{R}e(\omega)$ and $\mathcal{I}m(\omega)$, say those for $u = 0$, a minimizing procedure (e.g. a secant method) finds the appropriate values of $\mathcal{R}e(\omega)$ and $\mathcal{I}m(\omega)$ which result in $\mathcal{R}e(\Delta) = \mathcal{I}m(\Delta) = 0$ to within desired accuracy; (b) the value of u is increased by δu, say by 0.1, and using the $\mathcal{R}e(\omega)$ and $\mathcal{I}m(\omega)$ found in (a) as first approximations, the minimizing procedure determines the complex frequency for $u = 0.2$; and so on.

Clearly, this method has to be applied for each mode separately (for a given value of β), the locus to be followed depending on the initial trial value for $\mathcal{R}e(\omega)$.

(b) Second method

The fuller equation of motion (3.70) is nonhomogeneous, since the coefficients of derivatives of η are explicit functions of ξ and/or implicit ones of τ, because $u = u(\tau)$; hence, the foregoing method of solution is inapplicable. A solution for $u =$ const. is, however, readily possible via the Galerkin method and will be given here; the case for $u = u(\tau)$ is considered in Chapter 4. This is approximate, not only in the strict numerical sense, but also because of the finite number of terms utilized in the Galerkin expansion (Section 2.1).

Let

$$\eta(\xi, \tau) = \sum_{r=1}^{\infty} \phi_r(\xi)\, q_r(\tau), \tag{3.85}$$

where $q_r(\tau)$ are the generalized coordinates of the discretized system and $\phi_r(\xi)$ are the dimensionless eigenfunctions of a beam with the same boundary conditions as the pipe under consideration, and hence they are appropriate comparison functions (Section 2.1.3). It is presumed that the series (3.85) may be truncated at a suitably high value of r, $r = N$. Substitution of (3.85) into (3.70) with $\dot{u} = 0$, followed by multiplication by ϕ_s and

†Computers were then new and slow, and ἡ πενία τέχνας κατεργάζεται; i.e. poverty (necessity) develops ingenuity!

Table 3.1 The constants b_{sr}, c_{sr} and d_{sr}.

	Pinned–pinned pipes	Clamped–clamped pipes	Cantilevered pipes
$b_{sr}(s \neq r)$	$\dfrac{2\lambda_r\lambda_s}{\lambda_r^2 - \lambda_s^2}\{(-1)^{r+s} - 1\}$	$\dfrac{4\lambda_r^2\lambda_s^2}{\lambda_r^4 - \lambda_s^4}\{(-1)^{r+s} - 1\}$	$\dfrac{4}{(\lambda_s/\lambda_r)^2 + (-1)^{r+s}}$
b_{rr}	0	0	2
$c_{sr}(s \neq r)$	0	$\dfrac{4\lambda_r^2\lambda_s^2}{\lambda_r^4 - \lambda_s^4}(\lambda_r\sigma_r - \lambda_s\sigma_s)\{(-1)^{r+s} + 1\}$	$\dfrac{4(\lambda_r\sigma_r - \lambda_s\sigma_s)}{(-1)^{r+s} - (\lambda_s/\lambda_r)^2}$
c_{rr}	$-\lambda_r^2$	$\lambda_r\sigma_r(2 - \lambda_r\sigma_r)$	$\lambda_r\sigma_r(2 - \lambda_r\sigma_r)$
$d_{sr}(s \neq r)$	$\dfrac{4\lambda_r^3\lambda_s}{(\lambda_r^2 - \lambda_s^2)^2}\{1 - (-1)^{r+s}\}$	$\dfrac{4\lambda_r^2\lambda_s^2(\lambda_r\sigma_r - \lambda_s\sigma_s)}{\lambda_r^4 - \lambda_s^4}(-1)^{r+s} - \dfrac{3\lambda_r^4 + \lambda_s^4}{\lambda_r^4 - \lambda_s^4}b_{sr}$	$\dfrac{4(\lambda_r\sigma_r - \lambda_s\sigma_s + 2)}{1 - (\lambda_s/\lambda_r)^4}(-1)^{r+s} - \dfrac{3 + (\lambda_s/\lambda_r)^4}{1 - (\lambda_s/\lambda_r)^4}b_{sr}$
d_{rr}	$\frac{1}{2}c_{rr}$	$\frac{1}{2}c_{rr}$	$\frac{1}{2}c_{rr}$

integration over the domain [0,1] yields

$$\sum_{r=1}^{N} \left\{ \delta_{sr}\ddot{q}_r + \left[(\alpha\lambda_r^4 + \sigma)\delta_{sr} + 2\beta^{1/2}u \int_0^1 \phi_s\phi_r' \,\mathrm{d}\xi \right] \dot{q}_r \right.$$
$$+ \left[(\lambda_r^4 + k)\delta_{sr} + \{u^2 - \Gamma + \Pi\,(1 - 2\nu\delta) - \gamma\} \int_0^1 \phi_s\phi_r'' \,\mathrm{d}\xi \right. \tag{3.86}$$
$$\left. \left. + \gamma \int_0^1 \phi_s\phi_r' \,\mathrm{d}\xi + \gamma \int_0^1 \phi_s\xi\phi_r'' \,\mathrm{d}\xi \right] q_r \right\} = 0, \qquad s = 1, 2, \ldots, N,$$

in which the orthonormality of the eigenfunctions was utilized (i.e. the fact that $\int_0^1 \phi_s\,\phi_r\,\mathrm{d}\xi = \delta_{sr}$, δ_{sr} being Kronecker's delta), as well as the fact that $\phi_r'''' = \lambda_r^4\,\phi_r$, λ_r being the rth dimensionless eigenvalue of the beam. The definite integrals may be evaluated in closed form, defining the following set of constants:

$$b_{sr} = \int_0^1 \phi_s\phi_r' \,\mathrm{d}\xi, \qquad c_{sr} = \int_0^1 \phi_s\phi_r'' \,\mathrm{d}\xi, \qquad d_{sr} = \int_0^1 \phi_s\xi\phi_r'' \,\mathrm{d}\xi. \tag{3.87}$$

Their values for some sets of boundary conditions are given in Table 3.1, in which the σ_r are the constants associated with the ϕ_r [Bishop & Johnson 1960; cf. equation (2.28)]. The method for evaluating b_{sr}, c_{sr} and d_{sr} analytically is illustrated in Appendix B.

Equation (3.86) may be written in matrix form as follows:

$$\ddot{\mathbf{q}} + [\mathbf{F} + 2\beta^{1/2}u\mathbf{B}]\dot{\mathbf{q}} + \{\mathbf{\Lambda} + \gamma\mathbf{B} + [u^2 - \Gamma + \Pi(1 - 2\nu\delta) - \gamma]\mathbf{C} + \gamma\mathbf{D}\}\mathbf{q} = \mathbf{0}, \tag{3.88}$$

where $\mathbf{q} = \{q_1, q_2, \ldots, q_N\}^{\mathrm{T}}$, $\mathbf{F}$ and $\mathbf{\Lambda}$ are diagonal matrices with elements $(\alpha\lambda_r^4 + \sigma)$ and $(\lambda_r^4 + k)$, respectively, and $\mathbf{B}$, $\mathbf{C}$ and $\mathbf{D}$ are matrices with elements b_{sr}, c_{sr} and d_{sr}, respectively. This equation may be written in standard form,

$$[M]\ddot{\mathbf{q}} + [C]\dot{\mathbf{q}} + [K]\mathbf{q} = \mathbf{0} \tag{3.89}$$

cf. equation (2.1), Section 2.1. Its eigenvalues may be found in various ways; e.g. by transforming it into first-order form by the procedure leading from equation (2.15) to (2.17), and then to the standard eigenvalue problem of equation (2.18). The eigenvalues may be obtained numerically, e.g. by the IMSL library subroutines or those given by Press *et al*. (1992).

3.4 PIPES WITH SUPPORTED ENDS

3.4.1 Main theoretical results

We first consider the simplest possible system: a simply-supported (or 'pinned–pinned') horizontal pipe ($\gamma = 0$) with zero dissipation, and with $\beta = 0.1$, $\Gamma = \Pi = k = 0$ in equation (3.70). The dynamical behaviour of this system with increasing dimensionless flow velocity, u, is illustrated by the Argand diagram of Figure 3.9. It is recalled that $\mathfrak{Re}(\omega)$ is the dimensionless oscillation frequency, while $\mathfrak{Im}(\omega)$ is related to damping, the damping ratio being $\zeta = \mathfrak{Im}(\omega)/\mathfrak{Re}(\omega)$. The general dynamical features already remarked upon in Sections 3.2.1 and 3.2.3 are clearly seen: (i) since dissipation is absent in this example, the

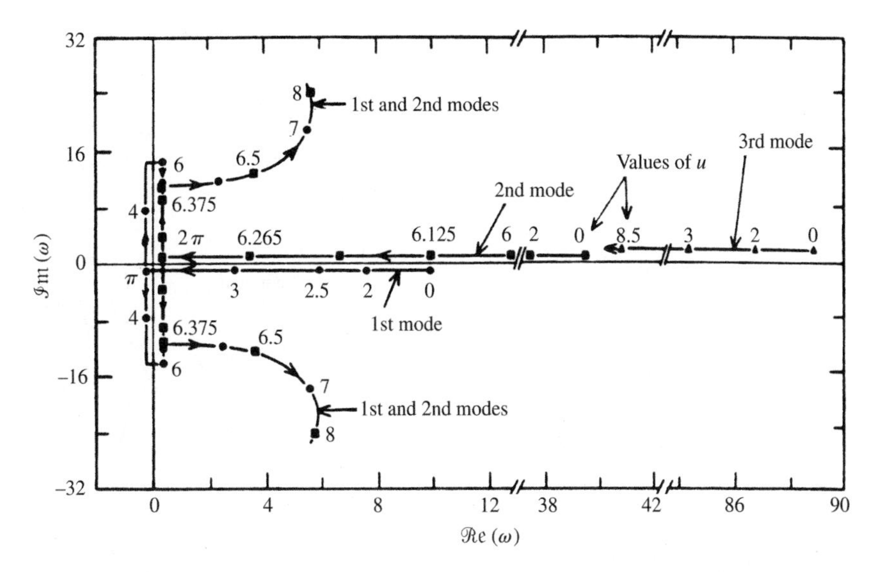

Figure 3.9 Dimensionless complex frequency diagrams for a pinned–pinned pipe; $\beta = 0.1$ and $\Gamma = \Pi = \alpha = \sigma = k = \gamma = 0$ [see equations (3.71) for meaning of symbols]. The loci that actually lie on the axes have been drawn slightly off the axes but parallel to them for the sake of clarity. — • — , first mode; — ■ — , second mode; — ▲ — , third mode; — ■ — • — ■ — , combined first and second modes (Païdoussis & Issid 1974).

eigenfrequencies are purely real and they are diminished with increasing u, for $0 \leq u < \pi$; (ii) at $u = u_{cd} = \pi$ the system loses stability in its first mode by divergence, via a pitchfork bifurcation, and thereafter the eigenfrequencies become purely imaginary — cf. Figure 3.4(a).

The dynamics of the same system but with clamped ends is illustrated in Figure 3.10, which also shows another way of presenting the results. In this case, $u_{cd} = 2\pi$, but the qualitative dynamics is similar to that in Figure 3.9; for $u < u_{cd}$ the eigenfrequencies are all purely real, whilst for $u > u_{cd}$ those associated with the first mode are, initially at least, purely imaginary.

The values of u_{cd} in Figures 3.9 and 3.10 may readily be found by the method of Section 3.3.6(a). By setting $\omega = 0$ in equation (3.82), one obtains $\alpha_{1,2} = 0$, $\alpha_{3,4} = \pm u$, and hence $\eta(\xi) = A_1 + A_2\xi + A_3 \exp(\mathrm{i}u\xi) + A_4 \exp(-\mathrm{i}u\xi)$, which is the appropriate form of (3.83) in this case. Then, application of boundary conditions (3.77) for pipes with simply-supported (pinned) ends leads to the characteristic equation

$$\sin u = 0, \tag{3.90a}$$

with roots $u = n\pi$, the first nontrivial one of which is $u = u_{cd} = \pi$. The second root, $u = 2\pi$, is associated with divergence of the second mode or restabilization of the first, as will be seen in the following. Proceeding in a similar way for clamped–clamped pipes, the characteristic equation is found to be

$$2(1 - \cos u) - u \sin u = 0, \tag{3.90b}$$

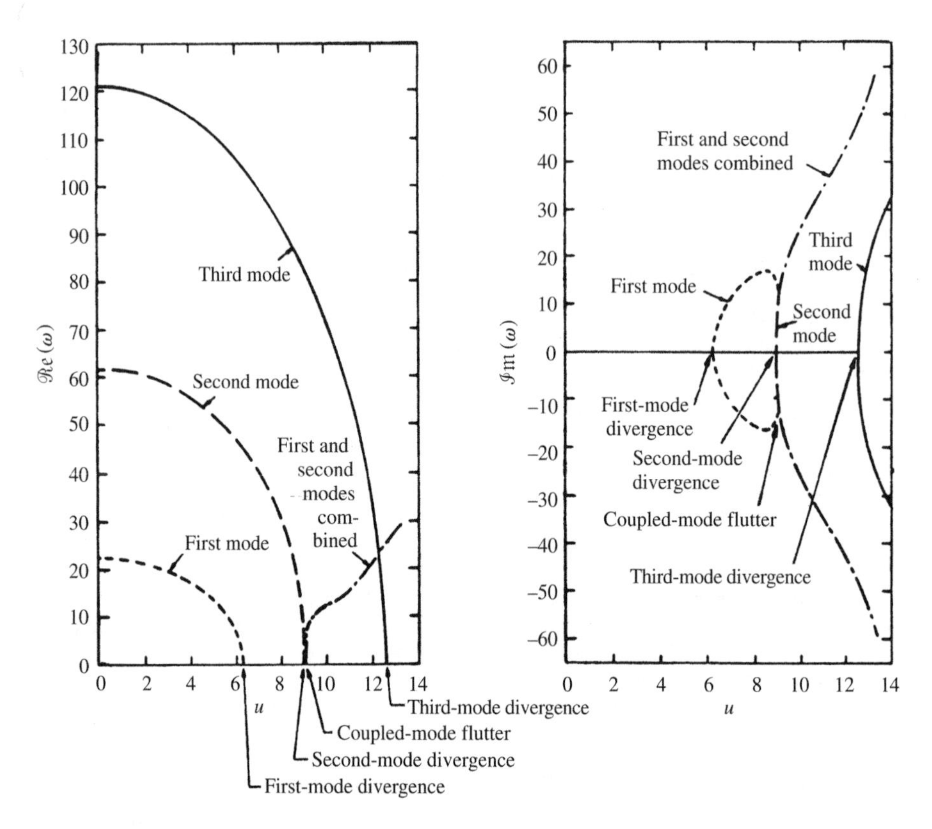

Figure 3.10 The real and imaginary components of the dimensionless frequency, ω, as functions of the dimensionless flow velocity, u, for the lowest three modes of a clamped–clamped pipe; $\beta = 0.1$, $\Gamma = \Pi = \alpha = \sigma = k = \gamma = 0$ (Païdoussis 1975).

with roots $u = 2\pi, 8.99, \ldots, 4\pi, \ldots,$ so that $u_{cd} = 2\pi$ as in Figure 3.10. For clamped–pinned ends, the characteristic equation is

$$u - \tan u = 0, \tag{3.90c}$$

which gives $u = u_{cd} \simeq 4.49$. Incidentally, this static analysis for the stability of conservative systems is known as *Euler's method of equilibrium* (Ziegler 1968).

The dynamics of a clamped–clamped system with $\beta = 0.5$ is illustrated in Figure 3.11. Once again, $u_{cd} = 2\pi$. In fact, u_{cd} is independent of β, as already seen in the results obtained by Euler's method; this is so because β is always associated with velocity-dependent terms in the equation of motion, while divergence represents a static loss of stability. Once more, the dynamics up to $u \simeq 8.99$ is similar to that in Figures 3.9 and 3.10.

The results presented here are based mainly on Païdoussis & Issid's (1974) work. Before embarking on the discussion of post-divergence dynamics, a historical parenthesis on the early, successful work on the dependence of ω on u and on the determination of u_{cd} is in order, some of it predating the computer era. Feodos'ev (1951) and Housner (1952) utilize Galerkin's method, essentially the method of Section 3.3.6(b), to examine stability and determine ω as a function of u. Li & DiMaggio (1964) use the method

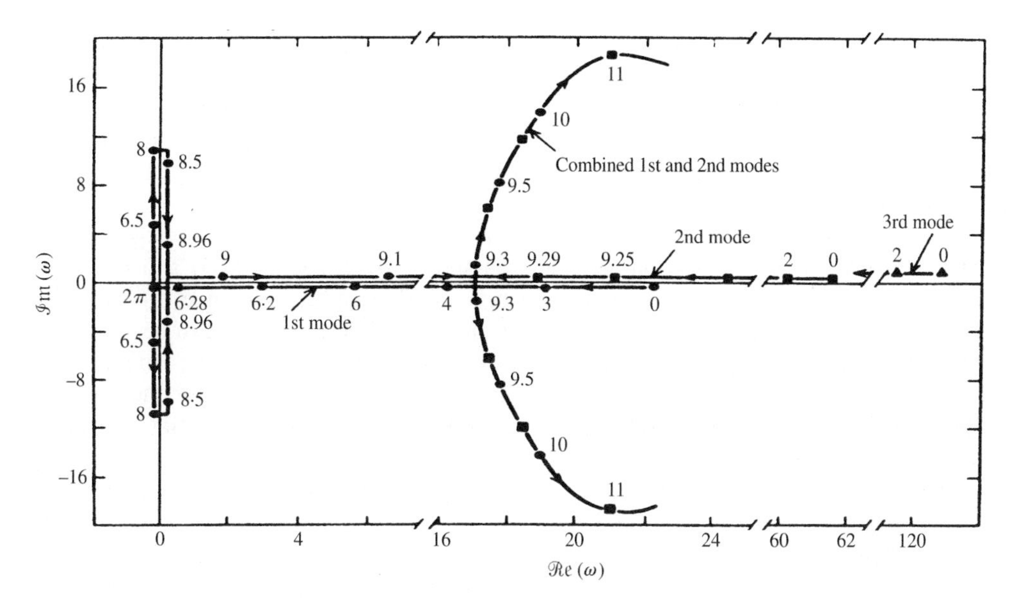

Figure 3.11 Dimensionless complex frequency diagram for a clamped–clamped pipe; $\beta = 0.5$ and $\Gamma = \Pi = \alpha = \sigma = k = \gamma = 0$. The loci that actually lie on the axes have been drawn slightly off the axes but parallel to them for the sake of clarity (Païdoussis & Issid 1974).

of Section 3.3.6(a) and obtain the full curve of the first-mode ω versus u, up to u_{cd}, by computer. However, more interesting methods have also been employed: the direct method of Lyapunov (Appendix F.1.3) by Movchan (1965), and the methods of integral equations by Jones & Goodwin (1971). Also, utilizing a perturbation method, Handleman (1955) determines the dependence of ω on u in the vicinity of $u = 0$ and $u = u_{cd}$. In all cases the simplest form of the equation of motions is considered, equation (3.1), and in all cases but the last for pinned–pinned pipes only. Finally, Niordson (1953) presents an elegant wave solution to the more general problem of a thin-walled pipe, modelled as a shell (Chapter 7); the required results for beam-like motions are then obtained by considering the $n = 1$ mode of the shell — see Figure 2.7(c).

The post-divergence dynamical behaviour of these systems, i.e. for $u > u_{cd}$, is of considerable interest. It should, however, immediately be remarked that strictly, linear theory is applicable only up to the first loss of stability. The reason for this is that, in the linear equation of motion, it is required that motions be small, in the vicinity of the equilibrium state, while for $u > u_{cd}$ the system has diverged away from that state.[†] However, in some cases (e.g. in Chapter 8), the buckled state is not so far away from the original stable equilibrium configuration, and then linear theory is capable of predicting the post-divergence dynamics of the system reasonably well. Hence, it is not pointless to examine the post-divergence dynamics as predicted by linear theory.

It is seen in Figures 3.9 and 3.10 ($\beta = 0.1$) that the simply-supported and clamped systems develop divergence in the second mode at $u = 2\pi$ and 8.99, respectively. Then, the loci of the two modes coalesce on the $\mathscr{I}m(\omega)$-axis and, at slightly higher u ($u \simeq 6.38$

[†]Of course, the stability of the original equilibrium as predicted by linear theory is always valid, but other states emerge once nonlinear effects are considered.

in Figure 3.9 and $u \simeq 9.0^+$ in Figure 3.10), they leave the axis, indicating the onset of Païdoussis-type coupled-mode flutter[†] as defined in Section 3.2.3 and by Figure 3.4(d).

The behaviour of Figure 3.11 ($\beta = 0.5$) is different. The $\omega = 0$ solution for $u \simeq 8.99$ does not correspond to a second divergence, but to *restabilization* of the system. This lasts to $u \simeq 9.3$, whereupon coupled-mode flutter occurs via a Hamiltonian Hopf bifurcation, as defined in Figure 3.4(c).

What is particularly interesting about this predicted coupled-mode flutter is its origination. As discussed in Section 3.2.1 and as shown by equations (3.5) and (3.6), for periodic motions there is no energy transfer between the fluid and the pipe. Hence, since the system is conservative, the question arises as to how the instability can be supported whilst the total energy of the system remains constant. As pointed out by Païdoussis & Issid(1974), the question is not quite like this, since the critical point for the onset of flutter, unlike for the nonconservative (cantilevered) system, is *not* a point of neutral stability; rather, it involves the coincidence of two real frequencies, and hence growing oscillations of the form $\eta(\xi, \tau) = \Re e[f(\xi)(a + b\tau)\exp(i\omega\tau)]$, with ω real. The source of energy is of course the flowing fluid, yet *how* some energy is channelled to generate the oscillatory state remains the question. A possible answer was provided, via an ingenious set of arguments, by Done & Simpson (1977) for a pipe with supported ends but with the downstream end free to slide axially [Figure 3.1(b)].

First, one may consider a two-mode Galerkin approximation of the system, namely

$$\ddot{\mathbf{q}} + \begin{bmatrix} 2\beta^{1/2}ub_{11} & 2\beta^{1/2}ub_{12} \\ 2\beta^{1/2}ub_{21} & 2\beta^{1/2}ub_{22} \end{bmatrix} \dot{\mathbf{q}} + \begin{bmatrix} \lambda_1^4 + u^2c_{11} & u^2c_{12} \\ u^2c_{21} & \lambda_2^4 + u^2c_{22} \end{bmatrix} \mathbf{q} = \mathbf{0}. \tag{3.91}$$

For clamped and pinned ends, $b_{rr} = 0$ and $b_{sr} = -b_{rs}$; for pinned ends, $c_{sr} = 0$ for all $r \neq s$, while the same applies to clamped ends for $r + s$ odd, which is the case here. Hence, equation (3.91) may be written as

$$\ddot{\mathbf{q}} + \begin{bmatrix} 0 & -2\beta^{1/2}ub_{21} \\ 2\beta^{1/2}ub_{21} & 0 \end{bmatrix} \dot{\mathbf{q}} + \begin{bmatrix} \lambda_1^4 + u^2c_{11} & 0 \\ 0 & \lambda_2^4 + u^2c_{22} \end{bmatrix} \mathbf{q} = \mathbf{0}. \tag{3.92}$$

It is of interest to remark that (i) the damping matrix is skew symmetric, which is a characteristic of the system being gyroscopic conservative, as already remarked, and (ii) by setting $\det[K] = 0$, $[K]$ being the stiffness matrix, one retrieves the zeros for static loss of stability $u_{cd} = \pi$ and $u = 2\pi$, exactly for simply-supported ends and approximately for clamped ends (since in this case the matrix is not fully diagonal for $N > 2$).

Then, solutions of the form $\mathbf{q} = \mathbf{q}_0 \exp(\lambda\tau)$ are considered, leading to the characteristic equation

$$p_4\lambda^4 + p_2\lambda^2 + p_0 = 0, \tag{3.93}$$

with $p_4 = 1$, $p_2 = [\lambda_1^4 + \lambda_2^4 - u^2(c_{11} + c_{22}) - 4\beta u^2 b_{21}^2]$, $p_0 = (\lambda_1^4 - u^2c_{11})(\lambda_2^4 - u^2c_{22})$. The condition of coalescence of two eigenfrequencies corresponds to two equal roots of (3.93), which occurs if $p_2^2 - 4p_4p_0 = 0$. The results for clamped ends are shown in Figure 3.12, where it is seen that all the critical points of Figures 3.10 and 3.11 are

[†]This flattering appellation, coined by Done & Simpson (1977), has been retained here for this particular form of coupled-mode flutter. This phenomenon, however, although analytically intriguing, was shown to be physically doubtful with the appearance of Holmes' (1977, 1978) work. This would have rendered any claim to fame by this book's author rather ephemeral, were it not for the fact that, luckily, the physical reality of the phenomenon is firmly established for another system (Chapter 8)!

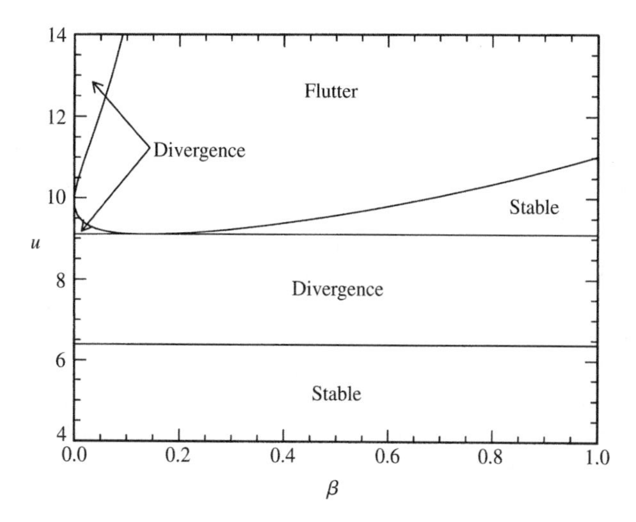

Figure 3.12 Map of different kinds of instabilities predicted by linear theory for clamped–clamped pipes with a sliding downstream end ($\delta = 0$), for varying β and $\Gamma = \Pi = \alpha = \sigma = k = \gamma = 0$, following Done & Simpson (1977). The first (lower) divergence zone is associated with the first mode; the second with the second mode. For $\beta < 0.139$ the coupled-mode flutter is of the Païdoussis type; for $\beta > 0.139$ it is via a Hamiltonian Hopf bifurcation.

reproduced quite well. It is also seen that the two types of coupled-mode flutter are neatly separated: Païdoussis flutter for $\beta < 0.139$, and flutter via a Hamiltonian Hopf bifurcation for higher β. (The results for pipes with pinned ends are quite similar, but the critical value of β is $\beta = 0.26$ in that case.)

Next, since the pipe is free to slide axially at $\xi = 1$, the total dimensionless 'contraction' (see Section 3.3.3) as a result of motions is given by

$$c = |u_L|/L = \tfrac{1}{2}\int_0^1 (w')^2\,\mathrm{d}\xi = \tfrac{1}{2}\int_0^1 \left[q_1(\tau)\phi_1'(\xi) + q_2(\tau)\phi_2'(\xi)\right]^2\,\mathrm{d}\xi, \tag{3.94}$$

where u_L is the axial contraction, defined by (3.17b), at $s = L$. The integral gives rise to quantities of the type $\int_0^1 \phi_s'\phi_r'\,\mathrm{d}\xi \equiv e_{sr}$ and, for the boundary conditions of interest, integrating by parts yields $e_{sr} = -c_{sr}$. Since the cross-terms ($r \neq s$, $r + s =$ odd) are zero as per Table 3.1, one is left with $e_{rr} = -c_{rr} = \lambda_r\sigma_r(\lambda_r\sigma_r - 2)$, which shows that $e_{rr} > 0$ for all r, for either clamped or pinned ends. Hence, c may be re-written as

$$c = \tfrac{1}{2}(e_{11}q_1^2 + e_{22}q_2^2), \tag{3.95}$$

a positive quantity. Consider now the particular case of coupled-mode flutter via a Hamiltonian Hopf bifurcation. At the onset of flutter, $q_1 = q_{10}\exp(\mathrm{i}\omega\tau)$ and $q_2 = q_{20}\exp(\mathrm{i}\omega\tau)$, while the ratio of q_{20}/q_{10} may be obtained from either of the two equations in (3.92), say the first, namely $q_{20}/q_{10} = [-\omega^2 + \lambda_1^4 - u^2c_{11}]/[\beta^{1/2}u\,b_{21}\omega\mathrm{i}]$, an imaginary quantity; hence the displacements in the two modes are in quadrature (90° out of phase), and one can write $q_1 = \overline{q}_1\cos\omega\tau$, $q_2 = \overline{q}_2\cos(\omega\tau + \tfrac{1}{2}\pi) = \overline{q}_2\sin\omega\tau$. Therefore, the axial shortening (contraction) over one or several periods of oscillation may be calculated through (3.95), giving a mean value of the contraction, $\overline{c}$, and an oscillating component of frequency 2ω [because of the quadratic nature of (3.95) and sinusoidal form of q_1 and

q_2] and amplitude $\tilde{c}$:

$$\bar{c} = \tfrac{1}{4}(e_{11}\bar{q}_1^2 + e_{22}\bar{q}_2^2) \qquad \text{and} \qquad \tilde{c} = \tfrac{1}{4}(e_{11}\bar{q}_1^2 - e_{22}\bar{q}_2^2). \tag{3.96}$$

Clearly, at no time in the course of the oscillation can the contraction become instantaneously zero. A similar argument may be made in the case of Païdoussis flutter; in this case, q_{20}/q_{10} is not purely imaginary but complex, and the phase angle is not neatly $\frac{1}{2}\pi$ but an angle ϕ. Nevertheless, the same conclusion may be reached with regard to the overall contraction never becoming zero during oscillation.

The implication of this is that the momentum flux of the fluid issuing from the sliding end of the pipe does work on the system in achieving a certain oscillation, MU^2 acting as a compressive load P as discussed in Section 3.1 and acting over a distance equal to the mean contraction, $\bar{c}$. No net work is required thereafter to maintain the oscillation, but there is an oscillatory flow of energy because of the axial motion of the downstream end of the pipe, which nevertheless is zero over a cycle of oscillation. This energy may be thought of as being carried in the form of travelling waves, as will be seen in Figure 3.13, with a node moving down to the pipe exit in half a cycle of oscillation. It is in this ingenious way, thanks to Done & Simpson, that the paradox of oscillation with no net energy expenditure may be explained!

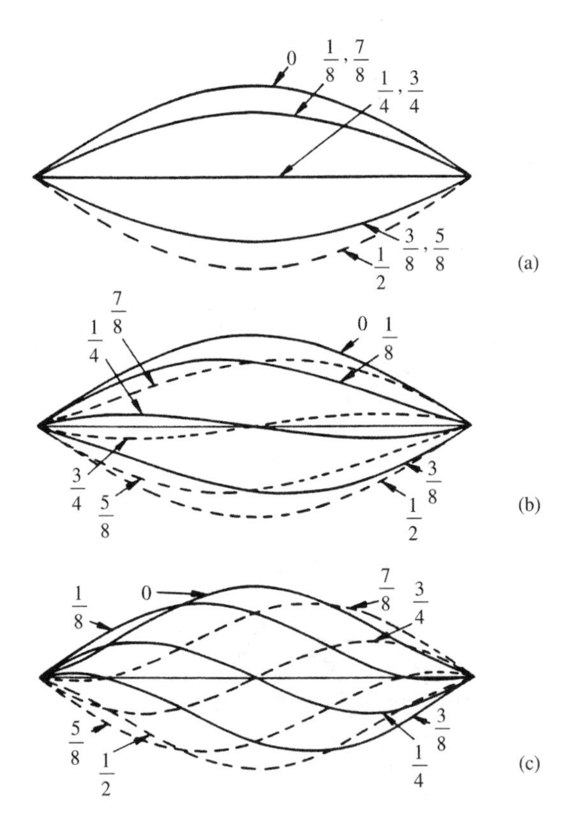

Figure 3.13 Variation of modal forms of the fundamental mode of a simply-supported pipe of vanishing flexural rigidity during a period of oscillation: (a) $u = 0$; (b) $u/u_c = 0.25$; (c) $u/u_c = 0.75$; the fractions denote fractions of the period (Chen & Rosenberg 1971).

It should be pointed out that the term $2\beta^{1/2}ub_{21}$ played an important role in all of the foregoing, not accidentally but because it is associated with the Coriolis term in the equation of motion, which in turn is what makes the system gyroscopic conservative, rather than just conservative. It is of interest that calculations with $\beta = 0$ show that, when the system is purely conservative, the only form of instability is divergence; coupled-mode flutter does not arise.

Another effect of the Coriolis forces — despite not doing any net work over a cycle of oscillation — is that they render classical normal modes impossible.[†] Thus, the modal displacement patterns contain both stationary and travelling-wave components, as seen in Figure 3.13(b,c). Physically, this is a consequence of the forward and backward travelling waves having different phase speeds (Chen & Rosenberg 1971) — see also Section 3.7. Contrast this to Figure 3.13(a), where $u = 0$ and the Coriolis forces vanish; in this case classical normal modes do exist.

The dynamics of the same system as in Figure 3.11 but with dissipation taken into account ($\alpha = 5 \times 10^{-3}$) is shown in Figure 3.14. It is seen that coupled-mode flutter of

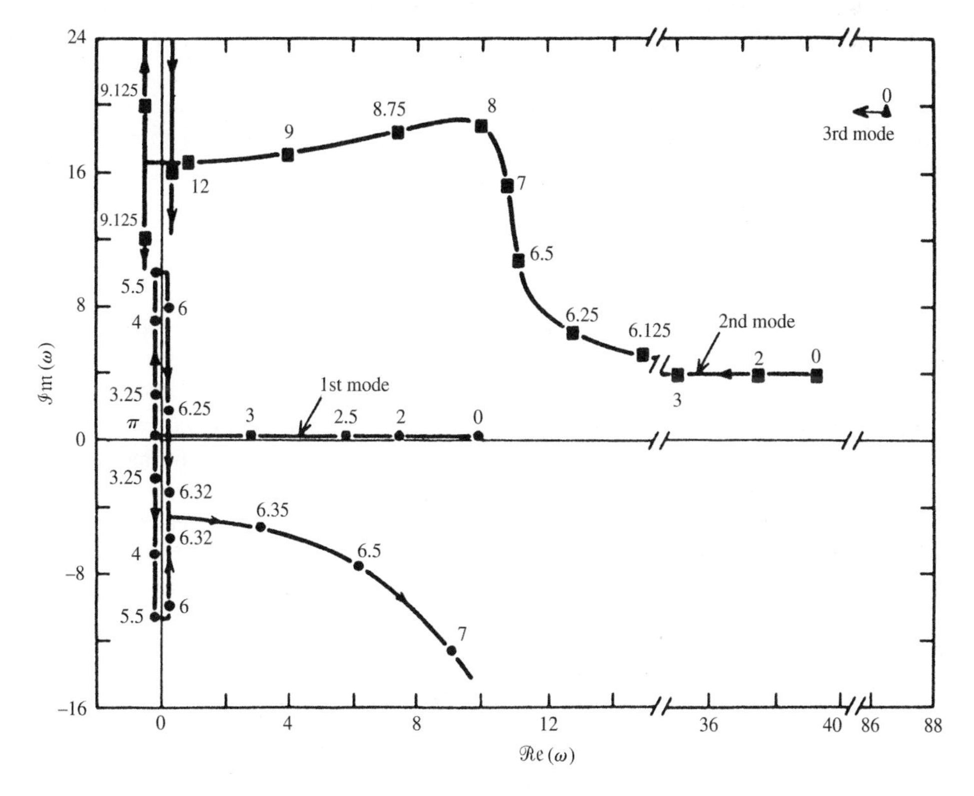

Figure 3.14 Dimensionless complex frequency diagram of a damped clamped–clamped pipe for $\beta = 0.5$, $\alpha = 5 \times 10^{-3}$, $\Gamma = \Pi = \sigma = k = \gamma = 0$. The loci that actually lie on the $[\mathcal{I}\mathrm{m}(\omega)]$-axis have been drawn off the axis but parallel to it for the sake of clarity (Païdoussis & Issid 1974).

[†]If the various parts of the system vibrate with the same phase and they pass through the equilibrium configuration at the same instant of time — as would be the case for a string or a beam — the normal modes (eigenmodes) are called *classical*. The necessary and sufficient conditions for their existence were investigated by Caughey & O'Kelley (1965) and others; see also Chen (1987; Appendix A).

another kind arises, at a slightly lower critical flow velocity, in which the two branches of the same mode are involved rather than two different modes. We shall continue calling this a coupled-mode flutter since, strictly speaking, the two branches on the $\mathcal{I}m(\omega)$-axis should be considered as being associated with different modes, from the left-hand (not shown) and right-hand sides of the complex ω-plane — see Figure 2.10(a).

The Done & Simpson argumentation for coupled-mode flutter may be extended to dissipative systems by supposing that, at the threshold of flutter, a sustained correction in the contraction c may be effected by the discharging axial momentum flux, so as to maintain a constant-amplitude motion. Thus, effectively, a sustained rate of work occurs through axial motion, whereas the dissipation occurs through lateral motion; note also that $\Delta W = 0$ in equation (3.95) in the undamped system applies to lateral motions.

It is important to stress, yet again, that both the restabilization of the system after divergence (e.g. in Figure 3.11) and the coupled-mode flutter are due to the gyroscopic nature of the system, i.e. to the Coriolis terms in the equation of motion. As pointed out by Shieh (1971) and Huseyin & Plaut (1974), purely conservative systems cannot be restabilized after divergence 'on their own', but gyroscopic forces *can* restabilize an otherwise conservative system, a fact known since Thomson & Tait's (1879) work. The possibility of coupled-mode flutter is a much newer 'discovery' which may be attributed to Shieh, who illustrated its existence with an example from gyrodynamics involving a shaft under an axial compression P, rotating with angular velocity Ω. The equations of motion are

$$\begin{aligned} EIy'''' + Py'' + M(\ddot{y} - 2\Omega\dot{z} - \Omega^2 y) &= 0, \\ EIz'''' + Pz'' + M(\ddot{z} + 2\Omega\dot{y} - \Omega^2 z) &= 0, \end{aligned} \tag{3.97}$$

in which y and z are mutually perpendicular deflections in a plane normal to the long axis; these equations clearly bear close similarity to that of the problem at hand — cf. equation (3.1).

Huseyin & Plaut (1974) discuss the dynamics of gyroscopic conservative systems in general, as well as the rotating shaft and pipe systems as examples. The latter will be discussed here briefly, partly (i) to introduce the concept of the 'corresponding nongyroscopic system' and (ii) to demonstrate the use of the so-called 'characteristic curves'. Huseyin & Plaut considered a two-degree-of-freedom discretization of the horizontal system, i.e. of equation (3.1), by using the beam eigenfunctions as suitable comparison functions. In the case of a clamped–pinned system, the results are shown in Figure 3.15 for three values of β;† also plotted are the results for $\beta = 0$, which is the *corresponding nongyroscopic system*, representing a column subjected to a load $\mathcal{P} = u^2$. The results are plotted in the form of *characteristic curves*, i.e. curves of loading versus ω^2, namely u^2 versus ω^2. Clearly, only $u^2 > 0$ is meaningful, but the extension of the curves to $u^2 < 0$ helps to show that the curves (full lines) are conic sections. In (a) it is seen that the system is initially stable ($\omega_1^2 > 0$, $\omega_2^2 > 0$), but for $u^2/\pi^2 = 2.05$ (at point A) corresponding to $u_{cd} = 4.49$ [cf. equation (3.90c)], the first-mode locus crosses to the $\omega^2 < 0$ half-plane, indicating divergence in the first mode. The system remains unstable with increasing u^2,

†These curves are not identical to Huseyin & Plaut's (1974), which are quantitatively in error (Plaut 1995); thus, the values of β for each of the three distinct types of behaviour are incorrect, and so is the value of u^2/π^2 for point B; otherwise, the results are qualitatively similar to those in Figure 3.15.

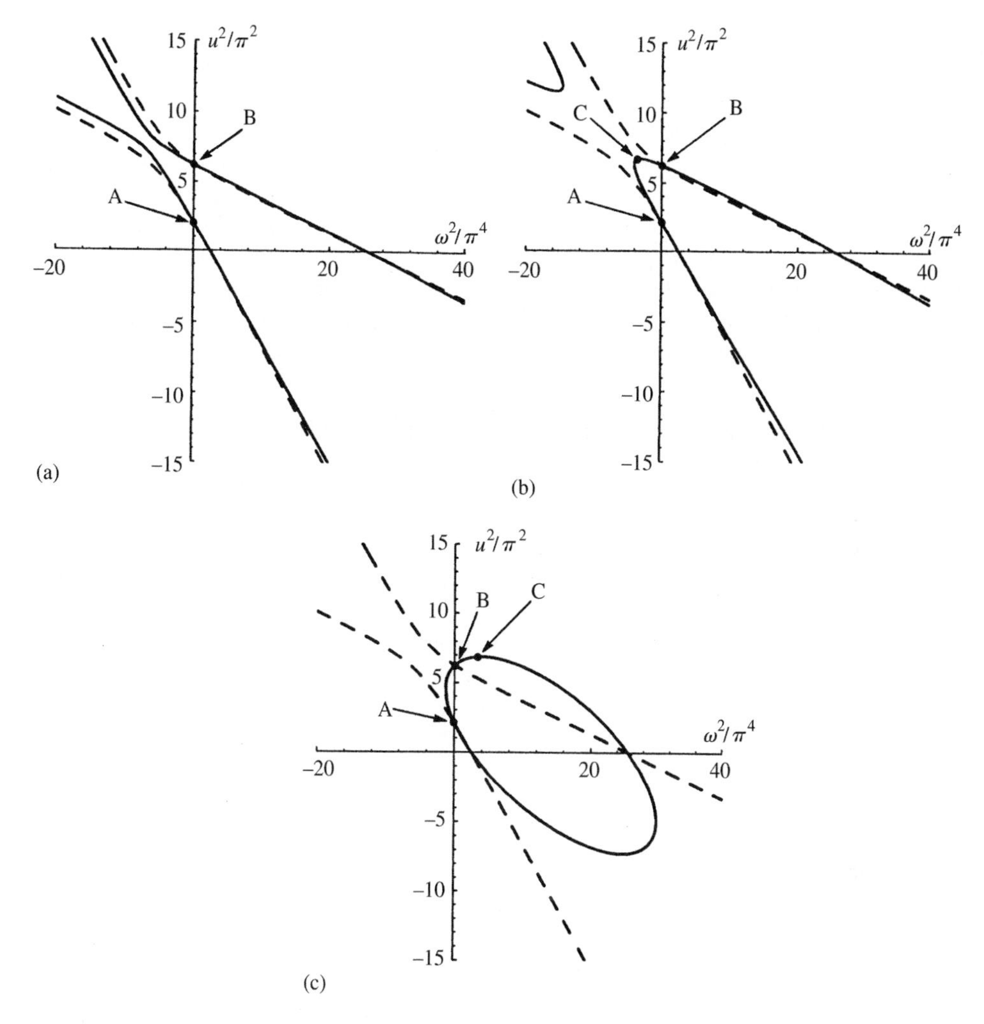

Figure 3.15 Stability behaviour of a clamped–pinned pipe ($\Gamma = \Pi = \alpha = \sigma = k = \gamma = 0$) in terms of 'characteristic curves' of u^2/π^2 versus ω^2/π^4 for (a) $\beta = 0.05$, (b) $\beta = 0.1$ and (c) $\beta = 0.7$: ——, the gyroscopic conservative system; – – –, the 'corresponding nongyroscopic system'.

but at point B ($u^2/\pi^2 = 6.24$)† divergence develops in the second mode also. In this case the dynamics is similar to that of the equivalent nongyroscopic system. In (b) it is seen that, after divergence at A and at B [for the same values of u^2 as in (a)], the ω_1^2 and ω_2^2 loci coalesce at point C, indicating the onset of Païdoussis-type coupled-mode flutter — i.e. directly from the divergent state. Thus, there is no post-divergence restabilization of the first mode for $u > u_{cd}$ in this case; coupled-mode flutter arises before it can materialize. In (c), after divergence at A, there *is* gyroscopic restabilization ($\omega_1^2 > 0$ again, at point B)

†An additional point of interest is that in this case, where the support conditions are asymmetrical, the stiffness matrix is not diagonal, unlike the case of simply-supported ends — refer to discussion on equation (3.92). Hence, this value differs considerably from that obtained from equation (3.90c).

at $u^2/\pi^2 \simeq 6.2$, followed by coupled-mode flutter at point C ($u^2/\pi^2 \simeq 7$), in this case via a Hamiltonian Hopf bifurcation. The value of u for restabilization at point B corresponds exactly to the point where the nongyroscopic system, or indeed the pipe systems in (a) and (b), develop divergence in their second mode.

In closing, the following two important points should be made. First, the results of Figures 3.9–3.11, 3.14 and 3.15 apply equally to pipes with a downstream end either free to slide axially or not [Figure 3.1(a,b)]: since linear theory cannot distinguish between the two, the same equation governs both; however, the foregoing explanation of the existence of coupled-mode flutter applies only to systems with a sliding end. Second, and as cautioned at the outset, the existence of coupled-mode flutter has to be decided by nonlinear theory (Chapter 5) and by experiments (Section 3.4.4).

3.4.2 Pressurization, tensioning and gravity effects

If dissipative and gravity effects are neglected and $\mathrm{d}U/\mathrm{d}t = 0$, equation (3.38) simplifies to

$$EI\,\frac{\partial^4 w}{\partial x^4} + \left[MU^2 + \overline{p}A(1-2\nu\delta) - \overline{T}\right]\frac{\partial^2 w}{\partial x^2} + 2MU\,\frac{\partial^2 w}{\partial x \partial t} + (M+m)\frac{\partial^2 w}{\partial t^2} = 0, \quad (3.98)$$

in which it is recalled that $\delta = 0$ if there is no axial constraint, so that axial sliding of the downstream end is permitted, and $\delta = 1$ if it is prevented. The case of $\delta = 1$ is shown in Figure 3.16(a), where p_f is the pressure expended in overcoming the frictional pressure

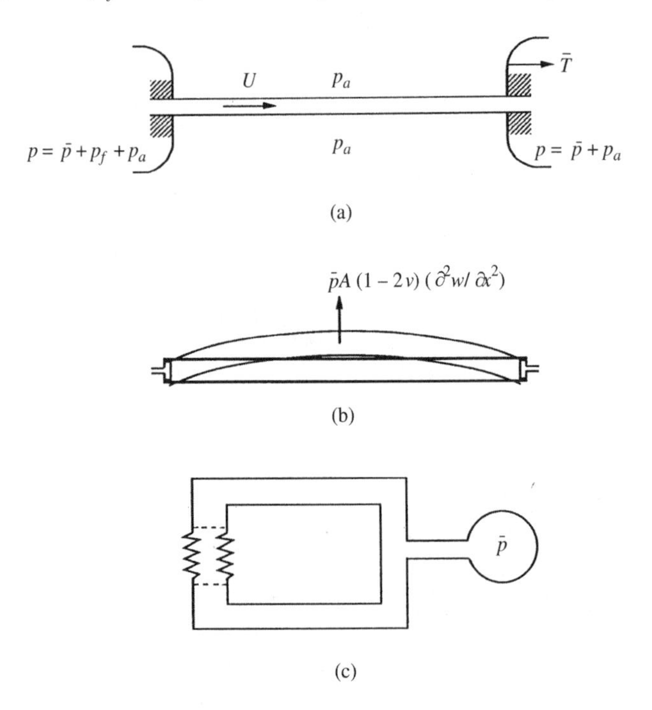

Figure 3.16 (a) A pipe subject to tensioning $\overline{T}$ and to pressurization $\overline{p}$, measured above the atmospheric pressure, p_a; (b) divergence due to presurization, represented as if the pipe were pressurized by floating pistons; (c) model experiment with bellows, to show pressurization-induced buckling.

drop and p_a is the atmospheric pressure, both of which do not enter the equation of motion (Section 3.3.4).

It is clear that the pressure term acts in the same way as the MU^2 term, and hence it is not surprising that, given a sufficiently high level of pressurization, divergence may be induced by pressure alone — just as it may do by compression alone, i.e. for $\overline{T} < 0$ and sufficiently large. Physically, one may think of the pressurization as being produced by floating pistons acting on both sides of a segment of the pipe, as shown in Figure 3.16(b). An easy experiment to demonstrate pressure-induced divergence consists in joining two rigid pipes with a straight rubber hose and then connecting the other ends of the rigid pipes to the same regulated pressure supply. As the pressure is increased, eventually the rubber hose buckles. The same effect may be obtained if, instead of a rubber pipe, bellows are used [Figure 3.16(c)].

The effect of pressurization may appear to be obvious and hence trivial. Nevertheless, consider the following two systems: (i) a pipe with an axially sliding end under pressurization $\overline{p}$ and tension $\overline{T}$, with zero flow [i.e. as in Figure 3.16(a) but with axial sliding permitted and $U = 0$; $\overline{T}$ being provided by a weight acting through pulleys], and (ii) a closed tube pressurized to $\overline{p}$. In both cases, the equation of motion is

$$EI\frac{\partial^4 w}{\partial x^4} + \overline{p}A\frac{\partial^2 w}{\partial x^2} - \overline{T}\frac{\partial^2 w}{\partial x^2} + (m+M)\frac{\partial^2 w}{\partial t^2} = 0. \tag{3.99}$$

In case (i), $\overline{p}$ and $\overline{T}$ are independent of each other, and $\overline{T}$ may possibly be zero. In case (ii), however, in the linear limit, $\overline{p}A = \overline{T}$ and the net effect of pressurization is nil. This, nevertheless, has not stopped an intrepid would-be inventor from obtaining a patent for stiffening hollow rotors against whirling by 'pressurization-induced tensioning', as shown in Figure 3.17(a,b), while conveniently forgetting about the destabilizing effect of pressurization illustrated in Figure 3.17(c). Thus, the inventor took into account the

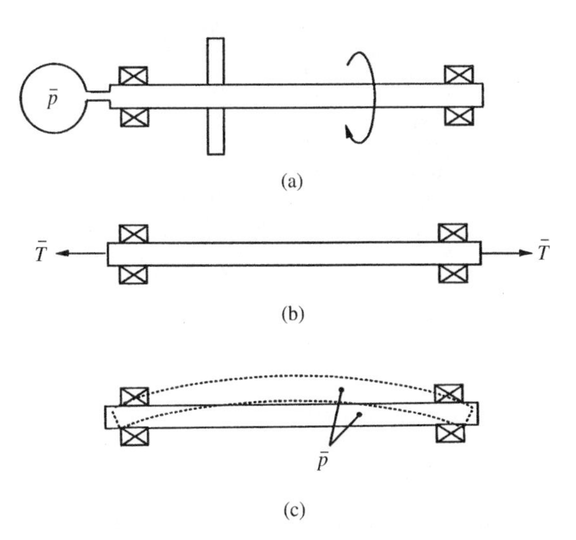

Figure 3.17 (a) The fallacious patent for delaying the onset of whirling through pressurization-induced tensioning; (b) the stabilizing effect of pressurization-induced tension, $\overline{T}$; (c) the destabilizing effect of pressurization, $\overline{p}$.

effect of the third term in (3.99), while ignoring the second, probably reasoning that since pressurization induces the tensioning, it need not be considered further — thus inventing the impossible! In reality, the net effect on whirling is zero.

The story of this fallacious patent is charmingly related by Den Hartog (1969), together with one on an earlier but similarly fallacious patent, this one for preventing buckling of drill-strings used in oil exploration. It is well known that the very long and slender drill-rods buckle under the compressive loading required for drilling and they touch the sidewalls in several places along the length. Then, as the drill-rod rotates and rubs against the sidewalls, up to 90% of the power is consumed for this non-useful work. The invention consisted of using a hollow drill-rod and a floating drill-bit, and pumping sludge down the drill-rod, which would rotate the drill-bit as a turbine, as depicted in Figure 3.18. Thus, it was thought, the removal of all compressive load from the drill rod would result in the elimination of all possibility of buckling. However, it should be realized that, to cause the drill-bit to press hard on the rock and to rotate against it, the pressure p_1, must be substantially larger than p_2. Hence, the truth emerges that the drill rod would buckle just the same due to pressurization, under much the same conditions as the original system — and perhaps earlier because of the flow effect.

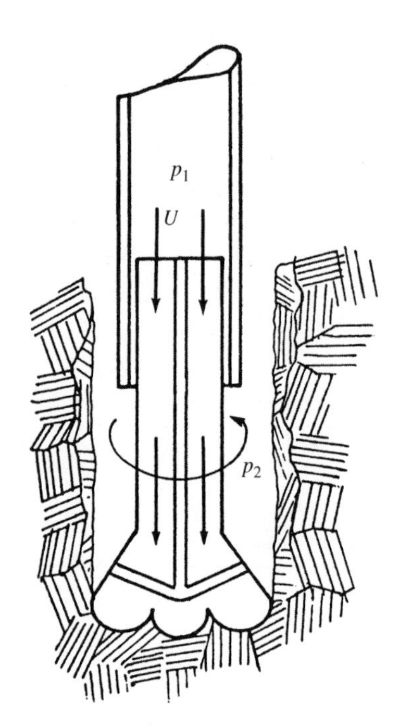

Figure 3.18 The fallacious patent for preventing buckling of drill-strings by the use of a floating drill-bit, rotating under the action of the flow (Den Hartog 1969).

Returning to a quantitative assessment of pressurization effects, equation (3.98) may be written in dimensionless terms as

$$\eta'''' + v^2\eta'' + 2\beta^{1/2}u\dot{\eta}' + \ddot{\eta} = 0, \qquad v^2 = u^2 + \Pi(1 - 2\nu\delta) - \Gamma. \tag{3.100}$$

Hence, it is clear that for pinned ends $v_{cd} = \pi$, while for clamped ends $v_{cd} = 2\pi$, since the Coriolis term is not involved in the divergence instability.

Gravity effects are considered next. If gravity is taken into account (i.e. if the system is vertical), but still taking $k = 0$ (no elastic foundation) in equation (3.70), the critical conditions are found to be as in Figure 3.19. Clearly, equations (3.90a,b) still apply, with v replacing u — with v as in the second of equations (3.100).

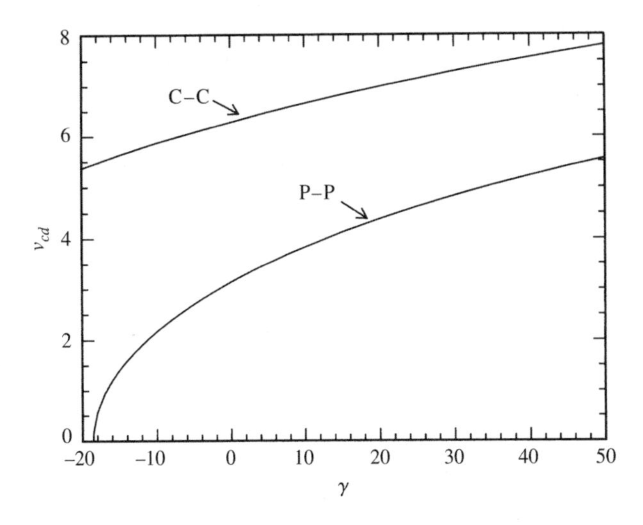

Figure 3.19 The critical value of v_{cd} for divergence of vertical pipes with supported ends ($\Gamma = \Pi = k = 0$), showing the effect of γ. P-P: pinned–pinned (simply-supported) pipes; C-C: clamped–clamped pipes; v is defined in the second of equations (3.100).

A value of $\gamma < 0$ signifies that gravity is in the opposite direction to the flow vector — i.e. upwards in Figure 3.1(b). Thus, for $\gamma < 0$ the pipe is under gravity-induced *compression*, while for $\gamma > 0$ it is under gravity-induced tension, which explains why u_{cd} for $\gamma < 0$ is smaller than for $\gamma > 0$; indeed, for γ sufficiently large and negative, the system diverges (buckles) under its own weight. [In the case of Figure 3.1(a), it is implicitly presumed that the pipe is hung before the downstream end is positively fixed; thus the pipe is subjected to the same gravity-induced tension/compression as in the case of Figure 3.1(b).]

It is also noted that, as γ increases, the ratio of v_{cd} for clamped and pinned pipes is diminished: $2\pi/\pi = 2$ for $\gamma = 0$ and $7.80/5.56 = 1.4$ for $\gamma = 50$. Physically, one may think of a larger γ as representing a longer pipe [equations (3.71)]; in the limit, the pipe will resemble a string rather than a beam, and hence will be less sensitive to boundary conditions. This breaks down for $\gamma < 0$, since in the case of pinned ends, as the critical γ is approached for divergence due to its own weight, v_{cd} is diminished very fast, while this is not yet true for clamped ends, for the range of γ in Figure 3.19. The critical γ values for divergence at $u = 0$ are $\gamma_{cr} = -18.55$ for pinned ends and -66.34 for clamped ones.

For other aspects and/or details of the effects of pressurization, the interested reader is referred to the work of Haringx (1952), Heinrich (1956), Hu & Tsoon (1957), Roth & Christ (1962), Naguleswaran & Williams (1968), Stein & Tobriner (1970) and Païdoussis

& Issid (1974), and for the effect of externally applied tension to Bolotin (1956) and Plaut & Huseyin (1975).

3.4.3 Pipes on an elastic foundation

An elastic foundation represents the distributed support provided to long pipes resting on a generally elastic medium, e.g. in the case of pipelines laid on the ocean floor. For pipes with supported ends the additional stiffness supplied by the elastic foundation simply renders the system stiffer [see equation (3.70)], and hence the qualitative effect on stability is predictable.

The critical flow velocity for divergence, u_{cd}, or more generally $\mathcal{v}_{cd}$ as per the second of equations (3.100), may be obtained by the method of Section 3.3.6(a) in a similar manner as used to obtain equations (3.90a–c); indeed, as first obtained by Roth (1964),[†]

$$\mathcal{v}_{cd} = \pi \left(1 + \frac{k}{\pi^4}\right)^{1/2}. \tag{3.101a}$$

However, if k is sufficiently large, e.g. $k = 10^3$, $\mathcal{v}_{cd}$ as given by (3.101a) is overestimated, because divergence can be associated with a higher mode at a lower value of $\mathcal{v}_{cd}$, obtained from

$$\mathcal{v}_{cd} = n\pi \left(1 + \frac{k}{(n\pi)^4}\right)^{1/2}, \tag{3.101b}$$

where the mode number n is identified with the beam eigenfunction $\sqrt{2}\,\sin(n\pi x/L)$.[‡] The mode to become unstable is that leading to the smallest $\mathcal{v}_{cd}$, and is thus associated with the *smallest* n satisfying

$$n^2(n+1)^2 \geq \frac{k}{\pi^4}; \tag{3.102}$$

e.g. for $k = 300$ one obtains $n = 1$, whereas for $k = 500$, $n = 2$. What happens physically is that the support provided by the elastic foundation can be thought of as providing added *supports* along the length, making the first divergence with one or more nodes within the span feasible.

For a clamped–clamped pipe, by Galerkin's method (Roth 1964), one obtains

$$\mathcal{v}_{cd} = 2\pi \left(1 + \frac{3k}{16\pi^4}\right)^{1/2} \qquad \text{for} \qquad k \leq (84/11)\pi^4$$

and (3.103)

$$\mathcal{v}_{cd} = \pi \left(\frac{n^4 + 6n^2 + 1}{n^2 + 1} + \frac{k}{\pi^4(n^2+1)}\right)^{1/2} \qquad \text{for} \qquad k \geq (84/11)\pi^4.$$

[†]Roth's excellent work, written in German, is unfortunately hardly ever cited in the English-language literature. The interested reader is encouraged to refer to Roth (1965a,b, 1966) also.

[‡]It is of interest that for all the solutions given by (3.101b), and also (3.103), the condition $\mathcal{v}_{cd}^4/k \geq 4$ is satisfied, so that the discriminant of (3.82) is positive (or zero, when $k = \pi^4$), and hence real values of the α_i are obtained.

The first equation is associated with $n = 1$; the second with $n \geq 2$, such that n is the smallest integer satisfying

$$n^4 + 2n^3 + 3n^2 + 2n + 6 \geq k/\pi^4, \tag{3.104}$$

e.g. $n = 2$ if $84/11 \leq k/\pi^4 \leq 54$, $n = 3$ if $54 \leq k/\pi^4 \leq 174$, $n = 4$ if $174 \leq k/\pi^4 \leq 446$, etc. Equations (3.102) and (3.104) differ from the criteria given by Roth, which can lead to a nonconservative value of v_{cd}. The Galerkin solutions (3.103) were compared to an exact solution and found always to overestimate the exact v_{cd}, but by less than 2%.

The values of v_{cd} versus k are plotted in Figure 3.20, showing the transition of divergence from $n = 1$ to higher n as k is increased.

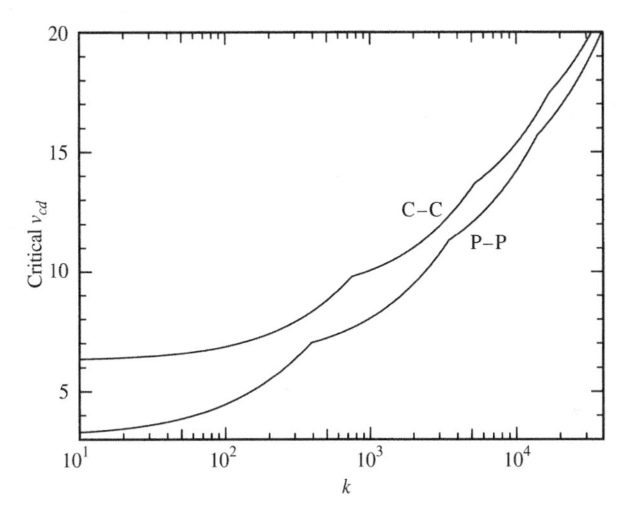

Figure 3.20 The critical values of v_{cd}, where $v^2 = u^2 + \Pi(1 - 2\nu\delta) - \Gamma$, for pinned–pinned (P-P) and clamped–clamped (C-C) pipes on an elastic foundation of dimensionless modulus k.

Some numerical results for a clamped–pinned pipe for divergence and coupled-mode flutter with $\beta = 0.9$ may be found in Lottati & Kornecki (1985).

Elastic foundations become particularly important for systems not otherwise supported, which in practice means that the end supports are very, or infinitely, far apart. They will be treated in Section 3.7.

3.4.4 Experiments

Experimental work on the dynamics of pipes conveying fluid commenced soon after Housner showed in 1952 that this system is subject to divergence (buckling) at sufficiently high flow velocity. The aim of the first set of such studies, implicitly at least, was the validation of the main theoretical findings: (i) that divergence does arise, (ii) that it occurs near the theoretical critical flow velocity, u_{cd}, and (iii) that the first-mode frequency, ω_1, varies with u parabolically, in the manner shown in Figure 3.10. Hence, for simplicity, in these studies (Long 1955; Dodds & Runyan 1965; Greenwald & Dugundji 1967; Yoshizawa *et al*. 1985, 1986) pressurization effects were not considered, by making the downstream end of the pipe free to slide axially [$\delta = 0$ in equations (3.37) and (3.38)].

Long's (1955) experiments involved simply-supported and clamped–clamped steel pipes conveying fluid; the downstream end was mounted on rollers. The simply-supported pipe had outer diameter $D_o = 25.4$ mm (1 in), wall thickness $h = 0.94$ mm (0.037 in), and span $L = 3.048$ m (120 in). Despite the length and hence relatively large flexibility of this pipe, $u_{cd} = \pi$ corresponds to $U_{cd} \simeq 52$ m/s (172 ft/s) — a high and difficult to achieve flow-rate, because of the pumping requirements implied: a high flow rate at a high pressure (to overcome the large pressure drop); indeed, beyond the capabilities of Long's apparatus. By means of strain gauges, Long measured the first-mode frequency and damping, and how they varied with u.[†] It should be recalled that $\mathcal{R}e(\Omega_1)$ is expected to decrease parabolically with u; also, since $\mathcal{I}m(\Omega_1)$ is approximately constant according to theory, $\delta_1 \simeq 2\pi\zeta_1 = 2\pi\mathcal{I}m(\Omega_1)/\mathcal{R}e(\Omega_1)$ is expected to *increase* parabolically. However, for $u < 1$, both $\mathrm{d}[\mathcal{R}e(\Omega_1)]/\mathrm{d}u$ and $\mathrm{d}\delta_1/\mathrm{d}u$ are small, and for the $u_{\max} \simeq 0.68$ achieved in these experiments the effect, if any, was judged to be within the margin of experimental error.[‡] Hence, these experiments were largely inconclusive.

A more effective experiment was conducted by Dodds & Runyan (1965), also with simply-supported pipes, as shown in Figure 3.21. The pipes were of aluminium alloy, with $D_o = 25.4$ mm, $h = 1.65$ mm, and an effective length $L = 3.812$ m (12.5 ft); the fluid was water. In this case, the critical flow velocity, $U_{cr} = 39.5$ m/s, was actually attained. Figure 3.22(a) displays the evolution of $\mathcal{R}e(\Omega_1)$ with u for two different pipes, and shows

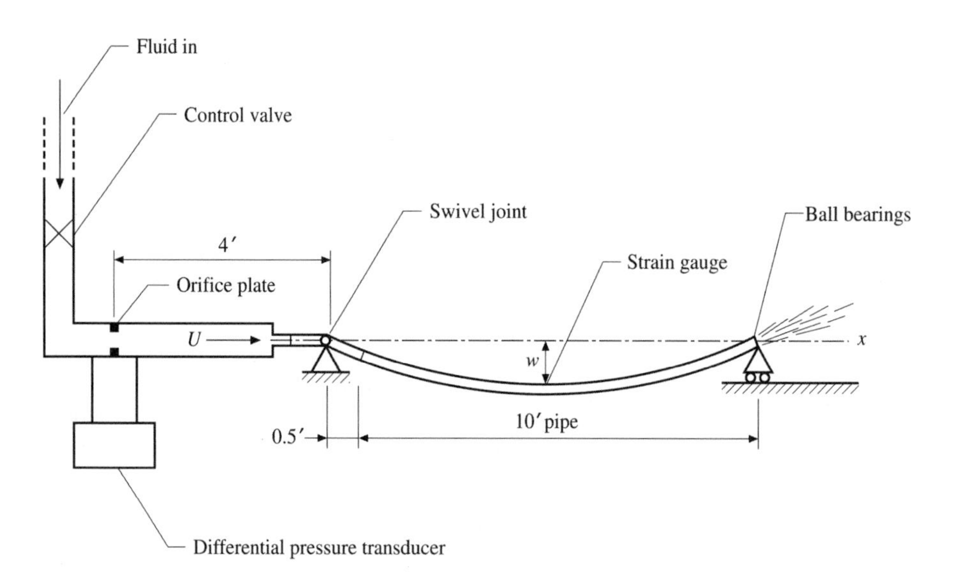

Figure 3.21 Schematic diagram of the experimental apparatus used by Dodds & Runyan (1965). All dimensions are in feet; 1 ft = 0.3048 m.

[†]Since various researchers have used different, and in some cases truly curious, schemes of nondimensionalization, wherever possible these have been converted to those used in this book, for the reader's convenience.

[‡]Long also reports on some experimental results by E. Ergin of Cal Tech, with a pipe 'similar to that used here', which show a clear quasi-parabolic Ω_1 versus u curve. However, there appears to be some error, at the very least in the nondimensionalization of u; for, whereas a $u_{\max} \simeq 6.0$ is shown, which greatly exceeds u_{cd}, the maximum reduction in $\mathcal{R}e(\Omega_1)$ is only 3.2%.

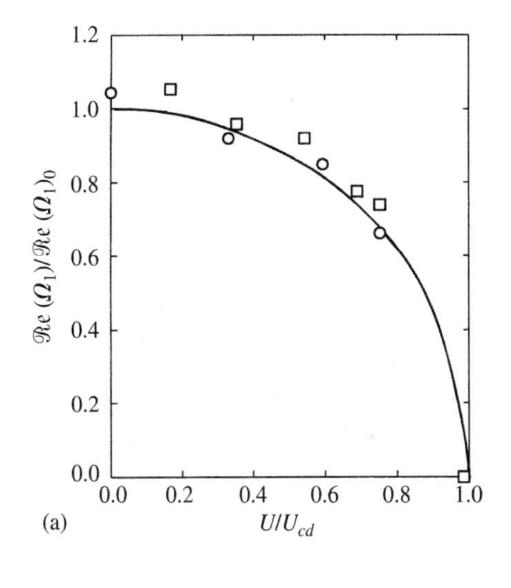

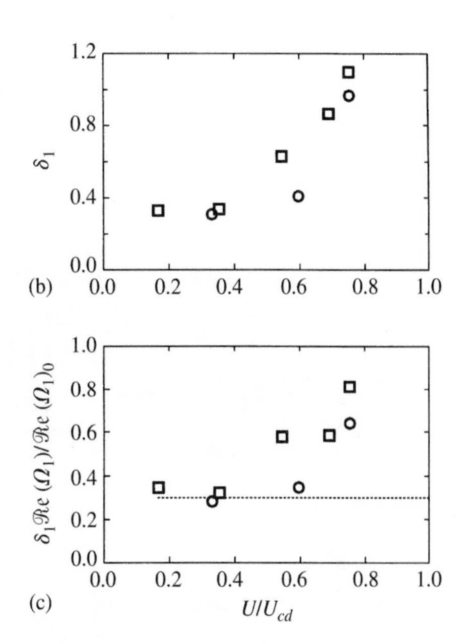

Figure 3.22 (a) The variation of the first-mode frequency $\mathcal{R}e(\Omega_1)$ with respect to U in Dodds & Runyan's experiments, respectively normalized by the zero-flow frequency $\mathcal{R}e(\Omega_1)_0$ and the flow velocity for divergence, U_{cd}, for two different pipes; (b) the variation of the first-mode logarithmic decrement, δ_1, with U/U_{cd} for the same two pipes; (c) the theoretically constant $\delta_1\mathcal{R}e(\Omega_1)/Re(\Omega_1)_0$. ——, Theory; ○, experiment with pipe 1; □, experiment with pipe 2. Data from Dodds & Runyan (1965).

near-perfect agreement with theory; $\mathcal{R}e(\Omega_1)_0$ is the value of $\mathcal{R}e(\Omega_1)$ at $U = 0$. However, agreement is likely not to have been as perfect as this figure would suggest, as may be appreciated from Figure 3.22(b,c), in which the authors' tabulated measurements of δ_1 as well as $\delta_1\mathcal{R}e(\Omega_1)/\mathcal{R}e(\Omega_1)_0$ have been plotted against u. This latter, being proportional to $\mathcal{I}m(\Omega_1)$, should theoretically be approximately constant with u, but in the experiments it increases substantially as u_{cd} is approached, reflecting most probably real effects at the supports as the pipe begins to bow. It is quite likely that these same effects involve an attendant stiffening of the pipe which neatly counterbalances any natural tendency of the pipe to buckle 'before its time' due to imperfections (e.g. initial curvature of the pipe, locked-in stresses, geometric and material nonuniformities), which, as is well known, would make the pipe diverge at a lower flow velocity than its perfect counterpart. This discussion is meant to provide physical insight into *some* of the real effects and difficulties encountered in experiments, and does not take away one iota of Dodds & Runyan's important achievement: to demonstrate convincingly the existence of divergence, as shown dramatically in Figure 3.23, and to validate items (i) and (ii) of the first paragraph of this section.

A more wide-ranging experimental and theoretical investigation was undertaken by Greenwald & Dugundji (1967), motivated by the same concern as Dodds & Runyan: the possibility of disastrous fluidelastic instabilities in the thin-walled propellant pipelines of liquid-fuel rocket engines. Experiments were conducted with clamped–pinned and cantilevered pipes. In contrast to previous studies, however, these were small-scale

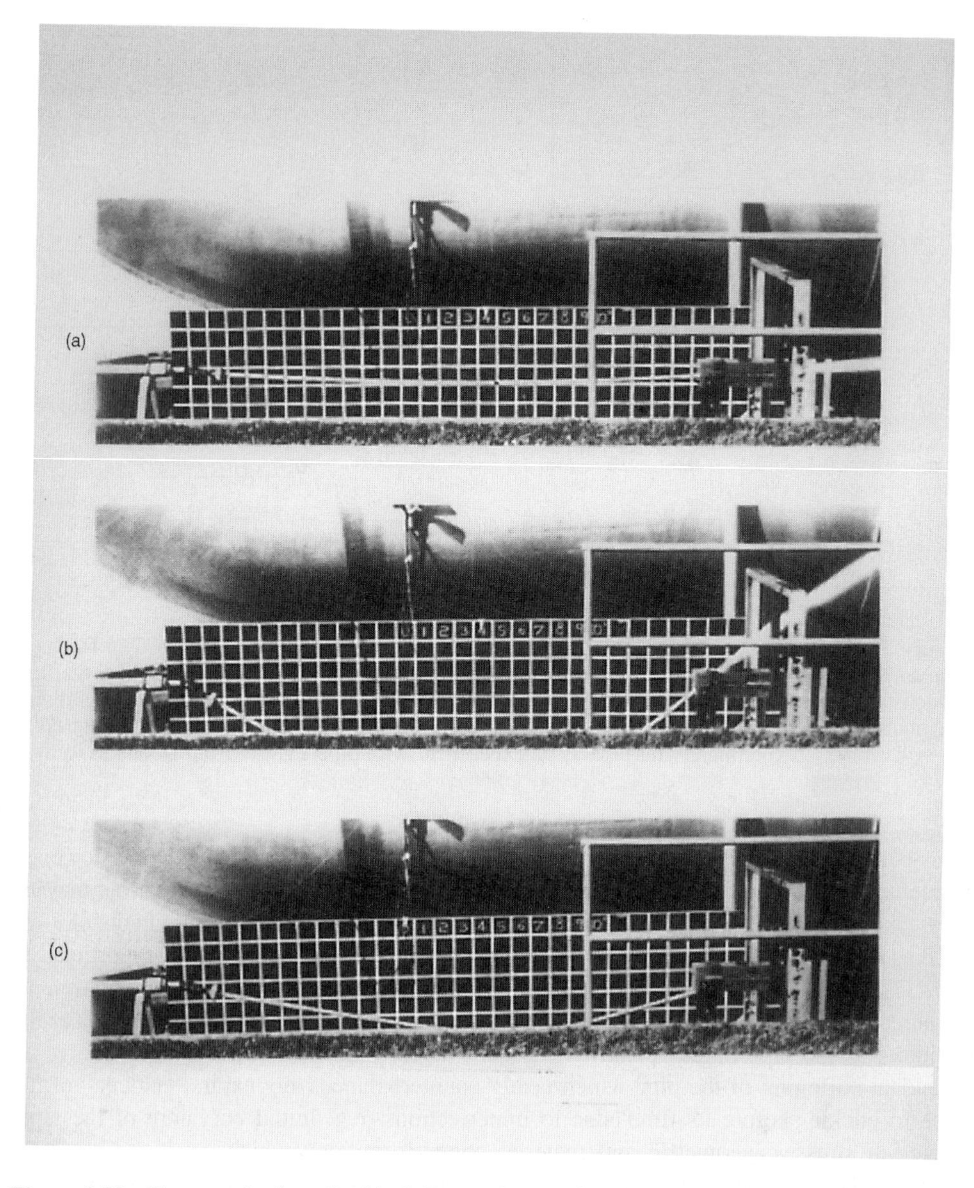

Figure 3.23 Photographs from Dodds & Runyan's experiments: (a) the pipe slightly curved just before divergence at $U = 36.6$ m/s (120 ft/s); (b) the pipe at divergence, with its middle part disappearing from view, at $U = 38.86$ m/s (127.5 ft/s); (c) curvature of the pipe after completion of the test, showing plastic deformation.

experiments (e.g. $D_o = 4.75$ mm, $h = 1.5$ mm, $L = 241.3$ mm) with elastomer and polyethylene pipes conveying water. These pipes have a Young's modulus more than two orders of magnitude smaller than metal pipes, and this results in a much simpler, low-pressure apparatus. However, there are at least three disadvantages in the use of such materials: (i) the damping characteristics are generally complex viscoelastic, and accurate representation requires at least a two-constant dissipation model; (ii) the cross-sectional area is generally a function of internal pressure; (iii) the pipe has an initial curvature as a result of being coiled during manufacture while still warm and of plastic set during storage. Of these, item (i) plays no role in the determination of u_{cd}, (ii) is not too important if the fluid discharges at $x = L$ so that the pressure is not too high at any point upstream, and (iii) was solved, according to the authors, by hanging the pipes vertically and pouring hot water through them.

In the clamped–pinned arrangement, the downstream support was provided quite simply by a greased steel rod in contact with the downstream end of the pipe. As the flow velocity was increased, the pipe began to bow slightly. At a certain critical speed the pipe was observed to statically diverge rapidly and to slide completely off the steel rod. This means that the measured u_{cd} was slightly higher than the real one. The experimental $u_{cd} = 4.70$ nevertheless compares favourably with the theoretical $u_{cd} = 4.49$ given by equation (3.90c).

A more recent, successful experiment for a clamped–pinned pipe, again with a sliding downstream end, was conducted by Yoshizawa *et al*. (1985, 1986) and is discussed in Section 5.5.3.

The main purpose of these studies was to validate items (i)–(iii) of the first paragraph of this section and it was partly achieved. It was also shown, by the way, that large flow velocities are necessary to induce divergence; hence, it is unlikely to arise in practice, except in specialized applications. Nevertheless, there is a high degree of idealization in the systems studied so far; certainly, systems of the type of Figure 3.21 are unlikely to be found in engineering applications. In more practical systems, the pipe would not discharge to atmosphere but would be connected to another component at a pressure higher than atmospheric [Figure 3.16(a)] — except after an accidental break (Section 4.7); moreover, axial sliding, if any, would not occur freely and destabilizing pressurization effects would come into play. In the next set of such studies, the dynamics under these more realistic conditions was considered.

A careful study of the effects of pressurization and tensioning was made by Naguleswaran & Williams (1968). Unfortunately, in the paper they do not give any of the dimensions and properties of their apparatus, nor any of their results in dimensional form. Nevertheless, Naguleswaran (1996) was kind enough to provide the approximate principal dimensions of the neoprene pipes used: $D_o = 15\,\mathrm{mm}$, $h = 2\,\mathrm{mm}$, and variable length, up to 880 mm. The pipe was attached on either side to rigid copper pipes, one of which was connected to the water mains and the other, after a certain length, discharged to atmosphere. The mean pressure in the whole system could be regulated, presumably by valves on the downstream end, so that pressurization was possible. Furthermore, axial tension could be applied by loading one of the copper pipe connections statically and then fixing it; thereafter, sliding was prevented ($\delta = 1$). The flow rate was determined by collecting and weighing the discharged water over a known time interval. Motions of the pipe were sensed at two locations along the span via capacitance transducers. The Poisson ratio, ν, of the pipe was determined in special tests by measuring the change in volume resulting from axial extension, and EI was determined from the natural frequency of a short cantilevered length of the pipe.

It was found that pressurization affected appreciably the first-mode natural frequency, to the extent that the pipe could be made to buckle quite readily without flow. For this reason, preliminary tests were made without flow. The variation of $\mathcal{R}\mathrm{e}(\Omega_1)$ with $\Pi/\Gamma \equiv \overline{p}A/\overline{T}$ is shown in Figure 3.24; $\mathcal{R}\mathrm{e}(\Omega_1)_0$ is the value for the pipe under $\overline{T}$ but for $\overline{p} = 0$. Since

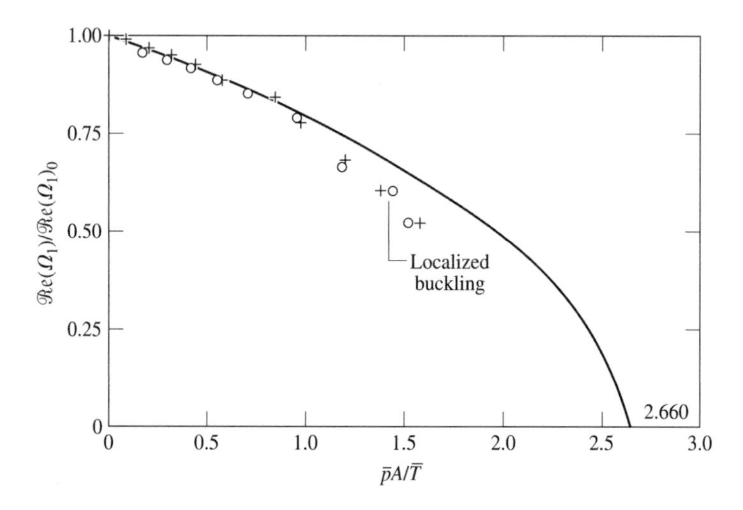

Figure 3.24 The effect of pressurization $\overline{p}$ on a tensioned pipe ($\overline{T} \neq 0$) for $u = 0$ according to Naguleswaran & Williams (1968): ——, theory with the measured value of $\nu = 0.312$; ∘, +, experimental data.

it was found that A varies appreciably with $\overline{p}$, the actual $A(\overline{p})$ were used in plotting the experimental points. The experimental values are compared with simplified theory, in which the pipe is assumed to be long enough for flexural effects to be less important than tensile ones; thus, by taking $v^2\eta'' \gg \eta'''' \to 0$ in equations (3.100), as well as $u = 0$ and $\delta = 1$, it is easy to find $\mathcal{R}e(\omega_1)/\mathcal{R}e(\omega_1)_0 = \mathcal{R}e(\Omega_1)/\mathcal{R}e(\Omega_1)_0 = [1 - \Pi(1 - 2\nu)/\Gamma]^{1/2}$.

It is seen in Figure 3.24 that the agreement is good for low enough Π, but as the buckling condition is approached (for $\Pi/\Gamma = 2.66$ for the experimental $\nu = 0.312$), there is considerable discrepancy, as a result of 'small irregularities, or kinks in the tube', i.e. imperfections, which lead to localized buckling. Furthermore, when Π is increased beyond that point, overall buckling (divergence) is never realized, because the axial length of the pipe is constrained and deflection of the pipe gives rise to increased tension.

Similar results are obtained with flow, as shown in Figure 3.25(a); since dimensional quantities are not given, the peculiar nondimensionalization of this study is retained. The experimental data are compared with (i) simple theory in which pressurization and dilatation of the pipe are ignored ($\Pi = 0$, $A =$ const.) and (ii) theory in which these effects are taken into account. As expected, agreement is far better with the latter. Figure 3.25(b) shows the phase difference in the motion at two locations ($\xi = 0.175$ and 0.815). Because of the opposite rotation of points with $\xi < 0.5$ and > 0.5 approximately, the Coriolis term is responsible for this phase difference, and it is seen that it increases nearly linearly with u, so long as the condition of divergence is not close; at $u = u_{cd}$, of course, ψ must be zero.

The condition of zero frequency (and zero phase) was never, indeed can never, be achieved for systems in which axial sliding is prevented, for the reasons already given: increased deflection generates an increase in tension and thus $\omega_1 = 0$ is unattainable. Thus, in this case there is a component of tension proportional to deflection, and the equation of motion becomes nonlinear. Hence, the dynamics of the system as the linear u_{cd} is approached (in this case, as seen in Figure 3.25, for $u > 0.5u_{cd}$ approximately) should be studied by means of nonlinear theory.

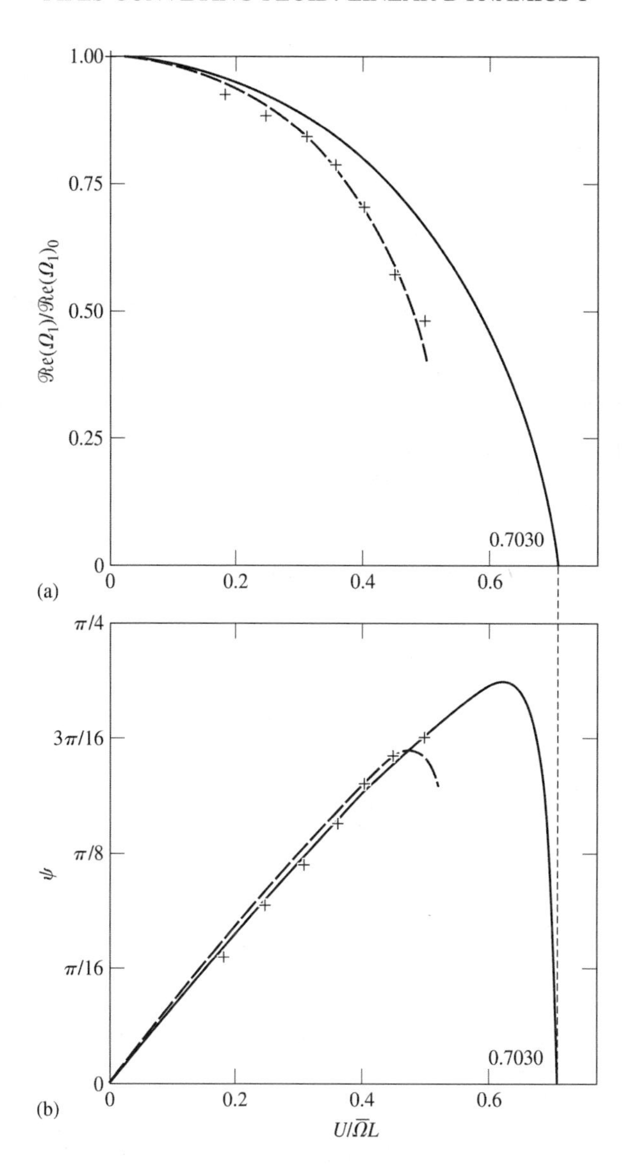

Figure 3.25 (a) The ratio of $\Re e(\Omega_1)/\Re e(\Omega_1)_0$ as a function of the dimensionless flow velocity $U/\overline{\Omega}L = u/\{\pi[\beta(\Gamma + \pi^2)]^{1/2}\}$ for a tensioned and pressurized pipe with $\{[\Gamma - \Pi(1 - 2\nu)]/[\Gamma + \pi^2]\}^{1/2}/\pi = 0.0636$ and $\beta = 0.4338$ at $u = 0$ and $\nu = 0.312$; (b) the phase difference in the displacement at $\xi = 0.175$ and $\xi = 0.815$ during vibration as a function of $U/\overline{\Omega}L$. +, Experiment; ——, theory with pressurization and resulting dilatation of the pipe ignored; — — —, theory with these effects taken into account (Naguleswaran & Williams 1968).

Another set of experiments was conducted by Liu & Mote (1974), aiming to study the effects of flow and tensioning on the dynamics of the system. They used small-diameter vertical aluminium pipes ($D_o = 6.375$ mm, $h = 0.559$ mm), 1.829 m (6 ft) long, conveying an oil–water emulsion circulated with the aid of a gear pump. The apparatus involved a shaker to excite the pipe and strain gauges to measure the vibration. The fluid was discharged to atmosphere at the downstream end (collected and recirculated), and axial sliding was permitted. Tension was applied via a pulley–weight mechanism. Typical results are shown in Figure 3.26 for two values of tension, $\Gamma = 0$ and 5, and nominally pinned ends. It is noted that the pinning is far from perfect: the first-mode measured frequency is 5.1 Hz for $\Gamma = 0$, whilst the theoretical one is 3.8 Hz; this is mostly due to the flexible coupling connecting the upstream end to the rest of the system, which when disconnected results in a measured frequency of 4.0 Hz, much closer to the theoretical one. Nevertheless, the normalized form of Figure 3.26 has the advantage of permitting the direct comparison of theoretical and experimental trends with increasing u and varying Γ.

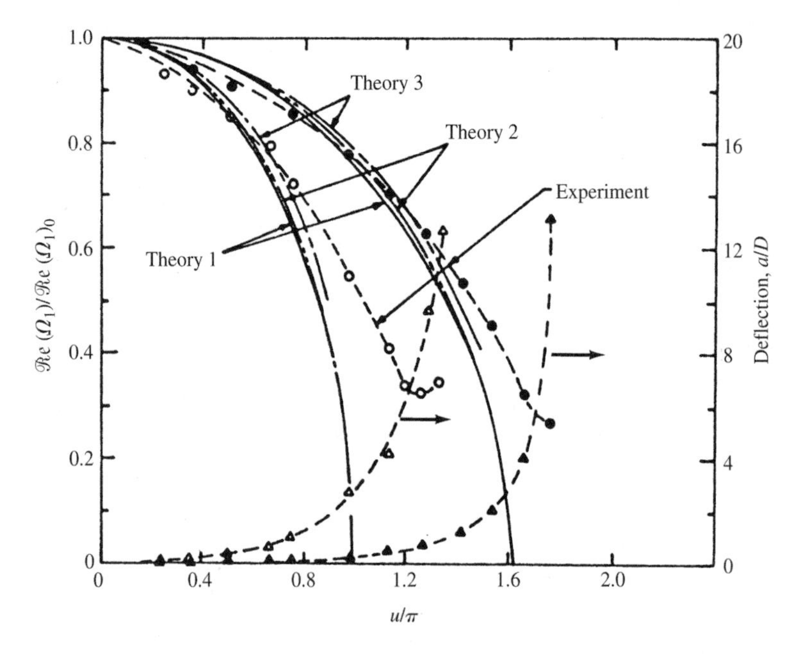

Figure 3.26 Fundamental resonance obtained from vibration measurements on a shaker-excited simply-supported pipe under tensioning, as a function of u/π. Theory 1 is the linear theory of Naguleswaran & Williams (1968); theory 2 and theory 3 are, respectively, Thurman & Mote's (1969b) linear and nonlinear theory. Experimental/theoretical reference frequencies $\mathcal{R}e(\Omega_1)_0 = 5.1/3.8$ Hz for $\Gamma = 0$, and 7.2/5.9 Hz for $\Gamma = 5$. The deflection has been nondimensionalized with respect to the pipe diameter (Liu & Mote 1974).

The measured frequencies decrease with u, initially as predicted by theory, but later the curves bottom out and the frequency begins to increase with u — an effect which is even more pronounced in some other of the authors' results. This is very perplexing, since for these conditions of support (with sliding permitted), the zero-frequency condition should have been attainable. Before proposing an explanation, it should be said that these experiments suffered from a number of weaknesses, as acknowledged by the authors: (a) the aforementioned nonzero bending moment (imperfect pinning) at $\xi = 0$; (b) a substantial and undesirable out-of-plane vibration, at times larger than the excited (and plotted) in-plane one; (c) an initial curvature (bow) and/or locked-in stresses in the pipe which gave a gradual and continuous increase in deflection with increasing u, rather than a precipitous one as u_{cd} was approached. Also, (d) there is a discrepancy of the theoretical results with regard to the Γ given in the figures. For instance, for $\Gamma = 5$, $u_{cd}/\pi = 1.23$ should have been obtained (cf. Section 3.4.2) and not 1.62.† However, since it is the *qualitative* nature of the frequency-versus-u variation that is perplexing, item (d) will be ignored here. On reflection, neither (a) nor (b) can provide a convincing explanation, but they do point out how demanding these deceptively simple experiments can be. Item (c), however, provides a likely explanation. As will be shown in Chapter 6, initially curved pipes in fact *do not* diverge. Thus, for semicircular pipes the reduction in frequency with flow is minimal; in that sense, strictly according to this hypothesis, this represents an intermediate system, behaving as a straight pipe for low u and as a curved one for higher u. Another outlook on this is provided by nonlinear theory. As discussed in Chapter 5, the pitchfork bifurcation is *structurally unstable* (in the mathematical sense), and the displacement-versus-u curve evolves more smoothly‡ in the presence of a small, or not-so-small, initial asymmetry. This corresponds physically to a gradual exaggeration of the asymmetry as u is increased (as observed), in contrast to the explosive divergence of the imperfection-free system. Furthermore, since neither the initial ($u = 0$) nor the 'final' state (for u larger than the theoretical u_{cd}) is associated with $\omega = 0$, the frequency in-between tends to bridge these two states without passing through zero.

Experiments were also conducted on clamped–clamped pipes by Jendrzejczyk & Chen (1985), with no sliding permitted. They found that divergence does not occur for the reasons already given; indeed the r.m.s. vibration amplitude was found to decrease as the theoretical critical u_{cd} is exceeded, which was attributed to deflection-induced tensioning.

A final comment is that in all these experimental studies there has been no reported observation of post-divergence coupled-mode flutter. Although this does not prove that it cannot exist — especially noting that the violence of the onset of divergence makes experimentation, at more than twice the critical flow rate, problematical — it would tend to support Holmes' finding, via nonlinear analysis, that pipes with supported ends cannot flutter, as discussed in Chapter 5.

3.5 CANTILEVERED PIPES

3.5.1 Main theoretical results

The essential dynamics of cantilevered pipes conveying fluid has already been outlined in Sections 3.1 and 3.2. Referring to the dimensionless equation of motion, equation (3.70), it is noted that, for cantilevered systems, $\Gamma = \Pi = 0$ always; furthermore, since the case of time-varying flow and elastic foundations will not be considered till later, $\dot{u} = k = 0$ as well. Hence, the only parameters that remain to be considered for the results to be presented in this section are the damping parameters α and σ, the mass parameter β, and the gravity parameter γ.

The simplest system is considered first, in which $\alpha = \sigma = \gamma = 0$ additionally, i.e. a horizontal system with the dissipation ignored, which thus depends only on β. In this case, solutions are possible via the First Method of Section 3.3.6(a) and involve no approximations (due to Galerkin truncation, for instance). Typical results are shown in Figures 3.27 and 3.28 for $\beta = 0.2$ and 0.295, respectively. It is seen that for small u ($u < 4$

†The effect of gravity was neglected by the authors ($\gamma = 0$), but in fact it is very small.
‡It is 'unfolded', in nonlinear terminology.

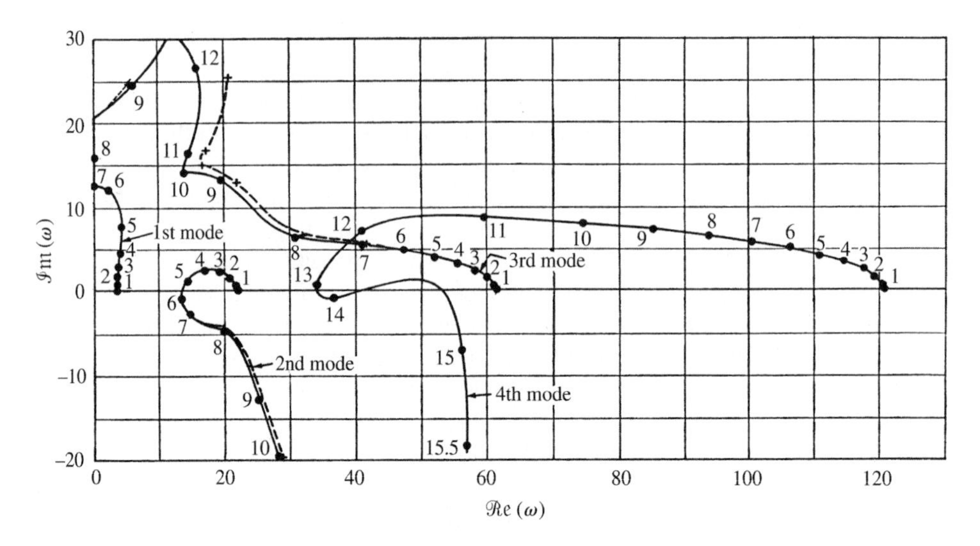

Figure 3.27 The dimensionless complex frequency of the four lowest modes of the cantilevered system ($\gamma = \alpha = \sigma = k = 0$) as a function of the dimensionless flow velocity, u, for $\beta = 0.2$: ——, exact analysis; – – –, four-mode Galerkin approximation (Gregory & Païdoussis 1966a).

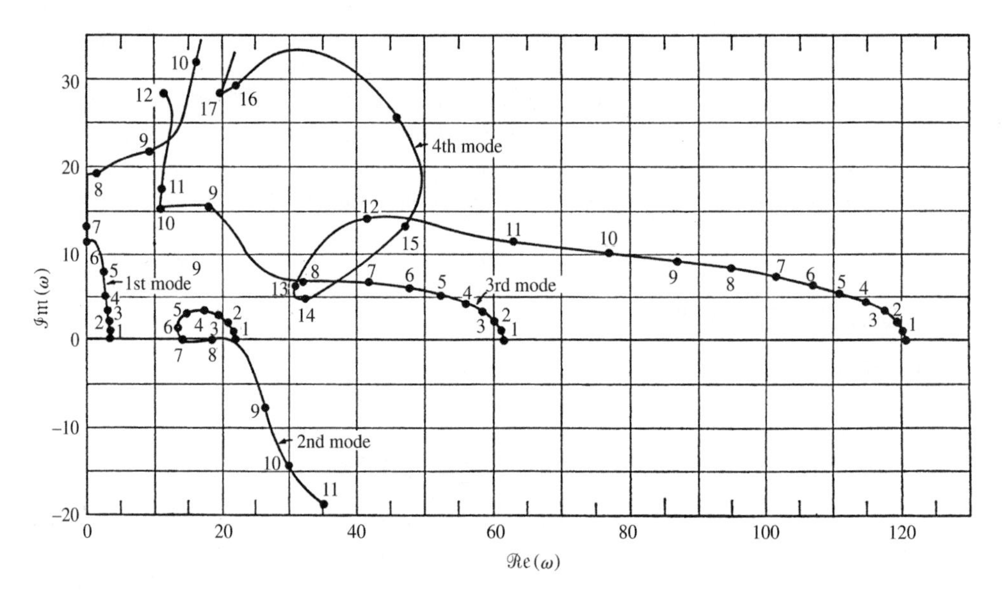

Figure 3.28 The dimensionless complex frequency of the four lowest modes of the cantilevered system ($\gamma = \alpha = \sigma = k = 0$) as a function of the dimensionless flow velocity, u, for $\beta = 0.295$ (Gregory & Païdoussis 1966a).

approximately), flow induces damping in all modes of the system; i.e. $\mathfrak{Im}(\omega) > 0$, or $\zeta = \mathfrak{Im}(\omega)/\mathfrak{Re}(\omega) > 0$. This is in line with the energy considerations of Section 3.2.2, in connection with equation (3.11). For higher u, $\mathfrak{Im}(\omega)$ in the second mode of the system begins to decrease and eventually becomes negative; thus, a Hopf bifurcation occurs at $u = u_{cf} \simeq 5.6$ and 7.0, for $\beta = 0.2$ and 0.295, respectively, and the system becomes unstable (in the linear sense) by flutter. For $\beta = 0.2$, there is also a fourth-mode oscillatory instability, via another Hopf bifurcation, at $u \simeq 13$.[†]

In the case of $\beta = 0.295$ and for $7 < u < 8.2$, the system loses stability, regains it and loses it again, as the locus meanders along the $\mathcal{R}e(\omega)$-axis. This cannot be seen very clearly in the scale of Figure 3.28, but it is similar to what is easily visible in Figure 3.27 for $13 < u < 15$ in the fourth mode.

Flutter does not always occur in the second mode of the system, as may be seen in Figure 3.29 for $\beta = 0.5$, where it is in the third mode that the system loses stability. It is of interest that (i) for $\beta = 0.2$ and 0.295, the second-mode locus bends downwards and crosses the axis to instability, while the third-mode locus moves towards higher $+\mathcal{I}m(\omega)$ values; (ii) for $\beta = 0.5$, the opposite takes place. This 'role reversal' or 'mode exchange' characteristic is a frequently occurring feature of the dynamics of the system. Thus, for $\beta = 0.2$ (Figure 3.27) the fourth mode leads to the higher-mode instability; in contrast, for $\beta = 0.295$ (Figure 3.28) the fourth-mode locus makes a loop, while the fifth mode (not shown) curves down to instability (cf. the third-mode locus of Figure 3.29). Another aspect of this behaviour is the closeness of the loci for some specific u, in Figure 3.29 for $u = 8.8125$; near the 'critical' β for which the mode exchange occurs, the two loci can be extremely close (Païdoussis 1969; Seyranian 1994).

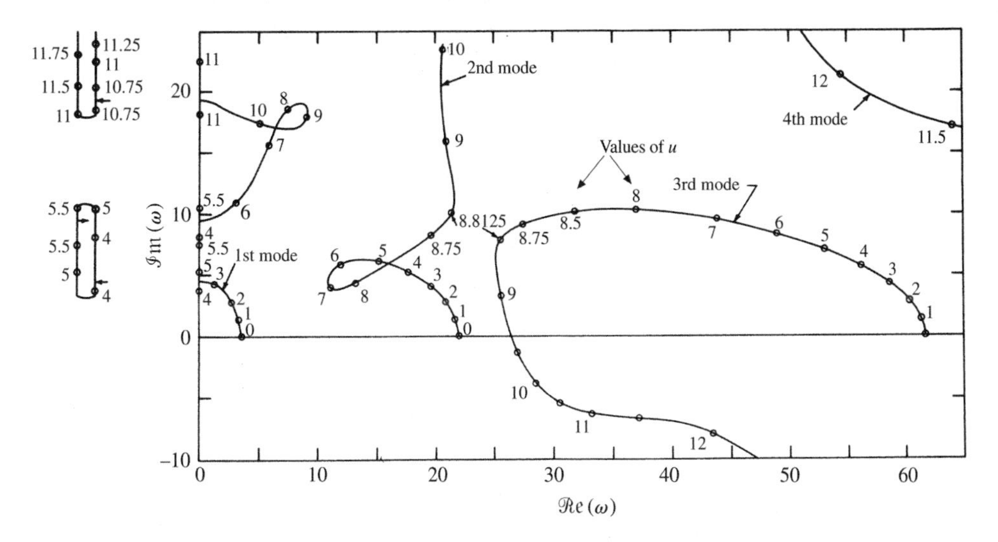

Figure 3.29 The dimensionless complex frequency as a function of u for a cantilevered system ($\gamma = \alpha = \sigma = k = 0$) for $\beta = 0.5$. The diagrams on the left-hand side of the figure display the behaviour of the loci while on the $\mathcal{I}m(\omega)$-axis (Païdoussis 1969).

With regard to the foregoing discussion, a very important point should be stressed. We have been talking about the 'second mode' and 'third mode', and so on, simply because they are part of the thus numbered loci. However, for $u \neq 0$, the *mode shapes* associated with these modes differ significantly from those at $u = 0$ (which are the classical beam modes), as first shown by Gregory & Païdoussis (1966b). Thus, for $u = 3$ or 4

[†]The *caveat* concerning the limitations of linear theory for predicting the dynamics beyond the first loss of stability applies here too.

the first-mode shape contains appreciable second-mode content, the second-mode shape third-mode content, and so on. Nevertheless, the present appellation is clearly a reasonable one. Another important point is that, similarly to the pipe with supported ends, these are not stationary, classical modes with fixed nodal points, but contain appreciable travelling wave components, to be discussed with the experiments in Section 3.5.6.

The critical flow velocity, u_{cf} as a function of β is shown in Figure 3.30;† there exists a similar curve for the corresponding frequency at $u = u_{cf}$, labelled ω_{cf} — see Figure 3.35. It is clear that, u_{cf} depends strongly on β. Furthermore, the u_{cf} and ω_{cf} curves contain a set of S-shaped segments. By referring to Figure 3.28 it is recognized that they are associated with the instability–restabilization–instability sequence discussed in the foregoing; hence, in Figure 3.30, the negative-slope portions of the curve correspond to thresholds of restabilization. If an experiment could be conceived in which the material damping is zero and β could be varied in very small steps, then around these points there would be 'jumps' in u_{cf}; e.g. for β in the vicinity of 0.69, from $u_{cf} \simeq 11$ to $u_{cf} \simeq 12.8$ for a very small increase in β. The values of β associated with these S-shaped segments of the stability curve (at $\beta \simeq 0.30$, 0.69, 0.92) will be found to be associated with a large number of perplexing linear and nonlinear characteristics of the system — in the sense of acting as separatrices for differing dynamical behaviour. Yet, the origin of their existence is not fully understood (see Section 3.5.4). As $\beta \to 1$, more and more S-shaped jumps are encountered. Mukhin has shown that for $\beta = 1$ no flutter solution may be possible, i.e. $u_{cf} \to \infty$ (Mukhin 1965; Lottati & Kornecki 1986).

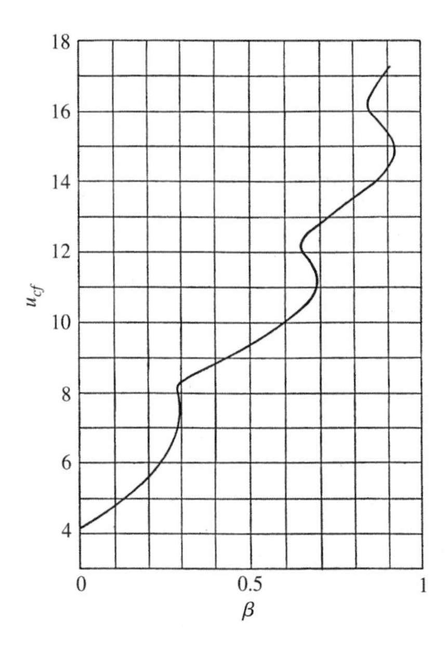

Figure 3.30 The dimensionless critical flow velocity for flutter, u_{cf}, of a cantilevered pipe conveying fluid, as a function of β, for $\gamma = \alpha = \sigma = k = 0$ (Gregory & Païdoussis 1966a).

Nevertheless, it appears that these jumps are related in some way to mode content. This is made clear by Figure 3.31 in which, in addition to results obtained by the method

†Numerically, such curves are computed by determining β for each assumed u_{cf}, rather than *vice versa*.

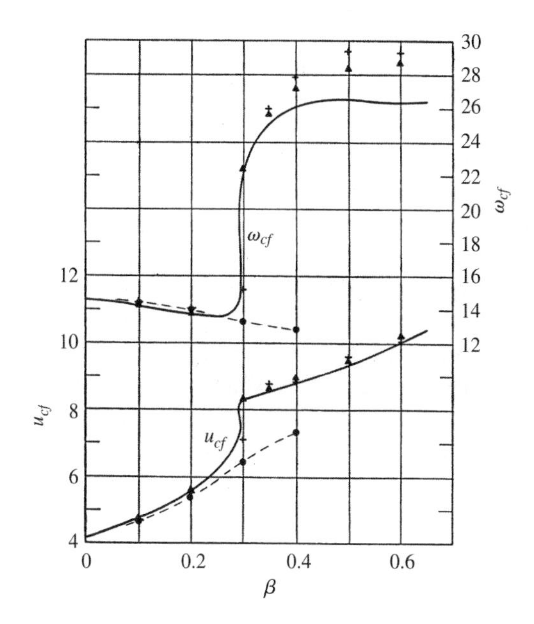

Figure 3.31 Comparison between u_{cf} and ω_{cf} obtained by the exact solution (——), cf. Figure 3.30, and Galerkin approximations: •, $N = 2$; +, $N = 3$, ▲, $N = 4$ (Gregory & Païdoussis 1966a).

of Section 3.3.6(a), some obtained by the Galerkin method of Section 3.3.6(b) are also presented, for $N = 2$, 3 and 4, N being the number of beam modes utilized. It is obvious that, although $N = 3$ and 4 may be adequate for predicting u_{cf} (see also Figure 3.27), the two-beam-mode approximation ($N = 2$) is not, failing to reproduce the S-shaped behaviour, as will be discussed further in Section 3.5.4; on the other hand, the $N = 2$ approximation is quite reasonable for $\beta \leq 0.2$, or even $\beta = 0.25$. In general, higher-N approximations become necessary to adequately represent the dynamics of the system as u and β are increased. This contrasts sharply to the inherently conservative system [cf. equation (3.92) of Section 3.4.1 and the attendant discussion], where $N = 2$ and even $N = 1$ Galerkin approximations can predict u_{cd} very well.

3.5.2 The effect of gravity

The motivation for investigating the effect of gravity ($\gamma \neq 0$) on the dynamics of the system comes from two sources. The first is to obtain theoretical results for comparison against measurements from experiments with pipes oscillating in a *vertical* rather than a horizontal plane, the former being easier to conduct. In this regard, recalling that $\gamma = (M + m)gL^3/EI$, it turns out that for metal pipes conveying fluid, unless L is very large, γ is small and its effect on the dynamics may well be negligible; for rubber or elastomer pipes, however, with which the majority of the experiments are conducted, because E is considerably lower, gravity effects should normally be accounted for. The second source of impetus was provided by Benjamin's (1961a,b) findings with articulated cantilevered pipes conveying fluid: that horizontal systems lose stability exclusively by flutter, whereas vertical ones can do so by divergence also (Section 3.8). Hence, since the continuously flexible system may be considered as the limiting case of an articulated one as the number of degrees of freedom $N \to \infty$, it is of interest to discover if divergence can arise in the vertical continuous system as well.

Extensive calculations of Argand and stability diagrams for $\gamma \neq 0$ were conducted by Païdoussis (1969, 1970), using the method of Section 3.3.6(b); it was found that $N = 9$ or 10 in the Galerkin series ensured accuracy of the eigenfrequencies to three significant figures. Similar calculations were done by Bishop & Fawzy (1976).

A summary of the results is presented in the form of a stability diagram in Figure 3.32. It is seen that the general dynamics of the system with $\gamma \neq 0$ is similar to that for $\gamma = 0$, but for $\gamma > 0$ the additional restoring force due to gravity causes u_{cf} to be higher. It is recalled that $\gamma < 0$ represents an up-standing system,[†] with the downstream free end *above* the clamped one. As expected, the system is less stable in this case. In contrast to the articulated system, no flow-induced divergence is possible in this one. It is seen in Figure 3.32 that each of the curves contains a number of S-shaped segments, indeed more of them as γ is increased.

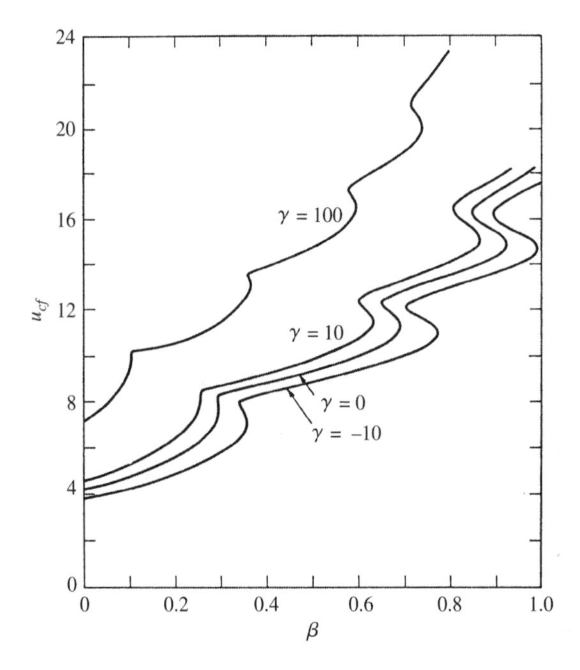

Figure 3.32 The dimensionless critical flow velocity for flutter, u_{cf}, of a vertical cantilevered pipe conveying fluid, as a function of β for varying γ, compared to the horizontal system, $\gamma = 0$; $\alpha = \sigma = k = 0$ (Païdoussis 1970).

More interesting dynamical behaviour is obtained if γ is negative and fairly large. In that case, corresponding to relatively long pipes, $\gamma < -7.83$ approximately, the cantilever buckles under its own weight at zero flow. The linear dynamics of the system is illustrated in Figure 3.33. Consider first a system with $\gamma = -10$ and $\beta = 0.2$, which is buckled under its own weight for $u = 0$: as u is increased (i.e. progressing vertically up in the

[†]Although this is hardly a symbol of moral rectitude!

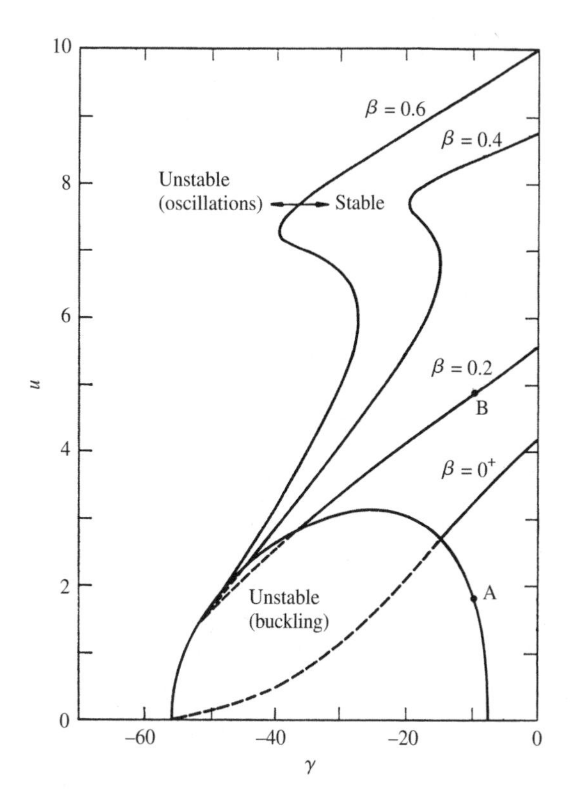

Figure 3.33 Stability map for a 'standing' cantilever ($\gamma < 0$), with the discharging free end of the pipe vertically above the clamped end, showing the effects of β and γ on stability (Païdoussis 1970). The dashed line corresponds to the onset of flutter, superposed on divergence. The dynamics for $\gamma < -55.9$ is more complex and is not detailed in the figure.

figure), the system is restabilized at $u \simeq 1.8$ (at point A), and then loses stability by flutter at $u \simeq 4.85$ (at point B). For $\gamma = -20$, $\beta = 0.2$, restabilization and flutter occur at $u \simeq 3.1$ and $u = 4.25$, respectively. Variants of this behaviour are represented by $\gamma = -20$, $\beta = 0^+$ or by $\gamma = -40$, $\beta = 0.2$; in such cases, again according to linear theory, the system develops flutter, while still under divergence. For $\gamma < -55.9$ approximately, the system buckles under its own weight in both its first and second modes and apparently remains unstable with increasing flow; this more complex behaviour is not detailed in Figure 3.33.

It is noted that the values of $\gamma \simeq -7.83$ and -55.9 agree well with those obtained by exact analysis of the static stability of an up-standing cantilever, corresponding to the first two zeros of the equation $\mathrm{J}_{-1/3}[\frac{2}{3}(-\gamma)^{1/2}] = 0$, where $\mathrm{J}_{-1/3}$ is the Bessel function of the first kind and order $-\frac{1}{3}$. The first and second zeros occur at $\frac{2}{3}(-\gamma)^{1/2} \simeq 1.87$ and 4.99, respectively, yielding the values of γ in question to within 0.5%.

A priori, whether any of this post-buckling behaviour materializes in practice is questionable, because in this linear theory the stability is considered for small motions about the *straight* equilibrium configuration, whereas the buckled system is certainly not in that state. Nevertheless, as will be seen in Section 3.5.6, the dynamics of the system as observed in experiments *is* substantially as just described.

3.5.3 The effect of dissipation

We next consider the effect of dissipation on stability. As first shown by Ziegler (1952) for the nonconservative system of a double pendulum subjected to a follower load, weak damping may actually *destabilize* the system. The same was found in the study of stability of a compliant surface over which there exists a flow (Benjamin 1960, 1963). Benjamin classified the various possible modes of instability into three distinct classes, according to the mode of energy exchange between fluid and solid. Benjamin shows that 'class A' waves are destabilized by damping, and Landahl (1962) has contributed to the discussion and clarification of this paradox; see also Section 3.5.5. It was in this same period that it was found that cantilevered pipes conveying fluid can also be destabilized by dissipation (Païdoussis 1963). Subsequently, a considerable amount of work has been done on this topic [e.g. by Gregory & Païdoussis (1966b), Nemat-Nasser *et al*. (1966), Bolotin & Zhinzher (1969), Païdoussis (1970), Païdoussis & Issid (1974)].

Figure 3.34(a,b) shows examples of a cantilevered pipe ($\beta = 0.65$, $\gamma = 10$) subjected to damping modelled (a) by a Kelvin–Voigt viscoelastic model (with $\alpha = 0.0189$), and (b) by a hysteretic or 'structural' damping model (with $\mu = 0.1$) — see equations (3.39) and (3.72). A number of interesting features of the system are displayed in this figure, as follows. (i) First, this is yet another example where it is not the second mode that is associated with flutter; here, after considerable peregrinations, it is the first, although the modal form is similar to that of the second mode by the time it crosses to the $-\mathfrak{Im}(\omega)$ half-plane.[†] (ii) By comparing the critical flow velocity for the undamped system ($u_{cf} = 12.88$) to that of the damped system [$u_{cf} = 9.85$ in Figure 3.34(a) and $u_{cf} \simeq 11$ in Figure 3.34(b)], it is clear that dissipation *destabilizes* the system. (iii) For the hysteretic system, the character of the equation of motion is quite different from that of the viscoelastically damped one in the following sense. For the viscoelastic system ($\alpha \neq 0$, $\mu = 0$), if iω is a root of the equation of motion, so is its complex conjugate, and the root loci are symmetric about the $\mathfrak{Im}(\omega)$-axis; hence, only the positive $\mathfrak{Re}(\omega)$ half-plane needs be shown, as in Figure 3.34(a). For the hysteretic system ($\mu \neq 0$, $\alpha = 0$), however, this is no longer true, and hence (partly) both sides of the plane have to be shown. It is of particular interest to note that for $u > 4$ there would appear to exist discontinuities in the values of $\mathfrak{Im}(\omega)$ as the $\mathfrak{Im}(\omega)$-axis is crossed, *if* only the positive $\mathfrak{Im}(\omega)$-plane were considered; in particular, in the vicinity of $u \simeq 5$ in the first mode and $u \simeq 11$ in the second. Finally, it must be recalled that, in accordance with the limitations to the validity of the hysteretic dissipation model referred to in Sections 3.3.2 and 3.3.5, only the portions of the loci near the $\mathfrak{Re}(\omega)$-axis have physical significance.

It is noted that, whereas hysteretic damping destabilizes the system for $\beta > 0.285$ approximately, it exerts a stabilizing influence for smaller values of β, as may be seen in Figure 3.35. This dependence of the dynamical behaviour on the mass ratio was also found by Benjamin (1963) in the stability of a compliant surface subjected to flow.

It should also be remarked that the values of α and μ utilized in these calculations are relatively high and representative of rubber and elastomer pipes (the values of $\alpha = 0.0189$ and $\mu = 0.1$ give identical logarithmic decrement, $\delta \simeq \pi\mu$, for the first mode at $u = 0$).

[†] As shown by Gregory & Païdoussis (1966b), the theoretical and experimental mode shapes associated with flutter, although displaying elements of higher beam modes with increasing β, in their essence retain the second-beam-mode 'dragging' form, despite changes in the numeration of the mode involved (Table 3.2).

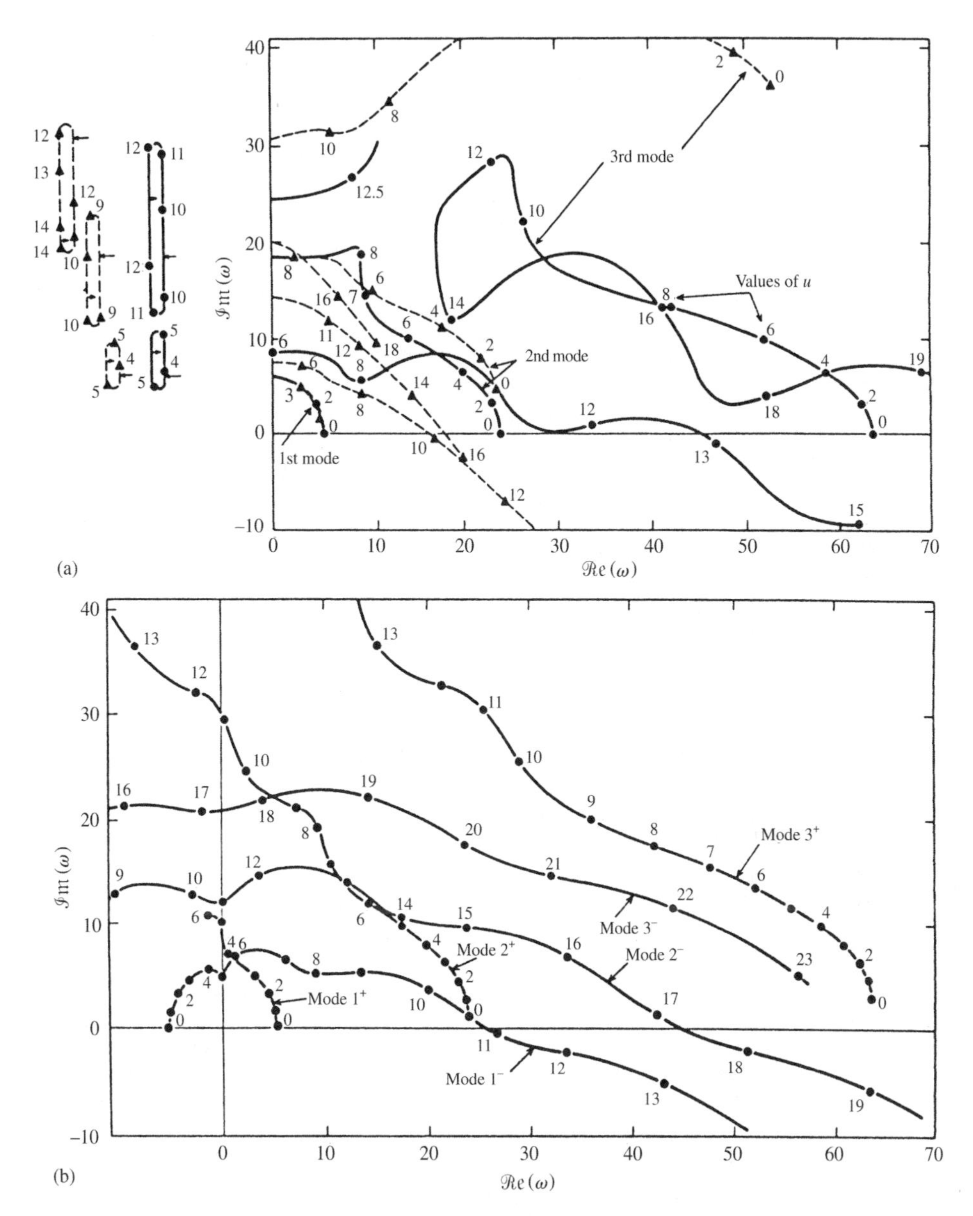

Figure 3.34 (a) Argand diagrams showing the effect of viscoelastic damping on stability of a cantilevered system: ——, undamped system ($\alpha = \sigma = 0$); – – –, viscoelastically damped system ($\alpha = 0.0189, \sigma = 0$); $\beta = 0.65$, $\gamma = 10$, $k = 0$. (b) Argand diagram showing the effect of hysteretic damping ($\mu = 0.1$, $\alpha = \sigma = 0$) on stability of otherwise the same system (Païdoussis & Issid 1974).

For metal pipes, typical values of μ would be $\mu = \mathbb{O}(10^{-3})$ or less [see, e.g. Snowdon (1975)].

We next turn our attention to the other source of dissipation in the system, namely to the damping introduced by friction with the surrounding air, characterized by σ [defined in (3.71)]. Especially for non-metallic pipes, this effect is negligible *vis-à-vis* damping

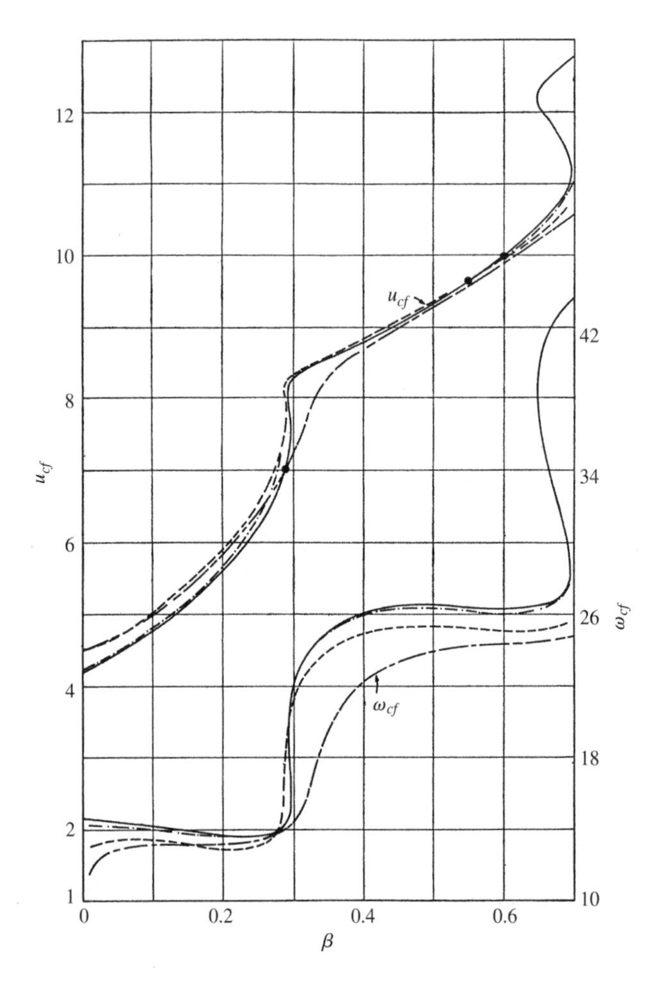

Figure 3.35 The effect of material damping, modelled by the hysteretic model, and of external viscous damping on stability of a cantilevered pipe conveying fluid ($\gamma = k = 0$): ——, undamped system; — - —, with hysteretic damping, $\mu = 0.065$; – – –, with viscous damping, $\sigma = 1.42$; — · —, with viscous damping, $\sigma = 0.23$ (Gregory & Païdoussis 1966b). The black dots mark the threshold values of β beyond which, for each of these three cases, the system is destabilized by damping.

due to dissipation in the material of the pipe. In this respect, it is useful to adapt the relationships of Section 2.2.3 to the work at hand. For unconfined fluid, the damping force is given by

$$F_d = C_d \Omega \rho A \frac{\mathrm{d}z}{\mathrm{d}t}, \qquad C_d = \frac{2\sqrt{2}}{\sqrt{S}} \tag{3.105}$$

from (2.153) and (2.157), respectively, where S$= \Omega R_o^2/\nu$, R_o being the outer radius of the pipe. In terms of the dimensionless parameters used here,

$$\sigma = \frac{2\sqrt{2}\rho\pi R_o^2 \Omega L^2}{[EI(M+m)\Omega R_o^2/\nu]^{1/2}}. \tag{3.106}$$

For a typical pipe used in the experiments (Païdoussis 1969, 1970) and $\Omega = 10\pi$, one finds $\sigma \simeq 10^{-2}$ or less. The effect of this is clearly very small as compared to, say, hysteretic damping with $\mu \sim \mathbb{O}(10^{-2})$, since in the equation of motion μ is multiplied by λ_i^4.[†] Nevertheless, the effect of an artificially large σ on stability as investigated by Païdoussis (1963) and Gregory & Païdoussis (1966b) is of interest; in these calculations the whole of the observed damping in the first and second mode of one of the pipes used in the experiments is assumed to be entirely due to σ (which, of course, cannot be so), yielding $\sigma = 0.23$ and $\sigma = 1.42$, respectively.[‡] As seen in Figure 3.35, viscous damping with $\sigma = 1.42$ destabilizes the system only for $\beta > 0.55$. With $\sigma = 0.23$ this occurs for $\beta > 0.60$, while for $0.3 < \beta < 0.6$ the critical flow velocity is less than 1% higher than for the undamped case. The critical frequency, ω_{cf}, is reduced in almost all cases.

The effect of very large values of σ is examined by Lottati & Kornecki (1986). Such large σ would arise if the pipe were immersed in water or a more viscous fluid (but in that case m, the pipe mass, must be presumed to include the fluid added mass). As shown in Figure 3.36, σ is stabilizing for $\beta \leq 0.5$ approximately, as in the foregoing, but for $\beta = 0.8$ it is destabilizing, with an interesting 'negative jump' in the curve.

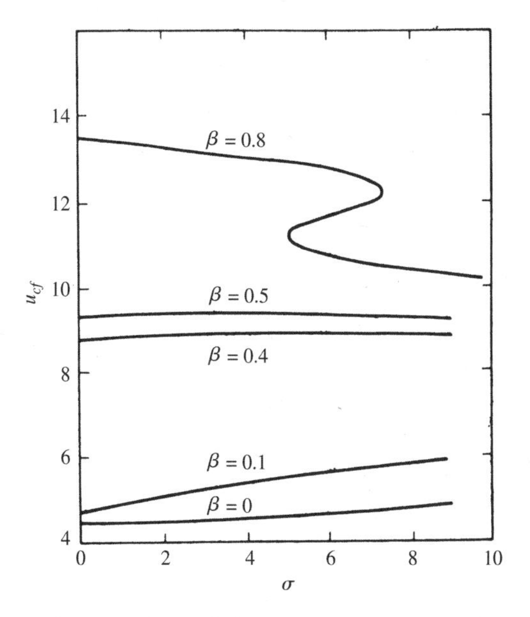

Figure 3.36 The effect of large values of viscous damping, σ, on the critical flow velocity for flutter, u_{cf}, of a cantilevered pipe for various β (Lottati & Kornecki 1986).

Another interesting dynamical feature of nonconservative systems is related to the non-smooth variation in the critical load as damping is varied from vanishingly small to zero, as first discussed in general terms by Bolotin (1963). This has been studied extensively for two-degree-of-freedom articulated columns [looking like the pipe system of Figure 3.1(d) but without flow] subjected to compressive follower loads (Herrmann & Bungay 1964; Herrmann & Jong 1965, 1966).[§] Such systems lose stability either by divergence or by

[†]See Sections 3.3.5 and 3.3.6.
[‡]These values correspond to $\mu = 0.065$ and are computed via $\sigma = \lambda_1^2\mu$ and $\sigma = \lambda_2^2\mu$, respectively.
[§]See also Section 2.1.5.

flutter, depending on the angle of the follower load to the last articulation and the ratio of the viscous damping at the two articulations, and they display several other interesting dynamical features. The destabilizing effect of vanishingly small damping as opposed to zero damping can be so large as to reduce the critical load by a factor of more than 6!

The investigation has been extended to cantilevered pipes conveying fluid by Nemat-Nasser *et al.* (1966), who examine the effect on stability of *all velocity-dependent forces*, as opposed to just damping: i.e. not only internal and external damping (α and σ), but also 'Coriolis damping' associated with β. They consider $\sqrt{\beta} = \nu\beta^*$, $\alpha = \nu\alpha^*$, $\sigma = 2\nu\sigma^*$, where ν is small, and then obtain solutions of the characteristic equation, neglecting terms of $\mathcal{O}(\nu^2)$ and higher. More specifically, they are concerned with the discontinuity in u_{cf} for $\beta = 0^+$ and $\beta = 0$ exactly: in the first case, as seen from Figure 3.30, $u_{cf} \simeq 4.21$; in the second case the problem reduces to Beck's (Section 3.3.2) for which $\mathcal{P}_c = 20.05$ and, since $\mathcal{P}_c$ is equivalent to u_{cf}^2, $u_{cf} = 4.48$. Thus, there is a jump up from $u_{cf} = 4.21$ to 4.48 if β is reduced from $\beta = 0^+$ to 0 (see insert in Figure 3.56).† The effect is greatly exaggerated when internal (material) damping is taken into account ($\alpha \neq 0$), as shown

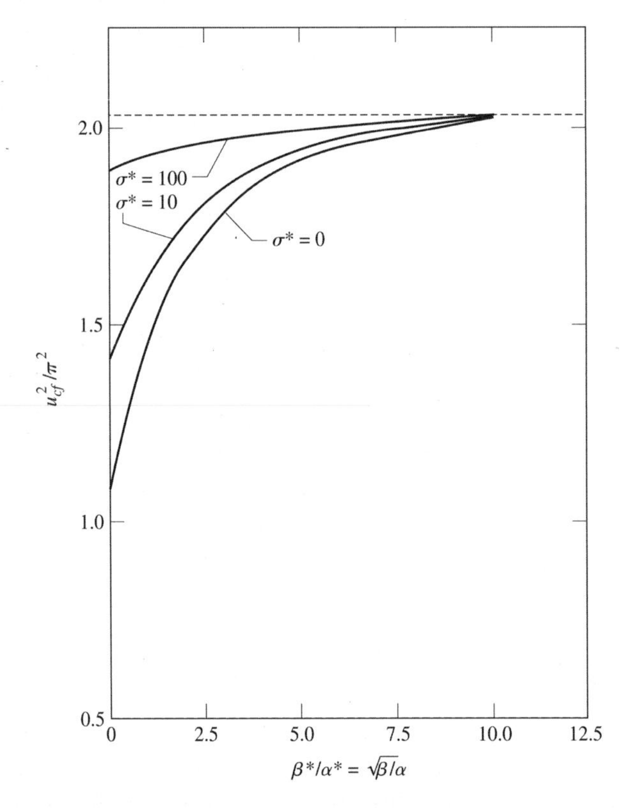

Figure 3.37 The critical flow velocity for small velocity-dependent forces acting on the cantilevered pipe system as a function of $\beta^*/\alpha^* = \beta^{1/2}/\alpha$ and σ^*, where $\sqrt{\beta} = \nu\beta^*$, $\alpha = \nu\alpha^*$ and $\sigma = 2\nu\sigma^*$, in which ν is a small parameter; - - -, u_{cf}^2/π^2 for Beck's problem, $\alpha = \beta = \sigma = 0$ (Nemat-Nasser *et al.* 1966).

†Of course, as we have already seen, the bifurcation leading to flutter is different: for $\beta = 0^+$ a Hopf bifurcation; for $\beta = 0$ a Hamiltonian Hopf, so that, in that sense, the discontinuity is not *too* surprising.

in Figure 3.37: it is seen that whereas $u_{cf} = 4.48$ for $\alpha = \beta = \sigma = 0$ (corresponding to $u_{cf}^2 = 20.05$ or $u_{cf}^2/\pi^2 = 2.03$, the dashed line in the figure), it can be as low as $u_{cf} = 3.31$ for $\beta = \sigma = 0$, $\alpha^* = 1$ ($u_{cf}^2/\pi^2 = 1.107$). Furthermore, agreeing with the results of Figure 3.35 for small β, external damping ($\sigma^* \neq 0$) stabilizes the system.

3.5.4 The S-shaped discontinuities

As already mentioned, the nature of the S-shaped discontinuities in the stability curves of u versus β, e.g. in Figures 3.30 and 3.32, is of interest not only *per se*, but also because the critical values of β at which these discontinuities occur are frequently associated with, or are separatrices for, distinctly different dynamical behaviour. The reader is referred to the discussion of Figures 3.63 and 3.68 in Section 3.6 and Figures 5.19–5.21 of Section 5.7, as well as to Païdoussis (1997).

An early attempt to reach some understanding of this matter was made in 1969. Specifically, it was attempted to link the occurrence of these S-shaped portions in the stability curves and the attendant jumps in u_{cf} to changes in the mode leading to flutter. Specifically, the mode in which the system becomes unstable is identified on either side of the jump, to see if there is a mode change (mode switching) across it. The results are shown in Table 3.2 (in the *conventional* ordering of the modes), and it is seen that this hypothesis fails. Thus, for $\gamma = 0$, there are two mode changes between the first and second jump ($0.4 \leq \beta \leq 0.65$), while the β versus u curve remains smooth (Figure 3.30); for $\gamma = 10$ there is no mode change across the first jump. The modes are then reordered, strictly in ascending order of magnitude of $\mathcal{Re}(\omega)$; for instance, in Figure 3.28 for $u = 9$–11, the second mode is now called 'third', and the third 'second'; in some cases [see Figure 3.34(a)] this causes very radical renumbering. The results of this reordering are also given in Table 3.2. The new scheme is partly successful, in the

Table 3.2 Relation between mode number of the mode becoming unstable and the 'jumps' in the u_{cf} versus β curves (Païdoussis 1969).

γ	Values of β tested	Nomenclature for mode becoming unstable		Range of β relative to 'jumps'
		Conventional	Reordered	
0	0.1, 0.2, 0.295	Second	'Second'	$\beta <$ 1st jump
	0.4, 0.5	Third	'Third'	1st jump $< \beta <$ 2nd jump
	0.6	Second	'Third'	
	0.65	First	'Third'	
10	0.1, 0.2	Second	'Second'	$\beta <$ 1st jump
	0.3	Second	'Third'	1st jump $< \beta <$ 2nd jump
	0.4, 0.5	Third	'Third'	
	0.65	First	'Third'	2nd jump $< \beta <$ 3rd jump
100	0.075, 0.1	Second	'Second'	$\beta <$ 1st jump
	0.113, 0.2	Third	'Third'	1st jump $< \beta <$ 2nd jump
	0.4, 0.5, 0.58	Fourth	'Fourth'	2nd jump $< \beta <$ 3rd jump
	0.65	First	'Fourth'	3rd jump $< \beta <$ 4th jump

sense that it imposes a systematic increase in the mode associated with instability as β is increased. Also, an improvement in the correspondence between jumps and mode changes is achieved: it works for the first jump in all cases; nevertheless, it fails for the second jump when $\gamma = 10$, and for the third jump when $\gamma = 100$. Clearly something more profound is involved.

A similar but more mathematical attempt was made more recently by Seyranian (1994), starting from the same observation that motivated Païdoussis' (1969) work: the 'drawing near' of two mode loci (e.g. in Figure 3.29 for the second and third modes at $u = 8.8125$) with increasing β, prior to switching of the mode leading to flutter, which often occurs as β is varied. Seyranian argues convincingly that this 'drawing near' of the loci implies actual frequency coincidence (a repeated root) at some nearby point in the parameter space — a 'collision of eigenvalues' in his terminology — if only an additional parameter (in this case, other than β and u) is varied at the same time. This may well be true, although Seyranian demonstrates it only for *nongyroscopic* nonconservative systems (e.g. for an articulated column with a follower load). As seen in the 'conventional' mode-ordering column of Table 3.2, however, there is not always a mode switch across an S-shaped jump, nor does mode switching necessarily imply an impending jump (Figure 3.29 *vis-à-vis* Table 3.2 being a case in point).

Either of these attempts, even if successful, would have given a mathematical explanation rather than physical insight into the nature of the S-shaped discontinuities. A more successful interpretation in this respect was provided by Semler *et al*. (1998), which also throws some light onto the destabilizing effect of damping.

Semler *et al*. (1998) consider a double pendulum under zero gravity, subjected to a follower load, P, as shown in Figure 3.38(a). The two rods are constrained by rotational springs of equal stiffness, k, and rotational dashpots, c_1 and c_2. The equations of motion are rendered dimensionless by introducing $\tau = t\sqrt{k/(mL^2)}$ for the time and the parameters

$$\gamma_1 = \frac{c_1}{\sqrt{kmL^2}}, \qquad \gamma_2 = \frac{c_2}{\sqrt{kmL^2}}, \qquad \mathcal{P} = \frac{PL}{k}. \tag{3.107}$$

Stability is lost via a Hopf bifurcation and the critical value of $\mathcal{P}$ for flutter, $\mathcal{P}_{cr} = f(\gamma_1, \gamma_2)$, may be derived in closed form. Figure 3.38(b) shows some results obtained for fixed γ_1 while γ_2 is varied. It is shown, for all γ_1, that increasing γ_2 from zero initially stabilizes the system (i.e. a higher $\mathcal{P}$ is required to cause flutter), but the trend is eventually reversed and then γ_2 becomes 'destabilizing'.

To understand the mechanism leading to this behaviour, the net energy gained by the system over a period of not necessarily neutrally stable oscillation, T, is considered,

$$\Delta E = W - D = \int_0^T \mathcal{P}\dot{\phi}\,\sin\chi\,\mathrm{d}\tau - \int_0^T [\gamma_1\dot{\phi}^2 + \gamma_2\dot{\chi}^2]\,\mathrm{d}\tau, \tag{3.108}$$

where $\phi \equiv \phi_1$ and $\chi = \phi_1 - \phi_2$ are an alternative set of generalized coordinates. ΔE has the same meaning as $E - \mathscr{E}_o$ in (C.6). Once the equations of motion are decoupled via modal analysis techniques (Section 2.1.2), it is possible to consider ΔE for each of the two modes separately. One can thus obtain the diagram of Figure 3.39(a). It is seen that mode 1, the 'stable mode' (i.e. *not* the one associated with flutter), becomes more and more stable as γ_2 is increased, being associated with progressively more negative ΔE. However, mode 2, the flutter mode, becomes less stable with increasing γ_2; eventually, for $\gamma_2 = 0.025$ [cf. Figure 3.38(b)] $\Delta E > 0$ is obtained, and hence amplified oscillations.

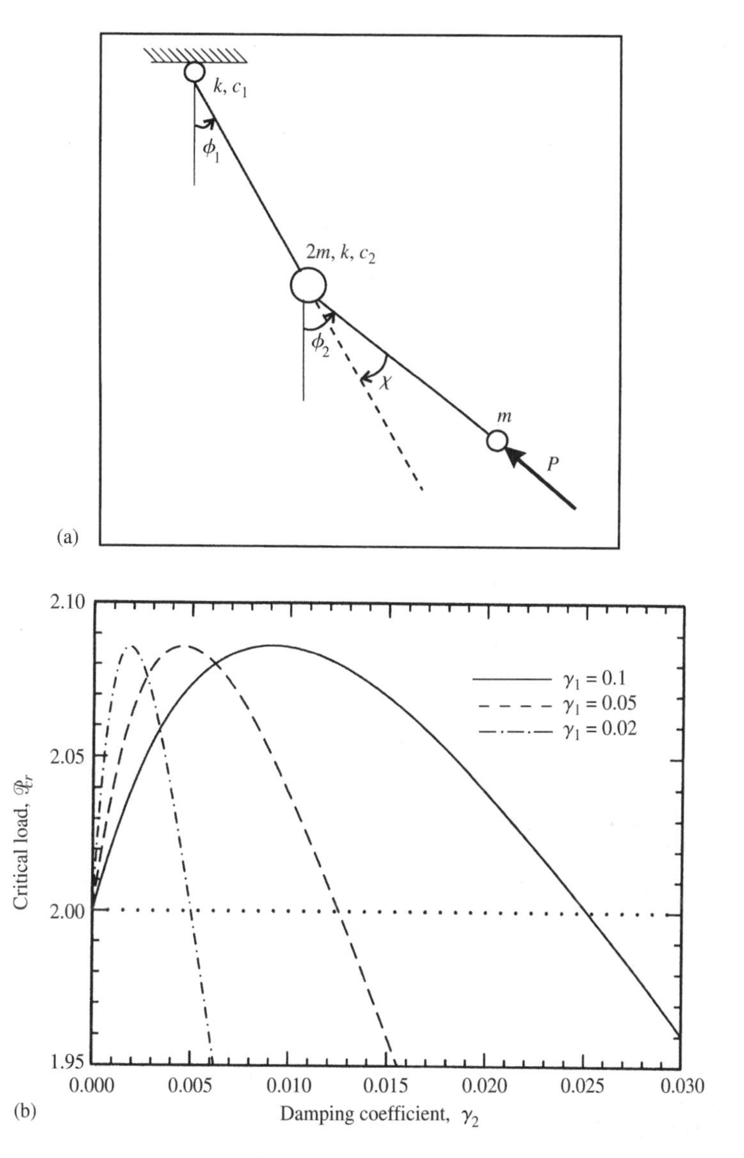

Figure 3.38 (a) Diagram of the 'double pendulum' system in zero gravity, subjected to a follower force, P. (b) The effect of increasing γ_2, while γ_1 is fixed, on the critical load for flutter, $\mathcal{P}_{cr}$ (Semler *et al*. 1998).

Insight into the dynamics of the system is obtained by looking at the relative amplitudes of the two generalized coordinates, ϕ and χ, when the response is periodic, i.e. at $\mathcal{P} = \mathcal{P}_{cr}$. It is noted that, whereas W is at most linearly dependent on χ, D is quadratically affected, and so a high χ-content in one of the modes means that it will be preferentially damped. The results are shown in Figure 3.39(b). It is seen that for low γ_2, the χ-content of mode 2 (which is the flutter mode) is higher, and hence this mode will be damped more than mode 1 which remains stable; hence, the effect of increasing γ_2 here is *stabilizing*. For larger γ_2, however, it is mode 1 that has the higher χ-content and hence *it* will be

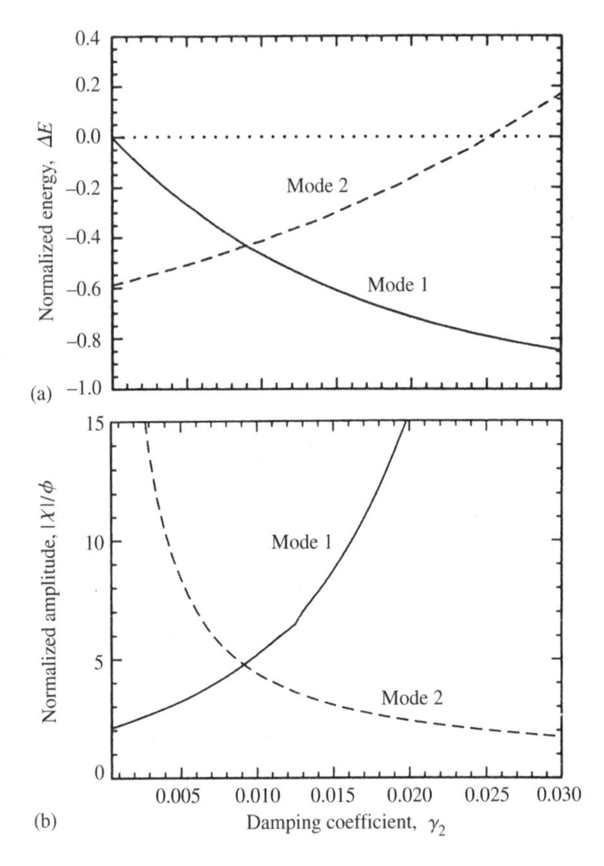

Figure 3.39 (a) The energy gained ($\Delta E > 0$) or lost by the double pendulum system in its two modes of vibration, normalized with respect to the initial energy, as a function of γ_2; $\gamma_1 = 0.1$, $\mathcal{P} = 2$. (b) The amplitude ratio of the generalized coordinates ϕ and χ, as a function of γ_2; $\gamma_1 = 0.1$ but $\mathcal{P}$ is varied (Semler *et al*. 1998).

preferentially damped, while the stable mode 2 is less damped; hence, increasing γ_2 is now *destabilizing*. The cross-over point occurs at $\gamma_2 = 0.095$, corresponding to the same point in Figure 3.38(b) where stabilization by γ_2 ceases and destabilization begins.

Moreover, not only the relative amplitude of the two generalized coordinates is important, but also the phase between them. On the stability boundary, where $\mathcal{P} = \mathcal{P}_{cr}$ and $\Delta E = 0$, the motion must be harmonic; since the amplitude is arbitrary, we can take $\phi = 1 \sin \omega\tau$, $\chi = B \sin(\omega\tau - \theta)$. Then, assuming ϕ and χ to be small and evaluating (3.108) with $\Delta E = 0$, one obtains $\mathcal{P}_{cr} \sin\theta = \omega(\gamma_1 + B^2\gamma_2)/B$. For $B > 0$, it is seen that θ must be positive for $\mathcal{P}_{cr}$ to exist, and the higher it is (but always $\theta < \pi$), the lower the value of $\mathcal{P}_{cr}$. Of course, $\mathcal{P}_{cr}$ also depends on γ_1, γ_2, B and ω, but the phase angle θ is of paramount importance.

Armed with these insights, the modal composition of the mode associated with instability is now considered in the pipe problem. The system of equation (3.76) is discretized by the Galerkin method [Section 3.3.6(b)], using the beam eigenfunctions, $\phi_r(\xi)$, as comparison functions and the associated generalized coordinates, $q_r(\tau)$. The system is then reduced to first order and decoupled by modal techniques — cf.

equations (2.16)–(2.19) — so that each eigenmode may be considered separately. It is the modal content of the mode leading to flutter, in terms of the amplitudes $\hat{q}_r$ and the phases between them, that is of interest. Here the q_r are equivalent to ϕ and χ in the foregoing. The centrifugal term ($\propto u^2$) plays the role of $\mathcal{P}$, and Coriolis damping ($\propto \beta^{1/2}u$) the role of the dissipative force due to γ_2.

Figure 3.40 shows the stability diagram constructed with a progressively higher number of modes in the Galerkin discretization. It is seen that not only does one not get the first 'jump' (at β_{S1}) with $N = 2$ and does so with $N = 3$ or higher (cf. Figure 3.31), but also $N = 4$ is required to obtain the second jump (at β_{S2}), $N = 5$ to obtain the third one, and so on! Thus, each jump is associated with the addition of another generalized coordinate, while the approximation prior to the jump is quite reasonable without it.

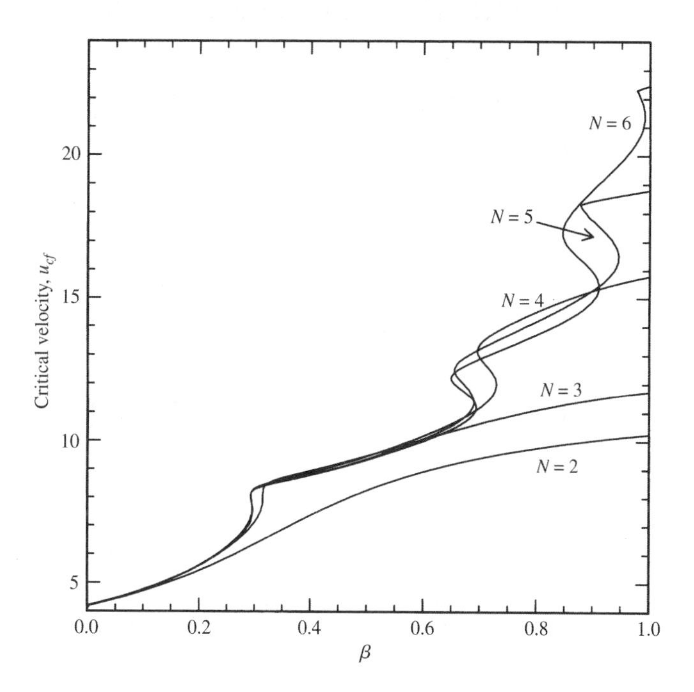

Figure 3.40 The stability diagram of u_{cf} versus β for Galerkin solutions of the undamped horizontal cantilevered pipe with an increasing number of comparison functions, N.

Figure 3.41(a) shows the evolution of the ratio of $\hat{q}_2/\hat{q}_1$ and $\hat{q}_3/\hat{q}_1$ with u. It is seen that around $u_{cf} \simeq 7.5$, which corresponds to the first jump, the $\hat{q}_2$ content reaches a minimum, while $\hat{q}_3$ begins to increase sharply — in which $\hat{q}_1 = 1$ was taken arbitrarily. It is noted that these variations with u are smooth, but when plotted versus β as in Figure 3.41(b), they become much more violent, generating jumps. Just beyond the jump, $\hat{q}_2$ increases once more, together with $\hat{q}_3$.

To interpret these results, and similar ones associated with β_{S2} *et seq.*, it must be recalled that work is done on the system by the centrifugal and the Coriolis forces, equal to

$$W_{\text{centrif}} = -u^2 \int_0^T \eta'(1,\tau)\dot{\eta}(1,\tau)\,\mathrm{d}\tau, \qquad W_{\text{Cor}} = -u\sqrt{\beta}\int_0^T \dot{\eta}^2(1,\tau)\,\mathrm{d}\tau, \tag{3.109}$$

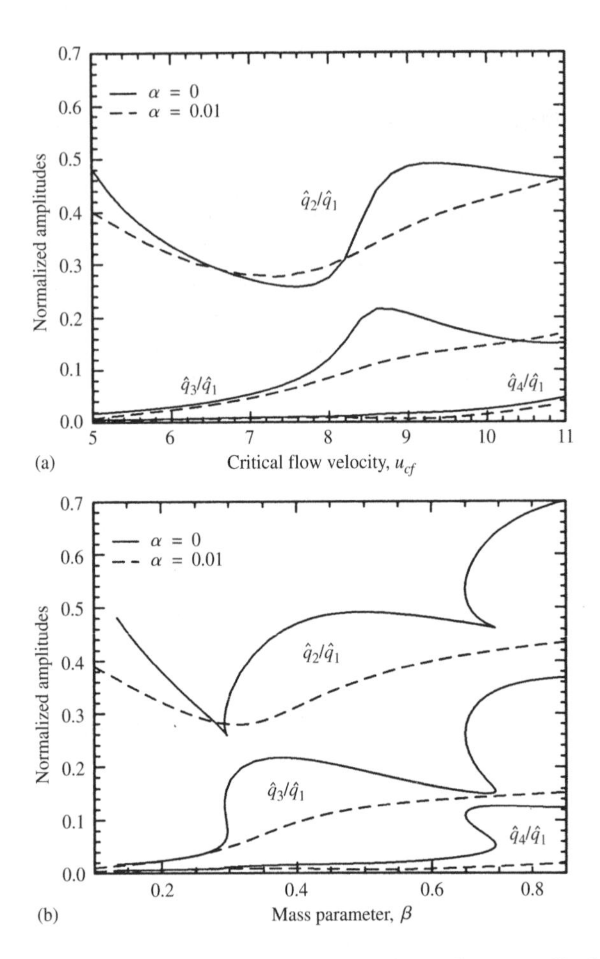

Figure 3.41 The evolution of the normalized generalized coordinate amplitudes, $\hat{q}_i/\hat{q}_1$, associated with the Galerkin discretization of the horizontal cantilevered pipe system: (a) as a function of u_{cf}; (b) as a function of 'the critical β' corresponding to $u = u_{cf}$ (Semler *et al*. 1998).

over a period of neutrally stable oscillation T.† For a three-mode Galerkin solution, taking $q_r = A_r \sin(\omega t - \theta_r)$, one obtains

$$W_{\text{centrif}} = u^2[13.62 A_1 A_2 \sin(\theta_1 - \theta_2) + 25.89 A_1 A_3 \sin(\theta_3 - \theta_1) + 12.27 A_2 A_3 \sin(\theta_2 - \theta_3)],$$

†The *physical* similarity to the follower-force system becomes even clearer if equation (3.108) is rewritten for *neutrally stable oscillations* (periodic motions) and small angles, giving

$$\Delta E = -\mathcal{P} \int_0^T \dot{\phi}_1 \phi_2 \,\mathrm{d}\tau - \int_0^T \left[\gamma_1 \dot{\phi}_1^2 + \gamma_2 (\dot{\phi}_1 - \dot{\phi}_2)^2\right] \mathrm{d}\tau = 0.$$

The first term simplifies to the form above since the integral of $\dot{\phi}_1 \phi_1 \equiv \mathrm{d}\left(\frac{1}{2}\phi_1^2\right)/\mathrm{d}\tau$ vanishes for purely periodic motions — the same conditions leading to (3.109). The similarity of this first term to W_{centrif} and of the term involving γ_1 to W_{Cor} now becomes very clear. The term involving γ_2 corresponds to viscoelastic damping in the pipe ($\alpha \neq 0$). Hence, it is obvious that positive work can be done on the system by the lateral component of the follower force $\mathcal{P}$ or, for the pipe problem, by the lateral component of the jet reaction, u^2 (of the jet emerging from the free end of the pipe); to the linear limit the axial component is conservative. Thus, the physical parallelism between the two cases of the follower-force system and the pipe system is therefore very close.

$$W_{\text{Cor}} = -4u\sqrt{\beta}\omega[A_1^2 + A_2^2 + A_3^2 - 2A_1A_2\cos(\theta_1 - \theta_2) + 2A_1A_3\cos(\theta_3 - \theta_1) - 2A_1A_3\cos(\theta_2 - \theta_3)].$$

Broadly, the centrifugal force imparts energy to the system, while the Coriolis force (involving β) absorbs energy, the balance between the two, in the absence of dissipation, giving rise to flutter (Section 3.2.2). However, as discussed in Section 3.5.6, the flutter mode shape remains broadly similar with varying β, though the mode content is clearly altered. Hence, as β is increased, the amplitude and phase of the q_r components have to be adjusted to provide a composite shape capable of absorbing energy from the fluid. For low enough β, q_1 and q_2 are quite sufficient. However, as $\beta \simeq \beta_{S1}$ is approached, the third component, q_3, has to come in to achieve the required modal mix; and similarly q_4 for β_{S2}, and q_5 for β_{S3}.

This leaves unanswered the question of why this adjustment in modal content is not gradual but rather abrupt. The answer is furnished by the phase information, Figure 3.42.

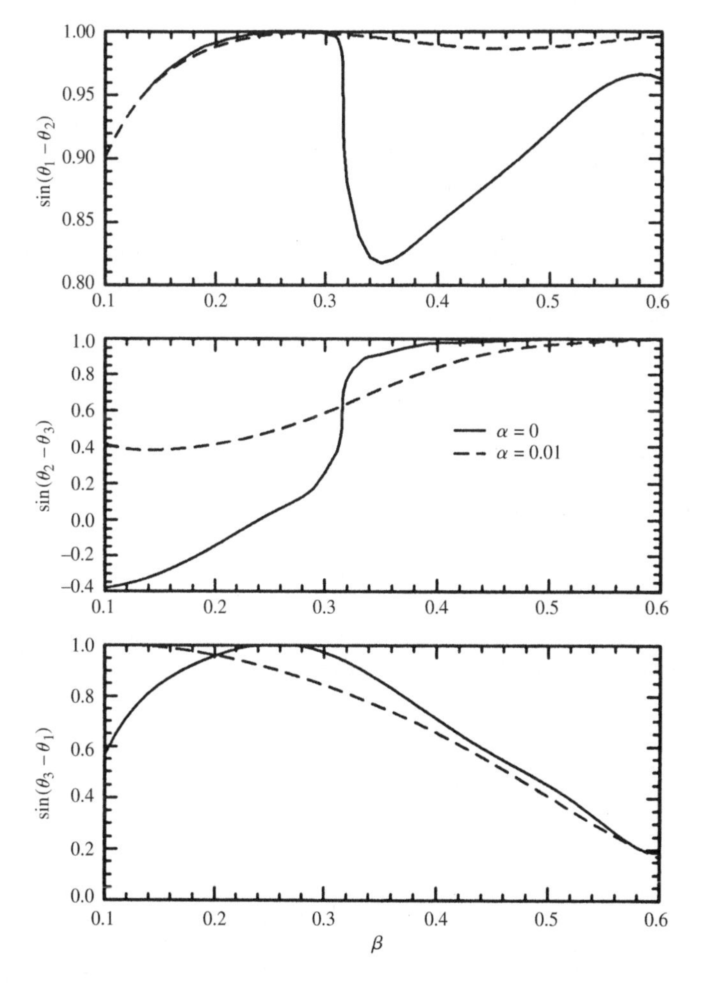

Figure 3.42 The evolution of phase differences between the generalized coordinates q_i associated with the Galerkin discretization of the horizontal cantilevered pipe system, as a function of β.

Even though each $\theta_i - \theta_j$ varies smoothly with β, as it crosses $\frac{1}{2}\pi$ and π, $\cos(\theta_i - \theta_j)$ and $\sin(\theta_i - \theta_j)$ respectively, change sign — with attendant abrupt changes in the energy expressions. For example, $\theta_2 - \theta_3 > \pi$ for low β, it crosses π at $\beta \simeq 0.24$, and then decreases sharply to $\sim \frac{1}{2}\pi$ near β_{S1}; hence, $\sin(\theta_2 - \theta_3)$ becomes positive for $\beta \geq 0.24$ and then increases precipitously near β_{S1}, while $\cos(\theta_2 - \theta_3)$ becomes small. Similarly, $\theta_3 - \theta_4$ crosses π at $\beta \simeq 0.6$ prior to dropping to $\sim \frac{1}{2}\pi$ at β_{S2} (not shown), and $\theta_4 - \theta_5$ does the same at $\beta \simeq 0.83$ and β_{S3}, respectively.

Thus, the answer to the existence of the jumps lies in the modal content of the flutter mode *and* the phase differences between its component parts. Moreover, at each jump, there is a transition zone in which three possible mixes of modes are feasible with different u_{cf}, one low, one middle (unstable), and the other high (e.g. at β_{S2}, for $0.65 < \beta < 0.69$ approximately), but as β is increased sufficiently, only the one with the higher u_{cf} survives.

As a cautionary note it should be mentioned that, in the foregoing, the travelling wave component in the mode shape was ignored, whereas in reality (see Section 3.5.6) $\theta_j \equiv \theta_j(\xi)$ generally. Clearly, this also must play a role.

3.5.5 On destabilization by damping

To those with a structural mechanics background the very statement that dissipation, i.e. energy loss, may make a stable system unstable might appear paradoxical. In gyrodynamics, however, this effect has been known for a long time (Den Hartog 1956; Crandall 1995a,b) — certainly since Thomson (Lord Kelvin) and Tait demonstrated in 1879 that damping in a 'gyroscopic pendulum' can be destabilizing. A gyroscopic pendulum is an 'up-standing', up-turned pendulum to which spin has been added so as to stabilize the statically unstable system. Stability can nevertheless be destroyed if damping is added, no matter what the spin-rate (Crandall 1995a).†

The effect is not surprising to fluid mechanicians either. For instance, they know of Reynolds' two hypotheses, formulated in 1883, stating that: (a) in some situations the inviscid fluid may be unstable, while the viscous one is stable, so that the effect of viscosity is purely stabilizing; (b) in other situations the inviscid fluid may be stable while the viscous one unstable, indicating that viscosity is destabilizing (Drazin & Reid 1981; Chapter 4). Examples may be found in shear flow instability (Tritton 1988; section 17.6), arising in 2-D velocity profiles with a discontinuity (e.g. a jet or a wake) or in profiles with no point of inflection (e.g. a pipe flow or a boundary layer with a favourable pressure gradient). In the first type of flow, viscosity is primarily stabilizing, preventing the Kelvin–Helmholtz‡ type instability at low Reynolds number ($\mathcal{R}e$). In the second type of flow this instability does not occur, but viscosity can cause instability of a different kind. Viscosity now plays a dual role: stabilizing at low Re, but destabilizing at high Re. In aeronautics the destabilizing effect of damping has been known for a long time, in relation to aircraft flutter, and has been carefully studied (Broadbent & Williams 1956; Done 1963; Nissim 1965); also, in satellite dynamics this untoward effect of dissipation is

†Crandall shows that, although ordinary damping is always destabilizing, 'rotating damping' is not, thus explaining how in practice such pendula are stabilized at high spin-rates.

‡The Kelvin–Helmholtz instability is the premier example of shear flow instability in profiles with a point of inflection. It may be demonstrated theoretically by a flow in which the upper half-plane has a uniform velocity to the right, and the lower half-plane to the left. If waviness develops in the interface, the pressures generated (via Bernoulli's equation for inviscid flow) tend to exaggerate the waviness, leading to instability.

now textbook material (Hughes 1986; Chapters 5 and 7). Nevertheless, for fluid–structure interaction phenomena, destabilization by dissipation is sufficiently perplexing to deserve further attention.

Several attempts have been made to understand the mechanism of destabilization. Of these, Benjamin's (1963) work, applying to all fluid–structure interaction systems, will be discussed first, followed by that of Bolotin & Zhinzher (1969) and Semler *et al*. (1998).

An attempt to explain the phenomenon in simple terms was made by Benjamin (1963) in connection with the stability of compliant surfaces in fluid flow. Specifically, considering a one-degree-of-freedom mechanical system, $m\ddot{q} + c\dot{q} + kq = Q$, where $Q = M\ddot{q} + C\dot{q} + Kq$ is associated with the fluid forces, and introducing the concept of an 'activation energy', Benjamin shows that (i) if $m > M$ and $k > K$, dissipation stabilizes the system (class B instability), while (ii) if $m < M$ and $k < K$, dissipation destabilizes it (class A instability). Since $-M$ is the added mass, $M < 0$ must hold for a physically meaningful system, and hence the condition $m < M$ is nonphysical. Benjamin recognized this and so considered next an infinitely long compliant surface, disturbed by a sinusoidal wave travelling along it. In this case, physically meaningful conditions are obtained for the existence of class A and B instabilities, once again with the aid of the activation energy [see also Yeo & Dawling (1987)]; as before, these conditions are dependent on the fluid/solid mass and stiffness ratios. This work is discussed in greater detail in Appendix C.

It was initially thought (Païdoussis 1969) that Benjamin's work could explain both dissipative destabilization and the stability curve jumps in the pipe problem. Certainly, for $\beta < \beta_{S1}$, where β_{S1} is the value for the first discontinuity, dissipation is stabilizing (Figure 3.35) and for $\beta > \beta_{S1}$ it is destabilizing. However, as seen in Figure 3.35, dissipation continues to be destabilizing across the second discontinuity at β_{S2}. Hence, Benjamin's work can only explain the destabilizing effect of damping for $\beta > \beta_{S1}$, but cannot explain the jumps themselves.

Another point of view was expressed by Bolotin & Zhinzher (1969), whose thesis may be summarized as follows: the very statement that 'damping is destabilizing' in a nonconservative system is flawed in that the analysis with zero damping gives a false indication of the stability region, a portion of which, if the analysis is properly conducted with *some* (even infinitesimally small) damping, is really unstable. Thus, the presence of purely imaginary eigenvalues on the imaginary axis merely indicates 'quasi-stability' rather than stability. This work is very important and it can explain the dynamics for $\beta = 0$ and $\beta = 0^+$ discussed at the end of Section 3.5.3; see also Section 3.7. However, it applies to nongyroscopic nonconservative systems and hence cannot help us, since the instability here is via a classical Hopf rather than a Hamiltonian Hopf bifurcation. For the pipe system, one not only obtains that nonzero dissipative forces are destabilizing *vis-à-vis* the undamped system, but also that in some cases (e.g. Figure 3.35 for $\sigma = 0.23$ and 1.42 and also Figure 3.43) increased dissipation *further* destabilizes the system. In this regard the dynamical behaviour is more closely related to Benjamin's system. Under conditions where dissipation-induced destabilization occurs (class A instability), the system must be allowed to do work against the external forces providing the excitation; i.e. the absolute energy level of the whole system must be reduced in the process of creating a free oscillation. The interested reader is also referred to Craik (1985) and Triantafyllou (1992) for a discussion of 'negative energy modes', requiring an energy *sink* in order to be excited.

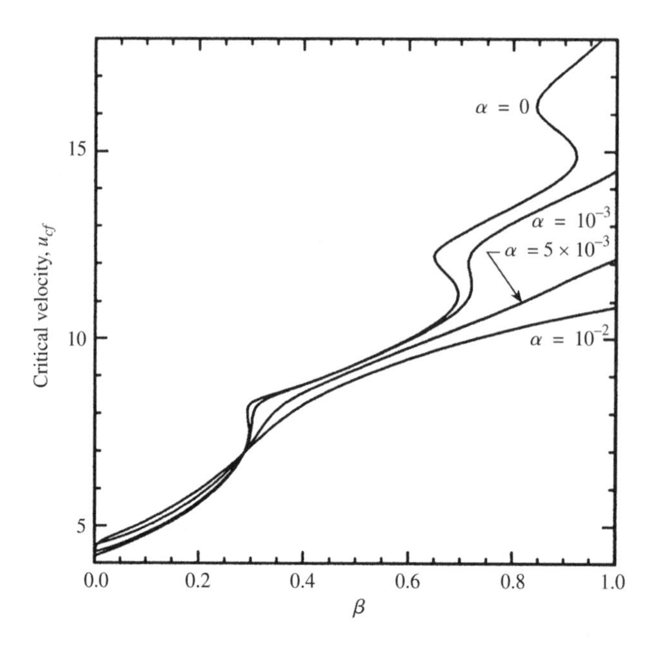

Figure 3.43 The stability diagram of u_{cf} versus β of the horizontal cantilevered pipe, for progressively higher values of the viscoelastic dissipation constant, α (Semler *et al*. 1998).

The question of destabilization by damping according to the Semler *et al*. (1998) thesis is considered next. It is recalled that, in viscoelastic or hysteretic damping, each generalized coordinate component, q_r, is damped proportionately to λ_r^4; so, the higher the value of r, the more is the corresponding q_r damped. Let us consider the first jump, at β_{S1}. The effect of $\alpha \neq 0$ is to damp q_3 more than q_1 and q_2, and to effectively wipe out all the higher components $q_r > q_3$. Now, it is evident from Figure 3.40 that, when it comes into play, q_3 has a stabilizing effect on the system, as manifested by the increase in u_{cf} at β_{S1}; hence, its diminution by α means that the system is effectively *destabilized*. As a result, this jump, which has been shown to be related to the emergence of q_3, can be entirely suppressed, as shown in Figure 3.43! One can similarly see how the other jumps can also be suppressed. Looking again at Figures 3.41 and 3.42 (the dashed lines), it is seen that both the amplitude ratios and phase differences of the q_r are significantly affected. Thus, it is seen that, with damping present, $\hat{q}_3/\hat{q}_1$ and $\hat{q}_2/\hat{q}_1$ increase more gradually with β beyond β_{S1}. Also, some of the 'saturation characteristics' of the phase differences disappear (e.g. for $\theta_3 - \theta_1$), and both $\theta_2 - \theta_3$ and $\theta_3 - \theta_1$ vary more gradually — thus making the discontinuous changes in u_{cf} with β unnecessary.

Another, physical way of looking at the problem is to realize that, in some circumstances, if the fluid pressure acting on an undamped oscillating body is completely in phase with its acceleration (out of phase with the displacement), there can be no interaction between fluid and solid. However, the introduction of dissipation in the solid would produce a phase shift in its oscillation, thereby enabling the fluid to do work on the solid or *vice versa*. In a situation where energy transfer occurs in any case, independently of dissipation, as for the pipe conveying fluid, one can say that the phase shift may either facilitate or hinder energy exchange, thus destabilizing or stabilizing the system as the case may be.

3.5.6 Experiments

The first set of experiments were conducted with horizontal cantilevers conveying air, water or oil, with rubber pipes, in some cases fitted with end-nozzles, and metal pipes (Païdoussis 1963; Gregory & Païdoussis 1966b). The apparatus for the experiments with rubber pipes is shown in Figure 3.44; the same apparatus was used for experiments with water flow (as shown) and with air flow (in which case a volumetric flow meter was inserted in the supply line); in the latter case, the air pressure was sufficiently low for compressibility effects to be neglected. The pipes were horizontal, hung from the ceiling by thin threads, so that motions were in a horizontal plane. In experiments with metal pipes, a different apparatus was used in basically the same arrangement, but the fluid was oil supplied by a suitably modified variable-speed hydraulic pump capable of delivering 66 cm^3/s (4 in^3/s) at up to 9.7 MPa (1400 psig).

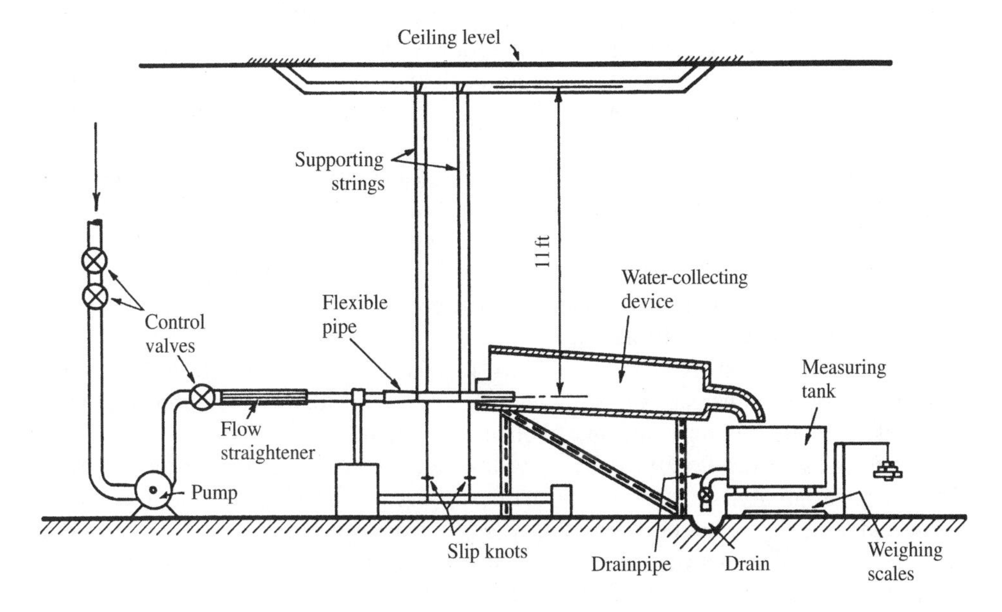

Figure 3.44 Schematic diagram of the apparatus used in Païdoussis' experiments with horizontal cantilevered rubber pipes conveying water or air; the apparatus for the metal pipe experiments was similar (Gregory & Païdoussis 1966b).

The rubber pipes were either pure latex rubber or of the type known as surgical quality rubber tubing; their inside diameter ranged from $D_i = 1.59$ to 12.70 mm ($\frac{1}{16}$–$\frac{1}{2}$ in), the wall thickness from $h = 0.79$ to 3.18 mm ($\frac{1}{32}$–$\frac{1}{8}$ in) and the length from 0.20 to 0.76 m. Although the pipes were carefully selected for uniformity and freedom from kinks and other flaws, all rubber pipes were found to have a permanent bow in one plane (cf. Section 3.4.4), countered by using pipes which, when supported by the strings with the bow in the vertical plane, would straighten out under their own weight together with that of the contained fluid. The two metal pipes were specially manufactured, stress-relieved and straightened by the suppliers. They were both of outer diameter $D_o = 1.59$ mm and 1.98 m (78 in) long; $h = 0.152$ and 0.193 mm. The supporting threads in this case were 6.1 m (20 ft) long.

All measurements were straightforward, except perhaps the measurement of EI for rubber pipes, required in the determination of the dimensionless flow velocity, u. The techniques utilized for this are summarized in Appendix D.

Some of the general observations on the dynamical behaviour of this system are worth giving in some detail; they are similar to those made by other researchers in the experiments to be discussed later.

Small flow velocities increased the damping of the system, and oscillations induced by light taps close to the free end, which were still of the general shape of the first cantilever mode, decayed much faster. At somewhat larger flow velocities the system became overdamped and any displacement of the pipe was followed by a return to rest without any oscillatory motion. In some cases physical contact of the free end of the pipe with the hand, momentarily transforming the system to one supported at both ends, caused the pipe to buckle by bowing out near the middle. When contact was broken suddenly, the pipe returned rapidly to its position of rest, but when the hand was removed only slowly, the pipe pressed against and followed the hand with the result that the timid observer was soon faced with a stream of fluid directed against himself (or nearby colleagues!) — this, as already remarked, being a demonstration of a negative-stiffness (divergence) instability.

At still higher flow velocities, light taps resulted in heavily damped oscillation with a form rather more like that of the second cantilever mode than the first. As the flow velocity increased further, the system became less heavily damped until at a certain critical velocity of flow the disturbance produced by lightly tapping the pipe grew into a self-supporting oscillation. If no outside disturbance was introduced, the system eventually became unstable spontaneously. This usually occurred at measurably higher flow velocities than were sufficient for 'induced' instability to take place, particularly in the case of rubber pipes. Further increase of the flow velocity beyond the stability limit resulted in an increase in both the amplitude and frequency of oscillation.

When decreasing the flow velocity, it was noted that oscillation persisted below the point where instability, spontaneous or 'induced', first occurred. This, and also the fact that in some cases the onset of instability depended on the amplitude of the applied disturbance, indicated that the experimental systems behaved in general nonlinearly.

The mode of deformation of the unstable system was recorded with a ciné-camera in a few selected typical cases and a number of successive frames of the film are shown in Figures 3.45(a–c). In general, for very small values of β, the modal form was essentially that of the first cantilever mode, with a small component of the second. For higher values of β, the second cantilever mode became more prominent, and for $\beta > 0.3$ approximately the third mode became apparent [e.g. see frame 8 of Figure 3.45(c)]. In all cases, the tangent to the free end of the pipe sloped backwards to the direction of motion of the free end over the greater part of a cycle of oscillation. This 'dragging' motion was predicted to be necessary for flutter, in conjunction with the energy considerations of Section 3.2.2.

Indeed, all observations described are in agreement with the theoretical predictions of Sections 3.5.1 and 3.5.2. However, two additional comments should be made. First, the dynamical behaviour of the system is, to some extent, nonlinear — as noted above — suggesting that the Hopf bifurcation is subcritical. Second, according to linear theory, once instability is developed, the amplitude should increase without limit; of course, once the amplitude becomes large, nonlinear forces come into play, and in this case evidently their net effect is to limit the amplitude, thereby establishing a limit cycle.

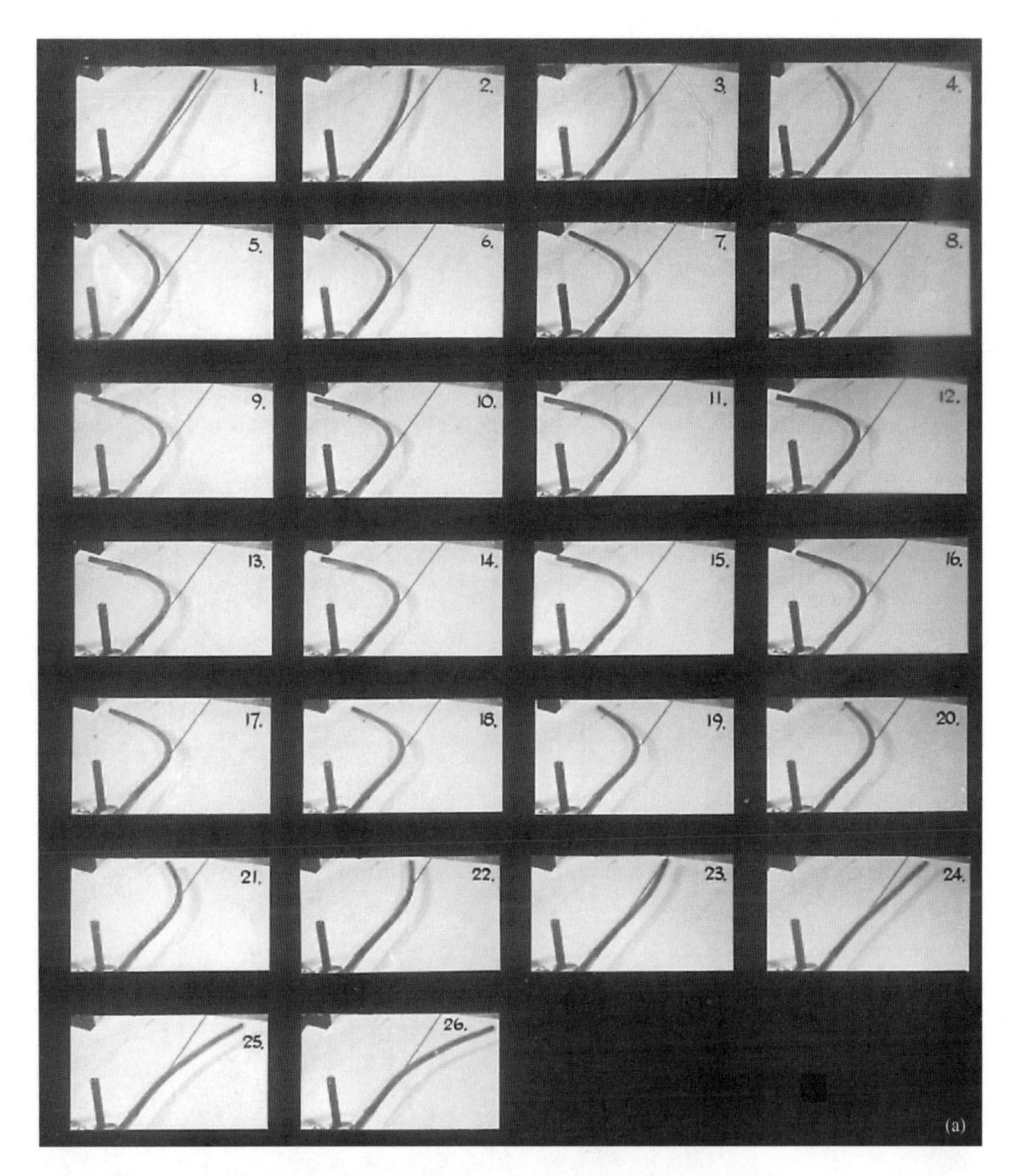

Figure 3.45 Ciné-film sequences showing the limit-cycle motion of cantilevered rubber pipes conveying fluid: (a) air, $L = 457$ mm, frequency $(f) = 1.56$ Hz, $\beta = 0.001$; (b) water, $L = 551$ mm, $f = 3.25$ Hz, $\beta = 0.479$; (c) water, $L = 724$ mm, $f = 1.82$ Hz, $\beta = 0.556$. The camera was located upstream and above the horizontal pipe; the straight black line (drawn on a board just below the flexible pipe) shows the equilibrium position of the pipe.

The limit-cycle amplitude of the free end at the onset of the flutter could be as large as $\frac{1}{4}L$ for the rubber pipes, but less than $\frac{1}{15}L$ for the metal pipes.

The dimensionless critical flow velocities, u_{cf}, and frequencies, ω_{cf}, are shown in Figure 3.46 for rubber pipes. Two sets of experimental points are presented: those of spontaneous instability and those of 'induced instability' — induced by the light taps on

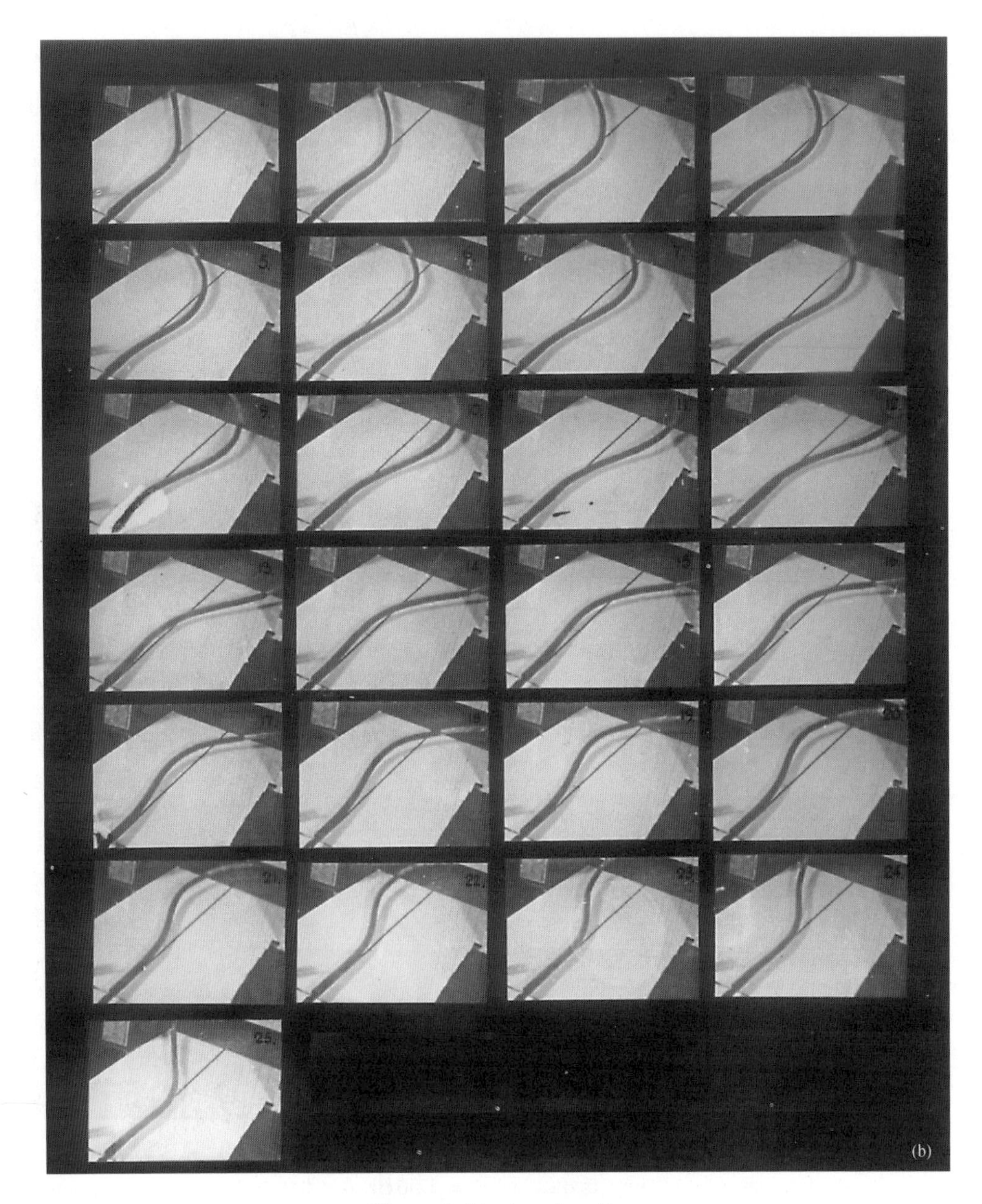

Figure 3.45 (*continued*).

the pipe referred to above. The data points within each set for the same value of β represent experiments with different lengths of the same pipe; experiments were conducted with a given initial length, subsequently shortened in steps by cutting off pieces of the pipe. The internal (material) damping used in the theory is an average for all the experiments, but nevertheless taking it into account improves agreement between theory and experiment,

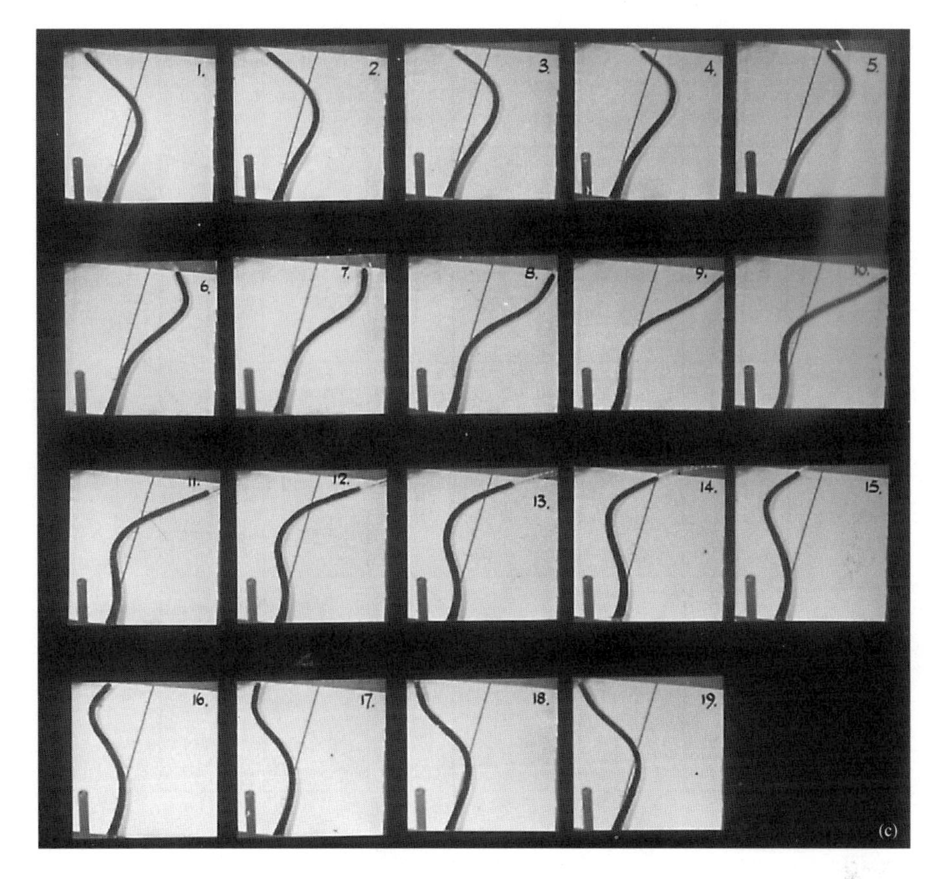

Figure 3.45 (*continued*).

as compared to theoretical results with zero damping [shown in Gregory & Païdoussis (1966b)]. In general, agreement is reasonable; it would have been better if the damping corresponding to each different pipe had been used (Appendix D).

For experiments with nozzles, a latex pipe 12.70 mm inside diameter and approximately 0.508 m long was used. The nozzles were machined in Perspex (Plexiglas) cylinders, 6.35 mm long and 12.70 mm in diameter, which were glued to the inside of the free end of the pipe with soluble glue. The nozzle cross-section converged smoothly over half the length from the diameter of 12.70 mm to the required exit diameter, which varied from 3.18 to 9.13 mm. After each test, the glue was dissolved and a new nozzle was inserted. The original β (without a nozzle) was $\beta \simeq 0.56$, and six experiments were conducted with nozzles in the range of $\beta \simeq 0.03$–0.30 [cf. equations (3.74)]. Experimental data are compared with theory in Gregory & Païdoussis (1966b); the degree of agreement is similar to that in Figure 3.46, but a little worse, possibly as a result of changes in the pipe cross-section due to pressurization of the pipe because of the constriction introduced by the end-nozzle.

The experimental data for the two experiments with metal pipes are compared with theory in Figure 3.47. The experimental values of u_{cf} corresponding to $\beta = 0.111$ and

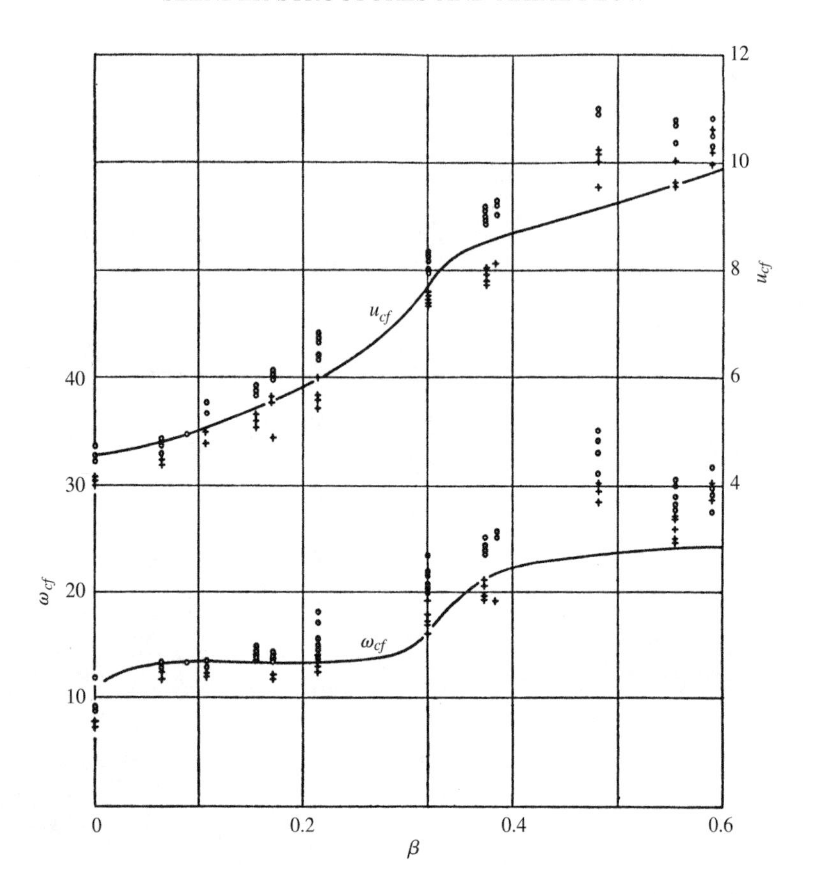

Figure 3.46 Comparison of the experimental values of u_{cf} and ω_{cf} for cantilevered rubber pipes with the theory taking into account internal (hysteretic) damping in the pipe material: •, measurements for spontaneous instability; +, measurements for 'induced' instability; ——, theoretical curves for hysteretic damping coefficient $\mu = 0.065$ (Gregory & Païdoussis 1966b).

0.170 are respectively 9% and 12% below the theoretical values. In this case damping is ignored, because it is quite small. The discrepancy between theory and experiment is likely caused by variations in the effective density and viscosity of the oil with pressure and temperature, as well as cavitation effects, all of which would generate a nonuniform flow velocity along the pipe. Nevertheless, the most significant point about these experiments is that they substantiate the theoretical prediction that frictional forces associated with pressure drop — even when of the order of 8.3 MPa (1200 psig) — do not affect the dynamics in any important way, as predicted in Section 3.3.4.

In Figure 3.46, noting that the values of u_{cf} for induced instability are generally substantially below those for spontaneous instability, it is tempting to conclude that the Hopf bifurcation is subcritical in all cases (see Section 2.3). However, as there was essentially no difference between spontaneous and induced instability thresholds in the case of metal pipes, this was thought to be related perhaps to the difference in material. Indeed, there is a property of carbon-black-'filled' rubbers known as 'stress softening', but latex rubbers should be free of that. As shown in Chapter 5, the difference may be related to the different ranges of L/D_i involved: subcritical Hopf bifurcation for relatively

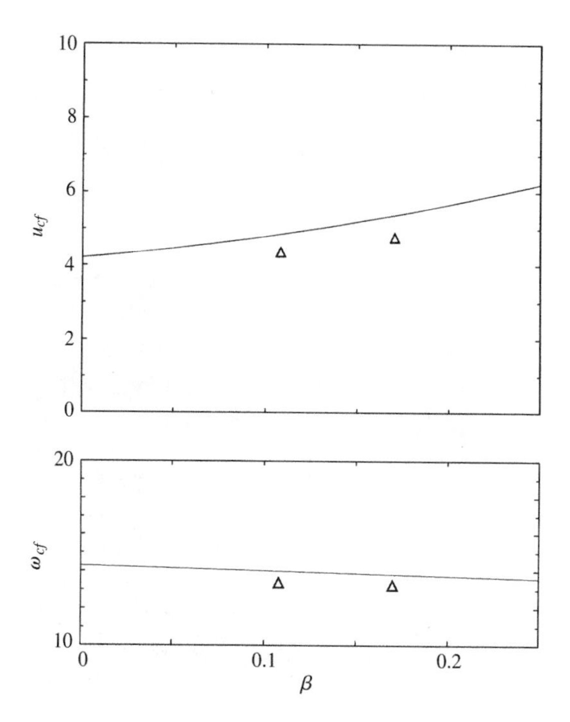

Figure 3.47 Comparison of the experimental values of u_{cf} and ω_{cf} for cantilevered metal pipes: Δ, measurements; ——, theory (Païdoussis 1963).

short pipes ($L/D_i \simeq 36$–350 for the rubber pipes) and supercritical Hopf bifurcation for long pipes ($L/D_i \simeq 1545$–1650 for the metal pipes).

Chronologically, the second set of experiments was conducted by Greenwald & Dugundji (1967) — see also Section 3.4.4. They conducted experiments with three elastomer pipes ($D_o = 3.00$–4.75 mm, $h = 0.86$–1.50 mm). The pipes were hung vertically and clamped at their upper end. The authors have made similar general observations to those discussed in the foregoing. A very nice photograph is shown in Figure 3.48, corresponding to a pipe with $\beta = 0.471$, which shows more clearly than Figure 3.45(a–c) the nonstationarity of the modes and the travelling wave component in the mode shapes. The measured critical flow velocities are compared with theory in Table 3.3. It is seen that agreement is at least as good as in Figure 3.46 when viscoelastic damping is taken into account. In the theory the authors neglected gravity; this is reasonable: from their data one finds $\gamma = 1.68$–2.89, which results in theoretical values of u_{cf} higher than those in Table 3.3 by less than 2%.

A more extensive set of experiments was conducted by Païdoussis (1970) with *vertical* pipes, either hanging or standing. This is the first instance when such pipes were cast by the researcher, and this allowed the manufacture of truly straight pipes for the first time, thus facilitating the experiments a great deal — not only for pipes conveying fluid, but also for experiments with shells and cylinders. The 'manufacturing techniques' are outlined in Appendix D.

The general observations of the dynamical behaviour of hanging pipes are much as described before and need not be repeated here. However, two additional points are useful

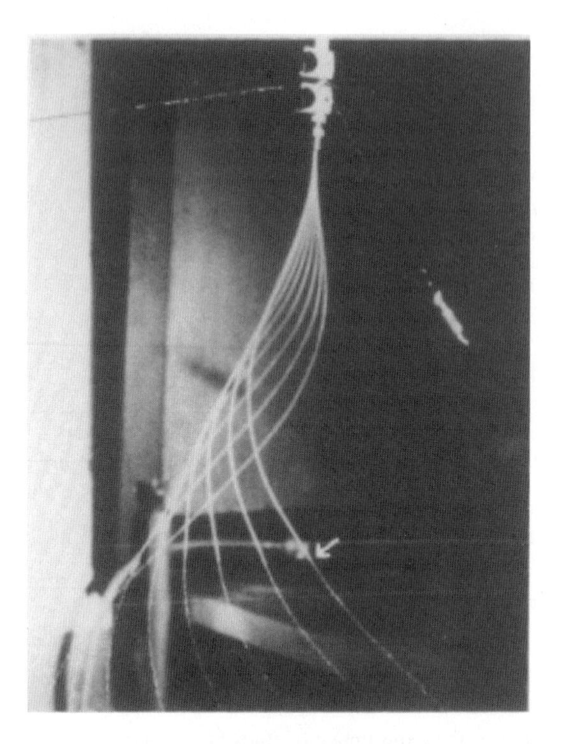

Figure 3.48 Photograph of a fluttering vertical pipe (pipe #3) from the experiments by Greenwald & Dugundji (1967). The arrow shows the end of the pipe; what is seen below that point is the free water jet.

Table 3.3 Comparison between the experimental and theoretical values of u_{cf} from Greenwald & Dugundji's (1967) experiments; the values have been scaled from their figure 13a.

β	Values of u_{cf}		
	Theory (no damping)	Experiment	Theory (with damping)
0.342	8.48	6.85	7.50
0.471	9.15	8.30	8.10
0.500	9.32	9.55	8.30

to make. The first is that, before the occurrence of flutter, in some cases, small movements of the cantilever away from its vertical position of rest were observed with increasing flow. These movements developed gradually with flow and never exceeded 6 mm (1–2% of the length); they could be made to vanish by suitable, slight circumferential adjustments of the tubular cantilever at its upstream support. Clearly, these could not be construed to be a buckling instability. They must be interpreted as 'localized' buckling resulting from small nonuniformities in the cantilever, or due to release of strains imposed by imperfect circumferential support at the clamped end. Similar occurrences of localized buckling were observed in experiments with horizontal cantilevers, and by Benjamin (1961b) in his experiments with articulated pipes. It is of interest that, if the tubular cantilever is initially (i.e. at zero flow) not substantially straight, flow can produce large lateral movements which are much larger than the initial departures from straightness. This can be observed by conducting an experiment using as the cantilever a piece of commercial rubber tubing, which normally has a set bow in it. Flow exaggerates the original bow, the shape of the tube continually changing with increasing flow velocity. Clearly, this could be misinterpreted as *buckling* of a *straight* pipe.

The second point of interest is that, in some of these experiments, it was possible to demonstrate the nonlinear dynamical behaviour displayed in Figures 2.12 and 2.13 about the origin. Over a very small range of flow velocities, it was found that: (i) weak taps to the pipe caused it to oscillate, but the oscillation decayed and the pipe returned to its equilibrium state; (ii) stronger taps induced the system to develop limit-cycle oscillation — thus demonstrating the existence of a small unstable limit cycle and a larger stable one.

Several experiments were conducted with different lengths (different γ) of a number of pipes with varying β. The pipes were all with $D_o = 15.5$ mm and $h = 2.79$–9.14 mm; the initial length was ~ 480 mm and experiments were conducted with $L = 230$–480 mm. Two different materials were used, Silastic A and Silastic B (Appendix D), the latter having a larger E and higher damping. In comparing with theory, the dissipation was modelled as a hysteretic effect, and average values were used: $\mu = 0.02$ for Silastic A and $\mu = 0.10$ for Silastic B.

Typical results for the experimental u_{cf} and ω_{cf} for spontaneous flutter of hanging cantilevers ($\gamma > 0$) are shown in Figures 3.49 and 3.50 for water flow and Table 3.4 for air flow, where they are compared with theory. It is clear that agreement between theory and experiment is reasonably good, especially when dissipation is taken into account. It is interesting that in some cases the measurements provide indirect experimental support to the theoretical prediction that damping may destabilize the system (e.g. for $\beta = 0.241$, $\gamma \simeq 16$ and for $\beta = 0.645$, $\gamma \simeq 8.6$).

In assessing agreement between theory and experiment, greater weight should be placed on the critical flow velocity than on the critical frequency, as the latter is measured after the limit cycle has been established, when nonlinear forces not taken into account in the theory have already come into play. Accordingly, the fact that taking into account dissipation seems to worsen agreement in the frequency between theory and experiment, in nearly all cases, cannot be interpreted as a weakness of the theory; rather, it should be viewed as being symptomatic of the limitations in the experimental procedure (in identifying the limit-cycle frequency with ω_{cf}).

As already remarked in Section 3.5.2, the impetus for these experiments was partly provided by Benjamin's (1961a,b) findings in connection with dynamical behaviour of articulated pipes conveying fluid. Benjamin found that divergence is sometimes possible in cases of vertically hanging articulated cantilevers conveying water; yet it does not occur if the conveyed fluid is air, the only form of instability possible in that case being flutter. However, in the case of continuous (hanging) cantilevers, it was found that divergence is not possible *at all* whatever the fluid conveyed, only flutter. This matter is clarified in Section 3.8.

We next consider the experiments with standing cantilevers conveying air only, for obvious reasons. The dynamical behaviour of the system was of three distinct types, which for ease of description will be categorized as applying to long, intermediate and short cantilevers.

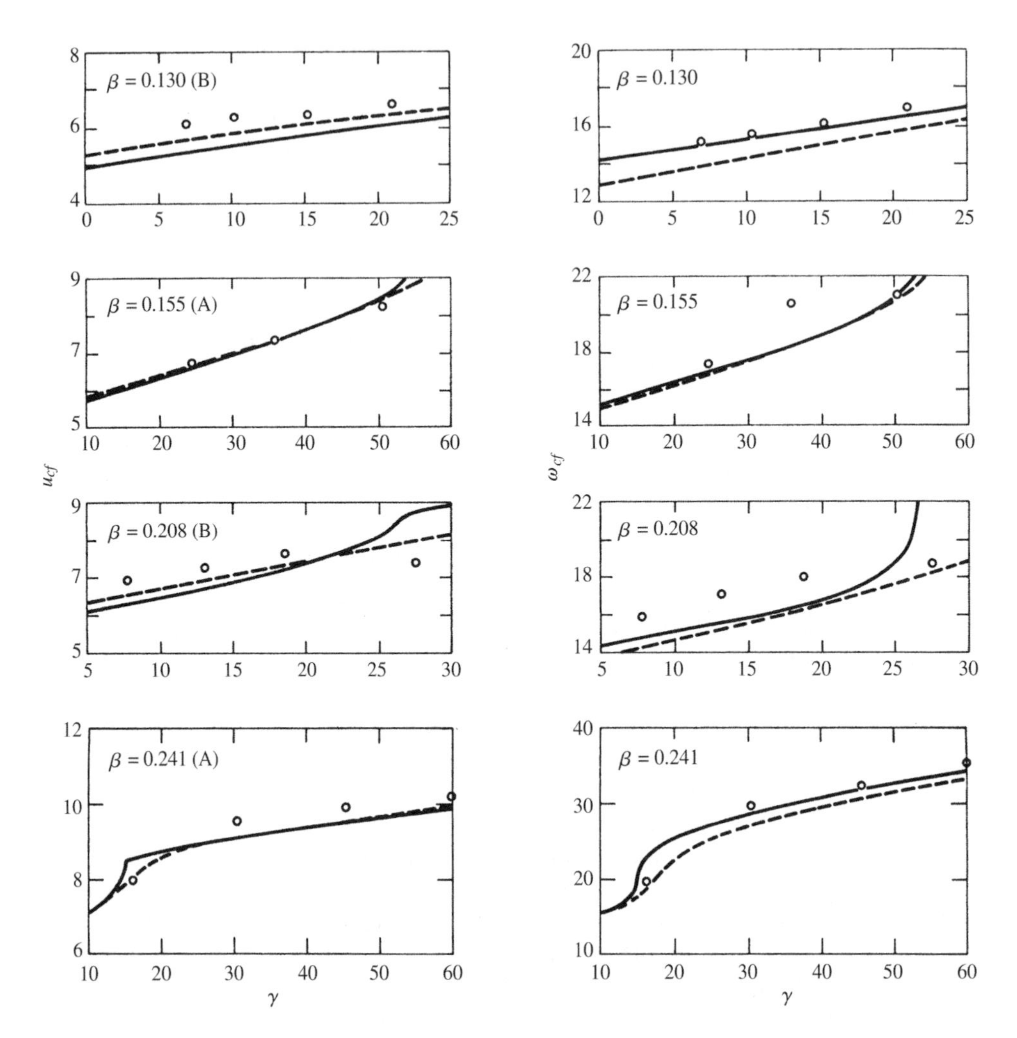

Figure 3.49 Comparison between theoretical and experimental values of u_{cf} and ω_{cf} for a number of vertical (hanging) cantilevered pipes conveying water with different β and lengths thereof (different γ): for $0.130 \leq \beta \leq 0.241$: o, experiment; ——, theory with no damping; – – –, theory with damping ($\mu = 0.02$ for Silastic type A rubber; $\mu = 0.10$ for Silastic type B); (Païdoussis 1970).

Long cantilevers were buckled under their own weight at zero flow velocity. The dynamical behaviour of the system was assessed by supporting the cantilever by hand in its unflexed shape, while the flow was incremented, and then releasing it. Long cantilevers ($\gamma < -23$) were unstable at all flow velocities. At low flows a long cantilever continued to be unstable by buckling; at higher flow velocities, oscillations were superposed on buckling, resulting in an erratic, thrashing motion.

Short cantilevers ($\gamma > -8$ approximately) did not buckle under their own weight at zero flow. Their behaviour with increasing flow was essentially as for hanging cantilevers; the system remained stable with increasing flow until, at a sufficiently high flow velocity, flutter developed spontaneously.

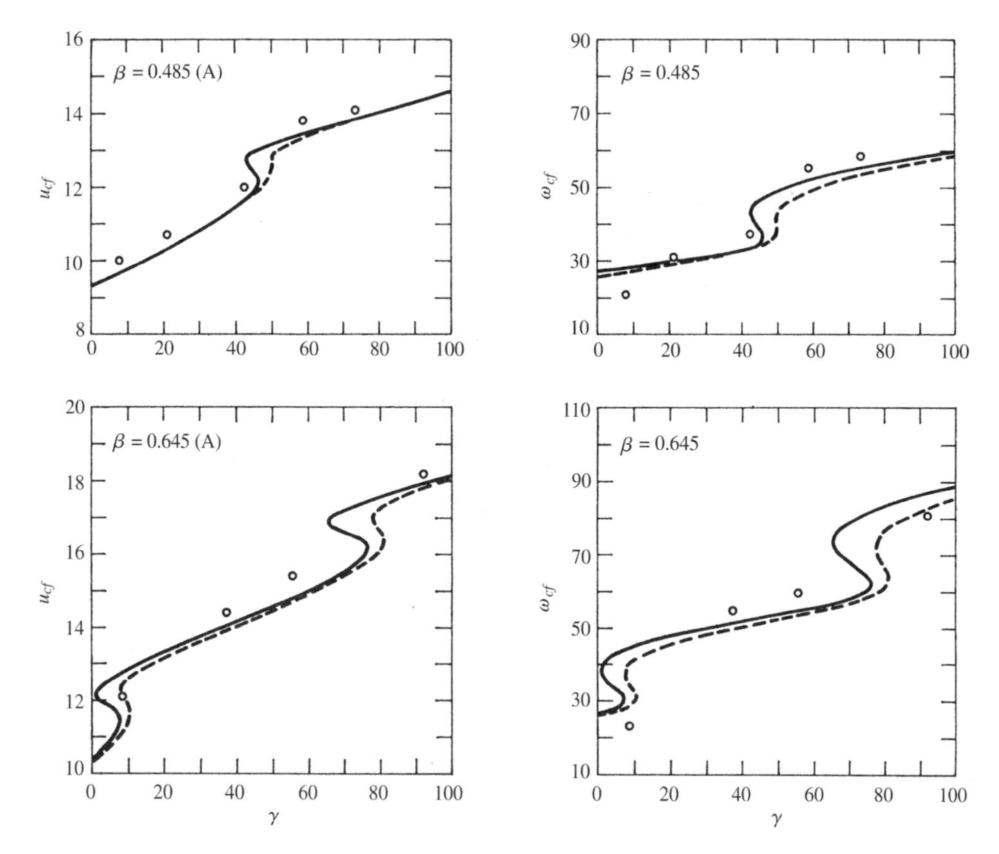

Figure 3.50 Continuation of the comparison as in Figure 3.49, for pipes with $\beta = 0.485$ and 0.645 (Païdoussis 1970).

Table 3.4 Comparison of experimental results with theory for hanging cantilevers conveying air (Païdoussis 1970).

$\beta \times 10^3$	γ	Values of u_{cf}			Values of ω_{cf}		
		Theory $\mu = 0$	Exp't	Theory $\mu = 0.02$	Theory $\mu = 0$	Exp't	Theory $\mu = 0.02$
0.23	61.1	6.15	6.33	6.33	21.3	18.0	19.7
2.03	42.4	5.62	6.05	5.79	19.5	19.5	18.1

Cantilevers of intermediate length, while unstable by buckling at zero and small flow velocities, were stable at a higher flow range. Thus, if the cantilever was supported and the flow increased to a certain point, upon release the cantilever retained its straight, undeformed shape. Further increase of flow, nevertheless, eventually resulted in the development of oscillatory instability.

Clearly these observations agree with the theoretically predicted behaviour of standing cantilevers, if one interprets increasing length as increasing negative γ.

Experimental results for standing cantilevers are shown, and compared with theory ($\beta = 1.1 \times 10^{-3}$, $\mu = 0.02$), in Figure 3.51. It is noted that measurements with $\gamma < -22.8$

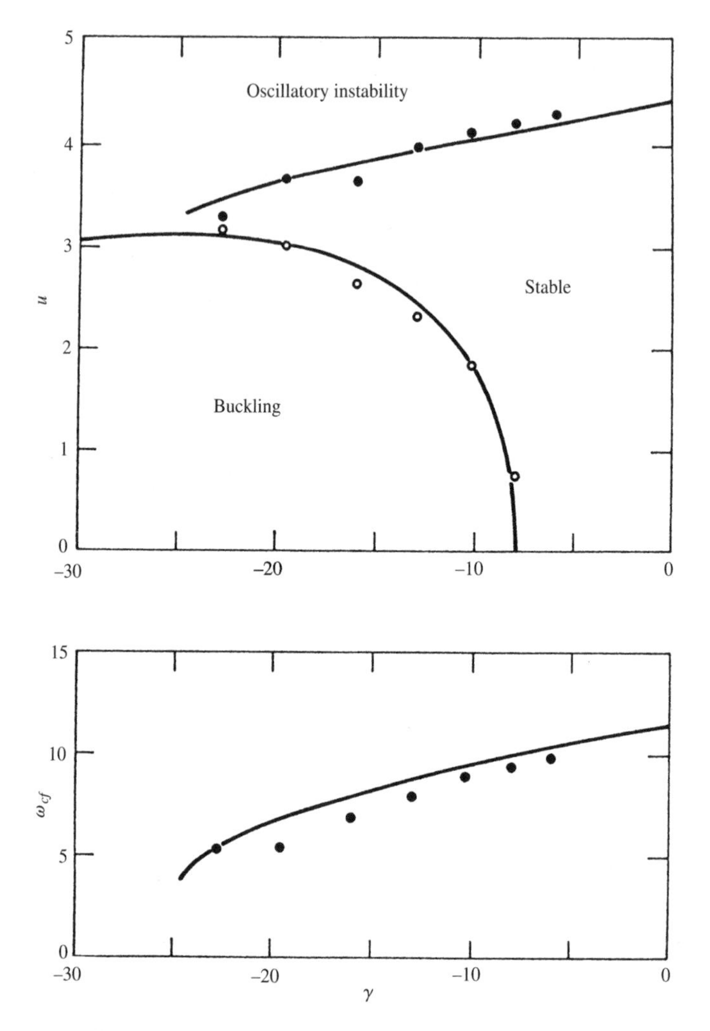

Figure 3.51 Comparison between theoretical and experimental values of u for (i) the threshold of restabilization of pipes buckled under their own weight and (ii) onset of flutter of 'standing' vertical cantilevers conveying air ($\beta = 1.1 \times 10^{-3}$): ∘, •, experiment; ——, theory (Païdoussis 1970).

approximately were not feasible, as the system then remained unstable at all flows; the transition from instability by divergence to instability involving both divergence and flutter proved to be very difficult to pin-point. It is also noted that the theoretical results in Figure 3.51 are quite different, for the given β and μ, from those in Figure 3.33 for $\beta = 0^+$, $\mu = 0$ — for the reasons discussed at the end of Section 3.5.3.

It is seen in Figure 3.51 that agreement between experiment and theory is quite good, particularly in the case of the dimensionless flow velocities, where in most cases the discrepancy is $< 5\%$, which is within the margin of experimental error. It is also remarked that, in this particular case, linear theory can predict the restabilization and second loss of stability of an initially unstable system quite well.†

†Provided that the system is first supported in more or less its equilibrium configuration and then released.

A successful experiment with a metal pipe was also conducted by Liu & Mote (1974) with their apparatus, described in Section 3.4.4. They used an end-nozzle, so as to reduce the effective β and achieve flutter with the available maximum flow rate (see Section 3.3.5); $\alpha_j = A/A_j$, the ratio of pipe flow area to terminal flow area, was 2.42. They obtained good agreement between the theoretical and experimental values of u_{cf}: 2.89 and 3.27, respectively, as well as between the theoretical and experimental frequency versus flow curves, as shown in Figure 3.52. In the absence of the nozzle, the agreement in frequency was less good, because the pipe was less straight. However, this is the second instance where the pressure drop in the pipe was very large, $\sim$ 3–10 MPa, yet the dynamics was essentially unaffected by it.

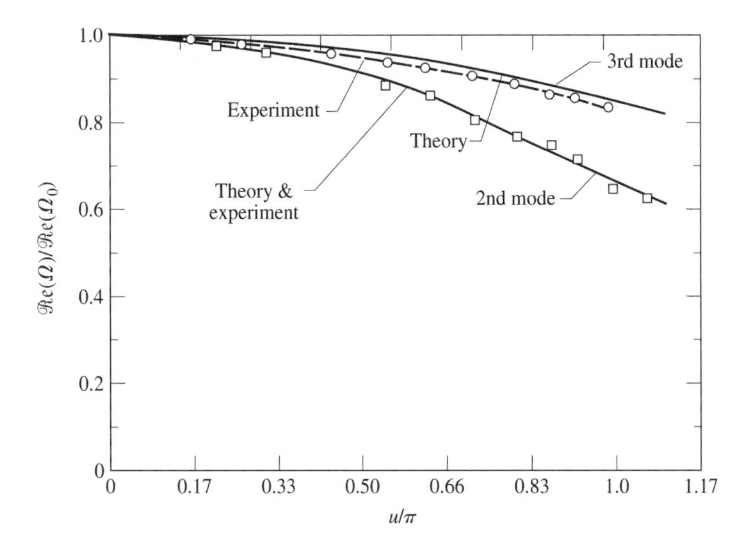

Figure 3.52 Variation of the second- and third-mode eigenfrequencies with increasing u for a metal cantilevered pipe conveying fluid. Experimental/theoretical reference frequencies (at $u = 0$): $\Re e(\Omega)_0 = 8.0/8.0$ Hz for the second mode and 25.5/21.7 Hz for the third (Liu & Mote 1974).

An important theoretical and experimental study, mainly on forced vibrations of vertical cantilevered pipes conveying fluid (see Section 4.6), was conducted by Bishop & Fawzy (1976). They also examined the free vibration characteristics, and a few words about that will be said here. The experiments were with surgical quality silicone rubber pipes conveying water. The authors studied extensively the static distortion from the stretched-straight state that they observed in their experiments and its evolution with flow. They concluded that it was due to lack of perfect straightness and residual internal stresses related to the manufacturing process, and *not* an instability (divergence) — in agreement with previous studies — even though some other researchers later misinterpreted this finding. Their experimental data and degree of agreement with theory were similar to those reported already. A typical set is shown in Figure 3.53 for experiments without ($\alpha_j = A/A_j = 1$) and with end-nozzles ($\alpha_j = 1.5$ and 2.49).

A simple but ingenious experiment was conducted by Becker *et al*. (1978) using drinking straws (of unspecified material) as pipes and air flow. The supported end was attached onto weighing scales and the flow-rate was determined from the reaction

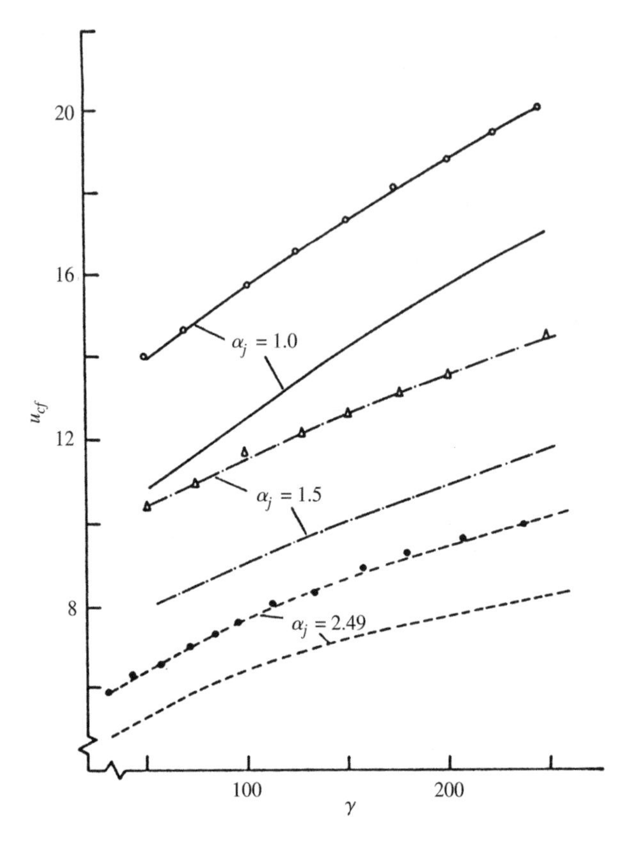

Figure 3.53 Theoretical and experimental values of u_{cf} for experiments with a vertical cantilevered rubber pipe conveying water ($\beta = 0.622$), fitted with different end-nozzles (different $\alpha_j = A/A_j$): curves with data points, experiments; curves without, theory. Dissipation modelled by viscoelastic model: $\alpha = 0.003$, 0.0025 and 0.002 for $\alpha_j = 1$, 1.5 and 2.49, respectively (Bishop & Fawzy 1976).

exerted thereon. The measured dimensionless critical flow velocity was found to be $u_{cf} = 4.45$ ($U_{cf} \simeq 150$ m/s), which is within 6% of the theoretical.

An extensive experimental programme (see Section 3.6) was undertaken by Jendrzejczyk & Chen (1985) and Chen & Jendrzejczyk (1985). They conducted two experiments with polyethylene cantilevered pipes ($D_o = 9.5$ and 12.7 mm, $h = 1.59$ mm and $L = 609.6$ mm) mounted vertically and conveying water. They obtained excellent agreement between theory and experiment, as illustrated by Table 3.5. The r.m.s.

Table 3.5 Jendrzejczyk & Chen's (1985) results for cantilevered polyethylene pipes conveying water.

Test no.	U_{cf} (m/s)		$\Omega_{cf}/2\pi$ (Hz)	
	Theory	Experiment	Theory	Experiment
1.1	25.0	24.9	12.9	12.0
1.2	30.7	31.4	16.0	14.5

amplitudes in two perpendicular planes are substantially equal over the whole flow range, as shown in Figure 3.54(a); this indicates that the plane of oscillation is not far from 45° to the two measurement planes — although, due to imperfections, the oscillation plane changed slightly with U. The power spectral densities (PSDs) at $U < U_{cf}$ are shown in Figure 3.54(b). It is seen that for low flow velocities the response of the pipe to flow turbulence is broad-banded; however, as U_{cf} is approached, the peak associated in this case with the second-mode frequency becomes dominant. There is an apparent discrepancy between the dominant frequency in Figure 3.54(b) for $U = 30.24$ m/s, a little before the onset of flutter, and the flutter frequency in Table 3.5, Test 1.2, at $U = 31.4$ m/s. This however, is explained by the fact that the establishment of the limit cycle (of amplitude $> \frac{1}{5}L$) is in this case accompanied by a drastic increase in frequency (Chen 1995), already referred to qualitatively in the foregoing.

An important set of experimental results on the onset of flutter and the evolution of limit-cycle oscillations was generated with a slightly longer sample of the smaller pipe ($\beta \simeq 0.45$, $L = 685.8$ mm, $L/D_o = 720$). In Figure 3.55(a), the data correspond to flutter induced by perturbing the pipe, while in (b) they correspond to spontaneously developed flutter. It is seen that, if the system is perturbed, the critical flow velocity is $U_c \simeq 22.3$ and the initial limit-cycle amplitude is $A/D_o \simeq 0.2$;† also, there is essentially no hysteresis (i.e.

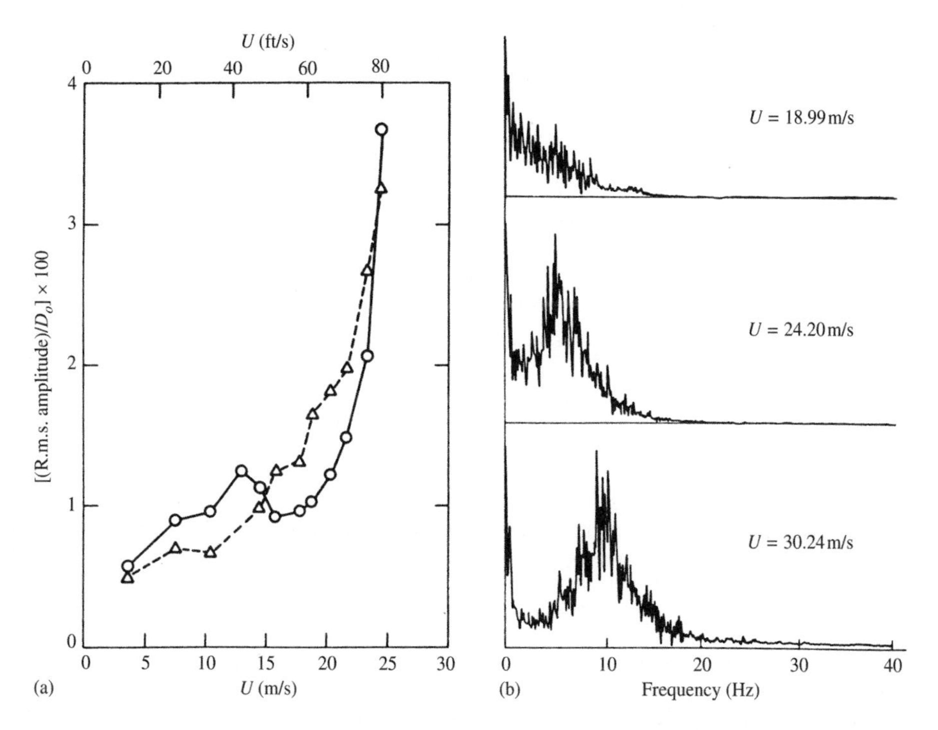

Figure 3.54 (a) The r.m.s. vibration amplitude of the pipe free end in two mutually perpendicular directions (○ and △) versus flow velocity for a vertical polyethylene pipe conveying water (Test 1.1); (b) PSDs from another pipe (Test 1.2) at three different flow velocities (Jendrzejczyk & Chen 1985).

†In this, A was measured at an unspecified point $x < L$; hence the apparent discrepancy between $A \simeq 0.2D_o$ here and the statement in the previous paragraph, pertaining to $x = L$, that $A \simeq 0.2L$ (Chen 1995).

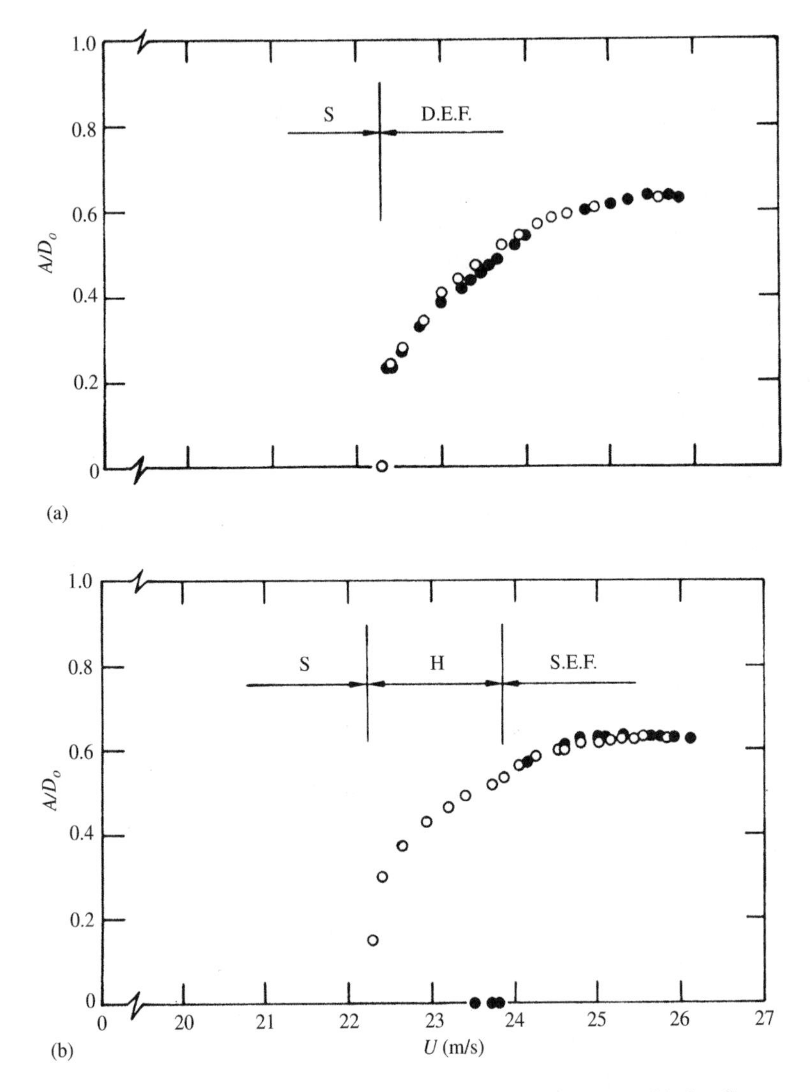

Figure 3.55 Limit-cycle amplitudes of a cantilevered pipe versus U: (a) for flutter excited by an external disturbance, (b) for spontaneously excited flutter. Region S corresponds to the stable region: region D.E.F. to disturbance-excited flutter; Region S.E.F. to spontaneously excited flutter; region H corresponds to the hysteresis region as U is decreased; •, increasing flow; ○, decreasing flow (Chen & Jendrzejczyk 1985).

no difference between amplitudes for increasing and decreasing U). For spontaneously excited flutter, however, $U_{cf} \simeq 23.8$ and $A/D_o \simeq 0.6$ at the onset; furthermore, there is a great deal of hysteresis. These results, taken together, suggest that the Hopf bifurcation in this case is subcritical [Figure 2.11(d)]; this gives quantitative substance to the earlier observations made by Païdoussis and discussed in conjunction with Figures 3.46 and 3.47.

The overall assessment is that the main features of the linear, free dynamics of this system have adequately been confirmed by experiment; this successful testing of theory is reinforced by some of the experiments described in Section 3.6.

3.5.7 The effect of an elastic foundation

Interest in the subject arises, in part, because of Smith & Herrmann's (1972) unexpected finding that for a cantilevered beam with a follower load the critical load (for coupled-mode flutter) is independent of the foundation modulus. This corresponds to the pipe system with $\beta = 0$.

For the pipe conveying fluid ($\beta \neq 0$), however, the effect of an elastic foundation is stabilizing, as shown by Lottati & Kornecki (1986), Figure 3.56. Thus, like gravity, the foundation provides an additional restoring force, which stabilizes the system.[†] The effect of foundation damping may be assessed from Figure 3.36, where the viscous damping may be considered to be associated with the foundation; thus, for high enough β, foundation damping is expected to be destabilizing.

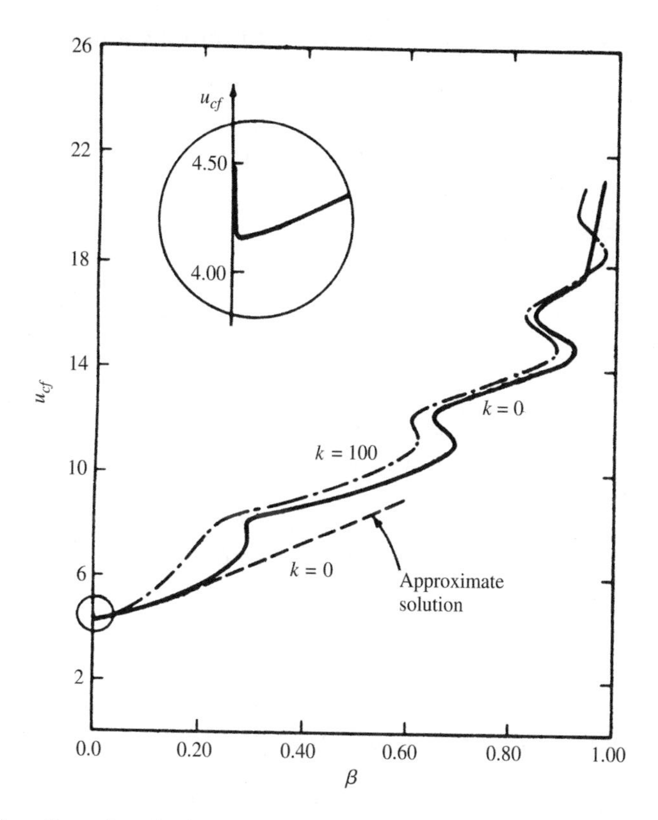

Figure 3.56 The effect of an elastic foundation with $k = 100$ on u_{cf} for the undamped cantilevered pipe (Lottati & Kornecki 1986).

Becker *et al*. (1978) studied the effect of a so-called Pasternak-type rotary foundation, in which the additional term $-c(\partial^2 w/\partial x^2)$ appears in the equation of motion, where c is the modulus of the rotary foundation — or the stiffness of distributed rotational springs

---

[†]Becker *et al*. (1978) obtain some results in which increasing k from zero to 10 is stabilizing, while further increasing it to 50 is slightly destabilizing, by less than 0.5%. However, these results are for $\beta = 10^{-3}$ and may be peculiar because of that — see discussion at the end of Section 3.5.3.

along the length of the pipe; a corresponding term appears in the shear boundary condition at the clamped end. The new term in the equation of motion opposes the centrifugal term, $U^2(\partial^2 w/\partial x^2)$, and this generates a strong stabilizing effect. Adding foundation damping in this case is destabilizing, although these results were confined to $\beta = 10^{-3}$ only, and hence should not be considered as general.

All the foregoing apply to uniform foundations. Unusual behaviour may be expected for nonuniform ones, however, in view of Hauger & Vetter's (1976) results for the system with the follower load (pipe with $\beta = 0$) and $k = k(\xi)$. Thus, if $k(\xi)$ is zero at $\xi = 0$ and 1 and triangularly distributed in-between, with a maximum at $\xi = \xi_m$ of $k_{\max}$, then the effect is destabilizing for all $k_{\max}$. The opposite is true if $k = 0$ at $\xi = \xi_m$, and $k_{\max}$ at $\xi = 0$ and 1.

3.5.8 Effects of tension and refined fluid mechanics modelling

The system shown in Figure 3.57(a) was studied by Guran & Plaut (1994), in which the compression P is conservative, i.e. it is constant and remains along the undeformed axis of the pipe. The equation of motion is equation (3.98) with $\overline{p} = 0$ and $\overline{T} = -P$, or $\Gamma = -\mathcal{P}$ in dimensionless form, where $\mathcal{P} = PL^2/EI$. The boundary conditions, in addition to $\eta(0, \tau) = 0$ and $(\partial^2\eta/\partial\xi^2)|_{\xi=1} = 0$ are

$$\frac{\partial^2\eta}{\partial\xi^2} - \kappa^* \frac{\partial\eta}{\partial\xi} = 0 \quad \text{at } \xi = 0 \quad \text{and} \quad \frac{\partial^3\eta}{\partial\xi^3} + \mathcal{P}\frac{\partial\eta}{\partial\xi} = 0 \quad \text{at } \xi = 1, \tag{3.110}$$

where $\kappa^* = CL/EI$. Clearly, in the limit of $\kappa^* = \infty$ the pipe becomes cantilevered.

Typical results for $\kappa^* = \infty$ are shown in Figure 3.57(b), and it is clear that they are similar to those of Figure 3.33. Indeed, the physical systems are similar: in the case of Figure 3.33 the pipe is subjected to conservative gravity-induced distributed compression, but in this case to conservative *uniform* compression along the length.

Results for $\kappa^* \neq \infty$ are similar. The influence of κ^* on stability is given in graphical form, for both divergence and flutter, in Guran & Plaut (1994). The condition for divergence is

$$u^2 + \mathcal{P} \cos v - (v\mathcal{P}/\kappa^*) \sin v = 0, \quad v = \sqrt{u^2 + \mathcal{P}}. \tag{3.111}$$

As suggested by Figure 3.57(b) and as may be verified numerically with equation (3.111), if the system is subjected to conservative *tension* it cannot lose stability by divergence, but only by flutter.

The effect of the small tension induced by the presence of a terminal nozzle on a cantilevered pipe — refer to the discussion associated with equations (3.40)-(3.42) — was taken into account by Bishop & Fawzy (1976), who found

$$T_L = \tfrac{1}{2}MU^2(\alpha_j - 1)^2 \tag{3.112}$$

by taking a force balance across the nozzle, where $\alpha_j = A/A_j$ is the ratio of pipe flow area to nozzle terminal flow area. This tension was neglected in the theoretical calculations to which the experiments of Gregory & Païdoussis (1966b) were compared. As seen in (3.112) this is not necessarily negligible for α_j substantially different from 1; it is properly taken into account in the comparison in Figure 3.53. A more refined treatment,

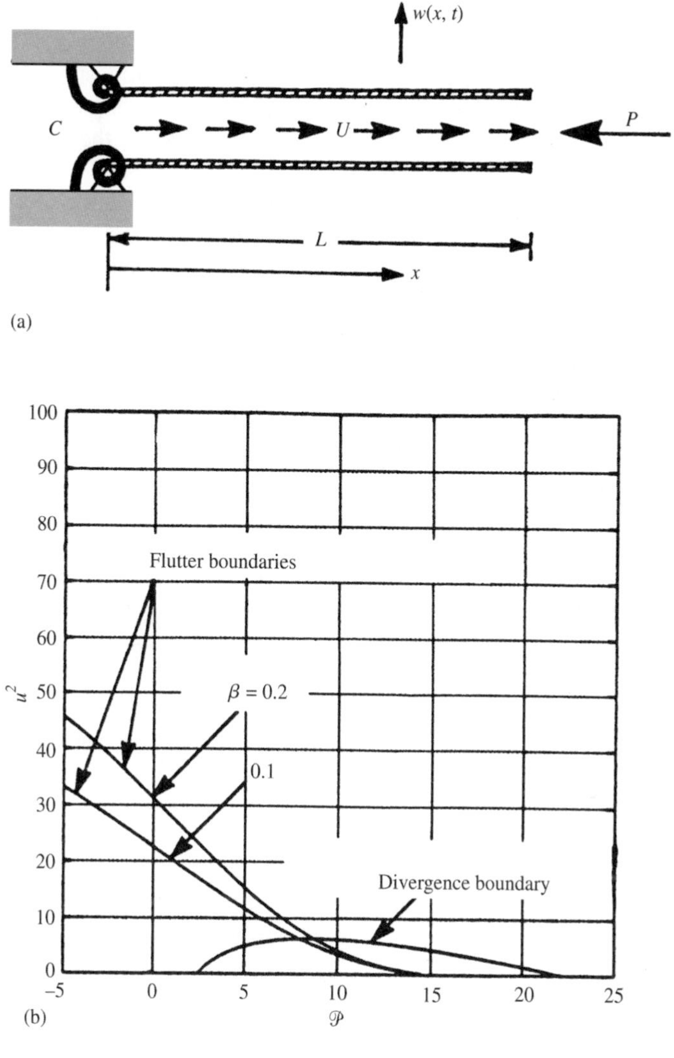

Figure 3.57 (a) Schematic of a pipe supported by a rotational spring of stiffness C at one end and free at the other, conveying fluid and simultaneously subjected to a conservative compression force, P; (b) stability diagram for the case $C = \infty$ ($\kappa^* = \infty$) (Guran & Plaut 1994).

taking into account the *vena contracta* that may arise beyond the free end, as well as frictional effects in the nozzle and air resistance, is given by Ilgamov *et al*. (1994).

In all of the foregoing it has tacitly been assumed that the jet issuing from the free end does not play any part in the dynamics of the system. This, despite the fact that the inverse is obviously untrue: as seen in some of the photographs of Figure 3.45(b) and in Figure 3.48, the jet continues the sinuous motion of the pipe well downstream of the free end before it breaks up. However, it may easily be confirmed experimentally that gross static or dynamic disturbances to the jet by the insertion of obstacles relatively close to the free end do not appear to have any significant effect on the dynamics of the pipe. It is partly thanks to this observation that it has implicitly been accepted that there exists a 'point relationship' between the fluid-dynamic force acting on the pipe at a particular point and the deflection at that point, resulting in equations (3.32) and (3.28) for instance. This makes the analysis much easier than if an 'integral relationship' were necessary, requiring, for instance, knowledge of the relationship between the unsteady pressure and streamwise position all along and beyond the end of the pipe for the force at any point x to be specified.

However, there is no guarantee in all of this that for sufficiently short pipes the jet behaviour beyond $x = L$ will not influence the dynamics of the pipe, or that this will be so in the case of shell motions. Furthermore, in some analyses, notably for short pipes and shell motions [Section 4.4 and Chapter 7], (i) three-dimensional potential flow theory is used for the formulation of the fluid-dynamic forces, which means that they are determined via integration of the unsteady pressure around the pipe circumference, and (ii) the generalized-fluid-dynamic-force Fourier-transform technique is employed which does require knowledge of the jet behaviour sufficiently far downstream of the free end — sufficiently far for the perturbation pressure to vanish; this, in effect, amounts to the specification of a downstream boundary condition for the fluid. As a result, a number of so-called *outflow models* have been proposed, starting with Shayo & Ellen (1978).

In most of these models (Shayo & Ellen 1978; Païdoussis *et al*. 1986, 1991b; Nguyen *et al*. 1993) the manner in which jet oscillations decay to zero is *prescribed*, based on more or less reasonable assumptions. A more physical approach, in which the dynamics of the free jet issuing from a vertical pipe with a terminal nozzle are coupled into the overall analysis, is adopted by Ilgamov *et al*. (1994).

Here some results obtained by Shayo & Ellen (1978) are presented, while other outflow models are discussed in Section 4.4 and Chapter 7. Shayo & Ellen proposed two such models. In the first, the so-called 'collector pipe model', it is supposed that there exists a collector pipe which is actuated by a sensor, so that its deflection matches that of the pipe outlet without touching it; the collector swallows up the fluid and discharges it at its other end, which is anchored on the undeformed x-axis, some distance downstream. In the analysis, the following extension to the cantilevered beam eigenfunctions $\phi_j(\xi)$, utilized as comparison functions, is introduced to describe the behaviour of the fluid for $\xi > 1$:

$$g_j(\xi) = \begin{cases} \phi_j(1)(l - \xi)/(l - 1), & 1 < \xi \leq l \\ 0, & \xi > l, \end{cases} \tag{3.113}$$

where l is chosen sufficiently large in the numerical calculations such that changes in its value have no effect on the fluid forces calculated. Thus, in this model it is presumed that the deflection dies out linearly to zero in a dimensional distance $(l - 1)L$, L being the length of the pipe. In the second, so-called 'free-flow model', it is supposed that the sinuous deflections persist in the fluid beyond $\xi = 1$, such that

$$g_j(\xi) = \phi_j(1) \exp\left[\mathrm{i}\omega L(\xi - 1)/U\right]. \tag{3.114}$$

Shayo & Ellen were concerned mostly with shell oscillations, but they also conducted calculations for beam-mode instabilities, albeit via the more complex three-dimensional potential flow theory (see Section 4.4.3 and Chapter 7) and shell theory for the pipe, instead of the simpler plug-flow Euler–Bernoulli beam theory. However, as shown by Païdoussis (1975) and discussed in Chapter 7, the results of the two theories converge for thin-walled slender pipes. Here, some of Shayo & Ellen's results are presented in Table 3.6, in terms of $\overline{U} = U/\{E/\rho_s(1 - \nu^2)\}^{1/2}$ for the given h/a and $\overline{\mu} = \rho a/\rho_s h$, where

Table 3.6 Values of the dimensionless critical flow velocity for flutter, $\overline{U}_{cf}$, for various L/a and $h/a = 0.0227$, $\overline{\mu} = 0.06$ and $\nu = 0.5$ (Shayo & Ellen 1978).

L/a	'Collector pipe' model	'Free-flow' model	'Long pipe' model
5	1.70	1.66	1.40
10	1.23	1.25	1.20
15	0.94	0.96	0.93
20	0.75	0.76	0.74

h is the wall thickness, a the internal radius, ρ_s the pipe wall density, ν the Poisson ratio, and the other symbols as before. These parameters are more appropriate for the analysis of shells than, say, β and u as used in the foregoing. The results for $\overline{U}_{cf}$ obtained with these two outflow models are compared with those of the 'long pipe model', in which the behaviour of the flow beyond $\xi = 1$ is ignored and the 'point relationship' between force and displacement [equation (3.28)] is utilized, as in most of the foregoing. It is seen that the results for length-to-radius ratio $L/a > 10$ are sensibly the same. Hence it must be concluded that, unless the pipe is very short, the use of a refined 3-D fluid dynamic model for the unsteady flow in the pipe, coupled with an outflow model, is not warranted. On the other hand, for very short pipes, $L/a \sim \mathbb{O}(5)$, the Euler–Bernoulli theory ceases being applicable and Timoshenko beam theory should be used instead. For this reason further discussion is deferred to Section 4.4.

3.6 SYSTEMS WITH ADDED SPRINGS, SUPPORTS, MASSES AND OTHER MODIFICATIONS

There has been a truly amazing array of studies of modified forms of the basic system discussed so far: e.g. cantilevers with one or more added masses at different locations, with intermediate supports, with different types of spring supports added at various locations, and so on. Some of these studies have been motivated by the interesting results obtained in similarly modified structural systems, notably columns subjected to follower loads; some by similarity to real physical systems; most, however, by pure curiosity: by the desire to know what the dynamical behaviour might be if this or that modification were introduced.

Since the analysis and dynamics of the basic systems have been discussed thoroughly in the foregoing, the treatment here will be more compact, concentrating on the differences *vis-à-vis* what has been described in Sections 3.2–3.5.

3.6.1 Pipes supported at $\xi = l/L < 1$

The system consists of a cantilevered pipe with an intermediate simple support, i.e. a support at $\xi = \xi_s = l/L < 1$, where $\xi = x/L$ and L is the overall pipe length, as shown in Figure 3.58(a). One would expect, therefore, the system to behave like a simple cantilevered pipe conveying fluid if l/L is sufficiently small, and like one with the two ends supported as $l/L \to 1$. This problem has been thoroughly studied, theoretically and experimentally, by Chen & Jendrzejczyk (1985), Edelstein & Chen (1985) and Jendrzejczyk & Chen (1985).

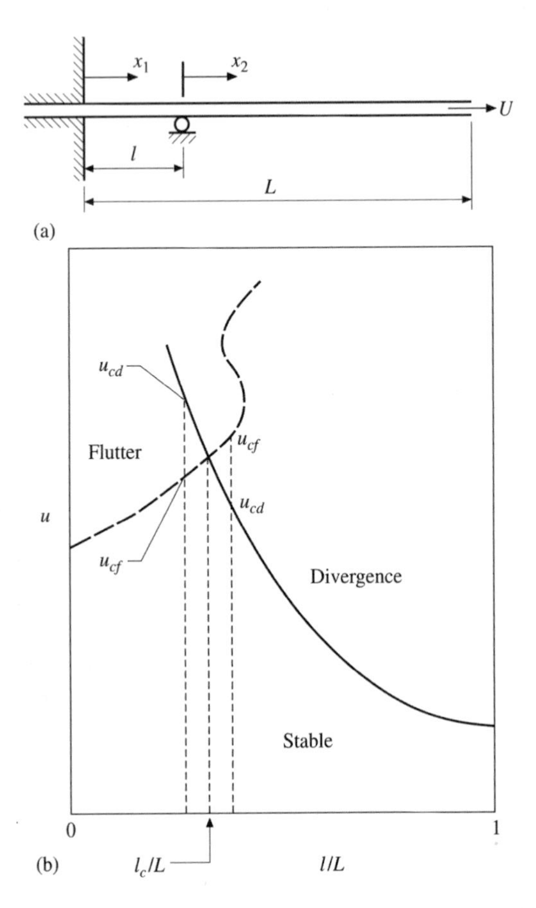

Figure 3.58 (a) Schematic of a pipe fixed at the upstream end and with a simple support at $\xi_s = l/L$; (b) qualitative stability diagram, showing the dimensionless flow velocities u versus l/L and the definition of l_c/L; u_{cd} and u_{cf} are the dimensionless critical flow velocities for divergence and flutter, respectively (Chen & Jendrzejczyk 1985).

The domain of the problem $\mathfrak{D} = [0, 1]$ is broken into two, $\xi_1 = [0, \xi_s]$ and $\xi_2 = [0, 1 - \xi_s]$ with $\xi_s = l/L$, wherein the dimensionless displacements of the pipe are η_1 and η_2, respectively. The equations of motion are then given by

$$\frac{\partial^4 \eta_i}{\partial \xi_i^4} + u^2 \frac{\partial^2 \eta_i}{\partial \xi_i^2} + 2\,\beta^{1/2} u \frac{\partial^2 \eta_i}{\partial \xi_i \partial \tau} + \frac{\partial^2 \eta_i}{\partial \tau^2} = 0, \qquad i = 1, 2, \tag{3.115}$$

cf. equation (3.1); the dimensionless quantities are the same as before, based on the overall length L. The boundary conditions are

$$\begin{aligned} \eta_1(0, \tau) &= \frac{\partial \eta_1}{\partial \xi_1}(0, \tau) = \eta_1(\xi_s, \tau) = 0, \\ \eta_2(0, \tau) &= \frac{\partial^2 \eta_2}{\partial \xi_2^2}(1 - \xi_s, \tau) = \frac{\partial^3 \eta_2}{\partial \xi_2^3}(1 - \xi_s, \tau) = 0. \end{aligned} \tag{3.116}$$

The system is completed by the compatibility conditions at $\xi_1 = \xi_s$ (or $\xi_2 = 0$), imposing the continuity of slope and bending moment at the pinned support:

$$\frac{\partial \eta_1}{\partial \xi_1}(\xi_s, \tau) = \frac{\partial \eta_2}{\partial \xi_2}(0, \tau), \qquad \frac{\partial^2 \eta_1}{\partial \xi_1^2}(\xi_s, \tau) = \frac{\partial^2 \eta_2}{\partial \xi_2^2}(0, \tau). \tag{3.117}$$

Solutions are obtainable via an obvious extension of the method of Section 3.3.6(a), eventually leading to an 8×8 determinant, in place of (3.84), which now is a function of ξ_s also (Chen & Jendrzejczyk 1985).

The qualitative dynamics of the system is illustrated in Figure 3.58(b). For $l/L < l_c/L$, l_c being a critical value depending on β, the system loses stability by flutter at a progressively higher flow velocity as l/L is increased, as compared to $l/L = 0$ which corresponds to the basic cantilevered system; theoretically at least, the system is also subject to divergence at higher flow velocity. For $l/L > l_c/L$, the system loses stability

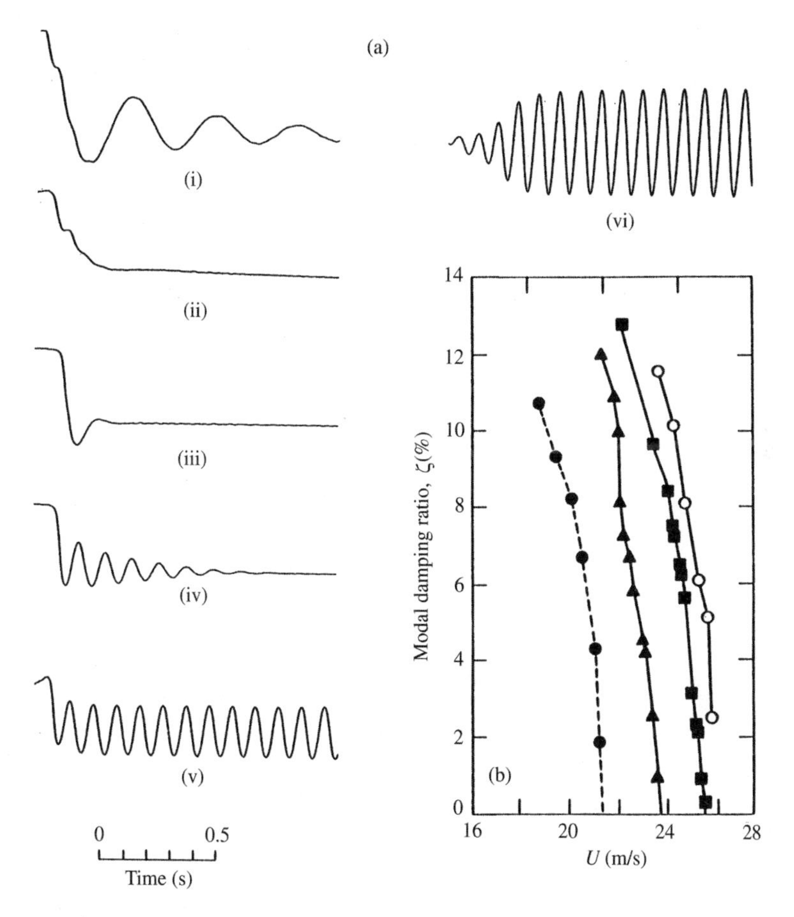

Figure 3.59 (a) Time histories of oscillation of a cantilevered pipe ($\beta = 0.48$) with an additional simple support at $l/L = 0.25$, at various flow velocities: (i) 0 m/s; (ii) 6.6 m/s; (iii) 19.0 m/s; (iv) 24.2 m/s; (v) 25.2 m/s; (vi) 26.5 m/s (Chen & Jendrzejczyk 1985). (b) The precipitously decreasing modal damping, ζ, towards zero as U_{cf} is approached for a similarly supported pipe ($\beta = 0.45$) and different values of l/L: •, $l/L = 0$; ▲, $l/L = 0.120$; ■, $l/L = 0.194$; ○, $l/L = 0.266$ (Jendrzejczyk & Chen 1985).

by divergence, much as a clamped–pinned system would, but is also subject to flutter at higher flow (generally single-mode flutter, not as the conservative system would). Finally, for $l/L = l_c/L$ the two critical flow velocities become coincident and Chen & Jendrzejczyk conjecture that this may lead to chaos (see Chapter 5).

The experiments were conducted with polyethylene pipes ($D_o = 9.5$ and 12.7 mm, wall thickness $h = 1.59$ mm, $L = 685.8$ mm) with a ring-type knife edge support at varying values of l/L. The corresponding values of β were 0.48 and 0.60 approximately, while $\gamma \simeq 2$ was sufficiently small for gravity effects to be neglected.

A great deal of high-quality data was obtained. Examples are shown in Figures 3.59 and 3.60. Some sample time traces for a pipe with $l/L = 0.25$ ($l_c/L \simeq 0.35$ in this case) are shown in Figure 3.59(a) and display dynamical behaviour similar to that of a simple cantilevered pipe as U is increased: (i) underdamped, (ii) and (iii) overdamped, (iv) again underdamped, (v) limit-cycle oscillation and (vi) larger amplitude limit-cycle oscillation. The oscillation in (i)–(v) was excited by perturbing the pipe, whereas in (vi) it developed spontaneously. Measurements of the modal damping ratio on a nominally identical pipe (but with $\beta = 0.45$) for varying l/L, shown in Figure 3.59(b), document its precipitous reduction as U_{cf} is approached.

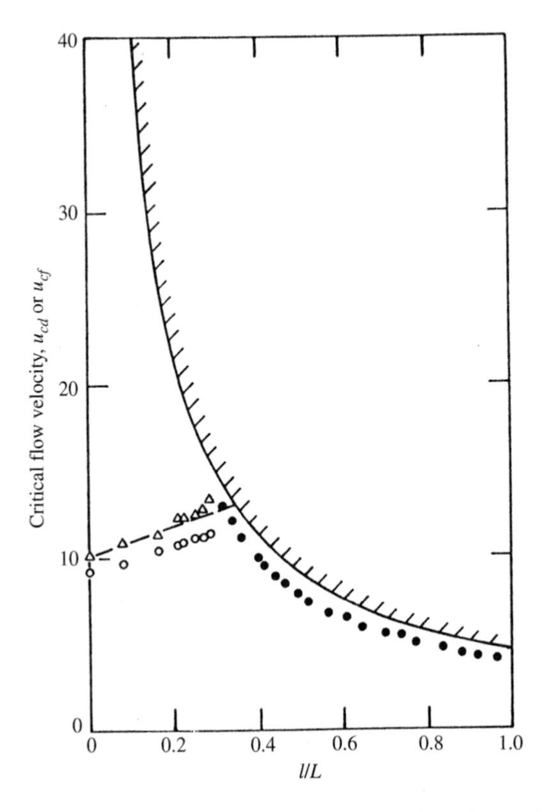

Figure 3.60 The critical flow velocities, u_{cd} or u_{cf}, for a pipe clamped at $\xi = 0$ and simply supported at $\xi = l/L$. Theoretical boundaries: ⊥⊥⊥⊥ for divergence; — — for flutter. Experimental data: •, divergence; ○, flutter induced by external disturbance; △, spontaneous flutter (Chen & Jendrzejczyk 1985).

Finally, l/L was varied systematically and the critical flow velocities for flutter or divergence was obtained and plotted versus l/L, as illustrated in Figure 3.60, where they are compared with theoretical results (apparently with dissipative forces ignored). It is seen that theory and experiment are in excellent agreement.

It is of interest that if the system lost stability by divergence, then, provided l/L was close to l_c/L [Figure 3.58(b)], flutter about the buckled state was observed to occur. On the other hand, if stability was lost by flutter, limit-cycle oscillation persisted at higher flows, 'and the tube does not buckle'; but it is not clear whether any asymmetry in the motion takes place which might be taken as evidence of a coexisting divergence.

3.6.2 Cantilevered pipes with additional spring supports

As we have seen in the foregoing, cantilevered pipes lose stability by flutter, whereas pipes supported at both ends do so by divergence. It was of interest, therefore, to study 'intermediate' support conditions, as initially done by Chen (1971a) [and later, apparently independently, by Becker (1979)], who examined the dynamics of the system of Figure 3.61(a). Physically, one would expect that for a very weak spring-constant K, the system would behave essentially as a cantilevered pipe; for sufficiently large K, however, the system would approach a clamped–pinned one. This, in fact, is what is obtained.

The dynamics of the system (neglecting gravity, dissipative effects, etc.) is governed by equation (3.1), or in dimensionless form by (3.76), and the same boundary conditions, except that the fourth, related to the shear at the downstream end, $EI(\partial^3 w/\partial x^3) = 0$, is

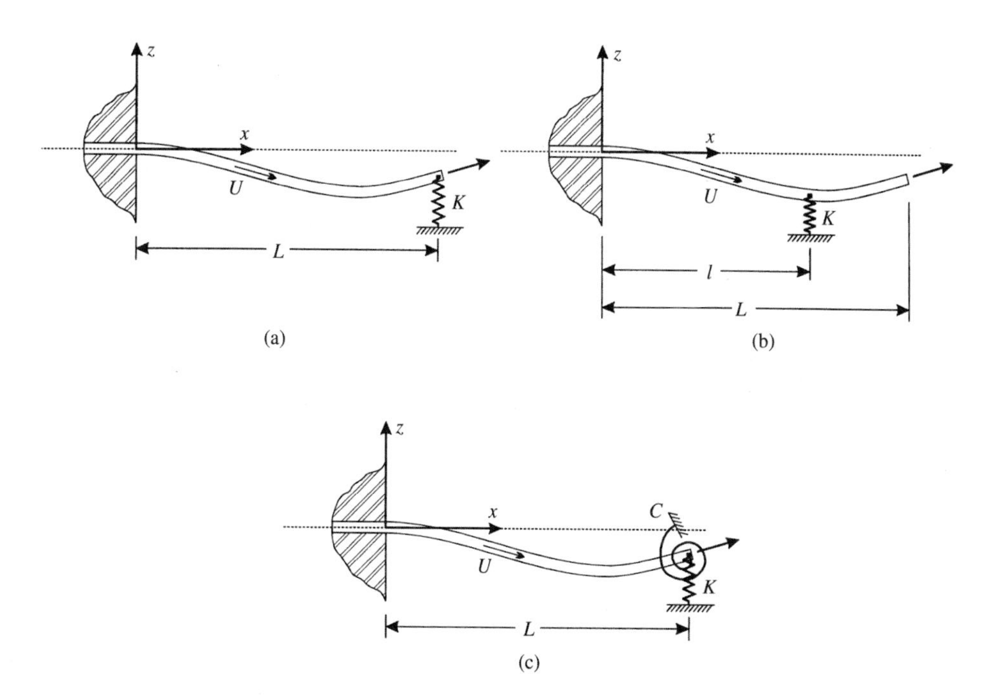

Figure 3.61 Various types of additional spring supports for cantilevered pipes conveying fluid: (a) translational spring at the downstream end, $x = L$; (b) translational spring at $x = l < L$; (c) translational and rotational springs at $x = L$.

now replaced by $EI(\partial^3 w/\partial x^3) - Kw = 0$, or in dimensionless form

$$\frac{\partial^3 \eta}{\partial \xi^3} - \kappa\eta = 0, \qquad \kappa = \frac{KL^3}{EI}. \tag{3.118}$$

Obviously, the method of solution of Section 3.3.6(a) may be utilized, except that the last line of determinant (3.84) is now replaced by $(\alpha_j^3 - \mathrm{i}\kappa)\exp(\mathrm{i}\alpha_j)$, $j = 1 - 4$. Moreover, working in a similar way as in Section 3.4.1, it is easy to find (Chen 1971a) that the condition for divergence, $u = u_{cd}$, is given by solutions of

$$u^3 + \kappa(\sin u - u\cos u) = 0. \tag{3.119}$$

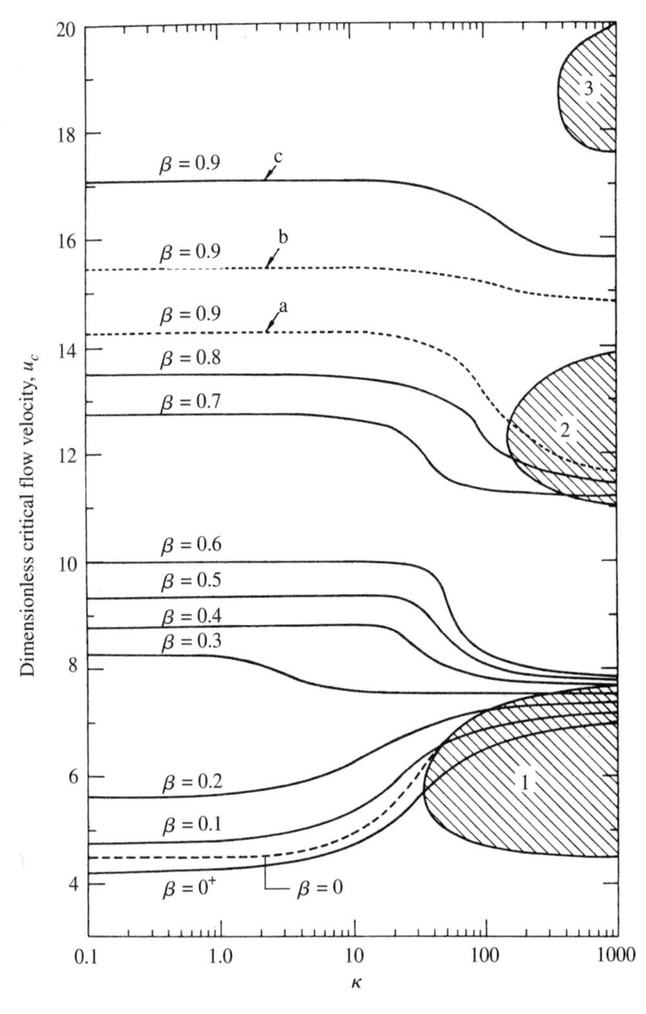

Figure 3.62 The dimensionless critical flow velocities of a cantilevered pipe with a spring support at $\xi = 1$ versus the dimensionless spring constant κ: ——, – – –, flutter boundaries; shaded areas are zones of divergence (Chen 1971a).

A typical Argand diagram is given by Chen (1971a) for $\beta = 0.6$, $\kappa = 100$. In this case the system loses stability by divergence at $u \simeq 4.7$, is restabilized at $u \simeq 7.2$, and then loses stability by single-mode flutter at $u \simeq 8.3$ — all in the first mode, but at $u \simeq 17.7$ flutter also occurs in the second mode. Thus, this system shares the characteristics of a cantilevered and a clamped–pinned pipe conveying fluid, with those of the latter being dominant. For smaller values of κ (e.g. $\kappa = 10$) the system behaves as a cantilever, and the only possible form of instability is flutter.

Figure 3.62 is the stability diagram in terms of the spring stiffness parameter κ. Several interesting observations may be made: (i) there is a critical value of κ, $\kappa_c = 34.81$, below which only flutter is possible; (ii) for sufficiently high κ, there is more than one divergence region, although the higher ones are of limited physical significance; (iii) for sufficiently high κ (say $\kappa > 200$), the values of u_{cf} (critical flutter velocities) become significantly less dependent on β than is the case for low κ (say $\kappa < 30$), as if the system tries to behave like a conservative one, but still loses stability by flutter: e.g. for $\beta = 0.4$, 0.5 and 0.6, and following the second S-shaped curve in the u_{cf} versus β curve (see Figure 3.30) for $\beta = 0.7$, 0.8 and 0.9; (iv) the three curves shown for $\beta = 0.9$ (two of which are dashed) correspond to loss, recovery and second loss of stability associated with the equivalent of the third of the S-shaped curves (Figure 3.30).

Another interesting result is shown in Figure 3.63. It is seen that the first S-shaped curve in the stability diagram marks a point of transition for the effect of κ on u_{cf}. Thus, for $\beta < 0.3$ approximately, κ stabilizes the system *vis-à-vis* $\kappa = 0$; for $\beta > 0.3$, however,

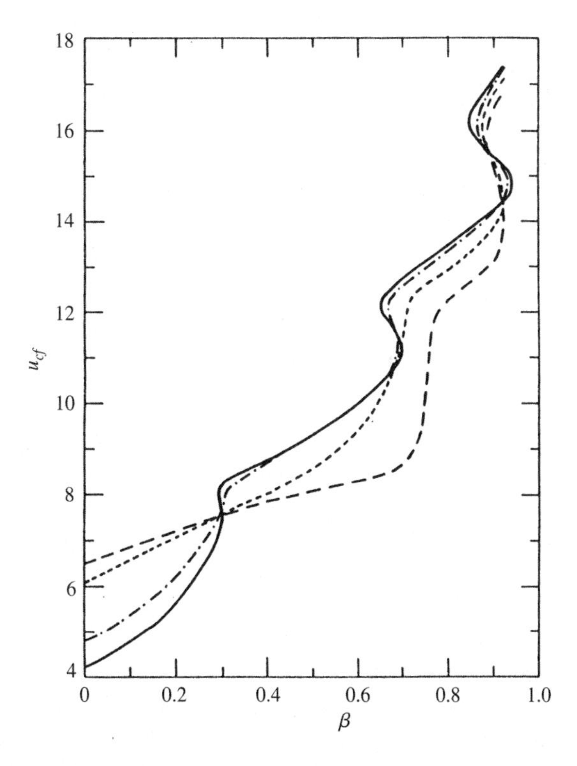

Figure 3.63 The dependence of u_{cf} on β for the system of Figure 3.61(a) with various values of κ: ——, $\kappa = 0$; - · - · -, $\kappa = 10$; - - -, $\kappa = 50$; — — —, $\kappa = 100$ (Chen 1971a).

κ *destabilizes* the system. This is the first of many unusual occurrences associated with these S-shaped curves, as we shall see.

Experimental verification of some of the foregoing is provided by Sugiyama *et al.* (1985a), whose work is described next. Figure 3.64 shows the theoretical and experimental critical flow velocities for three pipes (nominally with $\beta = 0.25$, 0.50 and 0.75) with varying κ. It is seen that agreement is reasonably good, in particular with regard to the critical value of κ at which transition from divergence to flutter occurs. It should be noted that, when comparing flutter velocities, the reader should consider only the curves for $\alpha = 0.02$, which corresponds to the average measured damping, whereas $\alpha = 10^{-3}$ represents some arbitrary minimal damping.

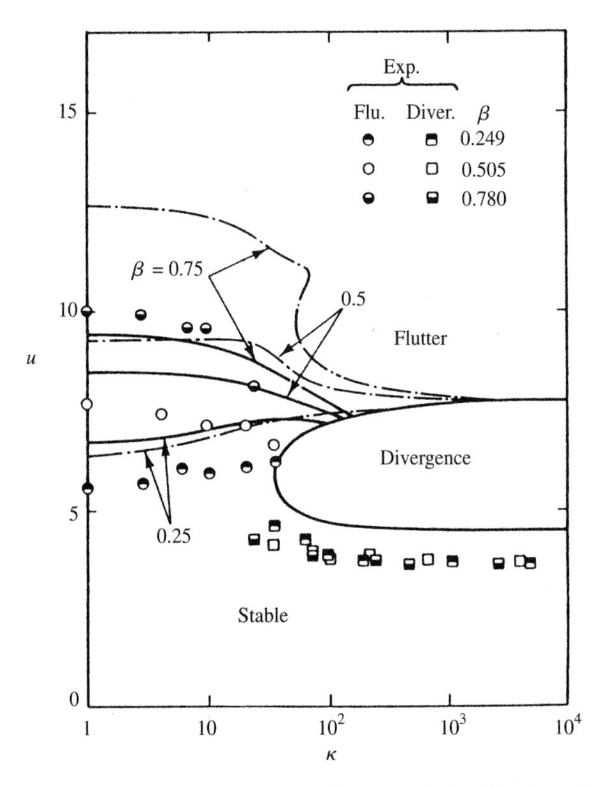

Figure 3.64 Comparison between theoretical stability boundaries (lines) and experimental points for flutter (circles) and divergence (squares) of a cantilevered pipe with an additional spring support at $\xi_s = 1$, as the spring stiffness κ is varied for the three values of β shown: $-\cdot-$, $\alpha = 0.001$; ——, $\alpha = 0.02$; the theoretical curves are for $\beta = 0.25$, 0.50 and 0.75, whereas the experimental values of β are as given in the legend (Sugiyama *et al.* 1985a).

Sugiyama *et al.* (1985a) examine the effect of an additional spring support at *any* location along the cantilever, as shown in Figure 3.61(b), both theoretically and experimentally. In this case the dimensionless equation of motion is modified by the addition of the term $\kappa\eta\delta(\xi - \xi_s)$, where δ is the Dirac delta function and $\xi_s = l/L$. Hence, the method of Section 3.3.6(b) may be utilized, with the cantilever beam eigenfunctions as comparison functions. It is found that as many as 14 such functions may be necessary to achieve convergence to three significant figures when $\xi_s = 1$, but that it is faster when there is some viscoelastic damping (even $\alpha = 10^{-3}$). This demonstrates the difficulty of the method in obtaining convergent results when the actual boundary conditions are different from those of the comparison functions; the use of the delta function to incorporate the spring forces into the equation of motion is a useful artifice, but it does have repercussions. However, with increasing dissipation (larger α), the higher mode content is damped out, and this is why convergence is easier to achieve.

In the Sugiyama *et al*. (1985a) experiments, elastomer pipes were cast by the authors in the manner described in Appendix C.[†] The pipes were supported by strings and oscillated in a horizontal plane as in Gregory & Païdoussis' experiments; water was used as the fluid. Experiments were conducted for $\xi_s = 0.25$, 0.50, 0.75 and 1.0 and many values of κ. Typical results are shown in Figures 3.64 and 3.65, wherein they are compared with theory. It should be stressed that experiments with springs are delicate, and hence the agreement achieved is quite reasonable. It is noted that in Figure 3.65 there are two flutter boundaries for $\beta = 0.50$ and 0.75 when $\alpha = 10^{-3}$; these correspond to the repeated loss of stability associated with S-shaped curves in the stability diagram, which do not exist at the higher value of dissipation (cf. Figure 3.35 for $\mu = 0.065$ and Figure 3.43).

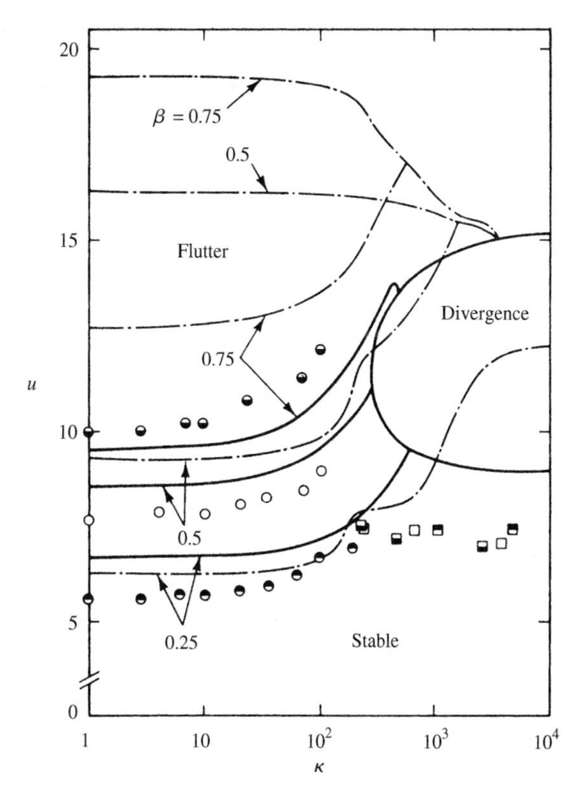

Figure 3.65 Comparison between theoretical and experimental stability thresholds of a cantilevered pipe with an additional spring support as in Figure 3.64, but now located at $\xi_s = 0.5$ (Sugiyama *et al*. 1985a).

[†]Three pipes were cast with slightly different $D_o \sim \mathcal{O}(10\,mm)$, $L \simeq 0.5$ m, yielding the values of β in Figure 3.64, slightly different from the nominal ones of $\beta = 0.25$, 0.5 and 0.75.

Two interesting features of the theoretical results are that: (i) the effect of κ is not dramatic until the critical value of κ for which divergence becomes possible is approached ($\kappa_c = 35$ for $\xi_s = 1$; $\kappa_c = 280$ for $\xi_s = 0.5$); (ii) as κ_c is approached, u_{cf} can be decreased with increasing κ if $\xi_s = 1$, which is surprising, but generally increases for $\xi_s = 0.5$. Furthermore, in the experiments for $\kappa \simeq \kappa_c$ it was found that after the onset of divergence, if the flow was increased slightly and the pipe was straightened by hand, it would remain straight upon release, so that the theoretical restabilization was actually observed; at higher flow, again as predicted, stability was lost once more by flutter.

This work has been extended to the case of several spring supports by Sugiyama *et al*. (1991).

Finally, Lin & Chen (1976) and Noah & Hopkins (1980) consider the case where the downstream end is supported simultaneously by a translational and a rotational spring [Figure 3.61(c)]. In this case the two boundary conditions at $\xi = 1$ are

$$\frac{\partial^2\eta}{\partial\xi^2} + \kappa^*\frac{\partial\eta}{\partial\xi} = 0, \qquad \frac{\partial^3\eta}{\partial\xi^3} - \kappa\eta = 0, \qquad \kappa^* = \frac{CL}{EI}, \qquad \kappa = \frac{KL^3}{EI}, \tag{3.120}$$

where C is the rotational and K the translational spring stiffness. The solution of the equation of motion, equation (3.76), subject to the boundary conditions $\eta = \partial\eta/\partial\xi = 0$ at $\xi = 0$ and equations (3.120), was obtained by Galerkin's method. The comparison functions, however, are obtained by solving the beam equation subject to these boundary conditions, yielding the following eigenfunctions (Noah & Hopkins 1980):

$$\phi_j(\xi) = \cosh(\lambda_j\xi) - \cos(\lambda_j\xi) - \sigma_j(\sinh(\lambda_j\xi) - \sin(\lambda_j\xi)), \tag{3.121a}$$

with

$$\sigma_j = \frac{(\kappa^*/\lambda_j)(\sinh\lambda_j + \sin\lambda_j) + \cosh\lambda_j + \cos\lambda_j}{(\kappa^*/\lambda_j)(\cosh\lambda_j - \cos\lambda_j) + \sinh\lambda_j + \sin\lambda_j}, \tag{3.121b}$$

which were shown to be orthogonal. The eigenvalues λ_j are solutions of

$$\frac{\kappa}{\lambda_j^3}(\tan\lambda_j - \tanh\lambda_j) + \frac{\kappa^*}{\lambda_j}(\tan\lambda_j + \tanh\lambda_j) + \frac{\kappa\kappa^*}{\lambda_j^4}\left(\frac{1}{\cos\lambda_j\cosh\lambda_j} - 1\right) + \left(\frac{1}{\cos\lambda_j\cosh\lambda_j} + 1\right) = 0. \tag{3.122}$$

In this way convergence, as the number of comparison functions is increased, is quite rapid (see Section 2.1).

If only a rotational spring is present and it is sufficiently stiff, then the system approaches a clamped–sliding system† and loses stability by divergence. However, the nonconservativeness of the system generates unexpected results when both κ and κ^* are present. Consider the following set of results obtained by Noah & Hopkins (1980) for $\beta = 0.125$, $\alpha = 10^{-3}$. (i) With $\kappa = \kappa^* = 0$ the system loses stability by flutter, but with $\kappa^* = 10$ it does so by divergence. (ii) If $\kappa = 25$ and $\kappa^* = 10$, however, the system loses stability by flutter once again, the divergence not occurring at all (not even at higher flow velocities). Thus, the addition of a translational spring, instead of aiding in the

†Sliding in the *transverse* direction, corresponding to the standard sliding support condition for a beam, with boundary conditions $w'(L) = 0$, $EIw'''(L) = 0$ (Bishop & Johnson 1960); not to be confused to axial sliding at an otherwise clamped or pinned end, as discussed in the foregoing.

development of divergence, makes it impossible. Finally, (iii) if κ is increased, so that $\kappa = 40$ and $\kappa^* = 10$, then the system loses stability by divergence once more. Yet, (iv) if $\kappa = 100$, the behaviour with $\kappa^* = 0$ and 10 is qualitatively similar: the system loses stability by divergence and at higher flow by flutter in both cases. Hence, κ and κ^* do not act synergistically; the dynamics of the system is affected not only by the values of the individual spring constants, but also by their relative magnitudes. Equivalent dynamical behaviour is found in aeroelasticity (Dowell *et al*. 1995; Section 3.6).

The foregoing peculiar stability behaviour follows the same pattern as in Figure 3.66, obtained by Lin & Chen (1976) for $\beta = 0$ — thus for a column subjected to a follower

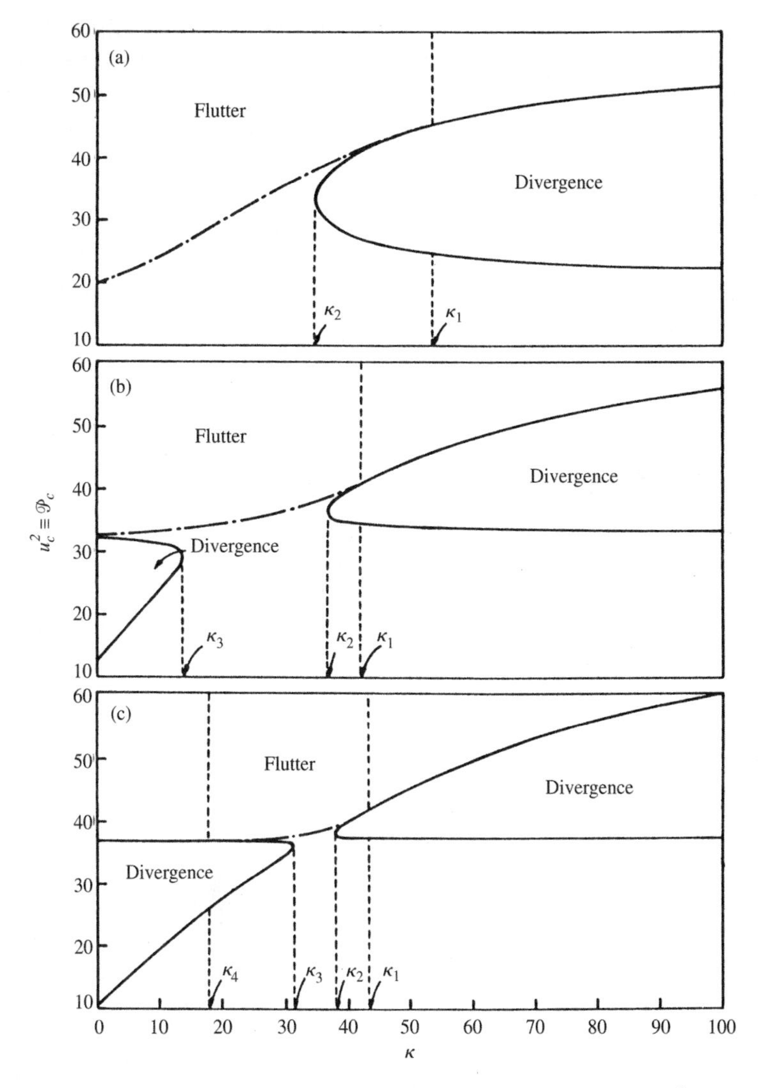

Figure 3.66 Stability of a cantilevered column subjected to a follower load $\mathcal{P}$ (or equivalently a pipe conveying fluid with $\beta = 0$, where $u^2 \equiv \mathcal{P}$), supported at the free end by a translational and a rotational spring of dimensionless stiffness κ and κ^*, respectively: (a) for $\kappa^* = 0$; (b) for $\kappa^* = 10$; (c) for $\kappa^* = 30$. At $\kappa = \kappa_1$ and κ_4 the divergence and flutter bounds coincide (Lin & Chen 1976).

load, rather than a pipe conveying fluid. It is seen that (i) for selected combinations of κ and κ^*, only flutter occurs; (ii) for a given $\kappa^* \neq 0$, if κ is small ($\kappa < \kappa_3$ in the figure), the system loses stability by divergence, and again if κ is relatively large ($\kappa > \kappa_2$). The physical mechanism must be that, for some combinations of (κ, κ^*), the eigenmodal shapes hinder the development of divergence, while being particularly propitious for flutter (Section 3.2.2), and so flutter develops rather than divergence.

3.6.3 Pipes with additional point masses

A very interesting study on the effect of lumped masses on the stability of cantilevered pipes conveying fluid has been made by Hill & Swanson (1970). The system is shown in Figure 3.67(a), and generally has several point masses at various locations, numbered as shown. In the equation of motion, equation (3.1), the term $m(\partial^2 w/\partial t^2)$ is now replaced by

$$\left[m + \sum_{j=1}^{J} m_j \delta(x - x_j)\right] \frac{\partial^2 w}{\partial t^2}. \tag{3.123}$$

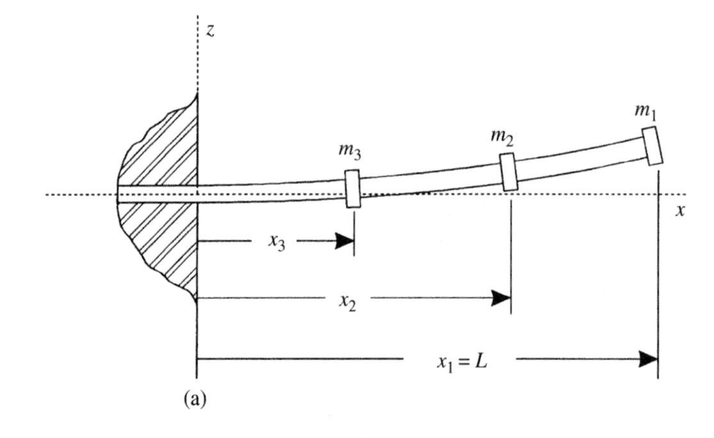

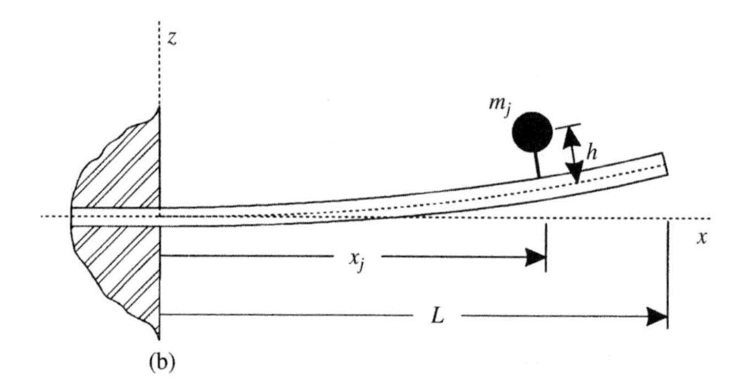

Figure 3.67 A cantilevered pipe conveying fluid (a) with three point masses added at x_1, x_2, x_3; (b) with one mass eccentrically located at x_j.

Thus, the method of Section 3.3.6(a) may be used to solve the problem. As shown in Section 2.1.4 and the discussion of Table 2.2, the method gives the correct results. In the dimensionless version of the equation of motion, the location and magnitude of the additional masses is expressed via

$$\xi_j = \frac{x_j}{L} \qquad \text{and} \qquad \mu_j = \frac{m_j}{(m+M)L}. \tag{3.124}$$

The theoretical results obtained are summarized in Figure 3.68, where they are compared to those of Gregory & Païdoussis (1966a) for a uniform pipe. It is seen at a glance that, in most cases, the additional masses *destabilize* the system. This is contrary to intuition, but nevertheless the effect is in the same sense as increasing the distributed mass m [i.e. decreasing β; recall that $\beta = M/(M+m)$]. However, on closer examination, a number of interesting and unusual features emerge, as follows.

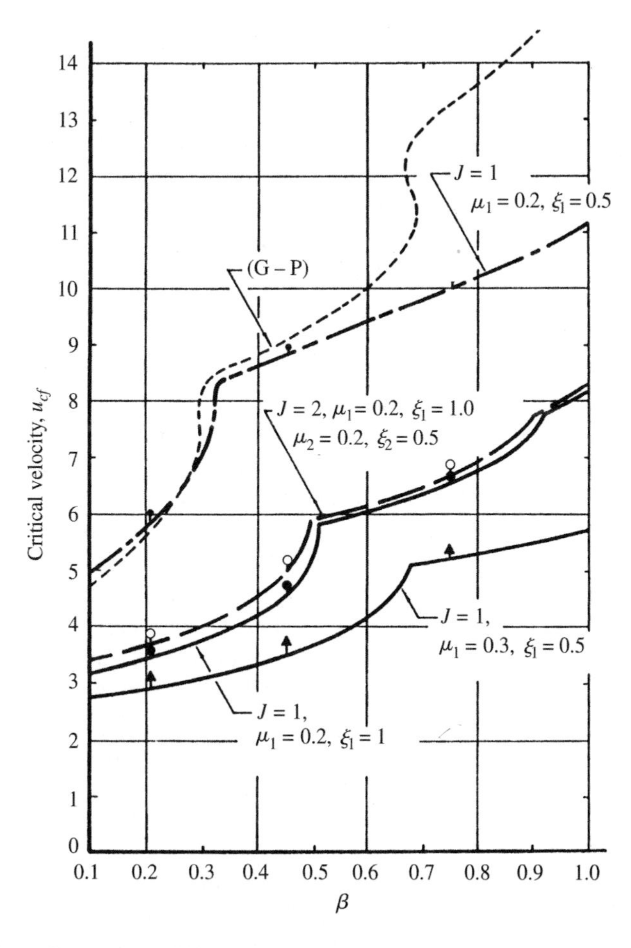

Figure 3.68 The effect of additional point masses, m_j, on the stability of a horizontal undamped cantilevered pipe conveying fluid, where J is the total number of the masses, $\mu_j = m_j/[(m+M)L]$, $\xi_j = x_j/L$. The stability curve marked as G-P represents Gregory & Païdoussis' results for $J = 0$, while the points represent experimental data (Hill & Swanson 1970).

First, consider the lowest three curves in the figure. Comparing the cases with a single mass at the downstream end ($J = 1$, $\xi_1 = 1$), it is seen that increasing the magnitude of the mass from $\mu_1 = 0.2$ to 0.3 destabilizes the system further, which agrees with the statement made just above. If, however, a second mass is added at mid-point ($J = 2$; $\mu_1 = 0.2$ at $\xi_1 = 1$; $\mu_2 = 0.2$ at $\xi_2 = 0.5$), the effect is to *stabilize* the system slightly, even though the combined additional mass is now higher than in either of the other two cases. On the other hand, if one starts with a mass at mid-point (top curve, other than G-P), then the addition of a second mass at the end (the next, lower curve) severely destabilizes the system.

Second, based on the foregoing, one might conclude that adding a mass at mid-point is always stabilizing. If, however, the mid-point mass is the only one added, as in the uppermost curve ($J = 1$; $\mu_1 = 0.2$ at $\xi_1 = 0.5$), the effect is stabilizing for $\beta \leq 0.27$ and destabilizing for larger β. This is another instance where the qualitative dynamical behaviour of the system is radically different on either side of the S-shaped bend in the stability curve — or, as is the case here, close to that bend (note that the transition at $\beta = 0.27$ is half-way between $\beta = 0.295$ for the system without an added mass and $\beta_{\text{equiv}} = M/[m + M + m_1/L] = 1/(1 + \mu_1) = 0.246$ for the system with one.)

Another point of interest in Figure 3.68 is the sharpness of the S-shaped bends, more like kinks here, in the lower stability curves. This is explained by Hill & Swanson as being due to sudden switches of the system from losing stability in one mode just below the β concerned, and in another mode just above it (and for a critical β two modes losing stability at the same u) — instead of the behaviour as in Figures 3.27 and 3.28 involving destabilization, stabilization and destabilization once more; thus, in this case there are real *discontinuities* in the values of ω_{cf}, as shown by Hill & Swanson, but not here for brevity.

Finally, the various data points (∘, •, etc.) correspond to experimental points obtained by Hill & Swanson, utilizing surgical rubber pipes conveying water in an apparatus similar to that of Gregory & Païdoussis (1966b). The agreement with theory is excellent, although if dissipation had been taken into account in the theory, it might have been less so.

Further studies on this problem have been made by Chen & Jendrzejczyk (1985) and Jendrzejczyk & Chen (1985) for a mass at the free end, and by Sugiyama *et al.* (1988a), who consider an additional mass together with a spring at some point $x < L$. Sugiyama *et al.* find that the u versus κ curve displays S-shaped discontinuities for selected combinations of κ and $\mu_1 = m_1/[(m + M)L]$. This means that there exists a region of restabilization between two critical values of u_{cf} for loss of stability (cf. Figure 3.28). It is of interest that the flutter mode is quite different at these two critical values, as shown in Figure 3.69: in both cases there are very strong travelling-wave components in the motion; however, in (a) the presence of the spring and mass at $\xi_1 = 0.25$ is hardly manifest, while in (b) there is a quasi-nodal point not far from ξ_1.

In another study, Silva (1979, 1981) examines the stability of pipes with attached valves — which can be quite massive relative to the pipe. Masses centred on the pipe or eccentric [overhanging, as shown in Figure 3.67(b)] are considered for both cantilevered and simply-supported pipes, but ignoring out-of-plane motions and possible coupling with torsional modes. As in the foregoing, the dynamical behaviour is affected by the value of μ_j (in this case $j = 1$ always) and also $\overline{h} = h/L$, where h is the distance of the point mass from the pipe centreline. As expected, the effect of the additional mass on the stability of simply-supported pipes is on coupled-mode flutter, rather than divergence which is a

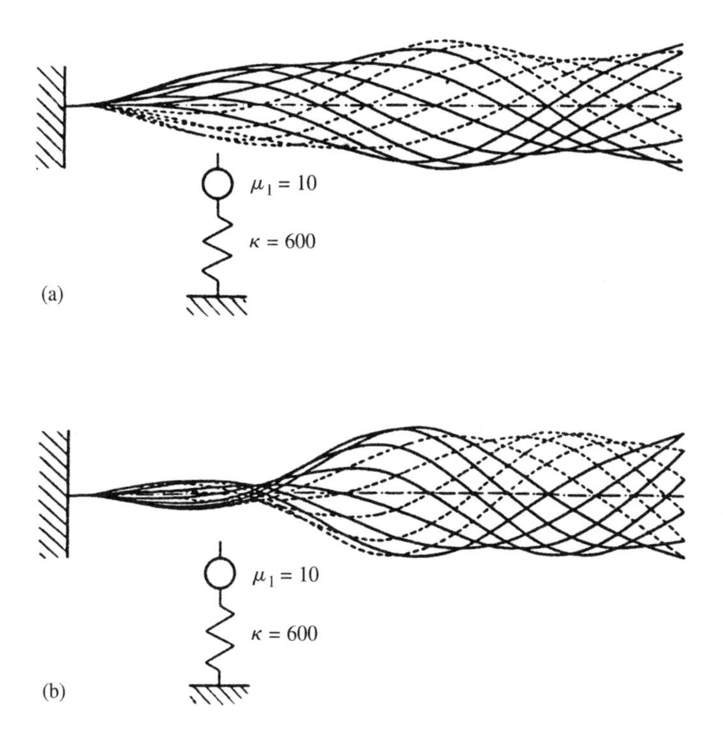

Figure 3.69 The flutter modes of a cantilevered pipe ($\beta = 0.50$, $\alpha = 0.02$) with an added mass and a translational spring at $\xi = 0.25$ ($\mu_1 = 10$, $\kappa = 600$): (a) at first loss of stability ($u_{cf} = 8.17$); (b) at the second loss of stability ($u_{cf} = 9.97$), after restabilization; obtained theoretically by Sugiyama *et al*. (1988a).

static phenomenon; it is found that this effect, both in terms of u_{cf} and the range of u over which coupled-mode flutter persists, can be affected a great deal, even if $\bar{h} = 0$. For cantilevered pipes it is found that eccentricity of an additional mass at the free end may further destabilize the system.

3.6.4 Pipes with additional dashpots

This problem has been studied theoretically and experimentally by Sugiyama *et al*. (1988b).[†] A dashpot is attached to a cantilevered pipe, located at some point $\xi_s = l/L \leq 1$, and sometimes also a mass, at the same point. The effect of the damper at the downstream end is generally destabilizing, with or without the mass, while at other locations it can sometimes be stabilizing (see Section 3.8.3), depending on β and μ_1 [equations (3.124)]. In some cases, multiple regions of flutter may exist.

The experiments were conducted in the same basic arrangement as for one or more additional springs, discussed in the foregoing. The dashpot was provided by attaching a thin flat plate to the pipe and immersing it in oil, with the motion parallel to the flat-plate surface. The experiments generally support theory quite well.

[†]The interested reader is also referred to Sugiyama *et al*. (1985b).

3.6.5 Fluid follower forces

This is a special class of problems involving beams subjected to fluid jets issuing tangentially from the beam, either within the span or at the free end. Hence, these are beams subjected to fluid follower loads, rather than pipes with flow all along.

Wiley & Furkert (1972) considered the system shown in Figure 3.70(a). The equations of motion may be written as follows:

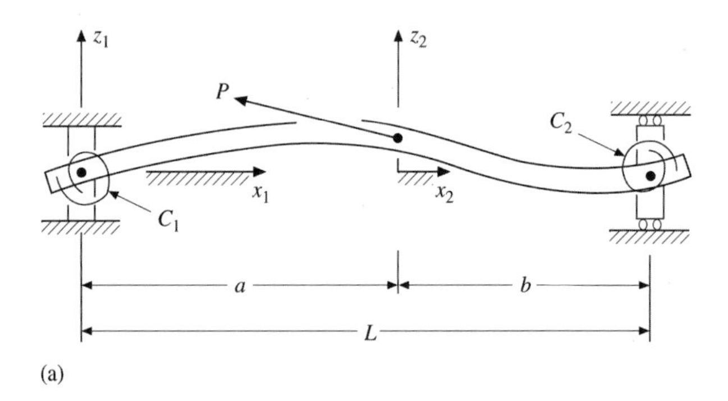

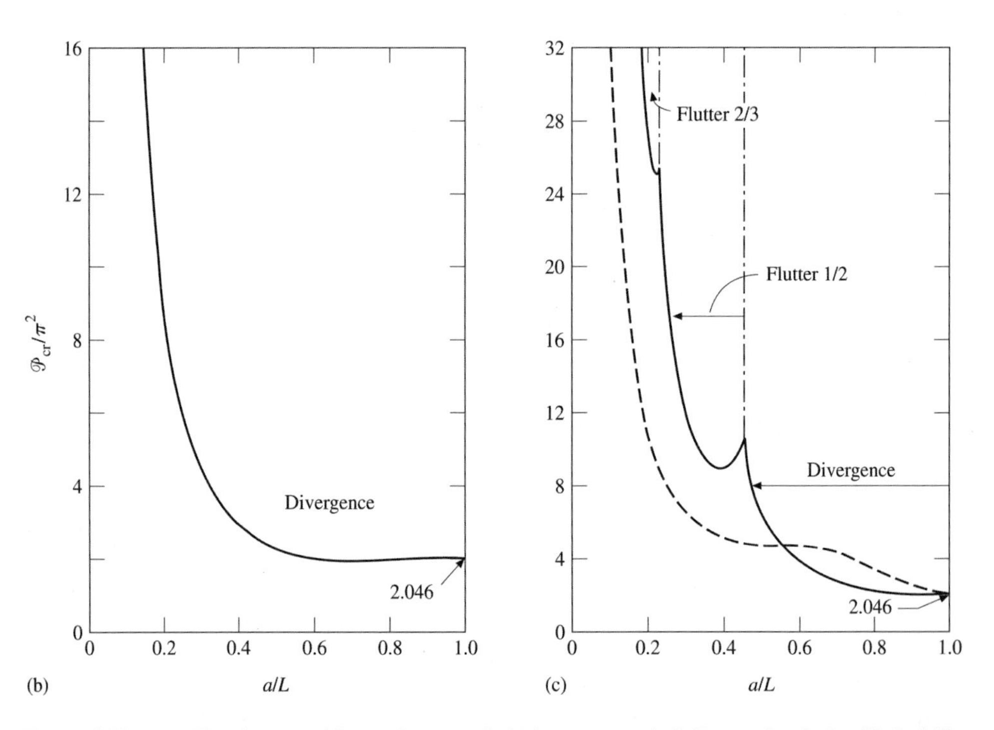

Figure 3.70 (a) The beam with an in-span fluid-jet-generated follower load P. (b) Stability boundary for the pinned–clamped system ($C_1 = 0, C_2 = \infty; \kappa_1^* = 0, \kappa_2^* = \infty$); (c) stability boundary for the clamped–pinned system ($\kappa_1^* = \infty, \kappa_2^* = 0$): ——, with tangential follower load; - - -, with horizontal, fixed-direction (conservative) load. *Flutter 1/2* stands for coupled-mode flutter involving the first and second modes; and similarly for *Flutter 2/3* (Wiley & Furkert 1972).

$$\begin{aligned} &EI \frac{\partial^4 w_1}{\partial x_1^4} + P \frac{\partial^2 w_1}{\partial x_1^2} + m \frac{\partial^2 w_1}{\partial t^2} = 0, \qquad 0 \leq x_1 \leq a; \\ &EI \frac{\partial^4 w_2}{\partial x_2^4} + m \frac{\partial^2 w_2}{\partial t_2} = 0, \qquad 0 < x_2 \leq b; \end{aligned} \tag{3.125}$$

subject to the boundary and compatibility conditions

$$\begin{aligned} &w_1 = 0, \quad EI\left(\frac{\partial^2 w_1}{\partial x_1^2}\right) - C_1\left(\frac{\partial w_1}{\partial x_1}\right) = 0 \quad \text{at} \quad x_1 = 0, \\ &w_1\Big|_a = w_2\Big|_0, \qquad \left(\frac{\partial w_1}{\partial x_1}\right)\Big|_a = \left(\frac{\partial w_2}{\partial x_2}\right)\Big|_0, \\ &w_2 = 0, \quad EI\left(\frac{\partial^2 w_2}{\partial x_2^2}\right) + C_2\left(\frac{\partial w_2}{\partial x_2}\right) = 0 \quad \text{at} \quad x_2 = b, \\ &\left(\frac{\partial^2 w_1}{\partial x_1^2}\right)\Big|_a = \left(\frac{\partial^2 w_2}{\partial x_2^2}\right)\Big|_0, \quad \left(\frac{\partial^3 w_1}{\partial x_1^3}\right)\Big|_a = \left(\frac{\partial^3 w_2}{\partial x_2^3}\right)\Big|_0. \end{aligned} \tag{3.126}$$

The system, once rendered nondimensional, may be solved by straightforward means (cf. Section 3.6.1). Its dynamics is governed by the following parameters:

$$\xi_s = a/L, \qquad \mathcal{P} = PL^2/EI, \qquad \kappa_1^* = C_1 L/EI, \qquad \kappa_2^* = C_2 L/EI. \tag{3.127}$$

It is noted that here $\beta = 0$, and hence there are no Coriolis terms, and $\mathcal{P} = -\Gamma$, Γ being the nondimensional tension, while $\mathcal{P}$ is a compression. By assigning to κ_1^* and κ_2^* the value of zero or infinity, a pinned or clamped end condition may be obtained at either end, or both, without change in the basic formulation.

Some typical and interesting results are presented in Figure 3.70(b,c). The stability boundary for a pinned–clamped system ($\kappa_1^* = 0$, $\kappa_2^* = \infty$) is shown in Figure 3.70(b). It is seen that the system loses stability by divergence throughout, with no coupled-mode flutter for higher values of $\mathcal{P}$ as would be the case for a pipe. The eigenfrequencies remain real until $\mathcal{P}_{cr}$ is reached, when they become imaginary; but they do not coalesce on either the real- or imaginary-frequency axis. Physically, it is clear that the follower force, once the beam is flexed as in Figure 3.70(a), cannot resist the moment generated by $\mathcal{P}$ and hence flutter cannot develop. Note also that in the absence of Coriolis forces there cannot be post-divergence restabilization. Hence, although the system is inherently nonconservative, it is *effectively* conservative, as is the case when both ends are pinned.

The behaviour of the clamped–pinned system ($\kappa_1^* = \infty$, $\kappa_2^* = 0$) is quite different, as seen in Figure 3.70(c); the conservative results (where the force remains parallel to the undeformed axis) are also shown. For these boundary conditions, for $0.2 < \xi_s < 0.45$ approximately, the system loses stability by coupled-mode flutter rather than divergence; this comes about through coalescence of two eigenfrequencies while on the real axis, either the first and the second or the second and third [cf. Figure 3.4(c)]. For lower ξ_s, progressively higher modes would be involved. For $\xi_s = 1$, the system becomes conservative.

Experiments were conducted by using a long aluminium blade ($L \simeq 1.2$ m, 50.8 mm wide in the vertical plane, and 5 mm thick), clamped at one end, and simply-supported and free to slide axially at the other, so as to oscillate in the horizontal direction. The compression was provided by an air jet, issuing from a pair of nozzles affixed to the beam at a slight angle, so as to avoid interaction with it; the compressive reaction force was towards the clamped end. The air was supplied via pairs of light rubber hoses, one vertically above and the other below the blade. Despite the obvious difficulties associated with minimizing the effect of the supply hoses, excellent qualitative and to some extent quantitative agreement with the theory of Figure 3.70(c) was obtained: for $\xi_s > 0.45$ divergence was observed, while for $\xi_s < 0.45$ flutter was observed.

In the case of a cantilevered beam with a tangential end-load at the free end, representing Beck's problem (Section 3.2.2), there is no simple way of minimizing the effect of fluid supply lines. Nevertheless, a successful experiment was conducted by Sugiyama *et al.* (1990, 1995) by attaching a solid-fuel rocket to the free end! The aluminium cantilever (section: 6×30 mm, $L = 800$–1400 mm) weighed 0.4–0.7 kg. The motor was much more massive, $\sim$ 14 kg, and could supply about 390 N force for 4 s. Hence, special techniques had to be developed for deciding whether a damped or amplified oscillation occurred from only a few cycles of oscillation in the period over which the rocket supplied full thrust. Also, not only the mass but the moment of inertia of the motor had to be taken into account. Agreement of experiment with theory is excellent, provided dissipation is ignored; once taken into account, viscoelastic damping in the column ($\alpha = 5 \times 10^{-4}$) is found to diminish the theoretical critical thrust by a factor of 2 as compared to the undamped system, thus rendering agreement apparently rather poor. However, once the criterion 'for stability in a finite time' (Leipholz 1970) is used, the two sets of theoretical results come very close to each other, thus leading to very good agreement with experiment.

3.6.6 Pipes with attached plates

One such system, depicted in Figure 3.71(a), is considered by Herrmann & Nemat-Nasser (1967) as part of a series of studies on the stability of nonconservative mechanical systems. It consists of a thin plate or I-section, with two pairs of flexible pipes attached to it and conveying fluid. This system can execute both flexural transverse motions and torsional motions [cf. Nemat-Nasser & Herrmann's (1966) work on the same structural system subjected to a follower load], and it is in the study of the latter that lies the main contribution of this work.

The equation of motion of the system for flexural transverse motion is the same as before, equation (3.1), except that $2M$ replaces M, which now is the mass per unit length for each *pair* of pipes. For torsional motions, adapting Benjamin's statement of Hamilton's principle to suit, Herrmann & Nemat-Nasser (1967) obtained the following equation of motion and boundary conditions:

$$EC_w \frac{\partial^4 \phi}{\partial x^4} + [2MU^2r^2 - GJ]\frac{\partial^2 \phi}{\partial x^2} + MUh^2 \frac{\partial^2 \phi}{\partial x \partial t} + (mr^2 + \tfrac{1}{2}Mh^2)\frac{\partial^2 \phi}{\partial t^2} = 0; \qquad (3.128)$$

$$\phi = \frac{\partial \phi}{\partial x} = 0 \qquad \text{at} \quad x = 0,$$

$$\frac{\partial^2 \phi}{\partial x^2} = 0, \qquad EC_w \frac{\partial^3 \phi}{\partial x^3} + [MU^2(2r^2 - \tfrac{1}{2}h^2) - GJ]\frac{\partial \phi}{\partial x} = 0 \qquad \text{at} \quad x = L; \qquad (3.129)$$

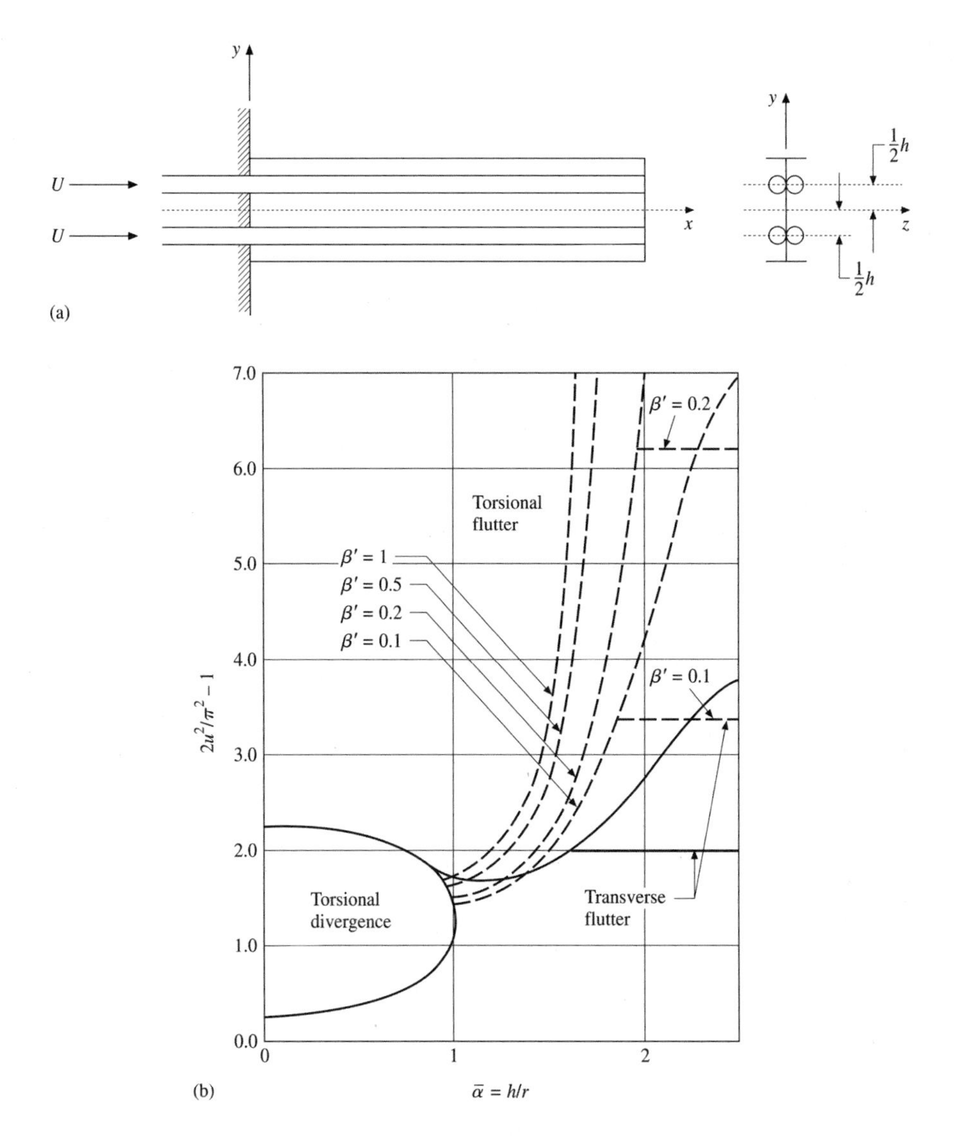

Figure 3.71 (a) The system of a cantilevered thin-plate structure with two pairs of flexible pipes attached, conveying fluid. (b) Stability diagram of $2u^2/\pi^2 - 1$ versus $\bar{\alpha} = h/r$ for $Ir^2/C_w = 1.5$; the solid lines are obtained with $\beta = 0$ (Herrmann & Nemat-Nasser 1967).

ϕ is the angle of twist about the x-axis, EC_w is the warping rigidity and GJ the torsional rigidity, E and G being Young's and the shearing modulus of elasticity, respectively, r is the polar radius of gyration of a section of the system, M is as just defined and m the mass of the structure per unit length. Here it should be recognized that an open section subjected to torsion also warps (Timoshenko & Gere 1961); e.g. an I-section, if subjected to torsion, will also sustain bending of the top and bottom flanges.†

†For example, for an I-section of height h, flange width b, flange thickness t_f and stem thickness t_w, $C_w = t_f h^2 b^3/24$ and $J = (2bt_f^3 + ht_w^3)/3$ (Timoshenko & Gere 1961; appendix).

Equations (3.128) and (3.129) may be rendered dimensionless by means of the following:

$$\xi = \frac{x}{L}, \qquad \tau = \left[\frac{EC_w}{mr^2 + \frac{1}{2}Mh^2}\right]^{1/2} \frac{t}{L^2}, \qquad \overline{\kappa} = \frac{GJL^2}{EC_w},$$
$$\overline{\alpha} = \frac{h}{r}, \qquad \beta' = \frac{M}{m}, \qquad \beta = \frac{\beta'}{1 + \frac{1}{2}\overline{\alpha}^2\beta'}, \qquad u = \left(\frac{Mr^2}{EC_w}\right)^{1/2} UL. \tag{3.130}$$

The solution of the equations of motion may be achieved by the method of Section 3.3.6(a). Typical results are shown in Figure 3.71(b), in terms of the parameter $2u^2/\pi^2 - 1$ for various values of β' and $Ir^2/C_w = 1.5$, for both transverse and torsional motions. The full lines correspond to results obtained for $\beta = 0$ and hence to a system without flow subjected to a follower force $\mathcal{P} = u^2$; however, since divergence is independent of β, the stability curve for $\beta = 0$ applies equally to cases with flow.

It is seen that three types of instability are possible: (i) torsional divergence for small enough $\overline{\alpha}$; (ii) torsional flutter (dashed curves) for intermediate $\overline{\alpha}$; (iii) transverse flutter (horizontal dashed lines) for high $\overline{\alpha}$. Thus, a system with $\beta' = 0.2$ would lose stability by divergence if $\overline{\alpha} = 0.5$, by torsional flutter if $\overline{\alpha} = 1.5$, and by transverse flutter if $\overline{\alpha} = 2.5$. Of course, according to linear theory, in the case of $\overline{\alpha} = 1.5$, transverse flutter would arise at higher flow (the horizontal lines for transverse flutter really extend across the figure), and so on.

It is of special interest that torsional divergence is possible, whereas transverse divergence is not. Equations (3.128) and (3.129) are similar in structure to those for transverse motion, with the torsional terms (proportional to GJ) playing the role of a conservative tensile load. However, it is known that tension does not induce divergence (Section 3.5.8). Hence, torsional divergence probably arises via the MU^2-related term in the boundary conditions (largest at small $\overline{\alpha}$) — cf. Chapter 8.

Another plate–pipe system, used for marine propulsion, is discussed in Section 4.7.

3.6.7 Concluding remarks

The main purpose of Section 3.6 is (i) to briefly document all these interesting studies in one place, and (ii) to show the veritable cornucopia of interesting dynamical problems that may be obtained with simple modifications to the basic system of a pipe conveying fluid — particularly the nonconservative case of a cantilevered pipe. This, despite the early scepticism on the practical value in studying the stability of such systems, as expressed for instance by Timoshenko & Gere (1961; section 2.21), regarding the critical load for a cantilevered column subjected to a tangential follower load: 'No definite conclusion can be made (as yet) regarding the practical value of the result, since no method has been devised for applying a tangential force to a column during bending'. Although a method *has* now been found, this is not really the important point. What *is* important in the study of these systems will emerge from the chapters that follow, and what is practically important from the pertinent sections on applications therein.

3.7 LONG PIPES AND WAVE PROPAGATION

If the pipe is very long between supports, or infinitely long, then the question of wave propagation becomes especially important. The main interest in this is for application to pipelines resting on the ground or on the ocean floor, or pipelines with many, periodically spaced supports. These two topics are treated here, after some preliminary discussion on wave propagation in simple systems.

3.7.1 Wave propagation

Some general characteristics of wave propagation will be reviewed here with the aid of some work by Chen & Rosenberg (1971) on 'pipe-strings' conveying fluid.

Consider first a totally unsupported very long, straight pipe of negligible rigidity, under tension — a very useful tutorial system. Since $EI = 0$ in (3.1), the equation of motion is rendered dimensionless by defining $\overline{u} = (M/\overline{T})^{1/2}U$, $\tau = [\overline{T}/(m+M)]^{1/2}t/L$, together with η, ξ and β as in (3.69) and (3.71), yielding

$$(\overline{u}^2 - c^2)\eta'' + 2\beta^{1/2}\overline{u}\dot{\eta}' + \ddot{\eta} = 0, \qquad c^2 \equiv 1. \tag{3.131}$$

The nondimensionalization gives $c = 1$; nevertheless, the equation is written like this to facilitate the physical interpretation of the results. Thus, if $\overline{u} = 0$, equation (3.131) is the wave equation and c is the dimensionless wave velocity.

Consider now a wave of the form $\eta = A\,\exp[\mathrm{i}\kappa(\xi - v_p\tau)]$, where κ is the *wavenumber* and v_p the *phase velocity*; $\kappa = 1/\lambda$, where λ is the wavelength. Substituting into (3.131), it is easy to see that the equation is of the hyperbolic type provided that $\overline{u}^2(1-\beta) < c^2$. In that case, either progressive or standing waves can exist, and the general solution is of the form (Morse 1948; Meirovitch 1967)

$$\eta = G(\xi - v_1\tau) + H(\xi + v_2\tau), \tag{3.132}$$

where

$$v_{1,2} = [c^2 - \overline{u}^2(1-\beta)]^{1/2} \pm \beta^{1/2}\overline{u}. \tag{3.133}$$

Considering the two component parts of (3.133), together with the form of (3.132), it is easy to show that (i) if $\overline{u} < c$, two waves propagate in the pipe-string, one in the downstream and the other in the upstream direction, with phase velocities v_1 and v_2, respectively, where $v_1 > v_2$; (ii) if $\overline{u} > c$, then both waves travel downstream; and (iii) if $\overline{u} = c$, there is one propagating and one standing wave. A disturbance, e.g. $\eta(\xi, 0) = \exp(-\xi^2)$, leads to waves travelling upstream and downstream without alteration in form. However, whereas for $\overline{u} = 0$ the two waves propagate with the same phase velocity and have the same form, for $\overline{u} \neq 0$ they do not: the wave with the larger phase velocity has smaller amplitude — unlike the classical string (Chen & Rosenberg 1971).

The case of a pipe-string of finite length and fixed ends is examined next. In this case, solutions of the form

$$\eta(\xi, \tau) = A_1\,\exp[\mathrm{i}(\kappa_1\xi + \omega\tau)] + A_2\,\exp[\mathrm{i}(\kappa_2\xi + \omega\tau)] \tag{3.134}$$

are considered, which satisfy (3.131) provided that

$$\kappa_{1,2} = \frac{\omega}{c^2 - \overline{u}^2}\left\{\beta^{1/2}\overline{u} \pm [c^2 - \overline{u}^2(1-\beta)]^{1/2}\right\}. \tag{3.135}$$

Applying the boundary conditions, the frequency equation is obtained, $\sin(\kappa_1 - \kappa_2) = 0$, and the dimensionless frequencies are found to be

$$\omega_n = \frac{n\pi(c^2 - \overline{u}^2)}{[c^2 - \overline{u}^2(1-\beta)]^{1/2}}, \qquad n = 1, 2, 3 \ldots \tag{3.136}$$

The corresponding mode shapes are given by

$$\psi_n = \sin n\pi\xi \, \cos[\kappa_n(\xi + v_p\tau + \theta_n)], \tag{3.137}$$

where the wavenumber and phase velocity are

$$\kappa_n = n\pi \frac{\beta^{1/2}\overline{u}}{[c^2 - \overline{u}^2(1-\beta)]^{1/2}}, \qquad v_p = \frac{c^2 - \overline{u}^2}{\beta^{1/2}\overline{u}},$$

and $v_p = \omega_n/\kappa_n$; v_p is related to its dimensional counterpart, V_p, via $v_p = [(M + m)/\mathcal{T}]^{1/2} V_p$. A number of useful observations can now be made. Wave propagation in this system is not frequency-dispersive, since the phase velocity is not a function of wavenumber (wavelength). Another manifestation of this is that the ratio of the frequency with flow to that without is independent of n. Finally, when $\overline{u} = 0$, v_p is infinite, and the system vibrates with the same phase, whereas for $\overline{u} \neq 0$, a finite v_p is obtained. This means that for $\overline{u} \neq 0$, no classical normal modes exist [cf. Section 3.4.1 and Figure 3.13]: various parts of the system pass through their equilibrium position at different times; i.e. the modal form contains a travelling wave component.

We next consider wave propagation in a beam ($EI \neq 0$), but taking $u = 0$ in equation (3.75), as discussed by Meirovitch (1967). In this case, the phase velocity, v_p, is a function of the wavenumber (wavelength): $v_p = \kappa$; hence, the beam is a *frequency-dispersive* medium. A general nonharmonic waveform may be thought of as a superposition of harmonic waves, $\eta(\xi, \tau) = \sum_n A_n \cos[\kappa_n(\xi - v_{pn}\tau)]$; since each component travels with different phase velocity, the wave form will change as the wave propagates along the beam, as a result of dispersion. If $u \neq 0$, the situation is further complicated. This is discussed next, for a pipe on an elastic foundation.

3.7.2 Infinitely long pipe on elastic foundation

This problem has been considered by Roth (1964) and Stein & Tobriner (1970), and what is presented here is a summary of some of their work.

A form of equation (3.70) is used for the pipe on a generally dissipative elastic foundation, but dissipation may also come from other sources, i.e.

$$\eta'''' + (u^2 - \Gamma + \Pi)\eta'' + 2\beta^{1/2}u\dot{\eta}' + k\eta + \sigma\dot{\eta} + \ddot{\eta} = 0; \tag{3.138}$$

no Poisson-ratio effects are considered, however, since no pressurization-induced tension can arise in the absence of end constraints. Since L could be infinite, in the dimensionless quantities a unit length could be used for L, or an appropriate length scale associated with the initial disturbance under consideration. Considering solutions of the form

$$\eta(\xi, \tau) = A\mathrm{e}^{\mathrm{i}\kappa\xi}\mathrm{e}^{\mathrm{i}\omega\tau}\mathrm{e}^{-\lambda\tau}, \tag{3.139}$$

it is found that equation (3.138) is satisfied if it is found that equation (3.138) is satisfied if

$$\begin{gathered}\omega^2 - \lambda^2 + 2\omega\beta^{1/2}u\kappa - [\kappa^4 - (u^2 - \Gamma + \Pi)\kappa^2 + k - \lambda\sigma] = 0,\\ 2\lambda(\omega + \beta^{1/2}u\kappa) - \omega\sigma = 0,\end{gathered} \tag{3.140}$$

which leads to

$$\omega_{1,2} = -\beta^{1/2}u\kappa \mp p, \qquad \lambda_{1,2} = \tfrac{1}{2}\sigma \pm q, \tag{3.141a}$$

$$p = \frac{1}{\sqrt{2}}\sqrt{\chi + |(\chi^2 + u^2\sigma^2\kappa^2)^{1/2}|}, \qquad q = \frac{1}{\sqrt{2}}\sqrt{-\chi + |(\chi^2 + u^2\sigma^2\kappa^2)^{1/2}|},$$

$$\chi = \kappa^4 - [u^2(1-\beta) - \Gamma + \Pi]\kappa^2 - k - \tfrac{1}{4}\sigma \tag{3.141b}$$

(Roth 1964). The similarity in the structure of $\omega_{1,2}$ in (3.141a) when $\sigma = 0$ to $\omega_{1,2} = v_{1,2}\kappa$ from (3.133) should be noted. Remarking that the form of equation (3.139) with κ replaced by $-\kappa$ and ω by $-\omega$ is also a solution, as easily seen from (3.140), one obtains for a general waveform the general solution

$$\begin{aligned}\eta(\xi, \tau) = &\sum_{n=0}^{\infty} \mathrm{e}^{-\lambda_{1n}\tau}[A_n \cos(\kappa_n\xi + \omega_{1n}\tau) + B_n \sin(\kappa_n\xi + \omega_{1n}\tau)]\\ &+ \sum_{n=0}^{\infty} \mathrm{e}^{-\lambda_{2n}\tau}[C_n \cos(\kappa_n\xi + \omega_{2n}\tau) + D_n \sin(\kappa_n\xi + \omega_{2n}\tau)].\end{aligned} \tag{3.142}$$

The arbitrary constants A_n to D_n are determined from the initial conditions. Thus, if $\eta(\xi, 0) = a(\xi)$, $\dot{\eta}(\xi, 0) = b(\xi)$ are periodic functions with $\kappa_n = n\pi$, the constants may be determined by the use of Fourier series, while a solution for a nonperiodic and spatially more general disturbance may be obtained with the aid of Fourier integrals (Roth 1964).

In solution (3.142) it is noted that the frequencies ω_{1n} and ω_{2n} are each associated with the phase velocities $v_1 = -\omega_{1n}/\kappa_n$ and $v_2 = -\omega_{2n}/\kappa_n$, for downstream- and upstream-travelling waves. For an observer travelling downstream with velocity $\beta^{1/2}u$, these waves propagate with wave speeds $\pm p_n/\kappa_n$, where p_n is as in (3.141a).

For stability of the pipe, λ_1 and λ_2 in solution (3.139) must be positive. This requires that

$$\sigma^2[\kappa^4 - (u^2 - \Gamma + \Pi)\kappa^2 + k] > 0, \tag{3.143}$$

which is true for all damping values, σ. The minimum of the function in square brackets occurs at $\kappa = [\frac{1}{2}(u^2 - \Gamma + \Pi)]^{1/2}$ and is equal to $\Delta(\kappa) = k - \frac{1}{4}(u^2 - \Gamma + \Pi)^2$. Hence, condition (3.143) is satisfied if $\Delta(\kappa) > 0$, or

$$v^2 \equiv u^2 - \Gamma + \Pi < 2\sqrt{k}. \tag{3.144a}$$

This result could be obtained also by the work leading to equation (3.101a). It is of interest that if $k = 0$, then $v = 0$, i.e. the pipe is unstable for all u in the case of $\Gamma = \Pi = 0$. This simply reflects that, in the absence of any support, a lateral displacement of the pipe is not opposed by any restraint.

Now, if the analysis is conducted with $\sigma = 0$ from the start, it is easy to show that in this case the condition of neutral stability, $\lambda_{1,2} = 0$, requires

$$u^2(1-\beta) - \Gamma + \Pi < 2\sqrt{k}. \tag{3.144b}$$

Clearly, since $\beta < 1$, this result is nonconservative; in particular, criterion (3.144b) predicts a system to be stable when, in fact, through (3.144a) it is unstable (Roth 1964)! This is a good demonstration of Bolotin & Zhinzher's (1969) thesis (Section 3.5.5).

It is Stein & Tobriner (1970) who consider wave propagation *per se*. They use the same equation as Roth, but with $\Gamma = 0$. They obtain a general solution to initial conditions $\eta(\xi, 0) = f(\xi)$ and $\dot{\eta}(\xi, 0) = g(\xi)$ by means of Laplace transforms in time (denoted by an overbar) and Fourier transforms in space (denoted by an asterisk), of the form

$$\overline{\eta}(\xi, \omega) = \int_0^\infty \eta(\xi, \tau)\mathrm{e}^{-\omega\tau}\,\mathrm{d}\tau, \qquad \eta^*(\alpha, \omega) = \frac{1}{\sqrt{2\pi}}\int_{-\infty}^\infty \overline{\eta}(\xi, \omega)\mathrm{e}^{\mathrm{i}\alpha\xi}\,\mathrm{d}\xi.$$

The Laplace transform over τ is applied first, and then the Fourier transform over ξ, on the resultant equation. After inversion, the general solution is

$$\begin{aligned}
\eta(\xi, \tau) = \frac{1}{\sqrt{2\pi}}\int_{-\infty}^\infty \Big\{ & f^*(\alpha)\mathrm{e}^{-\sigma\tau/2}[\cos\phi\,\cos\theta_1\,\cosh\theta_2 + \sin\phi\,\sin\theta_1\,\sinh\theta_2] \\
& + \frac{f^*(\alpha)}{2\sqrt{r}}\,\mathrm{e}^{-\sigma\tau/2}[2\beta^{1/2}u\alpha\,\cos(\phi-\mu) - \sigma\,\sin(\phi-\mu)\,\cos\theta_1\,\sinh\theta_2 \\
& + [2\beta^{1/2}u\alpha\,\sin(\phi-\mu) + \sigma\,\cos(\phi-\mu)]\,\sin\theta_1\,\cosh\theta_2] \\
& + \frac{g^*(\alpha)}{\sqrt{r}}\,\mathrm{e}^{-\sigma\tau/2}[\sin(\phi-\mu)\,\cos\theta_1\,\cosh\theta_2 - \cos(\phi-\mu)\,\sin\theta_1\,\cosh\theta_2]\Big\}\,\mathrm{d}\alpha,
\end{aligned} \tag{3.145}$$

where

$$\begin{aligned}
r &= \{[\alpha^2(\alpha^2 - (1-\beta)u^2 + \Pi) + k - \tfrac{1}{4}\sigma^2]^2 + [\beta^{1/2}u\sigma\alpha]^2\}^{1/2}, \\
\mu &= \frac{1}{2}\tan^{-1}\left\{\frac{\beta^{1/2}u\sigma}{\alpha^2[\alpha^2 - (1-\beta)u^2 + \Pi] + k - \frac{1}{4}\sigma^2}\right\}, \\
\theta_1 &= \tau\sqrt{r}\,\cos\mu, \qquad \theta_2 = \tau\sqrt{r}\,\sin\mu, \qquad \phi = \alpha(\beta^{1/2}u\tau - \xi).
\end{aligned} \tag{3.146}$$

Numerical results in the case of $\sigma = 0$ and $\eta = \exp[\mathrm{i}(\kappa\xi - \omega\tau)]$ are then considered. The characteristic equation in this case is

$$\omega^2 - 2\beta^{1/2}u\kappa\omega - \kappa^4 + (u^2 - \Pi)\kappa^2 - k = 0. \tag{3.147}$$

Thus, for each wavenumber (wavelength) there is an associated frequency and, in the absence of end constraints, all wavenumbers are permissible. The dominant wavelengths depend on the spatial distribution of the initial disturbance and the propagation characteristics of each of its Fourier components. Equation (3.147) may be solved for the phase velocity, $v_p = \omega/\kappa$, yielding

$$\frac{v_p}{u_c} = \beta^{1/2}\left\{\frac{u}{u_c} \pm \frac{1-\beta}{\beta}\left[\frac{(\kappa^4+\kappa_m^4)-\Pi}{\kappa^2(2\kappa_m^2-\Pi)} - \left(\frac{u}{u_c}\right)^2\right]^{1/2}\right\}, \tag{3.148}$$

where u_c is obtained from (3.144b) when it is transformed into an equality, while setting $\Gamma = 0$: $u_c^2 = [2k^{1/2} - \Pi]/(1-\beta)$; $\kappa_m = k^{1/4}$ is a critical wavenumber, which corresponds to the value of κ for which all positive roots of (3.148), whether $\kappa > \kappa_m$ or $< \kappa_m$, have phase velocities greater than that for κ_m. However, it is possible to obtain some positively travelling waves with $v_p < v_p(\kappa_m)$ from the negative roots of (3.148), namely for $u/u_c > (1-\beta)^{1/2}$. The dependence of v_p on u may be assessed from (3.148). For any κ, for the positively travelling wave, v_p increases with u; up to $u > v_p(\kappa_m)$, whereafter increasing u causes v_p to increase for some wavenumbers and to decrease for others. For the negatively travelling waves, v_p diminishes continuously with u, for all $u < u_c$.

One may retrieve from equation (3.148) Roth's result that for an observer travelling with a velocity $\beta^{1/2}u$,[†] upstream- and downstream-travelling waves would appear to have equal velocities; this would imply that the distribution of waves would always be symmetric about this translating axis for a symmetric disturbance about the origin. However, Stein & Tobriner show that this is true only asymptotically (in time), because the solution does not satisfy the boundary conditions in the limit as $\xi \to \infty$.

Some typical numerical results are shown in Figures 3.72–3.74 for a steel pipe conveying water, with zero dissipation ($\sigma = 0$); the larger foundation modulus, $k = 6.54$, is typical of crushed gravel. The initial disturbance is taken to be $\eta(\xi, 0) = \eta_0 \exp[-\frac{1}{4}(x/L)^2] \equiv \eta_0 \exp(-\frac{1}{4}\xi^2)$, with $L = 12.5$ ft (3.81 m); this same L is used in obtaining u, Π and k from the corresponding dimensional quantities.

Figure 3.72 shows the time evolution of the disturbance at $\xi = 0$ for $k = 0$ and $k = 6.3$, when $u = 0.160$ ($U = 30.48$ m/s or 100 ft/s) and $\Pi = 0$. In (a) it is seen that the system is unstable, as discussed, and the oscillations are amplified with time. In (b), condition (3.144b) is satisfied and hence the oscillation is stable ($u < u_c$); the amplitude of the oscillation at $\xi = 0$ is diminished with time as the disturbance energy is shared with progressively larger parts of the pipe, $|\xi| > 0$, as shown in Figure 3.73. Stein & Tobriner (1970) also show a case with $\Pi = 0.0256$ and $k = 2.43 \times 10^{-4}$, where $u = u_c$ and a neutrally stable oscillation at $\xi = 0$ is obtained.

In Figure 3.73(a) is shown the development of the initial disturbance when $u = 0$. It is seen that up- and downstream propagating waves are symmetric about the origin. It is also seen that the amplitudes for the lower k values are more severely attenuated than for the largest k. When $u > 0$, as in Figure 3.73(b), the symmetry about the origin is destroyed, and the waveform becomes symmetric with respect to an axis travelling at $\beta^{1/2}u$. In the figure this is visible only for large k; for the smaller k, this symmetry which occurs for large enough τ has not yet developed for the range of τ shown in the figure.

In Figure 3.74 we look at a particular point along the pipe, $\xi = 8$, versus time. It is seen that for a stiff enough foundation ($k = 6.54$), the wave retains its cohesion and propagates downstream as a 'wave packet', roughly at $\beta^{1/2}u$; the upstream-propagating component

[†]It is noted that $v_p/u = \beta^{-1/2}(V_p/U)$, where the capital letters are for the dimensional quantities, because of the different nondimensionalizing factors for v_p and u; thus, in *dimensional* terms, the observer travels with velocity βU.

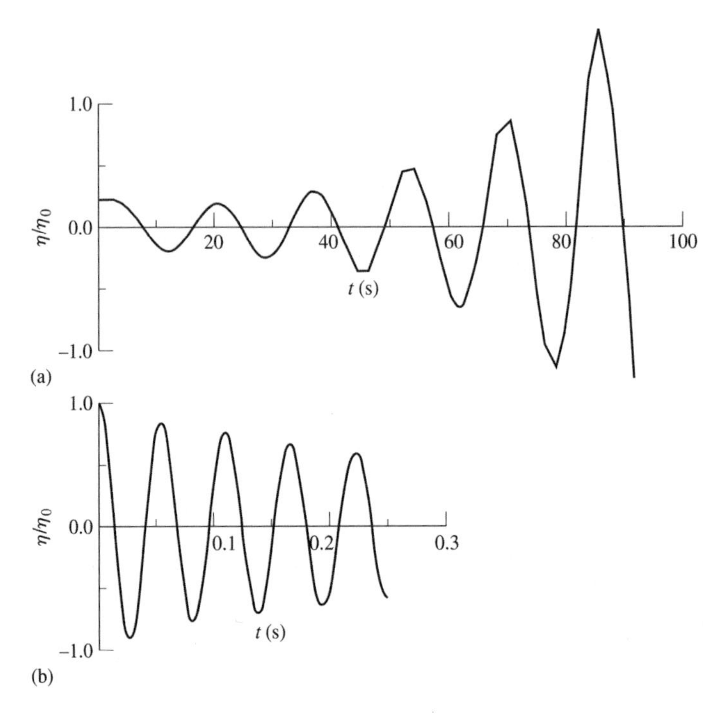

Figure 3.72 The time evolution of disturbance at $\xi = 0$ for an infinitely long pipe on an elastic foundation with (a) $k = 0$ and (b) $k = 6.30$, for $u = 0.160$, $\sigma = \Pi = 0$ (Stein & Tobriner 1970).

is much smaller. However, for smaller k (not shown here), neither a well-defined wave packet nor an axis of symmetry develops.

Finally, it is stressed that the calculations in Figures 3.72–3.74 have all been done with $\sigma = 0$. This should be borne in mind when considering wave propagation in real systems, in which dissipation is always present.

3.7.3 Periodically supported pipes

An excellent treatment of the subject was provided by Chen (1972a), an outline of the salient features of which is given in what follows.

Suppose that the pipe is simply-supported periodically at N supports, as shown in Figure 3.75(a), where N may be finite or tend to infinity. The equation of motion is

$$\eta'''' + (u^2 - \Gamma + \Pi)\eta'' + 2\beta^{1/2}u\dot{\eta}' + \ddot{\eta} = q\psi(\xi)e^{i\omega\tau}, \tag{3.149}$$

a version of (3.70); the term on the right side represents a possible forcing function.

Considering two neighbouring spans of the pipe on either side of the jth support, and denoting quantities on its left without a bar and those on its right with a bar, the boundary conditions to be satisfied are

$$\begin{aligned} &\eta(0) = \eta(1) = 0, \qquad \eta''(0) = -\alpha_{j-1}, \qquad \eta''(1) = -\alpha_j, \\ &\overline{\eta}(0) = \overline{\eta}(1) = 0, \qquad \overline{\eta}''(0) = -\overline{\alpha}_j, \qquad \overline{\eta}''(1) = -\overline{\alpha}_{j+1}, \end{aligned} \tag{3.150a}$$

$$\eta'(1) = \overline{\eta}'(0), \tag{3.150b}$$

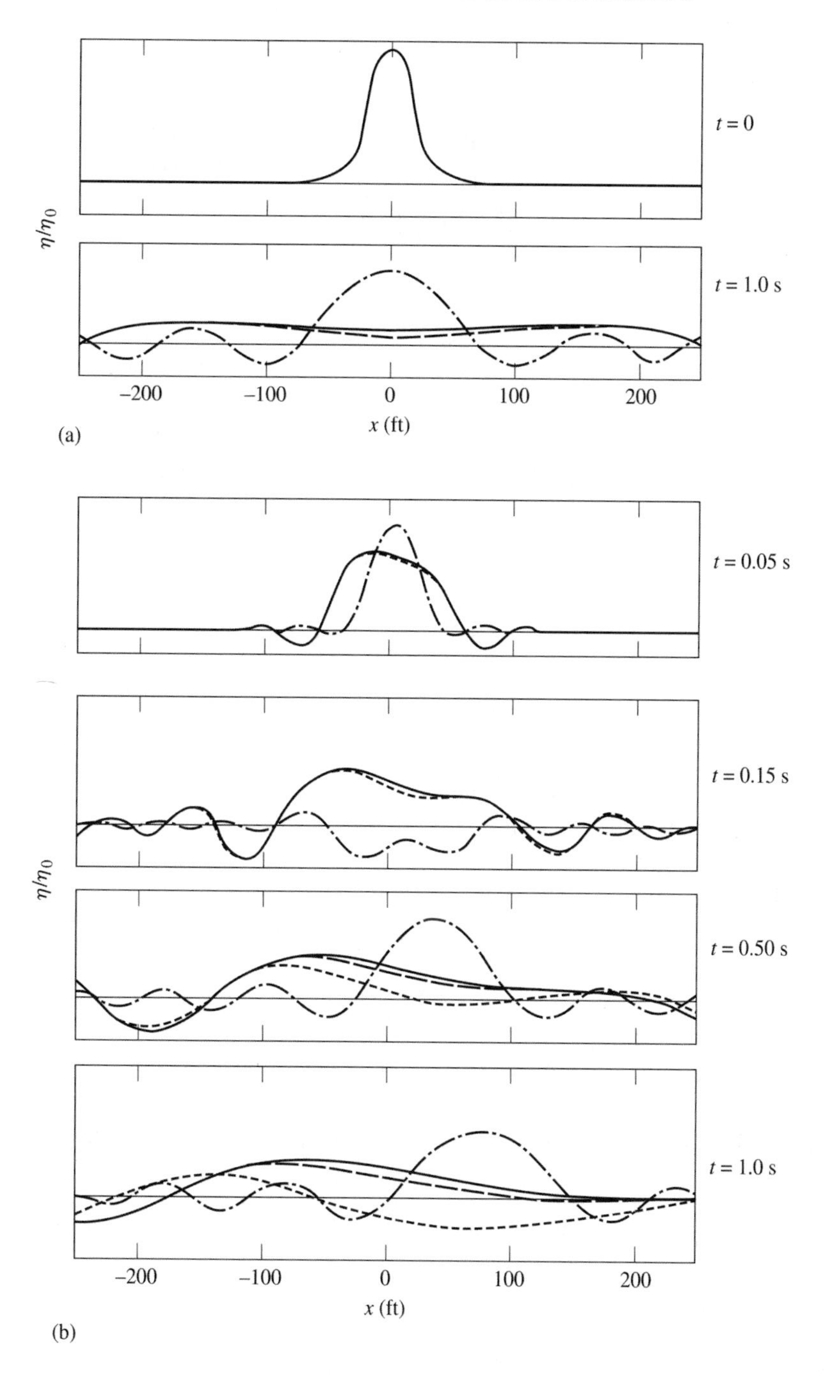

Figure 3.73 Displacement profile of the infinitely long pipe on an elastic foundation, for (a) $u = 0$ and (b) $u = 0.160$; $\Pi = \sigma = 0$: ——, $k = 0$; — — —, $k = 2.43 \times 10^{-4}$; - - -, $k = 2.75 \times 10^{-3}$; — - —, $k = 6.54$ (Stein & Tobriner 1970).

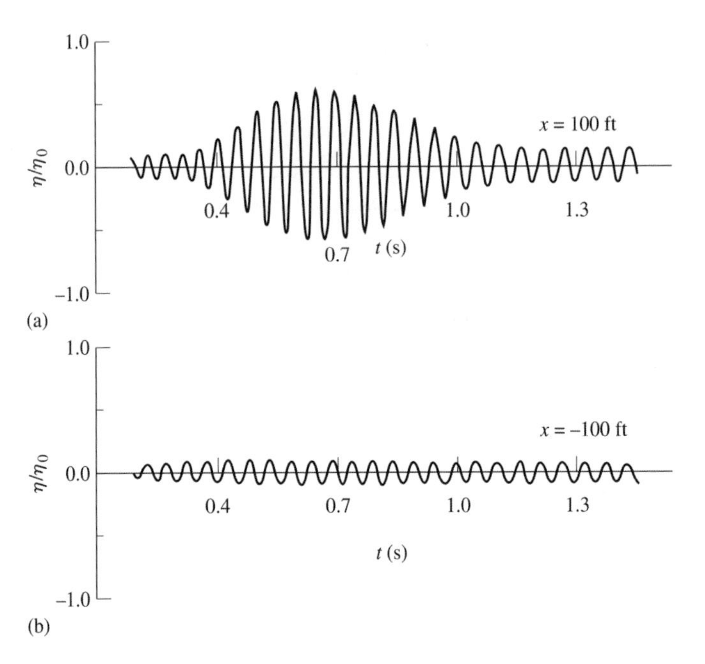

Figure 3.74 Propagation of a disturbance for a pipe on an elastic foundation, at (a) $\xi = 8$ ($x = 100$ ft) and (b) $\xi = -8$; $u = 0.320$, $\Pi = 2.56 \times 10^{-2}$, $k = 6.54$ (Stein & Tobriner 1970).

where $\alpha_j = \mathcal{M}_j l_j/EI$, $\overline{\alpha}_j = \mathcal{M}_j l_j/\overline{EI}$, and derivatives of the barred quantities are with respect to $\overline{\xi}$; $\mathcal{M}_{j-1}$, $\mathcal{M}_j$ and $\mathcal{M}_{j+1}$ are the bending moments at the supports (the same on either side of each support).

Now consider free vibration, as in Section 3.3.6(a). The general solution may be expressed as $\eta(\xi,\tau) = Y(\xi)\exp(\mathrm{i}\omega\tau) = \sum_{n=1}^{4} C_n \exp(\mathrm{i}\lambda_n\xi)\exp(\mathrm{i}\omega\tau)$. Substituting into (3.149) leads to an equation similar to (3.82), namely

$$\lambda_n^4 - (u^2 - \Gamma + \Pi)\lambda_n^2 + 2\beta^{1/2}u\omega\lambda_n - \omega^2 = 0; \tag{3.151}$$

hence, proceeding in the same manner but with two spans, one obtains

$$Y(\xi) = \sum_{n=1}^{4}\{a_n\alpha_{j-1} + b_n\alpha_j\}\mathrm{e}^{\mathrm{i}\lambda_n\xi}, \qquad \overline{Y}(\xi) = \sum_{n=1}^{4}\{\overline{a}_n\overline{\alpha}_j + \overline{b}_n\overline{\alpha}_{j+1}\}\mathrm{e}^{\mathrm{i}\overline{\lambda}_n\overline{\xi}}. \tag{3.152}$$

Then, with the aid of the continuity condition (3.150b), the following equation is obtained:

$$\sum_{n=1}^{4}\{(a_n\lambda_n\mathrm{e}^{\mathrm{i}\lambda_n})\alpha_{j-1} + (b_n\lambda_n\mathrm{e}^{\mathrm{i}\lambda_n})\alpha_j - (\overline{a}_n\overline{\lambda}_n)\overline{\alpha}_j - (\overline{b}_n\overline{\lambda}_n)\overline{\alpha}_{j+1}\} = 0. \tag{3.153}$$

For an infinite, uniform pipe with equispaced supports, $\overline{l} = l$ and $\overline{EI} = EI$, so that the bars in (3.153) may be removed. Equation (3.153) holds for all supports and may be viewed as a recurrence relationship between successive support moments. The general solution may be expressed as

$$\alpha_j = \alpha_{j-1}\exp(\mathrm{i}\mu), \tag{3.154}$$

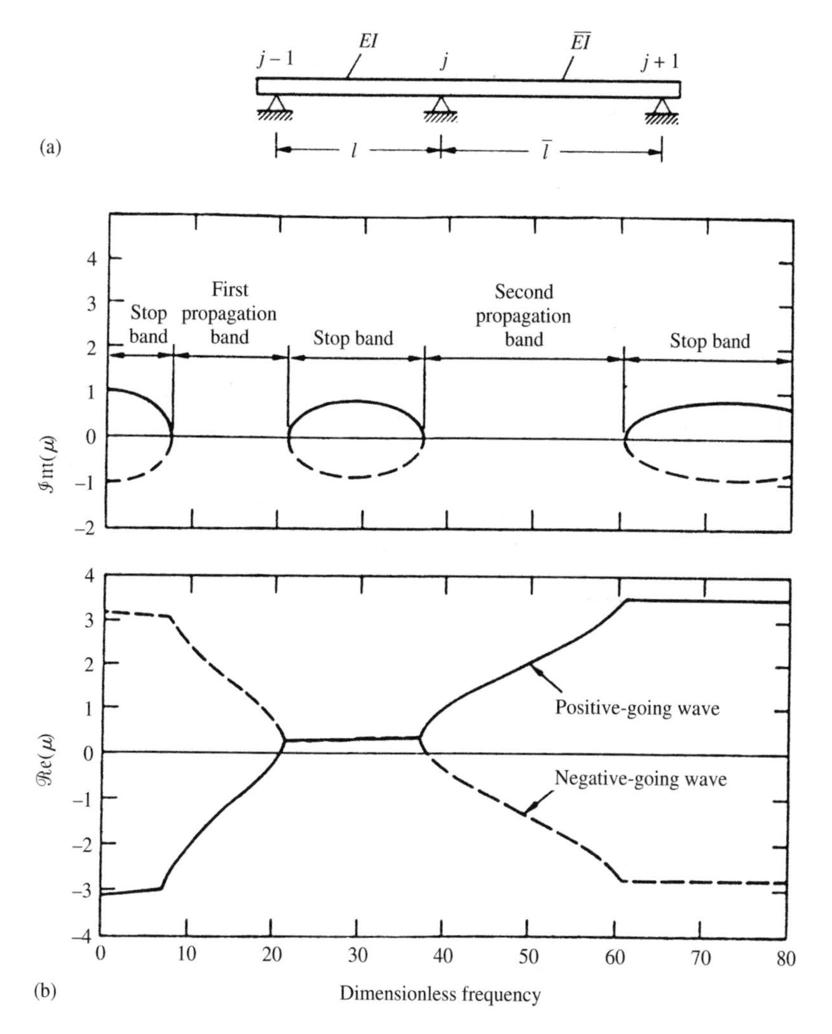

Figure 3.75 (a) A portion of a periodically supported infinitely long pipe, showing two spans of generally different length and flexural rigidity. (b) The propagation characteristics with varying ω for $\beta = 0.25$, $u = 2$, $\Pi = \Gamma = 0$ (Chen 1972a).

where μ is the *propagation constant*, which is generally complex; $\mathscr{R}e(\mu)$ represents the phase shift in the moments from one support to the next, while $\mathscr{I}m(\mu)$ represents the exponential decay. Clearly, unless $\mathscr{I}m(\mu) = 0$, the waves will decay to zero eventually, this being an infinite system. Hence, one may distinguish unattenuated *propagation bands*, where $\mathscr{I}m(\mu) = 0$, and nonpropagation *stop bands*, where $\mathscr{I}m(\mu) \neq 0$. Clearly, $\mu = \mu(u, \omega, \beta, \Gamma, \Pi)$.

A typical result is shown in Figure 3.75(b). It is seen that there is a succession of stop and propagation bands, each one beginning at the value of ω corresponding to one of the natural frequencies of a single span: $\omega = \pi^2, 4\pi^2$ *et seq.* for $u = 0$, and somewhat lower values for $u = 2$; the upper limit of each propagation band is the corresponding single-span eigenfrequency for a clamped–clamped pipe, $\omega = 22.37$, 61.67 *et seq.* for

$u = 0$, and a little lower for $u = 2$, for reasons to become evident two paragraphs hence. The propagation bands become wider with increasing u and, as the divergence limit is approached, $u = \pi$, the first propagation band reaches $\omega \to 0$. Also, from the $\mathscr{Re}(\mu)$ curves it is obvious that positively and negatively travelling waves have different phases and hence phase velocities, which again shows that the system does not possess classical normal modes (Section 3.7.1).

The case of a finite N follows the same pattern. One eventually obtains an $N \times N$ matrix equation giving N discrete frequencies for each propagation band, rather than a continuum. Thus, in the case of a pipe with $\beta = 0.25$, $u = \Pi = \Gamma = 0$ and $N = 8$, one obtains eight eigenfrequencies: $\pi^2 \leq \omega \leq 21.67\,(< 22.37)$ in the first band, and another eight $4\pi^2 \leq \omega \leq 60.52\,(< 61.67)$ in the second band.

To understand these results and those in Figure 3.75(b), it is important to realize that only modes with half-wavelength equal to or a submultiple of the single-span length can propagate: eight such modes when $N = 8$, and an infinite number for $N \to \infty$. The mode shapes can be visualized most easily for a three-span system ($N = 3$), as shown in Figure 3.76 for the first propagation band. The first mode obviously has the same frequency as the eigenfrequency of a pinned–pinned single-span pipe, while the other two have higher frequencies because of the additional strain at the supports where there is a change in slope. Clearly, however, the highest frequency in each band has to be lower than that of a single-span clamped–clamped pipe, approaching it only as $N \to \infty$. In the second propagation band, each mode has a second-mode shape within each span, and so on.

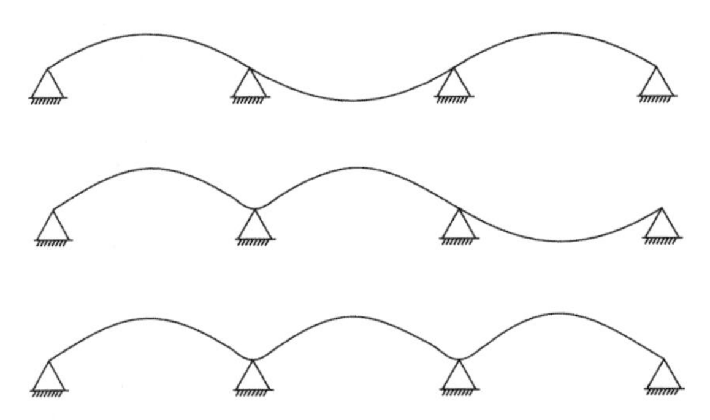

Figure 3.76 Schematics of the three modes in the first propagation band for a three-span pipe ($N = 3$).

If the pipe of the finite system is nonuniform, some new and interesting features develop. Chen considers the eight-span system with each span the same as all the others ($\beta = 0.25$, $u = \Pi = 0$), except that $\Gamma = 0$ for all spans but the fourth and fifth where $\Gamma = -4$. When the pipe is nonuniform, some eigenfrequencies exist in what would have been stop bands in the uniform pipe. Thus, the stop band of one portion of the piping system may be a propagation band in another portion; e.g. for $\Gamma = -4$, the propagation band is $7.51 \leq \omega \leq 9.91$, but for $\Gamma = 0$ waves are attenuated in that range of ω. The modes in that range are called *energy-trapping modes*, for obvious reasons: any energy that comes into the unattenuating part of the system is accumulated there, but dissipated elsewhere. For this example, two energy-trapping modes are found: $\omega = 9.00$ and $\omega = 38.81$. In these modes, the amplitudes of the fourth and fifth spans are much larger than those of the rest. This phenomenon is also known as *mode localization* — see, e.g. Pierre & Dowell (1987), Bendiksen (1987) and Vakakis (1994).

The response of the system to a convected pressure perturbation is considered next, of the form $q\,\exp[\mathrm{i}(\Omega\tau - \kappa\xi)]$, i.e. with $\psi(\xi) = \exp(-\mathrm{i}\kappa\xi)$ in (3.149). In this case, the solution is $\eta(\xi, \tau) = \sum_{n=1}^{4}[C_n \exp(\mathrm{i}\lambda_n\xi) + \Phi(\xi)] \exp(\mathrm{i}\omega\tau)$, where $\Phi(\xi)$ is the particular solution; for a uniform, infinitely long pipe,

$$\Phi(\xi) = q\mathrm{e}^{-\mathrm{i}\kappa\xi}[\kappa^4 - (u^2 - \Gamma + \Pi)\kappa^2 + 2\beta^{1/2}u\Omega\kappa - \Omega^2]^{-1}.$$

The solution follows the same pattern as before, but $\Phi(\xi)$ comes into the picture; i.e.

$$Y(\xi) = \sum_{n=1}^{4}\{a_n[\alpha_{j-1} + \Phi''(0)] + b_n[a_j + \Phi''(1)] + d_n\Phi(0) + e_n\Phi(1)\}\mathrm{e}^{\mathrm{i}\lambda_n\xi} + \Phi(\xi).$$

Hence, a more complex form of (3.153) results, involving $\Phi(0)$ and $\Phi(1)$. However, the form of the solution is the same, and taking $\alpha_j = \alpha_{j-1}\exp(\mathrm{i}\kappa)$, one eventually obtains

$$\mathcal{M}_j = qF(u, \beta, \Omega, \kappa, \Gamma, \Pi), \tag{3.155}$$

$$F = \frac{\displaystyle\sum_{n=1}^{4}[d_n\lambda_n(1 - \mathrm{e}^{-\mathrm{i}\lambda_n}) - a_n\lambda_n(1 - \mathrm{e}^{-\mathrm{i}\lambda_n})\kappa^2 - \kappa] + \sum_{n=1}^{4}[f(\lambda_n, \kappa, e_n, b_n)]\mathrm{e}^{-\mathrm{i}\kappa}}{\left\{\displaystyle\sum_{n=1}^{4}[a_n\lambda_n\mathrm{e}^{\mathrm{i}(\lambda_n-\kappa)} + b_n\lambda_n\mathrm{e}^{\mathrm{i}\lambda_n} - a_n\lambda_n - b_n\lambda_n\mathrm{e}^{\mathrm{i}\kappa}]\right\}\left\{\kappa^4 - v^2\kappa^2 + 2\beta^{1/2}u\Omega\kappa - \Omega^2\right\}},$$

where $v^2 = u^2 - \Gamma + \Pi$, and f is the same as the other expression in the numerator but involving e_n and b_n instead of d_n and a_n, and $+\kappa$ for the last term. The interesting part of this result is that F becomes infinite when either of the two bracketed expressions in the denominator vanishes. Comparing with (3.151), it is seen that the second bracketed quantity vanishes, if Ω coincides with one of the eigenfrequencies of the unsupported system: $\Omega = \omega$. This is the 'normal' resonance condition. Then, comparing the first bracketed expression to (3.153) with (3.154) substituted in it, it is clear that this too can vanish for $\kappa = \mu$, i.e. when the convection velocity of the pressure perturbation coincides with the phase velocity of free waves in the pipe, a 'new' type of resonance.

Similar work on wave propagation in periodically supported pipes (with an additional *rotational* stiffness present at each support) has been done by Singh & Mallik (1977). The interested reader should also refer to Mead (1970, 1973).

3.8 ARTICULATED PIPES

It is recalled that, essentially, the incredible saga of the dynamics of cantilevered pipes conveying fluid, in all its manifestations and variants, began with Benjamin's (1961a,b) work on articulated cantilevered pipes. Benjamin derived the correct statement of Hamilton's principle for an articulated system, equation (3.10), in much the same way as in Section 3.3.3, and in the process he discussed the incorrectness of previous derivations of the equations of motion of cantilevered pipes. He also examined the mechanisms of energy transfer and stability (Section 3.3.2), and illustrated the qualitatively predicted dynamical behaviour by sample calculations and model experiments. Further work on the subject was done by Païdoussis & Deksnis (1970), Bohn & Herrmann (1974a,b), Sugiyama & Noda (1981), Bajaj & Sethna (1982a), Sugiyama & Païdoussis (1982), Lunn (1982), Sugiyama (1984) and Sugiyama *et al*. (1986a,b) on linear aspects; a considerable amount of work was also done on the nonlinear dynamics of the system, which is discussed in Chapter 5.

The dynamics of the articulated system mirror those of the continuous one (which is treated first in this book), with the following difference: the *cantilevered* articulated system is not only subject to flutter but also to divergence, unlike the continuous system. The importance of this discrepancy should be viewed in the context of the popularity of low-dimensional (low-N) models for studying the dynamics of continuous systems (Herrmann 1967; Herrmann & Bungay 1964; Herrmann & Jong 1965, 1966). For columns subjected to axial loading, the dynamics is qualitatively the same in the discrete and continuous systems, and hence low-N models may be used without worry; however, this is not the case for pipes conveying fluid, as discussed in Section 3.8.2.

3.8.1 The basic dynamics

Consider the articulated system shown in Figure 3.1(d), oscillating in a vertical plane. The mass of the pipe per unit length is m and that of the fluid M, the length of the upper pipe l_1 and of the lower one l_2; the corresponding spring constants are k_1 and k_2, while the generalized coordinates are $q_1 = \theta$, $q_2 = \phi$. The equations of motion can be derived with the aid of (3.10) from the expressions for the kinetic and potential energies, correct to second order,

$$\begin{aligned} T &= \tfrac{1}{2}m\left[\left(\tfrac{1}{3}l_1^3 + l_1^2 l_2\right)\dot{\theta}^2 + l_1 l_2^2\dot{\theta}\dot{\phi} + \tfrac{1}{3}l_2^3\dot{\phi}^2\right] \\ &\qquad + \left\{\text{const.} + \tfrac{1}{2}M\left[\left(\tfrac{1}{3}l_1^3 + l_1^2 l_2\right)\dot{\theta}^2 + l_1 l_2^2\dot{\theta}\dot{\phi} + \tfrac{1}{3}l_2^3\dot{\phi}^2 + 2l_1 l_2 U(\phi - \theta)\dot{\theta}\right]\right\}, \quad (3.156) \\ V &= \tfrac{1}{2}\{k_1\theta^2 + k_2(\theta - \phi)^2 + \tfrac{1}{4}(m + M)g[(l_1^2 + 2l_1 l_2)\theta^2 + l_2^2\phi^2]\}, \end{aligned}$$

and $\mathbf{R} = (l_1\theta + l_2\phi)\mathbf{k} - \frac{1}{2}(l_1\theta^2 + l_2\phi^2)\mathbf{i}$ and $\boldsymbol{\tau} = \phi\mathbf{k} + \mathbf{i}$, where $\mathbf{k}$ and $\mathbf{i}$ are the unit vectors, respectively in the lateral z-direction and the axial x-direction. The equations of motion are rendered dimensionless by defining a dimensionless time $\tau = [3k_2/(M + m)l_2^3]^{1/2}t$ and the parameters $a = l_1/l_2$, $\overline{\beta} \equiv 3\beta = 3M/(M + m)$, $\kappa = k_1/k_2$, $u = [(M + m)l_2/3k_2]^{1/2}U$ and $\gamma = (M + m)gl_2^2/2k_2$.

For the system defined by $a = \kappa = 1$, $\beta = \frac{1}{2}$ and $\gamma = 0$ it is found that (i) the first mode remains stable, receding to even larger $\mathcal{I}\mathrm{m}(\omega)$ as u is increased, ω being the eigenfrequency; (ii) the system loses stability in its second mode at $u = 1.733$ by flutter; (iii) thereafter the second-mode locus reaches the $\mathcal{I}\mathrm{m}(\omega)$-axis and remains thereon, tending to $\omega \to 0$ as $u \to \infty$. For $\gamma \neq 0$, however, stability may be lost by divergence. No attempt was made by Benjamin to draw the map showing where divergence and where flutter would occur; this was done later, e.g. by Lunn (1982) — see Figure 5.13. Nevertheless, in the case where the restoring forces are due to gravity alone ($k_1 = k_2 = 0$), Benjamin shows that the system loses stability by divergence if $\beta > \frac{1}{6}$, and by flutter for lower β. Hence, in a typical experimental system, if the fluid conveyed is water the instability first observed will typically be divergence; if it is air, however, it will be flutter.

Benjamin (1961b) conducted a set of model experiments with articulated pipes made up of segments of brass or glass tubes (typically 12.7 mm in diameter, 0.20–0.62 m long), interconnected by joints made of short lengths of rubber tubing bound to the rigid tubes securely with wire. Care was taken to relieve stresses at the joints and to ensure a smooth flow passage from tube to joint and on to the next tube. Some experiments were conducted with $k_1 = k_2 \simeq 0$ by replacing each rubber joint by the neck of a toy balloon. The fluid was water ($\beta = 0.18, 0.31$ and 0.32). In some experiments, the pipe was vertical and in others horizontal (essentially as described in Section 3.5.6). In a few cases, both ends were supported.

Virtually all of the general qualitative observations made in Section 3.5.6 for flexible pipes have been noted earlier by Benjamin in his articulated pipe experiments: the violence of the divergence instability (which had to be limited by restricting its unimpeded growth, otherwise resulting in a broken joint), the destabilization of a cantilevered system by lightly touching the free end, limit-cycle motion, 'induced' versus self-excited flutter and hysteresis, etc.

Agreement between theoretical and experimental critical flow velocities was impressive: $U_{cd} = 0.34$ versus 0.36 m/s for divergence and $U_{cf} = 0.65$ versus 0.68 m/s for flutter are typical of a set of 18 experiments.

The 'mode exchange', already discussed in Section 3.5.1, also arises in the case of articulated systems, as demonstrated by Sugiyama & Noda (1981) and as shown in Figure 3.77, where it is seen that the mode loci come very close together before the switch actually takes place. The Argand diagram for $\beta = 0.50$ is identical to one of those originally obtained by Benjamin (1961a).

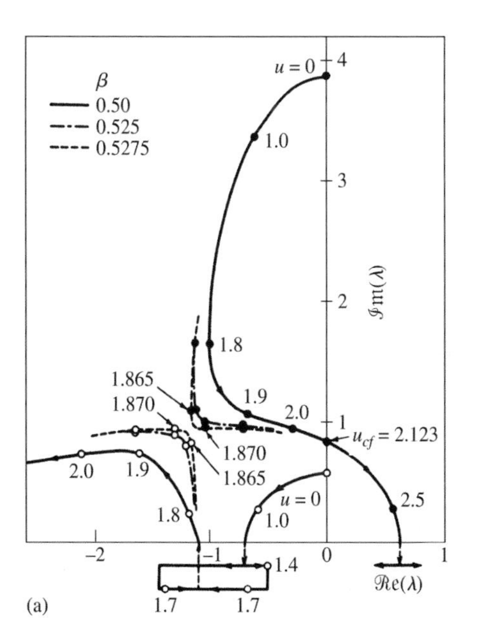

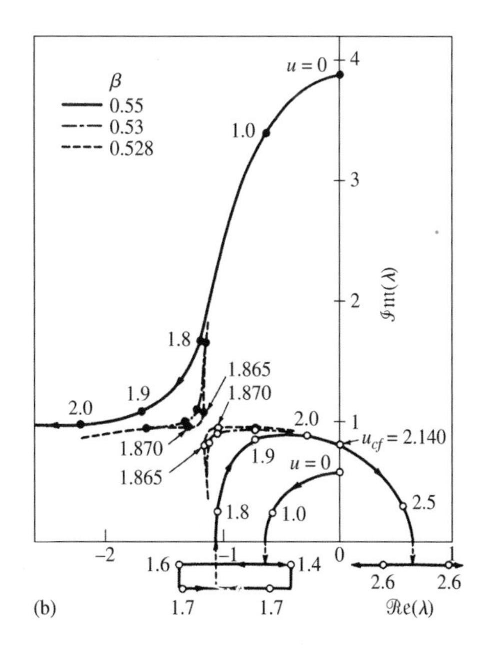

Figure 3.77 The 'mode exchange' from second- to first-mode flutter for an articulated cantilever with varying β for $\gamma = 0$: (a) for β increasing, starting with $\beta = 0.50$; (b) for β decreasing, starting with $\beta = 0.55$ (Sugiyama & Noda 1981).

Lunn (1982) studies the effect of dissipation on two-degree-of-freedom systems, in particular for $\beta \to 0$, obtaining similar behaviour to that discussed in Section 3.5.6 for continuous systems: i.e. severe destabilization due to internal damping (Figure 3.37). He also conducted a number of experiments, some of which are discussed in Chapter 5 (see Figure 5.14).

3.8.2 *N*-degree-of-freedom pipes

One major difference between the continuous and articulated cantilevered systems is that in the latter, if the pipe is vertical, divergence may occur, while for the continuous system it has been shown theoretically and confirmed experimentally that divergence is impossible. The resolution of this difference in behaviour was the motivation of the work Païdoussis & Deksnis (1970), dramatically entitled 'the study of a paradox'; 'paradox' simply because in the limit, as the number of articulations $N \to \infty$, one should expect the articulated system to approach the continuous one in every respect.

Païdoussis & Deksnis (1970) consider the vertical system of Figure 3.78, involving N articulations and N rigid tubes, of which $N - 1$ are of length l, while the last one is of length el, where $e \leq 1$. For reasons to be clarified later, a portion of the upstream immobile piping, $(1 - e)l$ long, is taken to be part of the system, so that the total length of the articulated pipe is $L = Nl$. The rigid tubes are interconnected by rotational springs of equal stiffness, k. Again, the masses per unit length of the pipe and of the fluid are m and M, respectively.

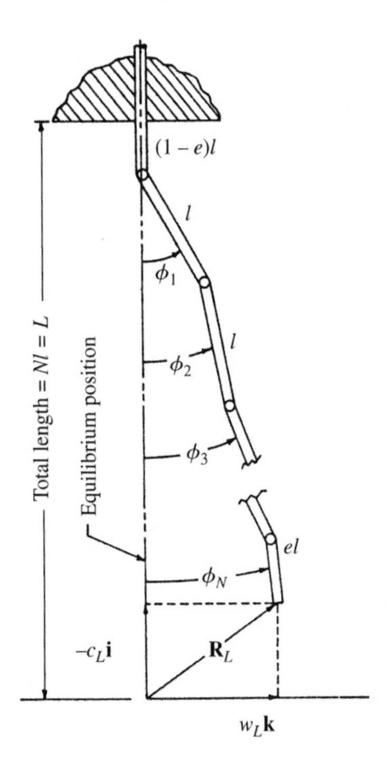

Figure 3.78 The N-degree-of-freedom articulated system, showing the lengths of the tubes, the generalized coordinates and the displacements of the free end. At each articulation there is a rotational spring of stiffness k, not shown (Païdoussis & Deksnis 1970).

Making allowance for the shorter tube at the free end, the kinetic energy of the pipe is

$$T_p = \tfrac{1}{2}ml^3 \sum_{p=1}^{N} \left\{ \tfrac{1}{3}c_{3p}\dot{\phi}_p^2 + c_{2p}\dot{\phi}_p \left(\sum_{q=0}^{p-1} \dot{\phi}_q \right) + c_{1p} \left(\sum_{q=0}^{p-1} \dot{\phi}_q \right)^2 \right\}, \tag{3.157}$$

where

$$c_{1p} = 1 + (e-1)\delta_{pN}, \qquad c_{2p} = 1 + (e^2-1)\delta_{pN}, \qquad c_{3p} = 1 + (e^3-1)\delta_{pN}, \tag{3.158}$$

and δ_{pN} is Kronecker's delta, while the kinetic energy of the fluid is

$$T_f = \tfrac{1}{2}Ml^3 \sum_{p=1}^{N} \left\{ \tfrac{1}{3}c_{3p}\dot{\phi}_p^2 + c_{2p}\dot{\phi}_p \left(\sum_{q=0}^{p-1} \dot{\phi}_q \right) + c_{1p} \left(\sum_{q=0}^{p-1} \dot{\phi}_q \right)^2 \right\}$$
$$+ \tfrac{1}{2}MUl^2 \sum_{p=1}^{N} \left\{ 2c_{1p} \left[\sum_{q=0}^{p-1} \dot{\phi}_q(\phi_p - \phi_q) \right] \right\} + \text{const}, \tag{3.159}$$

both correct up to the quadratic terms. The potential energy of the system is

$$V = \tfrac{1}{2}mgl^2 \sum_{p=1}^{N} \left\{ c_{1p} \left(\sum_{q=0}^{p-1} \phi_q^2 \right) + \tfrac{1}{2}c_{2p}\phi_p^2 \right\} + \tfrac{1}{2}k \sum_{p=1}^{N} (\phi_p - \phi_{p-1})^2. \tag{3.160}$$

The equations of motion are derived via Hamilton's principle, equation (3.10). In this case, $\mathbf{R} = w_L\mathbf{k} - c_L\mathbf{i}$, $\boldsymbol{\tau} = \phi_N\mathbf{k} + \mathbf{i}$, where

$$w_L = l \sum_{p=1}^{N} c_{1p}\phi_p \qquad \text{and} \qquad c_L = \tfrac{1}{2}l \sum_{p=1}^{N} c_{1p}\phi_p^2. \tag{3.161}$$

The equations of motion, in dimensionless form, are

$$\tfrac{2}{3}c_{3r}\ddot{\phi}_r + N\gamma c_{2r}\phi_r + 2N^2u^2c_{1r}(\phi_N - \phi_r)$$
$$+ \sum_{p=1}^{N} \left\{ c_{2p}\ddot{\phi}_p \left(\sum_{q=0}^{p-1} \delta_{qr} \right) + c_{2p}\delta_{pr} \left(\sum_{q=0}^{p-1} \ddot{\phi}_q \right) + 2c_{1p} \left(\sum_{q=0}^{p-1} \ddot{\phi}_q \right) \left(\sum_{q=0}^{p-1} \delta_{qr} \right) \right.$$
$$+ 2N\gamma c_{1p} \left(\sum_{q=0}^{p-1} \phi_q\delta_{qr} \right) + 2N^4(\phi_p - \phi_{p-1})[\delta_{pr} - \delta_{(p-1)r}]$$
$$\left. + 2N\beta^{1/2}uc_{1p} \left[c_{1r}\dot{\phi}_p + \dot{\phi}_p \left(\sum_{q=0}^{p-1} \delta_{qr} \right) - \delta_{pr} \left(\sum_{q=0}^{p-1} \dot{\phi}_q \right) \right] \right\} = 0,$$
$$r = 1, 2, 3, \ldots, N, \tag{3.162}$$

where

$$L = Nl, \quad \tau = \left(\frac{(M+m)L^3N}{k}\right)^{-1/2} t, \quad \beta = \frac{M}{M+m},$$

$$\gamma = \frac{(M+m)gL^2N}{k} \quad \text{and} \quad u = \left(\frac{MLN}{k}\right)^{1/2} U. \tag{3.163}$$

For $e = \frac{1}{2}$, one may consider the articulated system to be a *physically* discretized version of the continuous one, with the flexibility of the latter lumped at the mid-point of each l-length segment and equal to $k = EI/l$ — cf. Goldstein (1950; Chapter 11). It is the transition from the low-N discrete system to the continuous one that is the main concern of Païdoussis & Deksnis' work.

The dimensionless eigenfrequencies of the articulated system are compared with those of the continuous one,† first at $u = 0$, for increasing N. As expected, for $N = 2$ or 3, the two sets are appreciably different; with increasing N, however, they converge quite rapidly. Thus, for $N = 10$ the lowest five modes in the two sets are within 2%; for $N = 20$ within 1%, for $\gamma = 0$; and only slightly less close for $\gamma = 10$ [see table and figures in Païdoussis & Deksnis (1970)].

Then, the dynamical behaviour of the system with flow is investigated for various N. Figure 3.79(a,b) gives results for $\gamma = 10$ and 100. It is seen that for $\gamma = 10$ stability is lost by flutter, no matter what N is — although the Argand diagrams show that divergence is possible at $u > u_{cf}$. An interesting observation (cf. Sections 3.5.4 and 3.5.5) is that for sufficiently low N, no S-shaped jumps are manifested in the curves. Finally, from the results for $N = 8$ it is clear that, for sufficiently high N, the stability curve of the articulated system approaches that of the continuous one; since convergence in the lower eigenfrequencies is better than in the higher ones, agreement between the $N = 8$ discrete and the continuous system is better for lower β (cf. Section 3.5.4).

The situation depicted in Figure 3.79(b) for $\gamma = 100$ is more complex. It is seen that (i) for $N = 2$, the system loses stability by flutter only if $\beta < 0.195$, and by divergence for higher β; (ii) for $N = 3$ only flutter is possible; (iii) for $N = 4$ and 8, both divergence and flutter are possible but $u_{cd} > u_{cf}$, the difference between the two stability bounds being much larger for $N = 8$.

Indeed, observing the trend with increasing N in Figure 3.79(b), it is reasonable to suppose that $u_{cd} \to \infty$ as $N \to \infty$. This resolves the paradox that, whereas for the articulated system divergence is possible (and in some cases stability is lost that way), for the continuous system no divergence can occur. These same results explain the same paradox as expressed by Benjamin (1961b): that in some cases, divergence is possible with water-flow but not with air-flow. From Figure 3.79(b) we see that, for $N = 2$, stability is lost by divergence when $\beta = 0.2$ or higher and by flutter when $\beta \simeq 10^{-3}$, these two values of β being typical for water- and air-flow experiments respectively.

The non-occurrence of divergence for $N = 3$ is explained, phenomenologically at least, in Figure 3.80. For even values of N, there generally is a mode (typically the first), which crosses the origin from positive to negative $\mathcal{I}\mathrm{m}(\omega)$, the classical divergence path. In some

†The eigenfrequencies of the continuous system have themselves been obtained from a discretized (Galerkin) model, unless $\gamma = 0$ — see Section 3.3.6; however, the discretization in this case is *analytical* rather than physical.

cases, however, it is the locus of the mode giving rise to flutter that reaches the negative $\mathscr{I}\mathrm{m}(\omega)$-axis and then crosses the origin *from instability to stability*; these cases are shown as dashed lines for $N = 2$ and 4 and low γ. This appears to be the usual path for N odd, although for $N = 5$ and 7, over a range of γ, the mode locus recrosses to instability (the upper curve in each case). However, for $N = 5$ and large or small enough γ, there are no crossings of the origin at all, thus leading to the finite closed curve shown in the figure; for $N = 3$ the area of this closed curve is simply null. Thus, in this respect also, the transition from $N = 3$ to higher odd values of N may be considered to be 'smooth'.

Experiments were also conducted, similar to those of Benjamin's, with metal tubes (diameter = 9.5–12.7 mm, $L \simeq 0.6$–1.2 m), with connector-springs made of rubber tubing secured by jubilee clips, $N = 2$, 3 or 4, and water as the fluid. In some cases, in order to increase m and hence γ, the tubes were sheathed with larger diameter tubes. Typical results are given in Table 3.7. In all the cases in (a) stability is lost by flutter, while in (b) it is lost by divergence. The experiments in (b) were conducted with springs of negligible stiffness, in which case u and γ, as defined in (3.163) are meaningless; in that case, a new dimensionalization was made, in terms of the Froude number, $F = U/(gL)^{1/2}$. For $N = 2$, $e = \frac{1}{2}$, it is shown analytically that, for divergence, $F = F_{cd} = 1/(2\beta)^{1/2}$. As seen in the table, agreement between theory and experiment is reasonably good in all cases.

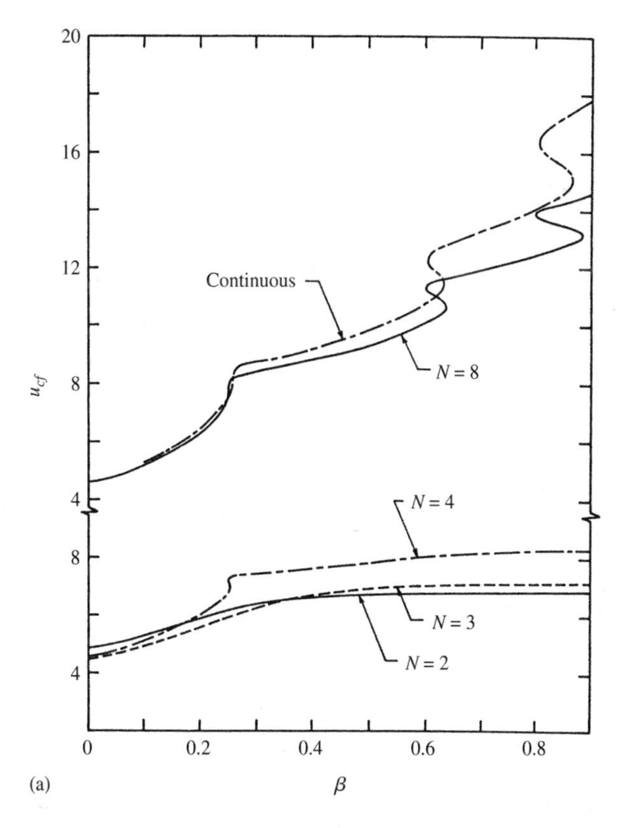

Figure 3.79 The dimensionless critical flow velocity u_{cd} for divergence and u_{cf} for flutter of the articulated cantilever for $N = 2$, 3, 4, 8 and for the continuous system ($N = \infty$) as a function of β: (a) for $\gamma = 10$; (b) for $\gamma = 100$; in both cases $e = \frac{1}{2}$ (Païdoussis & Deksnis 1970).

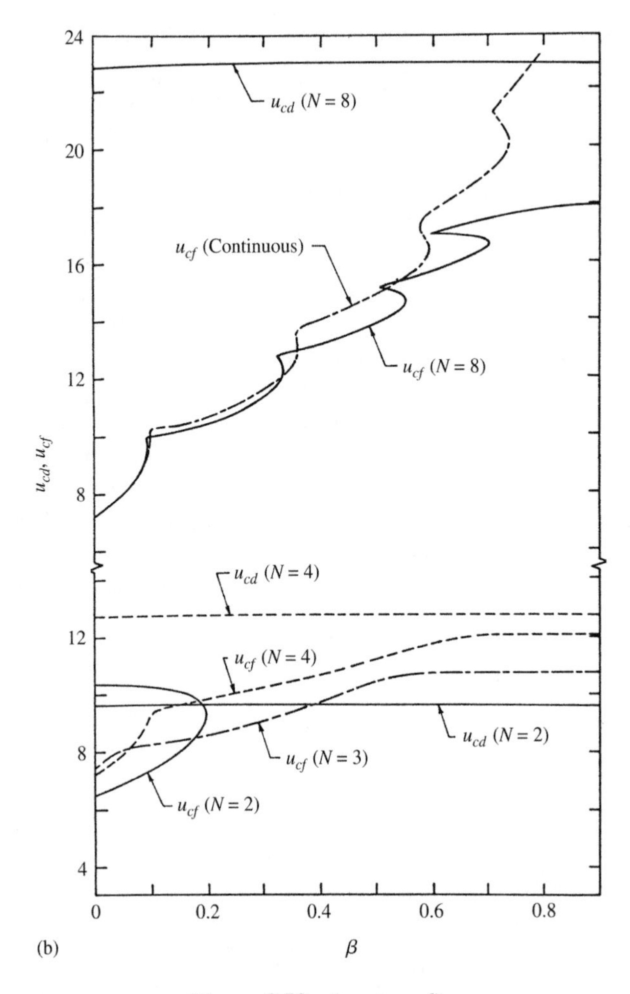

Figure 3.79 *(continued).*

Further theoretical and experimental results may be found in Païdoussis & Deksnis (1969, 1970). The final conclusions are the following. First, there is, after all, a smooth transition between the discrete system, as N is increased, and the continuously flexible one. Second, the low-N discrete system dynamics can be quite different from those of the continuous system [Figure 3.79(b)], and hence the popular two-degree-of-freedom articulated 'models', which work so well for Coriolis-free follower-force nonconservative systems, should be used with caution in the case of pipes conveying fluid if the results are meant to be extrapolated to those of the continuous system.

3.8.3 Modified systems

A very extensive and systematic study of various modified two-degree-of-freedom articulated cantilevered systems has been undertaken by Sugiyama and co-workers — 'modified' in a similar way as the continuous systems discussed in Section 3.6, by the addition of springs, masses, and so on.

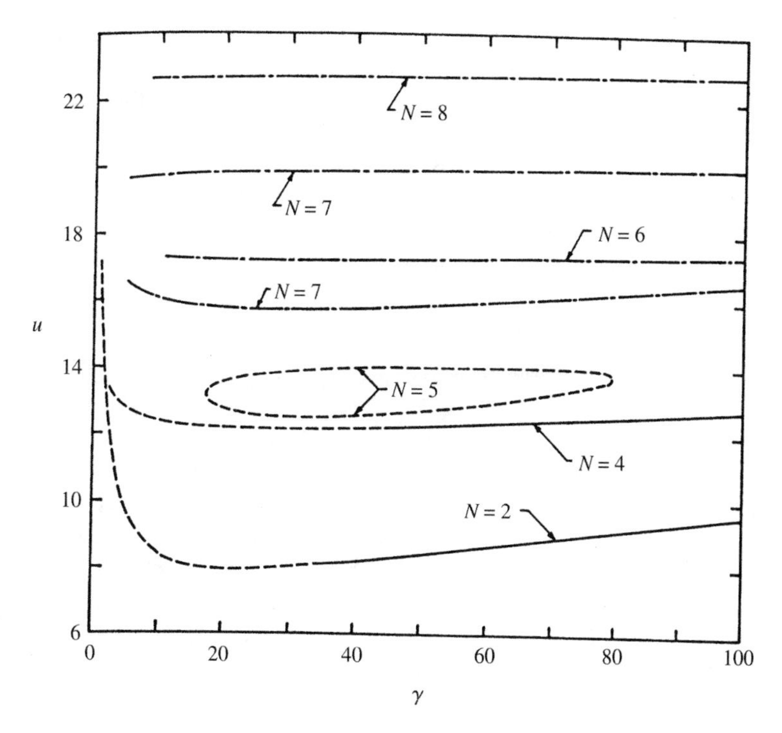

Figure 3.80 The values of u where the locus of one of the modes of the articulated system crosses the origin for $N = 2$–8 and $e = \frac{1}{2}$ as a function of γ; the meaning of the dashed lines is explained in the text (Païdoussis & Deksnis 1970).

Table 3.7 Conditions of stability: theory compared with experiments; in (a) k varied from one experiment to the next, in the range of $k = 0.33$–0.39 N m, while in (b) $k \simeq 0$ (Païdoussis & Deksnis 1970).

(a) **Flutter**

N	e	β	γ	u_{cf} (theory)	u_{cf} (exp.)
2	1	0.231	4.94	4.20	4.25
2	$\frac{1}{2}$	0.227	17.2	6.36	5.76
2	$\frac{1}{2}$	0.084	46.4	6.08	6.17
3	$\frac{1}{2}$	0.211	18.6	6.20	6.12
4	$\frac{1}{2}$	0.196	20.2	6.65	6.44

(b) **Divergence**

N	e	β	F_{cd} (theory)	F_{cd} (exp.)
2	$\frac{1}{2}$	0.258	1.39	1.50
2	$\frac{1}{2}$	0.258	1.40	1.34

The effect of some of the system parameters, e.g. the ratio of the stiffnesses and the associated damping at the joints, and the ratio of masses of the two articulations is investigated by Sugiyama & Païdoussis (1982), one aim being to find the configuration leading to the minimum value of u_{cf}.

The effect of an added lumped mass somewhere along the second segment and of damping at the articulations is examined by Sugiyama & Noda (1981), who find that the added mass virtually always destabilizes the system, as shown for example in Figure 3.81(a), both theoretically and experimentally. The notation in the figure is as

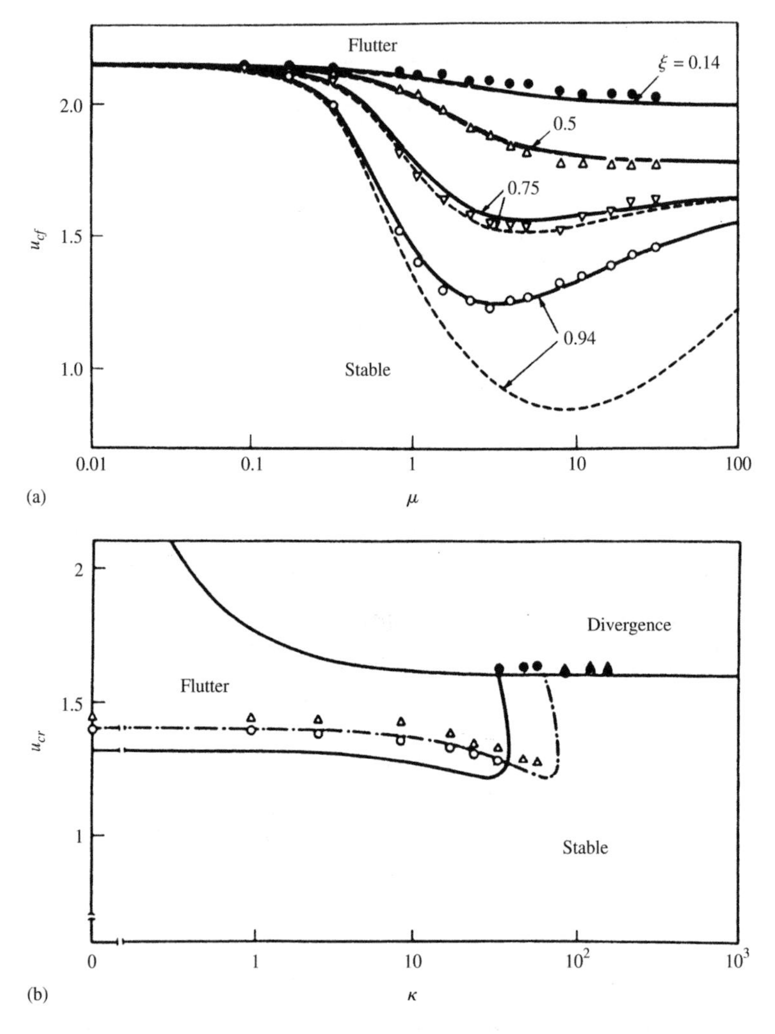

Figure 3.81 (a) The effect on stability of an added mass in the second segment of a two-segment horizontal articulated cantilever, for $\beta = 0.578$ and varying values of μ and ξ: ——, theory with measured damping; – – –, theory with no damping; •, △, ▽, ○, experiments (Sugiyama & Noda 1981). (b) The effect of an added mass–spring combination at $\xi = 0.94$ for $\beta = 0.299$ and dimensionless damping constant $\sigma = c/[k(m+M)l^3]^{1/2} = 0.0074$ for varying κ : ——, theoretical results for $\mu = 9.86$; — - — , theoretical results for $\mu = 18.5$; ○, △, corresponding experimental results for flutter; •, ▲, for divergence (Sugiyama 1984).

follows: $\mu = m_a/[(m+M)l]$ and $\xi = l_a/l$, where m_a is the added mass and l_a its location, measured from the beginning of the second segment of the system; $u = (Ml/k)^{1/2}U$.

The effect of an added spring–mass combination at a variable location in the second segment of the system is studied by Sugiyama (1984). Typical results are shown in Figure 3.81(b); u, μ and ξ are as just defined, while $\kappa = Kl^2/k$, K being the added spring stiffness, while k is the stiffness of the articulation joints. As seen in the figure, the system is generally subject to flutter for small κ and to divergence for higher κ (cf. Figures 3.64 and 3.65).

Finally, Figure 3.82(a) shows flutter of the system with an added dashpot, just before the second articulation joint, of the type discussed in Section 3.6.4. It is shown theoretically and experimentally (Sugiyama 1986a,b) that the dashpot is stabilizing if placed on the first segment of the system, but can be destabilizing if placed sufficiently far along the second segment, as shown in Figure 3.82(b); $\overline{\sigma} = c_a l/[k(m+M)l]^{1/2}$, where c_a is the added dashpot constant. The stabilization/destabilization mechanism is also discussed, and in the second case is shown to be related to a phase shift which facilitates energy transfer from the fluid to the pipe.

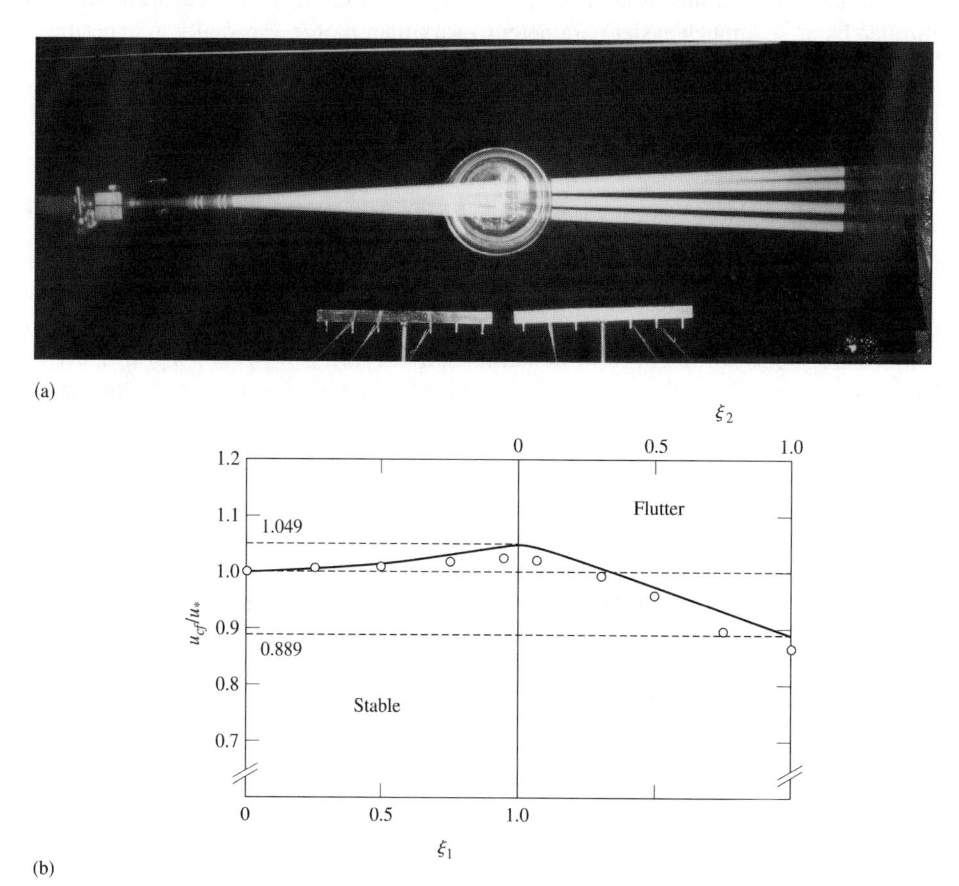

Figure 3.82 (a) The articulated cantilever with an added dashpot in flutter. (b) The effect of location of the dashpot on the first (at ξ_1) or the second (at ξ_2) segment, for $\beta = 0.575$, $\sigma = 1.7 \times 10^{-2}$, and $\overline{\sigma} = 0.59$ (Sugiyama *et al*. 1986a,b).

In all the experiments, the system was made of metal tubes interconnected by short rubber-pipe segments as in the foregoing, and it was suspended in a horizontal plane, much as in Figure 3.44. The design of the joints was much refined, however, and this is partly responsible for the excellent agreement with theory that has been achieved by Sugiyama and his colleagues.

A great deal of high-quality, interesting theoretical and experimental results have been obtained in all of this work, mostly anticipating those of the continuous system (Section 3.6). For that reason, it has been discussed here extremely briefly, but the interested reader is encouraged to refer to the original papers.

3.8.4 Spatial systems

A two-degree-of-freedom vertical articulated cantilever, with the lower tube out of plane by an angle ψ, is considered by Bohn & Herrmann (1974b), so that motions of the upper segment are constrained to occur in one plane and those of the lower one in another. The main advantage in this system is that the type of instability to occur turns out to be controlled by ψ — a much easier parameter to vary than β or γ, especially in experiments.

The equations of motion are again derived via equation (3.10). The linearized dimensionless equations are

$$\begin{aligned}(a+1)^3\ddot{\theta} + \tfrac{1}{2}(2+3a)(\cos\psi)\ddot{\phi} + (a+1)^2\bar{u}\dot{\theta} + (1+2a)(\cos\psi)\bar{u}\dot{\phi}&\\ + (a+1)^2\theta + \kappa_1\theta + (1+a\bar{u}^2/\bar{\beta})\phi = 0,&\qquad (3.164)\\ \tfrac{1}{2}(2+3a)(\cos\psi)\ddot{\theta} + \ddot{\phi} + \bar{u}(\cos\psi)\dot{\theta} + \bar{u}\dot{\phi} + (\cos\psi)\theta + (\kappa_2+1)\phi = 0,&\end{aligned}$$

where, since in most of the cases studied $k_1 = k_2 = 0$, the nondimensional quantities are slightly different from Benjamin's: $a = l_1/l_2$, $\bar{\beta} = 3\beta = 3M/(M+m)$, $\bar{u} = \bar{\beta}U/(\frac{3}{2}gl_2)^{1/2}$, $\tau = (\frac{3}{2}g/l_2)^{1/2}t$, and $\kappa_i = k_i/[\frac{1}{2}(M+m)l_2^2 g]$, $i = 1, 2$; the dot denotes $\mathrm{d}(\ )/\mathrm{d}\tau$.

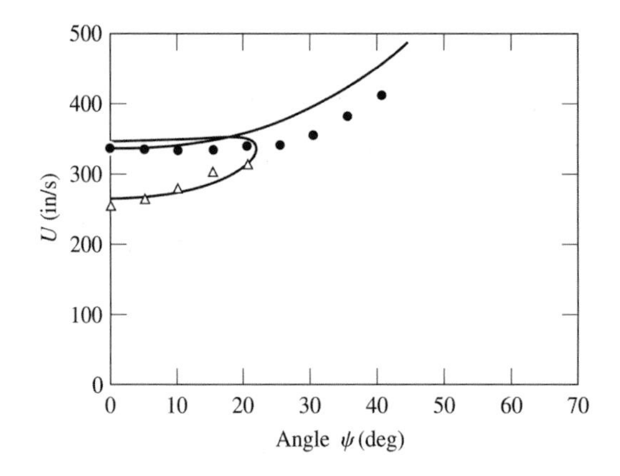

Figure 3.83 Theoretical and experimental results of U_{cr} (in/s; 1 in = 25.4 mm) for a 'spatially deformed' articulated system, by an angle ψ, for $\bar{\beta} = 0.328$: —— theory; •, experiment, divergence; △, experiment, flutter (Bohn & Herrmann 1974).

It is found that in the case of $\beta > \frac{1}{6}$ stability is lost by divergence — as found by Benjamin for $\psi = 0$ — no matter what the value of ψ. In the case of $\beta < \frac{1}{6}$, however, stability can be lost either by divergence of by flutter, depending on ψ. Typical results are shown in Figure 3.83, together with experiments. It is not clear from the text in Bohn & Herrmann (1974b) whether the theoretical results have been computed with the full theoretical model, i.e. taking into account k_1 and k_2 and also the damping at the joints (ignored here).

In the experiments, the joints were made of ball bearings and light brackets, attached as a 'backing' to the pipe system, while the flow was conducted from the upper to the lower tube by a latex-tubing connector. Great care had to be exercised in eliminating, to the extent possible, small eccentricities and the effect of a permanent bow in the latex tubing, as well as controlling and measuring the stiffening of this tubing with increasing internal pressure. Nevertheless, agreement between theory and experiment in Figure 3.83 is good, notably demonstrating the existence of a critical value of ψ, separating the domains of flutter and divergence. One unusual feature of the results in Figure 3.83 is that, apparently, divergence at a higher flow velocity than flutter materializes; unfortunately, this is not discussed by Bohn & Herrmann.

4
Pipes Conveying Fluid: Linear Dynamics II

4.1 INTRODUCTION

The linear dynamics of the basic system of a pipe conveying fluid has been considered in detail in Chapter 3, including, in more abbreviated form, the dynamics of some important modified systems (Section 3.6). A characteristic of all these systems, if they are continuously flexible, is that they are all governed by equations (3.38) and (3.70) and the dimensionless parameters of (3.71), or by simple variants thereof. Furthermore, solution of these equations may generally be achieved by one of the two methods of Section 3.3.6, or by straightforward extensions of these methods. The only 'unusual' system in this respect is that of articulated pipes, dealt with in Section 3.8.

The systems considered here, on the other hand, either are governed by substantially modified forms of the equations of motion or require different methods of solution. Specifically, the following topics are discussed. *Nonuniform pipes* are pipes with nonuniform cross-section and axially variable flow area. *Aspirating* or *sucking pipes* are pipes ingesting flow at a free end, rather than expelling it. *Short pipes* also require special treatment: from the solid mechanics side the use of Timoshenko beam theory, and from the fluid mechanics side the use of potential flow theory and the introduction of so-called 'outflow models' for the fluid discharging to atmosphere. *Pipes with harmonically perturbed flow velocity* are subject to parametric resonances and require special methods of solution; so does the treatment of *forced vibration* of pipes conveying fluid. Finally, the section on *applications* presents some expected and unexpected uses of the work discussed in Chapters 3 and 4.

4.2 NONUNIFORM PIPES

4.2.1 The equation of motion

The equation of motion will be derived for a pipe with a nonuniform flow passage and, generally, a nonuniform external form also. Variations in the shape of the pipe are axisymmetric, gradual and smooth with respect to the axial coordinate [see Figure 4.1(a)]. The pipe is immersed in air or water, so that hydrostatic, added-mass and damping effects associated with the external fluid need generally be taken into account.

In this derivation (Hannoyer & Païdoussis 1979a), the lateral dimensions of the flow passage will not *a priori* be considered to be negligible. However, the other assumptions made in Section 3.3.2 for uniform pipes are also made here, namely that motions are small, the flow is fully developed turbulent, the curvature of flow trajectories is small, etc. It is also assumed that (i) the profile of the axial component of the flow velocity, U_i, is uniform, and (ii) there are no significant secondary flows, other than that associated with changes in the cross-sectional flow area of the tubular beam. For simplicity, the flow velocity is assumed not to be time-varying. The subscript i, as in U_i, is added for two reasons: (a) since there is also an external fluid, to distinguish internal- and external-fluid properties, e.g. the densities ρ_i and ρ_e; (b) to facilitate the analysis in Chapter 8 (Volume 2) of the same system but with the outer fluid *flowing* with mean velocity, U_e.

In the following, the rate of change of the momentum of the flow associated with motions of the pipe will be derived first. This is then used in a Newtonian derivation of the equation of motion.

In the analysis, an inertial coordinate system $\{x, y, z\}$ is used, as shown in Figure 4.1(a). However, for convenience, a non-inertial frame $\{\xi, \eta, \zeta\}$ embedded in a cross-section of the pipe [Figure 4.1(b,c)] and centered at O in a cross-section of the pipe is also used. The conduit is assumed to be locally conical, with angle β_i sufficiently small for velocity terms of order β_i^2 to be negligible. On the centreline, the absolute velocity of the fluid, $\mathbf{V}_i$, is equal to the relative velocity on the centreline, $\mathbf{U}_i$, plus the velocity of the centreline, $\partial\mathbf{w}/\partial t$. Axial motion of the pipe is negligible (cf. Section 3.3.2); however, the effect of rotation needs generally to be taken into account. Thus, for a point off the centreline, the flow velocity relative to the pipe is $\mathbf{W}_i = \mathbf{U}_i + \mathbf{\Omega} \times \boldsymbol{\eta}$ [Figure 4.1(c)], where $\Omega = \partial^2 w/\partial x\, \partial t$ in the ζ-direction — obtained by assuming that the fluid essentially slips at the boundary and by neglecting second-order terms with respect to β_i.

The rate of change of the flow momentum is here derived via a control volume approach. In this case a convenient control volume, $\Delta\mathcal{V}$, is an elemental slice of the fluid in a cross-section of the pipe, of thickness $\delta\xi$. The rate of change of momentum in $\Delta\mathcal{V}$ may be expressed in terms of the material derivative of $\mathbf{V}_i$ as in equation (3.30). Alternatively and more conveniently, the rate of change of the flow momentum relative to the noninertial control volume attached to the tubular beam may be evaluated, and then the d'Alembert (apparent) body forces added to it, as follows:

$$\frac{\mathrm{d}}{\mathrm{d}t}\iiint_{\Delta\mathcal{V}} \rho_i \mathbf{V}_i\, \mathrm{d}(\Delta\mathcal{V}) = \iint_{\Delta S} \rho_i \mathbf{W}_i[\mathbf{W}_i \cdot \mathbf{n}\, \mathrm{d}(\Delta S)] + \frac{\partial}{\partial t}\iiint_{\Delta\mathcal{V}} \rho_i \mathbf{W}_i\, \mathrm{d}(\Delta\mathcal{V})$$
$$+ \iiint_{\Delta\mathcal{V}} \rho_i \left[\frac{\mathrm{d}^2\mathbf{R}}{\mathrm{d}t^2} + 2\mathbf{\Omega} \times \mathbf{W}_i + \mathbf{\Omega}\times(\mathbf{\Omega} \times \mathbf{r}) + \frac{\mathrm{d}\mathbf{\Omega}}{\mathrm{d}t}\times\mathbf{r} + \mathbf{a}_{\mathrm{rel}}\right] \mathrm{d}(\Delta\mathcal{V}), \quad (4.1)$$

where the surface integral represents the momentum flux across the surface ΔS of the noninertial control volume, the next integral represents the rate of change of momentum within the control volume, and the last integral the apparent (pseudo) body forces. $\mathbf{W}_i$ is the flow velocity of any point within $\Delta\mathcal{V}$, i.e. for any stream tube, not necessarily along the pipe centreline; $\mathbf{n}$ is the unit vector normal to the surface element $\mathrm{d}(\Delta S)$. $\mathbf{R}$ is the position vector of the origin O of the noninertial $\{\xi, \eta, \zeta\}$ frame *vis-à-vis* $\{x, y, z\}$, while $\mathbf{r}$ is the position vector of any point within $\Delta\mathcal{V}$ in the $\{\xi, \eta, \zeta\}$ frame; here, $\mathbf{r}$ is of the order of the pipe radius and therefore small, the pipe being slender; $\mathbf{a}_{\mathrm{rel}}$ is the fluid acceleration *vis-à-vis* the noninertial frame.

Each of the integrals in (4.1) will now be evaluated in turn. Because of the impermeability of the walls, the net momentum flux across ΔS is merely the difference between

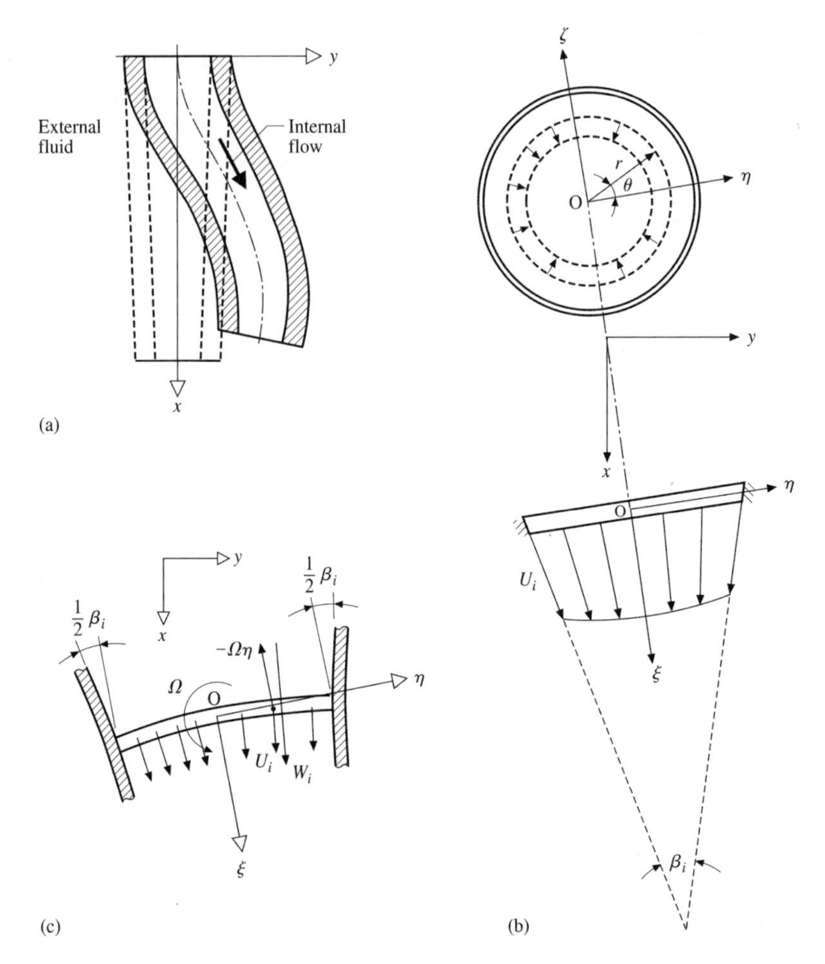

Figure 4.1 (a) Schematic of the system under consideration; (b) coordinate systems, a cross-section of the pipe, and an element of the fluid in a locally conical segment of the pipe; (c) flow velocities within the fluid element (Hannoyer 1977; Hannoyer & Païdoussis 1979a).

the fluxes across the flat surfaces in the flow direction, and it may be written as

$$\frac{\partial}{\partial x}\left\{\iint_{A_i} \rho_i \mathbf{W}_i(\mathbf{W}_i \cdot \mathbf{n}\, \mathrm{d}A_i)\, \mathrm{d}x\right\} = \rho_i\, \mathrm{d}x \iint_{A_i} \frac{\partial \mathbf{W}_i}{\partial x} W_i\, \mathrm{d}A_i, \tag{4.2}$$

by invoking continuity for each streamtube; A_i is the cross-sectional area of the flow conduit.

Since the control volume remains constant, the second integral on the right-hand side of (4.1) may be written as

$$s\frac{\partial}{\partial t}\iiint_{\Delta\mathcal{V}} \rho_i \mathbf{W}_i\, \mathrm{d}(\Delta\mathcal{V}) = \rho_i\, \mathrm{d}x \iint_{A_i} \frac{\partial \mathbf{W}_i}{\partial t}\, \mathrm{d}A_i \simeq \rho_i\, \mathrm{d}x \iint_{A_i} \frac{\partial \mathbf{W}_i}{\partial x}(-\Omega\eta)\, \mathrm{d}A_i; \tag{4.3}$$

in the last step the fact that $\mathbf{W}_i$ changes because of rotation of the control volume has been utilized, so that $(\partial \mathbf{W}_i/\partial t) = (\partial \mathbf{W}_i/\partial s)(\partial s/\partial t) \simeq (\partial \mathbf{W}_i/\partial x)(-\Omega\eta)$, where s here is the coordinate along a streamtube off the centreline (and should not be confused with the s used in Sections 3.3.1 and 3.3.2). Hence, the sum of (4.2) and (4.3) gives

$$\rho_i \,\mathrm{d}x \iint_{A_i} \frac{\partial \mathbf{W}_i}{\partial x}(W_i - \Omega\eta)\,\mathrm{d}A_i = \rho_i \,\mathrm{d}x \iint \frac{\partial \mathbf{W}_i}{\partial x} U_i \,\mathrm{d}A_i \simeq \rho_i A_i U_i \frac{\partial \mathbf{U}_i}{\partial x}\,\mathrm{d}x; \tag{4.4}$$

the intermediate result is obtained with the aid of Figure 4.1(c), while the last step is reached through neglect of second-order terms.

Since $\|\mathbf{r}\|$ is small and $\mathbf{a}_{\mathrm{rel}}$ is negligible, the last integral of (4.1) may be approximated as follows:

$$\iiint_{\Delta\mathscr{V}} \left[\frac{\partial^2 \mathbf{w}}{\partial t^2} + 2\boldsymbol{\Omega} \times \mathbf{W}_i\right] \rho_i \,\mathrm{d}(\Delta\mathscr{V}) \simeq \rho_i A_i \left(\frac{\partial^2 \mathbf{w}}{\partial t^2} + 2U_i \frac{\partial^2 \mathbf{w}}{\partial x\,\partial t}\right)\mathrm{d}x, \tag{4.5}$$

in which it is recalled that $\mathbf{w}$ is the vector displacement of the pipe centreline in the y-direction. The second term in (4.5) is obtained through the following sequence of operations: $2\boldsymbol{\Omega} \times \mathbf{W}_i = 2\boldsymbol{\Omega} \times \mathbf{U}_i(1 + \Omega\eta/U_i) \simeq 2\boldsymbol{\Omega} \times \mathbf{U}_i = 2\left[-(\partial^2 w/\partial x\,\partial t)\mathbf{k}\right] \times (U_i\mathbf{i}) = 2U_i(\partial^2 w/\partial x\,\partial t)\mathbf{j} = 2U_i(\partial^2 \mathbf{w}/\partial x\,\partial t)$, where $\{\mathbf{i}, \mathbf{j}, \mathbf{k}\}$ are unit vectors associated with $\{\xi, \eta, \zeta\}$. Throughout, the small inclination of the $\{\xi, \eta\}$-plane *vis-à-vis* the $\{x, y\}$-plane is utilized, subject to order-of-magnitude constraints. Hence, combining (4.4) and (4.5), the rate of change of fluid momentum is

$$\rho_i A_i \left[\frac{\partial^2 \mathbf{w}}{\partial t^2} + 2U_i \frac{\partial^2 \mathbf{w}}{\partial x\,\partial t} + U_i \frac{\partial \mathbf{U}_i}{\partial x}\right]\mathrm{d}x, \tag{4.6}$$

which yields components per unit length in the x- and y-direction, respectively equal to

$$\rho_i A_i U_i \frac{\mathrm{d}U_i}{\mathrm{d}x} \quad \text{and} \quad \rho_i A_i \left[\frac{\partial^2 w}{\partial t^2} + 2U_i \frac{\partial^2 w}{\partial x\,\partial t} + U_i \frac{\partial}{\partial x}\left(U_i \frac{\partial w}{\partial x}\right)\right]. \tag{4.7}$$

The second expression may be written in the compact form $\rho_i A_i \mathscr{D}^2 w$, where $\mathscr{D} = [(\partial/\partial t) + U_i(\partial/\partial x)]$, and

$$\mathscr{D}^2 w = \mathscr{D}[\mathscr{D}w] = \left[\frac{\partial^2 w}{\partial t^2} + 2U_i \frac{\partial^2 w}{\partial x\,\partial t} + U_i \frac{\partial}{\partial x}\left(U_i \frac{\partial w}{\partial x}\right)\right]. \tag{4.8}$$

It is instructive to note that there are no terms involving $\mathrm{d}A_i/\mathrm{d}x$ in (4.7), as there would have been if the lateral momentum change had erroneously been evaluated by a simplistic application of the formula $[(\partial/\partial t) + U_i(\partial/\partial x)]\{\rho_i A_i[(\partial w/\partial t) + U_i(\partial w/\partial x)]\}$!

Now, the next steps in the derivation of the equation of motion may be taken. Working in a similar way as in Section 3.3.2 (cf. Figure 3.6) by considering an element $\delta\xi$ of the pipe [Figure 4.2(a)], force balances in the x- and y-direction and a moment balance yield

$$\frac{\partial T}{\partial x} + F_{it} + F_{et} - (F_{in} + F_{en})\frac{\partial w}{\partial x} + \frac{\partial}{\partial x}\left(Q \frac{\partial w}{\partial x}\right) + mg = 0, \tag{4.9a}$$

$$\frac{\partial Q}{\partial x} + F_{in} + F_{en} + (F_{it} + F_{et})\frac{\partial w}{\partial x} + \frac{\partial}{\partial x}\left(T \frac{\partial w}{\partial x}\right) = m \frac{\partial^2 w}{\partial t^2}, \tag{4.9b}$$

$$Q + \frac{\partial}{\partial x}\left[\left(E^* \frac{\partial}{\partial t} + E\right) I \frac{\partial^2 w}{\partial x^2}\right] + \frac{\partial \mathscr{M}_f}{\partial x} = 0, \tag{4.9c}$$

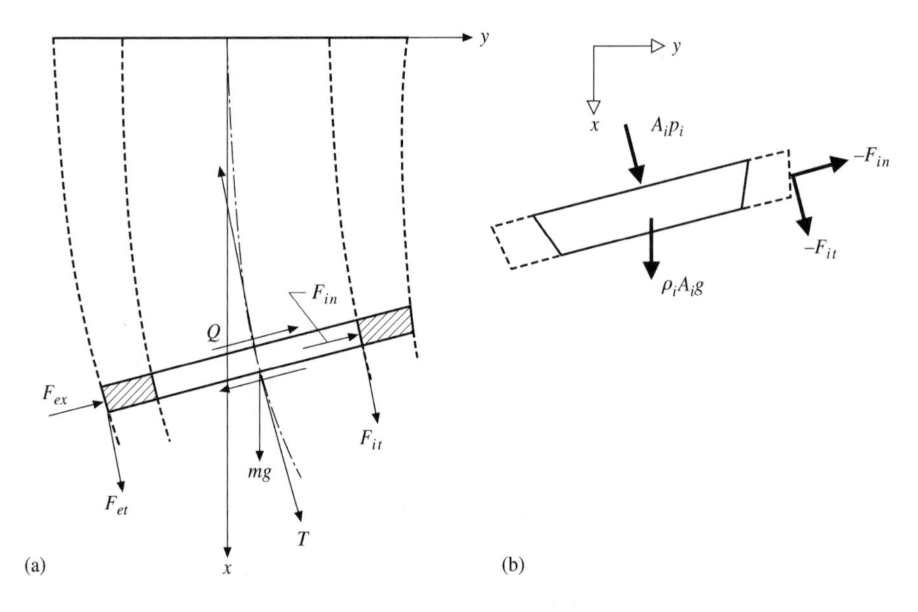

Figure 4.2 (a) An element of the pipe showing forces and moments acting on it; (b) an element of the contained fluid showing forces acting on it. Note that $Q + (\partial Q/\partial x)\,\delta x$ and $p_i A_i + (\partial/\partial x)(p_i A_i)\,\delta x$ on the lower surfaces have been omitted in these diagrams (Hannoyer 1977).

in which F_{it} and F_{in} are the tangential and normal components of the fluid–pipe interaction forces associated with the internal flow (equivalent to qS and F, respectively, in Figure 3.6), and F_{et} and F_{en} are the corresponding terms associated with the external, stagnant fluid;[†] T, Q, E, E^*, I, and m are the same as before, for uniform pipes: the tension, transverse shear force, modulus of elasticity, Kelvin–Voigt dissipation constant, area-moment of inertia, and mass per unit length, respectively. The term $\mathcal{M}_f$ is the fluid-related moment due to both internal flow and external fluid, which for a pipe of nonuniform cross-section may not tacitly be assumed to be zero.

Similarly, utilizing equations (4.7) and (4.8), x- and y-direction force balances on an element of the fluid [cf. equations (3.18) and (3.19) of Section 3.3.2 and Figure 4.2(b)] give

$$F_{it} - F_{in}\frac{\partial w}{\partial x} = -\frac{\partial}{\partial x}(p_i A_i) + \rho_i A_i g - \rho_i A_i U_i \frac{\mathrm{d}U_i}{\mathrm{d}x}, \tag{4.10a}$$

$$F_{in} + F_{it}\frac{\partial w}{\partial x} = -\frac{\partial}{\partial x}\left(p_i A_i \frac{\partial w}{\partial x}\right) - \rho_i A_i \mathfrak{D}^2 w, \tag{4.10b}$$

where $A_i = A_i(x)$, and $\rho_i A_i$ is what was previously called M, and $\mathfrak{D}^2 w$ has been defined in (4.8).

The external fluid, being stagnant, contributes only hydrostatic, inertial (added mass) and damping terms: respectively equal to the buoyancy force, $-\rho_e A_e g$, and to $-\rho_e A_e(\partial^2 w/\partial t^2)$ and $-\rho_e D_e U_v(\partial w/\partial t)$, where $U_v = (\mu_e C_D/\rho_e D_e)$ has the dimensions of

[†]The formal manner in which the external fluid forces are taken into account here is useful for later analysis, where F_{en} and F_{et} will be associated more generally with external *flow* (Chapter 8).

velocity, D_e is the external diameter of the pipe, μ_e is the dynamic viscosity, and C_D an empirical coefficient dependent on Stokes' number — see Section 2.2.1(g) and 2.2.3 and, for the viscous component, also Païdoussis (1973b) and Hannoyer & Païdoussis (1978). Hence, a balance of forces due to the external fluid gives

$$F_{et} - F_{en}\frac{\partial w}{\partial x} = -\rho_e A_e g + \frac{\partial}{\partial x}(p_e A_e), \tag{4.11a}$$

$$F_{en} + F_{et}\frac{\partial w}{\partial x} = \frac{\partial}{\partial x}\left(p_e A_e \frac{\partial w}{\partial x}\right) - \rho_e A_e \frac{\partial^2 w}{\partial t^2} - \rho_e D_e U_v \frac{\partial w}{\partial t}. \tag{4.11b}$$

The form of the pressure forces in equations (4.11a) and (4.11b) is clarified in Chapter 8; here one may simply accept it by similarity to the internal flow terms in equations (4.10a,b).

The evaluation of the $\partial \mathcal{M}_f/\partial x$ term in (4.9c) is quite tedious and will not be reproduced here. Suffice it to say that careful study (Hannoyer 1977) has shown that

$$-\frac{\partial \mathcal{M}_f}{\partial x} = \frac{\rho_i A_i}{2\pi}\frac{\mathrm{d}A_i}{\mathrm{d}x}\mathcal{D}^2 w + \frac{\rho_e A_e}{2\pi}\frac{\mathrm{d}A_e}{\mathrm{d}x}\frac{\partial^2 w}{\partial t^2}. \tag{4.12}$$

Equations (4.9a), (4.10a) and (4.11a) may be combined to give

$$\frac{\partial}{\partial x}[T + p_e A_e - p_i A_i - \rho_i (A_i U_i) U_i] = (\rho_e A_e - \rho_i A_i - m)g, \tag{4.13}$$

in which the fact that $A_i U_i$ is constant has been recognized. Then, by combining (4.9b,c), (4.10b) and (4.11b) and utilizing (4.13), the equation of lateral motion becomes

$$\begin{aligned}
&\frac{\partial^2}{\partial x^2}\left[\left(E^*\frac{\partial}{\partial t} + E\right) I \frac{\partial^2 w}{\partial x^2}\right] - \frac{\rho_i A_i}{2\pi}\frac{\mathrm{d}A_i}{\mathrm{d}x}\frac{\partial}{\partial x}(\mathcal{D}^2 w) - \frac{\rho_e A_e}{2\pi}\frac{\mathrm{d}A_e}{\mathrm{d}x}\frac{\partial}{\partial x}\left(\frac{\partial^2 w}{\partial t^2}\right)\\
&+ \rho_i A_i\left[\mathcal{D}^2 - U_i\frac{\mathrm{d}U_i}{\mathrm{d}x}\frac{\partial}{\partial x}\right] w + \rho_e A_e \frac{\partial^2 w}{\partial t^2} - (\rho_e A_e - \rho_i A_i - m) g \frac{\partial w}{\partial x}\\
&- (T + p_e A_e - p_i A_i)\frac{\partial^2 w}{\partial x^2} + m\frac{\partial^2 w}{\partial t^2} = 0;
\end{aligned} \tag{4.14}$$

it is important to note that, in the dominant term $\rho_i A_i[\mathcal{D}^2 - U_i(\mathrm{d}U_i/\mathrm{d}x)](\partial w/\partial x)$, the $U_i(\mathrm{d}U_i/\mathrm{d}x)(\partial w/\partial x)$ component cancels out once $\mathcal{D}^2 w$ is expanded — and this is true irrespective of magnitude considerations.

We next proceed to evaluate the only unspecified quantity in (4.14), namely that related to $T + p_e A_e - p_i A_i$. By integrating (4.13),

$$\mathcal{T}(x) = \mathcal{T}(L) - \rho_i (A_i U_i) U_i \Big|_x^L - \int_x^L (\rho_e A_e - \rho_i A_i - m) g \,\mathrm{d}x \tag{4.15}$$

is obtained, in which

$$\mathcal{T}(x) = (T + p_e A_e - p_i A_i); \tag{4.16}$$

it is recalled that T, p, A, U and m, unless otherwise denoted, are functions of x. Two cases will be analysed, separately, as follows.

(a) *Free or free-to-slide-axially downstream end.* In this case it is presumed that no externally imposed tensioning is possible; it is also assumed that the internal fluid discharges into the external fluid at $x = L$ and that $p_i(L) \simeq p_e(L)$, equal to the hydrostatic pressure at that point. Thus, $\mathcal{T}(L) = p(L)[A_e(L) - A_i(L)]$, which may be rewritten in terms of a drag coefficient

$$\mathcal{T}(L) = \tfrac{1}{2}\rho_i A_i U_i^2 C_{fi}; \tag{4.17}$$

it is recognized that, since $(A_e - A_i)_L$ is small, $\mathcal{T}(L)$ will be small and may alternatively be neglected.

(b) *Supported end with no axial sliding.* In this case,

$$\mathcal{T}(L) = \overline{T} + [T + p_e A_e - p_i A_i]_L, \tag{4.18}$$

where $\overline{T}$ represents a possible externally applied tension. The second term is evaluated by considering the flow-related terms by themselves and imposing the condition that the axial strain ε_x satisfy $\int_0^L \varepsilon_x \, \mathrm{d}x = 0$, as in the derivation of equation (3.37). It is noted that $\varepsilon_x = [\sigma_{xx} - \nu(\sigma_{rr} + \sigma_{\theta\theta})]/E$, in which $\sigma_{xx} = T(x)/A(x)$, where $A(x) = A_e(x) - A_i(x) \equiv (A_e - A_i)_x$, and ν is the Poisson ratio; furthermore, $\sigma_{rr} + \sigma_{\theta\theta} \simeq 2(p_i A_i - p_e A_e)/(A_e - A_i)_x$ by assuming that the tubular beam area variations are sufficiently gradual for the stress distribution applicable to a uniform tubular beam subjected to uniform internal and external pressure to hold true for each cross-section. Hence, one finds

$$\begin{aligned}(T + p_e A_e - p_i A_i)_L &\int_0^L \frac{\mathrm{d}x}{(A_e - A_i)_x} \\ &= (1 - 2\nu)\int_0^L \frac{(A_e p_e - A_i p_i)_x}{(A_e - A_i)_x}\,\mathrm{d}x + \rho_i (A_i U_i) \int_0^L \frac{U_i(L) - U_i(x)}{(A_e - A_i)_x}\,\mathrm{d}x, \end{aligned} \tag{4.19}$$

from which $(T + p_e A_e - p_i A_i)_L$ may be obtained if the form of $A_e(x), A_i(x)$ and the pressure distributions are known. In general, one may write

$$[T + p_e A_e - p_i A_i]_L = (1 - 2\nu)[p_e A_e - p_i A_i]_L f_1 + \rho_i (A_i U_i) U_i(L) f_2, \tag{4.20}$$

in which f_1 and f_2 must be obtained via (4.19). It is of interest to note that for a *uniform* tubular beam internally pressurized by $\overline{p}_i$ and immersed in a uniform ambient pressure, the second term in (4.20) vanishes while the first gives $-(1 - 2\nu)\overline{p}_i A_i$, thus retrieving the results of Section 3.3.2. It should also be noted that, unless pressurization effects exist, both f_1 and f_2 are very small terms which may be neglected for slightly tapered tubular beams.

Hence, the equation of small motions of the system, subject to all the assumptions and approximations made, is

$$\begin{aligned}\frac{\partial^2}{\partial x^2}\left[\left(E^* \frac{\partial}{\partial x} + E\right) I \frac{\partial^2 w}{\partial x^2}\right] &- \frac{\rho_i A_i}{2\pi}\frac{\mathrm{d}A_i}{\mathrm{d}x}\frac{\partial}{\partial x}\left\{\frac{\partial^2 w}{\partial t^2} + 2U_i \frac{\partial^2 w}{\partial x \partial t} + U_i^2 \frac{\partial^2 w}{\partial x^2}\right\} \\ - \frac{\rho_e A_e}{2\pi}\frac{\mathrm{d}A_e}{\mathrm{d}x}\frac{\partial}{\partial x}\left\{\frac{\partial^2 w}{\partial t^2}\right\} &+ \rho_i A_i \left\{\frac{\partial^2 w}{\partial t^2} + 2U_i \frac{\partial^2 w}{\partial x \partial t} + U_i^2 \frac{\partial^2 w}{\partial x^2}\right\} + \rho_e A_e \frac{\partial^2 w}{\partial t^2} \\ - \Big\{\mathcal{T}(L) + \rho_i A_i U_i [U_i - U_i(L)] &- \int_x^L (\rho_e A_e - \rho_i A_i - m) g\Big\} \frac{\partial^2 w}{\partial x^2}\end{aligned}$$

$$+ \rho_e D_e U_v \frac{\partial w}{\partial t} - [(\rho_e A_e - \rho_i A_i - m)g]\frac{\partial w}{\partial x} + m\frac{\partial^2 w}{\partial t^2} = 0, \tag{4.21}$$

in which $[\mathrm{d}A_i/\mathrm{d}x](\partial/\partial x)\{(\mathrm{d}U_i/\mathrm{d}x)(\partial w/\partial x)\}$ in the second, fluid-moment-related term has been neglected, as it is of second order for small taper angles; $\mathcal{T}(L)$ is given either by equation (4.17) or by (4.18)–(4.20). It is obvious that the second and third terms in equation (4.21), which are related to the fluid-related moment [equation (4.12)] are quite small as compared to, say, the fourth term; indeed, for sufficiently small $\mathrm{d}A_i/\mathrm{d}x$ and $\mathrm{d}A_e/\mathrm{d}x$, they may be neglected, and this is one of the reasons for not giving the derivation of $\partial \mathcal{M}_f/\partial x$ here in detail.

The boundary conditions are the same as for uniform tubular beams, e.g. equations (3.77) or (3.78).

The equations of motion and boundary conditions may be rendered dimensionless by the following set of nondimensional parameters:

$$\begin{aligned}
&\xi = x/L, \qquad \eta = w/L, \qquad \tau = [EI/(m + \rho_e A_e + \rho_i A_i)]^{1/2}_{\xi=0} t/L^2, \\
&\delta^2 = [A_i/A_e]_{\xi=0}, \qquad \sigma_e = A_e(\xi)/A_e(0), \qquad \sigma_i = A_i(\xi)/A_i(0), \qquad \epsilon = L/D_e(0), \\
&\nu_d = [I/\{E(m + \rho_e A_e + \rho_i A_i\}]^{1/2}_{\xi=0} E^*/L^2, \qquad \Theta = \mathcal{T}(L)L^2/EI(0), \qquad \Pi = \overline{T}L^2/EI(0), \\
&u_i = [\rho_i A_i/EI]^{1/2}_{\xi=0} U_i(0)L, \;\; c_v = [p_e A_e/EI]^{1/2}_{\xi=0} U_v L = [\rho_e A_e/EI]^{1/2}_{\xi=0}(\mu_e C_D/\rho_e D_e)L, \\
&\gamma = [\rho A_e/EI]_{\xi=0} g L^3, \qquad \gamma_e = 1 + \rho_e/\rho, \qquad \gamma_i = (\rho_i/\rho - 1)\delta^2,
\end{aligned} \tag{4.22}$$

where $\rho = m/(A_e - A_i)$. The equation of motion in dimensionless terms is then given by

$$\begin{aligned}
&\left(1 + \nu_d \frac{\partial}{\partial \tau}\right) \frac{\partial^2}{\partial \xi^2}\left\{\frac{\sigma_e - \delta^4\sigma_i}{1 - \delta^4}\frac{\partial^2 \eta}{\partial \xi^2}\right\} - \left\{\frac{\delta^2 u_i^2}{8\epsilon^2\sigma_i}\frac{\mathrm{d}\sigma_i}{\mathrm{d}\xi}\right\}\frac{\partial^3 \eta}{\partial \xi^3} \\
&+ \left\{\frac{u_i^2}{\sigma_i(1)} - \Theta - \int_\xi^1 \gamma[(2 - \gamma_e)\sigma_e + \gamma_i\sigma_i]\mathrm{d}\xi\right\}\frac{\partial^2 \eta}{\partial \xi^2} \\
&- \left\{\left[\frac{\delta^2 + \gamma_i}{\gamma_e + \gamma_i}\right]^{1/2} \frac{\mathrm{d}\sigma_i}{\mathrm{d}\xi}\frac{\delta^2 u_i}{4\epsilon^2}\right\}\frac{\partial^3 \eta}{\partial \xi^2\,\partial \tau} + \{\gamma[(2 - \gamma_e)\sigma_e + \gamma_i\sigma_i]\}\frac{\partial \eta}{\partial \xi} \\
&+ \left\{2\left[\frac{\delta^2 + \gamma_i}{\gamma_e + \gamma_i}\right]^{1/2} u_i\right\}\frac{\partial^2 \eta}{\partial \xi\,\partial \tau} - \left\{\frac{\delta^2 + \gamma_i}{\gamma_e + \gamma_i}\frac{\delta^2\sigma_i}{8\epsilon^2}\frac{\mathrm{d}\sigma_i}{\mathrm{d}\xi} + \frac{\gamma_e - 1}{\gamma_e + \gamma_i}\frac{\sigma_e}{8\epsilon^2}\frac{\mathrm{d}\sigma_e}{\mathrm{d}\xi}\right\}\frac{\partial^3 \eta}{\partial \xi\,\partial \tau^2} \\
&+ \left\{\left[\frac{\gamma_e - 1}{\gamma_e + \gamma_i}\right]^{1/2} \epsilon\sigma_e^{1/2} c_v\right\}\frac{\partial \eta}{\partial \tau} + \left\{\frac{\gamma_e\sigma_e + \gamma_i\sigma_i}{\gamma_e + \gamma_i}\right\}\frac{\partial^2 \eta}{\partial \tau^2} = 0.
\end{aligned} \tag{4.23}$$

4.2.2 Analysis and results

Some calculations have been conducted for conically tapered cantilevered tubular beams, i.e. either conical in outer form or with a conical flow passage. The notation 'cylindrical-conical' or 'conical-conical' is used here, the first denoting a cylindrical outer shape and a conical flow passage, while the second denotes conical outer and inner forms, as shown in Figure 4.3; the case of a 'cylindrical-cylindrical' pipe will simply be referred to as 'uniform'. In the case of conical passages, instead of σ_e and σ_i, it is more convenient to use the truncation factors α_e and α_i [see Figure 4.3(c)] or the cone angles β_e and β_i, defined by

$$\begin{aligned} \sigma_e &= (1-\alpha_e\xi)^2, \qquad \beta_e = \tan^{-1}\{[D_e(0)-D_e(L)]/L\} \simeq \alpha_e/\epsilon, \\ \sigma_i &= (1-\alpha_i\xi)^2, \qquad \beta_i = \tan^{-1}\{[D_i(0)-D_i(L)]/L\} \simeq \delta\alpha_i/\epsilon. \end{aligned} \tag{4.24}$$

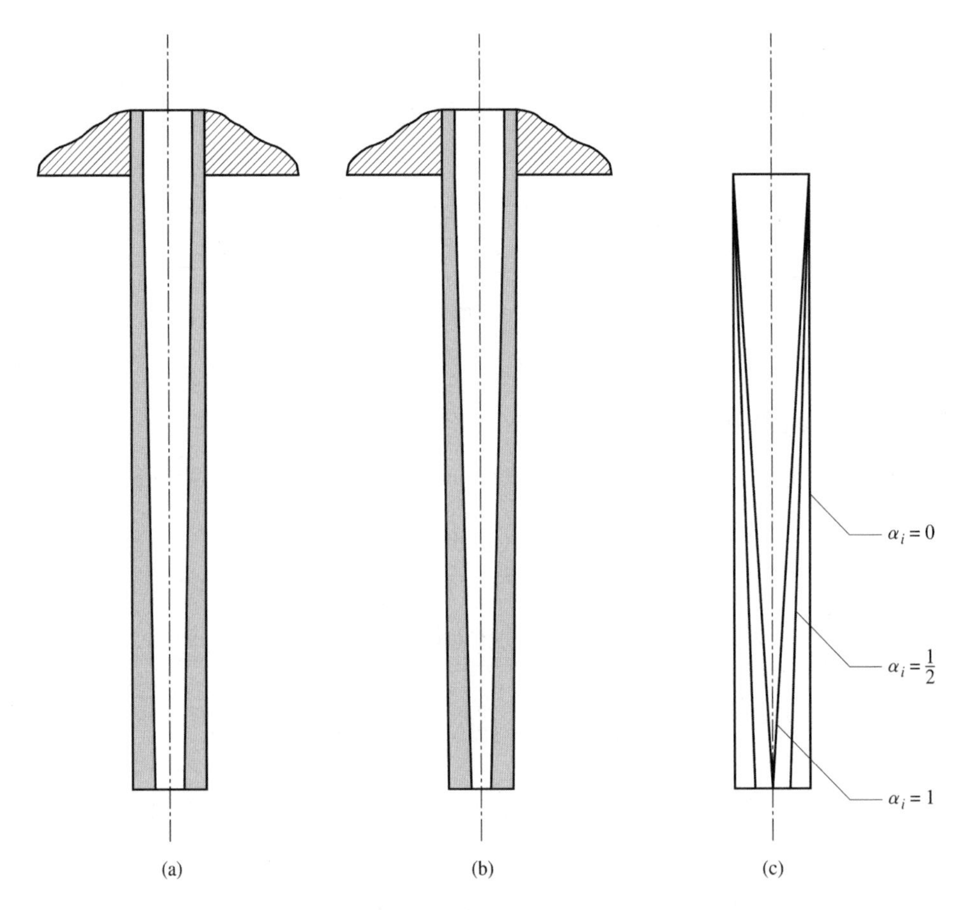

Figure 4.3 (a) A cylindrical-conical pipe, and (b) a conical-conical one. (c) Truncated cones representing possible internal conduit shapes, for the same ϵ ($\epsilon = 5$) and different α_i.

The method of solution, a modified Galerkin technique (Hannoyer 1972), is outlined in Chapter 8, where the system subjected concurrently to internal and external flow will be discussed.

In Figure 4.4(a), the dynamical behaviour with increasing u_i is compared for (i) a wholly uniform pipe and (ii) a cylindrical-conical one ($\alpha_e = 0$, $\alpha_i = 0.5$). It is seen that the dynamical behaviour is closely similar, but the critical flow velocity for the onset of flutter is considerably lower for the cylindrical-conical pipe ($u_{ic} \simeq 2.25$) than for the uniform

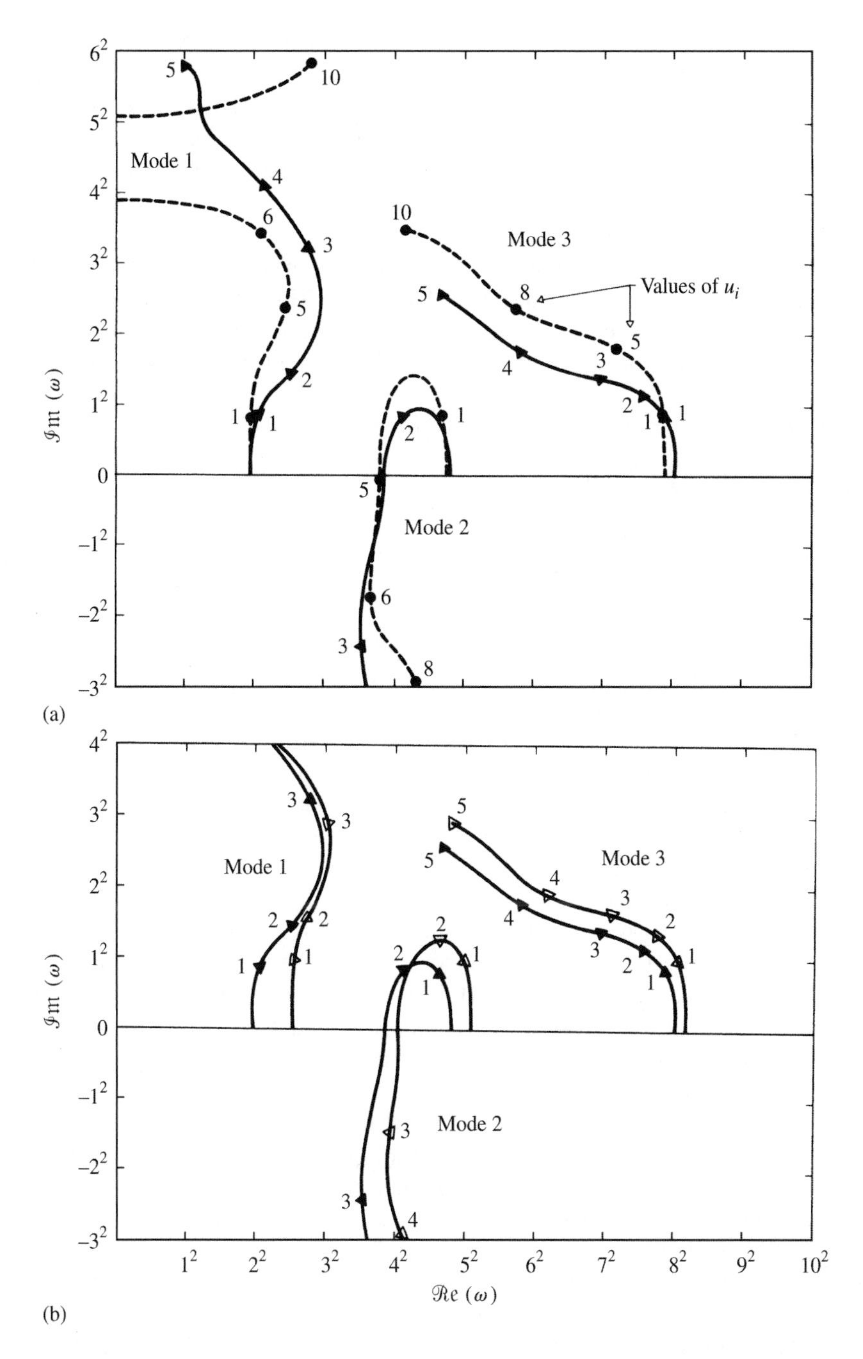

Figure 4.4 (a) Argand diagram of the complex eigenfrequencies of a tubular cantilever conveying fluid for a system with $\epsilon = 20$, $\delta = 0.5$, $\gamma = 20.05$, $\gamma_i = 0.03$, $\gamma_e = 1.9$, immersed in quiescent water, neglecting dissipation ($\nu_d = c_v = 0$): $- \bullet - \bullet -$, a uniform pipe ($\alpha_e = \alpha_i = 0$); —▼—▼, cylindrical-conical system ($\alpha_e = 0$, $\alpha_i = 0.5$). (b) Argand diagram of a similar cylindrical-conical system ($\alpha_e = 0$, $\alpha_i = 0.5$), with $\gamma_e = 1$ and all other parameters the same: ▼, immersed in water, ∇, immersed in air (Hannoyer & Païdoussis 1979a).

one ($u_{ic} \simeq 5.0$). By reconsidering the arguments originally made by Benjamin (1961a) and discussed in Section 3.2.2, flutter arises when the work done by centrifugal force $MU^2(\partial^2 w/\partial x^2) \equiv \rho_i A_i U_i^2(\partial^2 w/\partial x^2)$ overcomes that done by the Coriolis force. In the case of the nonuniform pipe, however, this term is equal to $[\rho_i A_i(x)U_i(x)U_i(L)](\partial^2 w/\partial x^2)$, where $\rho_i A_i(x)U_i(x) = \text{const}$. Hence, since $U_i(L) > U_i(0)$, the destabilizing force is higher at all points $x > 0$ in the cylindrical-conical system *vis-à-vis* the uniform one. In this case the ratio of critical flow velocities is $2.25/5.0 = 0.45$, which is close to the diameter ratio $(1 - \alpha_i)/1 = D_i(L)/D_i(0) = 0.5$. Similar calculations confirm that u_{ic} indeed decreases almost linearly with increasing 'truncation factor' α_i. Thus, the destabilizing effect of conicity of the flow passage is similar to that of mounting a convergent nozzle at the end of an otherwise uniform pipe [Sections 3.3.5 and 3.5.6 and Gregory & Païdoussis (1966b)].

Figure 4.4(b) shows the effect of density of the surrounding fluid on the dynamics of the cylindrical-conical pipe. The dimensionless frequency is defined, in terms of the dimensional circular frequency Ω, by

$$\omega = \left[\frac{m + \rho_e A_e + \rho_i A_i}{EI}\right]_{\xi=0}^{1/2} \Omega L^2. \tag{4.25}$$

Intuitively one would have supposed that when the surrounding fluid is water, the system would be more stable than when it is air. Yet, the opposite is found to be true. The increase in the surrounding fluid density acts in two ways: (i) to increase the effective inertia of the pipe through the added-mass effect and (ii) to decrease the gravity effect through buoyancy. Both have a destabilizing effect with increasing density of the external fluid, ρ_e. The latter is physically obvious. The former may be accepted by analogy to the case of uniform pipes where it was found that, as the mass ratio $\rho_i A_i/(\rho_i A_i + m)$ becomes smaller, the system is less stable (Section 3.5); the external stagnant fluid effectively adds $\rho_e A_e$ to m, producing the same effect.

In Figure 4.4(b) the real parts of the dimensionless frequencies $\mathfrak{Re}(\omega_j)$, j being the mode number, are lower for the pipe immersed in liquid than in air, which is reasonable in view of the added-mass effect; this is even more pronounced in dimensional terms — refer to equation (4.25). However, the $\mathfrak{Im}(\omega_j)$ are also lower in liquid than in air, which is contrary to physical intuition, as the added damping in liquid should be higher than in air. Nevertheless, it is recalled that the true measure of damping is $\zeta_j = \mathfrak{Im}(\omega_j)/\mathfrak{Re}(\omega_j)$, and this does show the expected behaviour. It may be shown by a perturbation analysis for small u_i that (a) $\mathfrak{Im}(\omega_j) = 2u_i[(\delta^2 + \gamma_i)/(\gamma_e + \gamma_i)]^{1/2}$ for all j, in the absence of dissipative forces, here taken to be zero for simplicity, and (b) the $\mathfrak{Re}(\omega_j)$ are approximately equal to their values at $u_i = 0$. These may be used to obtain estimates of ζ_i for small enough u_i.

Figure 4.5 shows the eigenfrequencies of some of the lowest modes of a conical-conical pipe in still water; internal dissipation has been taken into account[†] in one case. It is seen that the behaviour of the system in both cases is considerably different from that of the previous systems. First, the critical flow velocities are much lower, reflecting

[†] A modified viscoelastic dissipation model is utilized in this case to approximate the experimentally observed behaviour of silicone rubber, which exhibits hysteretical behaviour at high frequencies but is viscoelastic at low frequencies. This is achieved by replacing ν_d by $\nu_d[1 + (\nu_d/\mu_d)|\omega|]^{-1}$, where μ_d is the hysteretic damping coefficient as $\omega \to \infty$.

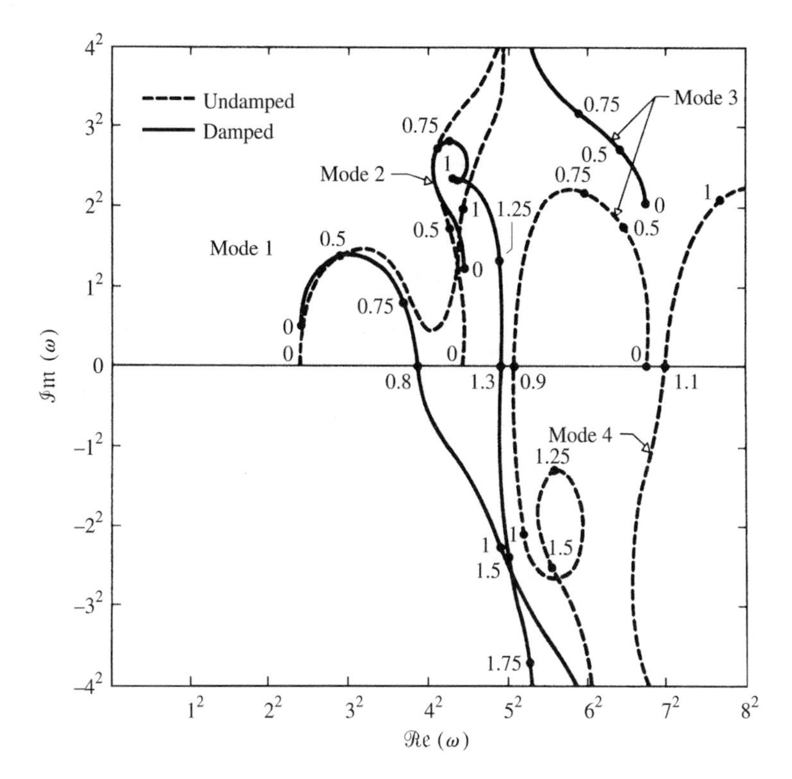

Figure 4.5 Argand diagram of the complex eigenfrequencies of a conical-conical cantilever conveying fluid and immersed in quiescent water, with and without dissipation in the pipe material taken into account ($\epsilon = 22$, $\delta = 0.5$, $\beta_e = 0.03$, $\beta_i = 0.016$, $\gamma = 16.47$, $\gamma_i = -0.08$, $\gamma_e = 1.7$, $c_v = 0$): ---, $\mu_d = \nu_d = 0$ (undamped); ——, $\mu_d = 0.20$, $\nu_d = 0.04$ (damped) (Hannoyer & Païdoussis 1979a).

the reduced flexural rigidity of conical-conical pipes and the diminished gravity effect ($\rho > \rho_e$ in the case presented). Second, there are two flutter instabilities close to each other (in terms of u_i). Comparing the undamped and damped systems, there is little evident similarity in the root loci. The differences are more apparent than real, however. Although different modes become unstable in the two cases, the critical flow velocities are not too different. It is recalled that this being a nonconservative system, dissipation can actually destabilize it.

Figure 4.6(a) shows that, for tubular cantilevers of constant cone angle β_e (and similarly for β_i), varying ϵ by cutting pieces off the free end entails variations in β_e (and similarly in β_i) — see equations (4.24). Figure 4.6(b) shows the effect of the slenderness ratio $\epsilon = L/D_e(0)$ on the critical flow velocity u_{ic} for a conical-conical pipe with constant β_i and β_e. (It is noted that as ϵ changes, the corresponding α_e and α_i also change.) It is seen that with increasing slenderness the system loses stability at a lower flow velocity. This contrasts with the case of uniform pipes where u_{ic} is almost independent of ϵ. Of course, the more slender the system, the lower is the *dimensional* critical flow velocity, in any case (*vide* definition of u_i: since $u_i \propto U_i(0)L$, as L increases, $U_i(0)$ decreases for a constant u_i); but in conical systems this effect is greatly amplified. Finally, the effect of the surrounding fluid is seen to be the same as for cylindrical-conical pipes.

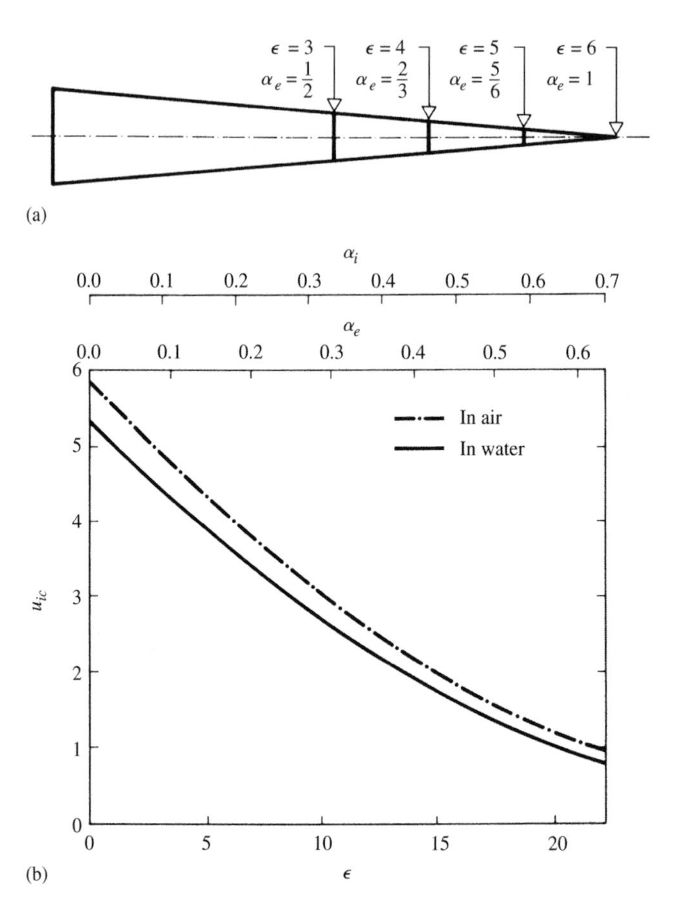

Figure 4.6 (a) Diagram showing that, for a cone of constant angle (here $\tan \frac{1}{2}\beta_e = \frac{1}{12}$, representing a possible exterior shape of the tubular cantilever), as ϵ is changed by truncating pieces from the free end, α_e changes also. (b) The effect of ϵ (and hence of α_i and α_e) on the critical flow velocity of a conical-conical cantilever of constant β_i and β_e conveying fluid [$\beta_e = 0.03$, $\beta_i = 0.016$, $\delta = 0.5$, $\gamma^* = (1 - \delta^4)\gamma/\epsilon^3 = 0.001\,45$ which is a version of γ independent of length, $\gamma_i = -0.08$, $\gamma_e = 1.70$, $\mu_d = 0.2$, $\nu_d = 0.04$, $c_v = 0$] (Hannoyer & Païdoussis 1979a).

4.2.3 Experiments

The validity of the theory was tested by experiments (Hannoyer & Païdoussis 1979b) with nonuniform elastomer tubular cantilevers conveying water. The pipes were centrally mounted in the vertical test-section of a water tunnel, so that external axial flow could also be imposed, as described in Chapter 8. Here we confine ourselves to experiments with internal flow, which was supplied from an external source through the supports of the upper end of the pipe. In the experiments the test-section was either empty or filled with stagnant fluid. The ratio of diameters of test-section and pipe was 200/25.4 mm $\simeq$ 8, so that the external fluid may be considered to be effectively unconfined.

Experiments were conducted with uniform, cylindrical-conical and conical-conical tubular beams (Figure 4.3), which were manufactured and their properties measured by variants of the methods described in Appendix D.

Experimentally determined values of the dissipative constants were used in the theory, using a mixed viscoelastic–hysteretic model, with corresponding coefficients ν_d and μ.

General observations

With increasing flow, externally induced beam motions become more heavily damped; however, beyond a certain flow the trend is reversed and, at sufficiently high flow, the stability limit is reached and flutter is precipitated.

Close to, but below, the critical flow for self-excited flutter, the system behaves as if it has a small unstable limit cycle within a larger stable one, so that external disturbances of a certain magnitude may precipitate flutter, yet small disturbances are damped. As the flow gets closer to the stability limit, the inner limit cycle becomes smaller, to the point where random, turbulence-induced disturbances are sufficient to propel the system beyond the confines of this limit cycle, precipitating amplified oscillation (flutter). These are clearly characteristics of a subcritical Hopf bifurcation [Figure 2.11(d)].

Limit cycles could generally be observed in the case of pipes hanging in air rather than water. The amplitude involved was larger for pipes with a uniform conduit than for those with a conical conduit. For flow velocities higher than those associated with the onset of instability, the amplitude of the limit cycle increased further. In contrast, for pipes in water, presumably because of buoyancy counteracting the stabilizing effect of gravity, the oscillations continued to grow until, in 10–20 cycles, the amplitude became large enough (i.e. about 8 pipe diameters) for the pipe to start hitting the walls of the test-section, whereupon the experiment was discontinued for fear of damage to the apparatus; thus, established limit-cycle motion could not actually be observed in this case.

Comparison between theory and experiment

The dimensionless critical flow velocities, u_{ic}, and the corresponding frequencies, ω_c, for flutter of a cylindrical-conical pipe in air and water are shown in Figures 4.7 and 4.8, respectively. Also shown is one experimental point for a cylindrical pipe, for comparison purposes.

It is seen that theoretical and experimental critical flow velocities agree very well — although the experimental values ought to have been a little lower than the theoretical ones, this being a subcritical Hopf bifurcation. The corresponding frequencies agree less well. However, this is not surprising, upon realizing that: (i) in the case of pipes in air, the measured frequencies were those of limit-cycle motion, rather than those associated with the onset of flutter; these two values could be quite different in the case of a subcritical Hopf bifurcation, since the initial limit cycle is of non-negligible magnitude; (ii) in the case of experiments in water, the frequency was measured during the first few cycles of motion, before the pipe started hitting the wall, and precision of measurement was not high.

The theoretically predicted reduction in dimensionless critical flow velocity with increasing slenderness (and hence the even more substantial reduction in dimensional flow velocity) is wholly supported by these experiments, as well as the theoretical finding that the system is less stable when immersed in water than in air.

Finally, the experimental frequencies for the cylindrical-conical pipes are lower than those of the uniform cylindrical ones, which is in agreement with theory.

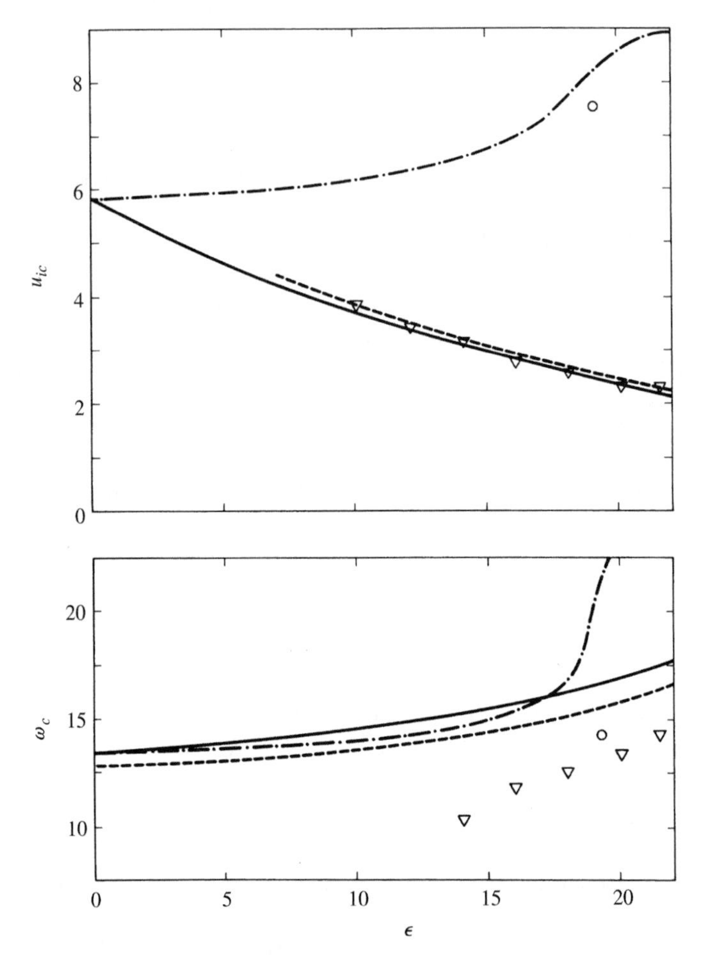

Figure 4.7 Theoretical and experimental critical flow velocities, u_{ic}, for flutter of cylindrical ($\beta_e = \beta_i = 0$) and cylindrical-conical ($\beta_e = 0$, $\beta_i = 0.014$) cantilevers conveying fluid, surrounded by still air; and the corresponding frequencies, ω_c. Other parameters: $\delta = 0.5$, $\gamma_e = 1.0$, $\gamma_i = -0.03$, $\gamma = 0.002\,51\epsilon^3$. Lines represent theoretical results and symbols are experimental data: —·— , ○, uniform cylindrical pipe; ——, cylindrical-conical pipe, undamped; – – –, ∇, cylindrical-conical pipe, damped ($\mu_d = 0.08$, $\nu_d = 0.02$); (Hannoyer & Païdoussis 1979b).

Figure 4.9 shows the corresponding case for a conical-conical pipe. In these experiments the pipe had fixed internal and external cone angles; changes in α_i, and hence α_e, were obtained by reducing the length of the pipe by cutting pieces off the free end — large α_e or α_i corresponding to fuller cones, and smaller values to more highly truncated ones — see Figures 4.3 and 4.6(a). As predicted by theory, it is seen that for the longer, more fully conical system, stability may be lost at very low flow velocity, many times smaller than for a cylindrical pipe. A change in the character of oscillation was observed at higher flows, but could not be recorded accurately enough to tell whether it is associated with the higher flutter instability predicted by theory or whether it corresponds to some other, secondary bifurcation.

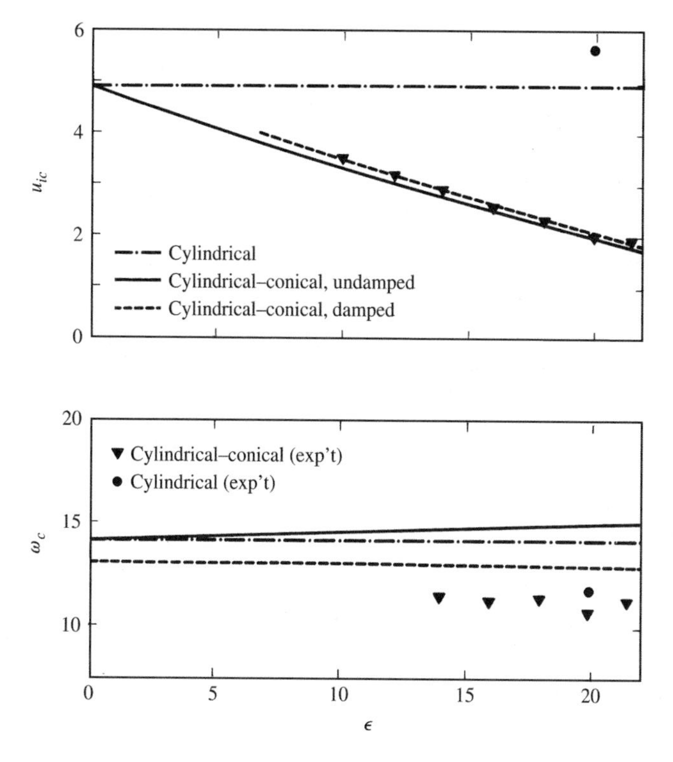

Figure 4.8 Theoretical and experimental values of u_{ic} and ω_c for the same system as in Figure 4.7, but immersed in stagnant water; all parameters are the same except $\gamma_e = 1.9$ (Hannoyer & Païdoussis 1979b).

It may be concluded, therefore, that these experiments validate the theoretical model. Both theory and experiments for nonuniform pipes subjected concurrently to internal and external axial flow are presented in Chapter 8 (Volume 2).

4.2.4 Other work on submerged pipes

Further work on the dynamics of *uniform* pipes immersed in fluid has been conducted, partly motivated by vibration of the inverted U-shaped pipe connecting the reactor vessel to the intermediate heat exchanger in a liquid-metal fast breeder reactor (LMFBR) [e.g. Inagaki *et al*. (1987), Sugiyama *et al*. (1996a)], and by more general applications in the marine and power-generating area [e.g. Shilling & Lou (1980), Langthjem (1995)].

The model utilized by Sugiyama *et al*. (1996a) is a variant of that in the foregoing, but modified to take into account immersion of only the lower part of the pipe. Similar results are obtained, but the effects of added mass, buoyancy and damping are studied more thoroughly through parametric calculations. The effect of partial immersion on stability is shown in Figure 4.10. The effect of immersion is generally destabilizing, for the reasons given following equation (4.25). However, partial immersion, as pointed out by Sugiyama *et al*. has a selective effect on mode shapes as well, mainly because of the discontinuous added-mass effect; see theoretical results for small l_o in Figure 4.10(c).

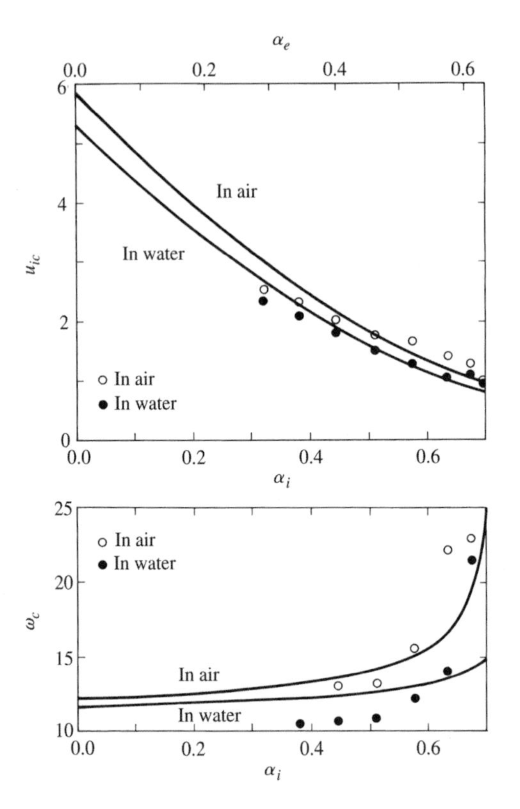

Figure 4.9 Theoretical and experimental critical flow velocities and frequencies for a conical-conical tubular cantilever conveying water; $\beta_e = 0.03$, $\beta_i = 0.016$, $\delta = 0.5$, $\gamma_e = 1.0$ (air) and $\gamma_e = 1.7$ (water), $\gamma_i = -0.08$, $\gamma = 0.001\,55\epsilon^3$, $\mu_d = 0.2$, $\nu_d = 0.05$ (air) and $\nu_d = 0.04$ (water); (Hannoyer & Païdoussis 1979b).

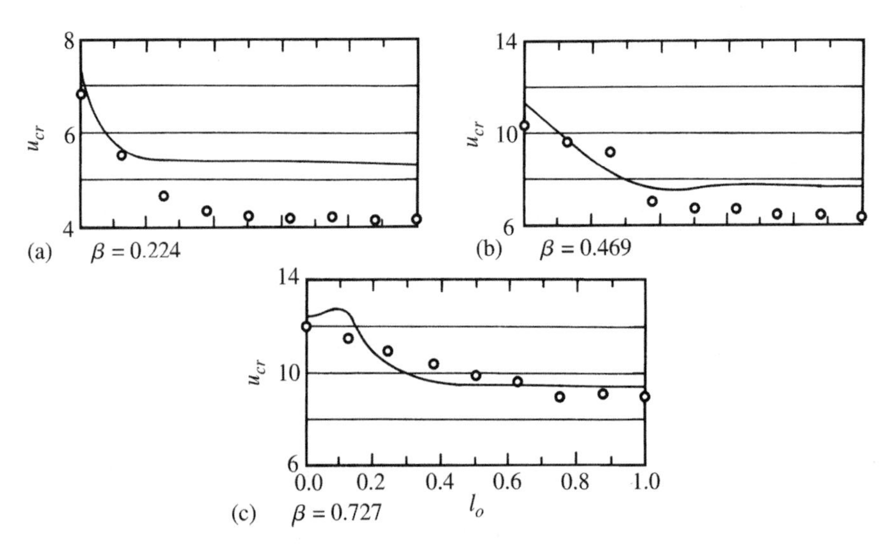

Figure 4.10 The effect of partial immersion of the lower portion of a cylindrical cantilevered pipe on stability; $l_o = 1$ represents total immersion (Sugiyama *et al*. 1996).

The same problem is studied by means of potential- rather than plug-flow theory and Timoshenko beam theory by Langthjem (1995) — see Section 4.4.10.

4.3 ASPIRATING PIPES AND OCEAN MINING

4.3.1 Background

In the discussion of energy transfer mechanisms for cantilevered pipes conveying fluid (Section 3.2.2) in conjunction with equation (3.11), it has generally been presumed that the flow velocity is 'positive', i.e. directed from the clamped towards the free end. However, it is obvious that if U is replaced by $-U$, all the arguments on stability and the predicted behaviour are reversed: for infinitesimally small U, and up to $|U_{cr}|$, the system would be unstable by flutter; then, for $|U| > |U_{cr}|$, it would regain stability! If dissipative forces were added, then perhaps 'infinitesimally small' would merely change to 'small'.

This intriguing possibility was explored experimentally by the author at the Chalk River Nuclear Laboratories in the mid-1960s, by immersing the lower end of an elastomer pipe in a barrel and connecting the upper end to a pump, as shown in Figure 4.11(a). The expected behaviour did not occur. However, a sort of amplified oscillation did occur, if the immersion was shallow; but the mechanism was soon discovered to be one of parametric resonance, involving the slurping of air-slugs into the pipe, sucked in at the extremes of the cycle of oscillation when the pipe end is closest to the free surface, as shown in Figure 4.11(b). Thus, the flow in the pipe has periodic density variations, with the optimum 2:1 parametric/natural frequency ratio (Section 4.5). Deeper immersion eliminated this mechanism of self-excitation. Attributing the non-occurrence of the expected 'regular' flutter at infinitesimal flow velocities to increased damping due to the water immersion, the flow rate was increased further, until a sufficiently large transmural pressure (external

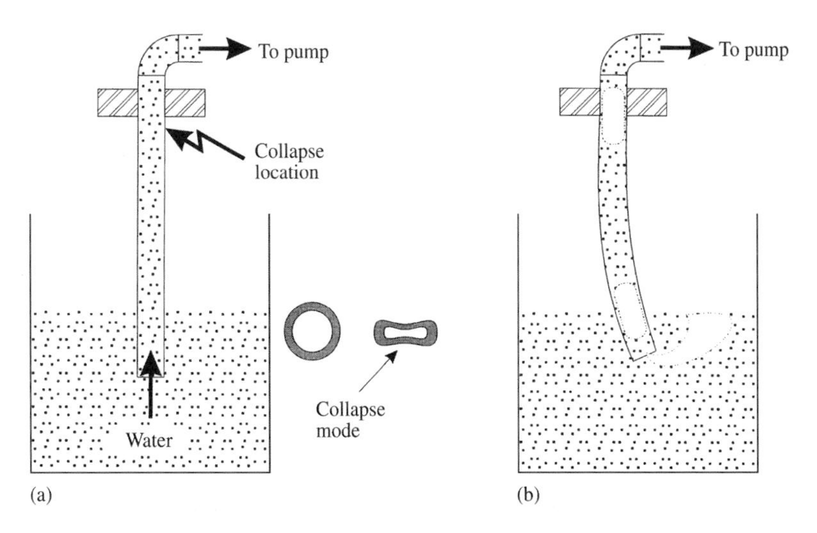

Figure 4.11 (a) Apparatus for experiments with water-aspirating pipes. (b) Diagram for understanding the mechanism of parametric resonance due to density pulsations occurring when the immersion is shallow.

ambient minus internal) caused local shell-type collapse of the pipe near the support. Reinforcing the pipe at that point simply postponed the collapse to a higher flow rate, at a lower point along the pipe; but there was still no sign of the elusive flutter! At the time, this was chalked up as due to 'experimental difficulties' and forgotten for a while.

Some time later, the author became aware of ocean mining and some aspects of research into the dynamics of such systems [e.g. Chung *et al*. (1981), Whitney *et al*. (1981), Felippa & Chung (1981), Koehne (1978, 1982), Chung & Whitney (1983), Aso & Kan (1986)], and work into the problem of sucking pipes received a new impetus. Ocean mining is basically the 'vacuuming' of minerals, notably of manganese nodules, which lie on the floor of the ocean, e.g. in the Northeast Pacific, at depths of the order of 5 km. The system involves a very long 'vacuum hose', with a massive 'vacuum head' which walks along the ocean floor and scours and sucks up nodule-rich sea-water, as shown in Figure 4.12(a). It occurred to the author that, the moment the bottom head loses contact with the sea floor, this becomes a cantilevered pipe with an end-mass, aspirating fluid and hence subject to flutter, as per equation (3.11). Therefore, it was decided that a more careful study of the problem was warranted.

4.3.2 Analysis of the ocean mining system

In most of the papers just cited, external flow and wave-related problems, as well as the dynamics of the long pipe itself, are the main concern. Only Koehne (1982) discusses briefly the modelling of the pipe with internal flow, but does not present any results. A systematic analysis of the general system of Figure 4.12(b) has been undertaken by Païdoussis & Luu (1985), which will be outlined briefly in what follows.

For simplicity, the pipe is assumed to be initially straight. Then, proceeding as in Section 4.2, the equation of motion is found to be[†]

$$\begin{aligned} EI\frac{\partial^4 w}{\partial x^4} &+ MUU_j\frac{\partial^2 w}{\partial x^2} - 2MU\frac{\partial^2 w}{\partial x\,\partial t} + (M+m+M_a)\frac{\partial^2 w}{\partial t^2} \\ &- \left\{(M+m)g(L-x) + \overline{M}g - \overline{F}_b - p_{oL}A_o\left(\frac{L/\alpha^2 - x}{L}\right)\right\}\frac{\partial^2 w}{\partial x^2} \\ &+ \{(M+m)g - p_{oL}A_o/L\}\frac{\partial w}{\partial x} + c\frac{\partial w}{\partial t} = 0, \end{aligned} \tag{4.26}$$

with boundary conditions

$$w = 0, \qquad \frac{\partial w}{\partial x} = 0 \tag{4.27a}$$

at $x = 0$, and

$$\begin{aligned} &EI\frac{\partial^3 w}{\partial x^3} - \overline{M}\,\overline{d}\frac{\partial^3 w}{\partial x\,\partial t^2} - (\overline{M}g - \overline{F}_b)\frac{\partial w}{\partial x} - (\overline{M} + \overline{M}_a)\frac{\partial^2 w}{\partial t^2} - \overline{c}\frac{\partial w}{\partial t} = 0, \\ &EI\frac{\partial^2 w}{\partial x^2} + (\overline{M}g - \overline{F}_b)\,\overline{d}\frac{\partial w}{\partial x} + \overline{M}\,\overline{d}\frac{\partial^2 w}{\partial t^2} + (\overline{J} + \overline{M}\,\overline{d}^2)\frac{\partial^3 w}{\partial x\,\partial t^2} = 0, \end{aligned} \tag{4.27b}$$

[†]The reader should consult the text in Section 4.3.3.

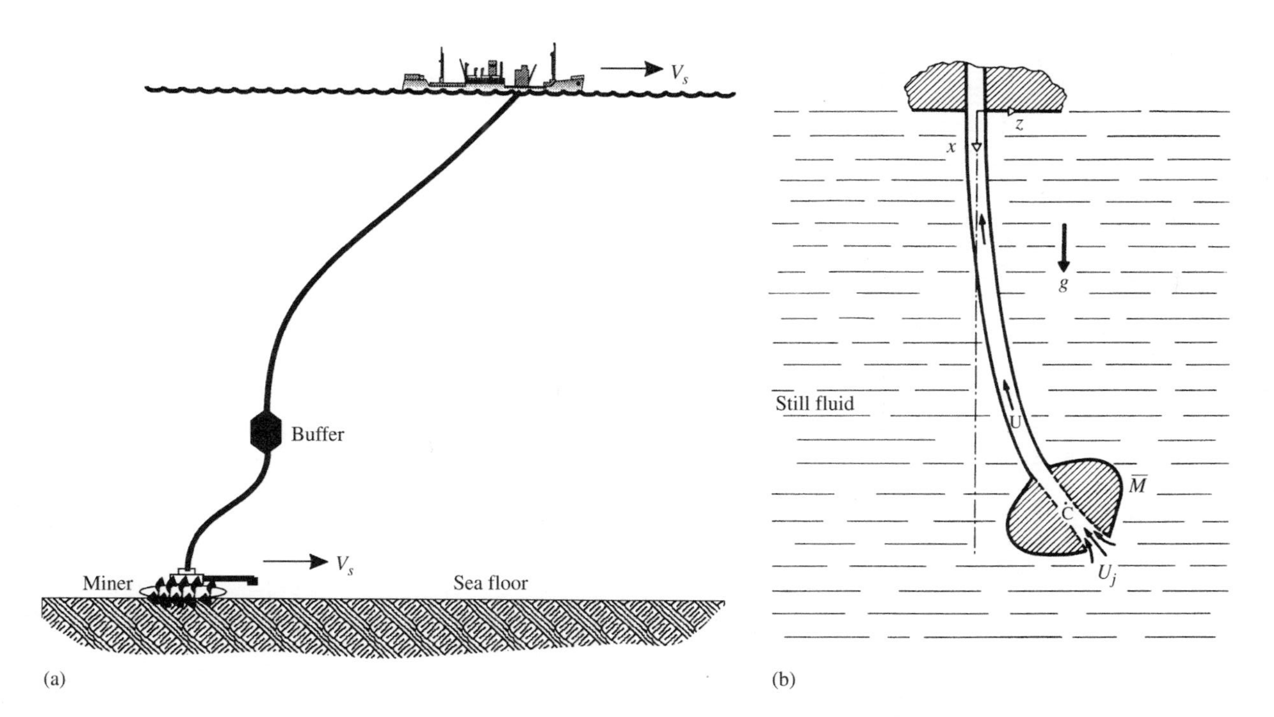

Figure 4.12 (a) The ocean mining system, after Chung & Whitney (1983); (b) the system modelled when the bottom mass loses contact with the ocean floor (Païdoussis & Luu 1985).

at $x = L$. All quantities are the same as in Section 3.3.2, but some new ones need be introduced: M_a is the added mass of the pipe per unit length; $\overline{M}$ is the end-mass, of mass-moment of inertia $\overline{J}$, and centre of mass at a distance $\overline{d}$ from the pipe end ($x = L$); $\overline{F}_b$ is the buoyancy force associated with $\overline{M}$; $\alpha^2 = A_o/A_j = U_j/U$ is the ratio of external pipe area to inlet jet area, with U_j as shown in Figure 4.12(b); p_{oL} is the external, hydrostatic pressure at $x = L$; other barred quantities have the same meaning as plain ones, but are associated with the end-mass. A form of expression (2.157) is used for c — see also equation (3.106). Furthermore, assuming a spherical form for $\overline{M}, \overline{c} = 6\pi\nu\rho\overline{d}$, where ν is the kinematic viscosity of the fluid.

It is stressed that in this formulation, in accordance with Figure 4.12(b), *a positive U corresponds to up-flow*, i.e. to what in Section 4.3.1 is called a *negative* flow velocity.

For very long pipes, a pipe-string approximation is normally used, i.e. the flexural rigidity is ignored; however, here flexural terms are retained. Because of the fact that the boundary conditions are frequency-dependent, the usual form of the Galerkin method is not applicable to this case (see also Section 4.6.2). A special hybrid Fourier–Galerkin method developed by Hannoyer (1972), outlined in Chapter 8, is used instead.

Some numerical calculations have been conducted for a system with parameters taken from Chung & Whitney (1983): a steel pipe ($E = 2 \times 10^8$ kN/m^2, $\rho_t = 7.83 \times 10^3$ kg/m^3) and $\overline{M} = 182 \times 10^3$ kg, $L = 1$ km, $D_i = 0.45$ m, $D_o = 0.50$ m, and $\alpha = 1$ for simplicity. Typical results are given in Figure 4.13 and Table 4.1.

The system loses stability by flutter at a very low flow velocity, $U_{cf} = 1.32$ m/s, corresponding to the dimensionless $u_{cf} = 1.129$ in Figure 4.13. As shown in Table 4.1,

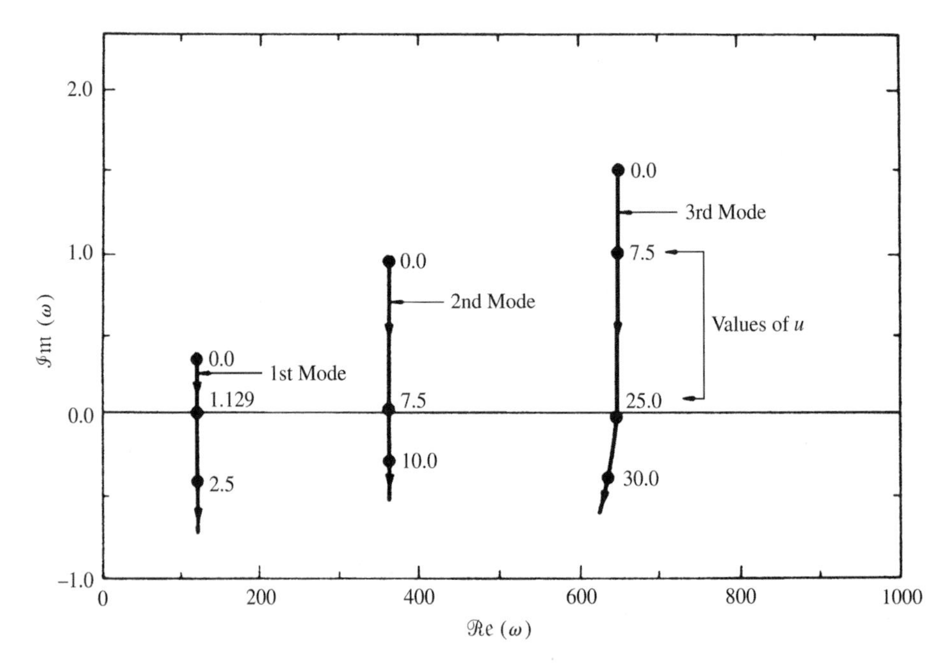

Figure 4.13 Dimensionless complex eigenfrequencies of the aspirating system of Figure 4.12(b) as functions of the up-flow dimensionless flow velocity, u, for $\overline{M} = 182 \times 10^3$ kg (Païdoussis & Luu 1985).

Table 4.1 The threshold flow velocity for flutter, $u_{cf} = (M/EI)^{1/2} U_{cf} L$, for various values of $\overline{M}$ and for zero dissipation (Païdoussis & Luu 1985).

$\overline{M}$ (kg)	Dissipation	u_c
182×10^3	Taken into account	1.13
1820	Taken into account	0.935
0	Taken into account	0.895
Any value	Neglected	0^+

the magnitude of $\overline{M}$ does not alter this value dramatically. If, however, the dissipative forces are taken to be zero, the system loses stability at $U = 0^+$.

Therefore, it would appear from these results that ocean mining designers and operators need to worry about flutter in their systems since, if a small safety factor were added, $U < 1$ m/s would be too small to live with — especially since, for the more realistic $L = 5$ km, one obtains $U_{cf} \leq 0.2$ m/s! Furthermore, the problem is of fundamental interest and hence work on experimental validation started anew.

A new apparatus was built at McGill in 1986, shown in Figure 4.14(a). This time the entire pipe, hung vertically, was immersed in water in a steel tank; water was supplied at the top of the tank, and was forced up the hanging pipe and out of the vessel. Compressed air was supplied at the top of the tank to achieve higher flows, but also to conduct experiments entirely with air up-flow. Several experiments were conducted, with thicker pipes to postpone the buckling collapse of Figure 4.11(a), and some with different-shaped inlet forms added, but the system remained unnervingly stable. The experiment was discontinued when, with ever-increasing air-pressure to force higher water flow up the pipe, the rubber hose leading the water to the drain burst free of its clamp, spraying water all over the laboratory and all over the instrumentation nearby, *and* giving the author an unwelcome cold shower! At that point, the author was certain that something was wrong with the theory; for one thing, the flow into the pipe is not exactly tangential, thus not replicating in reverse the outpouring jet in the case of down-flow. However, these negative results were not published,[†] precisely because they were negative and not fully understood — which is why the tale is worth telling.

Meanwhile, even without experimental verification, it was taken for granted that the Païdoussis & Luu flutter at infinitesimally small aspirating flow really does exist, and several more papers were published giving similar results [e.g. Sällström & Åkesson (1990)] and methods for suppressing the unwanted flutter [e.g. Kangaspuoskari *et al*. (1993)]. The only reference to absence 'of any physical evidence of this phenomenon' came out in the discussion by Dupuis & Rousselet (1991a), to which this author also contributed.

4.3.3 Recent developments

It was in 1995, during a visit by the author to Cambridge and upon recounting this paradoxial behaviour to Dr D.J. Maull, that the latter recalled reading 'something similar' in Richard Feynman's biography (Gleick 1992). It turns out that in 1939 or 1940,

[†]At least not until much later (Païdoussis 1997), when the reason why was much better understood.

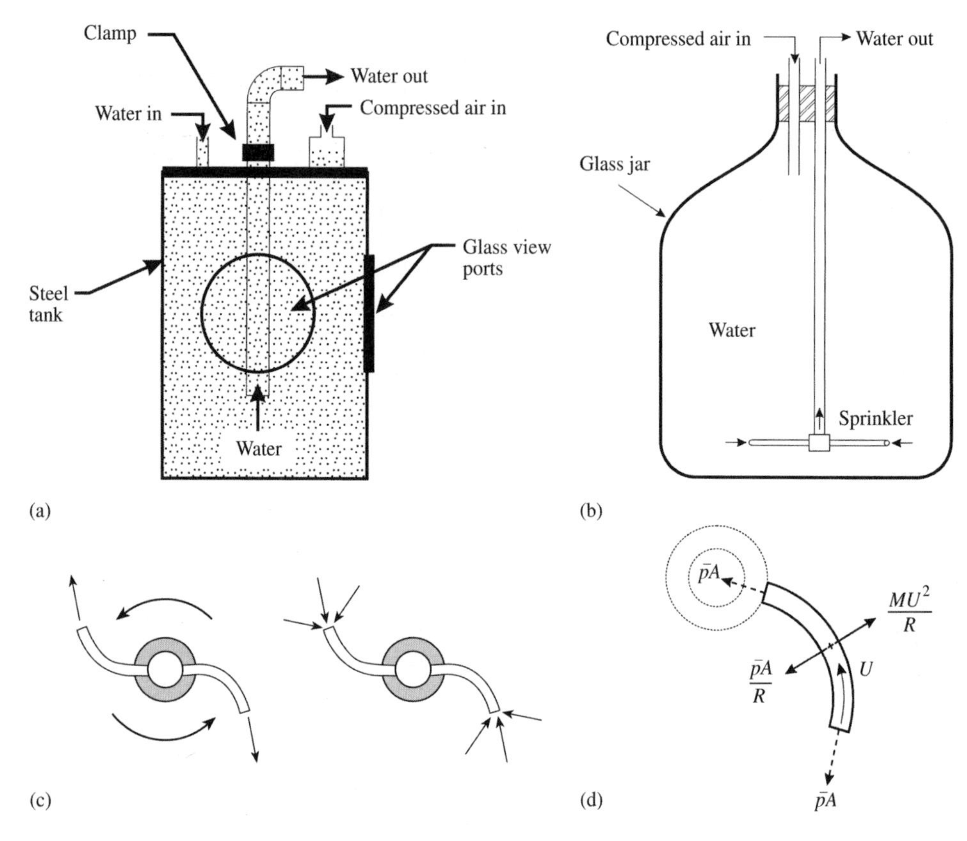

Figure 4.14 (a) New apparatus for forcing the fluid up the pipe in experiments by Païdoussis at McGill in 1980s; (b) Richard Feynman's apparatus for resolving the sprinkler problem at Princeton in late 1939 or 1940; (c) the sprinkler problem: which way does the sprinkler turn when aspirating fluid (Gleick 1992)? (d) 'negative pressurization' and centrifugal forces on one arm of the aspirating sprinkler.

Feynman's and most other physicists' tea-time conversation at Princeton and the Institute for Advanced Study was dominated by this problem: if a simple S-shaped lawn sprinkler were made to suck up water instead of spewing it out, Figure 4.14(c), would it rotate backwards or in the same way as for normal operation? (This problem was tied to the issue of reversibility of atomic processes!) Feynman could apparently argue convincingly either way.

Eventually, Feynman decided to do an experiment which, as shown in Figure 4.14(b), was remarkably similar to the author's. He immersed the lawn sprinkler into a glass jar filled with water, with an outlet connected to the sprinkler and a compressed air supply to force the water into the sprinkler and out. With increasing pressure and flow, the sprinkler refused to budge, up to the point where the glass jar exploded, spraying water all over. The result was that Feynman was banished from the laboratory thenceforth.[†]

[†]The author feels to be in good company with a Nobel prize winner, in retrospect, for even the accident in his laboratory is similar to Feynman's. More than that, however, he is thankful for his engineering training, to know not to do pressurized air experiments in glass jars — even if he did use a rubber hose!

Clearly the flow field is entirely different in 'forward' and 'reverse' flow through the sprinkler. This is the key that finally led the author to the resolution of the conundrum, for both the sprinkler and the pipe problem. Consider the stationary aspirating sprinkler, and imagine a flared funnel, not connected to it, channelling the flow in, thus modelling the sink flow. On reflection, the flow in the funnel is no different from that considered in Section 4.2 for nonuniform pipes. Hence, neglecting gravity, the axial balance of forces in the funnel is given by a form of equation (4.13),

$$\frac{\partial}{\partial x}\left[T + p_e A_e - p_i A_i - \rho_i (A_i U_i) U_i\right] = 0, \tag{4.28}$$

where x and U_i are directed as in Figure 4.12(b), and all quantities except ρ_i are functions of x. T is taken up by the imaginary funnel supports and may be ignored. Also, this expression may be simplified by taking $A_e \simeq A_i = A$, and by writing $U_i = U$ and $p_i - p_e = p$, and recalling that $\rho_i A_i U_i = MU =$ const. Then integrating from $x = \infty$, where $p \to 0$ and $U \to 0$, to $x = L$, the inlet of the sprinkler, we obtain $(pA)_L = -(MU^2)_L$. Hence, since MU^2 is the same for all $x < L$, one can write

$$\overline{p}A = -MU^2, \tag{4.29}$$

which clearly shows that at the sprinkler inlet, and hence throughout, there is a *suction* or *negative pressurization*, $\overline{p} = -\rho U^2 \equiv -MU^2/A$. Its effect is profound, as may be seen in Figure 4.14(d). The negative pressurization produces a lateral force $\overline{p}A/R = -MU^2/R$, R being the radius of curvature, which totally cancels the centrifugal force MU^2/R; hence, the sprinkler remains inert![†] Of course, these arguments do not hold once some rotation of the sprinkler takes place, but may be considered to be correct to first order.

The same applies to the pipe problem. Unlike the case of discharging fluid where the pressure at the free end (above the ambient) is zero, for the aspirating pipe there is a suction at the free end, equal to $-\rho U U_j$, and hence a negative pressurization equal to that, throughout the pipe (cf. Section 3.3.4). Therefore, a term $\overline{p}A(\partial^2 w/\partial x^2) = -MUU_j(\partial^2 w/\partial x^2)$ must be added to equation (4.26), which is incorrect as it stands. This cancels out the centrifugal force required for flutter (Section 3.2.2)!

Still, *seeing is believing*. Accordingly, an experiment was performed at McGill in 1997, in which two similar elastomer pipes were mounted as vertical cantilevers, immersed in a transparent water tank; at the free end of each pipe there was a light plastic 90° elbow. The clamped ends of the two pipes were interconnected via a pump. Once the pump was started, the pipe discharging fluid deformed in reaction to the emerging jet, as expected. The aspirating pipe, however, after a starting transient, returned to its original, no-flow configuration and thereafter remained limply straight.[‡] Therefore, it is now clear that aspirating pipes cannot aspire to flutter!

Before closing this section, it ought to be mentioned that there is another engineering application involving pipes aspirating fluid, namely the Ocean Thermal Energy Conversion (OTEC) plants. Shilling & Lou (1980) initially intended to conduct 'up-flow' experiments

---

[†] An alternative demonstration of this result may be made by control volume considerations and the fact that inlet and outlet vorticity is zero; however, some colleagues considered this less convincing.

[‡] The experiment was initially done with very flexible coiled Tygon tubing. In this case, there *was* steady-state flow-induced deformation, with the aspirating pipe coiling itself tighter. It was discovered, however, that this was due to the fact that, under suction, the pipe cross-section became oval, and the coiled pipe behaved like a Bourdon pressure gauge! This shows that there is no such thing as a simple experiment.

with this in mind but, because of 'existing equipment, measuring techniques and financial considerations', ended up doing regular down-flow experiments with mechanically forced excitation of the pipe (see Section 4.6).

4.4 SHORT PIPES AND REFINED FLOW MODELLING

In the foregoing (Chapter 3 and Sections 4.1–4.3), it has been assumed that (i) the pipe is sufficiently slender for Euler–Bernoulli beam theory to be adequate for describing the dynamics of the pipe, and (ii) that wavelength of deformation is sufficiently long for the plug-flow model to be acceptable, thus ignoring conditions upstream and downstream while determining the fluid-dynamic forces at a given point. If the pipe is sufficiently short, however, both assumptions become questionable, as will be discussed further in the following, and the use of Timoshenko beam theory and more elaborate fluid dynamics becomes necessary. In this section the necessary fundamentals are developed, by means of which (a) the limits of applicability of the Euler–Bernoulli plug-flow (EBPF for short) analytical model are determined, and (b) a theory for really short pipes conveying fluid is established.

Since stability is of primary concern, it is noted that short thin-walled pipes lose stability in their shell modes [$n \geq 2$; see Figure 2.7(c)] rather than in their beam modes ($n = 1$), as discussed in Chapter 7 (Volume 2). In what follows, however, it is presumed that the pipe is sufficiently thick-walled for its beam-mode dynamics to be of primary interest.

Timoshenko beam theory, where shear deformation and rotatory inertia are not neglected, was first applied to the study of dynamics of pipes conveying fluid by Païdoussis & Laithier (1976). This theory is applicable to articulated pipes in the limit of a very large number of articulations (Section 3.8), where the articulations permit substantial shear deformation. It is also applicable to continuously flexible short pipes, as well as for obtaining the dynamical behaviour of long pipes in their higher modes; in both these cases the necessity of utilizing Timoshenko, as opposed to Euler–Bernoulli beam theory, is well established (Meirovitch 1967). The equations of motion in Païdoussis & Laithier (1976) are derived by Newtonian methods, and solved by finite difference and variational techniques. They are rederived by Laithier & Païdoussis (1981) via Hamilton's principle — a nontrivial exercise. In terms of the fluid mechanics of the problem, however, the use of the plug-flow model is retained in both cases; this theory will be referred to as the Timoshenko plug-flow theory (TPF for short). Also, numerous finite element schemes based on TPF-type theory have been proposed and used for stability and more general dynamical analysis of piping conveying fluid (Sections 4.6 and 4.7), e.g. by Chen & Fan (1987), Pramila *et al.* (1991), Sällström & Åkesson (1990) and Sällström (1990, 1993).

It is nevertheless recognized that the applicability of the plug-flow model to short pipes — or indeed to the study of the high-mode dynamical behaviour of relatively longer pipes — is questionable, as discussed first by Niordson (1953) and also by others, e.g. Shayo & Ellen (1974): if the wavelength of deformation is not large, as compared to the pipe radius, the use of the plug-flow model for obtaining the fluid forces becomes invalid [Section 4.4.3(b)]. Hence, there is need for improvement of the fluid mechanics of the problem for studying the dynamics of this class of problem.

The dynamics and stability of short pipes conveying fluid are examined here by means of Timoshenko beam theory for the pipe and a three-dimensional fluid-mechanical model for the fluid flow, following closely the work by Païdoussis *et al.* (1986). This will be referred to as the Timoshenko refined-flow theory, or TRF for short. The pipes are either clamped at both ends or cantilevered; in the latter case, special 'outflow models' are introduced to describe the boundary conditions on the fluid exiting from the free end.

4.4.1 Equations of motion

The system under consideration consists of a tubular beam of length L, flexural rigidity EI_p, and shear rigidity GA_p, conveying fluid with an axial velocity which in the undeformed, straight pipe is equal to U. Here, with no loss of generality, the pipe is supposed to hang vertically, with the fluid flowing down, so that the x-axis is in the direction of gravity.

In contrast to the Euler–Bernoulli beam theory, the Timoshenko beam theory takes into account the deformation due to transverse shear. If ψ denotes the slope of the deflection curve by bending and χ the angle of shear at the neutral axis in the same cross-section (Figure 4.15), then the total slope $(\mathrm{d}w/\mathrm{d}x)$ is given by

$$\frac{\mathrm{d}w}{\mathrm{d}x} = \psi + \chi, \tag{4.30}$$

with

$$\frac{\mathrm{d}\psi}{\mathrm{d}x} = \frac{\mathcal{M}}{EI_p}, \tag{4.31a}$$

and

$$\chi = \frac{Q}{k'GA_p}, \tag{4.31b}$$

where $\mathcal{M}$ is the bending moment, Q the transverse shearing force, E Young's modulus and G the shear modulus; A_p is the cross-sectional area of the pipe (i.e. of the pipe material; as distinct from A_f, the flow area), and I_p the area-moment of inertia of the empty pipe cross-section; k' is the shear coefficient, which depends on the cross-sectional shape of the beam; for the circular cross-section of the tubular beam here under consideration, it is approximately given (Cowper 1966) by

$$k' = \frac{6(1+\nu)(1+\alpha^2)^2}{(7+6\nu)(1+\alpha^2)^2 + (20+12\nu)\alpha^2}, \tag{4.32}$$

in which ν is Poisson's ratio and α is the ratio of internal to external radius of the pipe.

In general, an element δx of the pipe is subjected to a fluid-dynamic force, the components of which, for steady flow and to first-order magnitude, are respectively zero and $F_A\,\delta x$ in the x and z directions (cf. Section 3.3.2). F_A, the lateral inviscid fluid-dynamic force (per unit length), the main concern of this work, is discussed in Sections 4.4.3 and 4.4.4; the subscript A denotes that it is related to the total acceleration of the fluid.

An element of the pipe and the forces and moments acting on it are considered next (Figure 4.15). By writing down the equations of dynamic equilibrium and neglecting terms of second-order magnitude, one can obtain the equations of motion of the system.

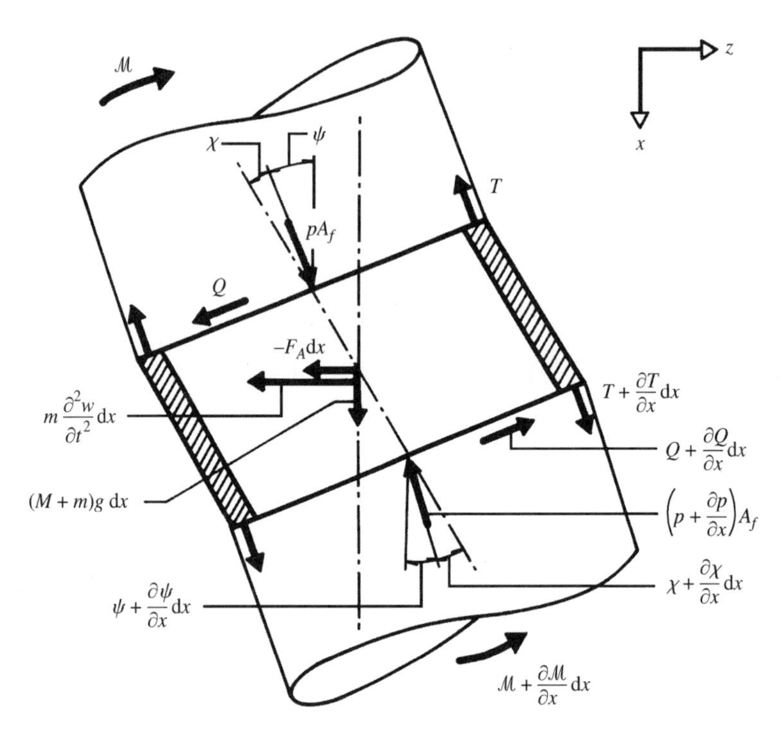

Figure 4.15 An infinitesimal element of the pipe and the enclosed fluid (under bending and shear), showing the forces and moments acting on it.

Projection of the forces on the x and z axes and consideration of moments, following a similar procedure to that in Section 3.3.2, gives (Païdoussis & Laithier 1976)

$$
\begin{aligned}
(M + m)g + \frac{\partial T}{\partial x} - A_f \frac{\partial p}{\partial x} &= 0, \\
F_A - A_f \frac{\partial}{\partial x}(p\psi) + \frac{\partial}{\partial x}(\psi T) + \frac{\partial Q}{\partial x} &= m \frac{\partial^2 w}{\partial t^2}, \\
\frac{\partial \mathcal{M}}{\partial x} + Q + \chi(pA_f - T) &= (\bar{I}_f + \bar{I}_p)\frac{\partial^2 \psi}{\partial t^2},
\end{aligned}
\tag{4.33}
$$

where p is the internal fluid pressure, above atmospheric, T is the tension in the pipe, A_f is the cross-sectional area of the enclosed fluid, M and m are the masses per unit length of the fluid and the empty pipe, respectively, and $\bar{I}_f$ and $\bar{I}_p$ are *mass*-moments of inertia per unit length of the fluid and the empty pipe, respectively.

If pressurization effects are neglected, then, by proceeding as in Section 3.3.2 and integrating the first of equation (4.33), the equivalent of equation (3.37) in this case is

$$
T - pA_f = (M + m)g(L - x) + \delta T(L), \tag{4.34}
$$

where $\delta = 0$ if the downstream end is free to slide axially and $\delta = 1$ if it is not, in which case $T(L) = \overline{T}$, where $\overline{T}$ is the mean tensioning. In the latter case, if additionally there is pressurization $\overline{p}$ *vis-à-vis* the outer ambient fluid, then $T(L) = \overline{T} - \overline{p}A(1 - 2\nu)$ approximately — see Section 3.3.2.

Using equations (4.30)–(4.31b), (4.33) and (4.34) and retaining ψ and w as variables, the following system of two differential equations may be obtained:

$$\begin{aligned} &F_A - m\frac{\partial^2 w}{\partial t^2} - (M+m)g\psi \\ &\quad + [(M+m)(L-x)g + \delta T(L)]\frac{\partial \psi}{\partial x} + k'GA_p\left(\frac{\partial^2 w}{\partial x^2} - \frac{\partial \psi}{\partial x}\right) = 0, \\ &EI_p\frac{\partial^2 \psi}{\partial x^2} + [k'GA_p - (M+m)g(L-x) - \delta T(L)]\left(\frac{\partial w}{\partial x} - \psi\right) \\ &\qquad\qquad - (\bar{I}_f + \bar{I}_p)\frac{\partial^2 \psi}{\partial t^2} = 0. \end{aligned} \tag{4.35}$$

It should be noted that equations (4.35) are not identical to those derived via Hamilton's principle. This is discussed in Appendix E.1. Here suffice it to say that the dynamical behaviour as obtained by the two sets of equations is sensibly the same for physically realistic conditions.

The system may be expressed in dimensionless terms by defining the following quantities:

$$\begin{gathered} \xi = x/L, \qquad \eta = w/L, \qquad \tau = [EI_p/(M+m)]^{1/2}t/L^2, \\ u = (M/EI_p)^{1/2}UL, \quad \beta = M/(M+m), \quad \gamma = (M+m)L^3g/EI_p, \\ \Lambda = k'GA_pL^2/EI_p, \qquad \sigma = (\bar{I}_f + \bar{I}_p)/[(M+m)L^2], \qquad \mathcal{T}_L = T(L)L^2/EI_p, \\ f_A = F_AL^3/EI_p, \qquad \varepsilon = L/2a, \end{gathered} \tag{4.36}$$

where a is the internal radius of the pipe. It is noted that for a given pipe material (i.e. for a given Poisson ratio, ν), Λ and ε are interrelated:

$$\Lambda = \frac{8k'\varepsilon^2\alpha^2}{(1+\alpha^2)(1+\nu)}, \tag{4.37}$$

where α, defined earlier, is equal to $a/(a+h)$, h being the wall thickness of the pipe. Substituting these terms into equations (4.35) gives the dimensionless equations of motion:

$$\begin{aligned} &f_A - (1-\beta)\frac{\partial^2 \eta}{\partial \tau^2} + \Lambda\frac{\partial^2 \eta}{\partial \xi^2} - \gamma\psi + [\gamma(1-\xi) + \delta\mathcal{T}_L - \Lambda]\frac{\partial \psi}{\partial \xi} = 0, \\ &\frac{\partial^2 \psi}{\partial \xi^2} + [\Lambda - (1-\xi)\gamma - \delta\mathcal{T}_L]\left(\frac{\partial \eta}{\partial \xi} - \psi\right) - \sigma\frac{\partial^2 \psi}{\partial \tau^2} = 0. \end{aligned} \tag{4.38}$$

It is noted that the equations of motion are not in their final form, as the fluid-dynamic force f_A is yet to be derived, in Sections 4.4.3 and 4.4.4. The parameter ε does not appear explicitly in equations (4.38), but it does in the expression for f_A in Section 4.4.4. It should also be noted that in equations (4.35) and (4.38), internal damping within the material of the pipe is neglected; if it is not, it may be modelled by a hysteretic damping model, wherefore Young's modulus E and the shear modulus G become complex: $E^* = E(1+\mathrm{i}\mu)$ and $G^* = G(1+\mathrm{i}\mu)$, with μ being the hysteretic damping constant.

The boundary conditions for a free end are $Q = \mathcal{M} = 0$; for a clamped end, $w = \psi = 0$. Thus, in dimensionless terms, we have

(i) for a clamped–clamped pipe:

$$\eta(0, \tau) = \psi(0, \tau) = \eta(1, \tau) = \psi(1, \tau) = 0; \tag{4.39a}$$

(ii) for a cantilevered pipe:

$$\eta(0, \tau) = \psi(0, \tau) = 0, \qquad \left.\frac{\partial \eta}{\partial \xi}\right|_{\xi=1} = \psi(1, \tau), \qquad \left.\frac{\partial \psi}{\partial \xi}\right|_{\xi=1} = 0. \tag{4.39b}$$

4.4.2 Method of analysis

The modal analysis method is utilized for the solution of the equations of motion. The motion being free, let

$$\eta(\xi, \tau) = \overline{\eta}(\xi)\mathrm{e}^{\mathrm{i}\omega\tau}, \qquad \psi(\xi, \tau) = \overline{\psi}(\xi)\mathrm{e}^{\mathrm{i}\omega\tau}, \tag{4.40}$$

where ω is a dimensionless frequency, related to the dimensional radian frequency of motion, Ω, by

$$\omega = \left(\frac{M+m}{EI_p}\right)^{1/2} L^2 \Omega. \tag{4.41}$$

Furthermore, the fluid-dynamic force f_A is assumed to vary temporally in the same manner, i.e.

$$f_A = \overline{f}_A \mathrm{e}^{\mathrm{i}\omega\tau}. \tag{4.42}$$

As in previous analyses, ω is generally complex, and the system is stable or unstable accordingly as the imaginary part of ω is positive or negative.

The modal analysis method proceeds by expressing $\overline{\eta}(\xi)$ and $\overline{\psi}(\xi)$ as the superposition of an infinite set of comparison functions (Galerkin's technique), i.e.

$$\overline{\eta}(\xi) = \sum_{n=1}^{\infty} a_n Y_n(\xi), \qquad \overline{\psi}(\xi) = \sum_{n=1}^{\infty} b_n \Psi_n(\xi), \tag{4.43}$$

where a_n and b_n are dimensionless generalized coefficients, and $Y_n(\xi)$ and $\Psi_n(\xi)$ are the eigenfunctions of a Timoshenko beam, with the appropriate boundary conditions, expressed in dimensionless form; this solution then inherently satisfies the boundary conditions. Substitution of equations (4.40), (4.42) and (4.43) into (4.38) and application of the

Galerkin procedure yields

$$\begin{aligned}
&\sum_{n=1}^{\infty} \{a_n(1-\beta)\omega^2 I_{kn}^{(1)} + a_n Q_{kn} + a_n \Lambda I_{kn}^{(3)} \\
&\qquad - b_n(\Lambda - \gamma - \delta\mathcal{T}_L) I_{kn}^{(4)} - b_n \gamma I_{kn}^{(5)} - b_n \gamma I_{kn}^{(6)}\} = 0, \\
&\sum_{n=1}^{\infty} \{b_n I_{kn}^{(7)} + a_n(\Lambda - \gamma - \delta\mathcal{T}_L) I_{kn}^{(8)} + a_n \gamma I_{kn}^{(10)} \\
&\qquad - b_n(\Lambda - \gamma - \delta\,\mathcal{T}_L) I_{kn}^{(9)} - b_n \gamma I_{kn}^{(11)} + b_n \sigma\omega^2 I_{kn}^{(9)}\} = 0,
\end{aligned} \tag{4.44}$$

where $k = 1, 2, ..., \infty$, and

$$\int_0^1 \overline{f}_A Y_k \, \mathrm{d}\xi \equiv \sum_{n=1}^{\infty} a_n Q_{kn}. \tag{4.45}$$

The constants $I_{kn}^{(i)}$, $i = 1, 2, ..., 11$, are defined as follows:

$$\begin{aligned}
I_{kn}^{(1)} &= \int_0^1 Y_n Y_k \, \mathrm{d}\xi, \quad I_{kn}^{(2)} = \int_0^1 Y_n' Y_k \, \mathrm{d}\xi, \quad I_{kn}^{(3)} = \int_0^1 Y_n'' Y_k \, \mathrm{d}\xi, \quad I_{kn}^{(4)} = \int_0^1 \Psi_n' Y_k \, \mathrm{d}\xi, \\
I_{kn}^{(5)} &= \int_0^1 \xi \Psi_n' Y_k \, \mathrm{d}\xi, \quad I_{kn}^{(6)} = \int_0^1 \Psi_n Y_k \, \mathrm{d}\xi, \quad I_{kn}^{(7)} = \int_0^1 \Psi_n'' \Psi_k \, \mathrm{d}\xi, \quad I_{kn}^{(8)} = \int_0^1 Y_n' \Psi_k \, \mathrm{d}\xi, \\
I_{kn}^{(9)} &= \int_0^1 \Psi_n \Psi_k \, \mathrm{d}\xi, \quad I_{kn}^{(10)} = \int_0^1 \xi Y_n' \Psi_k \, \mathrm{d}\xi, \quad I_{kn}^{(11)} = \int_0^1 \xi \Psi_n \Psi_k \, \mathrm{d}\xi.
\end{aligned}$$

The evaluation of these integrals in terms of the Timoshenko beam eigenfunctions is discussed in Appendix E.2.

The solution as expressed by equations (4.43) is then truncated at $n = N$, and equations (4.44) yield a vanishing determinant of order $2N$. This is solved to give the eigenfrequencies ω of the system, for different values of the dimensionless flow velocity u and of the other system parameters, β, Λ, γ, etc.

4.4.3 The inviscid fluid-dynamic force

Here the inviscid fluid-dynamic force, F_A, will be derived, first according to the plug-flow approximation and then in a more refined manner.

(a) The inviscid fluid-dynamic force for plug flow

This approximation, which applies to large length-to-diameter ratios, small displacements and, as we shall see, long wavelengths of deformation of the pipe as compared to its diameter, is what has been used in all of the foregoing. Thus, by using d'Alembert's principle, the force F_A is equal to the mass of the fluid per unit length multiplied by the reversed acceleration as given by equation (3.29), here with $\mathrm{d}U/\mathrm{d}t = 0$; hence,

$$F_A = -M\left\{\frac{\partial^2}{\partial t^2} + 2U\frac{\partial^2}{\partial x\,\partial t} + U^2\frac{\partial^2}{\partial x^2}\right\} w. \tag{4.46}$$

Expressed nondimensionally and in the form required by the modal analysis method (Section 4.4.2), the generalized fluid-dynamic force, $\overline{q}$, may be written as follows:

$$\overline{q} = \int_0^1 \overline{f}_A Y_k\,\mathrm{d}\xi = \sum_{n=1}^{\infty} a_n Q_{kn}, \tag{4.47}$$

with

$$Q_{kn} = \beta\omega^2 Q_{kn}^{(1)} - 2\mathrm{i}u\beta^{1/2}\omega Q_{kn}^{(2)} - u^2 Q_{kn}^{(3)}, \tag{4.48}$$

where

$$Q_{kn}^{(1)} = \int_0^1 Y_n Y_k\,\mathrm{d}\xi; \qquad Q_{kn}^{(2)} = \int_0^1 Y_n' Y_k\,\mathrm{d}\xi; \qquad Q_{kn}^{(3)} = \int_0^1 Y_n'' Y_k\,\mathrm{d}\xi.^\dagger \tag{4.49}$$

(b) The inviscid fluid-dynamic force for 3-D potential flow

The fluid is assumed to be inviscid and the flow irrotational, consisting of the mean flow $U\mathbf{i}$ along the pipe and a small perturbation $\mathbf{v}(r, \theta, x, t)$ associated with small motions of the pipe, which may be expressed in terms of a perturbation potential via $\mathbf{v} = \nabla\phi$. This potential must satisfy equation (2.73a), $\nabla^2\phi = 0$, which for this system is

$$\frac{\partial^2\phi}{\partial r^2} + \frac{1}{r}\frac{\partial\phi}{\partial r} + \frac{1}{r^2}\frac{\partial^2\phi}{\partial\theta^2} + \frac{\partial^2\phi}{\partial x^2} = 0, \tag{4.50}$$

as well as the compatibility and boundary conditions

$$\left.\frac{\partial\phi}{\partial r}\right|_{r=a} = \left(\frac{\partial w}{\partial t} + U\frac{\partial w}{\partial x}\right)\sin\theta, \qquad 0 \le x \le L, \qquad 0 \le \theta < 2\pi,$$
$$= 0, \qquad x < 0, \tag{4.51}$$

where motions are assumed to take place in the $\theta = \frac{1}{2}\pi$ plane and a is the internal pipe radius, and

$$\lim_{x\to\pm\infty} \phi = 0, \qquad \lim_{x\to\pm\infty} (\partial\phi/\partial x) = 0. \tag{4.52}$$

The force on the pipe is determined by integrating the pressure $p \equiv p(a, \theta, x, t)$ on the inner pipe boundary, which may be determined by substituting $\mathbf{v} = \nabla\phi$ and $\nu = 0$ in equation (2.67a), leading to

$$p = -\rho\left.\left(\frac{\partial\phi}{\partial t} + U\frac{\partial\phi}{\partial x}\right)\right|_{r=a}, \tag{4.53}$$

† Although these are equal to $I_{kn}^{(j)}$, $j = 1, 2, 3$, respectively, defined in conjunction with equation (4.44), they are denoted differently to indicate that they are related to the right-hand side of (4.45) or (4.47).

ρ being the fluid density. Assume now separable solutions of the form

$$\phi(r, \theta, x, t) = R(r) \sin \theta \exp[\mathrm{i}(\kappa x + \Omega t)], \tag{4.54}$$

where the form of the θ component has been suggested by (4.51), and the form of the x component emerges in the course of separating the variables. Substituting into (4.50) leads to

$$\frac{\mathrm{d}^2 R}{\mathrm{d}r^2} + \frac{1}{r}\frac{\mathrm{d}R}{\mathrm{d}r} - \left(\frac{1}{r^2} + \kappa^2\right) R = 0,$$

admitting solutions of the form

$$R(r) = C_1 \mathrm{I}_1(\kappa r) + D_1 \mathrm{K}_1(\kappa r), \tag{4.55}$$

where I_1 and K_1 are modified Bessel functions of the first and second kind of order 1, and where $D_1 = 0$ because ϕ must remain finite within the pipe. C_1 is determined by application of (4.51), and one finds

$$\phi(r, \theta, x, t) = \frac{\mathrm{I}_1(\kappa r)}{\kappa \mathrm{I}_1'(\kappa a)} \left(\frac{\partial w}{\partial t} + U\frac{\partial w}{\partial x}\right) \sin \theta, \tag{4.56}$$

in which $\mathrm{I}_1' = \mathrm{d}\mathrm{I}_1/\mathrm{d}(\kappa r)$. Then, utilizing the relation $\mathrm{I}_n'(x) = (1/x)[n\mathrm{I}_n(x) + x\mathrm{I}_{n+1}(x)]$ (Dwight 1961) for $n = 1$, one obtains from (4.53)

$$p = \frac{-\rho a}{1 + \kappa a \mathrm{I}_2(\kappa a)/\mathrm{I}_1(\kappa a)} \left[\left(\frac{\partial}{\partial t} + U\frac{\partial}{\partial x}\right)^2 w\right] \sin \theta. \tag{4.57}$$

From this, the force F_A is found to be

$$F_A = \int_0^{2\pi} pa \sin \theta \, \mathrm{d}\theta = \frac{-M}{1 + \kappa a \mathrm{I}_2(\kappa a)/\mathrm{I}_1(\kappa a)} \left(\frac{\partial}{\partial t} + U\frac{\partial}{\partial x}\right)^2 w, \tag{4.58}$$

where $M = \rho\pi a^2$ has been used. Comparing (4.58) to (4.46) it is clear that M is now replaced by $M/[1 + \kappa a \mathrm{I}_2(\kappa a)/\mathrm{I}_1(\kappa a)]$, where the denominator is generally larger than unity. Hence, for finite wavenumbers κa (and wavelengths of motion) the effective fluid-dynamic force is generally smaller than that given by the plug-flow approximation.

It is instructive to consider the case of κa small, i.e. motions of *large* wavelength. Utilizing the series expansion $\mathrm{I}_n(x) = (1/n!)(\frac{1}{2}x)^n[1 + \mathcal{O}(x^2)]$ (Dwight 1961), one obtains

$$\lim_{\kappa a \to 0} \frac{M}{1 + \kappa a[\frac{1}{2}(\frac{1}{2}\kappa a)^2/(\frac{1}{2}\kappa a)]} = M,$$

thus retrieving the form of F_A given by equation (4.46) and proving that it only holds true provided that the wavelength of motions is large compared to the pipe diameter.

However, for the analysis of short pipes the full form of (4.58) is retained. The pertinent forms of $\bar{q}$ and Q_{kn} — cf. equation (4.47) — are presented in the next section.

4.4.4 The fluid-dynamic force by the integral Fourier-transform method

It is noted that κ in equation (4.58) is not known *a priori*. Hence, there no longer exists a 'point relationship' between F_A and x as in most of the analyses of Chapter 3: F_A at any given x depends on the deformation all along the pipe. A powerful method for the solution of problems such as this was proposed by Dowell & Widnall (1966) — see also Widnall & Dowell (1967) and Dowell (1975) — the essence of which will become evident with its application in what follows.

We start by adapting what has just been obtained in Section 4.4.3(b) to a suitable form. We first redefine

$$\phi(r,\theta,x,t) = \psi(r,x)\sin\theta\, \mathrm{e}^{\mathrm{i}\Omega t}, \qquad w(x,t) = \overline{w}(x)\mathrm{e}^{\mathrm{i}\Omega t},$$
$$p(r,\theta,x,t) = \overline{p}(r,\theta,x)\mathrm{e}^{\mathrm{i}\Omega t}, \tag{4.59}$$

and define the Fourier transforms of $\psi(r,x)$ and $\overline{w}(x)$ by

$$\psi^*(r,\alpha) = \int_{-\infty}^{\infty} \psi(r,x)\mathrm{e}^{\mathrm{i}\alpha x}\,\mathrm{d}x, \qquad \overline{w}^*(\alpha) = \int_{-\infty}^{\infty} \overline{w}(x)\mathrm{e}^{\mathrm{i}\alpha x}\,\mathrm{d}x, \tag{4.60}$$

and similarly for $\overline{p}^*$ [see, e.g. Meirovitch (1967)]; the asterisk denotes the Fourier transform and α is the transform variable. The inverse transforms are

$$\psi(r,x) = \frac{1}{2\pi}\int_{-\infty}^{\infty} \psi^*(r,\alpha)\mathrm{e}^{-\mathrm{i}\alpha x}\,\mathrm{d}\alpha, \qquad \overline{w}(x) = \frac{1}{2\pi}\int_{-\infty}^{\infty} \overline{w}^*(x)\mathrm{e}^{-\mathrm{i}\alpha x}\,\mathrm{d}\alpha, \tag{4.61}$$

and similarly for $\overline{p}(r,\theta,x)$. Furthermore, we define

$$k = \frac{\Omega L}{U}, \qquad \overline{\alpha} = \alpha L, \qquad F(\overline{\alpha}) = \frac{2\epsilon}{\overline{\alpha}}\frac{\mathrm{I}_1(\overline{\alpha}/2\epsilon)}{\mathrm{I}_1'(\overline{\alpha}/2\epsilon)}, \tag{4.62}$$

where k is the so-called reduced frequency, $F(\overline{\alpha})$ is clearly the first part of (4.56) in the Fourier domain and $\epsilon = L/2a$, as already defined.

Proceeding with the analysis exactly as in Section 4.4.3(b) but in the Fourier domain, one finds for the perturbation pressure

$$\overline{p}^*(a,\theta,x) = \frac{\rho U^2 a}{L^3}(\overline{\alpha}-k)^2 F(\overline{\alpha})\overline{w}^*\sin\theta, \tag{4.63}$$

which inverted gives

$$\overline{p}(a,\theta,\xi) = \frac{1}{2\pi}\int_{-\infty}^{\infty}\frac{\rho U^2 a}{L^3}(\overline{\alpha}-k)^2 F(\overline{\alpha})\overline{w}^*\mathrm{e}^{-\mathrm{i}\overline{\alpha}\xi}\,\mathrm{d}\overline{\alpha}\sin\theta, \tag{4.64}$$

in terms of $\xi = x/L$. The inviscid fluid-dynamic force F_A is then found to be

$$F_A = MU^2\left(\frac{1}{2\pi L^2}\right)\mathrm{e}^{\mathrm{i}\Omega t}\int_{-\infty}^{\infty}(\overline{\alpha}-k)^2F(\overline{\alpha})\left\{\int_{-\infty}^{\infty}\overline{w}(\xi)\mathrm{e}^{\mathrm{i}\overline{\alpha}\xi}\,\mathrm{d}\xi\right\}\mathrm{e}^{-\mathrm{i}\overline{\alpha}\xi}\,\mathrm{d}\overline{\alpha}. \tag{4.65}$$

The physical domain of the problem is $[0, L]$; in terms of ξ, it is $[0, 1]$. However, this domain will be expanded, by taking in some additional space beyond the downstream end of the pipe, to $[0, l]$, where $l > 1$. This is necessary, particularly in the case of cantilevered pipes, as flow perturbations persist beyond the free end of the pipe, as discussed in Sections 3.5.8 and 4.4.5.† Accordingly, instead of the first of equations (4.43), the following form of the Galerkin expansion is adopted:

$$\begin{aligned} \overline{w}(\xi) &= \sum_{n=1}^{\infty} A_n Y_n(\xi), \qquad && 0 \le \xi \le 1 \\ &= \sum_{n=1}^{\infty} A_n G_n(\xi), && 1 < \xi \le l \\ &= 0, && \xi < 0 \text{ and } \xi > l; \end{aligned} \tag{4.66}$$

$Y_n(\xi)$ are the comparison functions associated with $\overline{\eta}(\xi)$ in equations (4.43), and $G_n(\xi)$ are the so-called 'outflow-model' functions which are associated with deflections of the fluid jet beyond the free end of a cantilevered pipe.

In the modal-analysis solution of the problem, the main interest is in the generalized fluid-dynamic force $\overline{q}$, rather than F_A, as defined by equations (4.47) and (4.48) or (4.45). In this case, $Q_{kn}^{(1)}$, $Q_{kn}^{(2)}$ and $Q_{kn}^{(3)}$ of (4.48) are given by

$$\begin{aligned} Q_{kn}^{(1)} &= \frac{1}{2\pi}\int_{-\infty}^{\infty} F(\overline{\alpha})\left\{\int_0^1 Y_n\, e^{i\overline{\alpha}\xi} d\xi + \int_1^l G_n\, e^{i\overline{\alpha}\xi}\, d\xi\right\}\left\{\int_0^1 Y_k\, e^{-i\overline{\alpha}\xi}\, d\xi\right\} d\overline{\alpha}, \\ Q_{kn}^{(2)} &= \frac{1}{2\pi i}\int_{-\infty}^{\infty} \overline{\alpha} F(\overline{\alpha})\left\{\int_0^1 Y_n e^{i\overline{\alpha}\xi}\, d\xi + \int_1^l G_n\, e^{i\overline{\alpha}\xi}\, d\xi\right\}\left\{\int_0^1 Y_k\, e^{-i\overline{\alpha}\xi}\, d\xi\right\} d\overline{\alpha}, \\ Q_{kn}^{(3)} &= \frac{-1}{2\pi}\int_{-\infty}^{\infty} \overline{\alpha}^2 F(\overline{\alpha})\left\{\int_0^1 Y_n e^{i\overline{\alpha}\xi}\, d\xi + \int_1^l G_n\, e^{i\overline{\alpha}\xi}\, d\xi\right\}\left\{\int_0^1 Y_k\, e^{-i\overline{\alpha}\xi}\, d\xi\right\} d\overline{\alpha}, \end{aligned} \tag{4.67}$$

with $F(\overline{\alpha})$ as given in (4.62)

4.4.5 Refined and plug-flow fluid-dynamic forces and specification of the outflow model

The lateral inviscid fluid-dynamic force derived by means of refined fluid mechanics and the integral-transform technique is intended to be used for short pipes. Nevertheless, in the limit of sufficiently long ones, it should give identical results to those obtained with the simpler, plug-flow model — for the reasons discussed already. In this section, a comparison is made of the generalized fluid-dynamic force components, $Q_{kn}^{(i)}$, $i = 1, 2, 3$, obtained by the refined fluid-mechanics model [equations (4.67)] and by the plug-flow model [equations (4.49)] — for clamped–clamped and cantilevered *long pipes*.

(a) Clamped–clamped pipes

For a pipe clamped at both ends, there is no need for an outflow model, insofar as the generalized fluid-dynamic force obtained by the integral (Fourier) transform method

†It is clear from (4.65), nevertheless, that the very nature of the Fourier-transform solution *requires* the specification of $\overline{w}(\xi)$ beyond $[0, 1]$, even if this means stating that $\overline{w}(\xi) = 0$ for $-\infty \le \xi < 0$ and $1 < \xi \le +\infty$.

is concerned, because the fluid discharging from the downstream end is assumed to enter a rigid pipe which experiences no deflection [Figure 4.16(a)]. Therefore, in expressions (4.67), $l = 1$ or $G_n(\xi) = 0$.

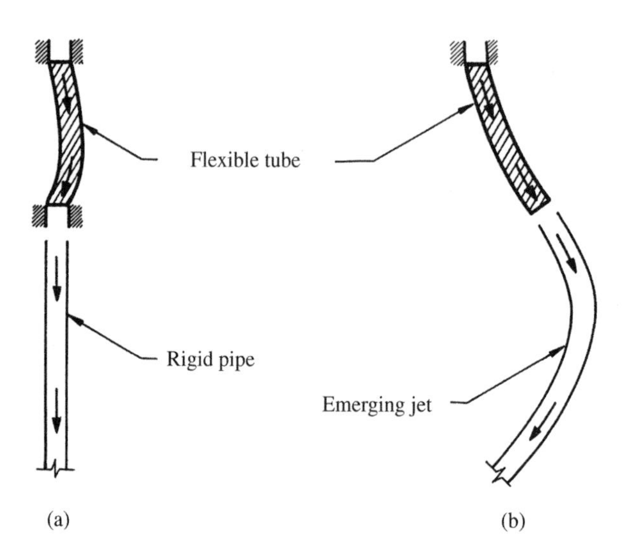

Figure 4.16 The physical form of the 'collector pipe' for a clamped–clamped pipe and the form of the free jet emerging from a cantilevered pipe (no collector pipe).

The two inner integrals of the expressions in (4.67) may be evaluated analytically (Luu 1983) or numerically, but the three outer integrals, which involve an infinite range of integration over $\overline{\alpha}$, have to be evaluated numerically; this is done by a two-point Gaussian numerical integration method. Based on a check on convergence for a long clamped–clamped pipe, calculations (throughout this work) of the generalized fluid-dynamic forces for a clamped–clamped pipe, either long or short, are done with the integration range $-100 \leq \overline{\alpha} \leq 100$ and the integration step $\delta\overline{\alpha} = 2$; they approximate the result for a larger range of $\overline{\alpha}$ (and hence $-\infty \leq \overline{\alpha} \leq \infty$) and a finer $\delta\overline{\alpha}$ very well.

The next step is to undertake a comparison between the results of the generalized hydrodynamic forces $Q_{kn}^{(i)}$, $i = 1, 2, 3$, for a long clamped–clamped pipe conveying fluid ($\Lambda = 10^{12}$, corresponding to $\varepsilon \equiv L/2a = 8.5 \times 10^5$)[†] obtained by (i) simple plug flow and (ii) refined fluid mechanics, where the $Y_j(\xi)$ used are the eigenfunctions of a clamped–clamped Timoshenko beam without internal flow, as given in Appendix E. For the first (lowest) three modes of the system ($k, n = 1, 2, 3$), the results are virtually identical: the largest discrepancy, associated with the $Q_{31}^{(3)}$ term, is only 0.023%. This, to some extent, validates the refined fluid mechanics model, which may now fairly confidently be used for short pipes clamped at both ends.

[†]This value of ε is clearly nonphysical, but has been dictated by the desire to obtain virtually *identical* results to those of the Euler–Bernoulli theory, to many significant figures. *Practically* identical results may be obtained for $\varepsilon \sim \mathcal{O}(10^3)$.

(b) Cantilevered pipes and outflow models

Unlike pipes with fixed ends, a cantilevered pipe discharges the fluid freely from its downstream end. The emerging jet continues its sinuous path in the ambient air, [Figure 4.16(b)], as briefly discussed in Section 3.5.8. The motion of the cantilever is therefore coupled with that of the downstream jet (at least in this kind of formulation) — as first discussed by Shayo & Ellen (1978). Thus, in a study of the flow-induced instability of a cantilevered pipe, it becomes necessary to construct an artificial 'outflow model' which describes the manner in which $\overline{w}(\xi)$ and hence the perturbations in the fluid are attenuated beyond the free end of the pipe.

For long pipes conveying fluid, the plug-flow model is fully expected to give reasonable approximations to the fluid-dynamic forces, and hence to predict reasonably well the dynamical behaviour of the system. Moreover, the results have been found to be in good agreement with experiments, and the plug-flow model may be considered to be quite adequate for long cantilevers conveying fluid. Therefore, the following approach is adopted: different outflow-model functions $G_n(\xi)$ and various values of l ($l > 1$) are tried and adjusted, so that the generalized fluid-dynamic forces (4.67) obtained by refined fluid mechanics agree with those obtained by simple fluid mechanics [plug-flow model with equations (4.49)] for a *long cantilevered pipe*. It is then assumed that the same outflow-model functions $G_n(\xi)$ and value of l would apply for short pipes — indeed to the very short cantilevered pipes which are the subject of this section. The validity of this assumption is tested, partially at least, by comparison with experimental measurements (Section 4.4.8). Following the mathematical formulation suggested by Shayo & Ellen (1978) for both the beam- and shell-mode dynamical behaviour of a cantilevered shell conveying fluid, three different downstream flow models are tried for the cantilevered tubular beam; their characteristics are summarized in Table 4.2, together with the 'no model' situation, in which the deflection of the perturbation in the fluid is supposed to vanish abruptly at $\xi = l = 1$; for the 'first', 'second' and 'third' models, the motion of the fluid beyond the free end is described by progressively higher-order polynomials of the fluid-jet deflection. The 'first model' is Shayo & Ellen's 'collector pipe model' (Section 3.5.8). The 'second model' is described mathematically by

$$G_n(\xi) = Y_n(1)\left[1 - \frac{(\xi-1)^2}{(l-1)^2}\right] + Y_n'(1)\left[(\xi-1) - \frac{(\xi-1)^2}{(l-1)}\right] \qquad \text{for } 1 < \xi \leq l,$$
$$= 0 \qquad \text{for } \xi > l. \tag{4.68}$$

The 'third model', which involves a cubic polynomial in ξ, is given in detail in Luu (1983); this model transcends physical reality by unjustifiably specifying a zero slope for the free jet far downstream (Table 4.2). Calculations done for a very long cantilever ($\Lambda = 10^{12}$) according to the various models of Table 4.2[†] show that the second model, equation (4.68), with $l = 2.8$ gives optimum results, as may be seen in Table 4.3. The second and third modes were also tested, and the second model with $l = 2.8$ again gives the best results. Hence, it

[†]In these calculations, the $Y_j(\xi)$ used in (4.67) are the eigenfunctions of clamped-free Timoshenko beam without flow; the integration range for $\overline{\alpha}$ is $[-150, 150]$ and the integration step $\delta\overline{\alpha} = 2$. These give convergent results and have been used throughout in calculations for cantilevered pipes.

Table 4.2 The characteristics of different outflow-model functions. The schematic presentation of outflow models is for the first beam mode only.

Characteristics	Type of outflow-model (0) No model	(1) 'First model' (1st-order polynomial)	(2) 'Second model' (2nd-order polynomial)	(3) 'Third model' (3rd-order polynomial)
Zero displacement at 'infinity' ($\xi = l$)	Yes	Yes	Yes	Yes
Continuity of displacement at outlet	No	Yes	Yes	Yes
Continuity of slope at outlet	No	No	Yes	Yes
Zero slope at 'infinity' ($\xi = l$)	Yes	No	No	Yes

Table 4.3 The results with different outflow models for the first terms of the generalized fluid-dynamic force, $Q_{kn}^{(i)}$ $(k = n = 1, i = 1, 2, 3)$ with $l = 2.8$, for a long cantilevered pipe ($\Lambda = 10^{12}$, $\varepsilon = 8.25 \times 10^5$).

Term	Plug-flow model	Refined fluid mechanics model: No model	1st model	2nd model	3rd model
$Q_{11}^{(1)}$	1.000	0.9915	1.000	1.000	1.000
$Q_{11}^{(2)}$	2.000	0.9941	1.984	2.000	2.000
$Q_{11}^{(3)}$	0.8582	−1.879	−2.873	0.8510	0.8222

has been adopted throughout this work for calculating the generalized fluid-dynamic forces for short cantilevered pipes.

4.4.6 Stability of clamped–clamped pipes

The calculations of the eigenfrequencies have been conducted by the methods of Section 4.4.2. Convergence of the eigenfrequencies by the modal analysis method is quite fast: for clamped–clamped boundary conditions, $N = 7$ yields convergent results.

Most of the calculations have been conducted for metallic pipes with $h/(a+h) = 0.10$, $\nu = 0.3$, $\sigma = 0$ and $\Lambda = 10^{12}$, 100 and 10, corresponding to $\varepsilon = 8.25 \times 10^5$, 8.25 and 2.61, respectively. $\Lambda = (k'GA_p/EI_p)L^2$, which is a measure of shear rigidity of the system, is very large for realistic systems, unless the pipe is quite short. For $\Lambda = 10^{12}$

shear deformation is minimal; it approximates $\Lambda \to \infty$ very well. The effect of σ (rotatory inertia) has been shown to be negligible for realistic systems (Païdoussis & Laithier 1976; Laithier 1979), and this is why the calculations have been conducted with $\sigma = 0$. The calculations are conducted according to the Timoshenko refined-flow (TRF), Timoshenko plug-flow (TPF) and the Euler–Bernoulli plug-flow (EBPF) theories and the results compared.

For the TRF theory, for each length-to-diameter ratio ε (and, correspondingly, for each Λ), the work involved consists of: (i) evaluating the generalized inviscid hydrodynamic forces Q_{kn} from (4.67); (ii) incorporating Q_{kn} into equations (4.44) to obtain the eigenfrequencies, and then (iii) constructing the corresponding Argand diagram of the system, to obtain the critical velocity u_{cd} for divergence (the system being conservative; Section 3.4) and the predicted post-divergence behaviour.

In Figure 4.17 are shown the first- and second-mode Argand diagrams of the system eigenfrequencies for the longest pipe ($\Lambda = 10^{12}$), as obtained by the TRF theory. The results obtained with the TPF and EBPF theories are virtually indistinguishable from those shown. This is as expected, since (i) as shown in the previous section, for a long clamped–clamped pipe the simple plug-flow model and the refined-flow model give the same values for the generalized fluid-dynamic forces, (ii) the dynamics of a very long pipe (here $\varepsilon = 8.25 \times 10^5$) are identical, whether analysed by Timoshenko or Euler–Bernoulli theory, at least in the low modes.

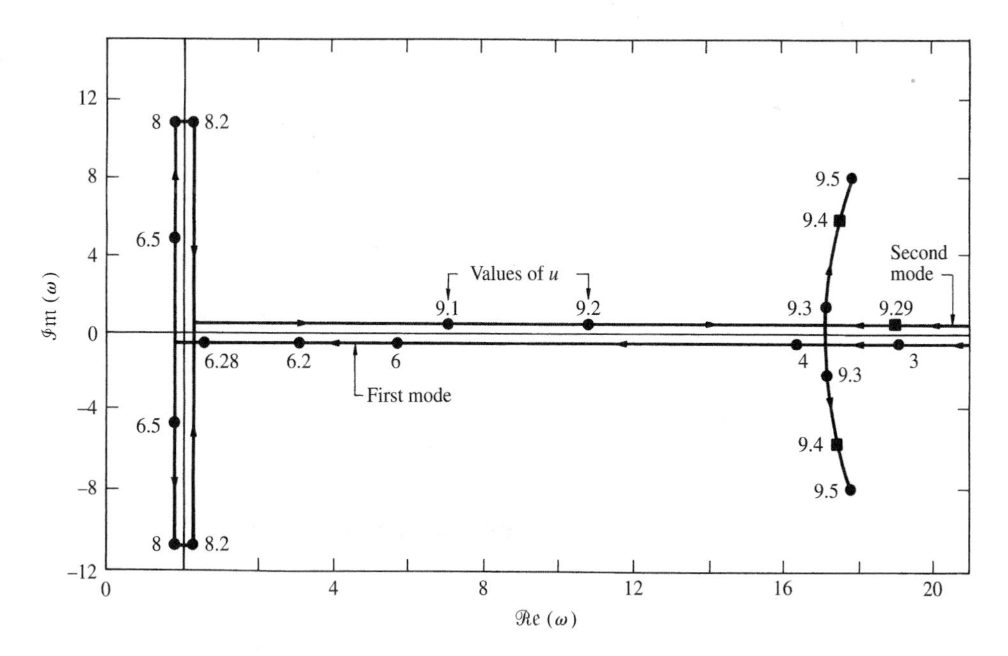

Figure 4.17 Dimensionless complex eigenfrequencies of an extremely long clamped–clamped pipe ($\beta = 0.5$, $\gamma = \mu = \sigma = 0$, $\Lambda = 10^{12}$, $\varepsilon = 8.25 \times 10^5$) as functions of the dimensionless flow velocity u, according to the Timoshenko refined-flow (TRF) theory: —●—, first mode; —■—, second mode; —●■—, combined first and second modes. The loci, which actually lie on the axes, have been drawn slightly off the axes but parallel to them for the sake of clarity (Païdoussis *et al*. 1986).

In Figure 4.18 is shown an Argand diagram for a shorter pipe ($\varepsilon = 8.25$, $\Lambda = 100$). The dynamical behaviour of the system is similar to that of a long pipe (Figure 4.17), but the eigenfrequencies obtained by TRF and EBPF theories are no longer coincident: the former are consistently lower than the latter. Moreover, the critical flow velocities, both for divergence and coupled-mode flutter, according to TRF theory are lower. These observations are reasonable since TRF theory correctly takes shear deformation into account; shear deformation renders the system effectively more flexible.

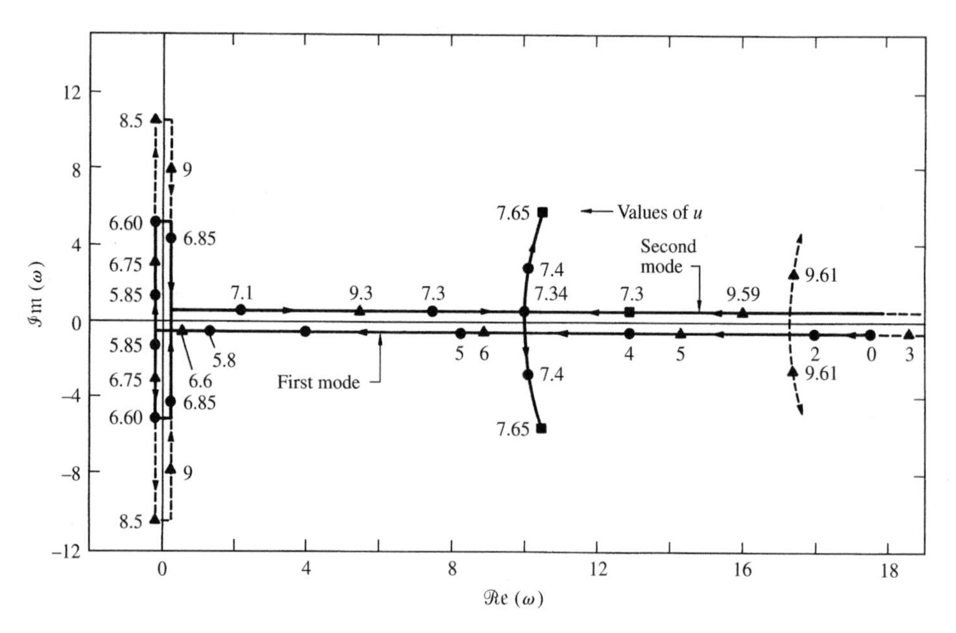

Figure 4.18 Dimensionless complex eigenfrequencies of a short clamped-clamped pipe ($\beta = 0.5$, $\gamma = 10$, $\mu = \sigma = 0$; $\Lambda = 100$, $\varepsilon = 8.25$), as functions of the dimensionless flow velocity u : •, 1st mode TRF (Timoshenko refined-flow theory); ■, 2nd mode TRF; ▲, 1st and 2nd mode EBPF (Euler–Bernoulli plug-flow theory). The loci, which actually lie on the axes, have been drawn slightly off but paralled to them for clarity (Païdoussis *et al*. 1986).

Similar observations can also be made for very short pipes ($\varepsilon = 2.61$). The trends referred to above are simply more pronounced in this case; hence, even lower dimensionless critical flow velocities are obtained.

Now, let us turn our attention to the differences in the results obtained by the simple and the refined fluid mechanics, and Timoshenko beam theory in both cases — i.e. let us compare the results of the TPF and TRF theories. The dimensionless eigenfrequencies of the first and second modes for $u = 0$ are shown in Table 4.4, and the critical flow velocities for divergence are shown in Figure 4.19.

At $u = 0$, the refined fluid mechanics model gives slightly higher values for the first-mode eigenfrequency than the simple, plug-flow one (Table 4.4). The difference is only noticeable for $\Lambda \leq 100$ and is larger for higher modes. The observed differences in eigenfrequencies are believed to arise from differences in the effective virtual mass per unit length. According to simple fluid mechanics, this mass is simply the enclosed mass of

Table 4.4 The eigenfrequencies of a heavy clamped–clamped short pipe ($u = 0$) for $\gamma = 10$, $\beta = 0.5$, $\mu = \sigma = 0$, by Timoshenko plug-flow (TPF) and Timoshenko refined-flow (TRF) theories.

Λ	Mode	ω by TPF theory	ω by TRF theory
100	1	19.552	19.599
	2	44.365	44.752
10	1	9.670	9.837
	2	19.342	20.493

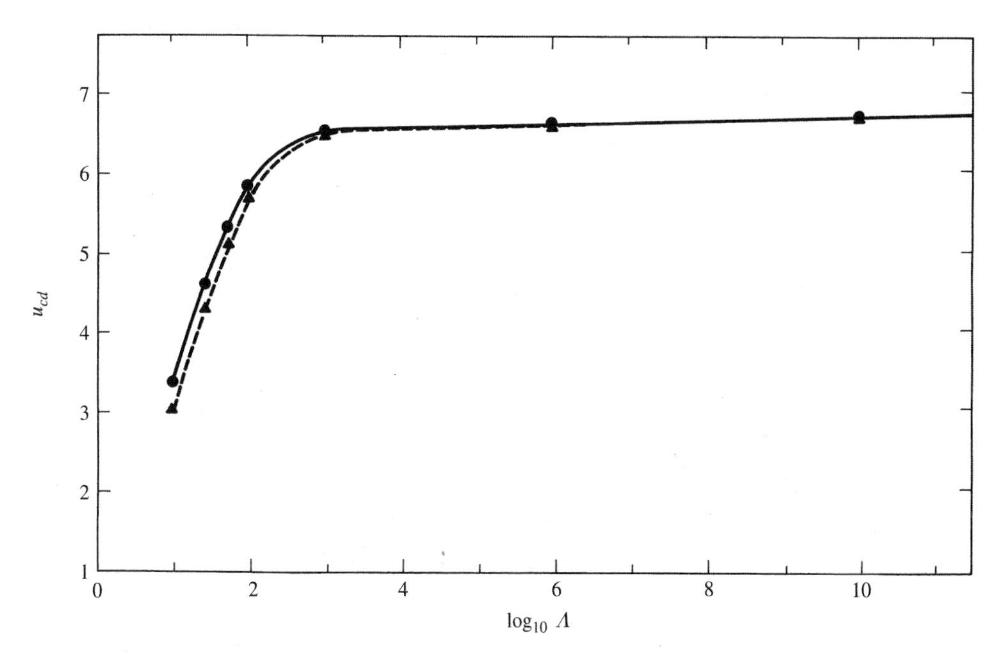

Figure 4.19 The critical dimensionless flow velocities for divergence, u_{cd}, of a pipe clamped at both ends, showing the effect of slenderness and related transverse shear [see equation (4.37)], for $\beta = 0.5$, $\gamma = 10$, $\mu = \sigma = 0$. —●—, Timoshenko refined-flow (TRF) theory; --▲--, Timoshenko plug-flow (TPF) theory (Païdoussis *et al*. 1986).

fluid per unit length — the 'slender body' approximation [cf. Section 2.2.2(e)(ii)] which in this case reduces to the plug-flow model. According to the refined model, however, this is smaller because of 'end effects' or departures from two-dimensionality [cf. equation (2.139) and the discussion of (4.58)], which are more important for short than for long pipes. Hence, the effective total mass per unit length is $m + M'$, with $M' < M$ where $M = \rho A_f$, and the values of ω [generally equal to (generalized stiffness)/(generalized mass)] are therefore larger.

Considering the critical flow velocities for divergence next (Figure 4.19), it is seen that the results for the TRF theory are indistinguishable from those obtained by the TPF theory for $\Lambda > 10^3$ approximately. On the other hand, for $\Lambda < 10^3$, TPF theory tends to underestimate u_{cd}. The argument of end effects just discussed may be invoked here also to explain these differences. The critical flow velocity for divergence depends principally on the excitation force $Q_{kn}^{(3)}$ — which in the plug-flow model is proportional to MU^2; this is smaller for refined fluid mechanics, since $M' < M$. Hence, this translates to $u_{cd}^{\text{TRF}} > u_{cd}^{\text{TPF}}$. For the EBPF theory the value of u_{cd} is independent of Λ and equal to $u_{cd}^{\text{EBPF}} = 6.66$ (cf. Figure 4.18), which is considerably higher than that obtained by the more appropriate TRF and TPF theories for $\Lambda < 1000$ or so.

4.4.7 Stability of cantilevered pipes

Calculations for cantilevered pipes are conducted, utilizing the outflow model developed in Section 4.4.5(b), i.e. the 'second' or quadratic model with $l = 2.8$. In this case $N = 7$, 8 and 9 terms in the modal expansions (4.43) are necessary for convergence in the first, second and third modes of the system, respectively. As in the previous section, the three theories (EBPF, TPF and TRF) are compared to one another for $\Lambda = 10^{12}$, 100 and 10.

Calculations for long pipes ($\Lambda = 10^{12}$) show that, similarly to the results of Figure 4.17 for clamped–clamped pipes, the eigenfrequencies obtained by EBPF, TPF and TRF theories are essentially identical (in the scale of the Argand diagram, not shown for brevity) in the lowest three modes. For shorter pipes, differences begin to become noticeable, as shown in Figures 4.20 and 4.21 for $\Lambda = 100$ and 10. The results of the EBPF theory are not shown, for clarity.

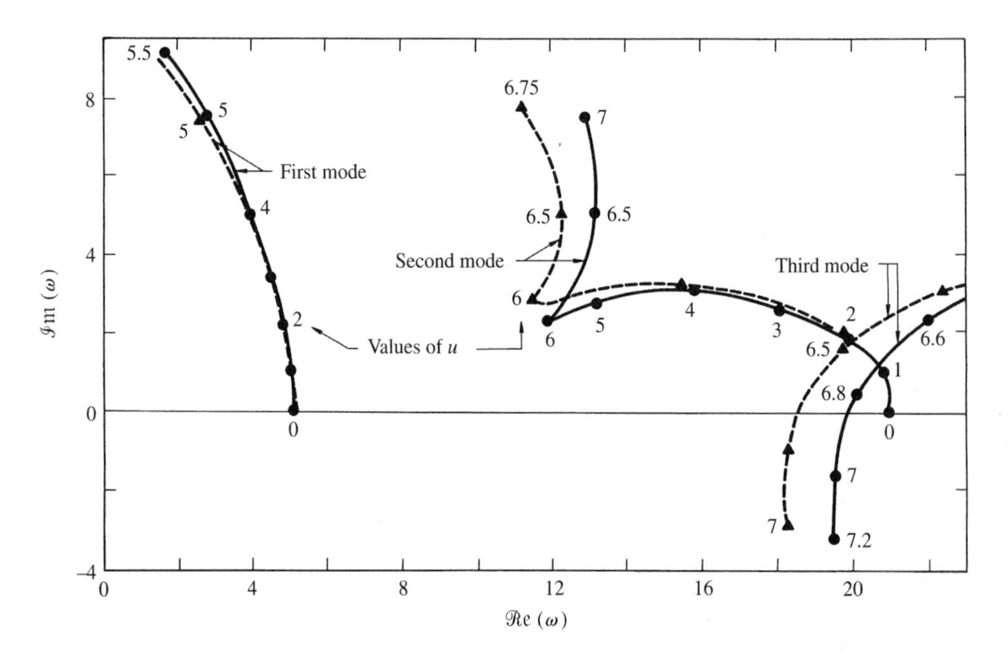

Figure 4.20 Dimensionless complex eigenfrequencies of a cantilevered pipe ($\beta = 0.3$, $\gamma = 10$, $\mu = \sigma = 0$, $\Lambda = 100$, $\varepsilon = 8.25$) as functions of the dimensionless flow velocity u, according to the two forms of the Timoshenko theory. Key as in Figure 4.19 (Païdoussis *et al*. 1986).

Considering Figure 4.20 first, it is noticed that the eigenfrequencies as given by TRF theory are higher than those obtained by TPF theory; the critical flow velocities for flutter obtained by refined fluid mechanics (TRF theory) are also higher. These observations are once again consistent with the concept of a smaller effective fluid mass per unit length, M', for the refined fluid mechanics, as compared to simple fluid mechanics. At the same values of flow velocity and mode number, the absolute value of the eigenfrequency obtained by the refined theory, $|\omega_{\text{ref}}|$, is always larger than that obtained by the simple theory, $|\omega_{\text{simp}}|$. Moreover, it is clear that M' becomes increasingly smaller than M for larger mode numbers (larger discrepancies in Figure 4.20); this is consistent with the fact that $M = \rho A_f$ applies only if the wavelength of deformation is long, as compared to the internal diameter of the pipe (Section 3.5.8) — which is not the case here for the second and third modes. In this connection it is recalled (Section 3.5.1) that the modal shapes for $u > 0$ contain components of higher zero-flow beam eigenfunctions, which reinforces the foregoing argument.

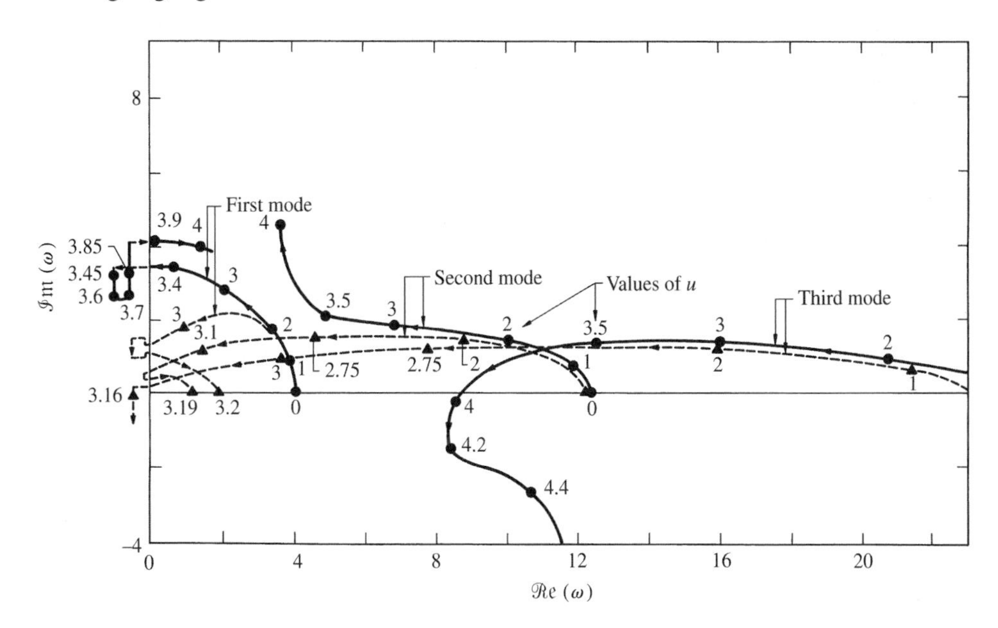

Figure 4.21 Dimensionless complex eigenfrequencies of a very short cantilevered pipe ($\beta = 0.3$, $\gamma = 10$, $\mu = \sigma = 0$, $\Lambda = 10$, $\varepsilon = 2.61$) as functions of the dimensionless flow velocity u, according to the two forms of the Timoshenko theory. Key as in Figure 4.19 (Païdoussis *et al.* 1986).

However, the extension of this argument to the question of stability of cantilevered pipes should be approached with caution, as loss of stability is not controlled by a single fluid-dynamic force term (as for clamped–clamped pipes), but by two — namely $Q_{kn}^{(2)}$ and $Q_{kn}^{(3)}$ of equations (4.49) and (4.67); it is a balance between these two forces which precipitates instability (Section 3.2.2). Indeed, as will be seen later, there are cases where u_{cf} according to TRF theory is lower than that obtained by TPF theory (plug-flow model), in contrast to the results of Figure 4.20.

A good deal of the foregoing discussion also applies to Figure 4.21. However, in a sense, this represents a very special case, since according to Timoshenko plug-flow (TPF) theory the system loses stability by divergence, at $u_{cd} \simeq 3.16$ — *vide* also Païdoussis & Laithier (1976). On the other hand, according to Timoshenko refined-flow (TRF) theory the system is shown to lose stability by flutter† at a higher value of u ($u_{cf} = 3.95$, $\omega_{cf} = 8.85$ for $\gamma = 10$; $u_{cf} = 3.63$, $\omega_{cf} = 8.40$ for $\gamma = 0$).

It is recalled that according to Euler–Bernoulli beam theory (and a simple plug-flow model) a cantilevered pipe conveying fluid can only lose stability by flutter. It is only in an earlier version of this work, in which the plug-flow model was used (Païdoussis & Laithier 1976), and here according to TPF theory, that loss of stability by divergence is predicted. On the other hand, once a more appropriate model for the fluid mechanics is used, flutter is predicted once again. Now, it cannot be said that the present TRF theory *never* predicts divergence for short cantilevered pipes, but simply that in some of the cases where TPF theory predicted divergence the present theory predicts flutter. In this connection, it is recalled that when the cantilevered pipe system is subjected to a second conservative force — other than the flexural restoring force — it sometimes loses stability by divergence. Examples are (i) the pipe–plate system of Section 3.6.6, subjected to warping as well as torsion, and (ii) the articulated pipe system of Section 3.8, subjected to gravity. Hence, there may be areas in the parameter space of the present system, also, where stability may be lost by divergence — according to TRF theory as well.

4.4.8 Comparison with experiment

The theory is compared with experimental results for cantilevered pipes, obtained by Laithier (1979). The pipes were made of silicone rubber, 15.60 mm in outside diameter and 6.35 mm in inside diameter. The fluid conveyed was water.

The pipes were specially moulded, with the upper end cast onto a special adaptor (Appendix D.2). The adaptor could be screwed directly to the piping supplying steady water flow. Special care was taken in designing the adaptor to ensure that (a) the upper support approaches the clamped condition as closely as possible, and (b) the entrance of the fluid to the supported part of the pipe is effected without disturbance (which in short pipes could have an important effect on their dynamical behaviour). The measured Young's modulus for these pipes was $E = 1.49 \times 10^6\ \mathrm{N/m^2}$, Poisson ratio $\nu = 0.45$, and the hysteretic damping coefficient $\mu = 0.02$. Utilization of equations (4.37) and (4.32) in this case gives $\Lambda = 0.538\varepsilon^2$. In the experiments, Λ was varied by progressively reducing the length of the pipe (by carefully cutting pieces off the free end), thus reducing ε; L was varied between 140 and 51 mm in one case, and 73 and 27 mm in another.

The flow velocity was measured by standard means. Oscillation was sensed by a fibre-optic sensor, measuring the lateral displacement close to the supported end of the pipe; the frequency of oscillation was measured from oscillation time-traces, recorded on a storage oscilloscope.

The critical flow velocities for flutter, u_{cf}, according to the three theories are compared with the experimental data in Figure 4.22(a) and the corresponding critical frequencies, ω_{cf}, in Figure 4.22(b); it is important to mention that the experimental values of ω_{cf} were measured at just the onset of instability and are *not* the limit-cycle values (which in this case are quite different), so that they should correspond better to those predicted by linear

†Surprisingly, this is the behaviour predicted by the Euler–Bernoulli theory, but at a very different critical flow velocity, $u_{cf} = 8.7$, and in the second mode.

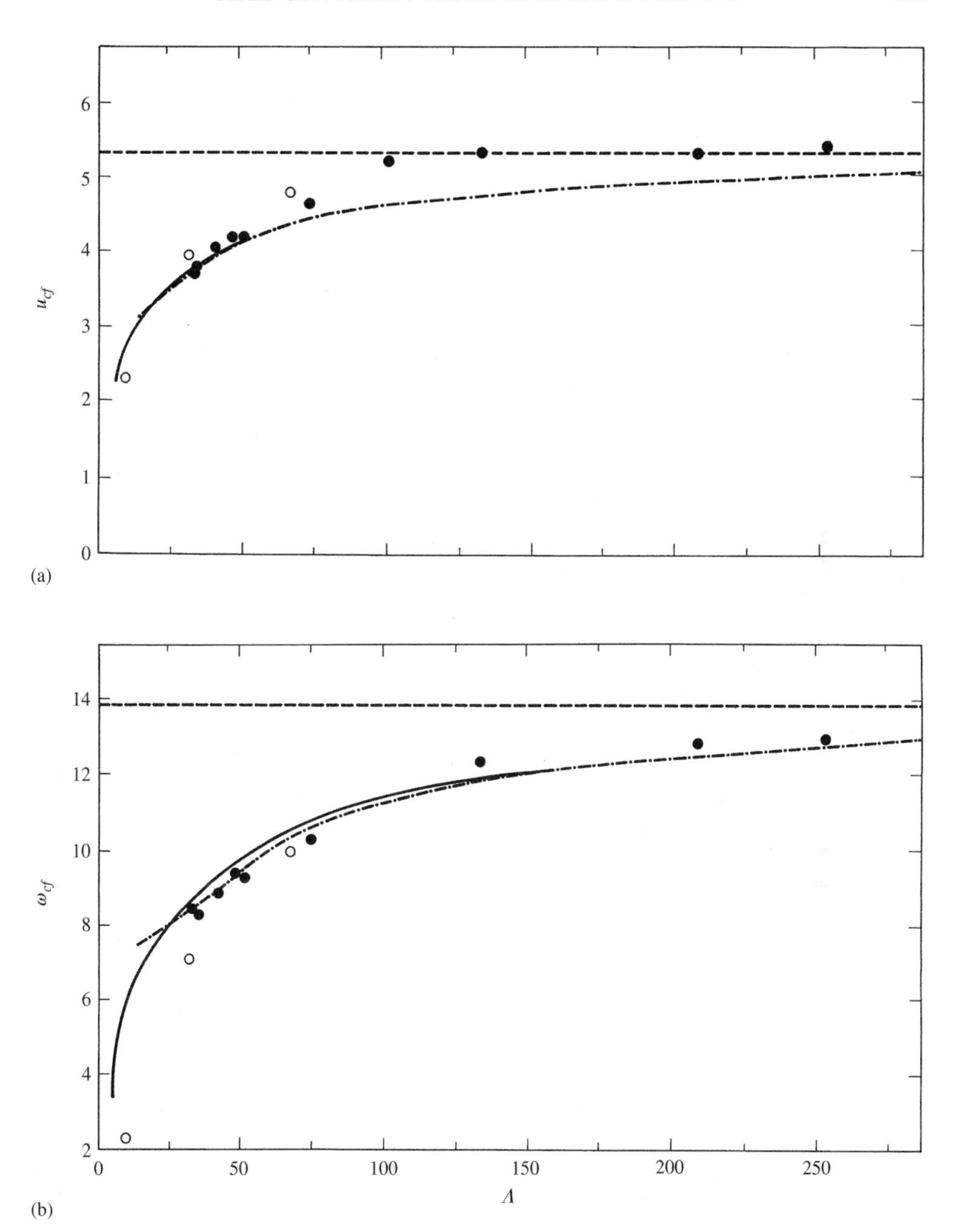

Figure 4.22 Comparison of dimensionless theoretical and experimental (a) critical flow velocities, u_{cf} and (b) the corresponding critical frequencies, ω_{cf}, for flutter of cantilevered pipes made of silicone rubber ($\beta = 0.155$, $\mu = 0.02$, $\sigma \simeq 0$). For $\Lambda > 50$ in (a) and $\Lambda > 125$ in (b), approximately, the results of the TRF theory coincide with those of TPF theory, in the scale of this figure. ——, TRF theory, $- \cdot - \cdot -$, TPF theory $- - -$, EBPF theory; •, ∘ experimental data.

theory. It is seen that both u_{cf} and ω_{cf} obtained by the Timoshenko theories agree better with the experimental data for $\Lambda < 75$ than the results obtained by the Euler–Bernoulli plug-flow theory; but surprisingly not for u_{cf} when $\Lambda > 100$. This last paradox may be explained in terms of nonlinear theory (Chapter 5).† Comparing the results obtained by the TPF and TRF theories, it is seen that they are very close. Nevertheless, for very short pipes, TRF theory displays superior agreement with the experimental data.‡

4.4.9 Concluding remarks on short pipes and refined-flow models

In general, for *short pipes clamped at both ends* the use of Timoshenko rather than Euler–Bernoulli beam theory results in lower critical flow velocities for divergence, u_{cd} — substantially lower for $\Lambda < 1000$ (Figure 4.19) — as a consequence of the pipe being effectively less stiff since it deforms not only by bending but also by transverse shear. The use of refined versus plug-flow fluid-dynamic modelling, on the other hand, has a less pronounced effect on the dynamics of the system: the refined model gives slightly higher values of the eigenfrequencies, as well as for the critical flow velocities for divergence. This is consistent with the concept of smaller-than-ideal virtual mass of the enclosed fluid, according to the refined three-dimensional fluid-mechanics model developed in this theory, as discussed in the foregoing. However, the differences in dynamical behaviour, both qualitative and quantitative, in terms of the refined and simple (plug-flow) Timoshenko theories are small; hence, from the practical point of view, down to $\Lambda = 10^2$, the simple (plug-flow) Timoshenko theory is good enough for predicting the dynamical behaviour of short clamped–clamped pipes conveying fluid.

In the case of short *cantilevered* pipes conveying fluid, the Euler–Bernoulli plug-flow model is adequate provided $\Lambda > 1000$ approximately. Once again, differences between refined and plug-flow Timoshenko theory are small, unless $\Lambda < 25$ approximately — an even lower Λ than for clamped–clamped pipes.

Finally, by comparison with experiments with cantilevered elastomer pipes, it was shown that the refined (TRF) theory is necessary for describing adequately the dynamical behaviour of short pipes ($L/D < 5$ approximately), although Timoshenko beam theory together with a plug-flow model (TPF theory) is quite satisfactory for relatively longer pipes; for 'long' pipes ($L/D > 15$), Euler–Bernoulli beam theory and the plug-flow model are perfectly adequate.

There is no question, however, that if one is interested in the dynamics of the system in its higher modes, e.g. for forced vibration analysis rather than stability (usually lost in one of the lower modes), then the differences between the three theories become larger, as may be appreciated from Figures 4.18, 4.20 and 4.21. Thus, although the first-mode behaviour is adequately predicted by EBPF theory down to $\Lambda = 1000$, third- and fourth-mode behaviour, and more so for higher modes, requires the use of Timoshenko theory and refined fluid mechanics (TRF theory) even at much larger values of Λ.

†The Hopf bifurcation for low Λ (hence low L/a) may be subcritical, while for higher Λ it is supercritical. Hence, for low Λ, the measured thresholds tend to be lower than would otherwise be the case. In this light, both the degree of excellence of the agreement with TRF theory for $\Lambda < 75$ and the better agreement with EBPF theory for $\Lambda > 100$ may be wholly fortuitous.

‡With the Timoshenko plug-flow theory, the shortest cantilevered pipe for which calculations have been conducted corresponds to $\Lambda = 13.07$. In the case of $\Lambda \leq 10\,(\beta = 0.155, \mu = 0.02, \gamma \leq 0.01)$, TPF theory, or at least the computer program utilized, fails to give a convergent solution.

4.4.10 Long pipes and refined flow theory

Despite what is said in the previous section regarding the superfluity of using Timoshenko or refined-flow theories except for really short pipes, there is no reason why they should *not* be used for longer pipes as well. This is particularly true in the case of general computational codes applicable to long and short pipes alike. An example is the work of Sällström & Åkesson (1990) and Sällström (1990, 1993), discussed in Section 4.7, which is based on Timoshenko beam theory.

Another example is a study by Langthjem (1995) on the dynamics of not necessarily short cantilevered pipes, partially or totally immersed in stagnant fluid, analysed by Timoshenko refined-flow (TRF) theory. Both the internal and external fluid dynamics are analysed by potential flow theory. Furthermore, it is argued that if the internal and external fluids are the same, e.g. liquid flow discharging into stagnant liquid, a turbulent jet develops and the flow is subjected to a velocity gradient at the free end; hence, yet another type of 'outflow model' is developed. It is found that, as a result of flow velocity reduction for $x > L$, the critical flutter speed (u_{cf}) may be diminished by 5–10%. Similar conclusions to those summarized in Section 4.4.9 are reached regarding the applicability of simpler theory down to very short pipes, and those in Section 4.2.4 and in Sugiyama *et al*. (1996a) regarding immersion effects. In particular, the destabilization when immersion is shallow, as compared to no immersion, is explained by noting that this enhances the 'dragging' form of the motion and hence optimizes energy transfer (Section 3.2.2).

Experiments with long elastomer pipes ($\varepsilon = 38$–60) conveying water support the theoretical findings, and agreement with theory in u_{cf} is within 10–15% — but not sufficiently close to validate the outflow model. It is of interest that, in one case, flutter was found to switch between planar and rotary motions in an unpredictable manner, suggesting that the oscillation may be chaotic. This should be compared to the physically similar case of a pipe with an additional end-mass (rather than immersion-related added mass) analysed by Copeland & Moon (1992) — see Section 5.8.3(b).

4.4.11 Pipes conveying compressible fluid

The dynamics of pipes conveying compressible fluid has been considered by Johnson *et al*. (1987), developing the theory initially formulated by Niordson (1953). Timoshenko beam theory is used for the pipe and a compressible potential flow for the fluid — in which $\nabla^2\phi = c^{-2}[(\partial^2\phi/\partial t^2) + 2U(\partial^2\phi/\partial x\,\partial t) + U^2(\partial^2\phi/\partial x^2)]$ is used instead of (4.50), c being the sonic speed. Results are given for the critical velocity for divergence, u_{cd}, in the first mode of a pinned–pinned pipe, obtained by Euler's method of equilibrium (Section 3.4.1).

The results are presented in a different and less physical manner than in Sections 4.4.1–4.4.9: the parameters ε and Λ, which are physically linked by equation (4.37), are varied *independently*; this, despite the fact that the only way of varying ε while keeping Λ constant is by changing the material constants and wall-thickness — which in practice cannot be varied widely. On the other hand, this allows the convenient separation of fluid-mechanical effects from structural ones (i.e. whether Timoshenko or Euler–Bernoulli theory is used). For example, some results are presented of u_{cd} versus ε,[†] while keeping Λ constant and infinite — what might be termed a Euler–Bernoulli refined-flow theory, which of course is physically impossible since the refined-flow effects come into play for small ε, when Euler–Bernoulli theory is not applicable. These results for incompressible flow (Mach number = 0) show that, as ε is decreased, u_{cd} is *raised*, because of the reduced effective virtual mass — contrary to the results in Figure 4.19, where both Λ and ε are varied together.

The above discussion is essential in understanding the results presented by Johnson *et al.*, the most notable of which are the following: decreasing the slenderness ε while keeping Λ constant (i) raises u_{cd} for subsonic flows, and (ii) lowers it for supersonic flows; also, reducing the sonic speed always diminishes u_{cd}.

Whereas the results obtained for low Mach numbers are probably sound, this is questionable in the case of near-sonic and supersonic, indeed hypersonic, flows because, as admitted by the authors, there are fundamental weaknesses in the model used, which supposes the fluid flow to be wholly isentropic. In the case of compressible flow, there are 'secondary effects' of fluid friction which generally cannot be ignored, e.g. causing choking; also, for the oscillating pipe, shock waves are generally inevitable. These real effects, which are difficult to model in a simple way, are not accounted for and their influence on the dynamics is unknown.

4.5 PIPES WITH HARMONICALLY PERTURBED FLOW

In all of the foregoing, except in the derivation of the equations of motion in Section 3.3, the mean flow has been taken to be steady ($\dot{u} = 0$). Here, the case of a harmonically perturbed flow is considered; i.e. it is supposed that a time-dependent harmonic component is superposed on the steady flow, such that

$$u = u_0(1 + \mu \cos \omega\tau), \tag{4.69}$$

where μ is generally small. This form of u may induce another type of instability, namely oscillations due to *parametric resonances*. These are akin to the parametric resonances experienced by, say, a pinned–pinned column subjected to a compressive end-load, $F = F_0(1 + \mu \cos \omega\tau)$. Especially in the case of $\omega/\omega_1 = 2$, where ω_1 is the first-mode natural frequency of the column, it is easy to appreciate physically that F pushes down most when the column end moves in the same direction 'naturally', at the two extremes of the oscillation cycle; hence, work is done on the column, resulting in amplification of the motion, i.e. in a parametric resonance (Bolotin 1964; Evan-Iwanowski 1976; Schmidt & Tondl 1986). Clearly, $\omega/\omega_1 = 1$ also leads to parametric resonance, although, as will be seen, other frequency ratios can also give rise to resonances. What renders the pipe conveying fluid particularly interesting and worth studying are the differences in dynamical behaviour *vis-à-vis* the column problem, because of the presence of gyroscopic terms and the fact that, in the case of a cantilevered pipe, the system is nonconservative.

The first to consider the problem, in terms of pressure pulsations arising from a pump, for example, was Roth (1964) and, in terms of a pulsating flow velocity, Chen (1971b) and Chen & Rosenberg (1971). However, as discussed in Section 3.3.2 in conjunction with equation (3.38), one of the terms in Chen's equations of motion is in error.

[†]In fact the results are presented in terms of the 'aspect ratio' a/L, the inverse of the slenderness ε.

Theoretical studies with the correct equations of motion were conducted by Ginsberg (1973) for pinned–pinned pipes and by Païdoussis & Issid (1974) and Païdoussis & Sundararajan (1975) for cantilevered and clamped–clamped pipes. Experiments were conducted by Païdoussis & Issid (1976). The work to be presented here is based mainly on these studies.

Further work was done by Ariaratnam & Namachchivaya (1986a) on pipes with supported ends, Bohn & Herrmann (1974a) on the articulated system, and Noah & Hopkins (1980) on an elastically supported cantilever [Figure 3.61(c)], to be discussed in Sections 4.5.4 and 4.5.5. Also, a great deal of work on the nonlinear dynamics of the system has been done in recent years, to be discussed in Section 5.9.

In what follows, we shall distinguish between *simple parametric* and *combination resonances*, which will be defined in due course.

4.5.1 Simple parametric resonances

For conservative systems, *simple* parametric resonances occur over specific ranges of ω in the vicinity of $2\omega_n/k$, $k = 1, 2, 3, \ldots$, where ω_n is one of the real eigenfrequencies of the system. ('Simple' is used in this book to differentiate parametric resonances associated with *one* eigenfrequency from combination resonances, defined in Section 4.5.2, involving two; however, they are often just referred to as 'parametric resonances' for simplicity.) As $\mu \to 0$, the resonances occur at $\omega/\omega_n = 2/k$, and for larger μ over a range of ω in the neighbourhood of these values. For nonconservative systems, there is a minimum value of μ below which parametric resonances are impossible, and for higher μ they occur in the vicinity of $\omega/\mathcal{Re}(\omega_n) = 2/k$, $\mathcal{Re}(\omega_n)$ being the frequency of oscillation associated with the nth complex eigenfrequency, ω_n.

One may distinguish *primary resonances*, corresponding to odd values of k, of which the *principal* one ($k = 1$ so that $\omega/\omega_n = 2$), a subharmonic resonance, is of particular importance, and *secondary resonances*, corresponding to even values of k. For the pipe problem, as the ω_n vary with u, it is expected that the ranges of ω necessary to induce parametric resonances will vary accordingly.

The easiest way of determining the regions of existence of parametric resonance is via a Fourier series solution approach, usually known as *Bolotin's method* (Bolotin 1964). To this end, the equation of motion, equation (3.70), is discretized by Galerkin's method, $\eta(\xi, \tau) = \sum_{r=1}^{N} \phi_r(\xi)\, q_r(\tau)$, where the ϕ_r are the beam eigenfunctions with the appropriate boundary conditions, leading to an equation similar to (3.86) but with the $\dot{u}$ terms retained; with (4.69) substituted therein, one obtains

$$\begin{aligned}\ddot{\mathbf{q}} &+ \{\mathbf{F} + 2\beta^{1/2}u_0(1 + \mu \cos \omega\tau)\mathbf{B}\}\dot{\mathbf{q}} \\ &+ \{\mathbf{\Lambda} + [u_0^2(1 + \mu \cos \omega\tau)^2 - \gamma - \beta^{1/2}u_0\mu\omega \sin \omega\tau - \Gamma \\ &+ \Pi(1 - 2\nu\delta)]\mathbf{C} + [\gamma + \beta^{1/2}u_0\mu\omega \sin \omega\tau]\mathbf{D} + \gamma\mathbf{B}\}\mathbf{q} = \mathbf{0},\end{aligned} \tag{4.70}$$

in which $\mathbf{q} = \{q_1, q_2, \ldots, q_N\}^{\mathrm{T}}$, $\mathbf{\Lambda}$ is the diagonal matrix with elements λ_r^4, λ_r being the rth dimensionless beam eigenvalue associated with ϕ_r, $\mathbf{F}$ is a diagonal matrix with elements $\alpha\lambda_r^4 + \sigma$, and $\mathbf{B}$, $\mathbf{C}$ and $\mathbf{D}$ are square matrices with elements b_{sr}, c_{sr} and d_{sr} defined in equation (3.87).

It is now presumed that there are regions in the $\{\omega, \mu\}$-plane where, for any given ω_n, there exist amplified oscillations or parametric resonances, and hence on the boundaries of these regions the oscillation is purely periodic. Since a periodic solution may be represented by a Fourier series, to obtain the *primary resonances*, $\mathbf{q}(\tau)$ may be expressed as

$$\mathbf{q} = \sum_{k=1,3,5,\ldots} \{\mathbf{a}_k \sin(\tfrac{1}{2}k\omega\tau) + \mathbf{b}_k \cos(\tfrac{1}{2}k\omega\tau)\}. \tag{4.71}$$

Substitution of equation (4.71) into (4.70) yields an infinite set of algebraic equations which, because of the presence of $\sin \omega\tau$, $\cos \omega\tau$ and $\cos^2 \omega\tau$ terms in (4.70) already, involves terms in $\sin(\frac{1}{2}m\omega\tau)$ and $\cos(\frac{1}{2}m\omega\tau)$, $m = k-4, k-2, k, k+2, k+4$ (Païdoussis & Issid 1974). Upon expanding this equation for $k = 1, 3, 5, \ldots$, and collecting terms in $\cos \frac{1}{2}\omega\tau$, $\sin \frac{1}{2}\omega\tau$, $\cos \frac{3}{2}\omega\tau$, etc., the coefficients of which must vanish independently, one obtains a matrix equation of the form

$$[\mathbf{G}]\left\{\begin{matrix}\mathbf{a}_j\\ \mathbf{b}_j\end{matrix}\right\} = \{\mathbf{0}\}, \tag{4.72}$$

or more explicitly

$$\begin{bmatrix} \cdots & \cdots & \cdots & \cdots & \cdots & \cdots \\ \cdots & G_{33} & G_{31} & G_{32} & G_{34} & \cdots \\ \cdots & G_{13} & G_{11} & G_{12} & G_{14} & \cdots \\ \cdots & G_{23} & G_{21} & G_{22} & G_{24} & \cdots \\ \cdots & G_{43} & G_{41} & G_{42} & G_{44} & \cdots \\ \cdots & \cdots & \cdots & \cdots & \cdots & \cdots \end{bmatrix} \left\{\begin{matrix} \vdots \\ a_3 \\ - - - \\ a_1 \\ b_1 \\ - - - \\ b_3 \\ \vdots \end{matrix}\right\} = \{0\}, \tag{4.73}$$

generally of infinite order. The $\mathbf{G}_{jk}$ are coefficients of $\mathbf{a}_k$ or $\mathbf{b}_k$ in the equations for $\sin(\frac{1}{2}j\omega\tau)$ or $\cos(\frac{1}{2}j\omega\tau)$. The odd j are associated with $\sin(\frac{1}{2}j\omega\tau)$ and the even j with $\cos[\frac{1}{2}(j-1)\omega\tau]$; while the odd k are associated with $\mathbf{a}_k$ and the even k with $\mathbf{b}_{k-1}$.

The equation for the boundary of the instability regions is obtained by setting the determinant of the matrix of the $\mathbf{G}_{jk}$ equal to zero. Of course the determinant is of infinite order, but it belongs to the class of normal determinants and is therefore absolutely convergent (Bolotin 1964). Hence, the boundaries of instability may be obtained approximately by equating to zero the determinant of the boxed matrix in (4.73); this is called the $k = 1$ approximation, which necessarily yields only the principal region of instability. A better approximation, as well as higher regions, would be obtained if the determinant involving all the terms shown explicitly in equation (4.73) is used; this is called the $k = 3$ approximation; and so on. Of course, the Galerkin series leading to equation (4.70) must be truncated at an adequately high N, which defines the order of the $\mathbf{G}_{jk}$.

Now, the *secondary resonances* may be obtained by expressing

$$\mathbf{q} = \sum_{k=0,2,4,\ldots} \{\mathbf{a}_k \sin(\tfrac{1}{2}k\omega\tau) + \mathbf{b}_k \cos(\tfrac{1}{2}k\omega\tau)\}, \tag{4.74}$$

which substituted into equation (4.70) leads once again to a matrix equation equivalent to equation (4.73) but with a vector $\{\ldots a_4, a_2, b_0, b_2, b_4, \ldots\}^{\mathrm{T}}$, and thence to a vanishing determinant which yields the boundaries of secondary resonances.

It has been found that, typically, the resonance boundaries associated with the first mode of pinned–pinned and clamped–clamped pipes can be determined with adequate precision by using the $k = 1$ approximation for the principal primary region and the $k = 2$ approximation for the main secondary one, and truncating the Galerkin series at $N = 2$. For cantilevered pipes, on the other hand, it was found necessary to use typically $k = 3$ and $k = 2$, respectively, and $N = 5$. Generally, the higher the flow velocity, or μ, or the mode number, the higher N and k have to be. Most of the calculations presented here have been performed with $k = 3$ or 2 and $N = 5$.

Typical results for a clamped–clamped pipe are shown in Figure 4.23, showing the effect of u_0 and damping (σ) on the principal resonance ($\omega \simeq 2\omega_1$) and the main secondary or 'fundamental' resonance ($\omega \simeq \omega_1$), where ω_1 is the first-mode eigenfrequency or the real part thereof for any given u_0. Resonance oscillations exist within the quasi-triangular regions, while the system is in its trivial equilibrium state outside. It is noted that, as the flow velocity is increased, the regions of parametric resonance are displaced downwards, which reflects the decrease in the first-mode eigenfrequency ω_1 with increasing u_0. Had ω/ω_1 been plotted in the figure instead of ω/ω_{01} — where ω_{01} is the zero-flow-velocity value of ω_1 — then all the curves would have been centred about $\omega/\omega_1 = 2$ and 1. For $\sigma = 0$, the parametric resonance regions exist even for $\mu = 0$, but at that point have zero width; for $\sigma > 0$, there is a minimum μ for each case, below which no resonance is possible.

It is also noticed that the resonance regions become broader with increasing flow, and that the effect of damping is correspondingly attenuated. Thus, at $u_0 = 3$, the primary resonance region is very little affected by fairly large dissipation ($\sigma = 0.5$), and the system is subject to parametric resonance even when $\mu < 0.1$.

Figure 4.24 shows the effect of β on parametric resonances. It is seen that with increasing β the resonance regions become broader; it is perhaps worth mentioning that with the equation of motion used by Chen the width of the unstable regions appears to be independent of β. The unstable regions are also displaced downwards with increasing β, which reflects the lowering of the natural frequencies as β is increased for this particular flow velocity; this is because ω_1 for $\beta^{1/2} = 0.8$ is lower than for $\beta^{1/2} = 0.2$[†] at $u_0 = 2$ (to the scale of this figure the difference is more clearly seen in the principal resonance regions).

It is noticed that the lower boundary of the principal primary resonance associated with $\beta^{1/2} = 0.8$ in Figure 4.24 is not straight; this is a characteristic of cases where the resonance boundary concerned is close to another resonance region (which is not shown here for the sake of simplicity). This also applies to the primary region for $u_0 = 3$ in Figure 4.23.

As seen in the foregoing, the dynamics of pipes with supported ends subjected to pulsating flow is, after all, not too dissimilar to that of columns with a harmonically perturbed conservative end-load. Nevertheless, a significant difference is that gyroscopic effects, operative in the case of pipes, and hence β, affect the location and particularly the *extent* of the resonance regions in the $\{\mu, \omega\}$-plane.

[†]Some of Païdoussis & Issid's (1974) calculations were conducted with specific values of $\beta^{1/2}$ rather than β, for direct comparison with Chen's (1971b).

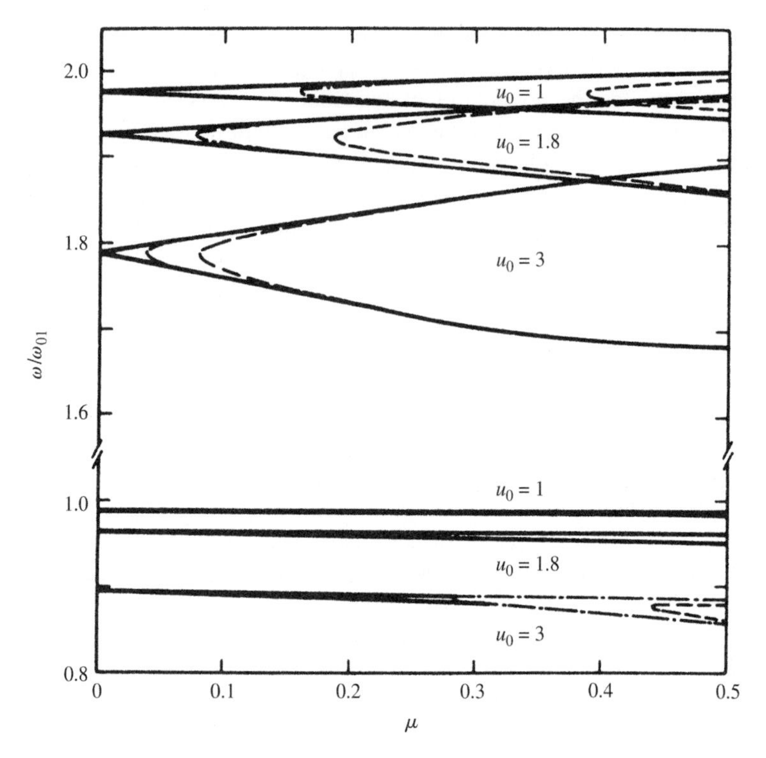

Figure 4.23 The effect of flow velocity, u_0, and viscous damping on the principal and fundamental parametric resonances associated with the first mode of a clamped–clamped pipe, centred respectively at $\omega \simeq 2\omega_1$ and $\omega \simeq \omega_1$, for $\Gamma = \Pi = \alpha = 0$, $\beta^{1/2} = 0.2$, $\gamma = 10$ and three values of u_0; ——, $\sigma = 0$; — - —, $\sigma = 0.2$; – – –, $\sigma = 0.5$; ω_{01} is the first-mode eigenfrequency for $u_0 = \sigma = 0$ (Païdoussis & Issid 1974).

Considering cantilevered pipes next, it is recalled that in this case free motions are damped by steady flow below the critical value for flutter; consequently, parametric resonances are not possible for all flow velocities. Moreover, the resonances are selectively associated with only some of the modes of the system, for reasons to become clear in what follows; thus, in all the calculations performed, at least for relatively low flow velocities, no resonances associated with the first mode have ever been found. To fully appreciate the results, it is necessary to give Argand diagrams for the particular systems considered, with steady flow, in Figure 4.25. Attention is drawn to the fact that for u_0 not too small (when the parametric excitation itself would be weak), $\mathcal{I}\mathrm{m}(\omega)$ and hence the flow-induced damping is smallest in the second mode for u_0 not too far from $u_0 = 6.0$ when $\beta = 0.2$, and from $u_0 = 6.0$–8.7 when $\beta = 0.3$. It is for these ranges of u_0 that parametric resonances should be most easily induced.

Figure 4.26 shows the parametric resonance regions for a cantilevered pipe with $\beta = 0.2$, $\gamma = 10$, $\alpha = \sigma = 0$ for $u_0 = 4.5$, 5.5 and 6.0 in the range $\omega/\omega_{02} < 2.4$. For $u_0 \leq 4$, no parametric resonances can occur, at least for the range of μ considered. The large resonance regions in the middle of the figure are the principal primary resonance regions associated with the second mode ($\omega \simeq 2\omega_2$), while at the bottom is the main secondary region (fundamental resonance, $\omega \simeq \omega_2$) which occurs for $u_0 = 6$ only. The small regions

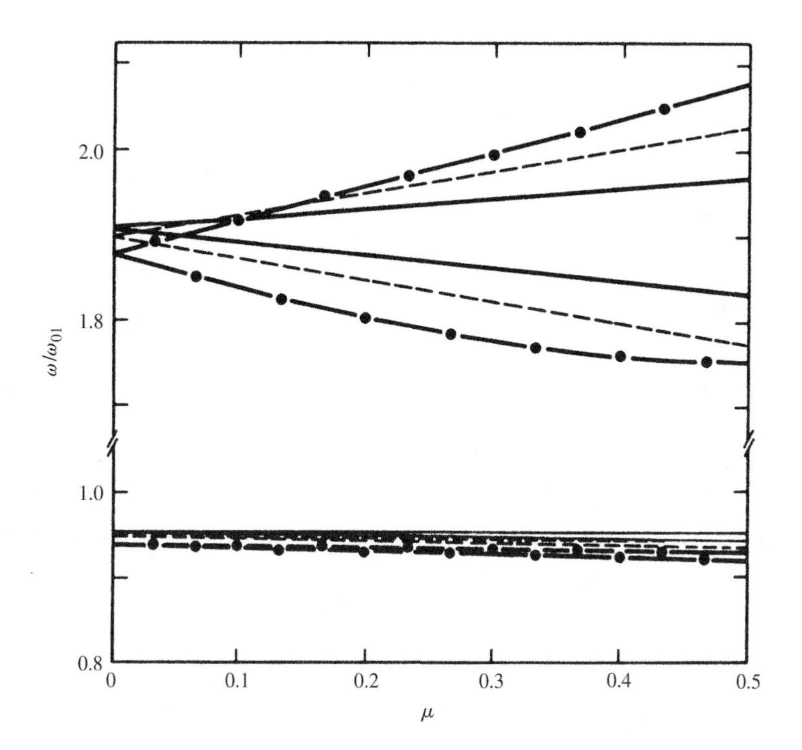

Figure 4.24 The effect of β on the principal and fundamental parametric resonances associated with the first mode of a clamped–clamped pipe ($\gamma = 10$, $\Gamma = \Pi = \alpha = \sigma = 0$, $u_0 = 2$): ——, $\beta^{1/2} = 0.2$; - - -, $\beta^{1/2} = 0.5$; •—•, $\beta^{1/2} = 0.8$ (Païdoussis & Issid 1974).

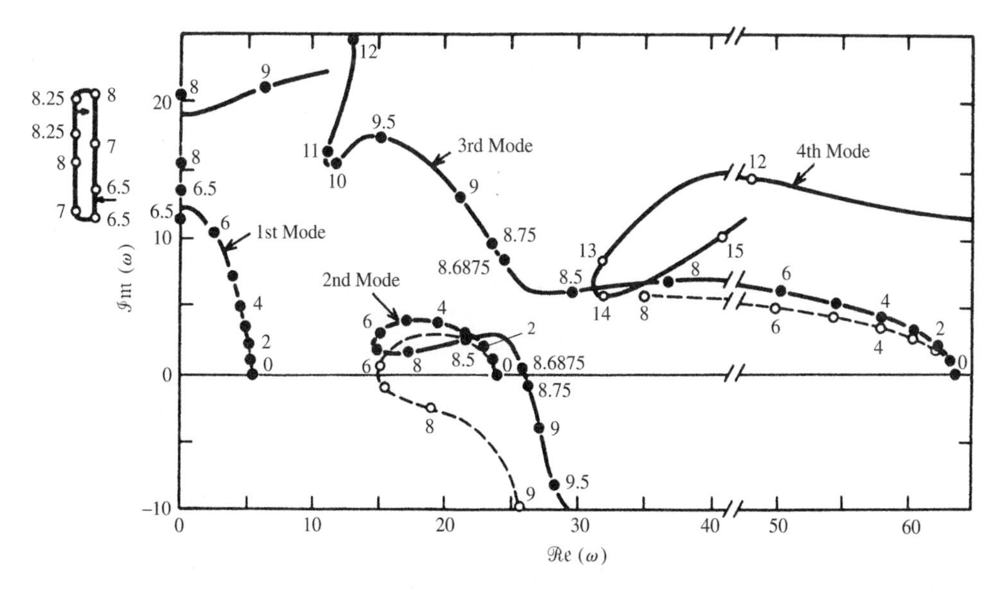

Figure 4.25 Argand diagrams for a cantilevered pipe with $\gamma = 10$, $\alpha = \sigma = 0$; - - -, $\beta = 0.2$; ——, $\beta = 0.3$ (Païdoussis & Issid 1974).

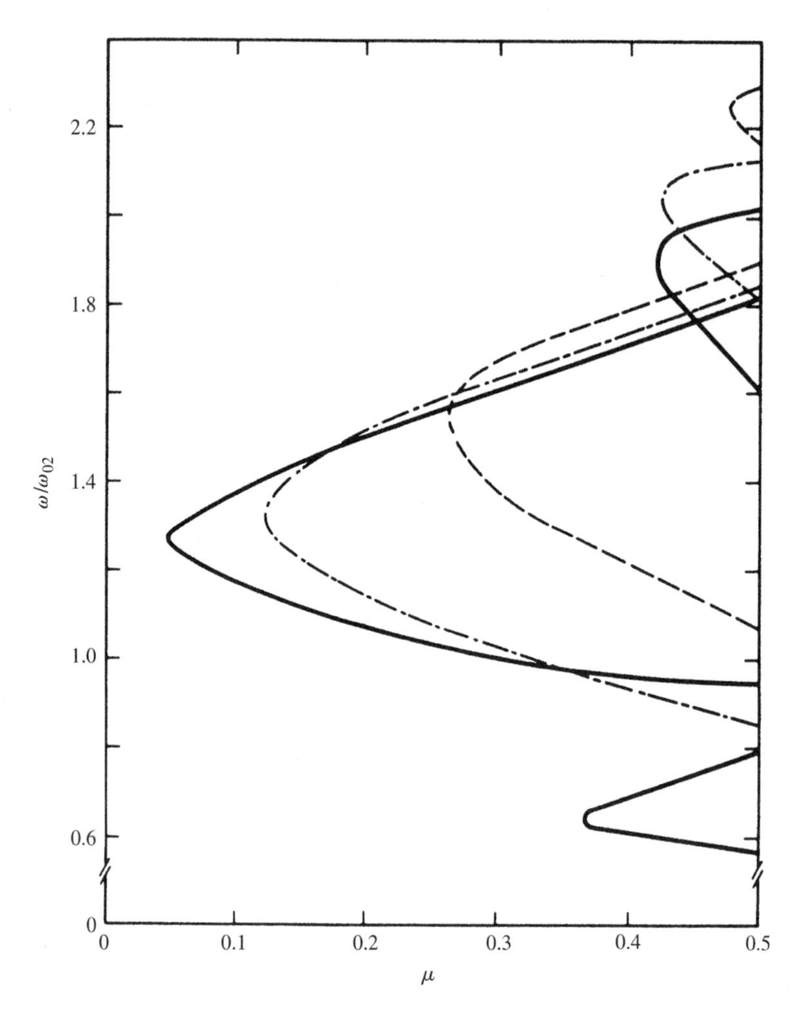

Figure 4.26 Parametric resonance boundaries for a cantilevered pipe (β=0.2, γ=10, $\alpha = \sigma = 0$); ---, $u_0 = 4.5$; — - — , $u_0 = 5.5$; ——, $u_0 = 6.0$ (Païdoussis & Issid 1974).

at the top are the main secondary regions associated with the third mode. It is noted that (i) a finite and quite substantial value of μ is generally necessary to induce parametric oscillations, (ii) this value of μ decreases with increasing flow velocity, and (iii) the resonance regions are more extensive at the higher flow velocities. These results are somewhat similar to those of Figure 4.23 for the damped clamped–clamped pipe with $\sigma = 0.5$. In contrast to the clamped–clamped pipe, however, damping in this case (arising by the action of the Coriolis forces) is intimately connected with the dynamics of the system; consequently, its effect on the parametric resonances is not uniform, nor easily predictable, as will be seen further below.

Figure 4.27(a,b) shows, respectively, the primary and secondary resonance regions, for the range of frequencies shown, of a system with $\beta = 0.3$, $\gamma = 10$, $\alpha = \sigma = 0$. The three uppermost resonance regions in Figure 4.27(a), for $u_0 = 6.0$, 7.5 and 8.0, are principal primary regions associated with the third mode, while the two large regions in the middle,

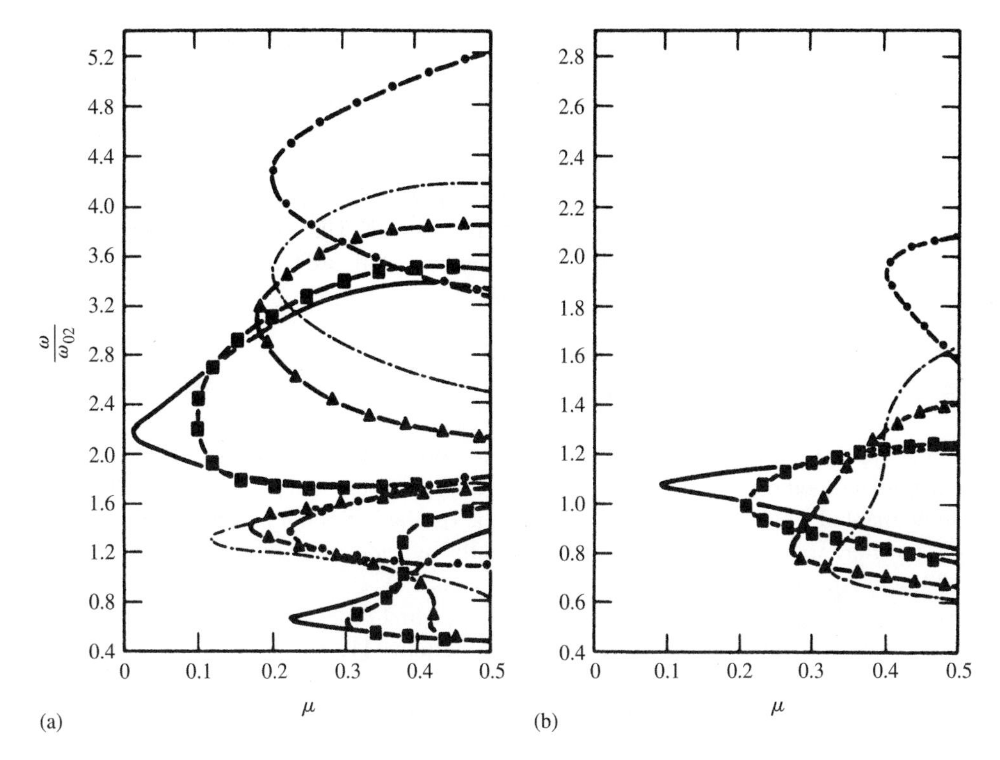

Figure 4.27 Parametric resonance boundaries for a cantilevered pipe ($\beta = 0.3$, $\gamma = 10$, $\alpha = \sigma = 0$). (a) Primary resonance regions; (b) secondary resonance regions. •, $u_0 = 6.0$; — - —, $u_0 = 7.5$; ▲, $u_0 = 8.0$; ■, $u_0 = 8.5$; ——, $u_0 = 8.6875$ (Païdoussis & Issid 1974).

for $u_0 = 8.5$ and 8.6875, are a mixture (i.e. a fusion) of principal primary regions associated with the second and third modes. By reference to Figure 4.25 it is noted that for $u_0 = 8.5$ and 8.6875 the real frequencies of oscillation of the second and third modes are relatively close.

Finally, the smaller regions at the bottom of Figure 4.27(a) may similarly be divided into the following two groups: (i) the regions for $u_0 = 6.0$, 7.5 and 8.0 are mixtures of principal primary regions associated with the second mode and of second primary regions associated with the third mode; (ii) the regions for $u_0 = 8.5$ and 8.6875 are mixtures of second primary regions associated with the second and third modes. This fusion of the regions of resonance is shown particularly well in the cases of $u_0 = 8.0$, 8.5 and 8.6875, where each of the regions is formed of two interlinked distinct zones, the upper of which is related to the second mode and the lower to the third mode.

In Figure 4.27(b) the upper region ($u_0 = 6.0$) is the fundamental secondary resonance region associated with the third mode, while the remaining regions are all mixtures of the fundamental resonance regions associated with the second and third modes. The upper areas of the latter are associated with the third mode and the lower areas with the second, except for $u_0 = 8.6875$ where no such distinction may be made; this is attributed to the proximity of $\mathfrak{Re}(\omega_2)$ and $\mathfrak{Re}(\omega_3)$ for $u_0 = 8.6875$ as shown in Figure 4.25.

One interesting aspect of the results of this section is that no parametric resonances occur at low values of u_0, where the damping effect of the mean flow is small. It is recalled, however, that both damping due to Coriolis forces and parametric excitation are proportional to u_0 for a constant μ; hence, for a given u_0 the magnitudes of the parametric excitation and of damping are predetermined, and as u_0 is decreased they both decrease proportionally.† Thus one cannot say *a priori* for what range of u_0 parametric resonances may occur, if at all, for a given range of μ.

Hence, the dynamics of cantilevered pipes with pulsating flow is quite different from that of columns with harmonically perturbed end-load. The combined effects of the nonconservative follower and Coriolis forces fundamentally affect both the existence and the extent of the parametric resonance regions.

4.5.2 Combination resonances

Another type of parametric oscillation is due to combination resonances, which occur in the neighbourhood of $\omega = (\omega_n \pm \omega_m)/k$, $k = 1, 2, \ldots$, where in this case $n \neq m$. They are not obtainable by Bolotin's method, because the oscillation is not periodic but *quasiperiodic*, but they may be obtained semi-analytically by Floquet analysis — see Meirovitch (1970) and Appendix F.1.2.

For the analysis, equations (4.70) are rewritten into first-order form, $\mathbf{u} = \{\mathbf{q}, \dot{\mathbf{q}}\}^{\mathrm{T}}$, and then integrated numerically with $2N$ different initial conditions: $\{1, 0, 0, \ldots\}^{\mathrm{T}}$, $\{0, 1, 0, \ldots\}^{\mathrm{T}}$, etc. The solutions thus obtained after one period, $\mathbf{u}^1(T), \mathbf{u}^2(T), \ldots, \mathbf{u}^{2N}(T)$ are used to construct the so-called fundamental or 'monodromy' matrix of the system,

$$[Y] \equiv [\mathbf{u}^1(T), \mathbf{u}^2(T), \ldots, \mathbf{u}^{2N}(T)]. \tag{4.75}$$

The characteristic or Floquet multipliers of the system are given by the eigenvalues of $[Y]$. If at least one of them has an absolute value greater than unity, the system is unstable (in the linear sense), giving rise to (i) a parametric resonance if this Floquet multiplier is real, and (ii) a combination resonance if it is complex. In this sense, combination resonances correspond to quasiperiodic motions — see Section 5.9.

A typical set of results for a clamped–clamped pipe are shown in Figure 4.28. A small amount of damping has been added ($\alpha = 10^{-3}$), so as to eliminate the profusion of very narrow resonance regions that would otherwise clutter up the figure. Regions of both simple and combination parametric resonance are seen to exist. The various resonances are well ordered and easily identifiable, as would for instance be the case for a column subjected to a harmonically perturbed end-load. Indeed, the analogy applies to the extent that combination resonances of the difference types do not materialize for these boundary conditions (Iwatsubo *et al.* 1974; Ariaratnam & Namachchivaya 1986a). In this figure, ω_{01}, the zero-flow first-mode undamped eigenfrequency, is used simply as a normalization factor throughout. As a result of this, however, the upper resonance regions in the figure appear to be more extensive than the lower ones; had ω/ω_{02} been used instead of ω/ω_{01} for the second mode, the same width as for the principal resonance of the first mode would have been obtained. The minimum value of μ below which resonance does not

---

†A different behaviour is obtained if the excitation force can be varied independently of damping. Calculations in which an external axial load $F_0 \cos \omega\tau$ is imposed and u is kept constant show that parametric resonances may then occur even for small u.

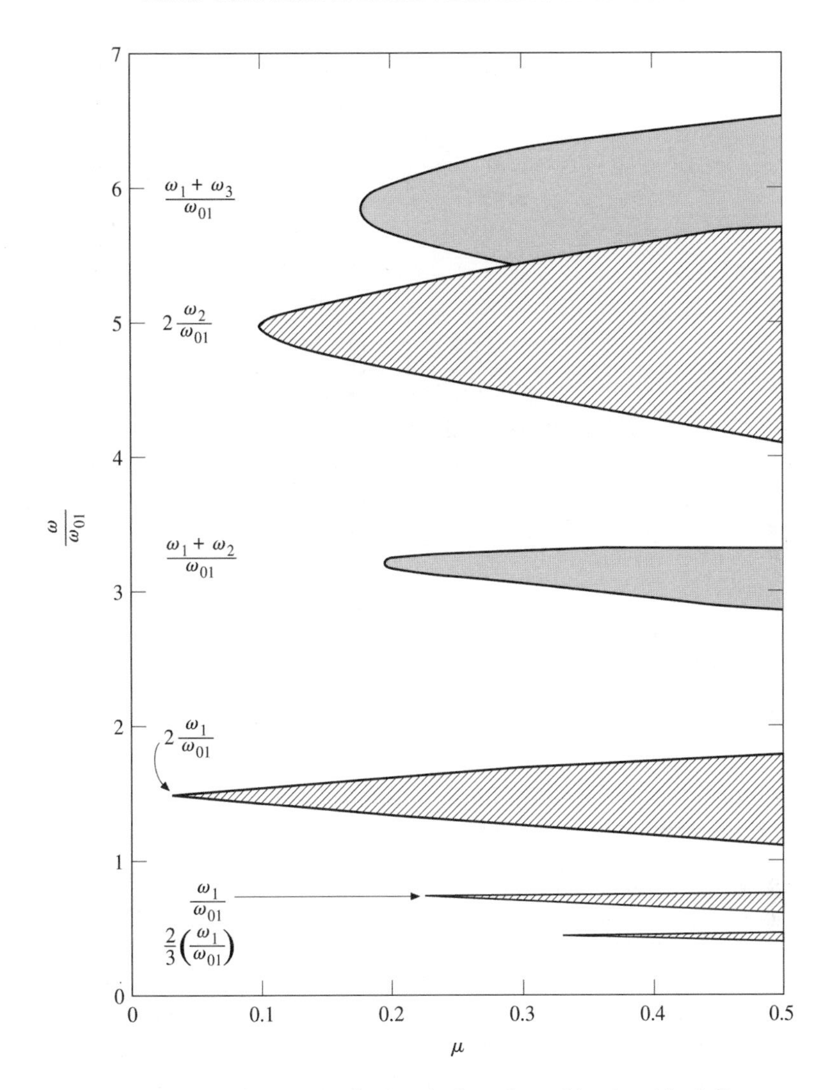

Figure 4.28 Regions of simple parametric (hatched) and combination (shaded) resonances for a pipe clamped at both ends ($\beta = 0.2$, $\gamma = \sigma = \Gamma = \Pi = 0$, $\alpha = 1 \times 10^{-3}$, $u_0 = 4$); $\omega_{01} = 22.3733$ (Païdoussis & Sundararajan 1975).

arise would not be affected, however; it is a function of the dissipation model used, and the minimum μ is higher for the higher modes.

It is noted that, as u_0 is reduced, all resonance regions become less extensive, and some are completely eliminated. In this respect, the regions of combination resonance suffer proportionately much more than those of simple parametric resonance.

Typical results for a cantilevered pipe are shown in Figure 4.29(a,b) for u_0 just below and just above the critical value for flutter when $\mu = 0$ ($u_0 = 6.34$); $N = 3$ and 4 represent the Galerkin–Floquet approximations with which the results have been computed. The results have been normalized by $\omega_{02} = 23.912$, and to fully understand them Table 4.5 is necessary.

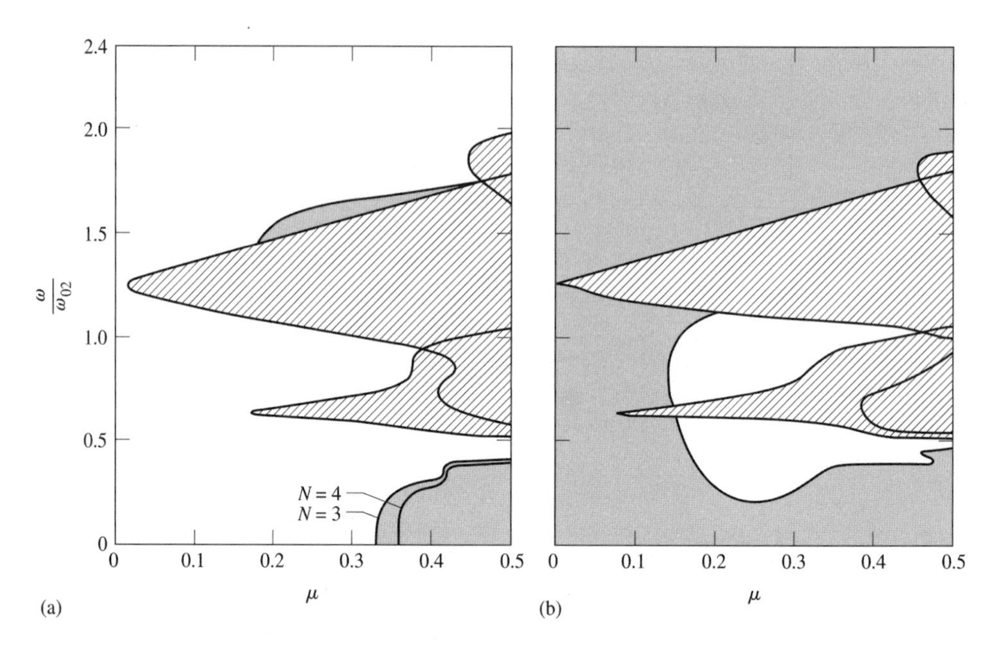

Figure 4.29 Regions of simple parametric (hatched) and combination (shaded) resonances for a cantilevered pipe conveying fluid ($\beta = 0.2$, $\gamma = 10$, $\alpha = \sigma = 0$; $\omega_{02} = 23.912$): (a) for $u_0 = 6.25$, with $N = 3$ and 4; (b) for $u_0 = 6.5$, with $N = 4$ (Païdoussis & Sundararajan 1975).

Table 4.5 Complex eigenfrequencies of the system of Figure 4.29.

u_0	ω_1	ω_2	ω_3
6.25	3.42 + 11.79i	15.02 + 0.17i	48.59 + 5.15i
6.50	3.36 + 12.82i	15.04 − 0.29i	47.15 + 5.29i

The large region of parametric resonance in the centre of the figures is the principal primary region ($k = 1$) associated with the second mode; significantly, near the 'nose' of the curve (small μ), the ratio of pulsation frequency to natural frequency is 2:1; see Table 4.5 ($\omega/\omega_{02} \simeq 1.25 \simeq 2\omega_2/\omega_{02}$). The lower bulge associated with this region corresponds to a higher-k primary region associated with the third mode. The lower simple resonance region is associated mainly with secondary resonance in the second mode ($k = 2$), while the uppermost region is also secondary, but associated with the third mode. The combination resonances at the bottom of Figure 4.29(a) involve the first and second (and perhaps other) modes of the system, while the upper region is associated with the second and third modes. In both cases the combination resonances appear to involve the differences, rather than the sums, of the natural frequencies; this is in agreement with some results obtained for columns subjected to periodic follower loads (Iwatsubo *et al*. 1974).

Calculations for $u_0 = 6.0$ show similar parametric resonances as in Figure 4.29(a), but smaller. Furthermore, the upper combination region disappears altogether. Calculations for $u_0 = 5.5$ show that only simple parametric resonances survive, and for $u_0 \leq 4$ all resonances vanish.

Following the trend just described in reverse, it is clear that one might expect the simple and combination parametric resonance regions to go on increasing in extent with increasing u_0 beyond u_{cf}. Nevertheless, the results in Figure 4.29(b) for $u > u_{cf}$ are both startling and interesting: the 'combination resonance region' corresponding to quasiperiodic motions has increased quite dramatically,† virtually covering all the previously stable area; nevertheless, it is of interest that a small region remains where the system is stable in pulsating flow, whereas in the absence of pulsation it would not be!

4.5.3 Experiments

Experiments were conducted with an apparatus and elastomer pipes similar to those in Païdoussis' (1970) steady-flow experiments, described in Section 3.5.6. However, the apparatus was modified to enable the addition of a harmonic perturbation component to the mean flow via a 'plunger pump' driven by a variable-stroke reciprocating mechanism, connected to a variable-speed drive, as shown in Figure 4.30. Thus both the amplitude and the frequency of the imposed harmonic perturbation could be varied; the frequency range was 1 to 16 Hz. Flexible bellows were inserted to isolate, as much as possible, the test pipe from vibration arising from the reciprocating mechanism and drive.

The flow velocity was measured just upstream of the elastic pipe by a hot-film anemometer. Traces of the periodically perturbed flow showed that the plunger pump gave almost truly sinusoidal perturbations to the flow, so that the flow velocity could be represented by $U = U_0(1 + \mu \cos \Omega t)$. Both μ and the mean flow velocity, U_0, were determined by the hot-film anemometer. Experiments were performed with clamped–clamped and cantilevered pipes. The lower clamped end in the former case was such as to permit axial sliding. The apparatus and the experiments are described in greater detail in Païdoussis & Issid (1976).

In general, the dynamical behaviour of clamped–clamped pipes is similar to that of columns subjected to periodic end-loading. The dominant resonances are associated with the first mode. The secondary parametric resonance was difficult to pin-point, particularly for small μ and for U far removed from U_{cd}; the main reason being that there was always a small-amplitude vibration of the pipe at the pulsation frequency, transmitted either mechanically or through the fluid, which proved impossible to eliminate completely. A stable region usually separated the secondary from the primary resonance, except for high μ and U_0 close to U_{cd}, where it was observed that the frequency of pipe oscillation changed directly from Ω to $\frac{1}{2}\Omega$. When the pulsation frequency was increased beyond the first-mode primary region, resonances associated with the second mode were observed. In some cases combination resonances or mixed resonance regions were encountered. In general, the observations are in qualitative agreement with theory.

A quantitative comparison between theory and experiment is made in Figure 4.31. It is evident that if the theoretical curves were shifted downwards, agreement with experiment would improve substantially; this would indicate that the theoretical frequencies may be incorrect and leads one to suspect that the lower sliding clamped support was not perfect; in fact, it was slightly loose to permit unimpeded axial movement. It is also noted that

†The notation of 'combination resonance' for $u > u_{cf}$ is inappropriate. 'Quasiperiodic' is much better to denote the presence of two incommensurate frequencies in the response.

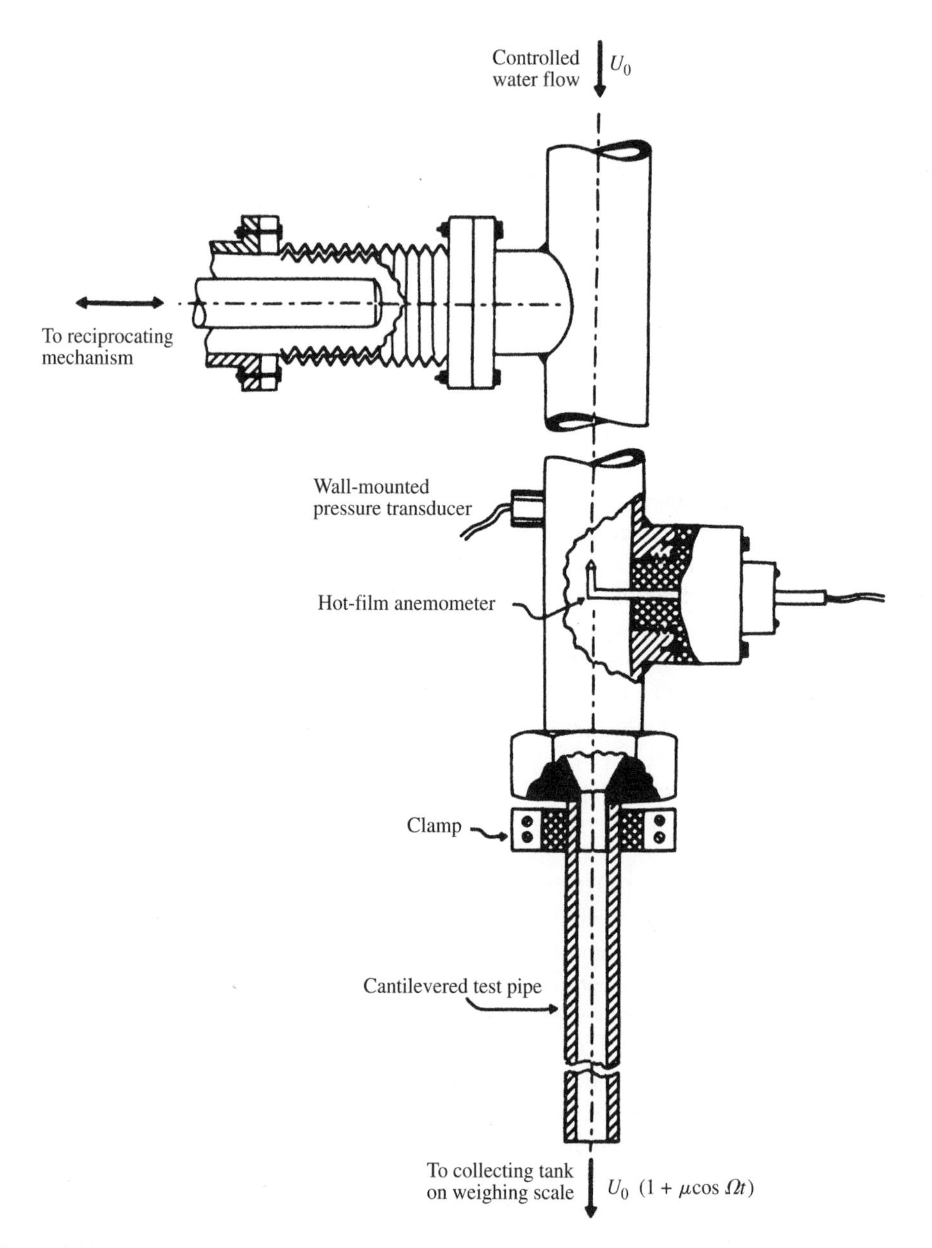

Figure 4.30 Schematic diagram of the apparatus used for parametric oscillations of pipes conveying fluid (Païdoussis & Issid 1976).

agreement between theory and experiment is particularly poor in the case of the secondary resonance. This is probably due to interference by forced vibration, transmitted through the apparatus, which resulted in effectively widening the region of resonance.

The dynamical behaviour in the case of cantilevered pipes is described next. It was found that unless the flow velocity was fairly close to the critical value, U_{cf}, at which the system loses stability by flutter in steady flow, the system remained stable, at least

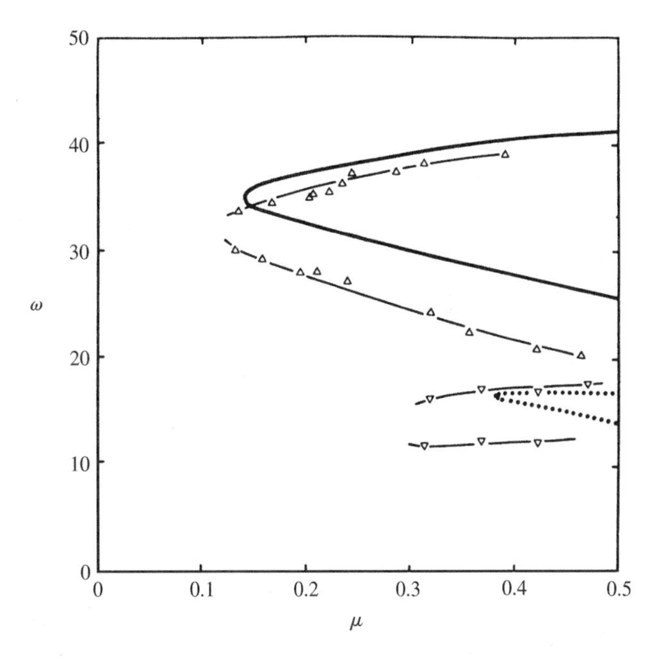

Figure 4.31 Experimental boundaries of the first-mode principal and fundamental parametric resonances compared with theory, for a clamped–clamped pipe with $\beta = 0.202$, $\gamma = 14.8$, $\alpha = 4.57 \times 10^{-3}$, $\sigma = 0$ and $u_0 = 4.55$. For primary resonance: ——, theory; -△-, experiment. For secondary resonance: ······, theory; -▽-, experiment (Païdoussis & Issid 1976).

for $0 < \mu < 0.5$ and a frequency range usually spanning the first three eigenfrequencies of the system.

If the flow velocity was close to the critical value mentioned in the foregoing, the following observations were made. At low frequencies, a secondary parametric resonance in the second mode of the system was observed, which was difficult to pin-point for the reasons already given. With increasing frequency, the amplitude of oscillation increased, to a maximum of typically five diameters, then decreased and finally ceased. The system remained stable with increasing frequency up to a certain value, where the principal primary instability was encountered, also in the second mode of the system, with the pulsation frequency equal to twice the oscillation frequency; the onset of this resonance was as easily pin-pointed as it was violent. With increasing frequency, a maximum amplitude of 20 pipe diameters was reached, then subsided and ceased.

Certain variations to the foregoing behaviour should be noted. In some cases, at low frequencies and high amplitudes of pulsation, quasiperiodic motion in a combination resonance region was encountered, involving the first and higher modes. In some cases, following the principal primary instability, either a combination resonance region or a region involving the superposition of more than one parametric resonance was observed, where the pipe seemed to be oscillating about a quasi-stationary deflected shape, thus displaying clearly nonlinear behaviour; these regions were difficult to decipher, and the mode shape transitions seemed to be gradual and difficult to pin-point. In yet other cases, the stable region between second-mode secondary and primary instabilities disappeared. In some cases, parametric resonances associated with the third mode were encountered, which were no different in character from those associated with the second mode. However, no simple parametric (as opposed to combination) resonances associated with the first mode were ever observed.

In one case, the flow velocity was increased sufficiently to cause flutter in steady flow. Interestingly, by adding a pulsatile component to the flow at certain frequencies and amplitudes, it was found possible to eliminate the flutter.

Once more, these general observations are in qualitative agreement with theory. Quantitative agreement may be assessed from Figures 4.32 and 4.33. In Figure 4.32 only the principal primary resonance region is shown, while in Figure 4.33(a) also the fundamental secondary one. In the latter case it is noted that no experimental points are shown for large μ; in that range, the resonance boundary was very difficult to define, as there was superposition of at least two resonance regions as shown in the theoretical results. Similarly, no experimental points are shown corresponding to the lower parts of the theoretical curves, which relate to lower subharmonics; in this case the experimentally observed resonance was of such small amplitude as to make it virtually impossible to define its boundaries.

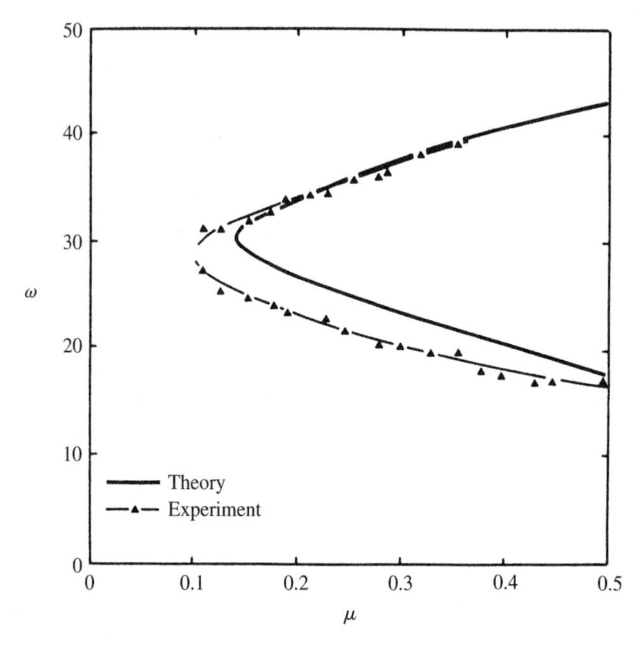

Figure 4.32 Experimental boundaries of the second-mode principal parametric resonance compared with theory, for a cantilevered pipe with $\beta = 0.205$, $\gamma = 8.22$, $\alpha = 3.75 \times 10^{-3}$, $\sigma = 0$ and $u_0 = 5.54$ (Païdoussis & Issid 1976).

It appears that theory generally underestimates the extent of the regions of resonance; moreover, it overestimates the value of μ_{cr}, the minimum value of μ necessary to cause parametric resonance. Agreement of experiment with theory is reasonable but not very good; plausible reasons for this are discussed by Païdoussis & Issid (1976), among them that certain assumptions in the theory are not quite true: e.g. that the flow-area of the pipe does not change with changing internal pressure and that the wave speed in the elastic pipe is essentially infinite.

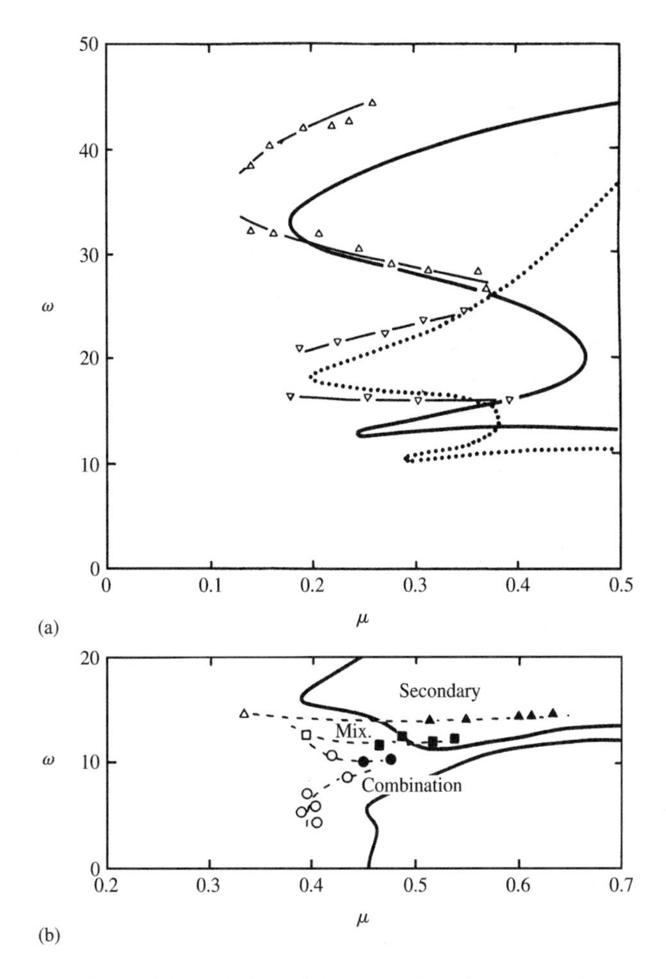

Figure 4.33 (a) Experimental boundaries of the second-mode parametric resonances compared with theory, for a cantilevered pipe with $\beta = 0.307$, $\gamma = 16.1$, $\alpha = 3.65 \times 10^{-3}$, $\sigma = 0$ and $u_0 = 7.86$. For primary resonances: ——, theory; -△-, experiment. For secondary resonances: ······, theory; -▽-, experiment. (b) Experimental boundaries of combination resonance and the lower boundary of simple secondary parametric resonance for a cantilevered pipe with $\beta = 0.203$, $\gamma = 13.3$, $\alpha = 4.54 \times 10^{-3}$, $\sigma = 0$ and $u_0 = 6.20$: ∘, •, combination resonance boundary; □, ■, combination-mixing transition; △, ▲, secondary parametric resonance threshold; ——, theory (Païdoussis & Issid 1976).

Theoretical and experimental combination resonance regions are compared in Figure 4.33(b) — some with $\mu > 0.5$, which is clearly beyond the theoretical assumption that μ is small. It is noted that theory underestimates the extent of combination resonance, but the shape of the left-hand boundary of the region is similar to that given by the experimental points.

Following a line of constant μ and increasing frequency (say, for $\mu = 0.5$), theory predicts that there should be a narrow region of stability separating the regions of combination and secondary parametric resonance. This was not observed experimentally; instead, the two regions were found to be separated by a 'mixing region', where, one might say, parametric resonance was trying to establish itself, but was confused by the presence of components of combination resonance.

4.5.4 Parametric resonances by analytical methods

The linear parametric resonance regions are frequently determined as special cases in *nonlinear* analyses of the system (Section 5.9). This is done for two reasons: to validate the more general nonlinear analysis in the linear limit, and to compare directly the linear and nonlinear dynamics of the system.

The first such analysis, unusually published as a paper wholly devoted to linear dynamics, is due to Ariaratnam & Namachchivaya (1986a) who study the principal (subharmonic) resonances associated with the first and second modes of pipes with supported ends ($\omega \simeq 2\omega_r$, $r = 1, 2$) and the corresponding combination resonances ($\omega \simeq \omega_2 \pm \omega_1$) by means of nonlinear analytical methods. These methods are in fact the same as those utilized to analyse the nonlinear dynamics of the system by Namachchivaya (1989) and Namachchivaya & Tien (1989a,b) and are described briefly in Section 5.9. Basically, the system is discretized into a two-degree-of-freedom one and transformed into first-order form while using an elegant Hamiltonian formulation; then the method of averaging is applied, via which the boundaries of the resonance regions are determined. The procedure is mathematically complex but very powerful: it yields analytical expressions for the resonance bounds, the minimum value of μ below which resonance does not occur, and so on. Also, by considering the stability of the solutions, it is shown which exist and which do not; specifically, it is shown that for pipes with both ends supported the difference combination resonance ($\omega \simeq \omega_2 - \omega_1$) does not exist (cf. Figure 4.28).

The analytical results are compared to numerical ones obtained by the authors and Païdoussis & Sundararajan (1975) — see the middle three regions of Figure 4.28. Agreement is quite good, despite the fact that the analytical method is meant to be valid only in the neighbourhood of the resonances, e.g. for $\omega = 2\omega_1 + \mathcal{O}(\epsilon)$; the discrepancy between analytical and numerical results becomes important for $\mu \geq 0.4$.

4.5.5 Articulated and modified systems

A two-segment articulated system hanging as a cantilever and conveying harmonically perturbed flow as in (4.69) has been examined thoroughly for parametric resonances by Bohn & Herrmann (1974a). The two pipe segments are of equal length, l, and no interconnecting springs are present, so that gravity is the only restoring force. Hence the following dimensionless parameters are used: $\overline{\beta} = 3\beta$, $\overline{u} = U/\left(\frac{3}{2}gl\right)^{1/2}$ and $\overline{\omega} = \left[l/\left(\frac{3}{2}g\right)\right]^{1/2}\Omega$. The resonance regions are determined by Bolotin's method and Floquet multipliers.

Basically, the dynamical behaviour of this system is similar to that described in Sections 4.5.1 and 4.5.2, for both simple and combination resonances. Of particular importance is the dynamical behaviour just above the critical flow velocity for instability *in steady flow*, as shown in Figure 4.34: (a) for $\overline{\beta} = 0.25$, when stability is lost by flutter, at $\overline{u}_{cf} = 2.632$; and (b) for $\overline{\beta} = 1.0$, when stability is lost by divergence, at $\overline{u}_{cd} = 1.732$. The dynamical behaviour in Figure 4.34(a) is similar to that in Figure 4.29(b), showing that what is a region of stability for $\overline{u} < \overline{u}_{cf}$ is essentially transformed into one of combination resonances (quasiperiodic oscillation) for $\overline{u} > \overline{u}_{cf}$; however, in this case also, there exists a region (albeit a very thin one) where the steady-flow flutter may be suppressed by flow pulsation. In Figure 4.34(b), whereas the parametric resonances almost fill the plane, there is quite a wide region where divergence is eliminated by pulsation.

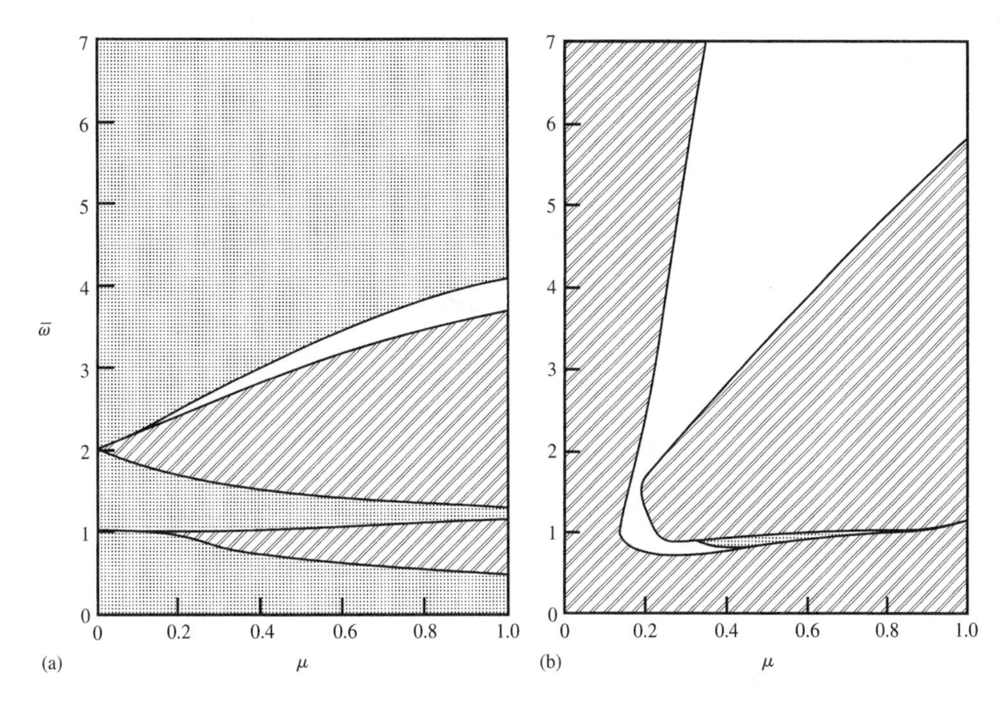

Figure 4.34 Regions of simple parametric (hatched) and combination (dotted) resonances for an articulated system with (a) $\overline{\beta} = 0.25$ at $\overline{u} = 1.05\overline{u}_{cf}$, and (b) $\overline{\beta} = 1$, $\overline{u} = 1.05\overline{u}_{cd}$ (Bohn & Herrmann 1974a).

A continuously flexible cantilevered system, modified by translational and rotational spring supports at the downstream end [Figure 3.61(c)], is analysed for parametric resonances by Noah & Hopkins (1980) — see Section 3.6.2 for the steady-flow dynamics. Typical results in Figure 4.35 show that, as for steady flow, the dynamics is intermediate between those for a cantilevered pipe and one with both ends supported, but more complex than either, and depends in an *a priori* unpredictable manner on the stiffness of the translational spring (κ) and the rotational one (κ^*). Of particular importance are that (a) parametric resonances related to the first mode can relatively easily be excited for some combinations of κ and κ^*, and (b) both sum- and difference-type combination resonances can arise in this case — both explainable in terms of the hybrid free but not totally free downstream end.

Finally, the analysis has also been extended to deal with periodically supported pipes by Singh & Mallik (1979), both by Bolotin's method and by a 'wave approach', which is particularly useful for pipes with a large number of spans and which is based on their earlier work with such pipes in steady flow (Singh & Mallik 1977). Unfortunately, their equations contain the same error as in Chen's work, referred to in the foregoing, and hence the results are quantitatively flawed, as are some of their conclusions — e.g. their

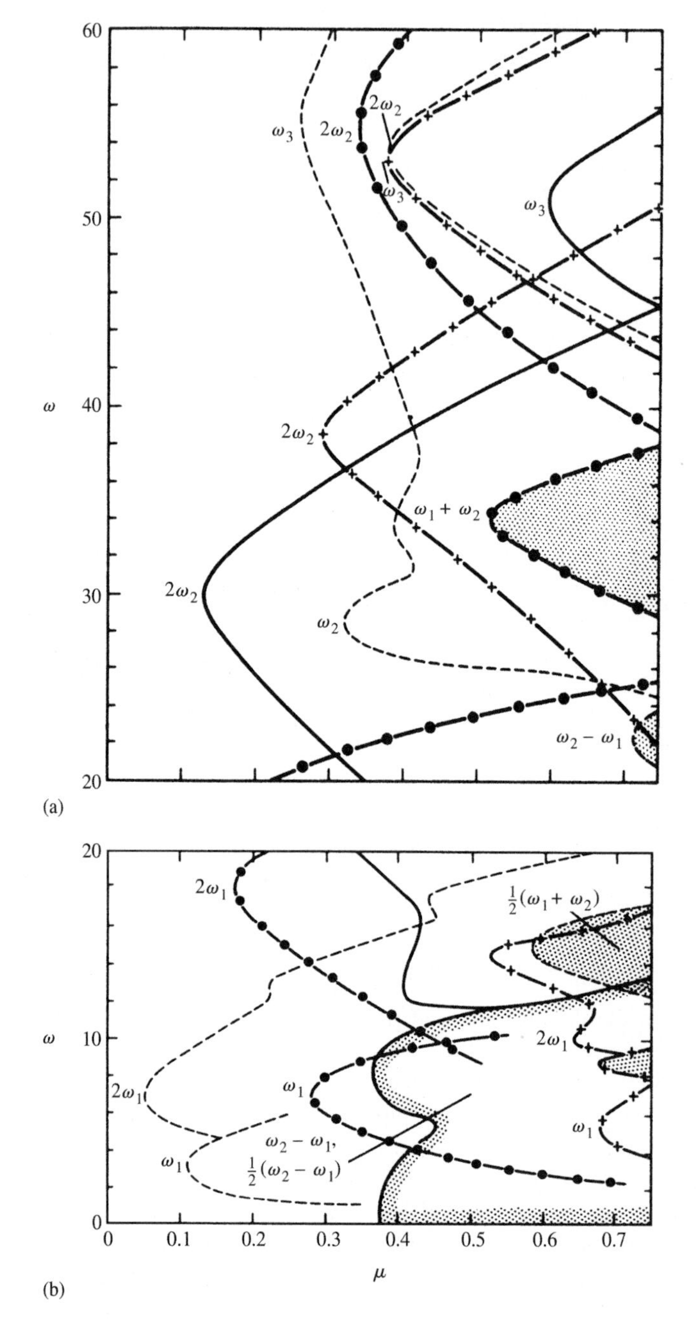

Figure 4.35 Regions of simple parametric (unshaded) and combination resonances (shaded) for cantilevered pipes with elastically supported downstream end, for $\beta = 0.125$, $\gamma = 0$, $\alpha = 10^{-3}$ and $u_0 = 4.5$: ——, $\kappa = \kappa^* = 0$; — + — , $\kappa = 25$, $\kappa^* = 0$; – – –, $\kappa = 100$, $\kappa^* = 0$; — • — , $\kappa = 100$, $\kappa^* = 10$. (a) For range of high forcing frequency, ω; (b) for lower range of ω (Noah & Hopkins 1980).

contention that parametric resonance regions calculated with $\beta = 0$ are representative of cases with effectively any value of β.

4.5.6 Two-phase and stochastically perturbed flows

As already mentioned, practical interest in this work has been associated with possible excitation of piping by pump-generated pulsations. A different and interesting application was studied by Hara (1977, 1980) in connection with *two-phase flow in piping*, in the 'slug-flow' regime, where the flow is essentially composed of alternating liquid and gaseous slugs.

Clearly, in this case it is not the velocity that is varying with time but the mass per unit length; moreover, the time variation is more like a square wave than a harmonic function. This is the phenomenon accidentally discovered in the experiments with piping aspirating flow, as shown in Figure 4.11(b). Hara's experiments, involving a 2.2 m long horizontal simply-supported pipe were in fact conducted with air–water mixtures simulating true two-phase flow. Parametric resonances were found for $\omega/\omega_n \simeq 0.65$, 0.95 and 1.94, where ω is the frequency associated with 'slug arrival times'; these ratios are remarkably close to the theoretically expected $\frac{2}{3}$, 1 and 2. The strongest excitation in this case was for $\omega/\omega_n = 1$ rather than 2; this is explained as being due to additional two-phase forced excitation when $\omega = \omega_n$.

Finally, the case of a pipe conveying *stochastically perturbed flow* was studied by Narayanan (1983), Ariaratnam & Namachchivaya (1986b) and Namachchivaya & Ariaratnam (1987). By assuming the intensity and correlation time of the stochastic perturbations to be small (broad-band spectrum), the problem is transformed into a Markov process, and solutions are obtained by stochastic averaging. It is found that the amount of damping necessary to ensure stability depends only on those values of the excitation PSD near twice the eigenfrequencies and near their sums and differences.

4.6 FORCED VIBRATION

There are two aspects of forced vibration of pipes conveying fluid worthy of discussion. The first is the physical aspect, which sheds further light onto the dynamics of the system, and the second is related to the analytical techniques which can be used to obtain the forced response of such systems. These will be dealt with separately in what follows.

4.6.1 The dynamics of forced vibration

Let us consider a pipe subjected to an arbitrary harmonic force field, such that it is governed by an equation of the form

$$\mathscr{L}(\eta) = f(\xi)\,\mathrm{e}^{\mathrm{i}\omega\tau}, \tag{4.76}$$

in which $\mathscr{L}(\eta)$ is given by equation (3.70). By means of Galerkin's method, this may be discretized into

$$\mathbf{M}\ddot{\mathbf{q}} + \mathbf{C}(u)\dot{\mathbf{q}} + \mathbf{K}(u)\,\mathbf{q} = \mathbf{F}\,\mathrm{e}^{\mathrm{i}\omega\tau}, \tag{4.77}$$

where the matrices **C** and **K** are functions of u. The steady response of the system is written as $\mathbf{q} = \mathbf{Q}\exp(\mathrm{i}\omega\tau)$ and hence equation (4.77) leads to

$$[\mathbf{K}(u) + \mathrm{i}\omega\mathbf{C}(u) - \omega^2\mathbf{M}]\mathbf{Q} \equiv \mathbf{S}(u, \mathrm{i}\omega)\mathbf{Q} = \mathbf{F}. \tag{4.78}$$

Hence, we may define the *direct receptance* (Bishop & Johnson 1960; Bishop & Fawzy 1976) at any generalized coordinate q_j as the generalized displacement at that coordinate due to a generalized force of unit amplitude and frequency ω applied at the same coordinate; then, application of Cramer's rule to equation (4.78) shows that the direct receptance α_{jj} is given by

$$\alpha_{jj} = \frac{Q_j}{F_j} = \frac{\text{cofactor}\, S_{jj}}{|\mathbf{S}|} = \frac{\displaystyle\prod_{a=1}^{2N-2}(\mathrm{i}\omega - \lambda_a)}{\displaystyle\prod_{e=1}^{2N}(\mathrm{i}\omega - \lambda_e)}, \tag{4.79}$$

where the λ_e are the $2N$ complex eigenvalues of **S** associated with *resonances* of the system. The λ_a are eigenvalues of **S** when coordinate q_j is locked and are associated with *antiresonances*. The treatment and the results to be presented are taken from Bishop & Fawzy (1976), in terms of plots of receptance and its inverse, the *inverse receptance*,[†] a form of mechanical impedance involving displacement rather than velocity. The motivation behind this study is to gain understanding useful in the dynamical testing of aircraft near the flutter boundary.

Typical results for a vertical cantilevered pipe fitted with an end-nozzle are shown in Figure 4.36, for the direct receptance at the free end, α_{11}, which relates the response at $\xi = 1$ to the excitation at the same point. The system is discretized by a Galerkin scheme with $N = 4$, and so four modes are involved; the critical flow velocity for flutter is $u_{cf} = 2.749$.

The four circular loops in Figure 4.36(a) correspond to the four degrees of freedom of the discretized system, which are traced by the solution as the forcing frequency ω is increased from zero (point P). The real parts of the eigenvalues of the system, $(-2.28 \pm 5.46\mathrm{i})$, $(-1.41 \pm 19.19\mathrm{i})$, $(-1.77 \pm 58.22\mathrm{i})$, $(-1.74 \pm 117.58\mathrm{i})$ are close to the minima of $\mathfrak{Im}(\alpha_n)$. Every point of a receptance diagram represents the sum of the responses in all the modes. This sum may be such that the curve intersects the positive real axis, as in Figure 4.36(a,b); at the frequency corresponding to such an intersection, no work is done by the driving force. Intersection with the negative real axis is also possible, again signifying no work done by the driving force, but for a different reason: this intersection occurs only for $u > u_{cf}$, at $\omega = \omega_{cf}$ — see Figure 4.36(b), where the intersection for $u = 3.0$ occurs at $\omega = 15$, beyond the confines of the figure.

It is also noted, in Figure 4.36(b), that as $u \to u_{cf}$ the diameter of the first loop of the receptance curve diminishes, while that of the second one, associated with the flutter mode tends to infinity; thus, as ω is increased, the receptance shoots off to infinity through the first quadrant, which seems reasonable on physical grounds. At u_{cf}, the pipe tends to

[†]Bishop developed the concept of receptance into a powerful tool for the analysis of all conceivable aspects of vibration of mechanical systems (Bishop & Johnson 1960). An anecdote making the rounds, in the U.K. at least, in the early 1960s is that one day the following sign was affixed (by a frustrated student, probably) on the door to this office: *No admittance without receptance!*

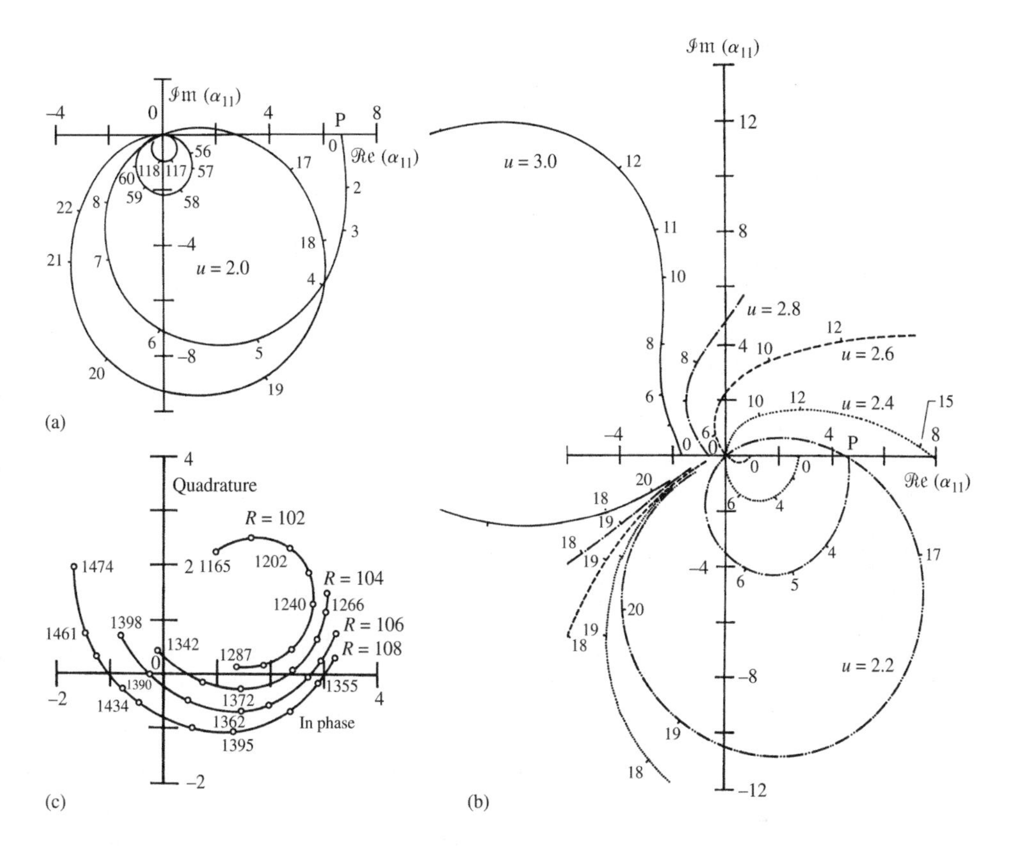

Figure 4.36 (a) Variation of the direct receptance α_{11} for a theoretical harmonically forced system with $u = 2$ and varying values of ω (nozzle area parameter $\alpha_j \equiv A/A_j = 3$, $\beta = 0.203$, $\gamma = 5$, zero dissipation, and $N = 4$). (b) Same as in (a), but $u = 2.2$, 2.4, 2.6, 2.8 and 3.0. (c) Inverse direct receptance at $\xi = 0.15$, close to the instability boundary of the experimental system ($\alpha_j = 1.27, \beta = 0.387, \gamma = 250$) for a number of values of the flow parameter, R, and varying frequency parameter, f, defined in the text (Bishop & Fawzy 1976).

oscillate at its critical frequency, $\omega_{cf} = 15$. When $\omega < \omega_{cf}$, energy flows from the pipe to the driving mechanism, and the displacement *leads* the excitation — which is not possible for passive systems. The phase lead continues until $\omega = \omega_{cf}$, when no energy flows to or from the driving mechanism, since all the energy required to achieve an infinite amplitude is supplied solely by the fluid. For $\omega > \omega_{cf}$, the pipe is forced to oscillate more rapidly, and the displacement *lags* behind the force; hence the receptance curve is now below the real axis.

A number of other, special and interesting features of these receptance curves are discussed by Bishop & Fawzy, among them: (i) the vanishing of the receptance at a finite ω; (ii) the migration of the starting point of the receptance curve along the real axis towards the origin. The first point suggests that the system may have some purely imaginary antiresonance eigenvalues — which means that these are the resonances of a system with the excitation point (at the downstream end of the pipe) constrained to zero, i.e. those of a clamped–pinned pipe. Indeed, it is known that the eigenvalues of the clamped–pinned system are purely imaginary up to a critical value (in this case

$u = u_{cd} = 2.7$). It is confirmed that the frequencies at which the receptance vanishes correspond to these clamped–pinned pipe eigenvalues — so that at those frequencies the pipe oscillates like a clamped–pinned one.

The second point, the migration with u of point P towards the origin, is at first sight paradoxical, since generally the eigenfrequencies *decrease* with increasing u (see Figures 3.27–3.29), implying a softening of the system, while the migration in Figure 4.36(a,b) indicates hardening! Recalling that the exciter is attached to the downstream end of the cantilevered pipe, the explanation is once more related to the characteristics of the clamped–pinned pipe, which is subject to divergence at $u_{cd} = 2.7$. Hence, at u_{cd} an infinite force is required to hold the free end in position. At $u > u_{cd}$, the tendency to buckle will cause the pipe to press against the support, and hence the displacement to be out of phase with the applied force; therefore, point P shifts to the negative $\mathcal{R}e(\alpha_{11})$ axis.†

Bishop & Fawzy tested the theory by conducting forced vibration experiments, using surgical quality silicone rubber tubes conveying water (cf. Section 3.5.6), excited sinusoidally via a carefully designed cross-head mechanism, based on the Scotch-yoke principle. The force was measured by a force transducer and the displacement by a displacement transducer, at the same point or elsewhere along the pipe. These experiments illustrate the difficulties in undertaking such experiments, especially near the flutter boundary.‡ Near u_{cf}, since $u_{cf} > u_{cd}$ for a clamped–pinned pipe in all cases when the system is excited at its lower end, experiments were practically impossible since 'it was extremely difficult to arrest the tube, let alone to oscillate it sinusoidally'. Hence, the pipe was excited at a point $x = 0.15L$–$0.4L$. A number of difficulties persisted, however. For example, for large force amplitudes, the system sometimes behaved as if composed of two subsystems joined together at the excitation point: a clamped–pinned beam and a pinned–free one — specifically for the forcing frequency close to that of the lower part of the pipe; this led to 'dynamic interference', beating and so on.

In the end, however, some successful experiments were performed, leading to several results of the type shown in Figure 4.36(c) — probably the first ever for an active system so close to the flutter boundary. Here the *inverse* direct receptance is plotted, so that at resonance the curve goes through zero. These curves are in terms of raw measurement quantities: a rotameter reading R, related to the flow velocity by $U = 1.185R \times 10^{-6}/A$ (m/s), where A (m^2) is the internal cross-flow area of the pipe; and a frequency factor, f, equal to 480 times the oscillation frequency in Hz. Unfortunately, the cross-sectional dimensions of the pipes are not given; hence, u and ω cannot be computed, and these results cannot be compared with the theory. The reason given for not presenting a comparison with theory is that dissipation, always present in the experiments, has been ignored in the theory.§

Nevertheless, for the experimental system, Figure 4.36(c) shows that for flutter, $104 < R_{cf} < 106$ and $1340 < f_{cf} < 1390$ approximately. This demonstrates that it *is* feasible

†The value of u_{cd} in this case is too close to the $u_{cf} = 2.749$ for the clamped–free system. Bishop & Fawzy present another calculation with $\alpha_j = 1$, $\beta = 0.203$ and $\gamma = 5$, for which $u_{cf} = 6.07$, while $u_{cd} = 4.74$. The receptance curve passes through the origin for u between 4.6 and 4.8, in agreement with the explanation given.

‡As expressed by the authors with exquisite British understatement: 'it has to be said that the study of a resonance test on an active system near an instability boundary is not easy'.

§One may nevertheless suspect, since *some* comparison, even with this limitation, would have been useful, that quantitative agreement cannot have been flattering.

to bracket the flutter conditions sufficiently closely, if this type of testing were used for the system motivating this study: the determination of the flutter boundary of aircraft via forced vibration testing.

Another set of forced vibration experiments was conducted by Shilling & Lou (1980) with a vertical cantilever (made of PVC and of $D_i = 17.6\,\text{mm}$) with several masses attached along the length [see Figure 3.67(a)]. The upstream support together with a suction pipe and a motor-pump unit were mounted on a horizontal track and oscillated by an exciter; the test pipe was partially or totally immersed in water, or totally surrounded by air. It was found that, with internal flow, the response is richer in higher harmonics. Also, immersion appears to greatly enhance the modal content of the vibration, but this clearly depends on the forcing frequency; see also Sections 4.2.4, 4.3.2 and 4.4.10.

4.6.2 Analytical methods for forced vibration

Consider a simplified form of equation (3.70) for a cantilever conveying fluid, subject to a forcing function,

$$\eta'''' + u^2\eta'' + 2\beta^{1/2}u\dot{\eta}' + \sigma\dot{\eta} + \ddot{\eta} = f(\xi, \tau), \tag{4.80}$$

with boundary conditions (3.78). By means of the Galerkin method (Section 2.1.6), this equation may be discretized into

$$[M]\{\ddot{q}\} + [C]\{\dot{q}\} + [K]\{q\} = \{Q\}, \qquad Q_j(\tau) = \int_0^1 \phi_j(\xi) f(\xi, \tau)\,\mathrm{d}\xi, \tag{4.81}$$

ϕ_j being the jth eigenfunction of a cantilever beam, and Q_j the corresponding element of $\{Q\}$. Here both $[C]$ and $[K]$ are nondiagonal, nonsymmetric matrices, functions of u. To decouple the system, the methods described in equations (2.16)–(2.19) may be utilized, in which the system is first transformed into one of first order,

$$[B]\{\dot{z}\} + [E]\{z\} = \{F\}. \tag{4.82}$$

The asymmetry of $[C]$ and $[K]$ means that $[B]$ and $[E]$ are also nonsymmetric. Hence, to decouple this system, one proceeds (Païdoussis 1973b; Section 4) to solve the eigenvalue problem $(\mu[B] + [E])\{u\} = \{0\}$ and its adjoint $(\mu[B]^{\mathrm{T}} + [E]^{\mathrm{T}})\{v\} = \{0\}$, from which the same set of eigenvalues μ_j are obtained, but two different sets of eigenvectors, χ_j and ψ_j, leading to modal matrices $[A]$ and $[N]$. Because of the weighted biorthogonality of the χ_j and ψ_j, $[N]^{\mathrm{T}}[B][A]$ and $[N]^{\mathrm{T}}[E][A]$ are diagonal, an easily proved result. Hence, introducing the transformation

$$\{z\} = [A]\{\zeta\} \tag{4.83}$$

into (4.82) and pre-multiplying by $[N]^{\mathrm{T}}$, one obtains an equation of the form

$$[J]\{\dot{\zeta}\} + [L]\{\zeta\} = [N]^{\mathrm{T}}\{F\} = \{\Phi\}, \tag{4.84}$$

in which $[J]$ and $[L]$ are diagonal; thus the system has been decoupled and hence is easily solvable. A particular example of excitation $f(\xi, \tau)$ due to a random pressure field (e.g. turbulence-induced excitation) is given in Païdoussis (1973b) for external axial flow which is very similar to the case of internal flow; as this is discussed in Chapter 8 (Volume 2), it will not be duplicated here.

An alternative method is to obtain the eigenvalues and eigenfunctions $\{\lambda_j, \chi_j(\xi)\}$ of the conservative part of (4.80), $\eta'''' + u^2\eta'' + \ddot{\eta} = 0$, as well as those of its adjoint, $\{\lambda_j, \psi_j(\xi)\}$; these are in fact the same as for problem (2.52), and are given by equations (2.59). Then, introducing $\sum \chi_j(\xi)q_j(\tau)$ into equation (4.80) and using the biorthogonality relation (2.57), another form of equation (4.81) is obtained, in which $[M]$ and $[K]$ are diagonal. The presence of the Coriolis term in (4.80), however, means that $[C]$ is not diagonal. Hence, even more than for the problem in Section 2.1.6, this method offers no special advantage, since it cannot diagonalize the nonhomogeneous problem 'in one step' as it would if this were an ordinary mechanical system.

Let us now turn our attention to the forced response of a cantilevered pipe with a *tip point mass*, $\mathcal{M}$, subjected to an arbitrary force field, $f(\xi, \tau)$. The dimensionless equations of motion and boundary conditions in this case are

$$\eta'''' + u^2\eta'' + 2\beta^{1/2}u\dot{\eta}' + \sigma\dot{\eta} + \ddot{\eta} = f(\xi, \tau); \tag{4.85a}$$

$$\eta(0) = \eta'(0) = 0, \qquad \eta''(1) = \eta'''(1) - \mu\ddot{\eta}(1) = 0, \tag{4.85b}$$

where $\mu = \mathcal{M}/[(M + m)L]$. An alternative way of formulating the problem leads to

$$\eta'''' + u^2\eta'' + 2\beta^{1/2}u\dot{\eta}' + \sigma\dot{\eta} + [1 + \mu\delta(\xi - 1)]\ddot{\eta} = f(\xi, \tau), \tag{4.86a}$$

$$\eta(0) = \eta'(0) = 0, \qquad \eta''(1) = \eta'''(1) = 0, \tag{4.86b}$$

in which $\delta(\xi - 1)$ is the Dirac delta function. As hinted in Section 2.1.4, the decoupling of the equations in this case poses some interesting problems, because the boundary conditions in (4.85a,b) are time-dependent. Three possible procedures immediately spring to mind, as follows:

Method (a): to utilize the eigenfunctions $\psi_j(\xi)$ of the problem $\eta'''' + \ddot{\eta} = 0$ subject to boundary conditions (4.85b) to discretize the system;

Method (b): to utilize these same eigenfunctions $\psi_j(\xi)$ but apply them to an 'expanded domain' of the problem (Friedman 1956), which effectively means that the time-dependent boundary conditions are added to the equation of motion, so that one obtains

$$\left[\int_0^1 \psi_i\psi_j \,\mathrm{d}\xi + \mu\psi_i(1)\psi_j(1)\right]\ddot{q}_j + \left[2u\beta^{1/2}\int_0^1 \psi_i\psi_j' \,\mathrm{d}\xi + \sigma\int_0^1 \psi_i\psi_j \mathrm{d}\xi\right]\dot{q}_j$$
$$+ \left[u^2\int_0^1 \psi_i\psi_j'' \,\mathrm{d}\xi + \int_0^1 \psi_i\psi_j'''' \,\mathrm{d}\xi - \psi_i(1)\psi_j'''(1)\right]q_j = \int_0^1 \psi_i f(\xi, \tau)\,\mathrm{d}\xi; \tag{4.87}$$

Method (c): to utilize the cantilever beam eigenfunctions, $\phi_j(\xi)$, directly to decouple equation (4.86a).

In principle, one can show directly which of these methods are correct or otherwise, but here we shall do so by means of sample computations. To simplify matters and since the main point of interest is the decoupling procedure, we consider the homogeneous undamped version of this system: $f(\xi, \tau) = 0$, $\sigma = 0$. The results are presented in Table 4.6, for two-mode discretization in all cases. For the same value of μ, the values of ω_1 and ω_2 for $u = 0$ are the same whether $\beta = 0$ or $\beta = 0.1$, and hence they are not

Table 4.6 The applicability of three possible decoupling schemes for a cantilevered pipe with an end-mass.

System	u	Method (a)	Method (b)	Method (c)
$\beta = 0$, $\mu = 0$	$u = 0$	$\omega_1 = 3.52$, $\omega_2 = 22.02$	$\omega_1 = 3.516$, $\omega_2 = 22.03$	$\omega_1 = 3.516$, $\omega_2 = 22.03$
	$u = u_H$	$u_H = 4.485$, $\omega_H = 11.2$	$u_H = 4.485$, $\omega_H = 11.6$	$u_H = 4.485$, $\omega_H = 11.6$
$\beta = 0.1$, $\mu = 0$	$u = u_H$	$u_H = 4.706$, $\omega_H = 14.49$	$u_H = 4.706$, $\omega_H = 14.49$	$u_H = 4.706$, $\omega_H = 14.49$
$\beta = 0$, $\mu = 0.3$	$u = 0$	$\omega_1 = 2.36$, $\omega_2 = 17.58$	$\omega_1 = 2.36$, $\omega_2 = 17.60$	$\omega_1 = 2.36$, $\omega_2 = 17.79$
	$u = u_H$	$u_H = 4.05$, $\omega_H = 8.0$	$u_H = 4.05$, $\omega_H = 8.0$	$u_H = 4.13$, $\omega_H = 7.9$
$\beta = 0.1$, $\mu = 0.3$	$u = u_H$	$u_H = 4.42$, $\omega_H = 10.45$	$u_H = 2.59$, $\omega_H = 15.31$	$u_H = 2.66$, $\omega_H = 15.37$

given in the second case. The values of u and ω at the Hopf bifurcation are denoted by u_H and ω_H.

It is clear from Table 4.6 that, in the absence of time-dependent boundary conditions ($\mu = 0$), any one of the three methods may be used. Any of the three methods may also be used provided that the system is nongyroscopic, e.g. for $\mu = 0.3$ when $\beta = 0$; the small differences in the results by method (c) are due to different rates of convergence (all results here are with $N = 2$). When, however, the system is nonconservative gyroscopic ($\beta \neq 0$) and has time-dependent boundary conditions, which is the last entry in the table, it is clearly seen that method (a) is incorrect. It is for this reason that the analysis in Section 5.8.3(a,c) is carried out with method (c), but that in Section 5.8.3(b) with method (b). The interested reader is also referred to Chen (1970) and Lin & Chen (1976).

Before closing this section, it should be mentioned that there now exist powerful general computational methods for solving free, transient and forced vibrations of this type of system, which are presented in Section 4.7.

4.7 APPLICATIONS

Virtually all of the research on the dynamics of pipes conveying fluid has been curiosity-driven, even though it sometimes was *inspired* by practical applications. Some attempts have been made to justify the effort by linking it, generally unconvincingly, to applications in oil pipelines, heat exchanger tubes, etc.; unconvincingly, because it has been known, certainly since the early 1950s, that the effect of internal flow on the dynamics of pipes conveying fluid begins to become interesting, let alone worrisome, at flow velocities at least ten times those found in typical engineering systems. That is the reason why most experiments have been done with elastomer rather than metal pipes, thus achieving the necessary dimensionless u with modest values of dimensional flow velocity, U.

Nevertheless, some applications do exist, as will be described in what follows. Most of them have emerged unexpectedly ten, twenty or thirty years after the basic work was done (Païdoussis 1993). Some have already been mentioned, sprinkled throughout the text, e.g. in connection with the stability of long pipelines on elastic foundations in Section 3.7.

The most enduring benefit of this research, however, is in developing the fundamentals and methods which are used in related topics involving axial-flow–structure interactions, which do have engineering applications. For example, the dynamics of cylindrical bodies in axial or annular flow and the dynamics of shells containing or immersed in axial flow, covered in Volume 2, can be understood *in simple terms*, modelled mathematically and solved by means of the work presented throughout Volume 1.

4.7.1 The Coriolis mass-flow meter

The principle of the Coriolis/gyroscopic mass-flow meter is familiar to most (Plache 1979; Smith & Ruesch 1991): the whole flow goes through a U- or Ω-shaped pipe which is attached to a T-shaped leaf-spring, as shown in Figure 4.37. Together they form a tuning fork which is excited electromagnetically close to its resonant frequency in the plane perpendicular to the paper. The resultant vibration (rotation vector $\mathbf{\Omega}$) subjects the fluid in the two legs of the U to Coriolis acceleration of opposing sign, generating a torque which periodically twists the pipe at the right-hand end in and out of the paper as shown. The twist angle θ is linearly related to the mass flow rate MU; it is usually measured optically, since deflections are generally very small. Alternatively, the phase of the vibration in the two legs of the U, which is 180° out of phase, may be measured instead. Many variants of the system described are now available, manufactured by different companies. A thorough analysis of the operation of the Coriolis mass-flow meter is provided by Raszillier & Durst (1991) and Raszillier *et al*. (1993); see also Sultan & Hemp (1989).

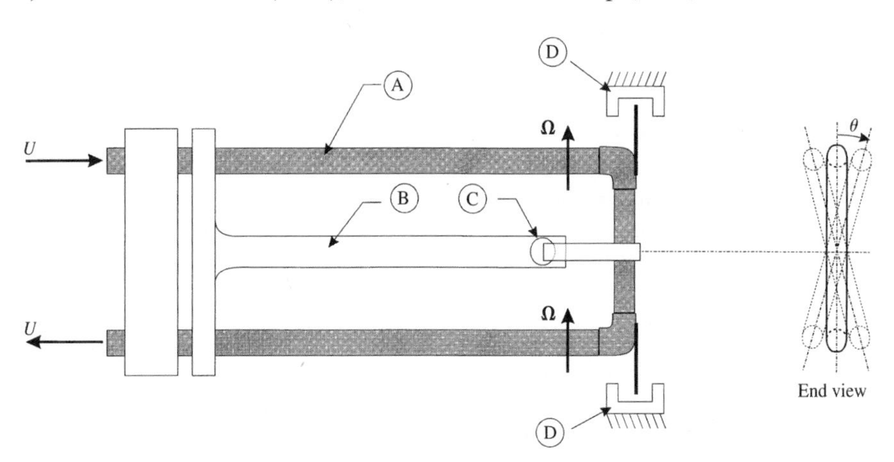

Figure 4.37 The operating principle of the Coriolis mass-flow meter. A, U-shaped pipe; B, T-shaped leaf spring; C, electromagnetic exciter; D, optical sensors; see Plache (1979).

It is not known to what extent the original invention was influenced by the fundamental work described in this book, but probably not much.† Nevertheless, when improvements to the original designs were contemplated, the manufacturers turned to the very researchers who contributed to the work in Chapters 3, 4 and 6 for consultation and further

†The first U.S. patent for a Coriolis-effect meter was issued in 1947.

research; alas, commercial confidentiality of products and methods precludes reporting much on this.

The work, however, continues. For instance, Tsutsui & Tomikawa (1993) propose a new straight-pipe mass-flow meter, on which is mounted an additional I-shaped oscillator. As described in conjunction with Figure 3.25(b), there is a phase difference in the two halves of the pipe when vibrating in its first mode; second-mode vibration actually produces a moment, and the operation of the new design is related to this effect and to the associated motion of the attached I-oscillator.

4.7.2 Hydroelastic ichthyoid propulsion

Noticing the similarity between the mode shapes of a fluttering cantilevered pipe (Figures 3.45 and 3.48) and a swimming slender fish, e.g. an eel as shown in Figure 4.38(a), a novel method of aquatic propulsion for watercraft was devised (Païdoussis 1976) and patented. It is recalled that for the cantilevered system, no classical modes exist: the limit-cycle motion envelope comprises standing and travelling wave components, the latter propagating from the clamped towards the free end, similarly to the anguiliform swimming motions of slender fish (Lighthill 1969; Triantafyllou *et al*. 1993).

By mounting a pair of Tygon pipes on either side of a straight thin brass plate as shown at the bottom of Figure 4.38(b), one can generate undulating motions of the plate (perpendicular to the plane of the paper) at sufficiently high flow rates in the pipe, beyond the flutter boundary. The system was tested by mounting this arrangement beneath a small vessel. The flow was generated by a motor-pump unit on board, powered in tram- or trolley-fashion by an overhead electrical conductor.

'Sea trials' were conducted in a long flume, approximately 0.9 m by 0.9 m in section and 15 m long. Propulsion of course occurs even without undulation of the plate, simply by the jet issuing from the twin pipes. Hence, two arrangements were tested: (i) one in which the plate was allowed to undulate, and (ii) another in which it was immobilized by attaching thin wooden stiffeners, shaped so as not to increase the drag. Typically, the forward speed was $V \simeq 1$ m/s, the wavelength of the motion $\lambda \simeq 0.6L$ and the frequency $\omega \simeq 15$ rad/s, so that the reduced frequency $\omega\lambda/V \simeq 10$. Allowing about 4 m for a constant speed to be reached, the motion of the vessel was timed over the next 8.5 m, establishing an average value of V.

It was found that 30% higher speeds, i.e. approximately 60% higher thrust, could be achieved with undulation as compared to without, provided that the downstream propagating wave velocity was faster than the forward speed of the vessel — alas, however, at considerably inferior efficiency to a propeller. Because of its similarity to fish motions, the name of *ichthyoid propulsion* was coined.

The experiments just described were, in effect, proving tests, with no attempt to optimize the system; in an optimized design, both the momentum flux and mass would be axially distributed in such a way as to give the most desirable wave-propagation characteristics, and hence propulsion efficiency.[†] This method of propulsion was put forward as a possible propulsion scheme for special purposes, e.g. where propellers are undesirable because of sealing (in great depths) or noise problems.

[†]See also Sugiyama & Païdoussis (1982).

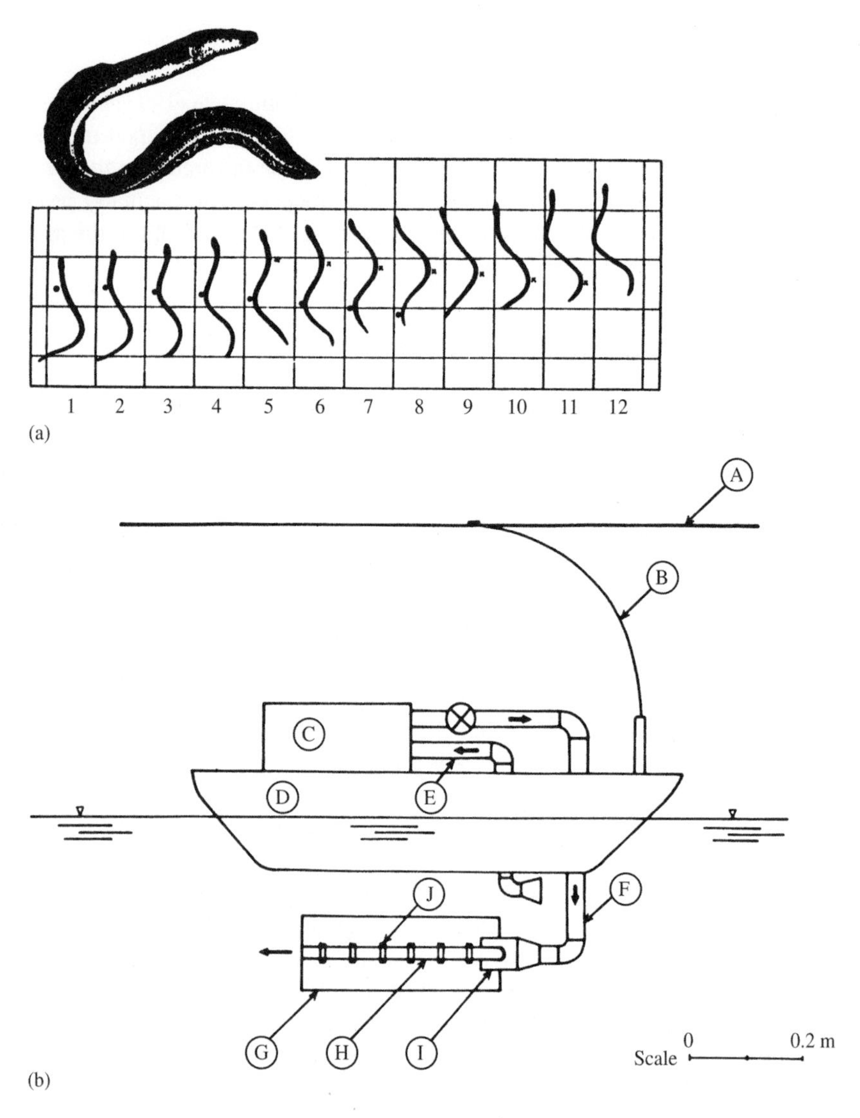

Figure 4.38 (a) Swimming motions of the common eel, *Anguilla vulgaris*, from photographs by Gray (1968), as given in Lighthill (1969). (b) Schematic elevation of the catamaran used to demonstrate ichthyoid propulsion, showing: A, overhead electrical conductor; B, 'trolley-type' conductor; C, motor-pump unit; D, catamaran; E, pump inlet; F, pump outlet; G, thin brass plate; H, Tygon pipes; I, flow adaptor; J, clips for attachment (Païdoussis 1976).

4.7.3 Vibration attenuation

Another application (Sugiyama *et al*. 1992, 1996b) involves one or more cantilevered pipes conveying fluid attached to a vibrating structure for the purposes of *damping* its vibration.

Looking at Figure 3.27, for instance, it is clear that for $u = 4$ there exists optimum damping of the pipe in all its modes. Hence, if this pipe were attached to a vibrating structure, the vibrations of the structure-cum-pipe system would be damped. An experiment with the set-up shown in Figure 4.39(a) demonstrates how the system can work when the 'structure' is the pipe itself. When the pipe is disturbed, it vibrates, and this is detected by a displacement sensor. If the vibration is above a predetermined threshold, the valve opens, admitting a fluid flow such as to give optimum damping. When the vibration level is reduced to below a given threshold, the valve closes, the pipe having accomplished its task. The vibration of the pipe without flow (the controller totally inoperative) and with control flow is shown in Figure 4.39(b,c).

The idea of such a vibration damper was also proposed by Lu *et al.* (1993).[†]

Of course, if the pipe is attached to a massive structure, the effective damping ratio will then be $\zeta_e = \frac{1}{2}c_p/[(m_s + m_p)(k_s + k_p)]^{1/2}$, where m_s and m_p are the modal masses of the structure and the pipe, respectively, in the mode concerned, k_s and k_p are the corresponding modal stiffnesses, while c_p is the modal damping of the pipe — neglecting c_s since presumably $c_s \ll c_p$. Hence, ζ_e will generally be considerably smaller than ζ for the pipe alone. This renders the application useful only for special cases, but no less interesting.

4.7.4 Stability of deep-water risers

Offshore risers are long pipes used in the exploration and production of oil and gas, connecting the sea-floor to an offshore floating or fixed platform or to a ship. With these activities moving to ever deeper waters, rigid-pipe risers have given way to flexible ones, such as shown in Figure 4.40. Sessions on riser dynamics are regular features of the annual *Offshore Technology Conference* (OTC), the ASME *Offshore Mechanics and Arctic Engineering Conference* (OMAE), ISOPE *Offshore and Polar Engineering Conference*, and other specialist conferences in the field, to the proceedings of which the interested reader is referred. All kinds of fluid–structure interactions are of concern, involving currents, waves and internal flow. Sample papers of interest here are by Sparks (1983), Vogel & Natvig (1987), Moe & Chucheepsakul (1988) and Moe *et al.* (1994).

Because of their great length, measured in kilometres, flexible risers may generally be considered to be hoses or *pipe-strings*, thus neglecting flexural restoring forces. As such, they are like any other string: effectively a limp strand of spaghetti, the configuration of which is solely determined by the imposed tension (applied by special tensioning devices and buoys), internal and external pressure, gravity and internal flow effects. The concept of an 'effective tension', incorporating tension and pressure effects, $T_{\text{eff}} = T + p_e A_e - p_i A_i$ as in equation (4.13) is widely used; cf. the 'combined force' Π in Chapter 6, defined in equations (6.46), and also refer to Section 3.4.2.

Elaborate computer codes exist for the calculation of the shape of, and stresses in, risers subject to given T_{eff} and to internal and external flow loading. Changes in pressure and flow, operational or accidental, give rise to transient motions and/or changes in configuration. Also, if the tensioning devices fail, loss of tension may give rise to 'instabilities' in the sense of large and sudden changes in configuration. The effects of

[†]In the oral presentation, when questioned as to possible applications, one of the authors proposed the 'damping of space structures'. An interesting idea, but the cost of transporting fluid into space and then sprinkling it all over the universe must be astronomical! The idea of damping wind-induced bridge vibrations is also a bit far-fetched. Nevertheless, the usefulness of the concept for special applications still stands.

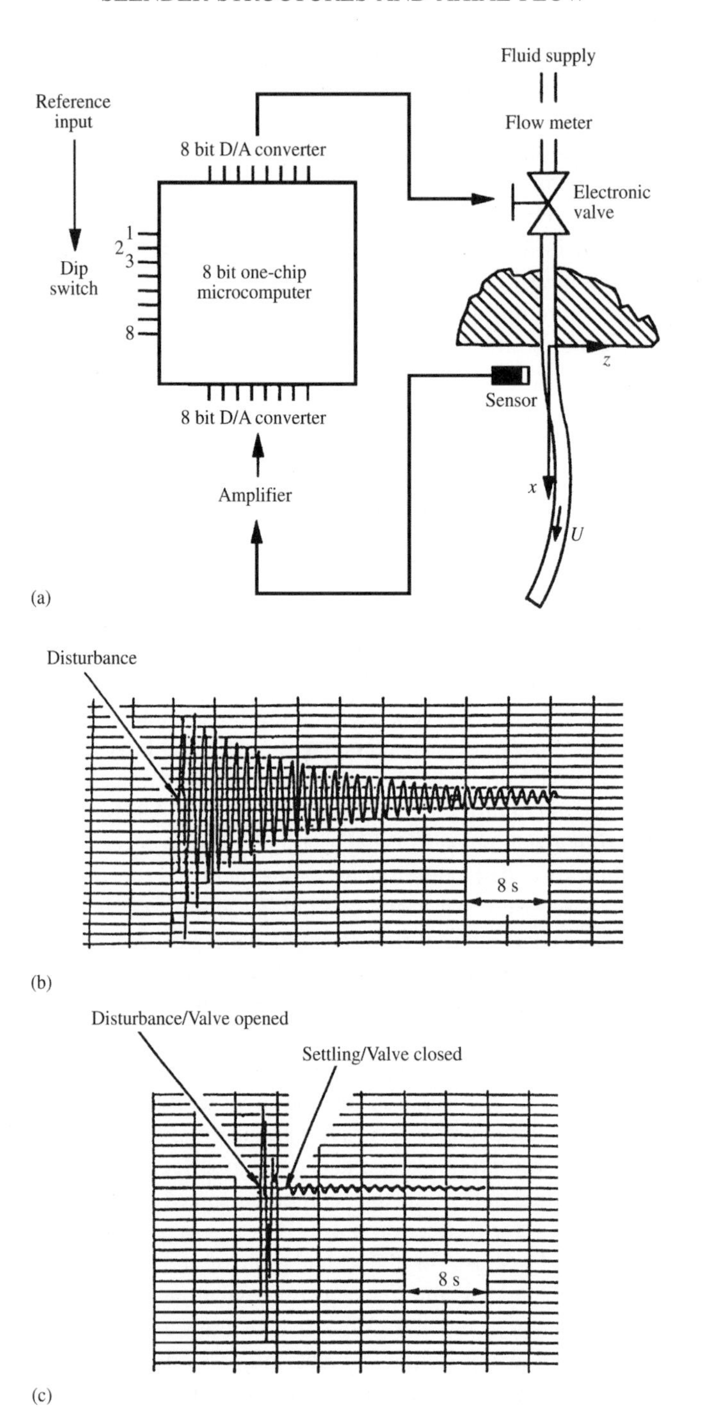

Figure 4.39 (a) Schematic diagram of the experimental vibration-suppression system. The response of the system to an impulse, (b) without fluid flow and (c) with controlled flow (Sugiyama *et al*. 1996b).

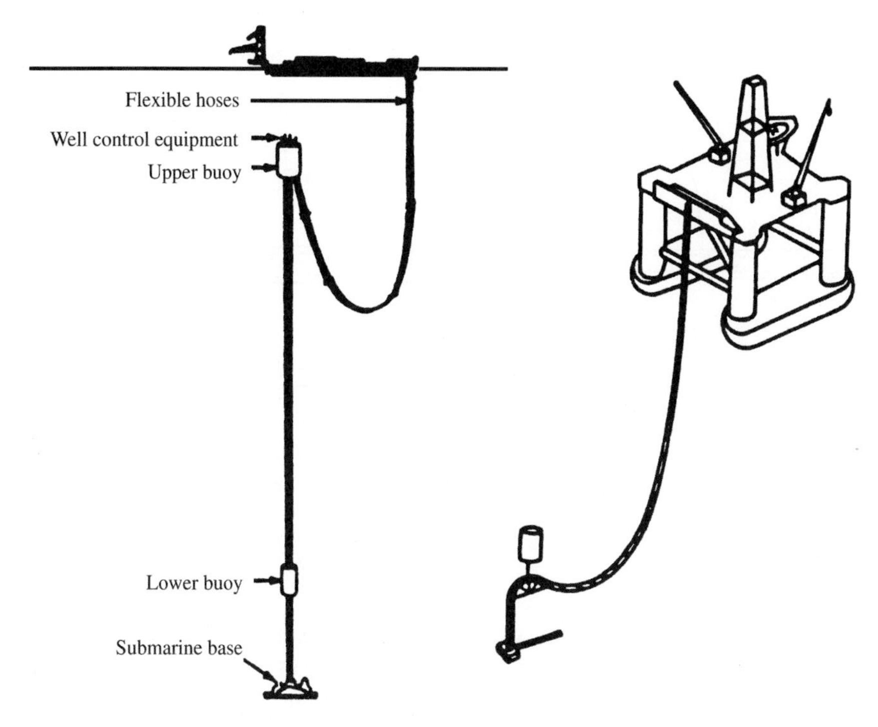

Figure 4.40 Diagrams of two possible deep-water flexible risers.

all such occurrences must be known at the design stage, predicted by the same computer codes. Therefore, understanding and modelling the effects of pressure and internal flow are essential building blocks in the development of these codes.

4.7.5 High-precision piping vibration codes

As has already been remarked, for most industrial applications the effect of steady internal flow in piping is not crucial. However, in specific applications, the piping is sufficiently flexible and failure sufficiently undesirable to make it important to develop high-precision computer codes for free, forced and transient vibration of pipes, taking internal flow effects into account. Such applications are those just discussed in offshore risers, ocean mining systems (Section 4.3), and special designs such as that discussed in Section 5.5.4, where long unsupported spans make the pipes effectively very flexible. Another example is a special low-cost condenser involving plastic tubes, designed by the French concern Ecopol, to be discussed in Chapter 9 (Volume 2). Other applications are in aircraft and rocket fuel lines, where the piping is very flexible because of weight considerations.

Computational tools for piping vibration have been developed, for example, by Ting & Hosseinipour (1983), Nakra & Kohli (1984), Dang *et al*. (1989), Piet-Lahanier & Ohayon (1990), Sällström (1990, 1993) and Sällström & Åkesson (1990). An example of the type of complex piping structures that can be handled is shown in Figure 4.41, analysed via an exact finite element formulation based on Timoshenko beam theory — i.e. using

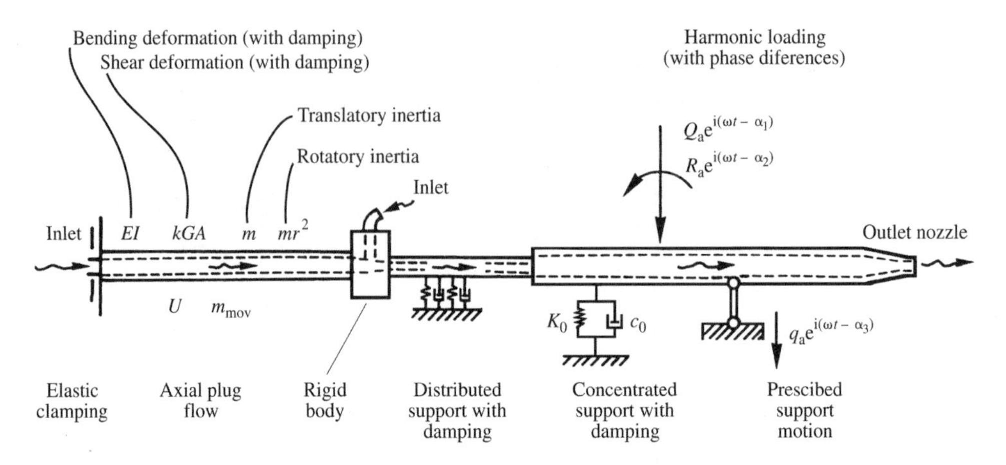

Figure 4.41 Example of a harmonically loaded piecewise uniform fluid-conveying piping structure in transverse vibration with angular frequency ω, showing how complex the structures analysed can be (Sällström & Åkesson 1990).

exact solutions of the equations for a uniform pipe rather than polynomial interpolation functions, thereby enabling the analysis of a complex system with very few finite elements (see also Chapter 7, Volume 2). The method has been found to be in excellent agreement with previous results and to be versatile, e.g. in handling systems such as that in Figure 4.41.

In contrast to steady flows, however, *unsteady flow*, e.g. due to pump-induced pulsation or acoustical effects, can and commonly does cause serious vibration problems (see Section 1.1), especially when light-gauge, low-damping piping is used, or in conjunction with flexible supports. Here the tools developed in Section 4.5 are of direct applicability.

The reader is also referred to the very extensive literature on the mainly unsteady fluid–structure interaction phenomena involving compressibility of the fluid and acoustical effects, including waterhammer, and more generally the effects of near-field and far-field noise which are not covered in this book (Wylie & Streeter 1978; Wiggert 1986, 1996; Moody 1990; Tijsseling 1996).

4.7.6 Vibration conveyance and vibration-induced flow

An unsigned 'focus' paper published in *Chemical Engineering* (March 1995, pp. 123–124) is entitled 'Pipes can't have "good vibrations".' Yet, as they say in Greek *μηδέν κακόν αμιγές καλοῦ*, i.e. nothing is bad without *some* good. An example of this is the turning of the tables on flow-induced vibration by *vibration-induced flow*.

Pipe vibration can be used to insert a long optical fibre into a long spatially curved, e.g. helical, steel pipe (Long *et al*. 1993, 1994). The optical fibre may be viewed as a 'plug-flow' model of a flowing fluid as in Bourrière's work (Sections 3.1 and 5.2.8). However, Jensen (1997) discovered recently that *real* vibration-induced flow is possible by nonlinear effects, as discussed in Section 5.10.

4.7.7 Miscellaneous applications

(a) Educational models

Some very interesting models for demonstrating the dynamics of nonconservative mechanical systems have been proposed by Herrmann *et al*. (1966), the central item in which is a cantilevered pipe conveying fluid. Every educator in the mechanics area should be aware of this publication.

(b) Buckling flow, turbulence and solar wind

In this case fluid-conveying pipes are used simply as conceptual devices in developing models for much more complex phenomena.

In a paper entitled 'Buckling flows: a new frontier in fluid mechanics', Bejan (1987) has assembled evidence to support the thesis that the buckling of flows is a generic phenomenon which may explain, among other things, the origins and structure of turbulence (Bejan 1989).

Bejan developed and systematized Taylor's (1969) and others' ideas and observations (e.g. Cruickshank & Munson 1981; Suleiman & Munson 1981), suggesting that there exists a characteristic wavelength/stream-thickness ratio to the undulations that may be seen in such diverse phenomena as the coiling of a honey (or maple syrup!) filament or the folding of a sheet of batter under gravity upon a solid surface, the sinuous shape taken by a jet of glycerine in quiescent water, the buckling of a falling sheet of toilet paper, a water jet hitting a free water surface, hot-air plumes, meandering rivers, etc. The interesting thing is that these phenomena are not confined to low-Reynolds-number flows. Although this wavelength-to-thickness ratio is different for each case, it remains in the range 1–10. The contention is that the large-scale structures in turbulent streams can be regarded as the 'fingerprint' of buckling.

Of interest here is that one of the examples cited by Bejan to support this thesis is the 'static buckling of a latex rubber hose' hanging vertically and conveying water — from his own experiments and those of Bishop & Fawzy (1976) and Lundgren *et al*. (1979). Of course, as discussed in Section 3.5.6, this is due to residual internal stresses; hence, in this context, the word 'imperfect' is required. However, this in itself is not deleterious to the thesis put forth by Bejan. Hence, this represents an unexpected use of the simple garden-hose problem towards modelling such a complex subject as turbulence!

Even more unexpected is the 'application' to an even more rarefied subject: solar wind modelling. *Solar wind* refers to the fast movements of plasma from the surface of the sun into space (all the way to earth), which, were it atmospheric air, would resemble wind. One of the early theories of the origin of solar wind (Axisa 1988) was that electromagnetically constricted 'tubes' of plasma develop, which are governed by fluid-dynamic equations (Dessler 1967; Parker 1963; Montgomery & Tidman 1964), and which move spirally into space. Bundles of such tubes of plasma could become unstable, similarly to fluttering cantilevered pipes, and then become intertwined, something like the snakes on Medusa's head, thus giving rise to turbulent mixing of the plasma. Alas, the real phenomenon is much more complex and such theories, though useful at the time, have long since been abandoned.

(c) Travelling bands and MAGLEV systems

The close similarity in the dynamics of travelling bands (band-saws, chains, magnetic tape, deploying antennas in space, etc.) have already been remarked upon on several occasions in this book; see, e.g. Mote (1968, 1972), Tabarrok *et al*. (1974), Wickert & Mote (1990). This is a case where cross-fertilization of ideas and swapping of techniques has been and is constantly taking place; a good example is given in Section 5.5.1. Another, among many areas benefitting from such cross-fertilization is magnetic levitation (MAGLEV) vehicle-guideway systems (Cai *et al*. 1992, 1996), where divergence and flutter instabilities can arise (Cai & Chen 1995).

(d) Severed pipes and pipe-whip

In many cases, a pipe may be acceptably 'stable' when connected the way it is meant to be, at both ends. If one end is accidentally disconnected or severed, however, which may sometimes mean the loss of tension which rendered it stable, the cantilevered pipe or pipe-string may well be unstable by flutter. Examples of such occurrences relate to fire-hoses and to life-lines used in space and underwater.

In the same family of accidents is the rupture in a pipe conveying high-pressure fluid, which causes the fluid to 'blow down' through the unevenly ruptured pipe into the surrounding fluid medium. As a result, the pipe may 'whip about', causing damage to surrounding structures. It is required practice for designers of power plants 'to postulate pipe ruptures and then perform analysis to determine what restraints or armor is required to prevent secondary failures' (Blevins 1990). A sample calculation of the motion of a pipe after rupture may be found in Blevins (1990, Chapter 10).

(e) Sprinkler system

A garden-variety type of application is the author's invention of a novel sprinkler for a McGill Open House circa 1976. It consists of an elastomer up-standing cantilevered pipe, mounted on a simple base and connected to the water mains. At the free end, a coarsely perforated stopper may be used. The hose performs a circular motion and waters a circular patch of lawn. Although it is not more effective than any other sprinkler, it is more aesthetically attractive than many, something like kinetic art!

4.8 CONCLUDING REMARKS

Even in an extensive treatment of the subject of the dynamics of straight pipes conveying fluid such as is given here in Chapters 3, 4 and 5, it is impossible to cite all the work, let alone discuss it. Hence, a great deal has been left out. Among that is the burgeoning effort on *control* of cantilevered pipes in flutter.

Control of the linear system has been studied by Takahashi *et al*. (1990), Kangaspuoskari *et al*. (1993), Cui *et al*. (1994, 1995), Tani & Sudani (1995), Lin & Chu (1996), Tsai & Lin (1997), Doki *et al*. (1998) and of the nonlinear chaotic system by Yau *et al*. (1995), utilizing a wide variety of control schemes. Some of the work, e.g. Tani & Sudani's (1995) and by Doki *et al*. (1998), is supported by experiments.

Some new work has began appearing also on *shape optimization* to maximize the critical flow velocity for flutter (Tanaka *et al*. 1993).

5

Pipes Conveying Fluid: Nonlinear and Chaotic Dynamics

5.1 INTRODUCTORY COMMENTS

One of the main reasons why the dynamics of pipes conveying fluid has remained of intense interest to dynamicists well into the 1980s and 1990s is the fact that (i) it displays interesting and sometimes perplexing *nonlinear* dynamical behaviour and (ii) it has become a handy tool in developing or testing modern dynamics theory.

In Chapters 3 and 4 we have mainly dealt with the stability of systems from the linear point of view, thus predicting loss of stability by divergence or flutter and, in some cases, a sequence of instabilities as the flow is increased beyond the onset of the first. As discussed in Section 2.3 with the aid of a one-degree-of-freedom model, divergence may arise via a *pitchfork bifurcation*, whereby the original equilibrium point becomes statically unstable, while two new equilibria are generated; on the other hand, flutter is often generated via a *Hopf bifurcation* and implies the generation of a limit cycle (Figures 2.10, 2.11 and 3.4). Additional bifurcations, e.g. a saddle-node and a period-doubling one, will be discussed in this chapter in due course.

Some key questions associated with these physical phenomena cannot be answered except by nonlinear theory, among them: (i) for the static instability, where are the new fixed points located for any value of the parameter being varied and are they foci (sinks) or saddles, and hence is the pitchfork bifurcation giving rise to the instability supercritical or subcritical? (ii) for the dynamic instability, is there an unstable 'inner' limit cycle in addition to a stable outer one (Figures 2.12 and 2.13), and hence is the Hopf bifurcation supercritical or subcritical (Figure 2.11); also, what is the amplitude of the limit cycle, and hence the amplitude of oscillation associated with flutter, and how does the frequency of oscillation vary with amplitude? Many of these questions are answered via the construction of appropriate *bifurcation diagrams* which, in compact form, display both (a) qualitative changes in the character of motion (bifurcations) and (b) the evolution in between some characteristic of the motion (typically the amplitude) with U. Also, the existence and nature of successive instabilities can be tackled by nonlinear theory, e.g. the question of post-divergence coupled-mode flutter, extensively discussed in linear terms in Chapters 3 and 4 by assuming implicitly that post-divergence oscillatory motions occur about the original equilibrium state; however, because the original equilibrium has become unstable after divergence, motions actually take place about the new equilibrium points, and hence stability has to be reassessed in this light.

Unfortunately, the methods required for the study of nonlinear dynamics (and for answering questions such as those in the previous paragraph) are much more complex than those for linear dynamics, and they are not in everyone's repertoire. Furthermore, they cannot easily be 'covered' in a book such as this, properly requiring one or more books of their own to do the job properly. One may distinguish 'classical' methods of nonlinear dynamics exemplified by the treatment in Minorsky (1962), Hayashi (1964) and Andronov *et al.* (1966), and 'modern' methods as exemplified by Guckenheimer & Holmes (1983), Sanders & Verhulst (1985) and Glendinning (1994). Other useful general references are Hirsh & Smale (1974), Nayfeh & Mook (1979), Hagedorn (1981), Rand & Armbruster (1987), Arnold (1988), Anosov *et al.* (1988), Arrowsmith & Place (1990) and Nayfeh (1981, 1993). An abbreviated treatment of some of the methods utilized extensively in this chapter and elsewhere in the book is given in Appendix F. This appendix, along with the specialized references cited therein, should be sufficient to guide the serious reader. The more casual reader may skip over the mathematics and concentrate on the physical interpretation of the results obtained in each case.

Most of the methods in Appendix F are concerned with *local analysis*, i.e. nonlinear behaviour in the vicinity of a fixed point or limit cycle; this is a requirement for the methods to work. More difficult is consideration of *global dynamics* aspects, e.g. the possibility of a large-amplitude limit cycle circumscribing two fixed points, which has not emanated from either. Although some aspects of global behaviour may sometimes be discerned from local analysis, the complete dynamical picture can only be provided by global analysis. The methods required for the latter will be discussed in *ad hoc* fashion and without too much detail.

The possibility of chaos in nonlinear systems has been known ever since Henri Poincaré at the turn of century, but it is fair to say that, for applied scientists, its existence lay dormant until the 1960s and the advent of Lorenz's work on thermal convection in the atmosphere. A good layman's introduction is given by Gleick (1987), and an excellent engineering treatment by Moon (1992). More sophisticated mathematical treatments are given by, among others, Thompson & Stewart (1986) and Wiggins (1988, 1990). Other useful references are Bergé *et al.* (1984), Devaney (1989), Parker & Chua (1989), Hao (1990) and Tsonis (1992), among others.

Basically, chaos arises when, over some ranges of parameters, the system ceases being predictable, in the sense that small changes in initial conditions may generate disproportionately large differences in the state of the system at any given time sufficiently long afterwards. The system is deterministic, but it behaves as if it were random — but with a most significant difference: its states are within specific regions of state-space, rather than all over, as would be the case for a truly random system. Thus, the trajectories of system response visit certain parts of the phase space, apparently randomly, but never others. A fractal nature in such plots is often revealed, whereby a small such region, when blown up, displays a similar character at a more microscopic scale.

Inevitably, specialized methods have been developed for the study of chaotic dynamics. These will be described in abbreviated form as necessary in the sections that follow.

5.2 THE NONLINEAR EQUATIONS OF MOTION

In many of the early papers on nonlinear dynamics of pipes conveying fluid (Holmes 1977; Ch'ng & Dowell 1979; Lundgren *et al.* 1979; Bajaj *et al.* 1980; Rousselet & Herrmann

1981), the equations of motion were derived *ab initio*, and hence several sets of different and certainly different-looking equations have come into existence. How different, and how complete and correct? The answering of these questions is not a trivial task, because of the different notations, approaches and assumptions involved in each of the derivations, and the relative obscurity of some. Hence, a definitive comparison was not undertaken until recently (Semler *et al*. 1994), but not before a number of *de facto* 'schools' had developed, followers of each utilizing the same basic assumptions and similar final forms of the equations.

In this section, following closely Semler *et al*. (1994), the equations of motion are derived via a Hamiltonian approach (while a Newtonian derivation is outlined in Appendix G) and, then, those of others' are discussed and their completeness and correctness assessed.

The system under consideration consists of a tubular beam of length L, internal cross-sectional area A, mass per unit length m and flexural rigidity EI, conveying a fluid of mass M per unit length with an axial velocity U, which may vary with time (Figure 5.1). The pipe is assumed to be initially lying along the x_0-axis (in the direction of gravity) and to oscillate in the (x_0, z_0) plane.

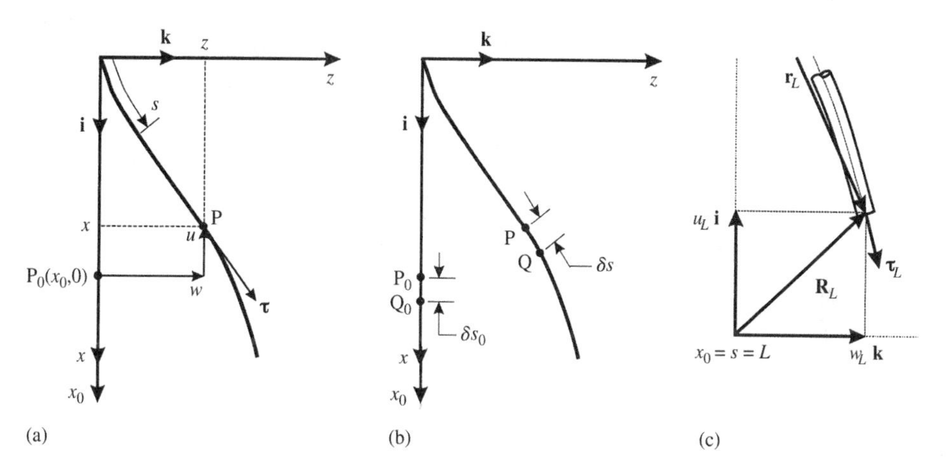

Figure 5.1 (a) The Eulerian (x, z) and Lagrangian (x_0, z_0) coordinate system and the coordinate s used when the centreline is considered to be inextensible; (b) diagram for the derivation of the inextensibility condition; (c) diagram defining terms for the statement of Hamilton's principle.

The basic assumptions made for the pipe and the fluid are as follows: (i) the fluid is incompressible; (ii) the velocity profile of the fluid is uniform (plug-flow approximation for a turbulent-flow profile); (iii) the diameter of the pipe is small compared to its length, so that the pipe behaves like an Euler–Bernoulli beam; (iv) the motion is planar; (v) the deflections of the pipe are large, but the strains are small; (vi) rotatory inertia and shear deformation are neglected; (vii) *in the case of a cantilevered pipe only*, the pipe centreline is inextensible.

5.2.1 Preliminaries

As in the derivation of the linear equations of motion (Section 3.3.1), two coordinate systems are utilized: the Eulerian (x, y, z) and the Lagrangian (x_0, y_0, z_0) — refer to

Figure 5.1(a,b). Then, as a material point of the pipe moves, its displacement may be defined by $u = x - x_0$ and $w = z - z_0$ for planar motions, where (u, w) may be expressed fully in either set of coordinates. Similarly, other quantities, such as the deformation gradients and strain tensors, can also be expressed in either set of coordinates. In infinitesimal deformation theory, which is the basis of the linear derivation, the distinction between Lagrangian and Eulerian strains disappears (Eringen 1987); however, this distinction must absolutely be made when nonlinear relationships are sought.

For a slender pipe with its initially undeformed state along the x_0-axis, undergoing motions in the (x_0, z_0) plane, we have $z_0 = 0$, so that $w \equiv z$. Here the Lagrangian representation is chosen, so that the position and deformation of any point of the pipe is expressed in terms of x_0, i.e. the position of that point in the undeformed state.

However, an exception is made in the case of a cantilevered pipe, the centreline of which may be assumed to be inextensible, in which case the coordinate s, measured along the centreline, is introduced (Figure 5.1), and all physical quantities, including the equations of motion, may be expressed in terms of (s, t).

The condition of inextensibility has already been defined in Section 3.3.1. Briefly, it states that the distance δs between two contiguous points P and Q, originally located at P_0 and Q_0 and δs_0 apart, satisfy $\delta s = \delta s_0 \equiv \delta x_0$; whereupon two forms of the inextensibility condition may be obtained, equations (3.14) and (3.15), repeated here for convenience:

$$\left(\frac{\partial x}{\partial x_0}\right)^2 + \left(\frac{\partial z}{\partial x_0}\right)^2 = 1, \qquad \left(1 + \frac{\partial u}{\partial x_0}\right)^2 + \left(\frac{\partial w}{\partial x_0}\right)^2 = 1. \tag{5.1}$$

For pipes fixed at both ends, δx_0 and δs are not identically equal, but they are related via equation (3.16), which may be expressed as

$$\frac{\partial x_0}{\partial s} = \frac{1}{1+\varepsilon} = \left[\left(1 + \frac{\partial u}{\partial x_0}\right)^2 + \left(\frac{\partial w}{\partial x_0}\right)^2\right]^{-1/2}, \tag{5.2}$$

where ε is the axial strain along the centreline. If $\varepsilon = 0$, the second of (5.1) is retrieved.

An exact expression for the curvature, κ, is useful in the derivations that follow, and is hence derived next. Depending on the choice of the coordinate system and the assumptions concerning the inextensibility of the pipe, the expression for κ differs. Let θ be the angle between the position of the pipe and the x_0-axis, and s the curvilinear coordinate along the pipe [Figure 5.1(b)]. For a pipe undergoing planar motion, extensible or inextensible, the curvature is given by

$$\kappa = \frac{\partial \theta}{\partial s}. \tag{5.3}$$

For simply-supported pipes, θ is defined by

$$\cos\theta = \frac{1 + \partial u/\partial x_0}{1 + \varepsilon(x_0)}, \qquad \sin\theta = \frac{\partial w/\partial x_0}{1 + \varepsilon(x_0)}. \tag{5.4}$$

In terms of the x_0-coordinate, equation (5.3) becomes

$$\kappa = \frac{\partial \theta}{\partial x_0}\frac{\partial x_0}{\partial s} = \frac{1}{1+\varepsilon}\frac{\partial \theta}{\partial x_0}. \tag{5.5}$$

The derivative in this expression may be obtained from (5.4),

$$\frac{\partial \theta}{\partial x_0} = \left[\frac{\partial^2 w}{\partial x_0^2}\left(1 + \frac{\partial u}{\partial x_0}\right) - \frac{\partial w}{\partial x_0}\frac{\partial^2 u}{\partial x_0^2}\right] \Big/ (1+\varepsilon)^2, \tag{5.6}$$

thus yielding the curvature (5.5) for pipes whose centreline may be extensible.

On the other hand, for cantilevered pipes whose centreline is assumed inextensible, expressions (5.3) and (5.4) still hold, except that $\varepsilon = 0$. In this case, $s = x_0$, and hence $\partial\theta/\partial x_0$ becomes

$$\frac{\partial \theta}{\partial x_0} = \frac{\partial x}{\partial s}\frac{\partial^2 z}{\partial s^2} - \frac{\partial z}{\partial s}\frac{\partial^2 x}{\partial s^2}. \tag{5.7}$$

Application of the inextensibility condition (5.1) leads to the following expression of the curvature:

$$\kappa = \frac{\partial^2 z/\partial s^2}{\sqrt{1 - (\partial z/\partial s)^2}}. \tag{5.8}$$

Alternatively, the curvature may also be defined as a vector:

$$\mathbf{b} = \frac{\partial^2 \mathbf{r}}{\partial s^2} \equiv \kappa\,\mathbf{n},$$

where $\mathbf{n}$ is the normal unit vector which is always perpendicular to the tangent direction of the pipe and $\mathbf{r} = (x, z)$ is the position vector along the pipe. Hence,

$$\kappa = \left|\frac{\partial^2 \mathbf{r}}{\partial s^2}\right| = \sqrt{\left(\frac{\partial^2 x}{\partial s^2}\right)^2 + \left(\frac{\partial^2 z}{\partial s^2}\right)^2}.$$

If the inextensibility condition is applied to the expression above, one obtains again equation (5.8).

Note that for a curve defined by $z(x)$ (Eulerian description) rather than $z(s)$, one has the familiar expression of curvature [e.g. Timoshenko (1955)],

$$\kappa = \frac{\partial^2 z/\partial x^2}{\left[1 + (\partial z/\partial x)^2\right]^{3/2}}. \tag{5.9}$$

Care must be taken as to which expression of κ is used, depending on the physical problem.

5.2.2 Hamilton's principle and energy expressions

A Hamiltonian derivation of the equations of motion is given here. It is based on the same statement of Hamilton's principle as in Section 3.3.3, namely

$$\delta \int_{t_1}^{t_2} \mathcal{L}\, \mathrm{d}t = \int_{t_1}^{t_2} \left\{ MU\left[\left(\frac{\partial \mathbf{r}_L}{\partial t}\right) + U\,\boldsymbol{\tau}_L\right] \cdot \delta \mathbf{r}_L \right\} \mathrm{d}t, \tag{5.10}$$

where $\mathcal{L}$ is the Lagrangian of the system ($\mathcal{L} = \mathcal{T}_f + \mathcal{T}_p - \mathcal{V}_f - \mathcal{V}_p$, $\mathcal{T}_p$ and $\mathcal{V}_p$ being the kinetic and potential energies associated with the pipe, and $\mathcal{T}_f$ and $\mathcal{V}_f$ the corresponding quantities for the enclosed fluid), and where $\mathbf{r}_L$ and $\boldsymbol{\tau}_L$ represent, respectively, the position vector and the tangential unit vector at the end of the pipe [Figure 5.1(c)].

However, before proceeding with the derivation of the equations of motion, some order-of-magnitude considerations are necessary. The lateral displacement of the pipe may be considered to be small relative to the length of the pipe, i.e.

$$z = w \sim \mathcal{O}(\epsilon), \qquad \epsilon \ll 1. \tag{5.11}$$

Large motions imply that terms of higher order than the linear ones have to be kept in the equation of motions. Because of the symmetry of the system, the nonlinear equations will necessarily be of odd order, and the derivation here will give a set of equations correct to $\mathcal{O}(\epsilon^3)$. However, the variational technique always requires a formulation correct to one order higher than that of the equation sought, so that all expressions under the integrand in statement (5.10) have to be at least of $\mathcal{O}(\epsilon^4)$. Finally, by considering the inextensibility condition, one can easily see that the longitudinal displacement u is

$$u \sim \mathcal{O}(\epsilon^2), \tag{5.12}$$

i.e. one order higher than w.

The total kinetic energy of the system is the sum of the kinetic energy of the pipe, $\mathcal{T}_p$, and the kinetic energy of the fluid, $\mathcal{T}_f$, defined by

$$\mathcal{T} = \mathcal{T}_p + \mathcal{T}_f = \tfrac{1}{2}m \int_0^L V_p^2 \, \mathrm{d}x_0 + \tfrac{1}{2}M \int_0^L V_f^2 \, \mathrm{d}x_0, \tag{5.13}$$

V_p and V_f being the corresponding velocities.

The potential energy comprises gravitational and strain energy components. In general, the gravitational energy depends on the distribution of mass (Fung 1969), and is written as $\mathcal{G} = \int \rho\phi(\xi)\, \mathrm{d}\mathcal{V}$, where ϕ is the gravitational potential per unit mass; in a uniform gravitational field it becomes $\mathcal{G} = \int \rho g \xi \, \mathrm{d}\mathcal{V}$, where g is the gravitational acceleration, ξ is a distance measured from a reference plane in a direction opposite to the gravitational field, and $\mathrm{d}\mathcal{V}$ is an elemental volume. Consequently, with the notation used here,

$$\mathcal{G} = -(m + M)\, g \int_0^L x \, \mathrm{d}x_0. \tag{5.14}$$

It is very important to define an exact form of the strain energy in the case of large deflections, correct to $\mathcal{O}(\epsilon^4)$. This problem is solved by Stoker (1968), with only one major (but not drastic) assumption: the strain is small even though the deflection can be large. His analysis finally leads to

$$\mathcal{V} = \tfrac{1}{2} \int_0^L [EA\varepsilon^2 + EI(1 + \varepsilon)^2 \kappa^2] \, \mathrm{d}x_0, \tag{5.15}$$

where x_0 represents the Lagrangian coordinate, A the cross-sectional area, I the moment of inertia and ε the axial strain.

5.2.3 The equation of motion of a cantilevered pipe

Consider a small segment of the pipe and the fluid. By definition, the velocity of the pipe element is

$$\mathbf{V}_p = \frac{\partial \mathbf{r}}{\partial t} = \dot{x}\mathbf{i} + \dot{z}\mathbf{k}, \tag{5.16}$$

and the velocity of the fluid element is $\mathbf{V}_f = \mathbf{V}_p + U\boldsymbol{\tau}$, where $U\boldsymbol{\tau}$ is the relative velocity of the fluid element with respect to the pipe element, $\boldsymbol{\tau}$ being the unit vector along s. For the cantilevered pipe, where the inextensibility condition is assumed to hold true, $\boldsymbol{\tau}$ has the form $\boldsymbol{\tau} = (\partial x/\partial s)\,\mathbf{i} + (\partial z/\partial s)\mathbf{k}$. Consequently,

$$\mathbf{V}_f = \left(\frac{\partial}{\partial t} + U\frac{\partial}{\partial s}\right)(x\mathbf{i} + z\mathbf{k}) \equiv \frac{\mathrm{D}\mathbf{r}}{\mathrm{D}t}, \tag{5.17}$$

where $\mathrm{D}/\mathrm{D}t$ is the material derivative of the fluid element (Section 3.3.3). By analogy, the accelerations of the pipe and of the fluid (used in Appendix G) are, respectively,

$$\mathbf{a}_p = \frac{\partial^2 \mathbf{r}}{\partial t^2}, \qquad \mathbf{a}_f = \frac{\mathrm{D}^2\mathbf{r}}{\mathrm{D}t^2}. \tag{5.18}$$

Hence, the total kinetic energy, $\mathcal{T}$, may be written as

$$\mathcal{T} = \tfrac{1}{2}m\int_0^L (\dot{x}^2 + \dot{z}^2)\,\mathrm{d}s + \tfrac{1}{2}M\int_0^L [(\dot{x} + Ux')^2 + (\dot{z} + Uz')^2]\,\mathrm{d}s, \tag{5.19}$$

where the dots and primes denote $\partial(\;)/\partial t$ and $\partial(\;)/\partial s$, respectively.

One important remark that ought to be made is that no variable term proportional to U^2 arises from expression (5.19) since, by expanding the integrand and by virtue of the inextensibility condition, one obtains only a constant term, $U^2x'^2 + U^2z'^2 = U^2$. This illustrates the importance of the right-hand side of statement (5.10), which will provide both linear and nonlinear components of the centrifugal force proportional to MU^2.

The variational operations on $\mathcal{T}$ lead to

$$\delta\int_{t_1}^{t_2}\mathcal{T}\,\mathrm{d}t = m\int_{t_1}^{t_2}\int_0^L(\dot{x}\,\delta\dot{x} + \dot{z}\,\delta\dot{z})\,\mathrm{d}s\,\mathrm{d}t + M\int_{t_1}^{t_2}\int_0^L[(\dot{x} + U\,x')(\delta\dot{x} + U\,\delta x')$$
$$+ (\dot{z} + Uz')(\delta\dot{z} + U\,\delta z')]\,\mathrm{d}s\,\mathrm{d}t.$$

Integrating by parts and noting that $x'\delta x' + z'\delta z' = 0$, one obtains

$$\begin{aligned}\delta\int_{t_1}^{t_2}\mathcal{T}\,\mathrm{d}t = &-\int_{t_1}^{t_2}\int_0^L[(m+M)\ddot{x} + M\,\dot{U}\,x' + 2MU\,\dot{x}']\,\delta x\,\mathrm{d}s\,\mathrm{d}t\\ &-\int_{t_1}^{t_2}\int_0^L[(m+M)\ddot{z} + M\,\dot{U}\,z' + 2MU\,\dot{z}']\,\delta z\,\mathrm{d}s\,\mathrm{d}t\\ &+MU\int_{t_1}^{t_2}[\dot{x}_L\,\delta x_L + \dot{z}_L\,\delta z_L]\,\mathrm{d}t,\end{aligned} \tag{5.20}$$

where $x_L = x(L)$ and $z_L = z(L)$ are the longitudinal and lateral displacements of the free end of the pipe.

The two components of the potential energy are derived next. Considering first the strain energy expression (5.15) with $\varepsilon = 0$, one can write

$$\delta \int_{t_1}^{t_2} \mathcal{V}\, \mathrm{d}t = \tfrac{1}{2} EI \iint \delta\,(\kappa^2)\, \mathrm{d}s\, \mathrm{d}t.$$

Utilization of the curvature expression (5.8) leads to

$$\begin{aligned}
\delta \int_{t_1}^{t_2} \mathcal{V}\, \mathrm{d}t &= \tfrac{1}{2} EI \int_{t_1}^{t_2}\int_0^L \delta[z''^2(1+z'^2)]\, \mathrm{d}s\, \mathrm{d}t + \mathcal{O}(\epsilon^5) \\
&= EI \int_{t_1}^{t_2}\int_0^L [(y''+z''\,z'^2)'' - (z''^2\,z')']\,\delta z\, \mathrm{d}s\, \mathrm{d}t + \mathcal{O}(\epsilon^5) \\
&= EI \int_{t_1}^{t_2}\int_0^L [z'''' + 4z'\,z''z''' + z''^3 + z''''\,z'^2]\,\delta z\, \mathrm{d}s\, \mathrm{d}t + \mathcal{O}(\epsilon^5). \qquad (5.21)
\end{aligned}$$

The gravitational potential (5.10) may be dealt with in a similar manner. However, since it will involve δx, a relationship between δx and δz needs to be found. This is done by taking variations of the first of (5.1), the inextensibility condition, yielding

$$\delta x' = -\frac{z'\,\delta z'}{\sqrt{1-z'^2}} = -z'\left(1+\tfrac{1}{2}z'^2\right)\delta z' + \mathcal{O}(\epsilon^4);$$

hence,

$$\delta x = -\int_0^s \left[z'\,\delta z' + \tfrac{1}{2}z'^3\,\delta z'\right] \mathrm{d}s. \qquad (5.22)$$

After integrating the right-hand side of (5.22) by parts and noting that $\delta z = 0$ at $s = 0$, one obtains

$$\delta x = -\left(z' + \tfrac{1}{2}z'^3\right)\delta z + \int_0^s \left(z'' + \tfrac{3}{2}z'^2 z''\right)\delta z\, \mathrm{d}s + \mathcal{O}(\epsilon^4). \qquad (5.23)$$

One can also prove quite easily (Semler 1991) that

$$\int_0^L g(s)\left(\int_0^s f(s)\,\delta z\, \mathrm{d}s\right)\mathrm{d}s = \int_0^L \left(\int_s^L g(s)\, \mathrm{d}s\right) f(s)\,\delta z\, \mathrm{d}s. \qquad (5.24)$$

Equation (5.24) is important, since terms of that form will arise from (5.23) in the process of relating δx to δz.

Now, using (5.23) and (5.24), the variation of the gravitational energy (5.14) is obtained:

$$\begin{aligned}
\delta \int_{t_1}^{t_2} \mathcal{G}\, \mathrm{d}t = -(m+M)\,g \int_{t_1}^{t_2}\int_0^L &\Big[-\left(z' + \tfrac{1}{2}z'^3\right)\delta z \\
&+ (L-s)\left(z'' + \tfrac{3}{2}z''\,z'^2\right)\delta z\Big]\, \mathrm{d}s\, \mathrm{d}t + \mathcal{O}(\epsilon^5). \qquad (5.25)
\end{aligned}$$

Applying next the variational procedure to the right-hand side (rhs) of Hamilton's principle, equation (5.10), leads to

$$\mathrm{rhs} = MU \int_{t_1}^{t_2} [(\dot{x}_L + U\,x_L')\,\delta x_L + (\dot{z}_L + U\,z_L')\delta z_L]\, \mathrm{d}t$$

$$= MU \int_{t_1}^{t_2} (\dot{x}_L\, \delta x_L + \dot{z}_L\, \delta z_L)\, \mathrm{d}t + MU^2 \int_{t_1}^{t_2} (x'_L\, \delta x_L + z'_L\, \delta z_L)\, \mathrm{d}t$$

$$= A + B. \tag{5.26}$$

The first term, A, cancels the last term in equation (5.20), while with the use of equations (5.1) and (5.23) B is found to be

$$B = MU^2 \iint \left[z'' + z'^2 z'' - z'' \int_s^L (z' z'')\, \mathrm{d}s \right] \delta z\, \mathrm{d}s\, \mathrm{d}t, \tag{5.27}$$

and hence contributes all the centrifugal-force terms.

Finally, after many transformations and manipulations, and recalling that $z = w$ as per equation (5.11), the general equation of motion is found to be

$$\begin{aligned}
&(m+M)\, \ddot{w} + 2MU\, \dot{w}'(1 + w'^2) + (m+M)\, g\, w' \left(1 + \tfrac{1}{2} w'^2\right) \\
&\quad + w'' \left[MU^2(1 + w'^2) + (M\dot{U} - (m+M)\, g)(L - s)\left(1 + \tfrac{3}{2} w'^2\right)\right] \\
&\quad + EI[w''''(1 + w'^2) + 4\, w'\, w''\, w''' + w''^3] \\
&\quad - w'' \left[\int_s^L \int_0^s (m+M)(\dot{w}'^2 + w'\, \ddot{w}')\, \mathrm{d}s\, \mathrm{d}s \right. \\
&\qquad \left. + \int_s^L \left(\tfrac{1}{2} M\dot{U} w'^2 + 2MU\, w'\, \dot{w}' + MU^2\, w'\, w''\right) \mathrm{d}s \right] \\
&\quad + w' \int_0^s (m+M)(\dot{w}'^2 + w'\ddot{w}')\, \mathrm{d}s = 0.
\end{aligned} \tag{5.28}$$

The Newtonian derivation of this equation is given in Appendix G.1.

5.2.4 The equation of motion for a pipe fixed at both ends

Here, as the inextensibility condition can no longer be applied, two equations are necessary: one in the x- and the other in the z-direction. Moreover, since both ends of the pipe are fixed, the right-hand side of expression (5.10) is now zero. Consequently, it is obvious that the contribution of the fluid forces is not the same as in the case of the cantilevered pipe.

When a bar is subjected to tension, the axial elongation is accompanied by a lateral contraction. Within the elastic range, the Poisson ratio ν is constant (Timoshenko & Gere 1961) and, for rubber-like materials, $\nu \simeq 0.5$. In the case where only a uniaxial load is applied to an elastic body, the change of unit volume is proportional to $1 - 2\nu$. Consequently, for rubber-like materials, the volume change due to uniaxial stress can be considered zero, i.e. they are incompressible. In the case of a pipe, for any initial volume of length $\mathrm{d}x_0$, this conservation of volume leads to $\mathrm{d}x_0\, S_0 = \mathrm{d}x_0(1+\varepsilon)\, S_1$, where S_1 represents the cross-sectional area of the pipe after elongation. For the incompressible fluid inside the pipe, one also has $U_0\, S_0 = U_1\, S_1$, with U_0 and U_1 being the flow velocities before and after elongation. Thus,

$$U_1(x_0) = U_0\, (S_0/S_1) = U_0\, (1+\varepsilon). \tag{5.29}$$

This shows that the *velocity of the fluid with respect to the pipe is no longer constant.* Hence, the absolute velocity is

$$\mathbf{V}_f = \mathbf{V}_p + U(x_0)\boldsymbol{\tau} = (\dot{x}\,\mathbf{i} + \dot{z}\,\mathbf{k}) + U(1+\varepsilon)\left(\frac{x'}{1+\varepsilon}\mathbf{i} + \frac{z'}{1+\varepsilon}\mathbf{k}\right),$$

where the prime denotes the derivative with respect to x_0. Consequently,

$$\mathbf{V}_f = \left(\frac{\partial}{\partial t} + U\frac{\partial}{\partial x_0}\right)\mathbf{r}. \tag{5.30}$$

Relationship (5.17) derived for a cantilevered pipe still holds, with the difference that the inextensibility condition is not valid here, so that U^2 terms in this case survive in the kinetic energy and are therefore not associated with the right-hand side of (5.10), which is zero! The total kinetic energy is given by

$$\mathcal{T} = \tfrac{1}{2}m\int_0^L (\dot{u}^2 + \dot{w}^2)\,\mathrm{d}x_0 + \tfrac{1}{2}M\int_0^L [(\dot{u} + U(1+u'))^2 + (\dot{w} + U\,w')^2]\,\mathrm{d}x_0. \tag{5.31}$$

For the case of non-rubber-like materials ($\nu \neq 0.5$), some additional words are necessary. The change of volume is no longer equal to zero, and $S_0/S_1 = 1/(1 - 2\,\nu\,\varepsilon)$. The fluid being incompressible, one obtains

$$U(x_0) = U_0\,(1 + 2\,\nu\,\varepsilon) = U_0(1+\varepsilon) + U_0\,\varepsilon\,(2\,\nu - 1) = U_1(x_0) + U_0\,\varepsilon\,(2\,\nu - 1),$$

i.e.

$$\frac{U(x_0) - U_1(x_0)}{U_0} = \varepsilon\,(2\,\nu - 1). \tag{5.32}$$

To fourth order, the strain ε is given by

$$\varepsilon = u' + \tfrac{1}{2}w'^2 + \mathcal{O}(\epsilon^4), \tag{5.33}$$

so that for a pipe of length $L = 1$, with $|u| \sim 0.01$, and $|w| \sim 0.1$, one obtains $|\varepsilon| \sim 1.5 \times 10^{-2}$. For $\nu = 0.4$ and 0.3, the error in the flow velocity associated with taking $\nu \simeq 0.5$ is 0.3% and 0.6%, respectively, which is of same order of magnitude as the error made by assuming the velocity profile to be uniform. Hence, equation (5.31) may still be considered valid.

The potential energy is considered next. To derive the strain energy, the axial strain is itself decomposed into two components: a steady-state strain due to an externally applied tension T_0 and pressurization P, and an oscillatory strain due to pipe oscillation. By reference to equation (5.15), this strain energy may be expressed as

$$\mathcal{V} = \tfrac{1}{2}EA\int_0^L \left(\frac{T_0 - P}{EA} + \varepsilon\right)^2 \mathrm{d}x_0 + \tfrac{1}{2}EI\int_0^L (1+\varepsilon)^2\,\kappa^2\,\mathrm{d}x_0.$$

By using (5.5), this is simplified to

$$\mathcal{V} = \tfrac{1}{2}EA\int_0^L \left(\frac{T_0 - P}{EA} + \varepsilon\right)^2 \mathrm{d}x_0 + \tfrac{1}{2}EI\int_0^L \left(\frac{\partial\theta}{\partial x_0}\right)^2 \mathrm{d}x_0. \tag{5.34}$$

Recalling that $u \sim \mathcal{O}(\epsilon^2)$, $w \sim \mathcal{O}(\epsilon)$, and using (5.6),

$$\left(\frac{\partial \theta}{\partial x_0}\right)^2 = w''^2 - 2\,w''^2\,u' - 2\,w''^2\,w'^2 - 2\,w'\,w''\,u'' + \mathcal{O}(\epsilon^5)$$

is obtained. Moreover, ε is given by equation (5.33).

The expression for gravitational energy is the same as in the case of the cantilevered pipe, i.e.

$$\mathcal{G} = -(m+M)\,g \int_0^L (x_0 + u)\,\mathrm{d}x_0. \tag{5.35}$$

The final equations of motions are obtained once again via application of variational techniques, this time with two independent variants, δu and δw. After many integrations by parts, one finally obtains

$$\begin{aligned}
&(m+M)\,\ddot{u} + M\dot{U} + 2MU\,\dot{u}' + MU^2\,u'' + M\dot{U}\,u' - EAu'' \\
&\quad - EI(w''''\,w' + w''\,w''') + (T_0 - P - EA)w'\,w'' - (m+M)g = 0, \qquad (5.36\text{a}) \\
&(m+M)\ddot{w} + M\dot{U}\,w' + 2MU\,\dot{w}' + MU^2\,w'' - (T_0 - P)w'' + EI\,w'''' \\
&\quad - EI(3\,u'''\,w'' + 4\,u''\,w''' + 2\,u'\,w'''' + w'\,u'''' + 2\,w'^2\,w'''' + 8\,w'\,w''\,w''' + 2\,w''^3) \\
&\quad + (T_0 - P - EA)\left(u''\,w' + u'\,w'' + \tfrac{3}{2}\,w'^2\,w''\right) = 0, \qquad (5.36\text{b})
\end{aligned}$$

where one now has two independent equations, instead of the one for a cantilevered pipe. A Newtonian derivation is outlined in Appendix G.2.

5.2.5 Boundary conditions

Using variational methods, it is straightforward to derive boundary conditions for the different cases considered. For the cantilevered pipe, the boundary conditions are the same as for the linear case: $w(0) = w'(0) = 0$ and $w''(L) = w'''(L) = 0$. For the pipe fixed at both ends, it is obvious that $u(0) = w(0) = u(L) = w(L) = 0$; in addition, if the pipe is simply-supported, one obtains $w''(0) = w''(L) = 0$, while for the clamped–clamped pipe, $w'(0) = w'(L) = 0$. Only two boundary conditions are necessary for u.

5.2.6 Dissipative terms

Dissipative terms have to be added to complete the equations. This can be done by assuming that the internal dissipation of the pipe material is viscoelastic and of the Kelvin–Voigt type (Snowdon 1968), i.e. that it is represented by $\sigma = E\,\varepsilon + E^*\,\dot{\varepsilon}$, where σ is the stress and ε the strain. Following then the approach used by Stoker (1968), the strain energy is modified, providing additional terms that can be written as

$$E \rightarrow E\left[1 + a\left(\frac{\partial}{\partial t}\right)\right], \tag{5.37}$$

where a is the coefficient of Kelvin–Voigt damping in the material. Therefore, in equations (5.28) and (5.36a,b), EI may be replaced by $EI(1 + a\,\partial/\partial t)$ and EA by $EA(1 +$

$a\,\partial/\partial t$). Moreover, for reasons of simplicity, and because, in any case, the Kelvin–Voigt dissipation is only an approximation, the dissipative terms are often assumed to be linear.

The damping associated with frictional forces due to surrounding air is neglected here.

5.2.7 Dimensionless equations

(a) Full equations for cantilevered pipes

Considering the cantilevered pipe first and introducing the same nondimensional quantities as in the linear case,

$$\xi = \frac{s}{L}, \qquad \eta = \frac{w}{L}, \qquad \tau = \left(\frac{EI}{m+M}\right)^{1/2} \frac{t}{L^2}, \qquad \alpha = \left(\frac{EI}{m+M}\right)^{1/2} \frac{a}{L^2},$$
$$\mathcal{U} = \left(\frac{M}{EI}\right)^{1/2} U\,L,^{\dagger} \qquad \gamma = \frac{m+M}{EI}\,L^3\,g, \qquad \beta = \frac{M}{m+M}, \tag{5.38}$$

one may rewrite equation (5.28), with (5.37) taken into account, in dimensionless form as follows:

$$\begin{aligned}
&\alpha\dot{\eta}'''' + \eta'''' + \ddot{\eta} + 2\,\mathcal{U}\,\sqrt{\beta}\,\dot{\eta}'(1+\eta'^2) \\
&\quad + \eta''\left[\mathcal{U}^2(1+\eta'^2) + (\dot{\mathcal{U}}\,\sqrt{\beta} - \gamma)(1-\xi)\left(1+\tfrac{3}{2}\,\eta'^2\right)\right] \\
&\quad + \gamma\,\eta'\left(1+\tfrac{1}{2}\,\eta'^2\right) + \left(1+\alpha\,\frac{\partial}{\partial t}\right)[\eta''''\,\eta'^2 + 4\,\eta'\,\eta''\,\eta''' + \eta''^3] \\
&\quad - \eta''\left[\int_\xi^1\int_0^\xi (\dot{\eta}'^2 + \eta'\,\ddot{\eta}')\,\mathrm{d}\xi\,\mathrm{d}\xi + \int_\xi^1 \left(\tfrac{1}{2}\dot{\mathcal{U}}\sqrt{\beta}\,\eta'^2 + 2\,\mathcal{U}\,\sqrt{\beta}\,\eta'\,\dot{\eta}' + \mathcal{U}^2\,\eta'\,\eta''\right)\mathrm{d}\xi\right] \\
&\quad + \eta'\int_0^\xi (\dot{\eta}'^2 + \eta'\,\ddot{\eta}')\,\mathrm{d}\xi = 0.
\end{aligned} \tag{5.39}$$

Of particular interest is the appearance in (5.39) of some nonlinear inertial terms which render the equations nonstandard and which, as discussed later, create difficulties in their solution.

(b) A modified approximate equation for cantilevered pipes

To circumvent difficulties associated with the nonlinear inertial terms, an approximate equation is obtained in which these terms are converted to equivalent stiffness and velocity-dependent terms via a perturbation technique (Li & Païdoussis 1994; Semler *et al.* 1994). To this end, equation (5.39) is rewritten as

$$L(\eta) + N_1(\eta) + N_2(\eta) = 0, \tag{5.40}$$

[†] Throughout Section 5.2, the dimensionless flow velocity is denoted by $\mathcal{U}$, to avoid confusion with the axial displacement, u. Later on in this chapter, however, for the sake of uniformity with the foregoing chapters, $\mathcal{U}$ will be replaced by u.

where $L(\eta)$ represents the linear terms, $N_1(\eta)$ the nonlinear terms not involving integrals, and $N_2(\eta)$ those that do and wherein all nonlinear inertial terms lie. The terms involving α are omitted for simplicity in the derivation.

It is noted that $L(\eta) \sim \mathcal{O}(\epsilon)$ and $N_1(\eta)$ and $N_2(\eta) \sim \mathcal{O}(\epsilon^3)$. After some manipulation of $L(\eta)$, starting with

$$\int_0^\xi \eta' L(\eta')\, \mathrm{d}\xi = \int_0^\xi \eta' [L(\eta)]'\, \mathrm{d}\xi \sim \mathcal{O}(\epsilon^2),$$

one obtains

$$\int_0^\xi \eta' \ddot{\eta}'\, \mathrm{d}\xi = -\int_0^\xi \Big[2\mathcal{U}\sqrt{\beta}\,\eta'\,\dot{\eta}'' + \eta'\,\eta'''\left(\mathcal{U}^2 + (\dot{\mathcal{U}}\sqrt{\beta} - \gamma)(1-\xi)\right) + \eta'\,\eta''(2\gamma - \dot{\mathcal{U}}\sqrt{\beta}) + \eta'\,\eta'''''\Big]\, \mathrm{d}\xi + \mathcal{O}(\epsilon^2), \tag{5.41}$$

which transforms one of the nonlinear inertial terms in $N_2(\eta)$. Integration of (5.41) from ξ to 1 yields the other nonlinear inertial term. These two terms are replaced in $N_2(\eta)$ to obtain, after some long but straightforward algebra,

$$\ddot{\eta} + 2\mathcal{U}\sqrt{\beta}\,\dot{\eta}' + \eta''\left(\mathcal{U}^2 + (\dot{\mathcal{U}}\sqrt{\beta} - \gamma)(1-\xi)\right) + \gamma\,\eta' + \eta'''' + N(\eta) = 0, \tag{5.42}$$

where

$$\begin{aligned} N(\eta) &= 2\mathcal{U}\sqrt{\beta}\,\dot{\eta}'\,\eta'^2 + \eta''\left(\mathcal{U}^2 + \tfrac{3}{2}(\dot{\mathcal{U}}\sqrt{\beta} - \gamma)(1-\xi)\right)\eta'^2 + \tfrac{1}{2}(\dot{\mathcal{U}}\sqrt{\beta} - \gamma)\eta'^3 \\ &\quad + 3\eta'\eta''\eta''' + \eta''^3 + \eta'\int_0^\xi \Big\{\dot{\eta}'^2 - 2\mathcal{U}\sqrt{\beta}\,\eta'\,\dot{\eta}'' - \eta'\,\eta'''\Big[\mathcal{U}^2 + (\dot{\mathcal{U}}\sqrt{\beta} - \gamma)(1-\xi)\Big] \\ &\quad + \eta''\,\eta''''\Big\}\, \mathrm{d}\xi - \eta''\int_\xi^1\int_0^\xi \Big\{\dot{\eta}'^2 - 2\mathcal{U}\sqrt{\beta}\,\eta'\,\dot{\eta}'' \\ &\quad - \eta'\,\eta'''\Big[\mathcal{U}^2 + (\dot{\mathcal{U}}\sqrt{\beta} - \gamma)(1-\xi)\Big] + \eta''\,\eta''''\Big\}\, \mathrm{d}\xi\, \mathrm{d}\xi \\ &\quad - \eta''\int_\xi^1 \Big\{(\dot{\mathcal{U}}\sqrt{\beta} - \gamma)\eta'^2 + 2\mathcal{U}\sqrt{\beta}\,\eta'\,\dot{\eta}' + \mathcal{U}^2\,\eta'\,\eta'' + \eta''\,\eta'''\Big\}\, \mathrm{d}\xi. \end{aligned} \tag{5.43}$$

Hence, the transformed equation is correct to $\mathcal{O}(\epsilon^3)$, as is equation (5.39). However, in view of the additional approximations introduced, this is a more approximate equation, albeit of the same order as the original.

(c) Equations for pipes with fixed ends

For pipes with both ends fixed, some additional nondimensional quantities need be introduced, as follows:

$$\Gamma = \frac{T_0 L^2}{EI}, \qquad \mathcal{A} = \frac{EAL^2}{EI}, \qquad \Pi = \frac{PL^2}{EI}, \tag{5.44}$$

together with the dimensionless displacements

$$\overline{u} = \frac{u}{L} \qquad \text{and} \qquad \overline{w} = \frac{w}{L}, \tag{5.45}$$

where $\overline{w} \equiv \eta$, but the barred quantity is used here for 'symmetry' with $\overline{u}$. Hence, equations (5.36a,b) may be written as follows:

$$\begin{aligned} &\ddot{\overline{u}} + \dot{\mathcal{U}}\sqrt{\beta} + 2\mathcal{U}\sqrt{\beta}\dot{\overline{u}}' + \mathcal{U}^2\overline{u}'' + \dot{\mathcal{U}}\sqrt{\beta}\overline{u}' - \mathcal{A}\overline{u}'' \\ &\quad - (\overline{w}''''\overline{w}' + \overline{w}''\overline{w}''') + (\Gamma - \mathcal{A} - \Pi)\overline{w}'\overline{w}'' - \gamma = 0, \end{aligned} \tag{5.46a}$$

$$\begin{aligned} &\ddot{\overline{w}} + \dot{\mathcal{U}}\sqrt{\beta}\overline{w}' + 2\mathcal{U}\sqrt{\beta}\dot{\overline{w}}' + \mathcal{U}^2\overline{w}'' - (\Gamma - \Pi)\overline{w}'' + \overline{w}'''' \\ &\quad - [3\overline{u}'''\overline{w}'' + 4\overline{u}''\overline{w}''' + 2\overline{u}'\overline{w}'''' + \overline{w}'\overline{u}'''' + 2\overline{w}'^2\overline{w}'''' + 8\overline{w}'\overline{w}''\overline{w}''' + 2\overline{w}''^3] \\ &\quad + (\Gamma - \mathcal{A} - \Pi)\left(\overline{u}''\overline{w}' + \overline{u}'\overline{w}'' + \tfrac{3}{2}\overline{w}'^2\overline{w}''\right) = 0. \end{aligned} \tag{5.46b}$$

Note that the dissipative terms have been omitted for clarity, and that the nonlinear inertial terms are not present in the current form of the equations. In fact, the only real penalty incurred for the absence of nonlinear inertial terms is that one has to deal with two equations, instead of just one.

5.2.8 Comparison with other equations for cantilevers

The nonlinear equations of motion obtained by different researchers are described and compared in some detail, here for cantilevered pipes and in Section 5.2.9 for pipes with fixed ends. In order to get a more 'comparable' set of equations, a standardization of the notation has been imposed.

(a) Bourrières' work

This work is very original, all the more so since it was written in 1939. Bourrières (1939) studied the case of planar motion of two interacting strings, one of them moving with respect to the other. The pipe and the fluid represented by the strings are assumed to be inextensible, and the string representing the fluid is supposed to be infinitely flexible. The equations of motion of the pipe and the fluid are derived via the force-balance method. The relationship between the shearing force Q and the bending moment $\mathcal{M}$, together with the condition of inextensibility, provides the nonlinear terms. Seven equations with nine parameters are obtained, two of which are independent, with coordinate s and time t as the two independent variables. After some algebraic manipulations, the fluid friction force is eliminated, yielding the following five equations:

$$\begin{aligned} &[(T+\Theta)x']' - (Qz')' - (m+M)\ddot{x} - 2MU\dot{x}' - MU^2x'' = 0, \\ &[(T+\Theta)z']' - (Qx')' - (m+M)\ddot{z} - 2MU\dot{z}' - MU^2z'' = 0, \\ &x'^2 + z'^2 = 1, \qquad Q = -\mathcal{M}', \qquad \mathcal{M} = EI(x'z'' - z'x''), \end{aligned} \tag{5.47}$$

where T and Θ represent the tension in the pipe and the negative of the pressure force in the fluid, respectively, and $(\ )' = \partial(\ )/\partial s$.

In his study, Bourrières considers only the linear case. However, his approach, if pursued far enough, would have led him eventually to expressions similar to those derived in Section 5.2.3. In fact, the only difference between equations (5.47) and those of Section 5.2.3 lies in the absence of gravity and time-varying flow terms, not considered by Bourrières; this makes his work irreproachable.

The remaining task would be to combine all the five equations of (5.47) into one, and to compare it with equation (5.39). This is not done here since it has effectively already been done by Rousselet & Herrmann (1981), to be discussed next.

(b) Rousselet and Herrmann's work

Rousselet & Herrmann (1981) derived the equations of motion in two different ways: by the force-balance method and the energy method. They obtain a set of equations, fairly close to the one found in Section 5.2.3, but with some minor differences. Their first method follows closely Bourrières' work. Two differences are simply due to the addition of gravity forces and the assumption that unsteady flow velocity effects may be present.

Considering an element of the system (see Figure G.1), the application of Newton's law leads to

$$\begin{aligned}
&\frac{\partial}{\partial s}[(T-P)\cos\theta] - \frac{\partial}{\partial s}(Q\sin\theta) + (m+M)g \\
&\quad = (m+M)\frac{\partial^2 x}{\partial t^2} + M\dot{U}\cos\theta - \frac{MU^2}{R}\sin\theta - 2MU\frac{d\theta}{dt}\sin\theta, \\
&\frac{\partial}{\partial s}[(T-P)\sin\theta] + \frac{\partial}{\partial s}(Q\cos\theta) \\
&\quad = (m+M)\frac{\partial^2 z}{\partial t^2} + 2MU\frac{d\theta}{dt}\cos\theta + \frac{MU^2}{R}\cos\theta + M\dot{U}\sin\theta,
\end{aligned} \tag{5.48}$$

where R is the local radius of curvature. In these equations, $(T-P)$ represents the tangential forces and Q the shear force, and $\sin\theta$ and $\cos\theta$ are related to x and z by

$$\sin\theta = \frac{\partial z}{\partial s}, \qquad \cos\theta = \frac{\partial x}{\partial s}. \tag{5.49}$$

By means of the inextensibility condition and the definition of the curvature κ, one can also prove that

$$\frac{1}{R}\frac{\partial x}{\partial s} = \frac{\partial^2 z}{\partial s^2}, \qquad -\frac{1}{R}\frac{\partial z}{\partial s} = \frac{\partial^2 x}{\partial s^2}. \tag{5.50}$$

Substituting (5.49) and (5.50) into (5.48), one obtains

$$\begin{aligned}
&\frac{\partial}{\partial s}\left((T-P)\frac{\partial x}{\partial s}\right) - \frac{\partial}{\partial s}\left(Q\frac{\partial z}{\partial s}\right) + (m+M)g \\
&\quad = (m+M)\frac{\partial^2 x}{\partial t^2} + 2MU\frac{\partial^2 x}{\partial s\partial t} + MU^2\frac{\partial^2 x}{\partial s^2} + M\dot{U}\frac{\partial x}{\partial s}, \\
&\frac{\partial}{\partial s}\left((T-P)\frac{\partial z}{\partial s}\right) + \frac{\partial}{\partial s}\left(Q\frac{\partial x}{\partial s}\right) \\
&\quad = (m+M)\frac{\partial^2 z}{\partial t^2} + 2MU\frac{\partial^2 z}{\partial s\partial t} + MU^2\frac{\partial^2 z}{\partial s^2} + M\dot{U}\frac{\partial z}{\partial s}.
\end{aligned} \tag{5.51}$$

In this form, the similarity with Bourrières' equations is self-evident. Note that κ and the condition of inextensibility have already been used implicitly. At this point, Rousselet & Herrmann proceed to reduce this set of equations into one. With the different relationships defined in Rousselet (1975), and after some manipulation, the nondimensional equation is obtained, which differs from (5.39) only in two nonlinear terms involving the unsteady velocity; they arise from an error in the use of the following relationship:

$$\int_0^L F(x)\left(\int_0^x (\tan\theta)'\right)\delta w \,\mathrm{d}x\,\mathrm{d}x = \int_0^L \left(\int_x^L F(x)\,\mathrm{d}x\right)(\tan\theta)'\,\delta w\,\mathrm{d}x. \tag{5.52}$$

This relationship is exact, but in the order analysis, if F is of order 0, then $\tan\theta$ must be approximated to the third order; this is not done by Rousselet & Herrmann. As explained in Section 5.2.1, this relationship [derived in Section 5.2.3, equation (5.24)] has to be rigorous to order $\mathcal{O}(\epsilon^4)$.

Rousselet & Herrmann also consider the effects of fluid friction or of the related pressure drop, and derive a flow equation,

$$P_0 - \alpha M U^2 + \int_0^L (M\,g\,x' - M\,\dot{U})\,\mathrm{d}s - \int_0^L M(\ddot{x}\,x' + \ddot{z}\,z')\,\mathrm{d}s = 0, \tag{5.53}$$

where P_0 is the compressive force acting on the fluid cross-section at $s = 0$, and αMU^2 is the sum of the friction forces between the fluid and the pipe (α is a constant which depends on the roughness of the pipe). The two partial differential equations are coupled through the nonlinear terms. Thus, instead of considering the flow velocity as constant, the upstream pressure (in a large reservoir) is assumed constant instead, as first proposed by Roth (1964).

(c) Sethna, Bajaj and Lundgren's work

Lundgren, Sethna & Bajaj (1979) and Bajaj *et al*. (1980) derived equations of motion by the Newtonian (force balance) method. The assumptions made are the same as in other work, but, from a mathematical point of view, every effort has been made to be as rigorous as possible. Their equations appear to be exact. They use the condition of inextensibility and the exact expression for curvature; in their derivation, all the nonlinearities come from the terms $(T_0 - P)$ and $EI\kappa^2$.

Lundgren *et al*. stopped their derivation at an early stage, without taking further advantage of the inextensibility condition. In their subsequent paper (Bajaj *et al*. 1980), some nonlinear terms are *apparently* missing, especially nonlinear velocity-dependent terms. In the form of an integrodifferential set of equations and neglecting, for the moment, the unsteady flow velocity, one may read [equation (5) in Bajaj *et al*. (1980)]

$$EI\,z'''' + 2M\,U\,\dot{z}' + M\,U^2 z'' + (m+M)\ddot{z} = \mathrm{NL}, \qquad \left(\frac{\partial x}{\partial s}\right)^2 + \left(\frac{\partial z}{\partial s}\right)^2 = 1, \tag{5.54}$$

where

$$\mathrm{NL} = -\tfrac{3}{2}EI\,\frac{\partial}{\partial s}[z'(x''^2 + z''^2)] - (m+M)\frac{\partial}{\partial s}\left(z'\int_s^L (x'\ddot{x} + z'\ddot{z})\,\mathrm{d}s\right).$$

At first glance these equations seem wrong (as no nonlinear velocity-dependent terms are present); however, if further simplification is carried out, equation (5.54) yields the correct form of governing equation in terms of z. The U and U^2 terms are actually 'hidden' in the nonlinear inertial term. Indeed, eliminating x through the condition of inextensibility leads to

$$\begin{aligned}(m+M)\ddot{z}(1-z'^2)+2MU\dot{z}'+MU^2z''+EI\left(z''''+3z'z''z'''+\tfrac{3}{2}z''^3\right)\\ +z'\int_0^s (m+M)(\dot{z}'^2+z'\ddot{z}')\,\mathrm{d}s\\ -z''\left(\int_s^L\int_0^s (m+M)(\dot{z}'^2+z'\ddot{z}')\,\mathrm{d}s\,\mathrm{d}s-\int_s^L (m+M)\ddot{z}z'\,\mathrm{d}s\right)=0.\end{aligned} \quad (5.55)$$

By multiplying by $(1+z'^2)$ throughout, keeping cubic nonlinear terms and replacing nonlinear inertial terms [cf. Section 5.2.7(b)], one may bring equation (5.55) with $z=w$ into the same form as (5.28).

Hence, this equation of motion is irreproachable. No nonlinear terms are missing, except for the gravity terms, since gravity has been neglected. However, the different steps from one equation to another are not very clear in the original derivation; also, Bajaj *et al*. (1980) use some implicit relationships of the curvature (Semler 1991), and the procedure for eliminating nonlinear inertial terms is not fully explained. Hence, verification is not easy.

Finally, similarly to Rousselet & Herrmann, Bajaj *et al*. also establish an equation for the flow, by considering a force balance on a fluid element, yielding

$$M\alpha(U_0^2-U^2)-M\int_0^L(\ddot{x}x'+\ddot{z}z')\,\mathrm{d}s-M\dot{U}L=0, \quad (5.56)$$

where U_0 is the constant flow velocity when the pipe is not in motion, α represents the resistance to the fluid motion (proportional to a friction factor) and αMU_0^2 represents the constant pressure force at the fixed end $s=0$ of the tube. It is found that α plays an important role in the dynamics, as discussed in Section 5.7.1.

(d) Ch'ng and Dowell's work

Ch'ng & Dowell (1979) obtained nonlinear equations of motion of a pipe conveying fluid by the energy method based on Hamilton's principle. An Eulerian approach is used to describe the dynamics of the system, and the flow is assumed to be steady. Using first only linear relationships, the well-known linear equation is found:

$$EI\,z''''+2MU\dot{z}'+MU^2z''-(M+m)g[(L-x)z']'+(m+M)\ddot{z}=0. \quad (5.57)$$

Ch'ng and Dowell then consider the nonlinear effects due to tension associated with the axial elongation of the pipe,

$$\int_0^L \mathrm{d}s=\int_0^L\sqrt{1+z'^2}\,\mathrm{d}x. \quad (5.58)$$

This relationship implies that the cantilevered pipe is extensible, which is an unusual but by no means erroneous assumption. By assuming the tube to be Hookean, an axial nonlinear force T is added to (5.57), giving rise to

$$-\left[\frac{EA}{2L}\int_0^L z'^2\,\mathrm{d}x\right] z''. \tag{5.59}$$

Because of the extensibility assumption, this equation cannot be compared with any of the previous ones. However, it should be mentioned that the strain is approximated to second order only, which does not fulfill the order considerations discussed in Section 5.2.1.

Additionally, Ch'ng and Dowell also consider a nonlinear relationship for the curvature. They use expression (5.9) for the curvature κ and the elastic strain energy

$$\mathcal{V} = \tfrac{1}{2}\int_0^L EI\,\kappa^2\,\mathrm{d}x, \tag{5.60}$$

and obtain additional terms, $-EI\,(3\,z'^2\,z'''' + 12\,z'\,z''\,z''' + 3\,z''^3)$. It is seen that expression (5.60) is not fully consistent with the strain energy derived by Stoker (1968), because the pipe is implicitly assumed to be extensible ($\varepsilon \neq 0$). Moreover, it is not obvious how the Eulerian description can be used with the energy method to derive nonlinear equations. Therefore, comparison cannot be made with other versions of the governing equations.

5.2.9 Comparison with other equations for pipes with fixed ends

In this section, two papers are discussed, representative of all the derivations for pipes fixed at both ends. Again, a standardization of the notation has been undertaken.

(a) Thurman and Mote's work

Thurman & Mote (1969b) were mainly concerned with the oscillations of bands of moving materials. They consider an axially-moving strip, simply-supported at its ends, in order to show how the axial motion could significantly reduce the applicability of linear analysis. This work is then extended to deal with pipes conveying fluid. The centreline being extensible, nonlinearities are associated with the axial elongation and the extension-induced tension in the tube. Therefore, the strain and the tension become

$$\varepsilon = \frac{T_0}{EA} + \sqrt{(1+u')^2 + w'^2} - 1, \qquad T = T_0 + EA\left(\sqrt{(1+u')^2 + w'^2} - 1\right). \tag{5.61}$$

Since a linear moment–curvature relationship and a linear approximation for the velocities are considered, the equations of motion obtained are

$$\begin{aligned} &EI\,w'''' - (T_0 - M\,U^2)w'' + 2M\,U\,\dot{w}' + (m+M)\ddot{w} \\ &\quad = (EA - T_0)\left(\tfrac{3}{2}w'^2\,w'' + u'\,w'' + u''\,w'\right), \\ &M\,\ddot{u} - EA\,u'' = (EA - T_0)\,w'\,w''. \end{aligned} \tag{5.62}$$

These are actually a simplified set of equations (5.36a,b). The differences come from the assumptions made: (i) no gravity forces, (ii) steady flow velocity, (iii) linear moment–curvature relationship, (iv) simple approximation of the fluid velocity; on the basis of these assumptions, the equations derived are correct.

(b) Holmes' work

Holmes (1977) was one of the first to develop the new tools of modern dynamics, and to introduce them into the study of fluid–structure dynamical systems; he was therefore less concerned with the derivation of the equations so much as he was with their *structure*. In that spirit, he considered only the major nonlinear terms associated with the deflection-induced tension in the pipe.

Starting from the linear equation obtained by Païdoussis & Issid (1974), Holmes adds the effect of the axial extension. To a first order approximation, the axial tension induced by lateral motions is

$$T = \sigma A = (E\,\varepsilon + E^*\,\dot{\varepsilon})A,$$

in which a Kelvin–Voigt viscoelastic material has been considered and where ε is the averaged axial strain defined by

$$\varepsilon = \frac{1}{2L}\int_0^L (z)'^2\,\mathrm{d}s.^\dagger$$

Thus, an axial force T is added to the linear equation, where

$$T = -\frac{EA}{2L}\int_0^L (z'^2)\,\mathrm{d}s + \frac{E^*A}{L}\int_0^L (z'\,\dot{z}')\,\mathrm{d}s. \tag{5.63}$$

The addition of this extra deflection-dependent axial force leads to one equation with two cubic nonlinear terms. This same axial force T (with $\eta = 0$) has also been obtained by Ch'ng & Dowell (1979) and by Namachchivaya (1989) through the energy method. In this case, however, attention must be paid to the order approximation, as already mentioned in Section 5.2.1.

It is noted that Holmes' version of the nonlinear equation is a single scalar one, as compared to the two equations derived in Section 5.2.7 and also by others. The implication is that, in Holmes' work, axial motion of the pipe is considered to be negligibly small and also that it is symmetric *vis-à-vis* the undeformed pipe shape.

5.2.10 Concluding remarks

The nonlinear equations of motion of a pipe conveying fluid have been derived in a simple manner, by both the energy and, in Appendix G, by the Newtonian method, following Semler *et al*. (1994). It is shown that the equations of motion of a cantilevered pipe and of a pipe fixed at both ends are fundamentally different. In the first case the pipe may be considered to be inextensible and nonlinearities are mainly geometric, related to the large curvature in the course of arbitrary motions. In the case of a pipe fixed at the ends, nonlinearities are mainly associated with stretching of the pipe and the nonlinear forces generated thereby.

Of the other derivations, some have been found to be absolutely correct, some correct for the purposes for which they are used, and some to contain errors or inconsistencies. Of the

† There are some errors in sign in a few intermediate steps in Holmes' derivation (1977); the final equation, however, is correct.

equations derived for *cantilevered pipes*, those by Lundgren *et al*. (1979) and Bajaj *et al*. (1980) are found to be absolutely correct, while those by Rousselet & Herrmann (1981) are found to be correct except for a small order-of-magnitude inconsistency. Furthermore, both sets contain a distinct refinement *vis-à-vis* those derived here: the flow velocity is not assumed to be constant; instead, the upstream pressure is taken to be constant, while the flow velocity generally varies with deformation. Of the equations derived for *pipes with fixed ends*, the set derived here is considered to be the only one available, correct to the same order as that for the cantilevered pipes. On the other hand, the simple equation derived originally by Holmes (1977) is correct as far as it goes and may be preferred in some cases because of its simplicity. It is of interest that the origin of the terms in the equations — even some of the linear terms — as well as the structure of the equations are distinctly different for pipes with both ends fixed as compared to cantilevered ones.

5.3 EQUATIONS FOR ARTICULATED SYSTEMS

Traditionally, articulated models of columns subjected to axial loading have been widely used as an aid in the study of their continuous, distributed parameter counterparts (Herrmann 1967). The same has occurred with pipes conveying fluid. For nonlinear dynamics this is particularly attractive, since many of the methods of nonlinear dynamics are best suited to low-dimensional discrete systems; with articulated systems, no questions need arise as to the adequacy of the Galerkin discretization of a continuous system: the *physical* system is discrete and may be low-dimensional to start with.

Most of the interesting dynamics is associated with cantilevered systems, and hence most of the research has been devoted to such systems. Furthermore, virtually all of that work has been confined to two-degree-of-freedom ($N = 2$) systems (Figure 5.2), and this despite the *caveat* (Section 3.8) that the dynamical behaviour of the $N = 2$ system is not generic with respect to N (Païdoussis & Deksnis 1970).

Two representative sets of equations are given here, both for cantilevered two-degree-of-freedom systems. The first set was derived by Rousselet & Herrmann (1977) by straightforward Newtonian methods via free-body diagrams and moment balances on the two segments of the pipe, yielding

$$\begin{aligned}
&(M+m)g\left(\tfrac{1}{2}l_1^2 + l_1 l_2\right)\sin\theta_1 \\
&\quad + (M+m)\left[-\tfrac{1}{2}l_1 l_2^2\dot{\theta}_2^2 \sin(\theta_2-\theta_1) + \ddot{\theta}_1\left(\tfrac{1}{3}l_1^3 + l_1^2 l_2\right) + \tfrac{1}{2}l_1 l_2^2\ddot{\theta}_2 \cos(\theta_2-\theta_1)\right] \\
&\quad + 2MU\left[\tfrac{1}{2}l_1^2\dot{\theta}_1 + l_1 l_2\dot{\theta}_2\cos(\theta_2-\theta_1)\right] + k_1\theta_1 - k_2(\theta_2-\theta_1) \\
&\quad + Ml_1 l_2\dot{U}\sin(\theta_2-\theta_1) + MU^2 l_1 \sin(\theta_2-\theta_1) = 0,
\end{aligned} \tag{5.64}$$

$$\begin{aligned}
&\tfrac{1}{2}(M+m)gl_2\sin\theta_2 + \tfrac{1}{3}(M+m)l_2^3\ddot{\theta}_2 + MUl_2^2\dot{\theta}_2 \\
&\quad + \tfrac{1}{2}(M+m)l_1 l_2^2\ddot{\theta}_1\cos(\theta_2-\theta_1) \\
&\quad + k_2(\theta_2-\theta_1) + \tfrac{1}{2}(M+m)l_1 l_2^2\dot{\theta}_1^2\sin(\theta_2-\theta_1) = 0,
\end{aligned} \tag{5.65}$$

where M and m are the mass of the fluid and of the pipe per unit length, and U the flow velocity; l_1 and l_2 are the lengths of the upper and lower segments of the pipe (Figure 5.2), and k_1 and k_2 are the stiffnesses of the interconnecting springs. The flow

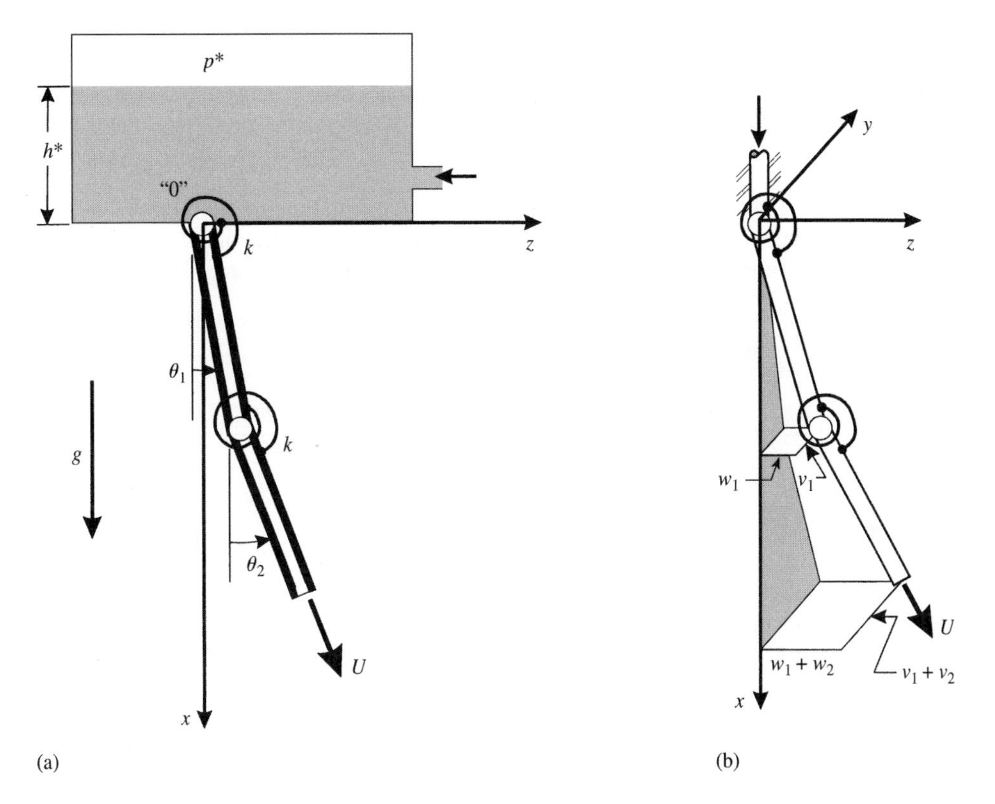

Figure 5.2 (a) A two-degree-of-freedom articulated pipe system conveying fluid, supplied by a constant-head tank and executing planar motions, as in Rousselet & Herrmann (1977); (b) an articulated system conveying fluid at a constant flow velocity U and executing three-dimensional motions, as in Bajaj & Sethna (1982a,b).

velocity is not assumed to be constant; rather, similarly to Roth (1964), the pressure is taken to be constant, at an upstream constant-head reservoir [Figure 5.2(a)]. Thus, a flow equation is also required, obtained by taking a force balance in the longitudinal (tangential) direction on a fluid element and subsequently integrating over the length of the pipe. This gives

$$p_0A_f - \overline{f}U^2 + Mg(l_1 \cos\theta_1 + l_2 \cos\theta_2) - M\dot{U}(l_1 + l_2)$$
$$+ M\dot{\theta}_1^2\left[\tfrac{1}{2}l_1^2 + l_1 l_2 \cos(\theta_2 - \theta_1)\right] - M\ddot{\theta}_1 l_1 l_2 \sin(\theta_2 - \theta_1) + \tfrac{1}{2}Ml_2^2\dot{\theta}_2 = 0, \quad (5.66)$$

where p_0A_f is the force due to pressure acting on the fluid at $x = 0$, A_f being the fluid area, and $\overline{f}U^2$ represents the force due to frictional losses along the pipe; in more conventional form, this term may be written as $(4fL/D_i)A_f(\frac{1}{2}\rho U^2)$, where f is a friction factor — see equation (2.98) and Massey (1979), for instance — which generally depends on wall roughness and Reynolds number, D_i is the internal diameter and L the total length. Equation (5.66) states that the pressure, as affected by the frictional losses and changing overall gravity head (the outlet pressure is always zero *vis-à-vis* the atmospheric), is equal to the mass × acceleration of the fluid (Massey 1979), this latter being equal to the longitudinal components of transverse and centrifugal accelerations of the pipe, plus the acceleration of the fluid relative to the pipe. The pressure p_0 is in turn related to the pressure in the reservoir p^* via

$$p_0 + \tfrac{1}{2}\rho U_0^2 = \rho g h^* + p^*, \tag{5.67}$$

where the subscript 0 refers to quantities at the entrance to the pipe; p^* and h^* are defined in Figure 5.2(a).

The same equations were derived by integration of the equations of motion of a continuously flexible pipe conveying fluid by Rousselet (1975). These equations may be rendered dimensionless through the following set of nondimensional parameters:

$$\begin{aligned} a &= \frac{l_1}{l_2}, \qquad \kappa = \frac{k_1}{k_2}, \qquad \beta \equiv \tfrac{1}{3}\overline{\beta} = \frac{M}{m+M}, \qquad \gamma = \left[\frac{m+M}{2k_2}\right] g l_2^2, \\ \tau &= \left[\frac{(m+M)l_2^3}{3k_2}\right]^{-1/2} \qquad t \equiv \overline{\Omega} t, \qquad u = \frac{U}{\overline{\Omega} l_2} = \left[\frac{(m+M)l_2}{3k_2}\right]^{1/2} U, \\ \lambda &= \frac{\overline{f}}{M}, \qquad \Pi_0 = \frac{p_0 A_f l_2}{k_2}, \qquad \overline{h}^* = \frac{h^*}{l_2}; \end{aligned} \tag{5.68}$$

and it is also noted that $\rho A_f = M$. These are more or less standard now [cf. Bajaj & Sethna (1982a,b)], but they are different from Rousselet & Herrmann's.

The second set of equations given here are Bajaj & Sethna's (1982a,b), who considered three-dimensional motions of the same system, i.e. motions in both the y- and z-directions, and a constant flow velocity [Figure 5.2(b)]. The Lagrangian procedure is utilized and hence Benjamin's equation (3.10). The generalized coordinates are the end-displacements of the two segments of the pipe, v_1 and $v_1 + v_2$ in the y-direction and w_1 and $w_1 + w_2$ in the z-direction. Hence, the kinetic and potential energies are given by

$$\begin{aligned} \mathcal{T} &= \tfrac{1}{6}(m+M)(l_1 + 3l_2)(\dot{v}_1^2 + \dot{w}_1^2 + \dot{u}_1^2) + \tfrac{1}{6}(m+M)l_2(\dot{v}_2^2 + \dot{w}_2^2 + \dot{u}_2^2) \\ &\quad + \tfrac{1}{2}MU^2(l_1 + l_2) + \tfrac{1}{2}(m+M)l_2(\dot{v}_1\dot{v}_2 + \dot{w}_1\dot{w}_2 + \dot{u}_1\dot{u}_2) \\ &\quad + MU(\dot{v}_1 v_2 + \dot{w}_1 w_2 + \dot{u}_1 u_2) \end{aligned} \tag{5.69}$$

and

$$\mathcal{V} = (m+M)g\left[\left(\tfrac{1}{2}l_1 + l_2\right)(l_1 - u_1) + \tfrac{1}{2}l_2(l_2 - u_2)\right] + \tfrac{1}{2}(k_1\phi_1^2 + k_2\phi_2^2), \tag{5.70}$$

where $u_1^2 = l_1^2 - (v_1^2 + w_1^2)$ and $u_2^2 = l_2^2 - (v_2^2 + w_2^2)$; ϕ_1 is the acute angle between the upper pipe and the x-axis, while ϕ_2 that between the two pipes,

$$\begin{aligned} \phi_1 &= \sin^{-1}[(v_1^2 + w_1^2)^{1/2}/l_1], \\ \phi_2 &= \cos^{-1}[\{v_1 v_2 + w_1 w_2 + [(l_1^2 - v_1^2 - w_1^2)(l_2^2 - v_2^2 - w_2^2)]^{1/2}\}/l_1 l_2]. \end{aligned} \tag{5.71}$$

Furthermore, the position vector $\mathbf{R}_L$ and the tangent vector $\boldsymbol{\tau}_L$, defined in Figure 3.1(d) are given by

$$\begin{aligned} \mathbf{R}_L &= (u_1 + u_2)\mathbf{i} + (v_1 + v_2)\mathbf{j} + (w_1 + w_2)\mathbf{k}, \\ \boldsymbol{\tau}_L &= (u_2\mathbf{i} + v_2\mathbf{j} + w_2\mathbf{k})/l_2, \end{aligned} \tag{5.72}$$

where $\mathbf{i}$, $\mathbf{j}$, and $\mathbf{k}$ are the unit vectors along the x-, y- and z-axes, respectively.

These equations are rendered dimensionless with the aid of relations (5.68) and the following additional ones:

$$\eta_1 = \frac{v_1}{l_1}, \quad \eta_2 = \frac{v_2}{l_2}; \qquad \zeta_1 = \frac{w_1}{l_1}, \quad \zeta_2 = \frac{w_2}{l_2}. \tag{5.73}$$

Substitution into (3.10) then leads to the following four dimensionless equations:[†]

$$\begin{aligned} a^2(a+3)\ddot{\eta}_1 &+ \tfrac{3}{2}a\ddot{\eta}_2 + a^2\overline{\beta}u\dot{\eta}_1 + 2a\overline{\beta}u\dot{\eta}_2 + a\overline{\beta}u^2(\eta_2 - \eta_1) + a(a+2)\gamma\eta_1 + \kappa\eta_1 + (\eta_1 - \eta_2) \\ &= -a^2(a+3)\eta_1(\dot{\eta}_1^2 + \dot{\zeta}_1^2 + \eta_1\ddot{\eta}_1 + \zeta_1\ddot{\zeta}_1) - \tfrac{3}{2}a\eta_1(\dot{\eta}_2^2 + \dot{\zeta}_2^2 + \eta_2\ddot{\eta}_2 + \zeta_2\ddot{\zeta}_2) \\ &\quad - 2a\overline{\beta}u\eta_1(\eta_2\dot{\eta}_2 + \zeta_2\dot{\zeta}_2) + \eta_1(\eta_1^2 + \zeta_1^2)\left\{-\tfrac{1}{2}a(a+2)\gamma - \tfrac{2}{3}\kappa - \tfrac{1}{2} + \tfrac{1}{2}a\overline{\beta}u^2\right\} \\ &\quad + \eta_1(\eta_2^2 + \zeta_2^2)\left\{\tfrac{1}{2} - \tfrac{1}{2}\overline{\beta}au^2\right\} - \tfrac{1}{6}(\eta_1 - \eta_2)\{(\eta_1 - \eta_2)^2 + (\zeta_1 - \zeta_2)^2\} \\ &\quad - \overline{\beta}ua^2\eta_1(\eta_1\dot{\eta}_1 + \zeta_1\dot{\zeta}_1) + \mathcal{O}(|\boldsymbol{\eta}|^5, |\boldsymbol{\zeta}|^5), \end{aligned} \tag{5.74}$$

$$\begin{aligned} a^2(a+3)\ddot{\zeta}_1 &+ \tfrac{3}{2}a\ddot{\zeta}_2 + \overline{\beta}ua^2u\dot{\zeta}_1 + 2a\overline{\beta}u\dot{\zeta}_2 + a\overline{\beta}u^2(\zeta_2 - \zeta_1) + a(a+2)\gamma\zeta_1 + \kappa\zeta_1 + (\zeta_1 - \zeta_2) \\ &= -a^2(a+3)\zeta_1(\dot{\eta}_1^2 + \dot{\zeta}_1^2 + \eta_1\ddot{\eta}_1 + \zeta_1\ddot{\zeta}_1) - \tfrac{3}{2}a\zeta_1(\dot{\eta}_2^2 + \dot{\zeta}_2^2 + \eta_2\ddot{\eta}_2 + \zeta_2\ddot{\zeta}_2) \\ &\quad - 2a\overline{\beta}u\zeta_1(\eta_2\dot{\eta}_2 + \zeta_2\dot{\zeta}_2) + \zeta_1(\eta_1^2 + \zeta_1^2)\left\{-\tfrac{1}{2}a(a+2)\gamma - \tfrac{2}{3}\kappa - \tfrac{1}{2} + \tfrac{1}{2}a\overline{\beta}u^2\right\} \\ &\quad + \zeta_1(\eta_2^2 + \zeta_2^2)\left\{\tfrac{1}{2} - \tfrac{1}{2}\overline{\beta}au^2\right\} - \tfrac{1}{6}(\zeta_1 - \zeta_2)\{(\eta_1 - \eta_2)^2 + (\zeta_1 - \zeta_2)^2\} \\ &\quad - \overline{\beta}ua^2\zeta_1(\eta_1\dot{\eta}_1 + \zeta_1\dot{\zeta}_1) + \mathcal{O}(|\boldsymbol{\eta}|^5, |\boldsymbol{\zeta}|^5), \end{aligned} \tag{5.75}$$

$$\begin{aligned} \tfrac{3}{2}a\ddot{\eta}_1 &+ \ddot{\eta}_2 + \overline{\beta}u\dot{\eta}_2 + \gamma\eta_2 + (\eta_2 - \eta_1) \\ &= -\eta_2[(\dot{\eta}_2^2 + \dot{\zeta}_2^2 + \eta_2\ddot{\eta}_2 + \zeta_2\ddot{\zeta}_2) + \tfrac{3}{2}a(\dot{\eta}_1^2 + \dot{\zeta}_1^2 + \eta_1\ddot{\eta}_1 + \zeta_1\ddot{\zeta}_1)] \\ &\quad - \tfrac{1}{2}(1+\gamma)\eta_2(\eta_2^2 + \zeta_2^2) + \tfrac{1}{2}\eta_2(\eta_1^2 + \zeta_1^2) + \tfrac{1}{6}(\eta_1 - \eta_2)\{(\eta_1 - \eta_2)^2 + (\zeta_1 - \zeta_2)^2\} \\ &\quad - \overline{\beta}u\eta_2(\eta_2\dot{\eta}_2 + \zeta_2\dot{\zeta}_2) + \mathcal{O}(|\boldsymbol{\eta}|^5, |\boldsymbol{\zeta}|^5), \end{aligned} \tag{5.76}$$

$$\begin{aligned} \tfrac{3}{2}a\ddot{\zeta}_1 &+ \ddot{\zeta}_2 + \overline{\beta}u\dot{\zeta}_2 + \gamma\zeta_2 + (\zeta_2 - \zeta_1) \\ &= -\zeta_2[(\dot{\eta}_2^2 + \dot{\zeta}_2^2 + \eta_2\ddot{\eta}_2 + \zeta_2\ddot{\zeta}_2) + \tfrac{3}{2}a(\dot{\eta}_1^2 + \dot{\zeta}_1^2 + \eta_1\ddot{\eta}_1 + \zeta_1\ddot{\zeta}_1)] \\ &\quad - \tfrac{1}{2}(1+\gamma)\zeta_2(\eta_2^2 + \zeta_2^2) + \tfrac{1}{2}\zeta_2(\eta_1^2 + \zeta_1^2) + \tfrac{1}{6}(\zeta_1 - \zeta_2)\{(\eta_1 - \eta_2)^2 + (\zeta_1 - \zeta_2)^2\} \\ &\quad - \overline{\beta}u\zeta_2(\eta_2\dot{\eta}_2 + \zeta_2\dot{\zeta}_2) + \mathcal{O}(|\boldsymbol{\eta}|^5, |\boldsymbol{\zeta}|^5). \end{aligned} \tag{5.77}$$

5.4 METHODS OF SOLUTION AND ANALYSIS

With a few rare exceptions,[‡] no *general* analytical solutions of the nonlinear equations of motion are possible. Therefore, recourse has to be taken to specialized analytical, semi-analytical and numerical techniques. Here, the classification proposed by Nayfeh

---

[†]With some corrections *vis-à-vis* Bajaj & Sethna (1982a) — Bajaj (1998).

[‡]For example, for the nonlinear equation of a simple pendulum, via elliptic integrals.

(1985) will be described in abbreviated form. However, before doing so, let us first distinguish between implicit and explicit forms of the equations to be solved.

Presuming that the equations to be solved are either discrete or discretized, they can be expressed as a set of second-order *implicit* nonlinear equations of the type

$$\mathbf{M}\ddot{\mathbf{x}} + \mathbf{C}\dot{\mathbf{x}} + \mathbf{K}\mathbf{x} = \mathbf{F}(\mathbf{x}, \dot{\mathbf{x}}, \ddot{\mathbf{x}}, t), \tag{5.78}$$

with appropriate initial conditions, $\mathbf{x}(0)$ and $\dot{\mathbf{x}}(0)$; $\mathbf{M}$, $\mathbf{C}$ and $\mathbf{K}$ are $N \times N$ matrices and all nonlinearities are included in $\mathbf{F}$. This equation is said to be implicit because of the presence in $\mathbf{F}$ of *nonlinear* inertial terms, i.e. terms involving $\ddot{\mathbf{x}}$, which cannot be removed or transformed.

In many cases, it is possible to express (5.78) as an *explicit* relation

$$\dot{\mathbf{y}} = \tilde{\mathbf{F}}(\mathbf{y}, t), \qquad \mathbf{y}(0) = \mathbf{y}_0, \tag{5.79}$$

which renders solution easier. However, when nonlinear inertial terms are present in $\mathbf{F}$, this transformation into (5.79) may not be possible, and means for the direct solution of (5.78) must be sought.

With this in mind, the various methods available for solving equations (5.78) or (5.79) will now be described.

Irrational analytical methods entail the simplification of the equations to be solved by neglecting or approximating various terms, e.g. by the use of small-deflection or small-angle approximations. Hence, such solutions are valid over a small range of parameters or for small deviations from the state of equilibrium.

Rational analytic methods, such as perturbation methods (Hagedorn 1981; Nayfeh & Mook 1979), the method of averaging (Hagedorn 1981; Nayfeh 1981) and its precursor the Krylov & Bololiubov method (Minorsky 1962; Nayfeh 1973), and the method of multiple scales (Nayfeh & Mook 1979; Nayfeh 1985), achieve solution by an asymptotic expansion or perturbation of the original set of equations, in terms of a small parameter ϵ ($\epsilon \ll 1$) which is either present in the equations *ab initio* or artificially introduced. Hence these methods are also known as 'small-parameter techniques', and they involve the sequential solution of simplified sets of equations, in which terms of order ϵ^{m+1} are disregarded while constructing the mth approximation. The method of averaging is described in Appendix F.4.

Numerical time-difference methods (Gear 1971; Lambert 1973; Press *et al*. 1992) are based on approximating the solution by its value at a sequence of discrete times. These methods have been developed mostly on the assumption that equation (5.78) may be rewritten as (5.79).† If this can be done, one can distinguish single-step and multistep methods of solution. The Runge–Kutta method is an example of the former; it requires the values of $\mathbf{x}$ and $\dot{\mathbf{x}}$ at time t_n, in order to compute the solution at t_{n+1}. Multistep or 'k-step' methods, e.g. that of Adams–Bashford–Moulton, accumulate information for the values of $\mathbf{x}$ and $\dot{\mathbf{x}}$ at $t_n, t_{n-1}, \ldots, t_{n-k}$ to proceed to the next step.

Combined analytical–numerical methods, such as the Rayleigh–Ritz, Galerkin (Meirovitch 1967) or harmonic balance methods (Hagedorn 1981; Nayfeh 1981) require an initial assumption as to the form of the approximate solution. The solution is typically expressed in the form of series, e.g. power, Taylor, Chebyshev, Fourier or Legendre series.

†The slightly more approximate form of the equation of motion of Section 5.2.7(b) has specifically been obtained to take advantage of this.

The assumed form involves coefficients determined by imposing minimizing conditions (Ritz–Galerkin method) or orthogonality conditions (Galerkin and harmonic balance methods), which effectively converts the nonlinear differential equation into a set of nonlinear algebraic equations, solved iteratively. The incremental harmonic balance (IHB) method (Lau *et al*. 1982; Ferri 1986), which has been found useful for the analysis of the cantilevered pipe system when the assumption of smallness of inertial nonlinearities is *not* made, is also of this class.

A few more words may be in order here regarding the difficulties encountered in solving equations with large inertial nonlinearities, e.g. equation (5.39). Several of the well known numerical and combined analytical–numerical methods for solving nonlinear differential equations fail, even though they work with large stiffness nonlinearities.[†] On the other hand, two finite difference methods (Houbolt's 4th order and an 8th order scheme) yield accurate results, but both introduce a phase shift and the 4th order scheme also some numerical damping. Only the IHB method (Lau *et al*. 1982, 1983; Lau & Yuen 1993; Semler *et al*. 1996) has proved to be totally satisfactory.

All of these methods of solution, despite some of them having been developed only recently, may be regarded as 'classical', at least in their outlook. Also considered classical is the use of the Lyapunov second method (Hagedorn 1981; Hahn 1963) to establish local, global or a symptotic stability[‡] — see Appendix F.1. Finally, also classical is the use of Floquet theory for assessing stability of limit cycles or the type of bifurcation emanating therefrom (Appendix F.1.2).

Another set of methods have come into prominence over the past 20 years or so, collectively referred to as the *modern methods of nonlinear dynamics*, which are at once more limited in scope and more powerful than the classical methods (Guckenheimer & Holmes 1983). Typically one starts from knowledge of the eigenvalues of the linearized system — which specify the linear behaviour — as well as their evolution as a given parameter (say, the dimensionless flow velocity, u) is varied, and one concentrates the investigation to the case where one of the eigenvalues has a *zero* real part. The *centre manifold method* then drastically reduces a nonlinear multidimensional system into a simpler low-dimensional subsystem, which nevertheless retains all the pertinent information on the bifurcating mode, and hence on the dynamics of the system, in the vicinity of $u = u_{cr}$ (Appendices F.2 and H.1). However, the equations on the centre manifold may still be too complex, and further simplification may be desirable. To this end, the *method of normal forms* (Appendices F.3 and H.2) provides a systematic way of simplifying a complex nonlinear system, by retaining only the essential nonlinear terms which decide its dynamical behaviour. Therefore, these two methods together (or alternatively the combination of the centre manifold and averaging methods) constitute a powerful tool for obtaining the simplest possible subsystem, capable of predicting the nonlinear dynamical behaviour for u not too far away from a particular u_{cr}. The use of *symbolic manipulation* computational software (e.g. MACSYMA, MAPLE, REDUCE

[†]For example, the Picard iteration scheme with Chebyshev polynomials fails for large inertial nonlinearities, not only for the pipe problem, but for the van der Pol type of equation $\ddot{x} + c\dot{x} + x = -x^2(\dot{x} + \ddot{x})$ when $c = -0.3$ but not when $c = -0.1$ (Semler *et al*. 1996).

[‡]While *local* stability applies to a solution for motions in some prescribed domain, *global* stability means that the solution is stable for all amplitudes. Similarly, *asymptotic stability* implies that the solution returns the system to its unperturbed state as $t \to \infty$, while mere *stability* means that it is returned to the neighbourhood of that state [see, e.g. Hagedorn (1981) and Appendix F.1 for more precise definitions].

and MATHEMATICA) renders these tools even more potent (Rand 1984; Rand & Armbruster 1987).

Another powerful set of *numerical* tools, developed relatively recently, after the concepts of bifurcation theory were established, are *continuation* or *homotopy methods*, nowadays available in computer packages. These methods, exemplified by AUTO (Doedel 1981; Doedel & Kernéves 1986), 'follow' a particular type of solution as it evolves in phase space as a result of varying a particular set of system parameters, and can detect the birth of a new type of solution via stability considerations. They are an invaluable tool in constructing *bifurcation diagrams*, which at a glance summarize the changes in dynamical behaviour occurring as the parameter in question is varied (Appendix F.5).

As noted in the foregoing, some of these methods are outlined in Appendix F. In what follows, similarly to the approach in Chapters 3 and 4, the methods used in each case are mentioned without much detail, and then the results are presented and discussed. In a few cases, however, e.g. in Sections 5.5.2 and 5.7.3(a,b), the analysis is outlined in fair detail, to give an appreciation of the power of these methods.

5.5 PIPES WITH SUPPORTED ENDS

5.5.1 The effect of amplitude on frequency

Perhaps the earliest study on nonlinear aspects of the dynamics of pipes with supported ends conveying fluid was conducted by Thurman & Mote (1969b), paralleling closely another on the nonlinear oscillation of axially moving strips (Thurman & Mote 1969a). In this study, motions in both the lateral and axial directions are taken into account, via equations (5.62). When these equations are rendered dimensionless, the additional nondimensional quantity $\mathscr{A} = AL^2/I$ emerges, which plays an important role in the dynamics of the system: all nonlinear terms are multiplied by $(\mathscr{A} - \Gamma)$.

The equations of motion are analysed by means of a hybrid method incorporating elements of Lindstedt's perturbation method and the Krylov–Bogoliubov method. The main finding is that the nonlinear natural frequencies prior to divergence are higher than the linear ones (i.e. the period of oscillation is lower). The discrepancy becomes progressively larger with increasing flow velocity, u,[†] as shown in Figure 5.3. To understand why, it is recalled that linear theory shows a very precipitous reduction in frequency with u close to the point of divergence (Figure 3.10), meaning that the effective stiffness of the pipe is diminished very rapidly; on the other hand, the nonlinear tension effects are not diminished. Hence, the *relative* discrepancy between linear and nonlinear analysis for a given tension increase is dramatically magnified with u. Finally, the flattening of the curves for each value of u/π in Figure 5.3 corresponds to a 'saturation' of the method of solution — carried to the second perturbation; beyond each local minimum, the accuracy of the result becomes doubtful. The important question of how these nonlinearities affect the transition to divergence was not addressed.

It should be remarked that in Thurman & Mote's work both axial motion and *variable* incremental tension associated with large deformations are taken into account — see equations (5.62). In contrast, the nonlinearity considered in Holmes' work (1977, 1978),

[†]It is recalled that the dimensionless flow velocity, $\mathscr{U}$ in Section 5.2, is now denoted by u throughout the rest of this chapter, for conformity with Chapters 3 and 4.

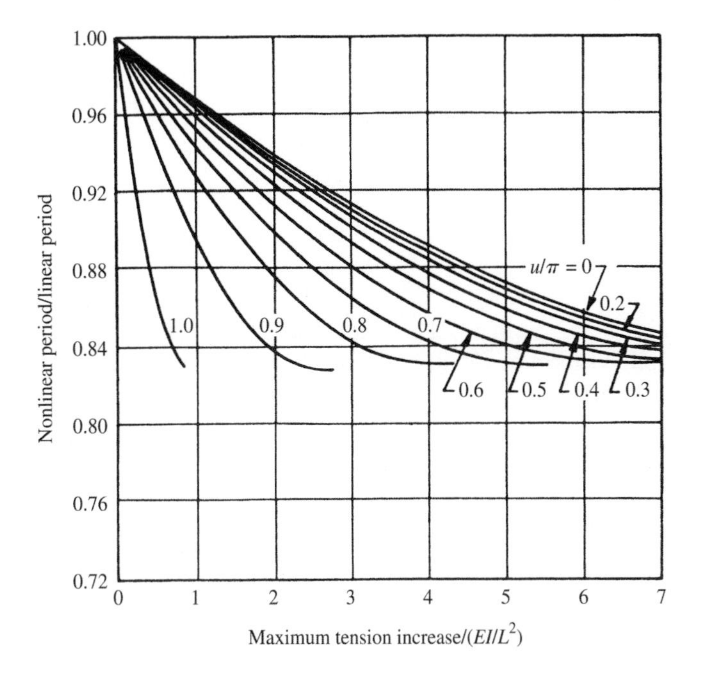

Figure 5.3 The variation of the fundamental period of oscillation versus motion (amplitude) related tension variation for a pinned–pinned pipe with $\beta = \frac{1}{3}$, $\Gamma = 1$ (Thurman & Mote 1969b).

to be discussed next, is related to the increase in *mean* tension due to moderate lateral deformations.

5.5.2 The post-divergence dynamics

The question of post-divergence coupled-mode flutter has already been discussed from the linear viewpoint in Section 3.4.1, where the paradox of how its existence may be reconciled with the fact of zero energy input was elucidated via the work of Done & Simpson (1977). However, there is no question that the existence or nonexistence of coupled-mode flutter has to be decided via nonlinear theory. This was done in two remarkable, authoritative studies by Holmes (1977, 1978), the latter of which is categorically entitled 'Pipes supported at both ends cannot flutter'.† Holmes was the first to use the modern tools of nonlinear dynamics for the analysis of two fluidelastic systems: the pipe conveying fluid and a panel in axial flow (Holmes 1977, 1978; Holmes & Marsden 1978). Some further work was done by Ch'ng (1977, 1978), Ch'ng & Dowell (1979) and Lunn (1982).

As discussed in Section 5.2.9(b), Holmes considered pipes with positively supported (non-sliding) ends, and obtained a nonlinear equation of motion by adding to Païdoussis & Issid's (1974) linear equation a nonlinear term representing the mean, deformation-induced tensioning — the principal nonlinearity. Thus, taking a component of $\overline{T}$ in equation (3.38) to be as in (5.63), the dimensionless form of the equation used, with $\mathcal{U} \equiv u$ and $\dot{u} = \Pi =$

†This is the ultimate in an executive summary: the main conclusion can be read in the title. In these busy times, this practice ought to be strongly encouraged!

0, is

$$\alpha\dot{\eta}'''' + \eta'''' + \left\{u^2 - \Gamma - \gamma(1-\xi) - \tfrac{1}{2}\mathscr{A}\int_0^1 (\eta')^2\,\mathrm{d}\xi\right\}\eta''$$
$$- \alpha\mathscr{A}\left\{\int_0^1 (\eta'\dot{\eta}')\,\mathrm{d}\xi\right\}\eta'' + 2\beta^{1/2}u\dot{\eta}' + \gamma\eta' + \sigma\dot{\eta} + \ddot{\eta} = 0, \tag{5.80}$$

with

$$\mathscr{A} = AL^2/I, \tag{5.81}$$

all other quantities being as in (3.71), (5.38) and (5.44). Most of the detailed work is done with a simplified form of (5.80) by taking $\Gamma = \gamma = 0$ and neglecting the nonlinear dissipative term, namely

$$\alpha\dot{\eta}'''' + \eta'''' + \left\{u^2 - \tfrac{1}{2}\mathscr{A}\int_0^1 (\eta')^2\,\mathrm{d}\xi\right\}\eta'' + 2\beta^{1/2}u\dot{\eta}' + \sigma\dot{\eta} + \ddot{\eta} = 0, \tag{5.82}$$

for a simply-supported (pinned–pinned) pipe — thus satisfying $\eta = \eta'' = 0$ at $\xi = 0, 1$.

Holmes considers the dynamics of the system in two ways, via (a) *finite dimensional analysis* and (b) *infinite dimensional analysis*, to be outlined in what follows. Then, some interesting work by Lunn (1982) is discussed in (c), leading to (d) the final conclusion.

(a) Finite dimensional analysis

For equation (5.82), a two-mode Galerkin discretization of the simply-supported pipe system is obtained via $\eta(\xi,\tau) = \sum[\sqrt{2}\,\sin(r\pi\xi)]q_r(\tau)$, $r = 1, 2$. Converting this to first-order form, leads to the following simple four-dimensional system:

$$\begin{aligned}
\dot{q}_1 &= p_1, \qquad \dot{q}_2 = p_2,\\
\dot{p}_1 &= -\pi^2(\pi^2 - u^2)q_1 - (\alpha\pi^4 + \sigma)p_1 + \tfrac{16}{3}\beta^{1/2}up_2 - \tfrac{1}{4}\mathscr{A}\pi^2(q_1^2 + 4q_2^2)q_1,\\
\dot{p}_2 &= -4\pi^2(4\pi^2 - u^2)q_2 - \tfrac{16}{3}\beta^{1/2}up_1 - (\alpha\pi^4 + \sigma)p_2 - \mathscr{A}\pi^2(q_1^2 + 4q_2^2)q_2.
\end{aligned} \tag{5.83}$$

It is seen that the nonlinearities are of the stiffening cubic type (of the same sign as the linear stiffness), which helps explain the global stability of the system.

It is useful here to refresh the reader's mind as to the linear dynamics of the system. Since damping is present ($\alpha, \sigma \neq 0$), the eigenvalues for $u = 0$ are complex conjugate pairs with negative real parts (Figure 2.10). Here the notation introduced by Holmes is utilized, in which such eigenvalues are denoted by the quartet $\lambda = \{-,-,-,-\}$, the signs being those of the real parts of the eigenvalues; a + means that one of the eigenvalues has positive real part, while 0 denotes a zero real part. As u is increased, the first bifurcation occurs at $u = \pi$ (Section 3.4.1); it is a pitchfork bifurcation, leading to divergence, as shown in Figure 5.4 (cf. Figures 3.9 and 3.14). Hence, for $u > \pi$ we have $\lambda = \{+,-,-,-\}$. The solution of equations (5.83) shows that two new fixed points are generated for $u > \pi$, located increasingly farther away from the origin. The numerical solutions for $\alpha = \sigma = 0.01$, $\beta = 0.2$, are shown in Figure 5.5, together with centre manifold predictions, which will be discussed next.

At the critical point where the bifurcation occurs, the four-dimensional system (5.83) is projected onto the centre manifold (Appendices F.2 and H.1), which in this case is

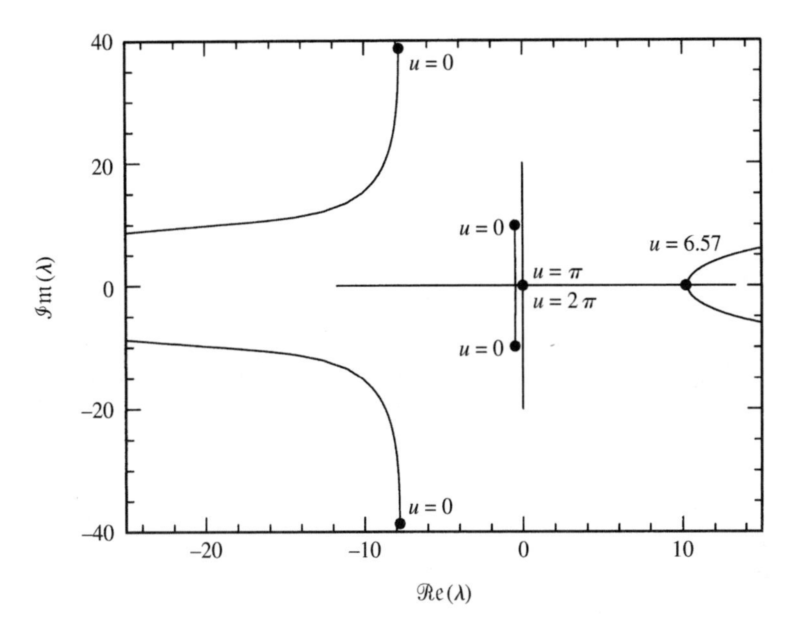

Figure 5.4 Argand diagram of the eigenvalues of the pinned–pinned pipe system ($\alpha = \sigma = 0.01$, $\beta = 0.2$, $\gamma = \Gamma = 0$).

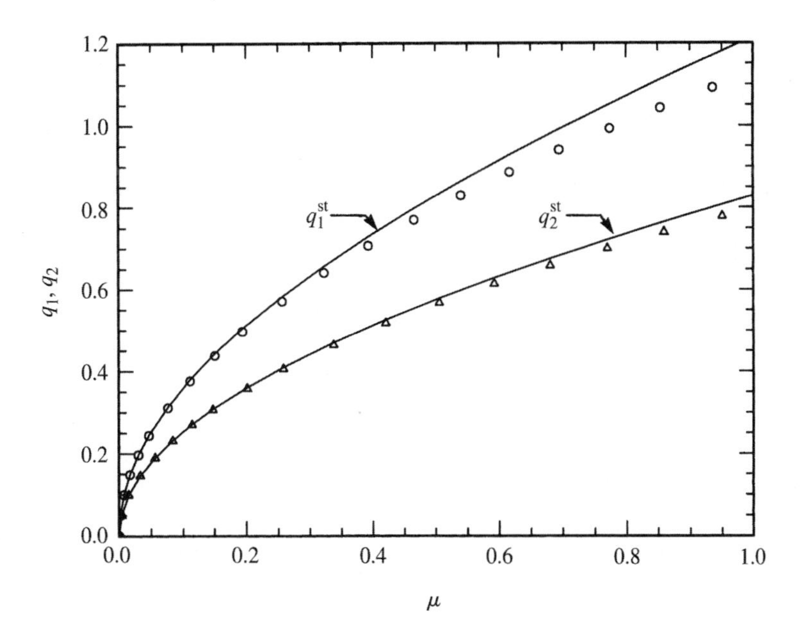

Figure 5.5 The location of the new stable fixed points (sinks), q_1^{st}, generated at $u = \pi + \mu$, and the unstable fixed points (saddles), q_2^{st}, at $u = 2\pi + \mu$, for the pinned–pinned system: ——, by numerical integration; ∘, △, by centre manifold theory. System parameters: $\alpha = \sigma = 0.01$, $\beta = 0.2$, $\mathscr{A} = 1$, $\gamma = \Gamma = 0$.

simply the centre eigenspace. The stable manifold is ignored, and attention is focused on the *centre* manifold; then the evolution of the system in this subspace, wherein the *interesting* aspects of the dynamics is happening, is examined. In the vicinity of $u = u_{cr} = \pi$, we have $\lambda = \{0, -, -, -\}$; re-writing $u = u_{cr} + \mu$, where $\mu \ll 1$, one eventually obtains (Appendix H.2.2):

$$\dot{x} = c_1 \mu x - c_2 \mathscr{A} x^3, \qquad \mu = u - \pi, \tag{5.84}$$

in transformed coordinates, where $c_1 = 63.015$ and $c_2 = 24.746$; this shows that this is a pitchfork bifurcation. Thus, putting $\dot{x} = 0$, it is obvious that there exist fixed points at $x_{st} = \pm[(c_1/c_2)(\mu/\mathscr{A})]^{1/2}$ and it is easy to show that they are stable;† i.e. the new fixed points are sinks (attractors in phase space). Transforming back to the original coordinates, one finds that these are located at

$$q_1^{st} = 1.596(\mu/\mathscr{A})^{1/2}, \tag{5.85}$$

i.e. there is a parabolic relationship between q_{st} and μ. As may be seen in Figure 5.5, agreement with numerical results is excellent, almost to $\mu = 0.4$, despite having specified $\mu \ll 1$.

This form of dependence of the post-divergence fixed points on u is also predicted in another way by Thompson & Lunn (1981), who develop an elegant 'static elastica' formulation of the nonlinear deformation of a pipe under equivalent static loading, effectively the equation of motion with all time-dependent effects ignored. Then, by similarity to the nonlinear behaviour of struts (columns) subjected to compressive loading (Thompson & Hunt 1973), they obtain a 'rising post-buckling path', the same as shown in Figure 5.5.

The question now is what happens for higher u, in particular for $u \geq 2\pi$. In this respect, it is instructive to look at the evolution of the linear system for the specific parameters in this example. As shown in Figure 5.4, because of the presence of dissipation, the mode loci evolve similarly to Figure 3.14 rather than to those typified by Figures 3.9–3.11. Dissipation renders restabilization of the first mode followed by a Hamiltonian Hopf bifurcation (cf. Figure 3.11) impossible, and it also prevents the pitchfork bifurcation associated with the second mode from happening (cf. Figure 3.9). Thus, at $u = 2\pi$ it is the second branch of the first mode that crosses to the unstable part of the complex frequency or the complex eigenvalue plane, rather than a branch of the second mode. At that point, one has $\lambda = \{+, 0, -, -\}$. However, the new fixed points originating at $u = 2\pi$ are saddles. The flow on the centre manifold in this case is governed by

$$\dot{x} = 31.81 \mu x - 24.99 \mathscr{A} x^3; \tag{5.86}$$

the origin in this case is unstable prior to the bifurcation, and so are the new fixed points. These fixed points, transformed back to the original coordinates, are also shown in Figure 5.5, where it is seen that, because the amplitude is smaller, they agree with numerically computed results even better than those for the stable fixed points.

Furthermore, it is shown that no further bifurcations occur; in particular, the only stable fixed points, those given by equation (5.85), do not give rise to Hopf or other bifurcations as u is increased.

---

†Basically, one perturbs (5.84) such that $x = x_{st} + \tilde{x}$, where x_{st} is given by (5.85) — or takes the Jacobian at $x = x_{st}$ — and eventually obtains $\dot{\tilde{x}} = -2c_1 \mu \tilde{x}$; from the solution $\tilde{x} = \tilde{x}_0 \exp(-2c_1 \mu \tau)$ it is clear that this solution branch is stable for $\mu > 0$ and unstable for $\mu < 0$.

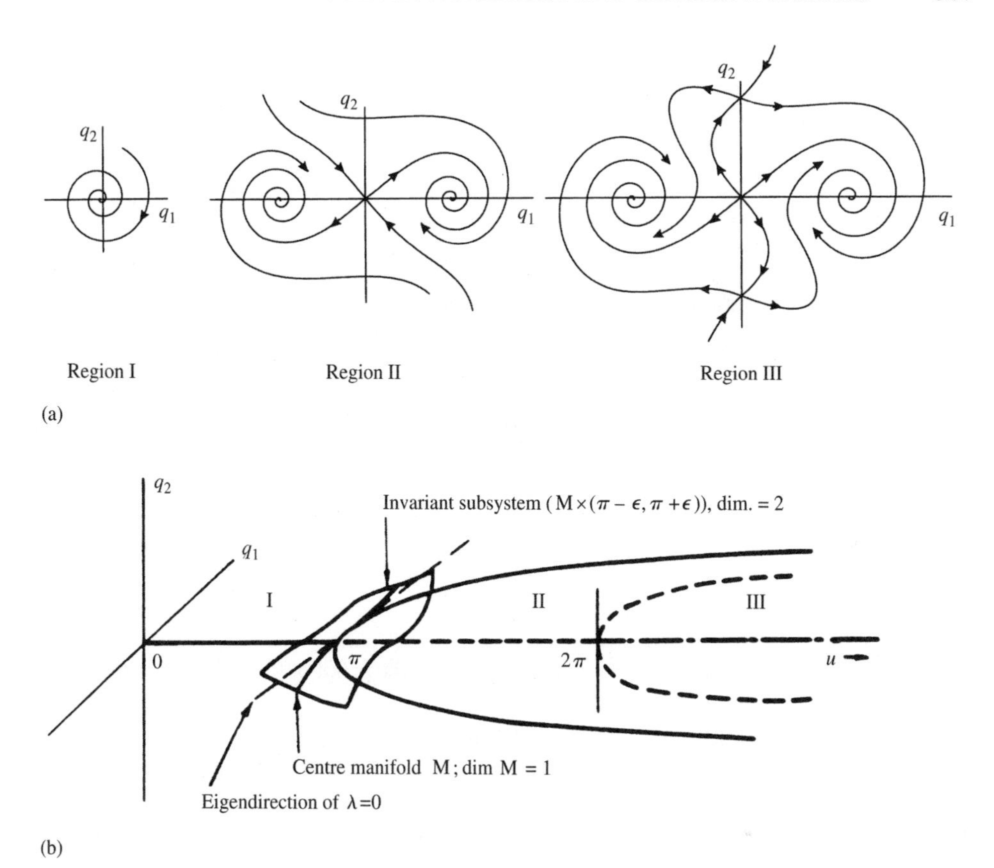

Figure 5.6 A qualitative picture of the bifurcations of the two-mode pinned–pinned pipe system (5.83) for $\alpha = \sigma = 0.01$, $\beta = 0.1$, $\gamma = \Gamma = 0$. (a) Vector fields projected on the $\{q_1, q_2\}$ plane; (b) evolution of the attractors in the $\{q_1, q_2, u\}$ space: ——, sink; – – –, saddle, $\lambda = \{+, -, -, -\}$; — - — , saddle, $\lambda = \{+, +, -, -\}$. After Holmes (1977), but the diagrams in (a) here are based on computed trajectories and are slightly different from Holmes' qualitative diagrams.

The dynamics may be summarized as in Figure 5.6. The four-dimensional, $\mathbb{R}^4$ vector field of (5.83) may be visualized by projecting solution curves onto the two-dimensional subspace $\{q_1, q_2;\ p_1 = p_2 = 0\}$; the resultant projection is shown diagrammatically in Figure 5.6(a), while the evolution with u is shown in Figure 5.6(b). For $\pi < u < 2\pi$ (region II), the flow along q_2 is stable, whereas for $u = 2\pi$ or just higher (region III) it is unstable — two new saddles having been generated along the q_2-axis; but the two sinks on q_1 still exist. Hence, in the flow range where coupled-mode flutter would exist according to linear theory (region III), Holmes concludes that (i) local amplified oscillatory motion can occur near the origin, but (ii) the system is eventually attracted by the sinks on the q_1-axis, since there exist no other attractors, as shown qualitatively in Figure 5.7(a). In fact, this diagram is typical of relatively high β and low α and σ (e.g. $\beta = 0.8$, $\alpha = \sigma = 10^{-3}$); for lower β and higher α, σ (e.g. $\beta = 0.1$ or 0.2 and $\alpha = \sigma = 10^{-2}$, as in the foregoing), the dynamical behaviour is much more like that in Figure 5.7(b). In any case, however, it is clear that with finite dimensional analysis of the problem, no limit-cycle oscillation is found to exist in this system. This,

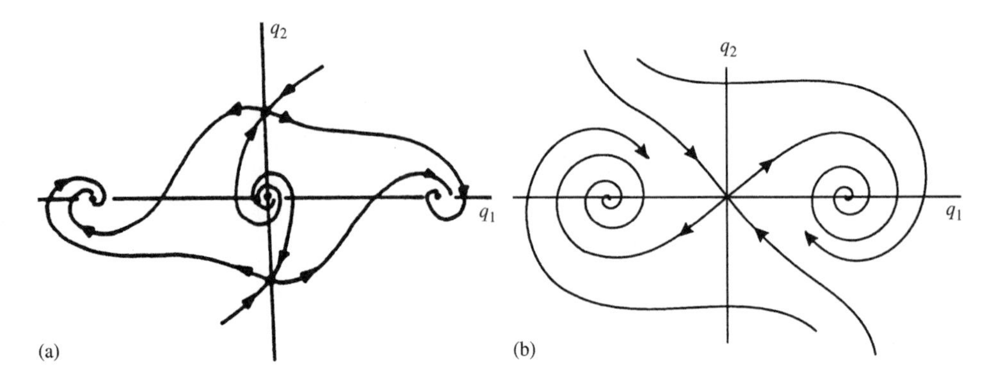

Figure 5.7 Schematics of the two-mode ($\mathbb{R}^4$) model projected onto the $\{q_1, q_2\}$ plane for $u > 2\pi$: (a) Holmes' (1978) original diagram, typical of high β and very low α and σ (e.g. $\beta = 0.8$, $\alpha = \sigma = 0.001$), showing that local oscillatory motion (coupled-mode flutter according to linear theory) about the origin does not lead to limit-cycle motion but eventually to the fixed points on the q_1-axis; (b) diagram based on computed solutions for small β and not very low α and σ, showing that trajectories are attracted by the stable fixed points with hardly any oscillation about the origin.

however, does not prove that a limit cycle *cannot* exist; the proof of that is given in subsection (b), via an infinite dimensional analysis of the system.

Before closing this section, a few words on the effect of symmetry on the pitchfork bifurcation are in order. The system here is symmetric. The mathematical manifestation of this is that the nonlinearities in (5.83) are cubic, so that if $\mathbf{q}_{\rm sol}$ is a solution, so is $-\mathbf{q}_{\rm sol}$. Hence, as is obvious from (5.85), there is another, mirror branch to the solution shown in Figure 5.5; the full 'picture' of the pitchfork bifurcation is as shown in Figure 5.8(a) — cf. Figures 2.11(a) and 5.6(b).

If, however, an imperfection (an initial deflection) is added to the system, so as to break the symmetry [e.g. by adding $+\varepsilon_0$ or $-\varepsilon_0$ to equation (5.84) or to the original system], then the bifurcation occurs as in Figure 5.8(b). This is an example of the generic form of the bifurcation (Holmes 1977), known also as the *canonical cusp* or *Riemann–Hugoniot catastrophe* of Thom (1972). This clearly is what happens in all experiments (Figures 3.22–3.26), since imperfections are always present: the deflection is not zero up to the bifurcation point, growing thereafter, initially with infinite slope, to a large value within a small interval Δu; rather, it is merely small before, and then grows to larger values, effectively more gradually. The fact that $\varepsilon_0 = \varepsilon_0(U)$ is a weakly increasing function as the threshold of divergence is approached makes the transition even more gradual.

(b) Infinite dimensional analysis

In this subsection, the stability of the system is reconsidered, this time by means of infinite dimensional analysis (Holmes 1978). Specifically, first the stability of the trivial equilibrium and then that of the nontrivial equilibria for $u > \pi$ is considered, and finally the possible existence of a limit cycle, independently of how it might emerge. The analysis is intricate and is here presented in greater detail than in Holmes' published work; hence, the casual reader may wish to skip over the details and go straight to the result.

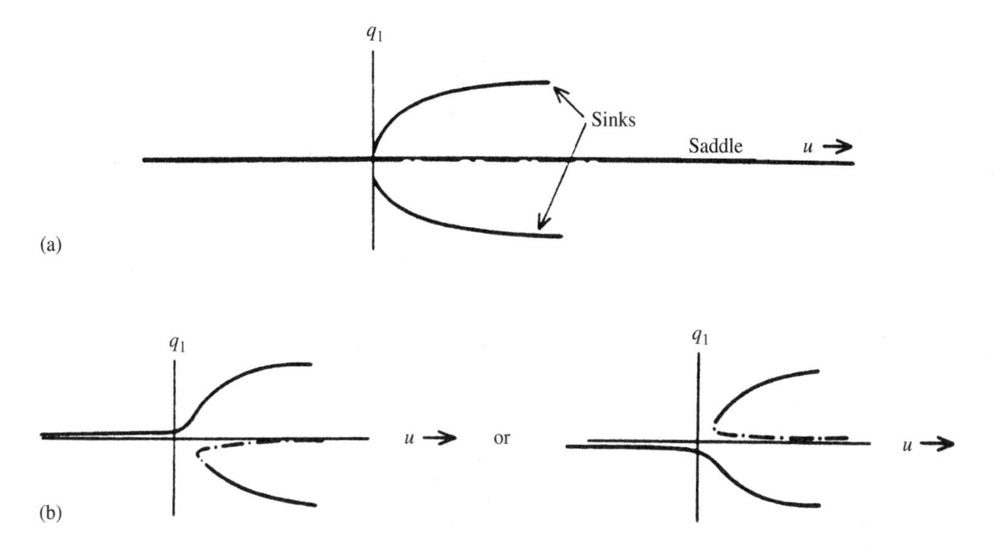

Figure 5.8 A qualitative picture of the occurrence of divergence via a pitchfork bifurcation (a) for a system with symmetry ($\varepsilon_0 = 0$), and (b) when the symmetry is broken ($\varepsilon_0 \neq 0$). Adapted from Holmes (1977).

In this case the system is not discretized. As known from linear analysis of equation (5.82), Section 3.4.1 and equation (3.90a) in particular, nontrivial equilibria for the pinned–pinned pipe arise for $u_j = j\pi$. Hence, from (5.82) it is seen that any nontrivial equilibrium point, $v_j \neq 0$, is an eigenfunction satisfying

$$v_j'''' + \lambda_j v_j'' = 0, \qquad |v_j'|^2 \equiv \int_0^1 (v_j')^2 \, d\xi = (u^2 - \lambda_j)/\tfrac{1}{2}\mathscr{A}, \qquad \lambda_j = j^2\pi^2, \tag{5.87}$$

where $|v_j'|$ denotes the norm; from this, it is clear that $u_c = \pi$, as found before. It is clear that no nontrivial equilibria exist for $u^2 \leq \lambda_1 = \pi^2$, where λ_1 is the first eigenvalue. If $u^2 > \lambda_1$, however, there are $2r$ distinct nontrivial equilibria occurring in pairs, corresponding to the r eigenvalues $\lambda_j < u^2$, the stability of which was examined by Holmes (1977, 1978).

To study the stability of a particular equilibrium position v_j [where it is understood that normally there exist a v_j^+ and a v_j^- because of (5.87)], consider a perturbation w about $v_j \equiv \overline{v}_j$, substitute $\overline{v}_j + w$ in (5.82) and then subtract the equation in $\overline{v}_j$, thus obtaining the equation for w:

$$\begin{aligned} w'''' + (u^2 - \tfrac{1}{2}\mathscr{A}|\overline{v}_j'|^2)w'' - \mathscr{A}(\overline{v}_j', w')\overline{v}_j'' + 2\beta^{1/2}u\dot{w}' \\ + \sigma\dot{w} + \ddot{w} - \tfrac{1}{2}\mathscr{A}\{2(\overline{v}_j', w')w'' + |w'|^2\overline{v}_j'' + |w'|^2 w''\} = 0, \end{aligned} \tag{5.88}$$

where $(a, b) = \int_0^1 a(\xi)b(\xi)\, d\xi$ is the inner product, and where $\alpha = 0$ is assumed without loss of generality.

The stability of $\overline{v}_j$ is studied via a generalization of the Lyapunov second (direct) method to partial differential equations (Movchan 1959; Parks 1967; Holmes & Marsden 1978) — see also Appendix F.1.

Consider first the trivial equilibrium position, $v_0 = 0$, i.e. the origin $\{\eta, \dot{\eta}\} = \{0, 0\}$, in which case we can use (5.82) rather than (5.88). The function

$$H_a = \tfrac{1}{2}|\dot{w}|^2 + \tfrac{1}{2}\{|w''|^2 - u^2|w'|^2\} + \tfrac{1}{8}\mathcal{A}|w'|^4 + \nu\left\{\tfrac{1}{2}\sigma|w|^2 + (w, \dot{w})\right\} \tag{5.89}$$

is a suitable Lyapunov function (Movchan 1965; Parks 1967), in which ν is to be chosen subsequently. It is noted that, essentially, (5.89) has the form of kinetic plus potential energy (the Hamiltonian), as related to (5.82); however, the extraneous last bracketed term is essential in rendering H_a a Lyapunov function. To prove that H_a is positive definite, we start with the inequality

$$H_a \geq \tfrac{1}{2}|\dot{w}|^2 + \tfrac{1}{2}\{\pi^4|w|^2 - u^2\pi^2|w|^2\} + \tfrac{1}{2}\nu\{\sigma|w|^2 + (w, \dot{w})\}, \tag{5.90}$$

in which (a) the first of equations (5.87) is used to show that $|w''|^2 = \pi^2|w'|^2$, and (b) the fact that $|\phi'|^2 \geq \pi^2|\phi|^2$ for any continuous function with $\phi(0) = \phi(1) = 0$ — as easily ascertained for trigonometric functions. Then, re-writing $|\dot{w}|^2 + 2\nu(w, \dot{w}) = |(\nu w + \dot{w})|^2 - \nu^2|w|^2$, inequality (5.90) may be written as

$$H_a \geq \tfrac{1}{2}[\nu(\sigma - \nu) + \pi^2(\pi^2 - u^2)]|w|^2 + \tfrac{1}{2}|(\nu w + \dot{w})|^2,$$

which is globally positive definite provided that $u < \pi$ and $0 \leq \nu \leq \sigma$. Therefore, for given ν and u, as $\{|w'|^2 + |\dot{w}|^2\}^{1/2}$ increases, so does H_a, monotonically.

Differentiating H_a with τ and using (5.82) with $\eta = w$, and then applying the boundary conditions in the resulting integrations by parts,

$$\frac{\mathrm{d}H_a}{\mathrm{d}\tau} = -(\sigma - \nu)|\dot{w}|^2 + 2\nu\beta^{1/2}u(w', \dot{w}) - \nu|w''|^2 + \nu u^2|w'|^2 - \tfrac{1}{2}\nu\mathcal{A}|w'|^4 \tag{5.91}$$

is obtained. Then, making use of the inequality $|\phi'|^2 \geq \pi^2|\phi|^2$ again, this may be written as

$$\frac{\mathrm{d}H_a}{\mathrm{d}\tau} \leq -\{(\sigma - \nu)|\dot{w}|^2 - 2\nu\beta^{1/2}u(w', \dot{w}) + \nu(\pi^2 - u^2)|w'|^2\} - \tfrac{1}{2}\nu\mathcal{A}|w'|^4. \tag{5.92}$$

Provided that $u < \pi$ and $\nu < \sigma$ this may be made negative definite if ν is chosen positive and sufficiently small. For example, letting $\nu = \sigma/k\beta u^2$, equation (5.92) is re-written as

$$\frac{\mathrm{d}H_a}{\mathrm{d}\tau} \leq -\left(\frac{\sigma}{k\beta u^2}\right)\{(k\beta u^2 - 1)|\dot{w}|^2 - 2\beta^{1/2}u(w', \dot{w}) + (\pi^2 - u^2)|w'|^2\}. \tag{5.93}$$

By utilizing the expansion of $|(\beta^{1/2}u\dot{w}/\sqrt{\pi^2 - u^2} - w'\sqrt{\pi^2 - u^2})|^2$ in a similar way as in the foregoing, inequality (5.93) may be re-written as

$$\frac{\mathrm{d}H_a}{\mathrm{d}\tau} \leq -\frac{\sigma}{k\beta u^2}\left\{(k\beta u^2 - 1)|\dot{w}|^2 + \left|\left(\frac{\beta^{1/2}u}{\sqrt{\pi^2 - u^2}}\dot{w} - w'\sqrt{\pi^2 - w^2}\right)\right|^2 - \frac{\beta u^2}{\pi^2 - u^2}|\tilde{w}|^2\right\}.$$

The bracketed quantity is clearly positive definite if $(k\beta u^2 - 1)(\pi^2 - u^2) \geq \beta u^2$; hence, if k is large enough (i.e. ν small enough), we have the required behaviour: $\mathrm{d}H_a/\mathrm{d}\tau < 0$ for all w' or $\dot{w} \neq 0$, i.e. $\mathrm{d}H_a/\mathrm{d}\tau$ is globally negative definite. Hence, the trivial equilibrium point $v_0 = 0$ is globally asymptotically stable if $u < \pi$. It is of interest that if $\sigma = 0$, ν must be set to zero also, and then it can only be proved that $\mathrm{d}H_a/\mathrm{d}\tau \leq 0$; hence only stability, but not asymptotic stability may be proved.

On the other hand, as $u^2 > \pi^2$ the system can no longer be proved to be stable; indeed, from linear theory and the finite dimensional analysis, we know that it is not.

The stability of the first pair of nontrivial equilibria, v_1^+ and $v_1^- = -v_1^+$ is assessed in the same way. A Lyapunov function related to equation (5.88) is now chosen, say for position v_1^+, namely

$$H_b = \tfrac{1}{2}|\dot{w}|^2 + \tfrac{1}{2}\left\{|w''|^2 - \pi^2|w'|^2\right\} + \tfrac{1}{2}\mathscr{A}(v_1^{+\prime}, w')^2 + \tfrac{1}{2}\mathscr{A}(v_1^{+\prime}, w')|w'|^2$$
$$+ \tfrac{1}{8}\mathscr{A}|w'|^4 + \nu\left\{\tfrac{1}{2}\sigma|w|^2 + (w, \dot{w})\right\}. \qquad (5.94)$$

Proceeding in a similar way, it is possible to prove that $H_b > 0$ and $\mathrm{d}H_b/\mathrm{d}\tau \leq 0$ in some neighbourhood of v_1^+ (Holmes 1978); thus, this equilibrium point (and similarly v_1^-) is locally asymptotically stable for all $u^2 > \pi^2$.

Similar forms as H_b but with $\lambda_j = j^2\pi^2$, $j \geq 2$, instead of $\lambda_1 = \pi^2$ as in the foregoing, are appropriate Lyapunov functions for the other points of equilibrium; but in this case the term $-\{|w''|^2 - \lambda_j|w'|^2\}$ appearing in the expression for $\mathrm{d}H_b/\mathrm{d}\tau$ cannot be proved to be negative definite. These points are unstable; in fact, they are saddle points.

The foregoing considerations, though important in the overall dynamical analysis, do not in themselves prove the existence or nonexistence of a limit cycle for $u > 2\pi$; indeed, a limit cycle could exist around v_1^+ (or more likely around both v_1^+ and v_1^-), but sufficiently far removed from it, since stability has only been proved in some neighbourhood of v_1^+; beyond that, it is conceivable that trajectories, also repelled by v_0, could be attracted by a stable limit cycle. The proof of nonexistence is provided by Holmes (1978) following a method developed by Ball (1973a,b) for the dynamic buckling of beams. This proof, outlined in what follows, makes use of the concept of a 'weak solution', which is introduced next.

A *weak solution* is a mathematical concept in functional analysis and topology [see, e.g. Oden (1979; Chapter 5) or, for a more accessible treatment, Curtain & Pritchard (1977)]. It signifies a generalized, nonclassical solution, e.g. one not satisfying the usual differentiability conditions. This concept allows the transformation of the problem from one involving differential operators, such as equation (5.82), to an equivalent problem involving continuous linear functionals, as in (5.95). For our purposes here, this enables us to reach some useful conclusions without first having to obtain a classical solution to equation (5.82).

A weak solution to equation (5.82) is a solution $\eta(\xi, \tau)$ which satisfies the equation

$$(\eta'', \phi'') + \{u^2 - \tfrac{1}{2}\mathscr{A}|\eta'|^2\}(\eta'', \phi) + 2\beta^{1/2}u(\dot{\eta}', \phi) + \sigma(\dot{\eta}, \phi) + (\ddot{\eta}, \phi) = 0, \qquad (5.95)$$

where the inner product is taken with a sufficiently differentiable function ϕ (Ball 1973a; Holmes & Marsden 1978). In (5.95), one can replace ϕ by $\dot{\eta}$ and integrate, thus obtaining an 'energy equation',

$$\tfrac{1}{2}|\dot{\eta}|^2 + \tfrac{1}{2}|\eta''|^2 - \tfrac{1}{2}u^2|\eta'|^2 + \tfrac{1}{8}\mathscr{A}|\eta'|^4 + \sigma\int_0^\tau |\dot{\eta}(\tau)|^2\,\mathrm{d}\tau = C_0, \qquad (5.96)$$

in which $(\dot{\eta}', \dot{\eta}) = 0$ has been utilized and C_0 is a constant. The similarity of (5.96) to (5.89) is obvious. This may be re-written as

$$E(\tau) \equiv \tfrac{1}{2}|\dot{\eta}|^2 + \mathscr{V}(\eta) = C_0 - \sigma\int_0^\tau |\dot{\eta}(\tau)|^2\,\mathrm{d}\tau, \qquad (5.97)$$

where $E(\tau)$, comprising the first four terms of (5.96), is the Hamiltonian (conservative) energy, i.e. the kinetic plus the potential energy

$$\mathcal{V}(\eta) = \tfrac{1}{2}\{|\eta''|^2 - u^2|\eta'|^2 + \tfrac{1}{4}\mathcal{A}|\eta'|^4\}; \tag{5.98}$$

thus, the integration constant, C_0, is equal to the initial energy supplied to the pipe. For $\eta = v_1^{\mp}$, with the aid of the first of equations (5.87), it is shown that $|\eta''|^2 = \pi^2|\eta'|^2$, and hence, by utilizing the second of (5.87), it is found that

$$\min \mathcal{V}(\eta) = -\tfrac{1}{2}(u^2 - \pi^2)^2/\mathcal{A}. \tag{5.99}$$

The integral term in (5.97) is strictly increasing with time and, since C_0 is constant, $E(\tau)$ must decrease unless $\dot{\eta} \equiv 0$. However, $\dot{\eta} = 0$ only at the equilibrium points; i.e. for $u > \pi$, at the saddle points where the minima of $\mathcal{V}(\eta)$ given by (5.99) occur. Therefore, these are also the minima of $E(\tau)$. Thus, the pipe will always approach an equilibrium point as $\tau \to \infty$. Consequently, by infinite dimensional analysis it has definitively been shown that no limit-cycle oscillation can exist in this system.

This completes the presentation of Holmes' work on this system, proving that 'pipes supported at both ends cannot flutter'. Or does it? The question of whether even this unequivocal statement has to be qualified is discussed next.

(c) Flutter in the Hamiltonian system

Lunn (1982) examined the equivalent problem to Holmes': a pin-ended pipe with one end free to slide, but constrained by an axially-disposed spring, so yielding equation (5.82) directly, with no approximation. On the other hand, the system was generalized by introducing an elastic foundation; hence, a term $k\eta$ appears in the equation of motion, where k is the dimensionless foundation stiffness — cf. equations (3.70) and (3.71). A two-mode Galerkin discretization is considered and, with the aid of centre manifold theory, similar conclusions to Holmes' are reached; but, in the process, a number of important contributions are made, as follows.

It is first observed that the region of gyroscopic stabilization, occurring for high enough β (Section 3.4; Figures 3.11 and 3.12) between the first and second critical flow velocities, is 'of purely "academic" concern' since, on first buckling, the deflections of the pipe would grow sufficiently to make the study of higher stability regions 'inapt'. Therefore, a system is sought which would remain stable up to the point of onset of linear coupled-mode flutter. This is achieved by a judicious choice of the elastic foundation (Section 3.4.3). For $k = 4\pi^4$, it is found that with zero dissipation the two eigenvalues reach zero in the Argand diagram at the same value of u, namely $u = \sqrt{5}\pi$ — cf. Figure 3.20; however, divergence does not develop thereafter, because of gyroscopic stabilization, and the eigenvalues remain purely imaginary up to $u_{cf} = 7.66$, where the system loses stability by Hamiltonian coupled-mode flutter. However, if even infinitesimally small dissipative forces are included, the gyroscopic stabilization is destroyed and hence coupled-mode flutter ceases to be the first instability to occur, divergence developing instead at $u = \sqrt{5}\pi$; which leads to qualitatively the same dynamics as discussed heretofore. This, however, raises the following question: Is it possible that the nonexistence of coupled-mode flutter in the nonlinear analysis is primarily due *not* to nonlinear effects but to dissipation?

To answer this question, Lunn reconsiders the nonlinear system, without any foundation but also without any dissipation. The startling result is illustrated in Figure 5.9(a), showing 'limit-cycle' motion of small amplitude about the origin! For $u > 2\pi$ the origin has become a higher-order saddle, resembling a potential energy 'hill'. The peculiar ornate character of Figure 5.9(a) derives from the nature of gyroscopic stabilization. The trajectory 'falls', moving away from the origin; gyroscopic forces then drive it at right angles to the instantaneous direction of motion, and eventually 'uphill'; when enough kinetic energy has been lost that way, the process begins anew. It is noted, however, that large-amplitude limit cycles are not possible, because of the attracting sinks.

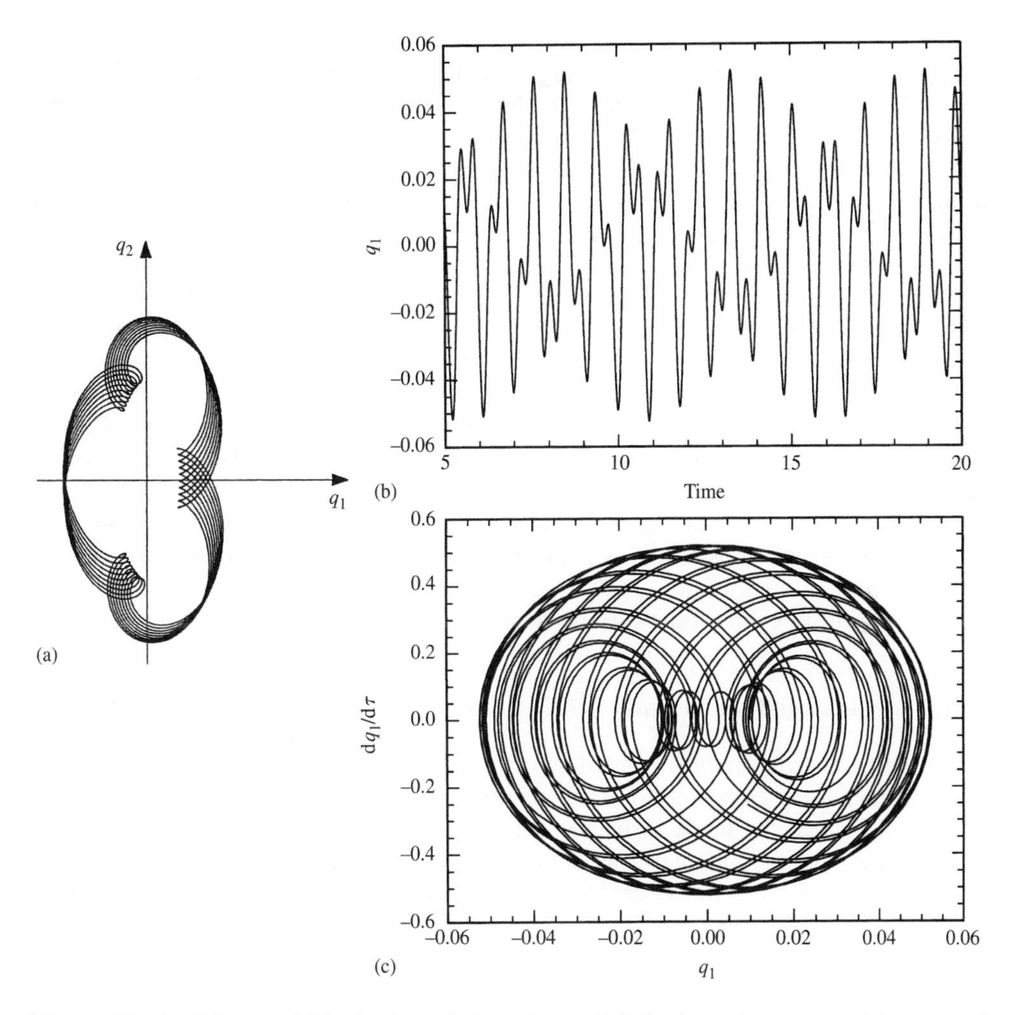

Figure 5.9 (a) 'Limit cycle' in the $\{q_1, q_2\}$ plane for $u = 2.025\pi$ for a pipe system with supported ends and $\beta = 0.694$, $\mathscr{A} = 0.4$ and $k = \alpha = \sigma = 0$ (Lunn 1982). (b) Time trace and (c) phase-plane plot of flutter of the Hamiltonian system ($\beta = 0.5$, $\alpha = \sigma = k = 0$, $\mathscr{A} = 1$, $u = 6.35$), as obtained numerically.

This result has been recalculated for $\beta = 0.5$ at $u = 6.35$, and is displayed as a time trace and a phase plane diagram in Figure 5.9(b,c). It is clear now that the motion is quasiperiodic (cf. Figure 2.4) and it involves two incommensurate frequencies. Hence, more strictly it is a motion on a torus, rather than a limit cycle which would imply a single frequency.

Therefore, *coupled-mode flutter does exist*, even in the nonlinear context. However, it is pathologically nonrobust: the slightest amount of damping destroys it utterly. On reflection, this is as unremarkable as 'finding' that periodic solutions can exist for a conservative system, but not when damping is included; what *is* remarkable, nevertheless, is that such periodic solutions are academically possible, *even after divergence*.

(d) Concluding remarks

The same conclusion as Holmes' with regard to the nonexistence of coupled-mode flutter for dissipative systems was reached by Ch'ng (1978) and Ch'ng & Dowell (1979), who utilize the same equation as Holmes, equation (5.82), discretize it and then integrate it numerically. [It is of interest that an error in an earlier attempt by Ch'ng (1977) led to the opposite conclusion. Holmes (1977) also admits that, in an earlier version of his work, a mistake led him too to the opposite conclusion. All this shows how sensitive this type of analysis can be.] On the other hand, Lunn's (1982) work shows that sustained oscillation, i.e. flutter, about the unstable initial equilibrium *is* possible, theoretically at least, provided that there exists *no* dissipation whatsoever. This, of course, is impossible in any *real* physical system.

In fact, as discussed in Section 3.4.4, no experimental evidence exists that pipes supported at both ends do flutter, whether axial sliding at the supports is permitted or not; in the former case violent divergence (buckling) develops and the $\omega = 0$ condition is obtained, while in the latter case this is not so. The main point here is that, for realistic systems, predictions of linear theory, beyond the onset of the first instability (divergence), do not materialize. This is not general, and in fact Holmes (1977) discusses another case, involving panel flutter, where post-divergence flutter does indeed materialize. This is also known to occur in cylindrical structures subjected to external axial flow (Chapter 8).

5.5.3 Pipes with an axially sliding downstream end

When a pipe has a laterally supported but axially free-to-slide downstream end, its equations of motion are essentially those of a cantilevered pipe (see end of Appendix G.2): the centreline may be taken to be inextensible, and the nonlinearities are mainly due to curvature effects, while the mean deformation-induced tension is zero. The nonlinear dynamics of such a system has been studied analytically, numerically and experimentally by Yoshizawa *et al*. (1985, 1986) up to and beyond the point of divergence.

The system considered is a clamped–pinned pipe, supplied by a constant-head tank, while the flow velocity is generally deformation-dependent. The equations derived are similar to Rousselet & Herrmann's [Section 5.2.8(b)]: (i) a 'flow equation' similar to equation (5.53), with a friction parameter α; (ii) an equation for the pipe motion involving both axial and transverse displacements, u and w, and the angle of deformation, θ — interrelated via $\sin\theta = \partial w/\partial s$, $\cos\theta = 1 + (\partial u/\partial s)$ as per equations (5.4) for an inextensible pipe.

The eigenfunctions of the subsystem $\eta'''' - \gamma[(1-\xi)\eta'' - \eta'] + u^2\eta'' + \ddot{\eta} = 0$ are determined and then the deflection of the pipe is approximated by a one-mode Galerkin scheme, $\eta(\xi,\tau) = \phi_1(\xi)q_1(\tau)$. Analytical solutions are obtained with this approximation, adequate for relatively modest deflections, as well as more accurate solutions for the post-divergence state by integrating numerically the time-independent version of the full equations of motion.

Experiments have also been conducted by Yoshizawa *et al*. (1985, 1986), utilizing vertical silicone rubber pipes ($D_o = 5$ mm, $D_i = 3$ mm, $L = 600$ mm) conveying water. Two stainless steel wires were attached to the pipe in one plane, to ensure that motions occur normal to that plane. The pinned, axially-sliding lower end was achieved by a short bar perpendicular to the pipe axis, in contact with the pipe.

Typical results are shown in Figure 5.10(a) for the variation of the first-mode eigenfrequency up to divergence, when theoretically $\omega_1 = 0$. The experimental frequencies are in excellent agreement with theory. Nevertheless, for obvious reasons, the experimental frequencies could not be measured all the way to divergence, the precise onset of which was difficult to pin-point.

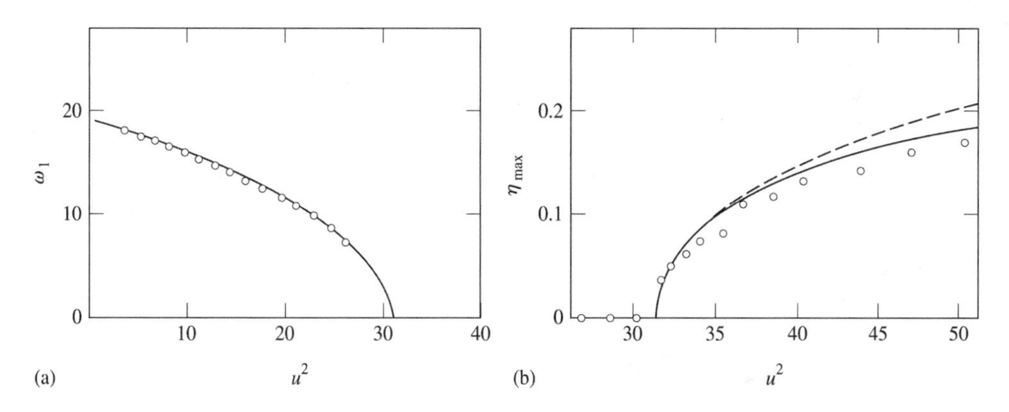

Figure 5.10 (a) Variation of the first-mode eigenfrequency with u^2 for a clamped-pinned pipe with an axially sliding downstream end ($\beta = 0.273$, $\gamma = 34.4$, $\alpha = 4.68$): ——, theory; ○, experiments. (b) The post-divergence maximum static pipe deflection, η_{max} versus u^2 ($\alpha = 5.56$): – – –, approximate analytical; ——, numerical; ○, experimental (Yoshizawa *et al*. 1985, 1986).

The variation of the maximum, steady post-divergence amplitude with u^2 is given in Figure 5.10(b). It is seen that (i) the approximate analytical and the numerical solutions agree for $\eta_{max} < 0.1$ approximately, and (ii) agreement with experimental values is very good overall, particularly for the more accurate numerical results.

5.5.4 Impulsively excited 3-D motions

An interesting, application-related study, examines the dynamics of inhibited flow/porous tubes (INPORTs) used to protect the inner wall of inertial confinement fusion (ICF) reactors† from X-rays, neutron bombardment and débris (Engelstad & Lovell 1985, 1995; Engelstad 1988). The porous, braided silicon-carbide-fibre tubes convey Li-Pb (molten lithium–lead), the fluid acting both as a coolant and a breeder; the tubes are porous, to ooze out a liquid film for protection from the same hazards. These very slender tubes

†The ICF is a precursor concept to the LIBRA fusion reaction chamber.

($D = 30\,\text{mm}$, $L = 10\,\text{m}$) are subjected to periodic blast waves transmitted through the gas in the reaction chamber.

The equations of motion are similar to Ch'ng & Dowell's and Thurman & Mote's (Sections 5.2.8 and 5.2.9) but more complete: three equations of motion are obtained for motion in two mutually perpendicular planes, and the flow velocity is generally harmonically perturbed as in equation (4.69) to account for pump-induced pulsations. Before analysis, however, the tension–gravity term is considerably simplified and the axial equation of motion is eliminated, so that each of the remaining equations becomes similar to Holmes' (Section 5.5.2); these two equations are coupled via the nonlinear tension terms. Furthermore, because tension–gravity effects are so large, flexural terms are neglected, so that the system becomes a pipe-string [cf. Copeland's work in Section 5.8.3(b)]. In the calculations presented, the flow velocity is steady and dissipation is taken into account. The equations of motion are discretized and then integrated numerically.

In the calculations, the pipe is excited by periodic impulses, introduced as initial conditions all along the length in the plane of the blast wave, and motion is monitored in both planes. With increasing frequency of impulses, the classical jump (down) phenomenon in the frequency response curve is obtained, characteristic of hardening nonlinear systems, and the associated jump (up) when the frequency is decreased.

If the pipe is perturbed in the plane perpendicular to that being excited, the oscillation either dies out or builds up to a steady limit-cycle motion, depending on the periodic impulse frequency; in the latter case, a generally oval whirling motion ensues with slow precession, which would be unacceptable in actual ICF operation.

5.6 ARTICULATED CANTILEVERED PIPES

Many of the methods for analysing nonlinear systems apply to ordinary differential equations, so that continuous systems must be discretized first before these methods can be applied. Furthermore, since most of these methods are practicable only for low-dimensional (low-D) systems (i.e. systems of only a few degrees of freedom), there is a natural tendency to study low-D discretizations of the continuous systems. This then opens the question, often left unanswered, of whether the low-D discretized model really captures adequately the essential dynamical features of the continuous system. This in turn provides the main impetus in the study of articulated systems: the very physical system is discrete and it can be chosen *a priori* to be low-D.

Most of the work on the nonlinear dynamics of articulated pipes conveying fluid has been done on the nonconservative cantilevered system (Figure 5.2). In many cases this serves as a preamble to the study of the same aspects of the *continuous* cantilevered system, discussed in Section 5.7; this is the reason for *this* section being where it is.

Before discussing cantilevered articulated pipes in Section 5.6.2, the case of a pipe with a constrained downstream end is treated first.

5.6.1 Cantilever with constrained end

No systematic study has been published on the nonlinear dynamics of the conservative system of articulated pipes with supported ends — perhaps because the work in Section 5.5 is considered to have settled all important issues.

Here, Thompson's (1982b) magic black box (Section 3.2.2, Figure 3.3) is discussed in detail. The system in the black box in Figure 5.11(a) is generally nonconservative. It consists of an articulated pipe, the downstream end of which is constrained by a string, supporting a weight $\overline{m}g$. Assuming equal spring stiffnesses, k, and lengths, l, neglecting gravity effects in the pipe system, and assuming small angular deflections, θ_1 and θ_2, Thompson (1982b) conducted an interesting static analysis of the system. In terms of statics, the fluid acts as a follower compressive load of magnitude MU^2 (Section 3.2.1). Taking moments about the joints, one obtains

$$\overline{m}gl + k(\theta_2 - \theta_1) = 0, \qquad 2\overline{m}gl + k\theta_1 + MU^2 l(\theta_2 - \theta_1) = 0. \tag{5.100}$$

The deflection at the end of the pipe system is $x = -l(\theta_1 + \theta_2)$, which, from the solution of (5.100), may be re-written as

$$x = -\overline{m}gl^2 \left(\frac{2MU^2 l}{k} - 5 \right) \Big/ k. \tag{5.101}$$

The flexibility may be defined as $x/\overline{m}$ (more usually $x/\overline{m}g$); its inverse, $\overline{m}/x$, is the stiffness of the system. It is clear that the stiffness is positive for small values of U, becomes infinite at the point of divergence ($2MU^2 l = 5k$), and then negative for larger values of U. This dynamical behaviour is illustrated in Figure 5.11(b) from experiments with Lunn's (1982) articulated pipes, involving Perspex or copper tubing and rubber joints, as described in Section 5.6.2. The observed behaviour is a little more complex than the linear relation between x and MU^2 in (5.101), but essentially the dynamics is as predicted. In particular, in the region of negative stiffness, when the weight $\overline{m}$ is doubled, x is *halved*, approximately; i.e. as the weight is increased, *it goes up* (Figure 3.3) — a graphic demonstration of 'paradoxial' mechanics due to negative stiffness.

The other interesting observation made by Thompson is this. For conventional structural systems (e.g. the inverted pendulum, loaded arches), the equilibrium path in the negative stiffness region is unstable under 'dead' load and, hence, can only be studied experimentally by using a suitable 'rigid' load, e.g. via a screw loading device (Thompson 1979). Here, however, we have a system in which the complete, stable load–deflection curve can be obtained, covering also the region of negative stiffness, by using dead weight loading — precisely because the system is nonconservative.

5.6.2 Unconstrained cantilevers

The main objective of virtually all nonlinear studies in this area is related to the characterization of the nature of the Hopf bifurcations leading to flutter. In the case of 2-D motions, this distinguishes subcritical from supercritical bifurcations in the parameter space. In the case of 3-D motions, however, the nonlinear analysis also defines whether the resulting flutter is planar or rotary. Of special interest is another set of studies, concerned with the dynamics of systems in the vicinity of a double degeneracy, characterized by two coincident bifurcations via which a rich variety of dynamical states may emerge, as discussed in part (c) of this section.

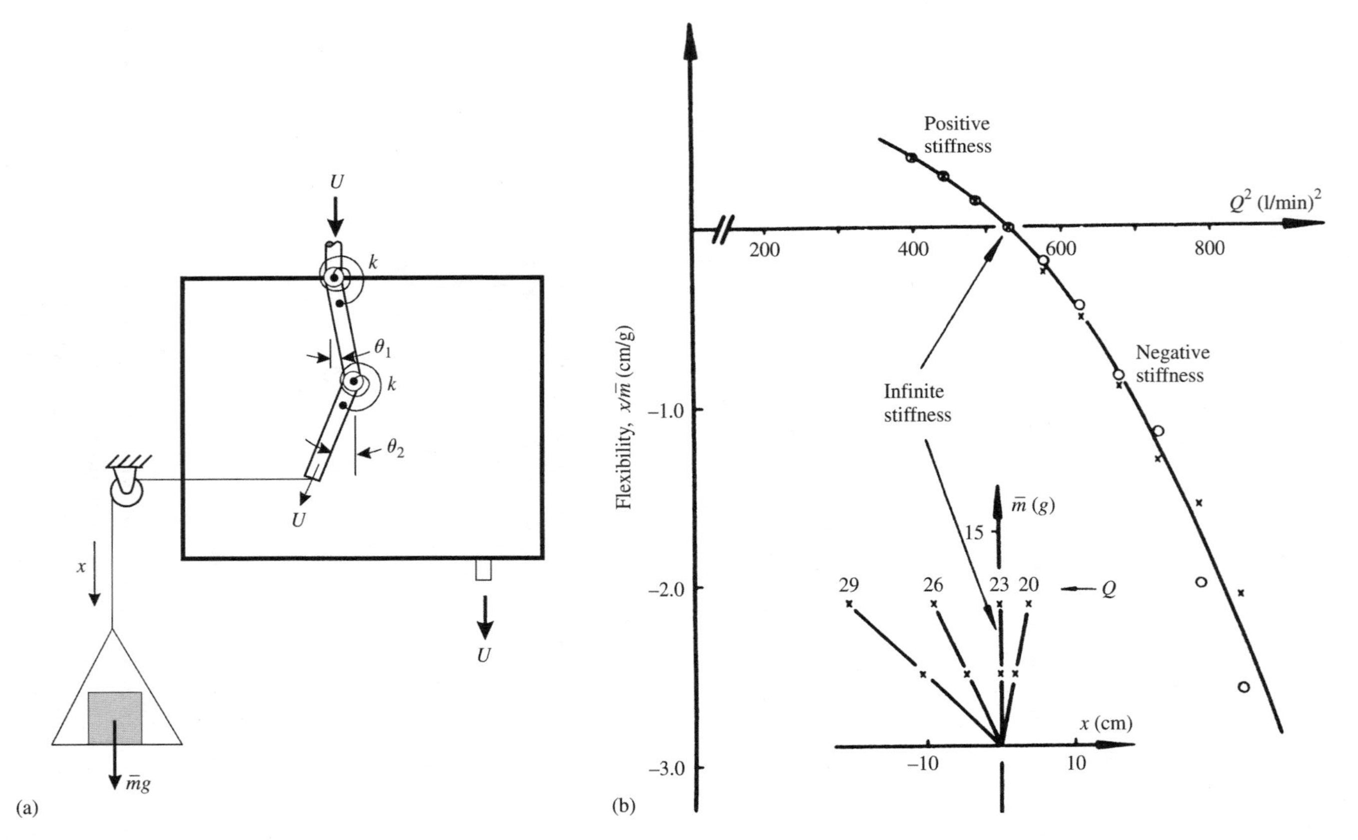

Figure 5.11 (a) The 'black box' containing Thompson's constrained articulated pipe system; (b) the experimental data showing the flexibility (inverse of stiffness) versus the square of the volumetric flow-rate, Q^2, where $Q = MU/\rho$, ρ being the fluid density (Thompson 1982b).

(a) 2-D motions

The first nonlinear study is due to Rousselet & Herrmann (1977), dealing with a system of two rigid pipes ($N = 2$) hanging vertically, with ideal articulations of zero stiffness and damping. The equations of motion are (5.64) and (5.65), but with $k_1 = k_2 = 0$. Hence, in the nondimensionalized equations, different parameters from (5.68) are utilized, e.g. $u = U/\sqrt{gl_2}$ and so on. Fluctuations in the flow velocity due to varying acceleration and gravity head are taken into account, as per equations (5.66) and (5.67); they are taken to be small, so that if u_0 is the mean flow velocity,

$$u = u_0 + \Delta u, \qquad \Delta u \ll 1. \tag{5.102}$$

The main objective is to obtain information on the dynamics in the vicinity of the Hopf bifurcation which leads to flutter; in particular, whether the predicted limit cycle is stable or unstable (supercritical or subcritical Hopf bifurcation), and what is its amplitude. With (5.102), (5.64) and (5.65), the equations of motion may be written in the form

$$[M]\{\ddot{\theta}\} + [C]\{\dot{\theta}\} + [K]\{\theta\} = \{F\}, \tag{5.103}$$

where $\{\theta\} = \{\theta_1, \theta_2\}^{\mathrm{T}}$; $\{F\} = \{F_1, F_2\}^{\mathrm{T}}$ contains all the nonlinear terms, i.e.

$$\{F_1, F_2\}^{\mathrm{T}} = \mathbf{f}(\theta_1, \theta_2, \dot{\theta}_1, \dot{\theta}_2, \ddot{\theta}_1, \ddot{\theta}_2, \Delta u, \Delta\dot{u};\ \text{system parameters}), \tag{5.104}$$

where the 'system parameters' are a, $\overline{\beta}$, γ, Π_0, λ and $\overline{h}^*$ — see equations (5.68). Similarly, (5.102), (5.66) and (5.67) lead to a 'flow equation',

$$\Delta\dot{u} = g(\Delta u, \theta_1, \theta_2, \dot{\theta}_1^2, \dot{\theta}_2^2;\ \text{system parameters}). \tag{5.105}$$

The ingenious procedure adopted to solve equations (5.103)–(5.105) (Rousselet 1975; Rousselet & Herrmann 1977) is outlined in what follows.

(i) The linear part of (5.103) is solved first, yielding the eigenvalues and eigenvectors, and hence also the critical value for flutter $u_0 = u_{0f}$, if it exists; it is noted that for $\overline{\beta} > 0.51$ it does not, and only divergence is then possible.

(ii) The equation of motion is then transformed into first-order form,

$$[B]\{\dot{z}\} + [E]\{z\} = \{\Phi\}, \tag{5.106}$$

$\{z\} = \left\{\{\dot{\theta}\}, \{\theta\}\right\}^{\mathrm{T}}$ — cf. equations (2.16) and (2.17). Then, the homogeneous form of (5.106) and its adjoint [see equation (2.20)] are solved simultaneously for $u_0 = u_{cf}$, yielding the same eigenvalues but different eigenvectors from those of the original system. The use of the biorthogonality property [equation (2.21)] then allows the decoupling of the system. Attention is thenceforth devoted exclusively to the mode associated with the Hopf bifurcation, ignoring the other (stable one), thus reducing the fourth-order system in (5.106) to one of order 2. Nowadays, the same would have been accomplished via the centre manifold method (Appendices F and H). The resulting equation has the form

$$\ddot{x} + (\alpha^2 + \omega^2)x = 2\alpha\dot{x} + f(F_1, F_2, u_0, \alpha, \omega; \text{system parameters, modal form}), \tag{5.107}$$

where $\lambda = \alpha + \mathrm{i}\omega$, $|\alpha| \ll 1$, it being understood that u_0 is close to u_{cf}; 'modal form' signifies the eigenvector information for the mode undergoing the Hopf bifurcation.

(iii) In equation (5.107), F_1 and F_2 contain the still unknown Δu and $\Delta \dot{u}$. These are determined by solving the differential equation in Δu, equation (5.105), after the 'modal form' of interest has been substituted in its right-hand side.

(iv) Now that all terms on the right-hand side of the reduced form of (5.103), equation (5.106), are known, the nonlinear equation is solved by the Krylov–Bogoliubov method, a form of averaging (Appendix F.4), keeping only the first term in the asymptotic expansion, $x = \frac{1}{2}\Theta \sin\psi = \frac{1}{2}\Theta \sin(\omega t + \phi)$, eventually leading to

$$\dot{\Theta}_{\rm avg} = \frac{\omega\alpha\Theta_{\rm avg} + K_1\Theta^3_{\rm avg}}{\omega}, \qquad (\Theta\dot{\phi})_{\rm avg} = \frac{-K_2\Theta^2_{\rm avg}}{\omega}, \tag{5.108}$$

where K_1 and K_2 are lengthy algebraic expressions involving the parameters in (5.107). For a limit cycle, $\dot{\Theta}_{\rm avg} = 0$; hence one obtains the limit-cycle amplitude

$$\Theta_{\rm LC} = \left(\frac{-\omega\alpha}{K_1}\right)^{1/2}. \tag{5.109}$$

It is clear from (5.108) that the origin becomes unstable for $\alpha > 0$; furthermore, if $K_1 < 0$ the emerging limit cycle is stable. On the other hand, if $\alpha < 0$ and $K_1 > 0$, the limit cycle is unstable.

Typical results are shown in Figure 5.12. It is seen that for $\overline{\beta} < 0.30$ the limit cycle is unstable, which suggests a subcritical Hopf bifurcation. However, the upper, stable branch of the solution cannot be predicted, since polynomial expansions to only fourth order are included in the analysis (cf. Section 2.3, Figures 2.12 and 2.13). For $\overline{\beta} > 0.30$, the limit cycle is stable and the Hopf bifurcation supercritical. For $\overline{\beta} = 0.30$, an infinite amplitude is obtained, but this should be interpreted as meaning that the effect of nonlinearities (to the order to which the Krylov–Bogoliubov analysis and the polynomial expansions have been carried out) is null — a higher order degeneracy.

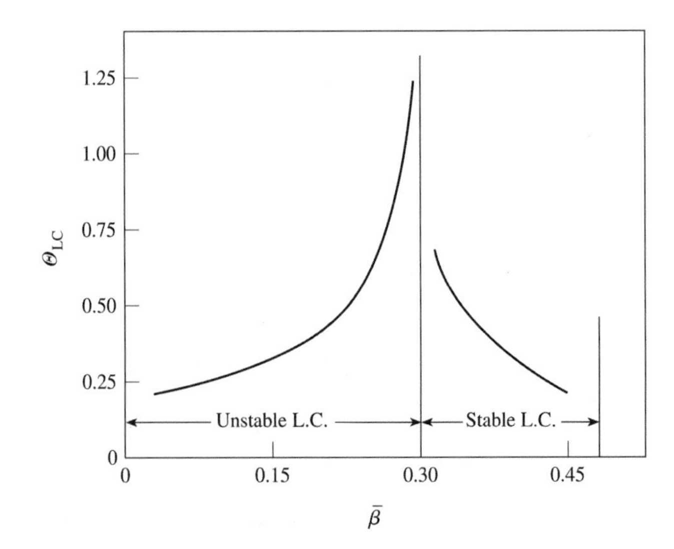

Figure 5.12 Limit-cycle amplitude, $\Theta_{\rm LC}$, versus the mass ratio $\overline{\beta}$, for an articulated cantilevered system with $c_1 = c_2 = 0, a = 1, \lambda = 0.5, |\Delta u_c| = 0.1$ ($\simeq 3\%$); (Rousselet & Herrmann 1977).

An important result obtained by Rousselet & Herrmann, with repercussions to most other analyses, is that an amplitude of angular motion of at least 10° is required to perturb the flow velocity by a few per cent. This justifies the assumption made in most other analyses that the flow velocity is independent of motion.

Finally, in a qualitative experiment involving a system with $\overline{\beta} = 0.216$, Rousselet & Herrmann found that the Hopf bifurcation is indeed subcritical: (i) for small disturbances, the oscillations die out and the system returns to equilibrium, and (ii) for larger disturbances, the oscillations grow, until the motion reaches a steady state (stable limit cycle).

The same problem, but with joints of nonzero stiffness and simplified by considering that the flow velocity is motion-independent, has been studied by Lunn (1982). In the nonlinear equations, only cubic nonlinearities are retained. The dynamics in the vicinity of the critical points is studied with the aid of centre manifold theory for both pitchfork and Hopf bifurcations, in the latter case also making use of the multiple scales perturbation technique (Nayfeh & Mook 1979; Nayfeh 1981).

Figure 5.13 shows a stability map in the $\{\beta, \gamma\}$-plane for the occurrence of divergence or flutter for a system with $a = \kappa = 1$ [i.e. $k_1 = k_2$ and $l_1 = l_2$; equations (5.68)]. It is seen that, for small enough β, the system loses stability by flutter: for very small β via a subcritical, and for larger β by a supercritical Hopf bifurcation (cf. Figure 5.12 and Rousselet & Herrmann's findings for $\gamma = 0$). For low enough γ, divergence is impossible. For higher γ and not too small β, stability is lost via a subcritical pitchfork bifurcation, indicating a 'falling post-buckling path' (Figure 5.14), and is therefore unstable. One can presume that there may be a stable solution branch at larger amplitude — which nevertheless cannot be determined except by a higher-order analysis or by numerical integration.

Lunn also conducted experiments with Perspex or copper tubing for the pipes and short pieces of rubber tubing for the spring-like joints (Section 3.8) conveying water.

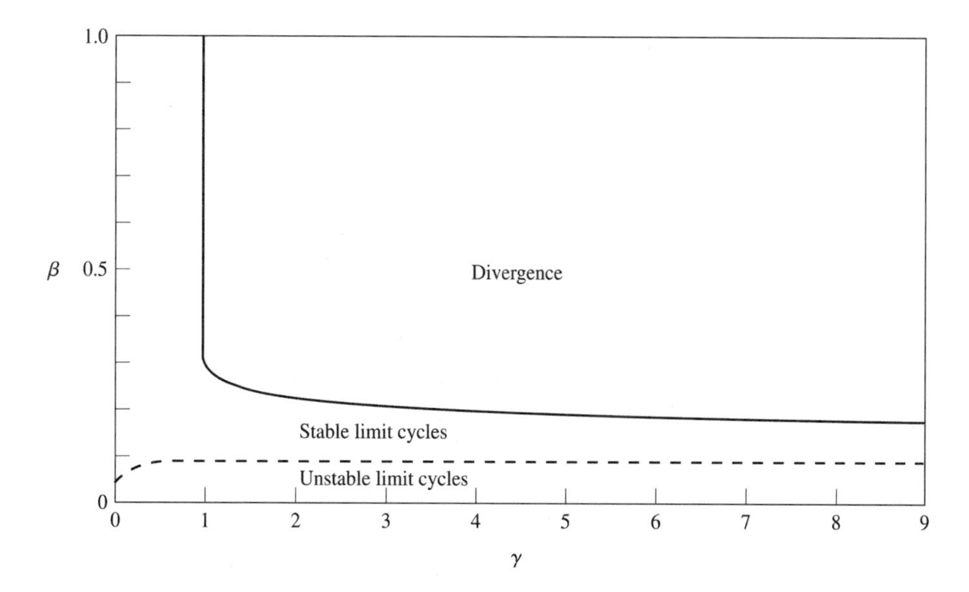

Figure 5.13 Stability map for an articulated cantilevered pipe ($N = 2, k_1 = k_2, l_1 = l_2$) in terms of β and γ, showing regions of loss of stability by a sub- or supercritical Hopf bifurcation or by a subcritical pitchfork bifurcation (Lunn 1982).

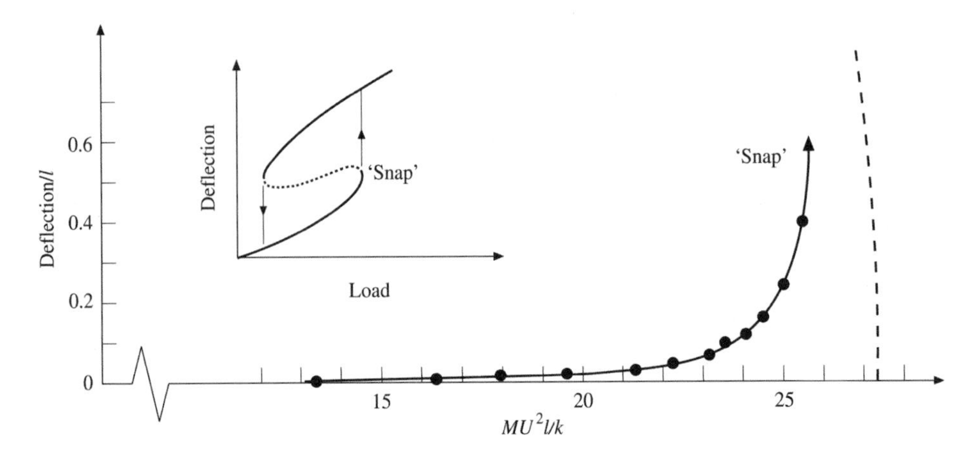

Figure 5.14 The development of divergence for a pipe with $N = 2, k_1 = k_2 = k, l_1 = l_2 = l$; $\beta = 0.549$ and $\gamma = 7.384$, showing the development of the lateral departure from equilibrium, $\delta = (\text{deflection})/l$ versus MU^2l/k: ----, theory; —•—, experiment (Lunn 1982). The inset diagram shows what the development beyond 'snap' might be: ——, stable solution; – – –, unstable solution.

Light caliper-type hinges at the joints ensured planar motions. The pipes were of different diameters, of $\mathcal{O}$ (10 mm), and 0.2–1.0 m long, thus varying both β and γ. Typical results for a pipe losing stability by divergence are shown in Figure 5.14. 'Snap' indicates the point where any further increase in the flow would 'cause deflections to grow so large that something would probably break — this was not attempted!' Hence, it can only be theorized that 'snap' corresponds to the point where the system would snap to the larger, stable solution branch, as sketched in the inset diagram. If that is so, then theoretical and experimental paths towards divergence agree remarkably well. Agreement between theory and experiment is less good for pipes theoretically losing stability by a supercritical Hopf bifurcation: in one case the experimental observations indicate that the bifurcation is subcritical (though the critical flow velocities agree very well); and in another, stability is lost by divergence in the experiment. These discrepancies are attributed to imperfections and peculiarities of the rubber joints.

An important recent study is due to Champneys (1991), in which a two-degree-of-freedom system is considered, modified as follows: (i) the interconnecting springs are nonlinear and (ii) the straight configuration does not correspond to the unstrained-spring case, the two articulations being at an angle ψ. Hence, this is a case where, as the flow velocity u is increased, the equilibrium configuration is altered continuously. The bifurcational behaviour beyond the Hopf bifurcation is tracked, and two kinds of homoclinic orbits are found to exist: so-called E-homoclinic orbits involving tangency to a stationary (equilibrium) point, and P-homoclinic orbits, bi-asymptotic to periodic orbits. Because the system is autonomous, AUTO (Doedel & Kernéves 1986) could be used to trace all the bifurcations as parameters are varied. The system dynamics is investigated by varying u and ψ. The system loses stability through a Hopf bifurcation for sufficiently high u. Thereafter, depending on the values of ψ, further increase in u could lead to period-doubling, reverse period-doubling, as well as homoclinic bifurcations (both of E- and P-type). Among the interesting and unusual dynamical features obtained are *isolas* in the global branches of some of the orbits, and *towers*, which are sequences of period-doubling and saddle-node bifurcations. Champneys goes on to show the existence of chaotic regions in this and in a subsequent paper (Champneys 1993) where the system asymmetry was removed — as discussed in Section 5.8.5.

(b) 3-D motions

A very sophisticated analysis of three-dimensional motions of the $N = 2$ system — effectively generalizing to 3-D the foregoing analysis, but utilizing entirely different techniques — has been conducted by Bajaj & Sethna (1982a,b). The equations of motion used are equations (5.74)–(5.77); hence the flow velocity is motion-independent. The springs are considered to be so designed as to allow both planar motions in two directions and rotational motion with zero torsional stiffness — a challenging design problem if attempted experimentally.

The particular problem investigated is the loss of stability by flutter. Because of the rotational symmetry of the system, a *double* pair of complex eigenvalues crosses simultaneously the imaginary axis from negative to positive. The nonlinear phenomena in this case are more complicated than those associated with simple Hopf bifurcations (Figure 2.11); e.g. supercritical bifurcations do not necessarily imply a stable system in this case (Iooss & Joseph 1980).

After considering the linear dynamics, the problem is transformed into Jordan canonical form. Then, periodic solutions of the nonlinear equations are analysed by the method of alternate problems (Hale 1969; Bajaj 1981, 1982), which is similar in spirit to the Lyapunov–Schmidt method (Appendix F, Sections F.6.2 and F.6.3). Two independent sets of periodic solutions are found to exist: clockwise or counterclockwise rotary motions about the x-axis and planar transverse motions. Their stability is determined by the Floquet exponents of the corresponding variational equations, leading finally to the following set of interesting results:

(i) both supercritical and subcritical solutions of both the rotary and planar kinds are generally possible for $0 < \overline{\beta} < 3$ $(0 < \beta < 1)$ and for given ranges of a, κ and γ, as defined by (5.68); as already mentioned, these are associated with double pairs of eigenvalues crossing the imaginary axis;
(ii) if both planar and rotary motions are supercritical, then the one with the larger amplitude is stable and the other unstable, whereas normally one would expect all supercritical solutions to be stable;
(iii) for a given $\overline{\beta}$, if either of the solutions (rotary or planar) is subcritical, both solutions are unstable.

Typical results are shown in Figure 5.15, where $1/|\mu_{20}|^{1/2}$ is a measure of the amplitude. In Figure 5.15(a) it is seen that both rotary and planar supercritical solutions exist. For $\overline{\beta} < 0.51$, the latter being larger, the planar oscillations are stable and hence should physically materialize; for $\overline{\beta} > 0.51$, it is the rotary motions that are stable. When gravity is present, as in Figure 5.15(b), the situation is more complex. Thus, for $\overline{\beta} < 0.45$, we once again have stable planar motions, while for $0.45 < \overline{\beta} < 1.19$ we have stable rotary motions. For $\overline{\beta} > 1.19$, however, there also exist subcritical solutions, not present in Figure 5.15(a); since one of the solutions is subcritical, both are unstable. This does not imply that there are no stable periodic solutions for $\overline{\beta} > 1.19$; it merely reflects the

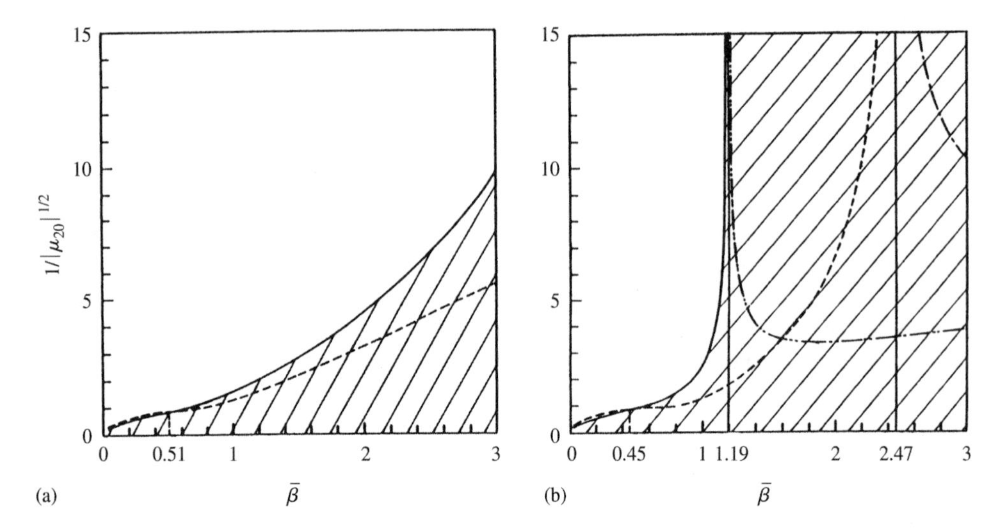

Figure 5.15 Amplitude of periodic solutions in the vicinity of u_{cf}, for 3-D motions of a two-degree-of-freedom articulated cantilevered system with $a = 2, \kappa = 1$ and (a) $\gamma = 0$ and (b) $\gamma = 0.25$: ——, rotary supercritical; — - - —, rotary subcritical; – – –, planar supercritical; — - —, planar subcritical (Bajaj & Sethna 1982b).

analysis having been carried out to only the first approximation level. The same applies to the 'infinite' amplitude at $\bar{\beta} = 1.19$.

Finally, Bajaj & Sethna (1982b) discuss the effect of a small asymmetry in the system, by making the spring stiffness in one plane slightly larger than in the other. It is found that the circular rotary motions become elliptical, but the dynamics in the foregoing are otherwise robust.

The foregoing analysis is restricted to solutions in the neighbourhood of the straight, vertical equilibrium. The situation when this restriction is removed has been studied by Sethna & Gu (1985), where the 'limiting configurations' as $u \to \infty$ are examined: (i) does the system perform a rotary motion in a horizontal plane with the two pipe segments at right angles to each other, or (ii) does it take on an S-shape in a vertical plane, or (iii) some other configuration? The authors examine five such generic shapes, all of the type in which the equations are invariant under rotation about the vertical axis. The stability of these generic shapes is studied by either a linear approach or by utilizing centre manifold theory (in the case of global circular motions). It is found that apart from shapes (i) and (ii) above, all other shapes eventually become unstable via a (secondary) Hopf bifurcation. The analytical results are complemented by numerical simulations.

(c) Double degeneracy

Two studies into the dynamics of articulated cantilevers near a point of double degeneracy are discussed here.

In the first such study, by Sethna & Shaw (1987), codimension-three bifurcations are considered of a two-segment articulated system vibrating in a plane; the double degeneracy in this case is associated with a pitchfork and a Hopf bifurcation occurring simultaneously for a special set of parameter values. It has already been mentioned previously that for different values of γ and β (or $\overline{\beta}$) the system may lose stability by divergence (pitchfork bifurcation) or flutter (Hopf bifurcation); for the right combinations of (γ, β), these two bifurcations may occur simultaneously, i.e. at the same u. Codimension-three refers to three parameters being used to 'unfold' the bifurcations in the vicinity of this double degeneracy — i.e. to develop the evolution of the bifurcations gradually as one or more parameters are varied. This is normally a codimension-two problem (Guckenheimer & Holmes 1983), but here a third parameter corresponding to imperfection-related asymmetries is added.

The system is first transformed into Jordan canonical form, and then through centre manifold reduction and averaging reduced to the deceptively simple set of equations

$$\dot{r} = \mu_1 r + r(-r^2 - bz^2), \qquad \dot{z} = \mu_4 + \mu_2 z + z(-cr^2 + z^2), \tag{5.110}$$

where $\mu_i, i = 1, 2$ and 4, are the unfolding parameters which are related to variations of the original system parameters from the critical point. By comparing these to the original nonlinear equations [cf. the planar version of equations (5.74)–(5.77)] involving more than 10–15 terms each, it is clear that a very dramatic simplification has been achieved. Yet, equations (5.110) are capable of capturing the essential dynamics of the system, as is illustrated, for instance, by comparison with simulations from the full form of the equations. The r equation is similar to the averaged equation for the classical van der Pol oscillator, with r representing the amplitude of oscillatory response, in this case due to the Hopf bifurcation, while the z equation represents pipe response due to the pitchfork bifurcation. The results are illustrated in Figure 5.16 for the case of no asymmetries ($\mu_4 = 0$), which in fact corresponds to codimension-two bifurcations — cf. Section 5.7.3(d) where, for a similar system, the analysis is outlined in greater detail. The parameter μ_1, for $\mu_1 > 0$, gives rise to a pitchfork bifurcation (only one side of which is shown) and to a new equilibrium point q_1 (on the r-axis) for the averaged system (5.110); in the original system this corresponds to the amplitude of periodic motions. For $\mu_2 < 0$, we additionally have a static subcritical pitchfork bifurcation, and the point q_2 (on the z-axis) is unstable. Of particular importance is line 2, on which there exists a heteroclinic cycle, across which the character of the solutions and the stability of the new fixed point q_3 change. According to Smale's horseshoe theory (Guckenheimer & Holmes 1983; Moon 1992), it is known that homoclinic and heteroclinic tangles may lead to complex dynamics and chaos.

The three-parameter, codimension-three case is very much more complex and will not be discussed here, even in outline. In addition to periodic motions, amplitude-modulated oscillations, i.e. motions on a 'two-torus' in four-dimensional space, are generally also possible. These manifest themselves as periodic orbits in r — only on line 2 for the system of Figure 5.16, but more widely for the asymmetric system — cf. Section 5.7.3(d). In total, in this remarkable study, 23 distinct open sets are found in the three-dimensional (μ_1, μ_2, μ_4) parameter space, each corresponding to qualitatively different dynamics!

Yet another type of double degeneracy in articulated pipes was studied by Langthjem (1995): the case of two Hopf bifurcations occurring simultaneously. This, though impossible for $N = 2$ and 3 systems, can occur for the $N = 4$ system. Hence, to study this system, Langthjem derives the nonlinear equations for the four-degree-of-freedom system, considering planar motions and a deflection-independent flow velocity; the connecting springs are taken to be nonlinear, with a linear and a cubic component.

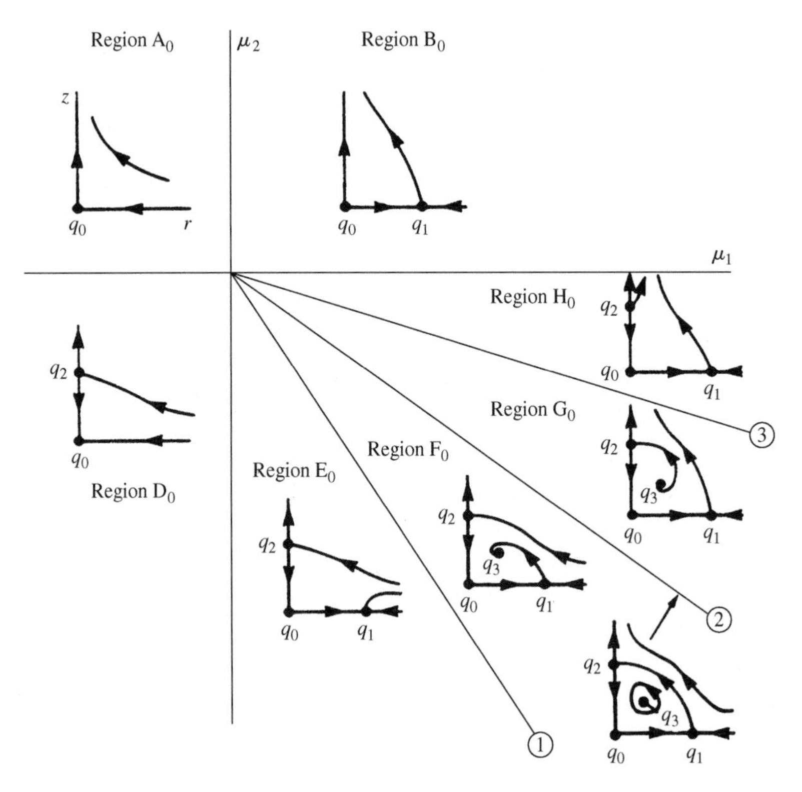

Figure 5.16 Two-parameter unfolding of bifurcations for the symmetric case ($\mu_4 = 0$) of an $N = 2$ vertical articulated cantilever near a point of double degeneracy (Sethna & Shaw 1987).

Analysis of the linear system yields the double-degeneracy conditions, corresponding to simultaneous Hopf bifurcations in the second and fourth modes, for instance. The system is transformed to Jordan canonical form, and then centre manifold and normal form theory are employed to study the dynamics in the neighbourhood of the double degeneracy. The reduced subsystem on the centre manifold is found to be governed by the amplitude equations

$$\dot{r}_1 = r_1(\delta_1 + r_1^2 + br_2^2) + \mathbb{O}(|r|^5), \qquad \dot{r}_2 = r_2(\delta_2 + cr_1^2 + dr_2^2) + \mathbb{O}(|r|^5), \tag{5.111}$$

and similar equations for the phase angles θ_1 and θ_2. In (5.111), δ_1 and δ_2 are the increments in the real parts of the eigenvalues (equal to zero at the critical points), which can be related to the bifurcation parameters μ and χ, associated with u and γ, respectively (Appendix F.5). Thus, this is a codimension-two analysis. System (5.111) has been analysed by Guckenheimer & Holmes (1983), and nine topologically different classes of solutions are found to be possible.

Langthjem found several of these solutions, involving different kinds of periodic and quasiperiodic motions. Sample results are shown in Figure 5.17: (a,c) in the $\{r_1, r_2\}$-plane and (b,d) in the $\{\phi_1, \dot{\phi}_1\}$ phase plane, where ϕ_1 is the angular deflection of the first articulation. In Figure 5.17(a) we see a fixed point on the r_1-axis, $(r_1, r_2) = (\sqrt{-\delta_1}, 0)$; in the physical system it corresponds to periodic oscillations at frequency

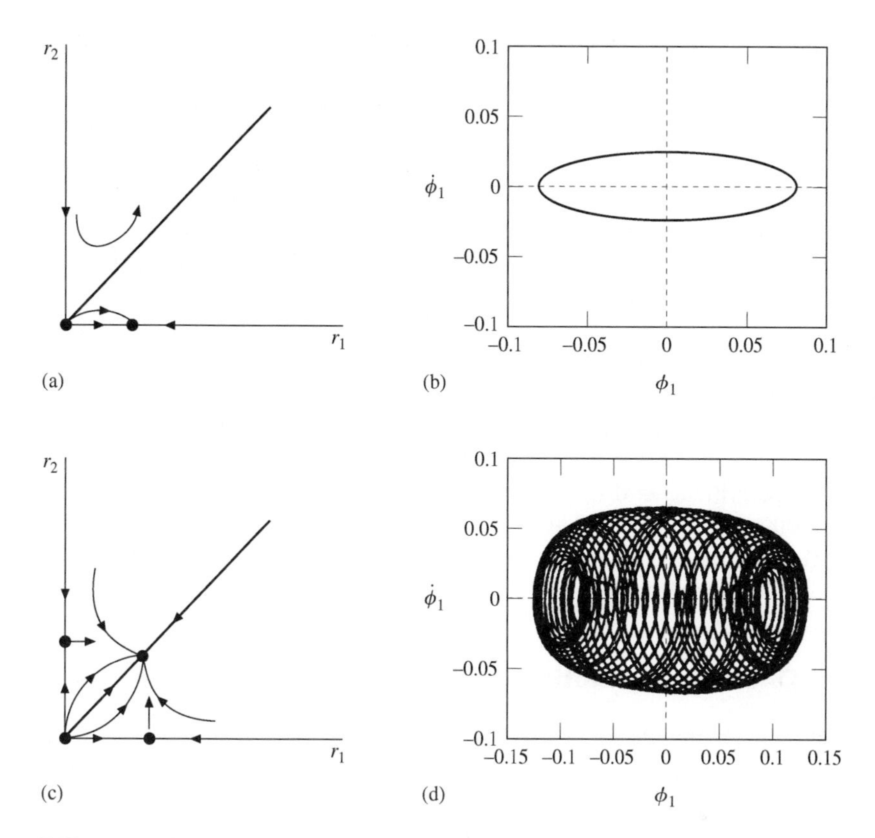

Figure 5.17 (a) A fixed point in the $\{r_1, r_2\}$ plane and (b) corresponding physical-coordinate phase-plane plot; (c) a fixed point on the invariant line $r_2 = 1.2599r_1$ and (d) the corresponding phase-plane plot; for the $N = 4$ articulated system near a point of double degeneracy involving two Hopf bifurcations (Langthjem 1995).

$\dot{\theta}_1$. In Figure 5.17(c) the fixed point lies on the invariant line $r_2 = 1.2599r_1$; physically this corresponds to quasiperiodic oscillations with two incommensurate frequencies. The phase-plane diagrams of Figure 5.17(b,d), obtained numerically by integrating the *full* equations verify the centre manifold predictions. This verification, by the way, is something that is rarely done but *should* be done, wherever possible.

5.6.3 Concluding comment

The paradoxical dynamics in Thompson's magic box, the prediction and confirmation of both subcritical and supercritical Hopf bifurcations in addition to divergence (Figures 5.13 and 5.14), the discovery of rotary as well as planar limit-cycle motions (Figure 5.15), and the existence of quasiperiodic motions, heteroclinic cycles and chaos in the vicinity of double-degeneracy conditions, all this shows that the nonlinear system is dynamically very rich and even more interesting than the linear one. This realization has added to the impetus for nonlinear analysis of the continuous counterpart of this system, to be discussed next.

5.7 CANTILEVERED PIPES

A rapid scan of publication dates will convince the reader that most of the activity on the dynamics of pipes conveying fluid has in recent years concentrated on the nonlinear dynamics of *continuous cantilevered pipes*, to be discussed here, or articulated cantilevers, covered in Section 5.6; this is more striking if one includes the work on chaotic dynamics, presented in Section 5.8. The reason is that these systems display as varied and fascinating nonlinear dynamical behaviour as the cornucopia in the linear dynamics domain already discussed in Chapters 3 and 4.

In keeping with the rest of this chapter, most important findings are discussed to a greater or lesser extent, but the mathematical methods and analytical details are only skimmed; only in one case, in Section 5.7.3, are they presented in fair detail.

As for articulated cantilevers, the character of the Hopf bifurcations leading to flutter is a question of considerable interest. For 2-D motions, this defines whether the bifurcation is sub- or supercritical, as discussed in Section 5.7.1. For 3-D motions, this additionally decides whether the flutter is planar or three-dimensional, as discussed in Section 5.7.2. Again as for articulated cantilevers, it is of interest to examine the dynamics near different types of double degeneracy conditions; this is done in Section 5.7.3.

5.7.1 2-D limit-cycle motions

Planar limit-cycle motions of a vertical cantilever in the vicinity of the critical flow velocity where the Hopf bifurcation arises were studied by Rousselet & Herrmann (1981). A constant-pressure tank is assumed to feed the flow into the pipe [see Figure 5.2(a)], while the flow velocity is pipe-deformation dependent, as discussed in Section 5.6.2 for the articulated counterpart of this system. Hence, there are two coupled governing equations, which are solved iteratively, in a manner similar to that described in Section 5.6.2: (i) the homogeneous, linear solution to the equation of motions is obtained first; (ii) the solution is substituted into the 'flow equation' which yields Δu and $\Delta \dot{u}$, the flow velocity and acceleration due to pipe deformation; (iii) the homogeneous solution together with Δu and $\Delta \dot{u}$ are substituted into the full nonlinear equation of motion, which is solved by the Krylov–Bogoliubov averaging method, to first order, yielding the averaged amplitude, $\overline{A}$, and the corresponding phase. Of course, at the very threshold of the Hopf bifurcation there is zero damping. A small amount of positive or negative external viscous damping, $\pm\sigma\dot{\eta}$, is added as a control parameter, used to achieve purely real eigenfrequencies for $u = u_c \pm \Delta u_c$, where u_c is the critical value; an equal and opposite amount of 'damping', $\pm\sigma\dot{\eta}$, is added to the nonlinear part of the equation, thus cancelling the total added damping. Also added to the nonlinear part of the equation are the gravity terms, which are small. These addenda to the nonlinear part of the equation are significant in the discussion of the results.

Unfortunately, not all parameters are defined in the results obtained. It is mentioned that, 'for a steel or plastic pipe $\frac{1}{4}$ in. in diameter and 2 ft in length' and $\beta = 0.5$, the gravity parameter is $\gamma = 0.01$ and 0.3, respectively; so, presumably γ is in that range for the calculations in Figure 5.18. The dimensions also give a clue as to the likely value for the fluid friction constant α in equation (5.53), which can play an important role in the dynamics (Bajaj *et al*. 1980).

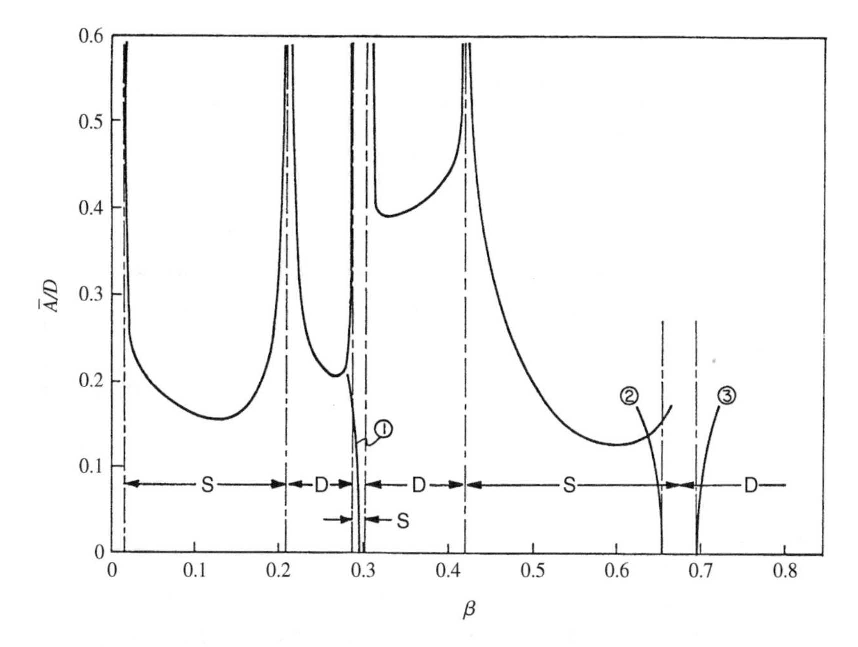

Figure 5.18 Limit cycle amplitude of free end of a cantilevered pipe, $\overline{A}/D$, versus β for $|\Delta u_c| = 0.1$; curves 1, 2, 3 represent solutions with $|\Delta u_c| = 1$ and unusually small amplitude (Rousselet & Herrmann 1981).

The main result obtained, Figure 5.18, shows the amplitude of the limit cycle as a function of β for $|\Delta u_c| = 0.1$ or 1. Also indicated (as S or D) is whether the effect of nonlinearities is stabilizing or destabilizing. It is here, however, that it must be recognized that the 'nonlinearities' include the linear gravity and reversed damping terms; as a result, the regions in which 'nonlinearities' are destabilizing is not necessarily associated with a subcritical Hopf bifurcation, since the damping which has been added artificially (the $\pm\sigma\dot{\eta}$ term) can also lead to destabilization (Section 3.5.3).

It is observed in Figure 5.18 that $\overline{A}/D > 0.15$ in most cases, where $\overline{A}$ is the limit-cycle amplitude, indicating that a relatively large amplitude of motion is required to compensate for the small amount of positive or negative damping associated with $|\Delta u_c|$.[†] This shows that the 'nonlinearities', as expected, do not have a strong effect on the system, except in the vicinity of the 'jumps' or S-shaped curves in the stability diagram, at $\beta \simeq 0.295$ and 0.67 (see Figures 3.30 and 3.32).[‡] $|\Delta u_c| = 0.1$ corresponds to approximately only 1% of u_c, $|\Delta u_c| = 1$ to about 10%. The infinite amplitudes correspond to effectively zero effect of the 'nonlinearities', at least in terms of the first-order averaging approximation.

A variant of the system was considered by Lundgren *et al.* (1979): a horizontal pipe, fitted with an inclined nozzle at the free end (at angle θ_j and terminal flow velocity U_j), as shown in Figure 5.19(a), and subjected to a deformation-independent flow velocity. As a result of the sideways load by the exiting fluid jet, the shape of the pipe in the plane of the nozzle is changed continually as u is increased. The static equilibrium of the pipe is

[†]To a certain extent, 'small' and 'large' are dependent on the nondimensionalization.

[‡]The dynamics in the vicinity of $\beta = 0.295$ is much more complex, and several solution curves are determined by Rousselet & Herrmann (1981).

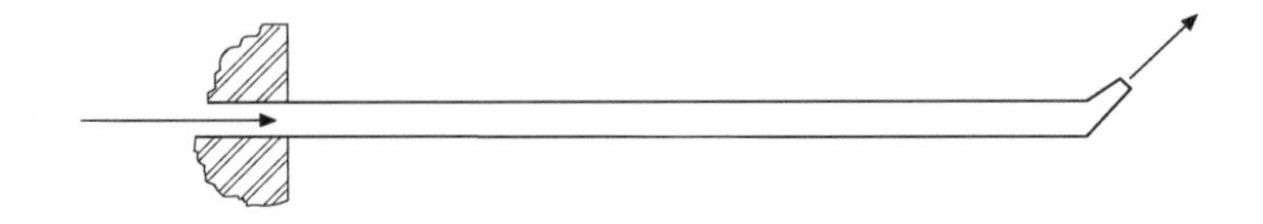

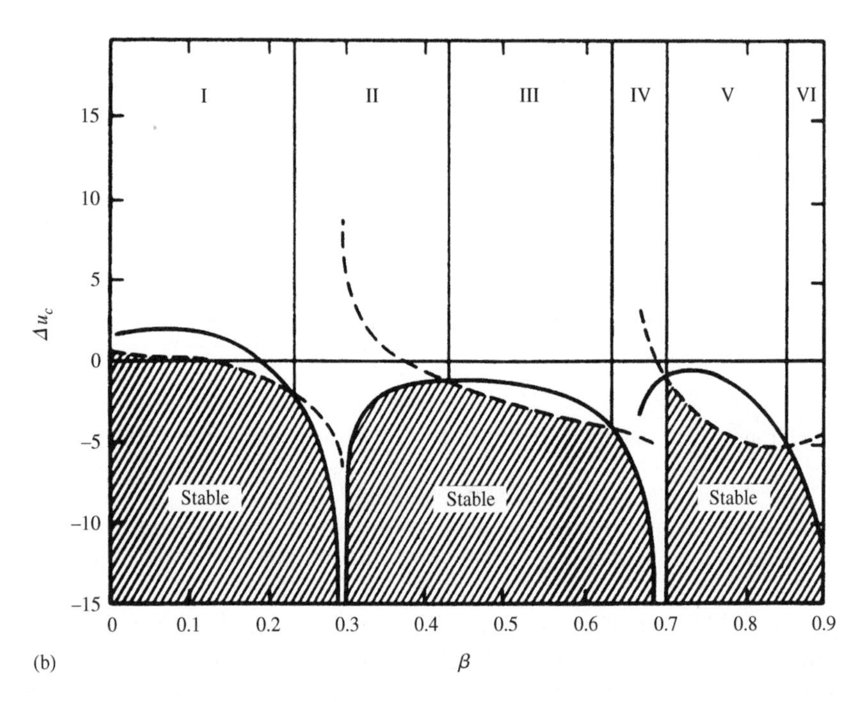

Figure 5.19 (a) The pipe fitted with a nozzle at the free end, inclined at an angle θ_j in the plane of the paper; planar motions, both in this plane ('in-plane') and perpendicular to it ('out-of-plane'), are considered. (b) Stability of the system for small θ_j; Δu_c stands for the change in u_c *vis-à-vis* linear theory: ——, stability boundary for in-plane motions; – – –, stability boundary for out-of-plane motions (Lundgren *et al*. 1979).

determined for any given $u = (MUU_j/EI)^{1/2}L$ — cf. equations (3.74) — by solving the time-independent form of the nonlinear equations via elliptic functions. For small θ_j, the equations are linearized and the solution becomes

$$\frac{w}{L} = \frac{\theta_j}{u}\,[\sin u(1-\xi) + u\xi\cos u - \sin u]\,, \tag{5.112}$$

where $w(\xi)$ is the lateral deflection along the length. This solution has a constant and a linear term in ξ, causing a tilt of the pipe from the axial ξ-axis, while the trigonometric part indicates a wavy shape. The number of periods of the waves increases with u and 'often this number becomes surprisingly large before the system develops a dynamic instability by flutter'. The equations are then linearized about the static equilibrium configuration, the stability of which is examined by obtaining perturbation solutions for small θ_j, for motions both in the plane of the nozzle ('in-plane') and perpendicular to it ('out-of-plane'). The results are shown in Figure 5.19(b). It is interesting that the zones where the system loses stability out-of-plane (zones I, III and V), correspond nearly exactly to the regions where Rousselet & Herrmann (1981) predict that the system is stabilized by 'nonlinearities'! Suppressing the small region around $\beta = 0.295$ in Figure 5.18 (where dissipation is known to have a destabilizing effect — Section 3.5.3), the comparison is made in Table 5.1. Evidently, the nozzle affects stability in a similar way as 'nonlinearities': when the nozzle has a stabilizing effect in its own plane, the pipe loses stability out-of-plane; and when the effect is destabilizing, then in-plane flutter is obtained.

The results were tested experimentally for varying values of β, and accord between theory and experiment, as to whether in-plane or out-of-plane limit cycles arise, is quite good, considering especially that the experiments were with vertical pipes, whereas in the theory gravity effects are neglected.

The definitive study into the nature of the Hopf bifurcations leading to flutter, for the same system as that examined by Rousselet & Herrmann, is due to Bajaj *et al.* (1980), who conducted a sophisticated analysis, utilizing the tools of modern dynamics theory. The equation of motion analysed involves a constant upstream pressure and a deformation-dependent flow velocity. The partial differential equation is transformed to vector form, and then the linear problem and its adjoint are solved. The solution procedure is nonstandard, but is based on the ideas of centre manifold theory and averaging [refer also to the discussion of Bajaj's (1987b) work in Section 5.9.2]. A periodic solution with undetermined amplitude and phase is assumed, and the equations are eventually reduced to the following normal form:

$$\dot{r} = \epsilon r(\mu + ar^2) + \mathcal{O}(\epsilon^2), \qquad \dot{\psi} = \omega_o + \epsilon[\mu c + br^2] + \mathcal{O}(\epsilon^2). \tag{5.113}$$

Depending on the sign of a, the emerging limit cycle is stable ($a < 0$, supercritical Hopf) or unstable ($a > 0$, subcritical Hopf). The results are shown in Figure 5.20, and depend on the parameter $\alpha = fL/\sqrt{A}$, f being the friction factor and $A = \frac{1}{4}\pi D_i^2$ the flow area; thus, $\alpha \propto L/D_i$, represents the slenderness of the pipe. It is seen that for sufficiently long (slender) pipes, hence large α, the bifurcation is always supercritical, whereas for short enough pipes[†] it can be subcritical. Significantly, for any given α, the regions of sub- and supercritical Hopf bifurcations in Figure 5.20 do not correspond to those of destabilization and stabilization due to 'nonlinearities' in Table 5.1, and hence contradict the results of Rousselet & Herrmann (1981); the reason for this is likely the fact that 'artificial damping' is included among the 'nonlinearities' in Rousselet & Herrmann's

Table 5.1 Effect of 'nonlinearities' (Rousselet & Herrmann 1981) and plane of flutter for a pipe fitted with an inclined end-nozzle (Lundgren *et al.* 1979) for different ranges of β; in-plane means in the plane of the nozzle, and out-of-plane perpendicular to the plane of the nozzle.

β	Effect of 'nonlinearities'	β	Plane of flutter
0.02–0.21	Stabilizing	0.00–0.23	Out-of-plane
0.21–0.42	Destabilizing	0.23–0.42	In-plane
0.42–0.66	Stabilizing	0.42–0.63	Out-of-plane

[†]Still presuming that they are long enough for Euler–Bernoulli theory to hold.

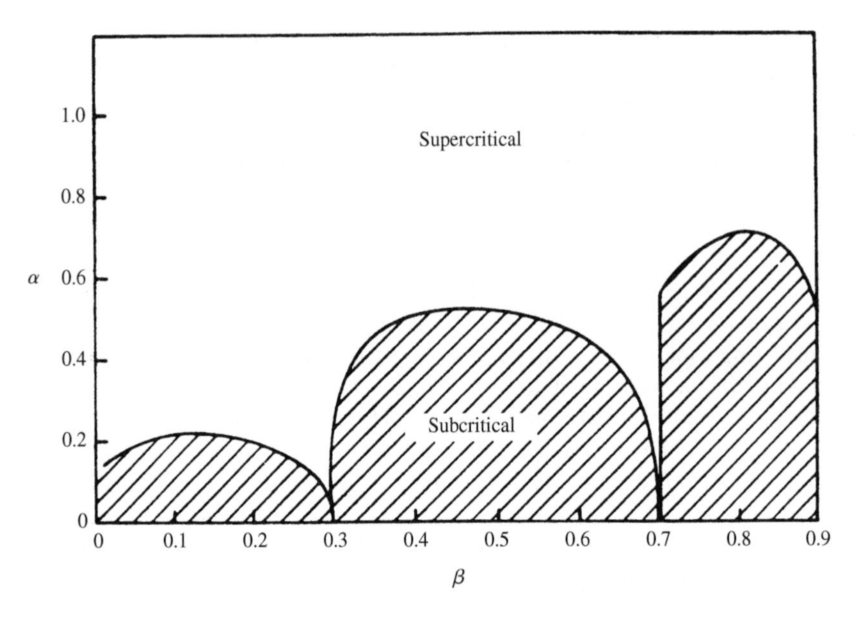

Figure 5.20 The different types of Hopf bifurcation of a cantilevered pipe depending on β and α, where α is a friction factor multiplied by $(4/\pi)^{1/2}L/D_i$, L/D_i being the pipe slenderness (Bajaj *et al*. 1980).

work, as discussed in the foregoing. One experiment by Bajaj *et al*. with three lengths of the same pipe ($\beta = 0.342$) showed supercritical loss of stability for the longer pipes and a subcritical one for the shorter pipe, in full agreement with theory.

The same topic was studied via simulation by Ch'ng & Dowell (1979), using two variants of their nonstandard equations of motion [Section 5.2.8(d)]. It is of interest that a subcritical Hopf bifurcation is found in a case with $\beta = 0.5$: for small initial conditions the oscillation dies out, while for large enough initial conditions the solution converges to a limit-cycle motion (implying the existence of an unstable limit cycle in-between, cf. Figure 2.11(d)), as found experimentally by Païdoussis (Section 3.5.6). This result by Ch'ng & Dowell is cited mainly to make the following point. In the semi-analytical studies, nonlinearities of only up to $\mathcal{O}(\epsilon^3)$ are generally retained, since to go to $\mathcal{O}(\epsilon^5)$ would make the analysis unwieldy. As a result, in all the foregoing, the existence of the outer, stable limit cycle in the case of a subcritical Hopf bifurcation could only be surmised — cf. equations (2.165) and (2.166) and Figures 2.12 and 2.13. In the simulations, however, this problem does not arise.

Another numerical study is due to Edelstein *et al*. (1986), utilizing the Lundgren *et al*. (1979) equations of motion (with no end-nozzle) and solving them by means of a finite element method and a 'penalty function' technique.† In the calculations, for a pipe studied

†In Lundgren's *et al*. equations one obtains two equations of motion involving $T - pA$, which should be eliminated to obtain just one equation. In the penalty function approach, $e = (T - pA)L^2/EI$ is defined and equated to $\lambda[(\partial x/\partial x_0)^2 + (\partial z/\partial x_0)^2 - 1]$, in which the penalty parameter λ is generally a large number $\sim \mathcal{O}(10^5)$ and the last bracketed expression is zero because of the inextensibility condition, equation (5.1). Thus, by allowing a small amount of 'mathematical extensibility' one can both eliminate e from the equations of motion and satisfy the inextensibility condition.

experimentally by Chen (see Section 3.5.6), the limit cycle appears to be supercritical, but this is not made absolutely clear, although in the experiments it is definitely subcritical. Nevertheless, agreement with the experimental stable limit-cycle amplitude and flutter frequency is reasonably good.

Before ending this section, the singular behaviour of the system in the vicinity of $\beta \simeq 0.3$ and 0.69 in Figures 5.18–5.20 should be remarked upon; i.e. at the same values of β where S-shaped discontinuities in the u_c versus β plot occur (Figure 3.30), and with which so many interesting features of linear dynamics are associated (Sections 3.5 and 3.6). Furthermore, the 'stabilizing/destabilizing' and in-plane/out-of-plane ranges of β in Table 5.1 straddle these same S-shaped discontinuities. Referring to the work presented in Section 3.5.4, this is hardly surprising: the modal content of the flutter mode and the energy transfer mechanism both experience radical alterations about these critical values of β. Hence, it is reasonable that the nature of limit-cycle motions should also be modified across these same values of β.

5.7.2 3-D limit-cycle motions

As already mentioned, the main objective in the case of 3-D motions of the pipe is the prediction of whether the flutter motions are planar or three-dimensional (orbital). The necessary analysis, using a simplified set of equations with a deflection-independent flow velocity, was done by Bajaj & Sethna (1984). As in Bajaj & Sethna's (1982a,b) similar analysis of articulated pipes (Section 5.6.2), stability in this system is lost via *two* pairs of complex eigenvalues crossing the imaginary axis simultaneously, i.e. via a generalized Hopf bifurcation. This system is said to have O(2) symmetry, possessing both rotational (about the x-axis) and 'reflective' symmetry across that axis.

The original governing PDE is reduced directly to a set of ODEs on the centre manifold without introducing Galerkin expansion, similarly to Bajaj *et al*. (1980). The discretized equations are then projected onto the centre manifold, and periodic solutions close to the critical flow velocity are sought. The resulting equations are then brought into normal form by the method of averaging. Eventually, equations similar to (5.113) are obtained, but in this case two different amplitude parameters, a_1 and a_2, are involved; for rotary motions $a_1 = a_2$. The final results are shown in Figure 5.21. Similarly to the articulated system (Section 5.6.2(b)), whenever two supercritical bifurcations occur for the same system, that with the larger amplitude is the stable one, while the other is unstable. Once more, the singularities in this figure, namely the values of β where the limit-cycle amplitude can be zero, correspond to the location of the S-shaped curves in the linear stability diagram (Figure 3.30). However, the points where limit-cycle motions switch from planar to circular (rotary) and *vice versa* do not correspond to any of the ranges in Table 5.1.

The dynamics of the same system was also studied (Bajaj & Sethna 1991) in the presence of small imperfections, representing different bending stiffnesses and additional viscous damping in two mutually perpendicular directions, imposed to break the rotational symmetry. Thus, the linear flexural rigidity term in one equation is multiplied by $(1 + \epsilon\delta)$ and in the other by $(1 - \epsilon\delta)$, while the damping coefficients are ϵc_1 and ϵc_2, respectively, ϵ being a small parameter. The rotational symmetry is therefore broken by the parameters δ and $(c_1 - c_2)$. Similar analytical techniques as in Bajaj & Sethna (1984) are used. At

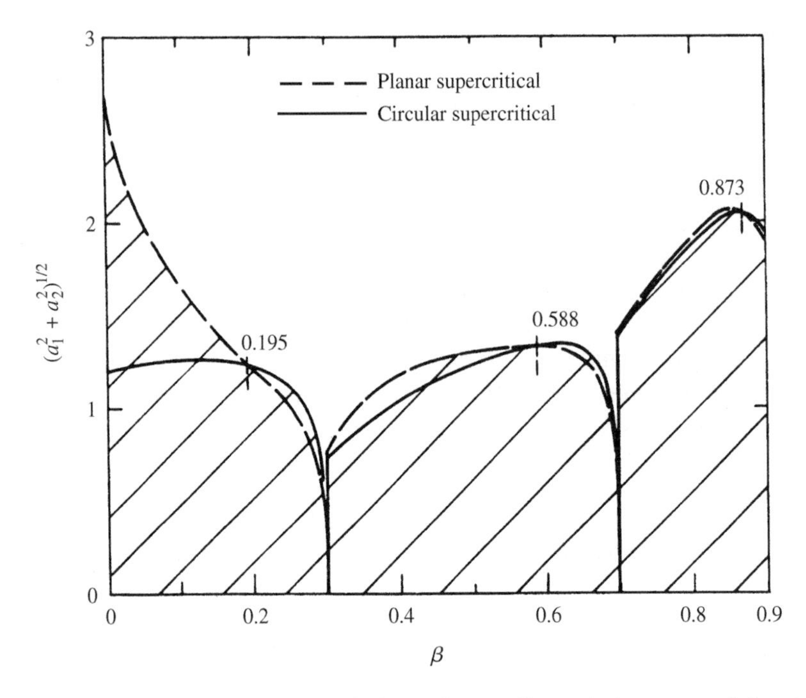

Figure 5.21 The amplitude of periodic solutions of a cantilevered pipe versus β for planar and circular (rotary) supercritical Hopf bifurcations, of which that with the larger amplitude is stable: – – –, planar motions; ——, circular motions (Bajaj & Sethna 1984).

flow rates past a critical value, two primary branches of 'standing waves', i.e. motions restricted to a plane, are found. Depending on β, the standing wave may undergo a pitchfork bifurcation into a 'travelling wave' (i.e. rotary pipe motions, which in this asymmetric case are *elliptic*), or it may coexist with travelling waves arising from a saddle-node bifurcation. Secondary bifurcations and codimension-two bifurcations to modulated waves are also considered. A very rich dynamical behaviour is displayed, a sample being shown in Figure 5.22.

It may be seen in that figure that, as a result of the breach of symmetry, the system now loses stability by planar ('standing wave', SW) motions at different flow rates in the two mutually perpendicular planes, differing by a phase angle ($\theta = 0$ or π); the critical values of the flow parameter are $\lambda = -0.5211$ and $\lambda = +0.5211$.[†] The SW_π solution, emerging via a supercritical Hopf bifurcation is associated with the weaker of the two planes and is stable, while the SW_0 solution is subcritical and unstable. Rotary ('travelling wave', TW) motions do not arise until a higher flow rate is reached ($\lambda > \lambda_{min} = 2.5113$), whereupon two such solutions emerge, one stable and the other unstable. The SW_π solution becomes unstable by a secondary Hopf bifurcation at $\lambda = \lambda_h = 4.600$, signifying possible modulated oscillations. However, this solution branch is unstable (the

[†]The 'flow parameter' λ is defined as $\lambda = \mu\beta_{1r} + \frac{1}{2}e_{1i}(c_1 + c_2)$, where μ is the flow bifurcation parameter, and ϵc_1 and ϵc_2 represent the damping in the two mutually perpendicular planes of motion, $\epsilon \ll 1$; β_{1r} and e_{1i} are constants dependent on the flow velocity and frequency at the critical point, on β, and on the deformation (Bajaj & Sethna 1991; equation (21)).

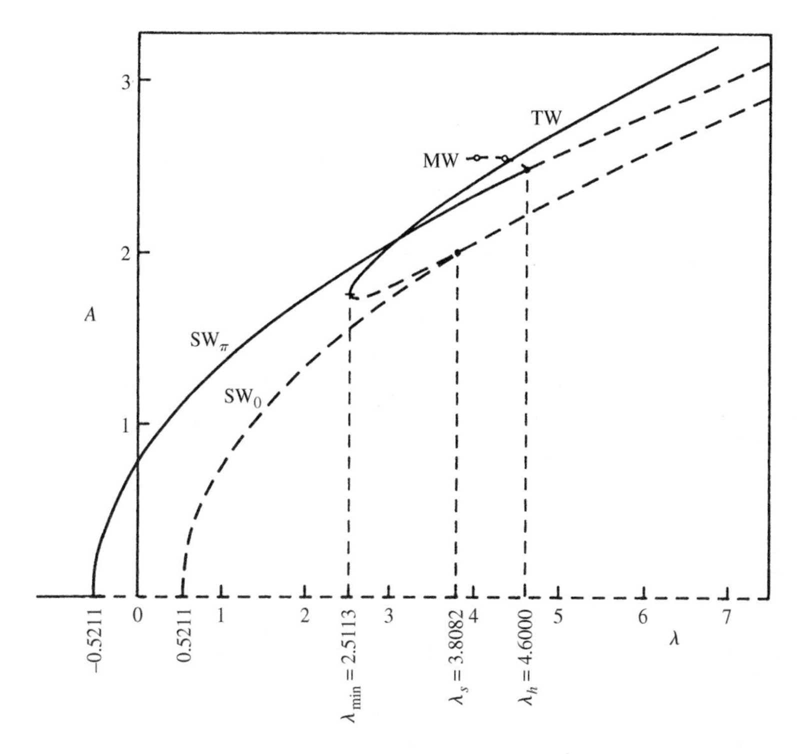

Figure 5.22 Bifurcation diagram of the perturbed cantilevered pipe system with a small stiffness asymmetry, showing the amplitude, A, of planar motions ('standing waves', SW), rotary motions ('travelling waves', TW) and modulated waves (MW), as a function of the flow parameter λ for $\beta = 0.25$ and $\delta = 0.1$ (Bajaj & Sethna 1991).

bifurcation is subcritical): for small perturbations the solution converges to SW$_\pi$, and for larger ones to TW — the two coexisting for $\lambda_{\min} < \lambda < \lambda_h$. For $\lambda > \lambda_h$, only rotary (TW) motions are possible.

These theoretical predictions are qualitatively supported by experiment, involving slender, vertical cellulose acetate butyrate pipes ($L = 1.83$ m, $D_i = 6.48$ mm, $D_o = 7.95$ mm). The pipes were fitted with straight nozzles at the free end, with area ratios of $A_j/A = 0.3$ and 0.4, resulting in $\beta = 0.173$ and 0.231 [cf. Section 3.5.6 and equations (3.74)] — the latter case being close to $\beta = 0.25$, as in Figure 5.22. Experiments were conducted with a round pipe and then with flats machined on diametrally opposite sides of its outer surface, resulting in $\delta \simeq 0.05$ (cf. $\delta = 0.1$ in Figure 5.22). The following remarkable set of observations are made.

(i) For the round, nearly perfectly symmetric pipe with $\beta = 0.173$, the initial limit-cycle motion is confined to one plane — as determined by whatever imperfections are present. With further increase in the flow rate, the amplitude of motion increases but it remains planar, in the same plane. This is as predicted in Figure 5.21, where the planar motion is stable, while the circular one is unstable.

(ii) For the same configuration but $\beta = 0.231$, the initial limit-cycle motion is again planar. For a slightly larger flow rate, however, the motion becomes circular (rotary), which is the situation predicted by theory (Figure 5.21).

(iii) The $\beta = 0.173$ pipe with flats behaves essentially as in (i); this, also agrees with Bajaj & Sethna's theoretical predictions.

(iv) The most interesting case is the $\beta = 0.231$ pipe with flats. The initial limit-cycle motion is again planar (SW). This motion persists to a certain flow rate, u_h, qualitatively corresponding to λ_h in Figure 5.22, whereupon the pipe motion suddenly develops 'complex spatial transients' and then settles down to a large elliptical limit cycle (TW). As the flow rate is increased, the ellipse becomes more nearly a circle. Changes in the initial conditions result in circular motion in the opposite direction. On reducing the flow rate, circular motion (TW) persists to below u_h, with reduced amplitudes and a more sharply elliptical shape (large ratio of major to minor axis). At a lower flow still, corresponding to $\lambda_{\min}$ in Figure 5.22, spatial transients develop again, and then motion settles down to a planar limit cycle (SW). These observations agree qualitatively remarkably well with theory (Figure 5.22). In particular, for flow rates between $\lambda_{\min}$ and λ_h, both planar (SW) and rotary (TW) stable motions are found to exist and, by carefully controlling the initial conditions, either can be achieved.

5.7.3 Dynamics under double degeneracy conditions

The main reason for studying doubly or multiply degenerate systems lies in the fact that, in the presence of 'competing attractors', the dynamics becomes particularly interesting. In 'unfolding the bifurcations' a variety of different dynamical behaviour is found, and regions where chaos may arise can be identified.

Some double degeneracies, e.g. those associated with 3-D motions and generalized Hopf bifurcations (which are discussed in Section 5.7.2, but which could equally well have been covered here), are *inherent* in the system — whatever the system parameters — provided that it is perfectly symmetric. Some others, such as those treated here, involving different types of coincident bifurcations (pitchfork and Hopf) arise for *particular* sets of system parameters, irrespective of symmetry. Studies of this type were conducted by Païdoussis & Semler (1993b), Li & Païdoussis (1994) and Steindl & Troger (1988, 1995). In all cases, these studies build upon Sethna & Shaw's (1987) pioneering work on articulated pipes.

Specifically, three particular studies into the dynamics of cantilevered pipes near a point of double degeneracy are presented in this section: (a, b) the 2-D and (c) the 3-D dynamics of a pipe with an intermediate spring support, and (d) the 2-D dynamics of an 'up-standing' cantilever in a gravity field. The case of 2-D dynamics of the pipe–spring system is the only one in this book where the analysis is presented in reasonable detail, starting with the nonlinear equations of Section 5.2.7; hence, the presentation is broken into two: (a) the general analysis, and (b) the analysis under double degeneracy conditions.

(a) 2-D motions of pipe–spring system; general analysis

The planar dynamics of a vertical cantilever with an intermediate, linear spring support (Figure 5.23) has been studied by Païdoussis & Semler (1993b). Equations (5.42) and (5.43) constitute the equation of motion used, but with the linear spring term

$$\kappa\eta\delta(\xi - \xi_s) \tag{5.114}$$

added, where $\kappa = kL^3/EI$ is the dimensionless spring constant and $\xi_s = x_s/L$ is the location of the spring. The linear dynamics of the system has been discussed in

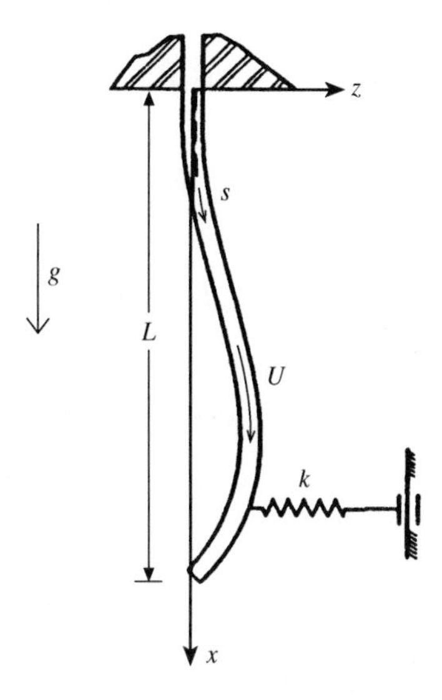

Figure 5.23 Schematic of the vertical cantilevered pipe with an intermediate linear spring support, capable of double degeneracy conditions (Païdoussis & Semler 1993b).

Section 3.6.2 (Figures 3.64 and 3.65), where it is shown that both divergence and flutter are generally possible.

The nonlinear equation of motion is discretized by Galerkin's method, yielding

$$\ddot{q}_i + C_{ij}\,\dot{q}_j + K_{ij}\,q_j + \epsilon(\alpha_{ijkl}\,q_j\,q_k\,q_l + \beta_{ijkl}\,q_j q_k\,\dot{q}_l + \gamma_{ijkl}\,q_j\,\dot{q}_k\,\dot{q}_l) = 0, \tag{5.115}$$

where C_{ij} and K_{ij} are defined as

$$C_{ij} = \alpha\lambda_i^4\delta_{ij} + 2\beta^{1/2}u b_{ij}, \qquad K_{ij} = \lambda_i^4\delta_{ij} + \kappa\phi_i(\xi_s)\phi_j(\xi_s) + u^2 c_{ij} + \gamma(b_{ij} - c_{ij} + d_{ij}),$$

$$\alpha_{ijkl} = u^2 a_{ijkl} + \gamma b_{ijkl} + c_{ijkl}, \qquad \beta_{ijkl} = 2u\beta^{1/2} d_{ijkl}, \tag{5.116a}$$

$$\gamma_{ijkl} = \int_0^1 \phi_i\phi_j'\left\{\int_0^\xi \phi_k'\phi_l'\,\mathrm{d}\xi\right\}\mathrm{d}\xi - \int_0^1 \phi_i\phi_j''\left\{\int_\xi^1\int_0^\xi \phi_k'\phi_l'\,\mathrm{d}\xi\,\mathrm{d}\xi\right\}\mathrm{d}\xi,$$

and

$$a_{ijkl} = \int_0^1 \phi_i\phi_j''\phi_k'\phi_l'\,\mathrm{d}\xi - \int_0^1 \phi_i\phi_j'\left\{\int_0^\xi \phi_k'\phi_l'''\,\mathrm{d}\xi\right\}\mathrm{d}\xi$$
$$+ \int_0^1 \phi_i\phi_j''\left\{\int_\xi^1\int_0^\xi \phi_k'\phi_l'''\,\mathrm{d}\xi\,\mathrm{d}\xi\right\}\mathrm{d}\xi - \int_0^1 \phi_i\phi_j''\left\{\int_\xi^1 \phi_k'\phi_l''\,\mathrm{d}\xi\right\}\mathrm{d}\xi,$$

$$b_{ijkl} = -\tfrac{3}{2}\int_0^1 \phi_i\phi_j''\phi_k'\phi_l'(1-\xi)\,\mathrm{d}\xi - \tfrac{1}{2}\int_0^1 \phi_i\phi_j'\phi_k'\phi_l'\,\mathrm{d}\xi$$
$$+ \int_0^1 \phi_i\phi_j'\left\{\int_0^\xi \phi_k'\phi_l'''(1-\xi)\,\mathrm{d}\xi\right\}\mathrm{d}\xi$$

$$
-\int_0^1 \phi_i\phi_j''\left\{\int_\xi^1\int_0^\xi \phi_k'\phi_l'''(1-\xi)\,\mathrm{d}\xi\,\mathrm{d}\xi\right\}\mathrm{d}\xi+\int_0^1 \phi_i\phi_j''\left\{\int_\xi^1 \phi_k'\phi_l'\,\mathrm{d}\xi\right\}\mathrm{d}\xi,
$$

$$
c_{ijkl}=3\int_0^1 \phi_i\phi_j'\phi_k''\phi_l'''\,\mathrm{d}\xi+\int_0^1 \phi_i\phi_j''\phi_k''\phi_l''\,\mathrm{d}\xi-\int_0^1 \phi_i\phi_j''\left\{\int_\xi^1\int_0^\xi \phi_k''\phi_l''''\,\mathrm{d}\xi\right\}\mathrm{d}\xi
$$

$$
+\int_0^1 \phi_i\phi_j'\left\{\int_0^\xi \phi_k''\phi_l''''\,\mathrm{d}\xi\right\}\mathrm{d}\xi-\int_0^1 \phi_i\phi_j''\left\{\int_\xi^1 \phi_k''\phi_l'''\,\mathrm{d}\xi\right\}\mathrm{d}\xi, \tag{5.116b}
$$

$$
d_{ijkl}=\int_0^1 \phi_i\phi_j'\phi_k'\phi_l'\,\mathrm{d}\xi-\int_0^1 \phi_i\phi_j'\left\{\int_0^\xi \phi_k'\phi_l''\,\mathrm{d}\xi\right\}\mathrm{d}\xi
$$

$$
+\int_0^1 \phi_i\phi_j''\left\{\int_\xi^1\int_0^\xi \phi_k'\phi_l''\,\mathrm{d}\xi\,\mathrm{d}\xi\right\}\mathrm{d}\xi-\int_0^1 \phi_i\phi_j''\left\{\int_\xi^1 \phi_k'\phi_l'\,\mathrm{d}\xi\right\}\mathrm{d}\xi,
$$

where the λ_i are the dimensionless eigenvalues associated with the beam eigenfunctions ϕ_i, used here as comparison functions in the discretization. The coefficients b_{ij}, c_{ij} and d_{ij} are computed from the integrals of the eigenfunctions (Section 3.3.6, Table 3.1) while a_{ijkl}, b_{ijkl}, c_{ijkl} and d_{ijkl} are computed numerically (Semler 1991; Li & Païdoussis 1994).

The repeated indices in (5.115) implicitly follow the summation convention. The nonlinear terms have been multiplied by ϵ, used here as a book-keeping device to indicate that they are small. Equation (5.115) may be re-written in first-order form:

$$
\begin{Bmatrix}\dot q\\ \dot p\end{Bmatrix}=\begin{bmatrix}0 & I\\ -K & -C\end{bmatrix}\begin{Bmatrix}p\\ q\end{Bmatrix}+\epsilon f(q,p),
$$

in which $p=\dot q$; or

$$
\dot y=[A]y+\epsilon f(y), \tag{5.117}
$$

where I, K, C are the identity, stiffness and damping matrices, and q, p and y are understood to be vectors. Although all this is applicable to any order of discretization, all numerical results (and the centre manifold calculations in subsection (b)) are confined to a two-term Galerkin discretization, $N=2$.

In the remainder of this subsection, a linear and then a nonlinear stability analysis is undertaken, in the latter case supplemented by simulation. A typical Argand diagram of the eigenvalues of $[A]$ as u is increased is shown in Figure 5.24(a). The system loses stability by divergence at $u=11.47$ in its first mode. Then, according to linear analysis, it is subject to flutter at $u=12.48$ in its second mode. The two branches of the divergent mode merge, and that mode regains stability, at $u=15.07$.

A nonlinear analysis of the dynamics in the vicinity of the fixed points is conducted next. The fixed points are given by

$$
K_{ij}q_j^0+\alpha_{ijkl}\,q_j^0q_k^0q_l^0=0, \tag{5.118}
$$

where the superscript 0 denotes the fixed point. Then, considering perturbations about a fixed point, $q_i=q_i^0+u_i$, $p_i=v_i$, the perturbation equations are obtained from (5.115) and (5.117):

$$
\begin{Bmatrix}\dot u_i\\ \dot v_i\end{Bmatrix}=\begin{bmatrix}0 & I_{ij}\\ -K_{ij}-(\alpha_{ijkl}+\alpha_{ikjl}+\alpha_{ilkj})\,q_k^0q_l^0 & -C_{ij}-\beta_{ijkl}\,q_j^0q_k^0\end{bmatrix}\begin{Bmatrix}u_j\\ v_j\end{Bmatrix}. \tag{5.119}
$$

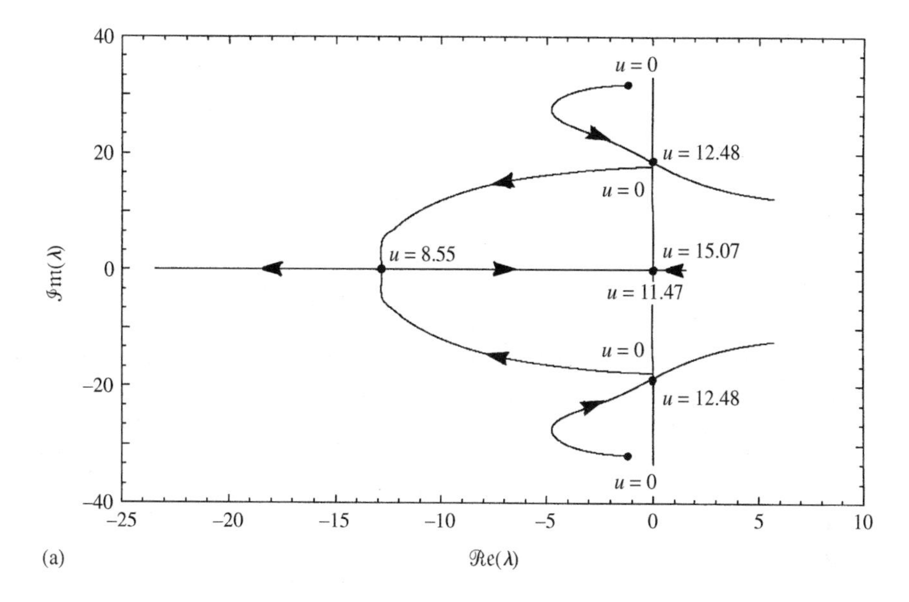

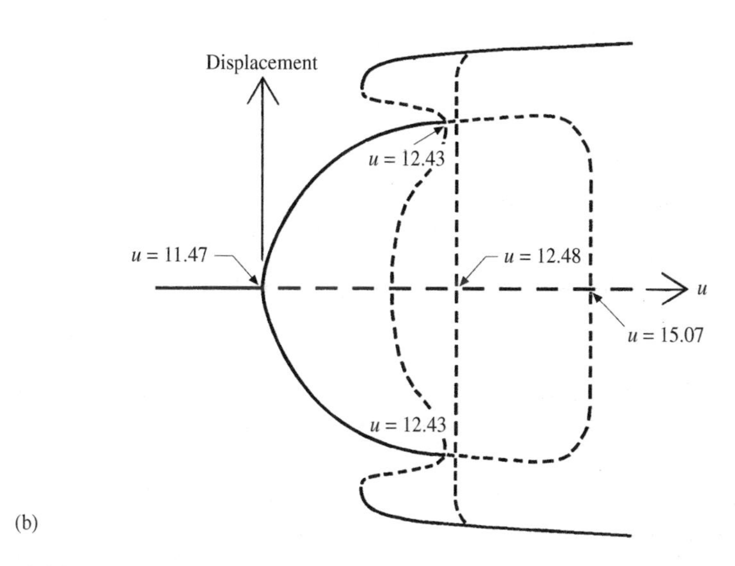

Figure 5.24 (a) Argand diagram of the eigenvalues of the linearized system of Figure 5.23 and (b) diagram of the tip displacement of the nonlinear system, for $\beta = 0.18$, $\gamma = 60$, $\alpha = 5 \times 10^{-3}$, $\kappa = 100$, $\xi_s = 0.8$ (Païdoussis & Semler 1993b).

Hence, stability is assessed from the eigenvalues of the matrix in (5.119). The results are shown in Figure 5.24(b). The same notation for the eigenvalues as in Section 5.5.2(a) is used here. It is seen that the origin {0} becomes unstable through a pitchfork bifurcation, $\lambda_{\{0\}} = (0, -, -, -)$ at $u = 11.47$, which corresponds to what is seen in Figure 5.24(a). Two stable equilibria appear: the fixed points $\{\pm 1\}$ with eigenvalues $\lambda_{\{\pm 1\}} = (-, -, -, -)$, where $\{\pm 1\}$ simply denotes the *first* set of fixed points. They remain stable until $u = 12.43$, when Hopf bifurcations occur and $\lambda_{\{\pm 1\}} = (+, +, -, -)$. This probably leads to stable limit-cycle motion, since there is no other stable equilibrium state. At $u = 12.48$, it is the origin {0} that undergoes a Hopf bifurcation. The three fixed points {0} and {±1} coalesce at $u = 15.07$ [$\lambda_{\{\pm 1\}} = \lambda_{\{0\}} = (+, +, 0, -)$]. A numerical investigation confirms the results found: limit-cycle oscillations are found to exist before the first Hopf bifurcation occurring at $u = 12.43$, showing that these oscillations are due to the subcritical bifurcation of the {±1} fixed points. For u a little less than 12.43, e.g. at $u = 12.35$, the orbit can be attracted either by one of the stable fixed points or by the attracting periodic limit-set.

It is of particular interest that, in this case, post-divergence flutter does materialize, although not in the manner predicted by linear theory: i.e. it emerges from the new stable fixed points associated with first-mode destabilization, rather than from the second mode. From this and other similar calculations, it is clear that the nonlinear dynamics of the system can be substantially different from linear predictions. Thus, the stability map in Figure 5.25(a), obtained by linear analysis, can only be relied upon for the *first* loss of stability: by divergence for $-56 < \gamma < 71.9$ and by flutter for $\gamma > 71.9$ approximately for the particular set of parameters given in the caption; the other predicted instabilities, beyond the first, do not necessarily materialize.

The region of 'global oscillations' in Figure 5.25(a) cannot be obtained by linear or even local nonlinear analysis, but was found numerically. 'Global' is used here to indicate that the oscillations circumnavigate more than one, in this case three, fixed points. For $u = 7.5$, it is seen in Figure 5.25(b) that the origin has become a saddle, but two new stable equilibria exist. For $u = 13.1$, however, the dynamics is more complicated, as shown in Figure 5.25(c). The origin {0} is a saddle, as well as the second pair of fixed points, {±2}; for clarity, not all the stable and unstable manifolds have been drawn in this figure, and the existence of only one fixed point of the second pair at ~ 0.1 is revealed by the trajectories shown. The first pair {±1} at ±0.2, is 'weakly' attracting. Flows with initial conditions close to the equilibrium are attracted by one of the fixed points {±1}. However, other attracting sets also exist: one may observe either limit-cycle oscillations around one of the equilibria or global limit-cycle oscillations around the five equilibria. Those oscillations do not come from local bifurcations. For Duffing's equation, for example, solutions lie on level curves of the Hamiltonian energy of the system, and these solutions are closed orbits representing a global stability state (Guckenheimer & Holmes 1983).

(b) 2-D motions of pipe–spring system; double degeneracy conditions

Here the dynamics of the system is discussed in the vicinity of the double degeneracy, in this case due to coincidence of a pitchfork and a Hopf bifurcation. Figure 5.26 shows appropriate combinations of β, γ and κ for which such a double degeneracy is obtained.

Appendix H shows how the nonlinear dynamics in the vicinity of a Hopf or a pitchfork bifurcation may be analysed by means of centre manifold theory and either normal form or averaging analysis. Here the same type of analysis is done under conditions of double degeneracy.

In this case, introducing an appropriate modal matrix $[P]$ and the transformation $y = [P]x$, equation (5.117) can be put into 'standard form', defined by

$$\dot{x} = [\Lambda]x + \epsilon[P]^{-1} f([P]x), \tag{5.120}$$

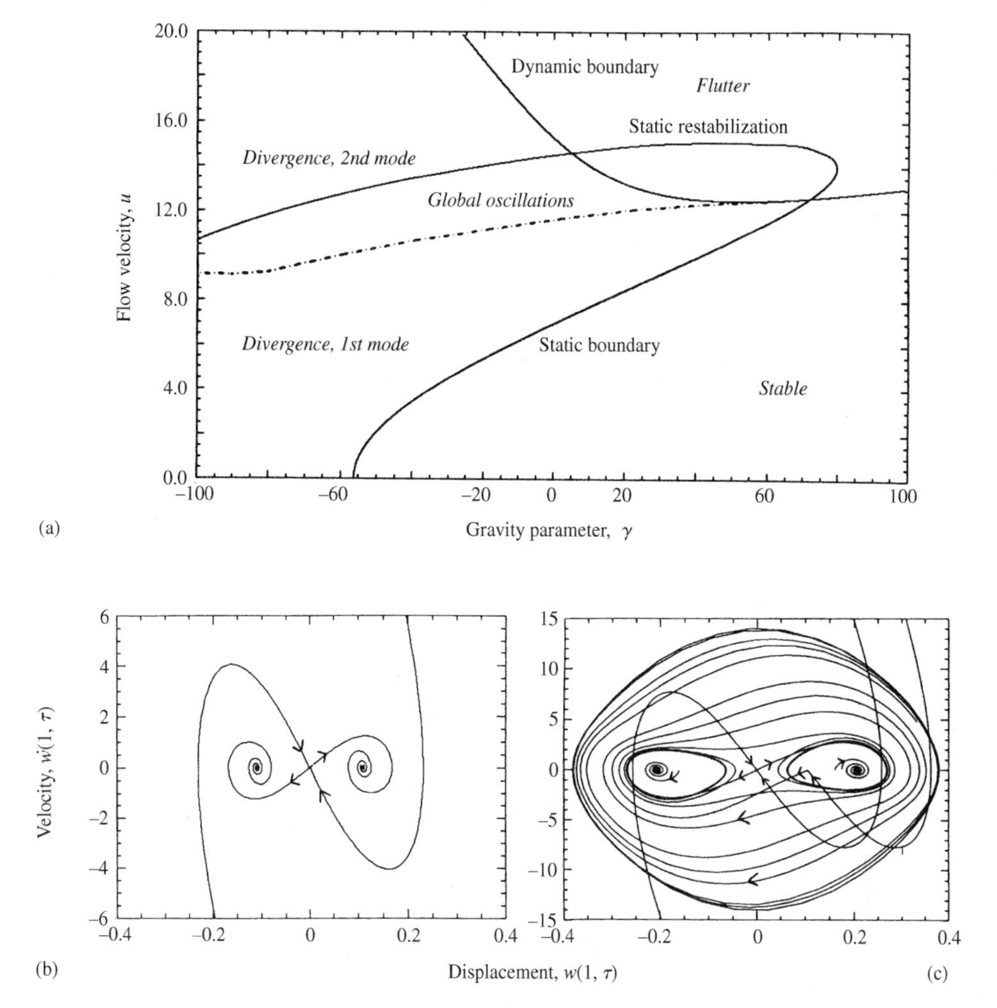

Figure 5.25 (a) Stability boundaries for the system of Figure 5.23 obtained by direct eigenvalue analysis, except for the 'global oscillations' region obtained numerically from the equations of the nonlinear system; (b) phase portrait of the system showing the saddle node at {0} and the two stable equilibria {±1} for $\gamma = -60$, and $u = 7.5$; (c) three saddles {0} and {±2}, two stable equlibria {±1} and global oscillations for $\gamma = -60$ and $u = 13.1$. In all cases $\beta = 0.18$, $\alpha = 5 \times 10^{-3}$, $\kappa = 100$, $\xi_s = 0.8$ (Païdoussis & Semler 1993b).

where

$$[\Lambda] = \begin{bmatrix} \begin{bmatrix} 0 & -\omega_0 & 0 \\ \omega_0 & 0 & 0 \\ 0 & 0 & 0 \end{bmatrix} & [0] \\ [0] & [M] \end{bmatrix}; \tag{5.121}$$

$[M]$ is the matrix of the eigenvalues with negative real parts; for this 4-D system the last row of $[\Lambda]$ involves a scalar.

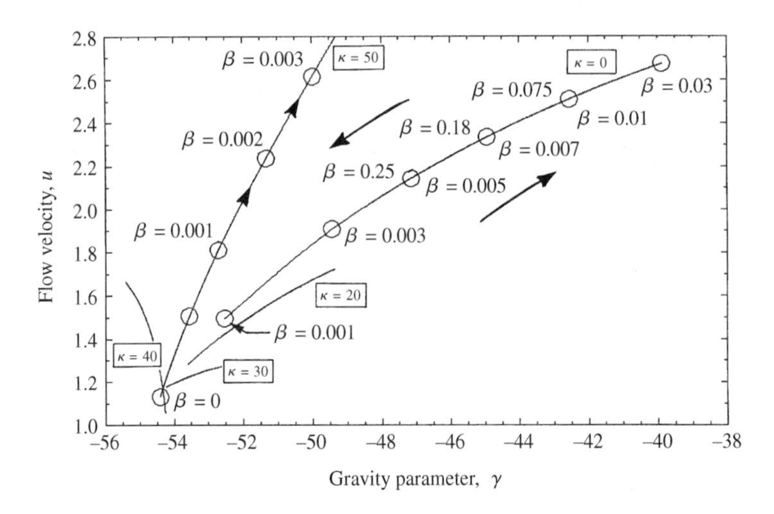

Figure 5.26 Double degeneracy conditions for the system of Figure 5.23 for different values of κ. The control parameter varied is β. It is noted that for $\kappa = 0$ essentially the same curve is retraced backwards as β is incremented past $\beta = 0.03$; this is not the case for higher κ, e.g. $\kappa = 50$, where the curve continues on to positive values of γ (Païdoussis & Semler 1993b).

Ignoring the stable eigendirection associated with $[M]$, see Appendix H, the flow in the vicinity of the double degeneracy on the centre manifold may be described by the subsystem

$$\dot{x} = \begin{bmatrix} 0 & -\omega_0 & 0 \\ \omega_0 & 0 & 0 \\ 0 & 0 & 0 \end{bmatrix} x + \begin{bmatrix} \epsilon\mu_1 & -\epsilon\mu_3 & 0 \\ \epsilon\mu_3 & \epsilon\mu_1 & 0 \\ 0 & 0 & \epsilon\mu_2 \end{bmatrix} x + \epsilon f(x), \tag{5.122}$$

where f is a third-order polynomial in x, different from f in (5.120). In the dynamics vocabulary, μ_1 and μ_2 are called unfolding parameters, and they represent the deviations of the real parameters from their critical values (see Appendix F). In the case of a double degeneracy, two such parameters are necessary to unfold the dynamics of the problem (codimension-two bifurcation).

The next step is to follow the strategy of normal forms, in which all the nonessential nonlinear terms of f are eliminated ('nonessential' meaning that they do not affect the qualitative dynamics), as described in Appendix F.3. In the case of the double degeneracy with certain symmetry properties, the normal form is shown to be

$$\begin{aligned} \dot{r} &= \epsilon[\mu_1 r - (a_{11}r^2 + a_{12}z^2)]r + \mathcal{O}(\epsilon^2), \\ \dot{z} &= \epsilon[\mu_2 z + (a_{21}r^2 + a_{22}z^2)]z + \mathcal{O}(\epsilon^2), \\ \dot{\phi} &= \omega_0 + \mathcal{O}(\epsilon), \end{aligned} \tag{5.123}$$

where $r^2 = x_1^2 + x_2^2$ (Takens 1974; Guckenheimer & Holmes 1983). In physical terms, r represents the amplitude of oscillatory motions of the pipe, z represents the buckled positions of the pipe, and $\mathrm{d}\phi/\mathrm{d}t$ the frequency of oscillations. It is interesting to note that the first two of equations (5.123) and the third one are decoupled, providing immediately

$$\phi = \omega_0 t + \phi_0 + \mathcal{O}(\epsilon).$$

A rescaling procedure can transform the first two equations to their usual form (Guckenheimer & Holmes 1983; pp. 396–411),

$$\dot{r} = r(\mu_1 + r^2 + bz^2), \qquad \dot{z} = z(\mu_2 + cr^2 + dz^2), \qquad d = \pm 1. \tag{5.124}$$

This system has been studied by Takens (1974) who found nine topologically distinct equivalence classes. Results obtained from three different sets of parameters are presented for comparison:

$$\begin{array}{lllll} \text{Case 1:} & u = 2.245 & \gamma = -46.001 & \beta = 0.20 & \kappa = 0, \\ \text{Case 2:} & u = 12.598 & \gamma = 71.941 & \beta = 0.18 & \kappa = 100, \\ \text{Case 3:} & u = 15.111 & \gamma = 46.88 & \beta = 0.25 & \kappa = 100. \end{array} \tag{5.125}$$

The location of the linear spring is constant, $\xi_s = 0.8$, and in all three cases $d - bc \neq 0$. Table 5.2 shows the coefficients found and the corresponding equivalence class (last column) defined in Guckenheimer & Holmes (1983; p. 399). Starting from system (5.124) and referring to Figure 5.27, the classification of the different unfoldings can be undertaken. For example, one can easily show that pitchfork bifurcations occur from $\{0\}$ on the lines $\mu_1 = 0$ and $\mu_2 = 0$, and also that pitchfork bifurcations occur from $\left(r = \sqrt{-\mu_1}, z = 0\right)$ on the line $\mu_2 = c\mu_1$, and from $\left(r = 0, z = \sqrt{\mu_2}\right)$ on the line $\mu_2 = -\mu_1/b$. The behaviour of the system remains simple, as long as Hopf bifurcations do not occur from the new fixed point. This is the case when $d - bc < 0$. Hence, in case 2, no Hopf bifurcation can occur, while it is possible in cases 1 and 3. The bifurcation sets, and the associated phase portraits can be constructed for the different unfoldings; it is evident that in case 2 [Figure 5.27(b)] no global bifurcations are involved, while in the other two cases a heteroclinic loop (or 'saddle loop') emerges [Figure 5.27(a)].

To get a physical understanding of the motions of the pipe from the phase portraits of Figure 5.27, it may be useful to recall that (a) a fixed point on the z-axis represents a static equilibrium position; (b) a fixed point with $r \neq 0$ represents a periodic solution because of the angular variable ϕ; (c) a closed orbit represents amplitude-modulated oscillatory motions. By integrating numerically the equations of motion, some of the results obtained here analytically can be verified. For example, it is possible to find (i) the stable fixed point $\{0\}$; (ii) the stable fixed point $\{\pm 1\}$ corresponding to the buckled state; (iii) oscillatory motions around the origin $\{0\}$. However, attempts to obtain some of the more unusual features of the system shown in Figure 5.27(a), such as amplitude-modulated motions,

Table 5.2 Normal form coefficients and equivalence class for the three cases defined in (5.125).

	d	c	b	$d - bc$	Class
Case 1	$-1 < 0$	$-1.52 < 0$	$3.954 > 0$	+	VIa
Case 2	$-1 < 0$	$-0.07 < 0$	$-24.3 < 0$	−	VIII
Case 3	$-1 < 0$	$-3.39 < 0$	$1.656 > 0$	+	VIa

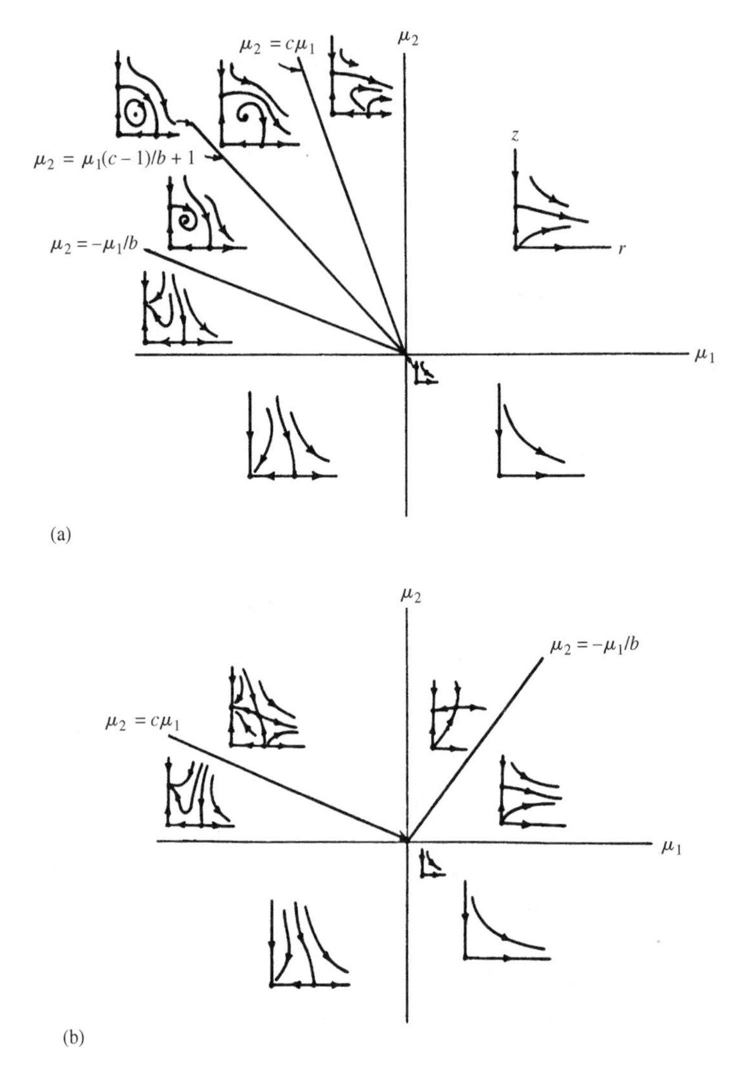

Figure 5.27 Codimension-2 bifurcation diagrams for the doubly degenerate system of Figure 5.23: (a) for cases 1 and 3, defined in (5.125), from Guckenheimer & Holmes (1983); (b) for case 2 (Païdoussis & Semler 1993b).

have not been successful. In Figure 5.27 it is seen that most of the limit sets are unstable. On the other hand, by numerical integration of the equations it is possible to find only the stable hyperbolic sets.

The heteroclinic orbit on the line $\mu_2 = \mu_1(c-1)/b + 1$ is of special interest. It is known that if perturbed, it may give rise to heteroclinic tangles and chaos (Guckenheimer & Holmes 1983; Moon 1992). This is discussed in more detail in Section 5.8.

(c) 3-D motions of the doubly degenerate pipe–spring system

Two studies on the three-dimensional motions of the system were conducted by Steindl & Troger (1988, 1994, 1995), utilizing the equations of motion of Lundgren *et al*. (1979).

Gravity in this case is neglected (or inoperative) and the spring is positively fixed (not as in Figure 5.23), so that associated geometric nonlinearities have to be taken into account. In a way, these studies represent an extension to Bajaj & Sethna's (1984, 1991) work to the case when there is an additional, intermediate spring support. This is a very complex problem, and unfolding the various bifurcations is an arduous task.

The wholly symmetric system is considered first. To achieve this, so far as the spring force is concerned, it is found that either the springs in the two planes should be very long or an array of many springs radiating from the support should be used. Although not much information is given on the method of discretization, centre manifold and normal form theory are used as in the foregoing to simplify the system in the vicinity of the principal bifurcation associated with loss of stability: (i) by a zero root, (ii) a zero and a purely imaginary pair, (iii) two purely imaginary pairs, *in all cases with multiplicity two*, because of the symmetry. Two unfolding parameters are used: μ, associated with $u - u_c$, and ν, associated with $\kappa - \kappa_c$. A typical case of a system with a double pair of zero roots and κ slightly above κ_c is shown in Figure 5.28(a). The system starts from the trivial stable state (TS), then becomes subject to divergence (static buckling, SB) and, after a secondary bifurcation, develops planar oscillation about the buckled state (SW_3) — cf. Figure 5.24(b). With increasing flow, the amplitude grows so that the oscillation crosses the origin and changes into planar oscillation about the origin (SW_2). All solution branches associated with a positive real part of an eigenvalue of the locally linearized system, marked with a + and drawn as dashed lines, are unstable. In this case, TW solutions, corresponding to rotary pipe motions, and MW solutions, corresponding to rotary pipe motions with a superposed radial oscillation, are unstable.

Another case of a system with a zero root and a purely imaginary pair with multiplicity of two is shown in Figure 5.28(b) — cf. Golubitsky & Stewart (1986). As seen in the figure, there are eight solution branches associated with: (i) the trivial equilibrium state, TS; (ii) the statically buckled state, SB; (iii) planar oscillations, SW_2, about TS; (iv) planar oscillations, SW_3, about SB in the plane of SB; (v) the same, but $SW = SW_4$, perpendicular to SB; (vi) rotary pipe oscillations, TW; (vii) modulated motion, MW; (viii) SB with superposed TW (SB/TW). As shown in the figure, most of these solution branches are unstable. The system, after buckling, develops planar oscillatory motions (SW_4) about the buckled state, in a plane perpendicular to that of buckling. More complicated motions are possible in the case of two pairs of purely imaginary roots.

The case of broken symmetry is considered next, in three ways: (i) via a small geometric imperfection, yielding a constant term in the bifurcation equations — cf. Figure 5.8(b); (ii) imperfect springs breaking rotational symmetry; (iii) imperfect loading breaking reflectional symmetry (e.g. by immersing the end of the pipe in a swirling fluid). Dynamical behaviour similar to that of Figure 5.22, but richer, is predicted.

(d) Planar motions of doubly degenerate up-standing cantilever

The planar dynamics of another type of system, that of the 'up-standing cantilever' (Section 3.5.2) where the free end is located above the clamped one, without any intermediate support, is considered next; this was studied under double degeneracy conditions by Li & Païdoussis (1994).† It is recalled that, generally, this system buckles

†This study though published later, was actually conducted prior to that by Païdoussis & Semler (1993b), which owes a great deal to the Li & Païdoussis study.

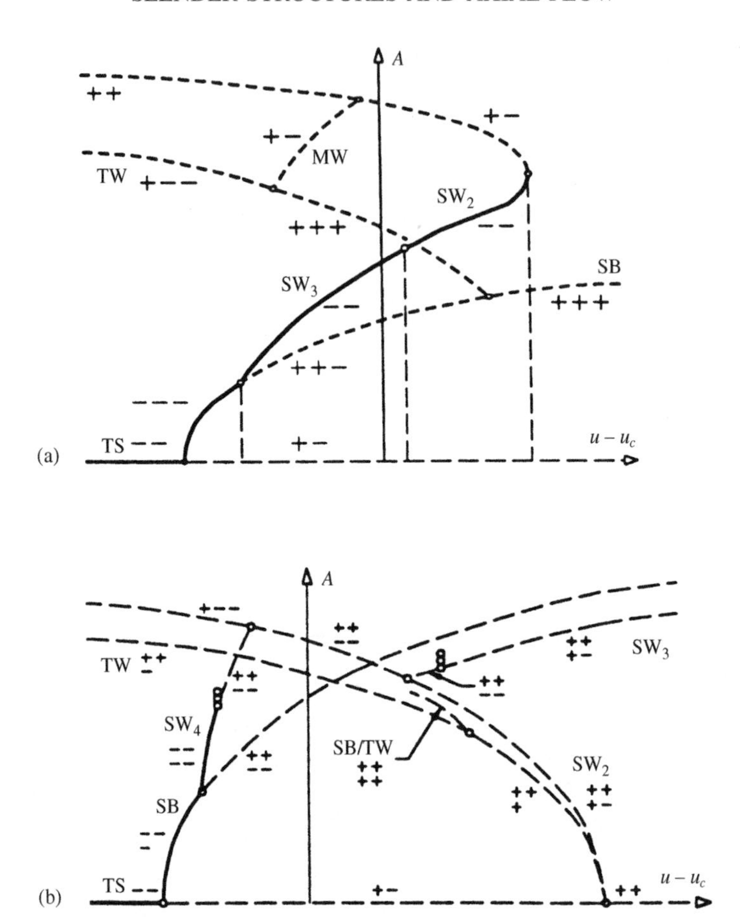

Figure 5.28 Amplitudes of solutions, A, versus $u - u_c$ for 3-D motions of the cantilevered pipe–spring system, with κ slightly higher than κ_c: (a) for a system with a double set of zero roots, which starts from a trivial stable state (TS), then develops buckling (SB), and then planar oscillations about the buckled state (SW_3) and about the origin (SW_2); (b) for a system with a double set of one zero root and one purely imaginary pair, which starts from the trivial state (TS), develops buckling (SB), and then planar oscillations perpendicular to the buckling plane (SW_4). Solution branches drawn as dashed lines are unstable (Steindl & Troger 1995).

under its own weight; as the flow velocity is increased, the system is restabilized via a reverse pitchfork bifurcation, and then loses stability by flutter via a Hopf bifurcation at higher flow. For the special sets of parameters shown in Figure 5.26 for $\kappa = 0$, these two bifurcations become coincident. The system is studied analytically and numerically, starting again with equation (5.42) and a two-degree-of-freedom discretization thereof. The system is first projected on the centre manifold and then the unfolding parameters μ_1, μ_2 and μ_3 are computed, as shown explicitly in Appendix F.5, equations (F.58)–(F.62). For $\beta = 0.2$, for which $u_c = 2.246$, $\gamma_c = -46.001$, computing the derivatives in (F.61) numerically and eliminating μ_3, the following relationships are obtained:

$$\mu = -0.467\mu_1 - 0.240\mu_2, \qquad \chi = -5.427\mu_1 - 2.647\mu_2. \tag{5.126}$$

The system is then further simplified by the method of normal forms, following Guckenheimer & Holmes (1983; exercise 7.4.1) and Sethna & Shaw (1987) — see Appendices F and H. Suffice it to say that the final form of the perturbation equations on the centre manifold is given by

$$\dot{r} = \mu_1 r - (r^2 + bz^2)r + \text{ h.o.t.}, \qquad \dot{z} = \mu_2 z + (cr^2 + z^2)z + \text{ h.o.t.}; \tag{5.127}$$

h.o.t. stands for higher order terms. For $\beta = 0.2$, $b = 1.518$ and $c = 3.954$.

A local bifurcation analysis shows that, generally, there are four equilibrium points:

$$\begin{gathered}(r, z) = (\sqrt{\mu_1}, 0) \quad \text{for } \mu_1 > 0, \qquad (r_1, z) = (0, \sqrt{-\mu_2}) \quad \text{for } \mu_2 < 0; \\ (r, z) = \left(\sqrt{-\frac{\mu_1 + b\mu_2}{bc - 1}}, \sqrt{\frac{c\mu_1 + \mu_2}{bc - 1}}\right), \qquad (r, z) = (0, 0);\end{gathered} \tag{5.128}$$

for the values of b and c just given, the third equilibrium exists only for $\mu_1 + b\mu_2 < 0$ and $c\mu_1 + \mu_2 > 0$. Topological features of the system near these equilibria can be determined by the eigenvalues of the matrix of the linearized autonomous version of (5.127). For an equilibrium point (r_0, z_0), this matrix is

$$[A] = \begin{bmatrix} \mu_1 - 3r_0^2 - bz_0^2 & -2br_0 z_0 \\ 2cr_0 z_0 & \mu_2 + 3z_0^2 + cr_0^2 \end{bmatrix}. \tag{5.129}$$

Figure 5.29 shows phase flows emanating from the equilibria in the fourth quadrant of the parameter plane. Within the segment defined by the dashed lines, the equilibrium of the third equation (5.128) emerges, and it may be a spiral sink or a source; all other equilibria are saddles. A bifurcation curve exists, where the sink and the source collapse

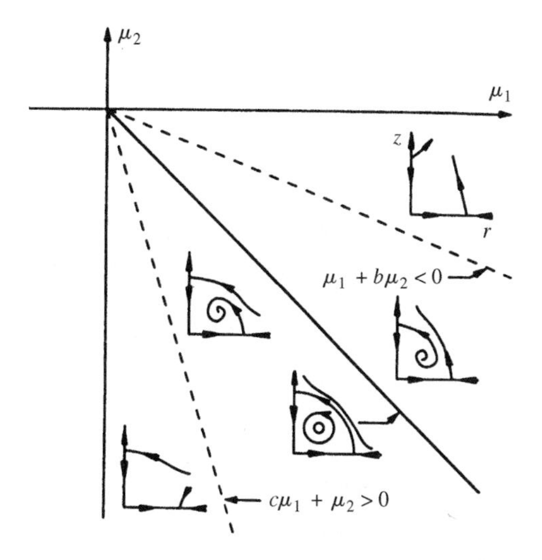

Figure 5.29 Bifurcation diagram for the doubly degenerate up-standing cantilevered pipe on the centre manifold, showing typical phase flows in the fourth quadrant (Li & Païdoussis 1994).

into a centre, when $\mathrm{tr}([A]) = 0$, where tr stands for trace, which gives

$$(1 + c)\mu_1 + (1 + b)\mu_2 = 0. \tag{5.130}$$

Along this line, a heteroclinic loop emerges. This in turn may give rise to the possibility of heteroclinic tangles and chaos, as discussed in Section 5.8.

5.7.4 Concluding comment

The work presented in this section is representative of a larger set, not fully discussed for the sake of brevity. The reader is also referred to the next section, primarily devoted to chaotic dynamics, but also containing work of interest here (e.g. on the dynamics of pipes fitted with an additional mass at the free end).

It is hoped that the work in Section 5.7 has shown that, similarly to the linear dynamics of cantilevers conveying fluid (Chapters 3 and 4), the nonlinear dynamics is equally fascinating. Of special interest are the results in Figures 5.19–5.22, where the regions of sub- and supercritical Hopf bifurcations are defined, as well as whether motions are three-dimensional or planar and, in the case of an inclined nozzle, in which plane. Once more, the special importance of the 'critical values of β', associated with S-shaped discontinuities in the u_c versus β plot, emerges; thus, it is seen that, for nonlinear dynamics also, these values of β are either separatrices or backbones of peculiar behaviour. Also of importance are the codimension-2 and -3 bifurcation sets emanating from the vicinity of double degeneracies, samples of which are given in Figures 5.27–5.29. Of special interest in this regard is the existence of conditions leading to heteroclinic loops which are often associated with chaotic behaviour (Section 5.8).

Finally, it is also hoped that the material in Sections 5.5, 5.6 and 5.7 has made abundantly clear — not only by the substance of the work, but also by the authors' names — that this system has served both as an example of a physical system on which the various modern methods of nonlinear dynamics could be demonstrated and as a system through which these methods could be further developed.

5.8 CHAOTIC DYNAMICS

With the rapidly developing, and deserved, fascination with chaos in engineering systems, it was inevitable that its possible existence in fluidelastic systems would be explored. As is well known, however, chaos is usually associated with strong nonlinearities (Moon 1992); hence, the first set of studies into chaotic dynamics involved modifications to the system so as to introduce strong nonlinear effects. Three such systems are discussed: (i) the pipe with loose lateral constraints (Section 5.8.1); (ii) with magnets added; (iii) with an added mass at the free end. Then, the existence of chaos under more particular conditions, e.g. near double degeneracies, is discussed in Sections 5.8.4 and 5.8.5.

5.8.1 Loosely constrained pipes

In contrast to other parts of this chapter, the presentation here is chronological as well as paedagogical in tone. The reason for this is that, in addition to showing how chaos can arise in loosely constrained pipes conveying fluid, there is another objective also: to show that the interpretation of the dynamics is very sensitive, in a way that nonchaotic dynamics can never be, particularly when judging the success or otherwise of analytical modelling. This will become evident in the course of the presentation, and is discussed in the two paragraphs preceding the last two of this section.

The system, first studied experimentally (Païdoussis & Moon 1988), is shown in Figure 5.30. It consists of an elastomer vertical pipe conveying fluid. As already discussed in Chapter 3 and Section 5.7, at sufficiently high flow velocity, a Hopf bifurcation leads to a stable limit cycle. As the amplitude of motion increases with flow, for appropriately positioned motion constraints (typically metal bars), the pipe bangs on one constraint, rebounds from it, and then generally on the other, back and forth, without making permanent contact. Thus, at location x_b, a very large restraining force is operative on impact, while during 'free flight' between restraints there is none. It should be remarked that, even when the constraints are rigid, the impact involves local deformation of the pipe, and hence the constraint may be modelled by a strongly nonlinear spring, involving a spring constant which varies discontinuously with displacement (see Figure 5.33).

Experiments were conducted with several pipes, e.g. those listed in Table 5.3, mostly conveying water, but in a few cases conveying air. For these pipes the limit-cycle motions are planar — in a plane defined by minute imperfections in the pipe. At impact, however, the motion tends to deteriorate into a three-dimensional one. In order to keep the dynamics as simple as possible for analytical modelling, the oscillation was restricted to a plane,

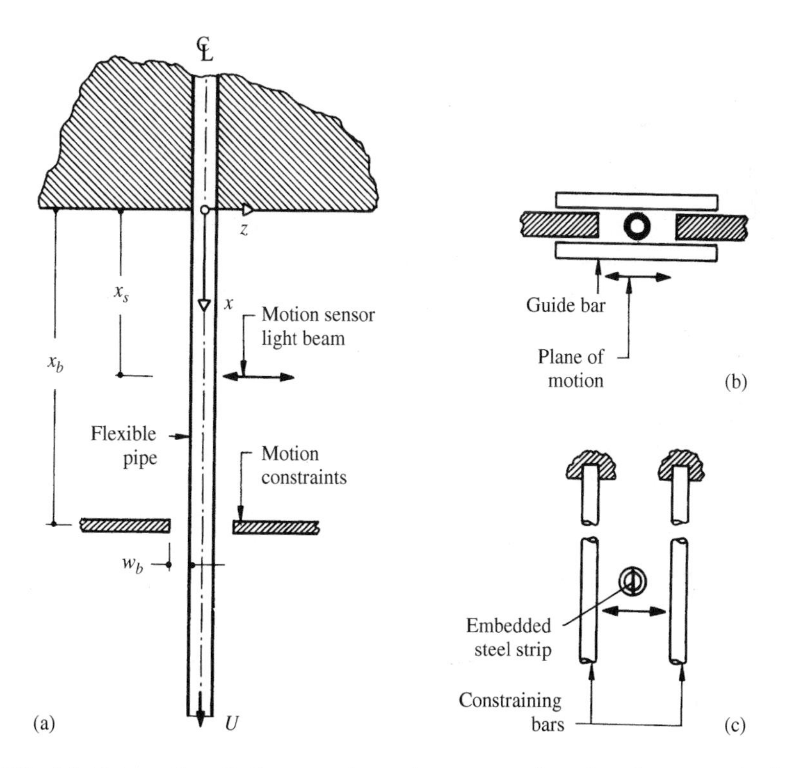

Figure 5.30 (a) Schematic of the experimental system of a loosely constrained vertical cantilevered pipe; (b) scheme of achieving planar motions by guide-bars; (c) refined scheme for planar motions, with steel strip embedded in the pipe, also showing motion-constraining bars.

Table 5.3 Pipe parameters for Païdoussis & Moon's (1988) experiments. D_o, D_i are the outer and inner pipe diameters, L and m are the pipe length and mass per unit length, respectively, EI the flexural rigidity, and δ_j the modal logarithmic decrement of damping in the jth beam mode. Note: for pipe 9, $\delta_2 = 0.081$ and $\delta_3 = 0.144$ were also measured.

Pipe number	D_o (mm)	D_i (mm)	L (mm)	m (kg/m)	$EI \times 10^3$ (N m^2)	δ_1 (–)
8	15.5	7.94	350	0.190	21.7	0.17
1	15.2	6.35	234	0.174	5.05	0.090
9	15.88	7.94	441	0.182	7.28	0.028

either by guiding bars [in early experiments, Figure 5.30(b)] or by embedding a thin metal plate all along the length of the pipe in the process of casting it [in later experiments, Figure 5.30(c); see also Appendix D].

The flow velocity was made to be as uniform and steady as possible by using a large accumulator tank to remove pump pulsations, and a long straight pipe upstream of the test-pipe, fitted with screens and honeycombs; a 36:1 smooth area-contraction in the piping just upstream of the test-pipe reduced incident turbulence. The flow rate was determined by measuring the time taken to collect a certain weight of water in the collecting tank beneath the test-pipe, or via standard rotameters in the airflow tests.

Experiments were conducted with fairly rigid motion constraints [Figure 5.30(b,c), with bars made of metal], or with more pliable ones, which deformed appreciably under impact [Figure 5.30(c), with polycarbonate plastic bars, or with leaf-type springs on metal bars]. The constraint location and the gap were varied: $\xi_s = x_s/L = 0.62 - 0.65$ and $w_b/L = 0.025 - 0.055$ for the water experiments, and $\xi_s = 0.84$ and $w_b/L = 0.130$ for the air-flow experiments.

The vibration of the pipe was monitored by non-contacting sensors: either a Fotonic fibre-optic sensor or an optical tracking system. In both cases, it was ensured that the measuring system operated in its linear range. The optical tracking probe was trained at a point $x_s/L = 0.22$ typically (Figure 5.30); the fibre-optic sensor, when it was used, was considerably closer to the fixed end, $x_s/L < 0.1$ typically.

The signal was processed in various ways: (i) it could be fed into an FFT signal analyser, to provide auto- or power spectra (PS), autocorrelations, or probability density functions (PDFs) of the system, and/or (ii) into a digital storage oscilloscope to generate phase-plane portraits and Poincaré maps; the signal could also be recorded by an instrumentation tape recorder for later processing.

The Poincaré map is a collection of points obtained by collecting and storing a single point of the trajectory of the system in phase space for each cycle of motion, with consistent timing (Moon 1992). In the present experiment, a circuit was triggered at impact with one of the two constraint bars, which was instrumented with strain gauges [i.e. when displacement $w(x_b, t) = w_b$, see Figure 5.30, and $\dot{w}(x_b, t) > 0$], which caused the displacement and velocity to be stored for that value of t. Thus, if the motion is periodic (period-1), the Poincaré map consists of but a single point in a $(\dot{w}, w)$-plot; it consists of two points for period-2 motion. A cloud of points in some defined pattern would suggest chaotic motion.

It is also recalled that the PDF of a periodic signal (period-1) displays two prominent peaks at the extremes of the displacement, where motion is slow, and hence the probability of finding the vibrating system there is high. Departures from this double-masted, suspension-bridge shape indicate departures from regularity in the motion; a Gaussian-like distribution corresponds to random or random-like (chaotic) motion. The autocorrelation of a periodic signal is periodic with time, whereas an aperiodic signal has a 'damped-response versus time' characteristic, showing loss of memory after a few cycles of motion — which is characteristic of random or more generally stochastic signals.

Typical results with pipe 9 (Table 5.3), water flow and the polycarbonate impact bars at $w_b/L = 0.055$ are presented in Figures 5.31 and 5.32: power spectra (PS), phase plots, probability density functions (PDF), autocorrelations and Poincaré maps. In Figure 5.31(a), the limit-cycle amplitude is sufficiently large to allow impacting, preferentially on one of the two bars. Therefore, the motion at this stage is asymmetric, biased towards one of the bars.† The profusion of harmonics of the main oscillation frequency ($f \simeq 2.6$ Hz) is due to the impacting. The double-masted shape of the PDF, the essentially constant-amplitude autocorrelation and the single-loop phase-plane plot all indicate periodic (period-1) motion. In Figure 5.31(b) it is seen that the motion is still periodic, but the strong subharmonic at $\frac{1}{2}f$ in the PS plot and the double-loop phase-plane plot indicate period-2 motion; physically, a typical sequence of motions is this: the pipe impacts on the bar, then executes a complete 'free' cycle of oscillation, before impacting again. The corresponding Poincaré map, Figure 5.32(a) shows two fuzzy 'points', indicating that a small chaotic component may be in existence already, but indicating a predominantly period-2 motion.

The motion is considerably more erratic for $U = 7.48$ m/s (not shown), but still periodic (period-2); there is a slight reduction of the autocorrelation with time and the trough of the PDF is considerably more filled out, with a clear double peak on the right side, corresponding to the period-2 trajectory. This perhaps is the limit of periodic or almost periodic motion. In Figure 5.31(c), the two subharmonics of the main frequency are at $\frac{1}{3}$ and $\frac{2}{3}$, signifying a period-3 motion (found in a few other instances also); however, the low-frequency content of the signal is wide-banded and erratic and hence the oscillation should be considered chaotic. Significantly, the PDF has become less concave in the centre region, almost convex, and the autocorrelation decays fairly rapidly with time, with beating (more easily visible if displayed over a longer time period). Finally, the motion is quite chaotic in Figure 5.31(d), as shown by the PS, the PDF and the autocorrelation equally. The phase portraits in Figure 5.31(a,b) correspond to the pipe impacting on one motion-constraint bar, whereas in Figure 5.31(c,d) it is impacting on both.

The Poincaré maps of Figure 5.32 correspond to (a) conditions just after the period-doubling bifurcation, with two attractors, as already remarked; (b) more erratic motion, with a suggestion of more complex attractors; (c) where the motion is more wide-band chaotic. In the last case, although the Poincaré map does not display the artistic merit of that in (b) and even less of the 'fleur de Poincaré' and some other remarkable examples (Moon 1992), still it does seem to have some structure. In this connection, the point should be made that these Poincaré maps represent two-dimensional sections of a multidimensional attractor, which may indeed have a great deal of structure that cannot be discerned in the planar sections taken; the construction of double-Poincaré maps (Moon 1992) might have been more successful in rendering any such structure more conspicuous. It is also

†This occurs soon after — in terms of increasing u — impacting begins to take place.

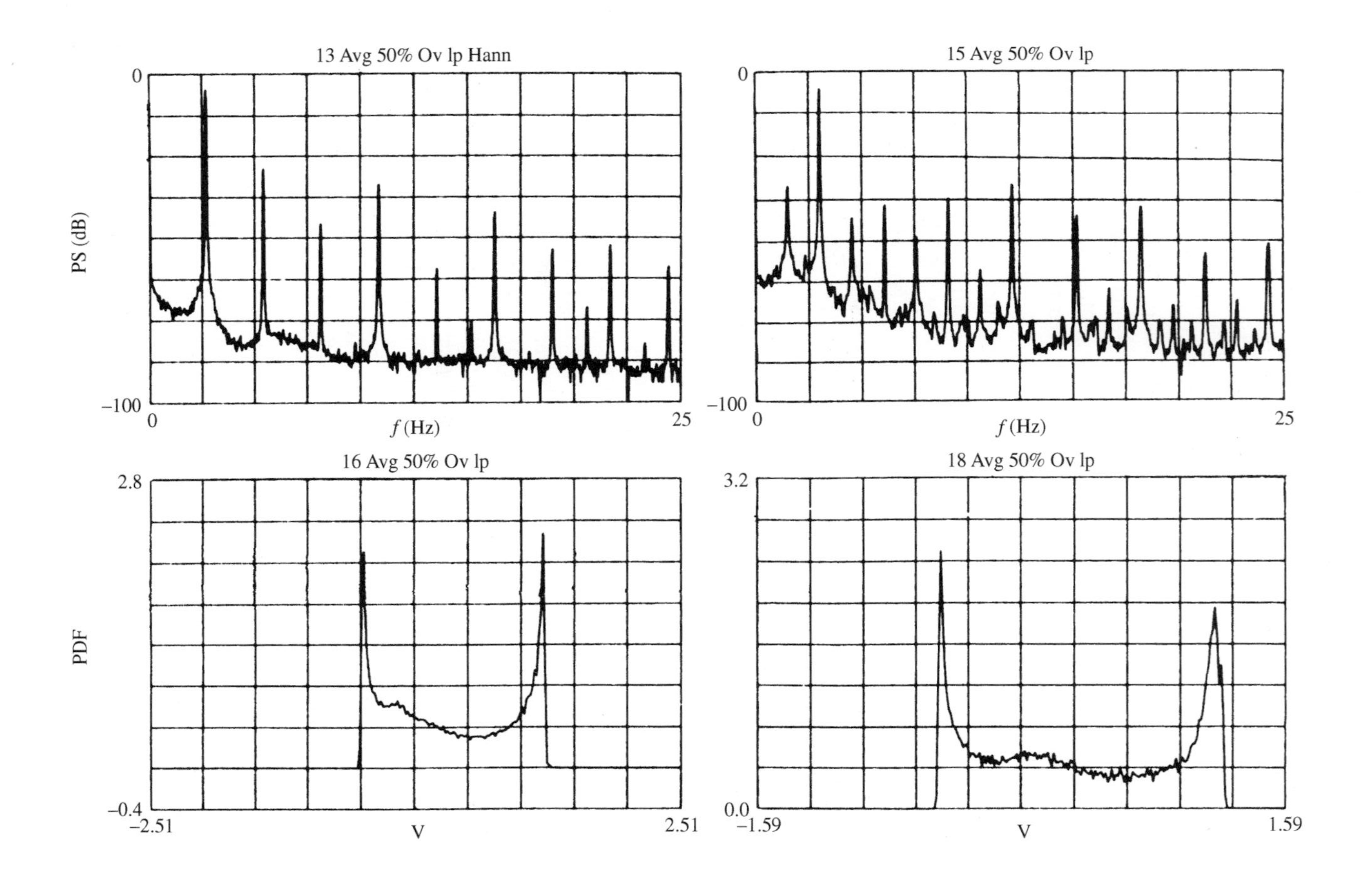

13 Avg 50% Ov lp Hann
15 Avg 50% Ov lp
PS (dB)
0
−100
0
25
f (Hz)
16 Avg 50% Ov lp
18 Avg 50% Ov lp
PDF
2.8
−0.4
−2.51
2.51
V
3.2
0.0
−1.59
1.59

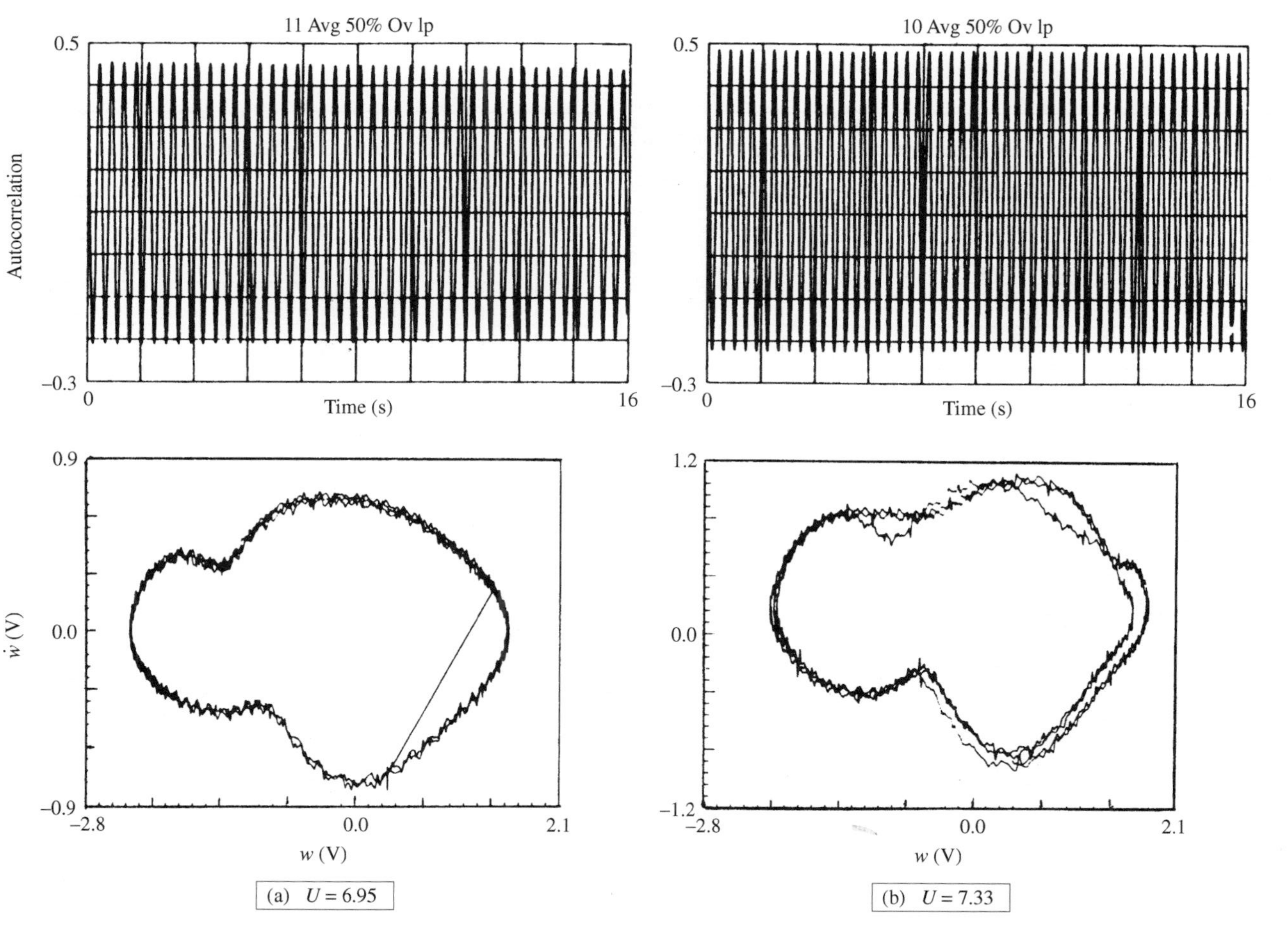

Figure 5.31 Experimental vibration spectra (PS), probability density functions (PDF), autocorrelations, and phase-plane plots for pipe #9 with the polycarbonate constraining bars [Figure 5.30; $\xi_b = 0.65$, $w_b/L = 0.055$ ($w_b/D_o = 1.52$)]: (a) $U = 6.95$ m/s ($u = 8.03$); (b) $U = 7.33$ m/s ($u = 8.47$).

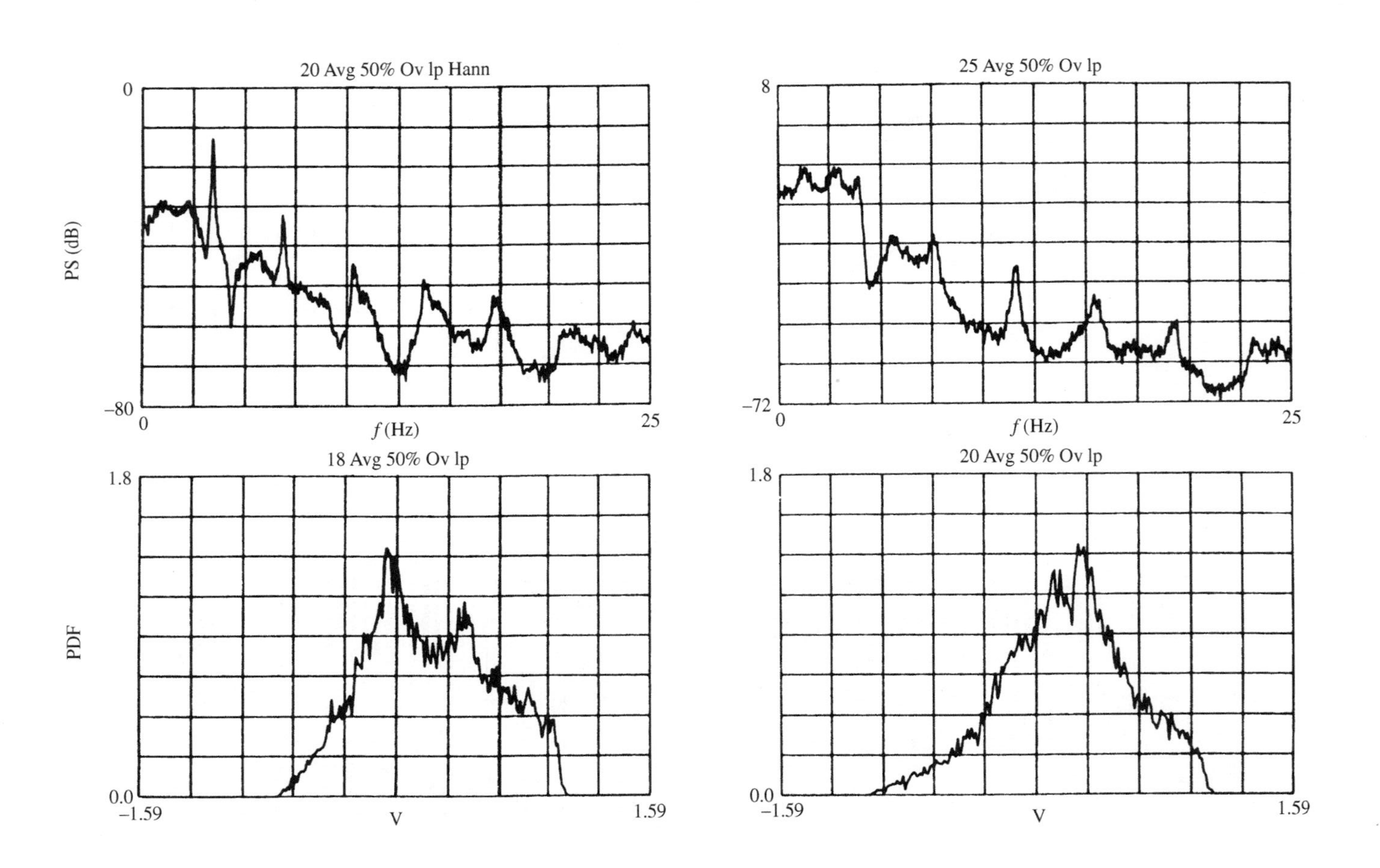

20 Avg 50% Ov lp Hann
PS (dB)
0
−80
0
f (Hz)
25
25 Avg 50% Ov lp
8
−72
0
f (Hz)
25
18 Avg 50% Ov lp
PDF
1.8
0.0
−1.59
V
1.59
20 Avg 50% Ov lp
1.8
0.0
−1.59
V
1.59

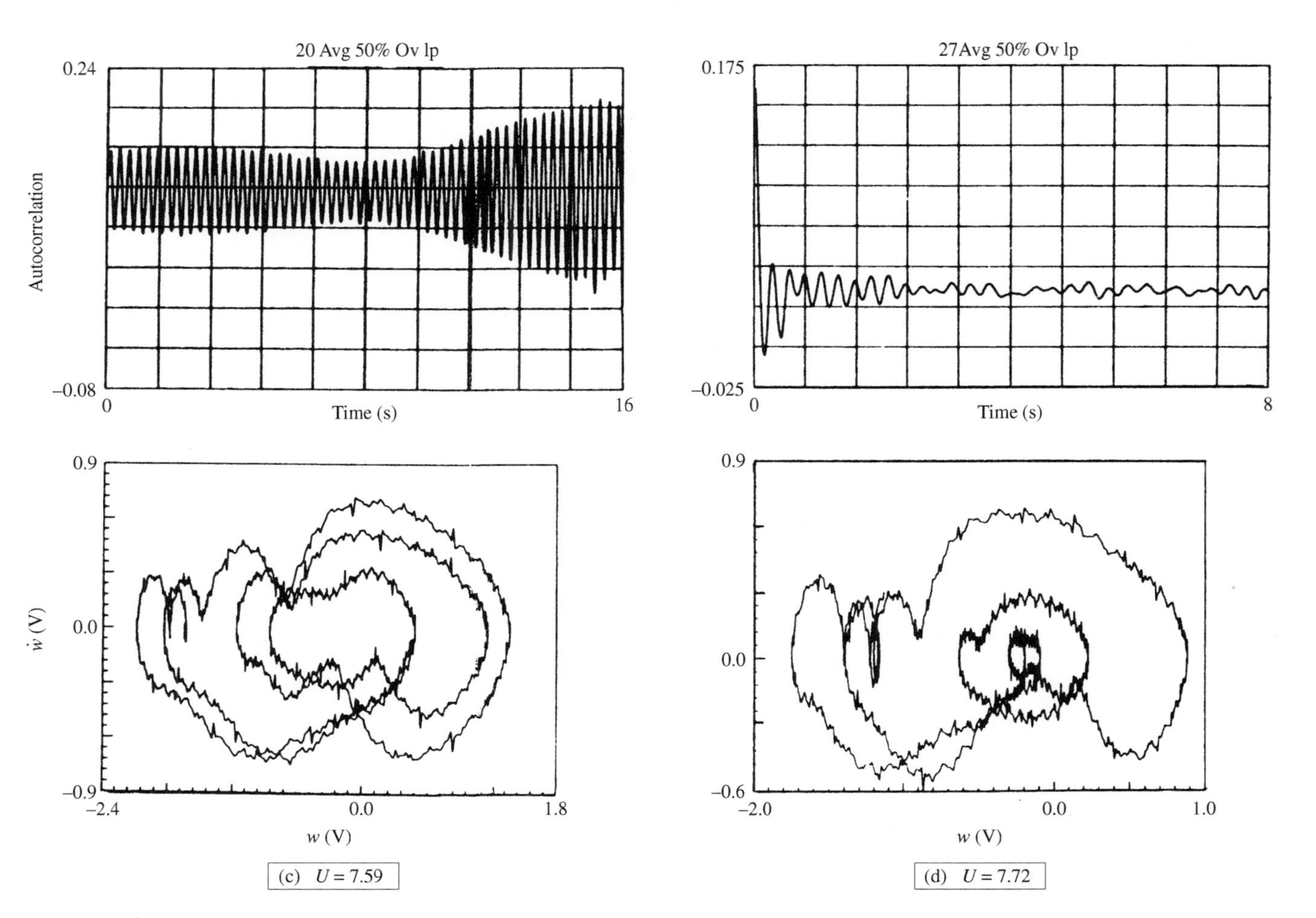

Figure 5.31 (*continued*) (c) $U = 7.59$ m/s ($u = 8.77$); (d) $U = 7.72$ m/s ($u = 8.92$); (Païdoussis & Moon 1988).

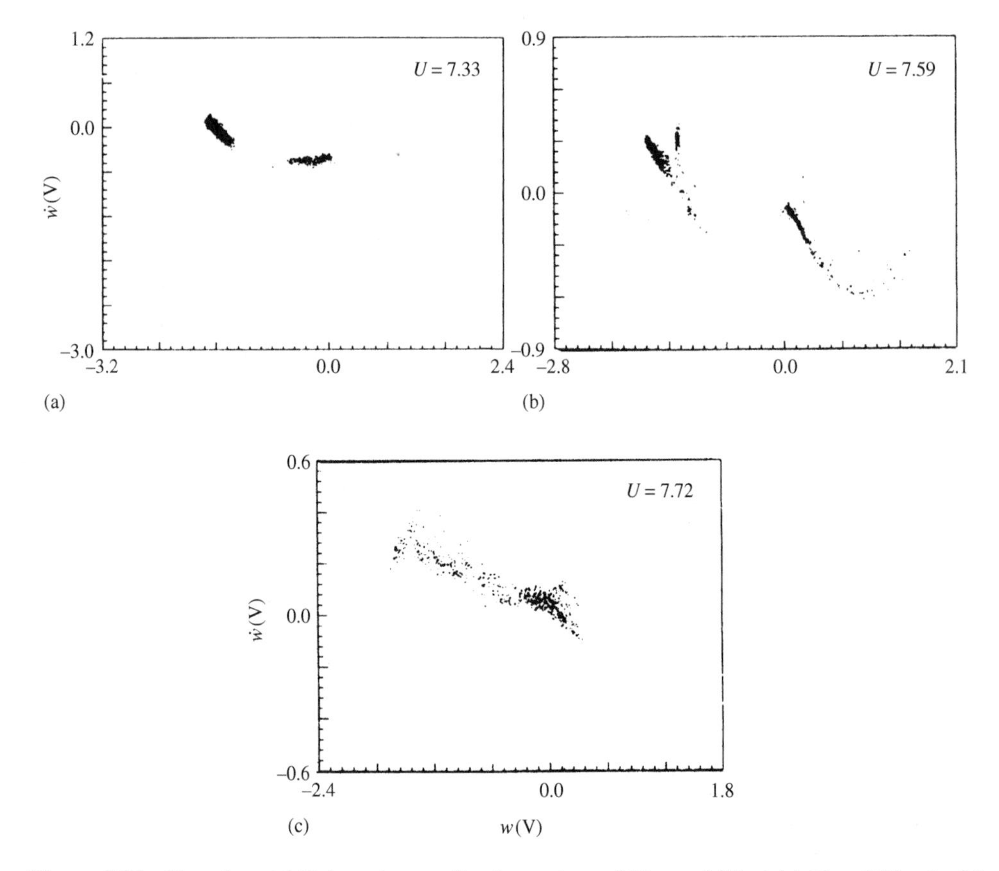

Figure 5.32 Experimental Poincaré maps for the system of Figure 5.30 at (a) $U = 7.33$ m/s, (b) $U = 7.59$ m/s, and (c) $U = 7.72$ m/s, corresponding to (b), (c) and (d) in Figure 5.31 (Païdoussis & Moon 1988).

noted that noise (due to random unsteadiness in the flow, for instance) tends to smudge some of the finer structure.

One significant conclusion that emerges from the results of Figures 5.31 and 5.32 is the importance of utilizing more than one measure in deciding on the existence of a strange attractor and chaos, especially in experimental systems, where some unsteadiness in one or more of the system parameters and random noise, no matter how small, are nevertheless ubiquitous. Thus, the PS and PDF in Figure 5.31(c) and the corresponding Poincaré map, Figure 5.32(b), suggest chaotic motions, whereas the autocorrelation is inconclusive; similarly, the Poincaré map of Figure 5.32(a) may be thought to indicate chaos, whereas all the other corresponding measures in Figure 5.31(b) show the motion to be periodic. Other cases of 'conflicting' conclusions by some of the measures of vibration are presented in Païdoussis & Moon (1988).† The wise experimenter would therefore do

†These difficulties are partly associated with the inevitable presence of random noise in the signal, associated with ubiquitous if minute unsteadiness in various experimental quantities, e.g. the flow velocity; also in 'extraneous' vibration transmission through supports, the ambient air, etc. A more quantitative measure of the

well not to rely on just one measure in deciding that motions are chaotic or otherwise. Professor F.C. Moon made the point eloquently in a course he gave at Cornell, through a parable, paraphrased here from memory. 'If you see something that *looks* like a duck, it does not necessarily mean that it *is* a duck. If, however, it also walks like a duck, it swims like a duck, and it quacks like a duck, then it is much more likely to *be* one!'[†]

The experimental system was also studied analytically, initially by the simplest possible model (Païdoussis & Moon 1988; Païdoussis *et al*. 1989). As motions are relatively small because of the constraining bars, the linear version of equation (5.39) is used as a first approximation — i.e. equation (3.70) — apart from the forces associated with impacting. A good model for the latter is a trilinear spring: zero stiffness when no contact is made, and a large stiffness once it is. For analytical convenience, this can be approximated by a cubic spring (Figure 5.33); hence, the following term is added to the dimensionless equation of motion: $\kappa\eta^3\delta(\xi - \xi_b)$, where ξ_b is the dimensionless axial location of the constraints, κ the dimensionless cubic-spring stiffness, $\kappa = kL^5/EI$, k being the dimensional value, and δ is Dirac's delta function. Thus, the equation of motion is

$$\left[1 + \alpha\left(\frac{\partial}{\partial\tau}\right)\right]\eta^{\text{iv}} + [u^2 - \gamma(1-\xi)]\eta'' + 2\beta^{1/2}u\dot{\eta}' + \gamma\eta' + \kappa\eta^3\delta(\xi - \xi_b) + \ddot{\eta} = 0. \quad (5.131)$$

The system is discretized by Galerkin's method into a two-degree-of-freedom ($N = 2$), four-dimensional (4-D) model. Solutions are obtained by numerical integration via a fourth-order Runge–Kutta integration algorithm.

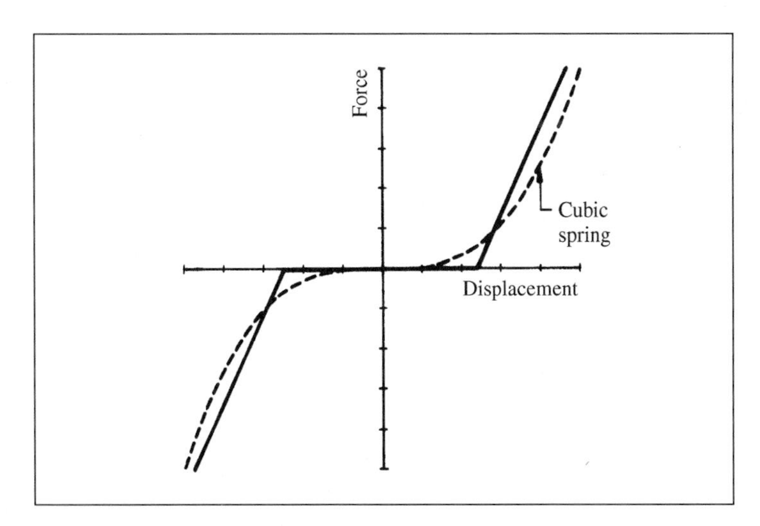

Figure 5.33 Diagrammatic view of the idealization of the trilinear (or 'bilinear') motion constraint (——) by a cubic spring (– – –) (Païdoussis & Moon 1988).

threshold of chaos is provided by the calculation of the Lyapunov exponents, discussed later. For this problem, however, such calculations were confined to the analytical model, although they are also possible, but not at all easy, for experimental data (Moon 1992).

[†]The author, having recently discovered the excellent Belgian beer *Kwak*, feels that this argument is further reinforced, since thirsty humans may just as plausibly emit 'Kwak, Kwak' as itinerant ducks.

This model gives reasonable qualitative agreement with experimental observations, as exemplified by the bifurcation diagram of Figure 5.34 and the phase-plane plots of Figure 5.35. Beyond the Hopf bifurcation (at $u = u_H$), the maximum and minimum amplitudes of the ensuing limit cycle are shown. At u_{pf}, a pitchfork bifurcation takes place, destroying the symmetry of the limit cycle [see also Figure 5.35(a)], which in the experiments corresponds to the motion biased towards one of the bars; this is followed at u_{p2} by period-2 [cf. Figure 5.35(b)], period-4 [Figure 5.35(c)], period-8, etc. bifurcations, leading to chaos [Figure 5.35(d)]. Thus Figures 5.34 and 5.35 clearly establish that the *period-doubling route* to chaos is followed in this case (Moon 1992): a theoretically infinite sequence of successive period-doubling bifurcations leading to an ever-increasing period of oscillation, aperiodicity and chaos. An outstanding feature of this scenario is that it can be represented by a very simple map (Feigenbaum 1978) and that successive values of these bifurcations, u_j, obey the following rule:

$$\frac{u_{j+1} - u_j}{u_{j+2} - u_{j+1}} \longrightarrow \mathrm{Fe} = 4.669\,2016, \tag{5.132}$$

where Fe is the Feigenbaum number. In practice, this value of Fe is approached by the third or fourth period doubling (Moon 1992). Indeed, in this case, taking the first three period-doubling bifurcations into account and then the second-to-fourth sequence, we obtain Fe = 4.124 and 4.613, the latter being quite close to the value in (5.132).

The quantitative agreement with experiment for the critical values of u associated with key bifurcations in Figure 5.34 is quite reasonable (within $\sim 20\%$). However, this agreement is achieved by grossly straining (relaxing) the values of some parameters, as follows:

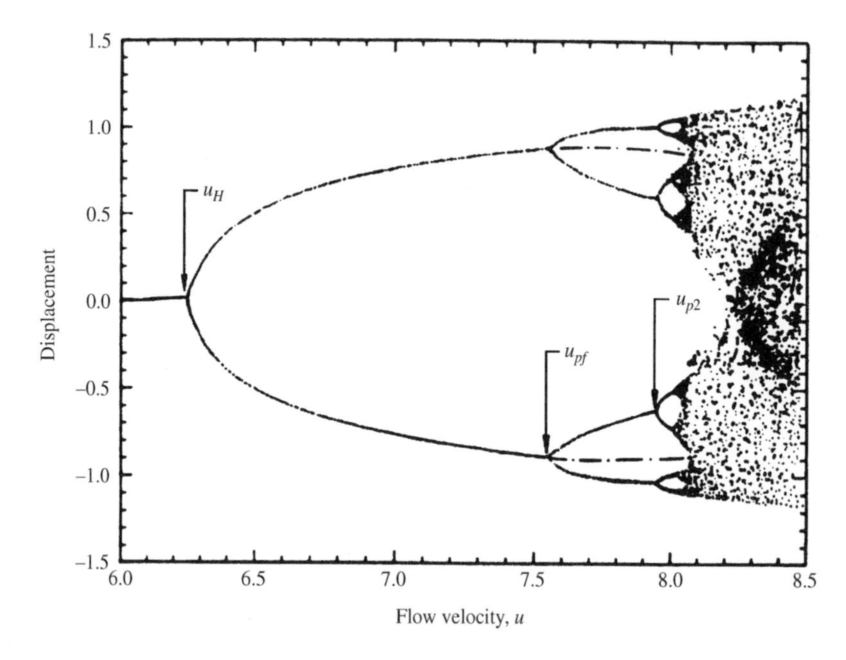

Figure 5.34 Analytical bifurcation diagram of the tip-displacement of the loosely constrained cantilevered pipe ($N = 2$, $\alpha = 5 \times 10^{-3}$, $\beta = 0.2$, $\gamma = 10$, $\xi_b = 0.82$, $\kappa = 100$), showing the onset of chaos through a cascade of period-doubling bifurcations (Païdoussis *et al*. 1989).

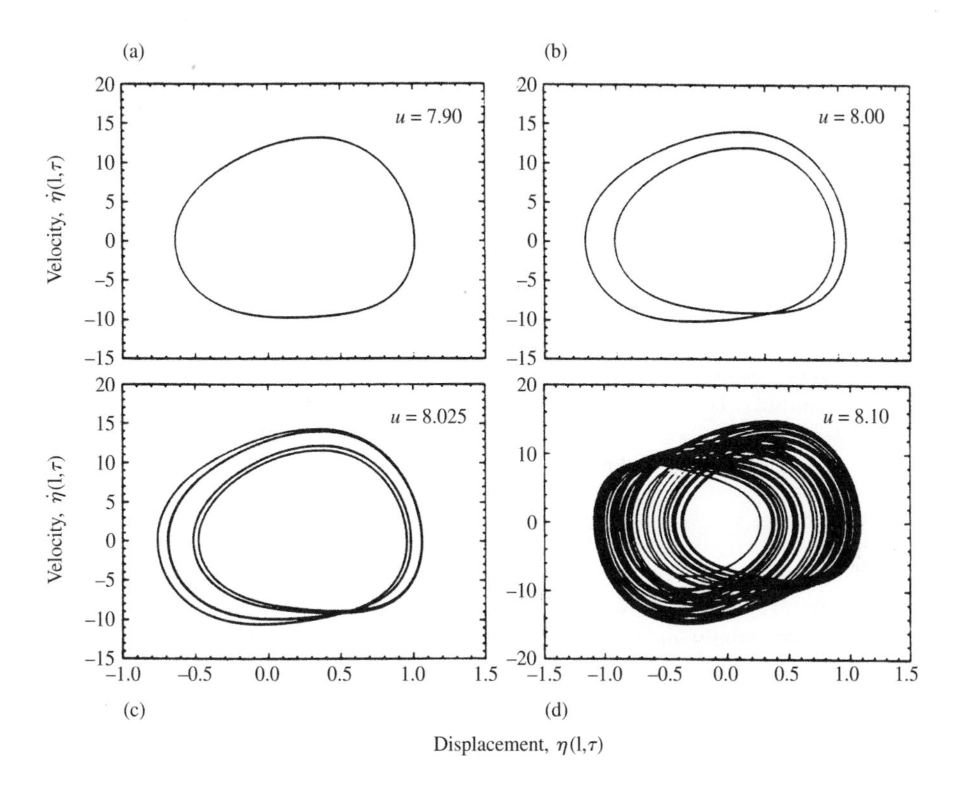

Figure 5.35 Phase-plane plots corresponding to selected u in Figure 5.34: (a) $u = 7.9$; (b) $u = 8.00$; (c) $u = 8.025$; (d) $u = 8.100$ (Païdoussis *et al*. 1989).

(i) the constraining bars are located at $\xi_b \equiv x_b/L = 0.82$ and (ii) the cubic spring stiffness is $\kappa = 10^2$–10^3 in the model, while $\xi_b = 0.65$ and $\kappa \sim \mathbb{O}(10^5)$ in the experiments; without such straining, the numerical solutions diverge. Furthermore, the predicted amplitudes of motion are unrealistically large (Figure 5.35). Hence, improvements to the analytical model are necessary, and they are discussed farther on.

A few explanatory words may be appropriate on how bifurcation diagrams are constructed. The amplitude of the free-end displacement, $\eta(1, \tau) = \phi_1(1)q_1(\tau) + \phi_2(1)q_2(\tau)$ is recorded and plotted when $\dot{\eta}(1, \tau) = 0$, $\phi_i(1)$ being the beam eigenfunctions at $\xi = 1$ and $q_i(\tau)$ the corresponding generalized coordinates; thus, both positive and negative amplitudes are recorded in Figure 5.34. Once the symmetry of the solutions is broken, at u_{pf}, only one branch of the solution shown in the figure is normally obtained. However, by conducting simulations with two 'opposite' sets of initial conditions, $q_1(0) = q_2(0) = \pm 0.1$ and $\dot{q}_1(0) = \dot{q}_2(0) = 0$, the two different solution branches (four, in effect: two positive and two negative, as explained above) are obtained.

The bifurcation diagram in Figure 5.34 and the phase portrait of Figure 5.35(d) may be considered to be sufficient evidence of the existence of chaotic regions in the parameter space of the system. However, even more definitive discriminators are the Lyapunov exponents (Guckenheimer & Holmes 1983; Moon 1992), discussed next.

Here also, an explanatory paragraph may be useful to the reader. Consider the n-dimensional system $\dot{y} = f(y)$ with a solution $\phi(\tau)$ corresponding to a set of initial conditions $\phi(\tau_0) = \phi_0$. Of concern here is the stability of $\phi(\tau)$. Suppose that $\phi_1(\tau)$ is another trajectory corresponding to different initial conditions. Then, defining the variational vector function $u(\tau) = \phi_1(\tau) - \phi(\tau)$ such that $\|u\| \ll 1$, the stability of $\phi(\tau)$ is governed by the equation

$$\dot{u} = \mathrm{D}f(\phi)u, \tag{5.133}$$

where $\mathrm{D}f(\phi)$ is the Jacobian matrix function for the vector field $f(y)$ along $\phi(\tau)$. Clearly, if $\phi_1(\tau)$ approaches $\phi(\tau)$, then $u(\tau)$ will tend to zero; while if it diverges away from it, then $u(\tau)$ will tend to grow. This may be expressed as

$$\|u(\tau)\| \sim \mathrm{e}^{\sigma\tau}. \tag{5.134}$$

As the system is bounded, however, the exponential behaviour indicated by (5.134) cannot continue indefinitely. Hence, in the computations the vector function $u(\tau)$ is renormalized from time to time and the calculation process reinitialized in that sense. The so-called Lyapunov exponent may be defined as

$$\sigma = \lim_{\tau\to\infty} \frac{1}{\tau} \ln\left(\frac{\|u(\tau)\|}{\|u(0)\|}\right). \tag{5.135}$$

Hence, the two trajectories in question, $\phi_1(\tau)$ and $\phi(\tau)$, may be considered to converge or diverge exponentially *on the average*, according as σ is negative or positive, with $\sigma = 0$ corresponding to neutral orbital stability (the case of a stable periodic orbit). Note that in an n-dimensional space there exist n Lyapunov exponents. However, the largest one dominates the dynamics of the system. Given an arbitrary initial condition $u(0)$, the solution $u(\tau)$ will converge to the direction of most rapid growth, which is associated with the largest Lyapunov exponent. A chaotic trajectory is defined as one with at least one positive Lyapunov exponent (Parker & Chua 1989; Moon 1992).

The problem at hand being represented by a fourth-order system, there will be four Lyapunov exponents, only the largest of which is computed for the purposes of defining the dynamical behaviour of the system. The largest one for the case corresponding to Figures 5.34 and 5.35 is shown in Figure 5.36. It is seen that, for $u < 8.027$, $\sigma_{\max} \simeq 0$, indicating stable periodic orbits, while for $u > 8.027$, $\sigma_{\max} > 0$, indicating chaotic behaviour. (For $u < u_H$, i.e. below the Hopf bifurcation, $\sigma_{\max} < 0$ is obtained.) It is also noted that there are so-called 'periodic windows' at $u \simeq 8.18$ and 8.19, where periodic motion is once more obtained over small ranges of u.

Simultaneously to all of the foregoing, the *fractal dimension* of the experimental system was determined (Païdoussis *et al*. 1992). This will help answer whether the $N = 2$ discretization is adequate or not. There are several measures of the fractal dimension (Moon 1992), of which the *correlation dimension*, as developed by Grassberger & Proccacia (1983a,b) is used here. The method is outlined in Appendix I, with experimental data from another run of the same system as in Figures 5.30–5.32.

The final results for three flow velocities, showing the correlation dimension, d_c, as a function of the embedding dimension, m, are shown in Figure 5.37, while the preliminary work leading to these figures is given in Figures I.1–I.3. In Figure 5.37(a,b,c), corresponding to Figures I.1, I.2 and I.3, the oscillation is periodic, 'fuzzy period-2' and chaotic, whereby 'fuzzy' is signified a periodic oscillation with a small chaotic component. The value of $d_c = 1.03$ in Figure 5.37(a) is sufficiently close to the ideal $d_c = 1$ for periodic oscillation; $d_c = 1.53$ in Figure 5.37(b) is substantially different from $d_c = 1$ because of the fuzzy nature of the oscillation. Finally, Figure 5.32(c) gives the most

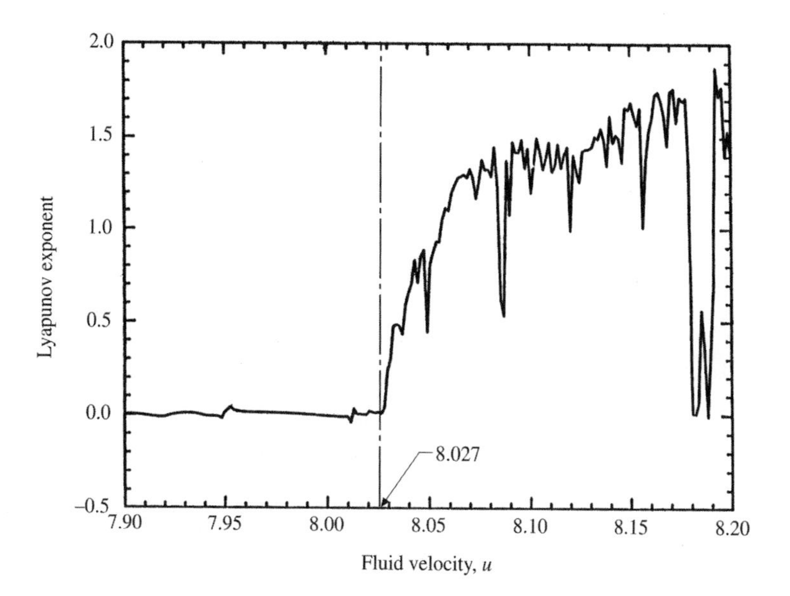

Figure 5.36 The largest Lyapunov exponent of the system of Figure 5.34 versus u (Païdoussis & Moon 1988).

significant result: that the correlation dimension in the chaotic regime is $d_c = 3.20$. Here, it is recalled that noninteger values for dimension are perfectly normal for, and indicative of, chaotic systems.

As shown by Mañé (1981), if d_c is the correlation dimension of the system, the *minimum* number of state variables, M, required for modelling the system is given by $d_c \leq M \leq 2\,d_c + 2$, where M corresponds to the dimension of the system, i.e. to twice the number of degrees of freedom N; therefore, this may be written as

$$d_c \leq 2N \leq 2\,d_c + 2. \tag{5.136}$$

Both M and N have to be integers and M must be even for an autonomous system. Therefore, for $d_c = 3.20$, the important result is obtained that the number of degrees of freedom required to capture the essential dynamics is $2 \leq N \leq 4$ or 5, depending on exactly how the inequality is interpreted. Hence, it is not surprising that the analytical results of Figures 5.33–5.36 obtained with $N = 2$ are in qualitative agreement with the experimental observations. On the other hand, it is clear that, to capture the *quantitative* aspects of the dynamics adequately, an $N = 4$ or $N = 5$ model may be necessary.

The first attempt to improve the analytical model aimed at (i) studying the dynamics of the system for $N > 2$ and (ii) conducting calculations with a more realistic model of impacting with the constraining bars, by using their true stiffness and location — rather than the strained ('relaxed') values used in the foregoing, resorted to solely to obtain convergent solutions. The measured stiffness of the constraining bars is very close to the trilinear model, as shown in Figure 5.38. With this model, the cubic spring term in equation (5.131) is replaced by

$$f(\eta)\delta(\xi - \xi_b), \qquad f(\eta) = \kappa\{\eta - \tfrac{1}{2}(|\eta + \eta_b| - |\eta - \eta_b|)\}. \tag{5.137}$$

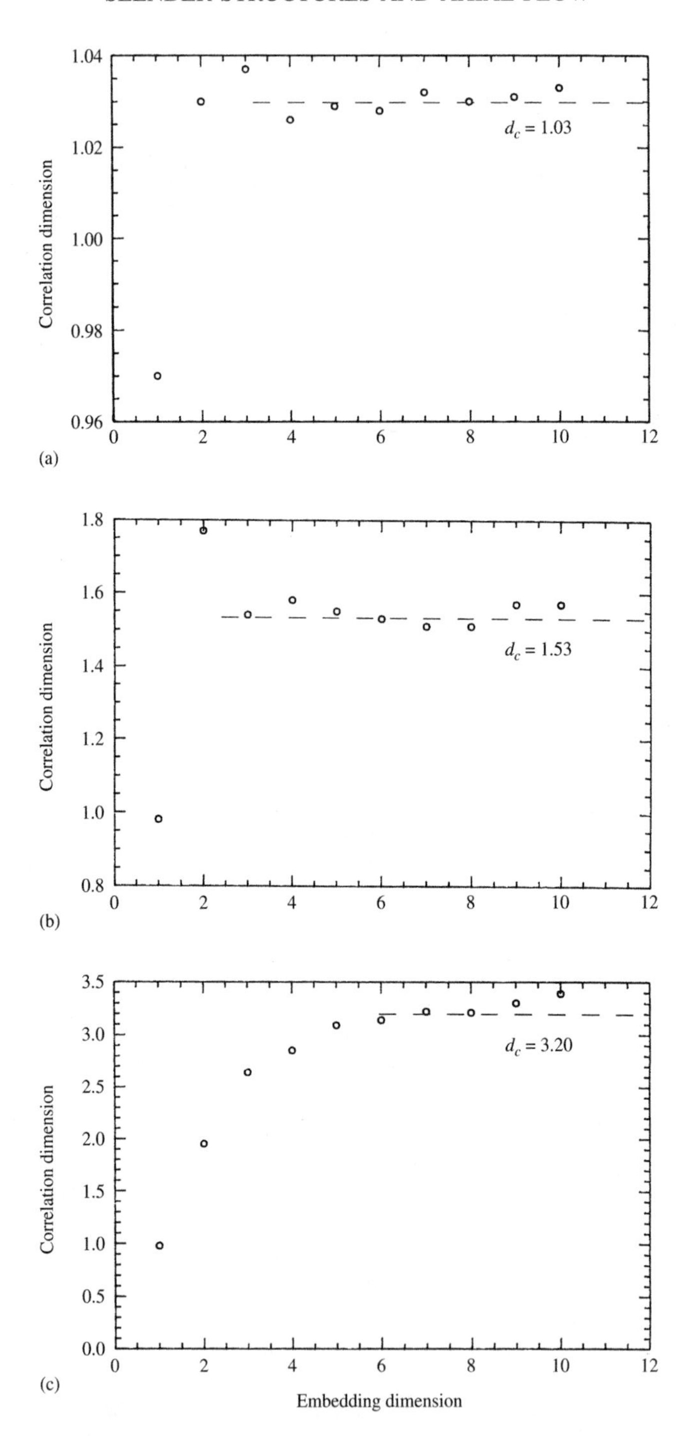

Figure 5.37 Correlation dimension, d_c, versus embedding dimension, m, defined in Appendix I, for the system of Figure 5.30 at (a) $U = 6.77$ m/s; (b) $U = 7.27$ m/s; (c) $U = 7.47$ m/s. The measurement error (68% confidence limit — see Appendix I) for convergent m (i.e. $m \geq 4$) was 0.004 (Païdoussis, Cusumano & Copeland 1992). Note that $u = 1.15[U(\text{m/s})]$.

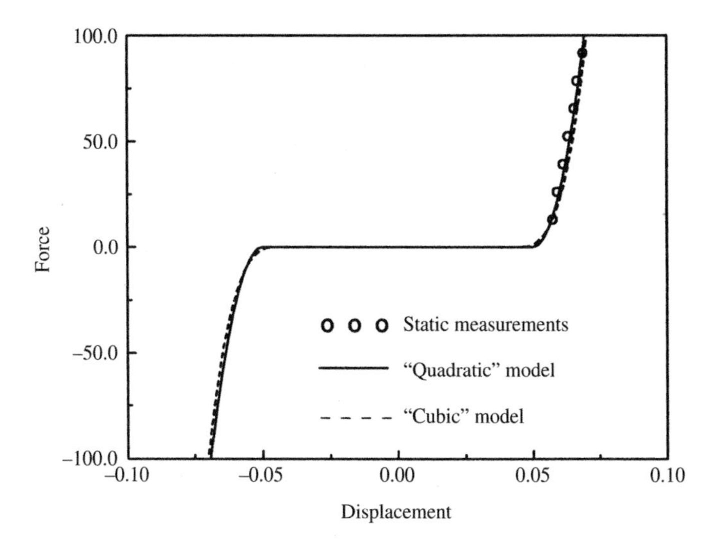

Figure 5.38 The static force–displacement measurements to determine the stiffness of the constraining bars (Figure 5.30) and the fitted smoothened-trilinear model for $n = 2$ and 3 in equation (5.138), labelled 'quadratic' and 'cubic' (Païdoussis, Li & Rand 1991a).

This model may be improved further, to account for the pipe itself being deformable at impact (Hunt & Crossley 1975), suggesting that a more gradual application of the full force following impact is closer to reality.† Hence, 'smoothened' versions of this model are used, as follows:

$$f(\eta) = \kappa_n\{\eta - \tfrac{1}{2}(|\eta + \eta_{bn}| - |\eta - \eta_{bn}|)\}^n, \qquad \text{with} \qquad n = 2, 3 \text{ or } 5; \qquad (5.138)$$

least-squares fitting gives the following set of values: for $n = 2$, $\kappa_2 = 2.7 \times 10^5$, $\eta_{b2} = 0.050$; for $n = 3$, $\kappa_3 = 5.6 \times 10^6$, $\eta_{b3} = 0.044$; for $n = 5$, $\kappa_5 = 1.0 \times 10^9$, $\eta_{b5} = 0.031$. These models will henceforth be referred to as 'quadratic', 'cubic', and 'quintic', for short; in the quadratic, clearly $(\eta - \eta_b)^2 = (\eta - \eta_b)|\eta - \eta_b|$ to preserve functional oddness. The resulting approximations to the dynamical restraint stiffness are shown in Figure 5.38 for $n = 2$ and 3; the curve for $n = 5$ is very close to those shown and is therefore omitted to preserve clarity.

It is found that with this impact model and $N \geq 3$ it is now possible to obtain convergent results, while using the correct stiffness κ_n and the impact location $\xi_b = 0.65$ as in the experiments. In all cases [$n = 2, 3, 5$ in equation (5.138)], the bifurcation diagrams, phase portraits, PSDs and so on are qualitatively similar to those already shown, confirming that the route to chaos is via a cascade of period-doubling bifurcations (Païdoussis *et al.* 1991a). Of special interest is the convergence of the various bifurcations with increasing N. As shown in Figure 5.39(a,b), the critical values of u for the Hopf and first period-doubling bifurcations have essentially converged when $N = 4$ or 5; for $N \geq 5$, the values of u_{pd} differ by less than 3.5%.

Furthermore, the degree of agreement with experiment for some of the key bifurcations, as shown in Table 5.4 (left side) is now excellent — discrepancies being of $\mathcal{O}(5\%)$, which

†Besides, numerical convergence problems continue to arise with the experimentally determined $\kappa \sim \mathcal{O}(10^5)$.

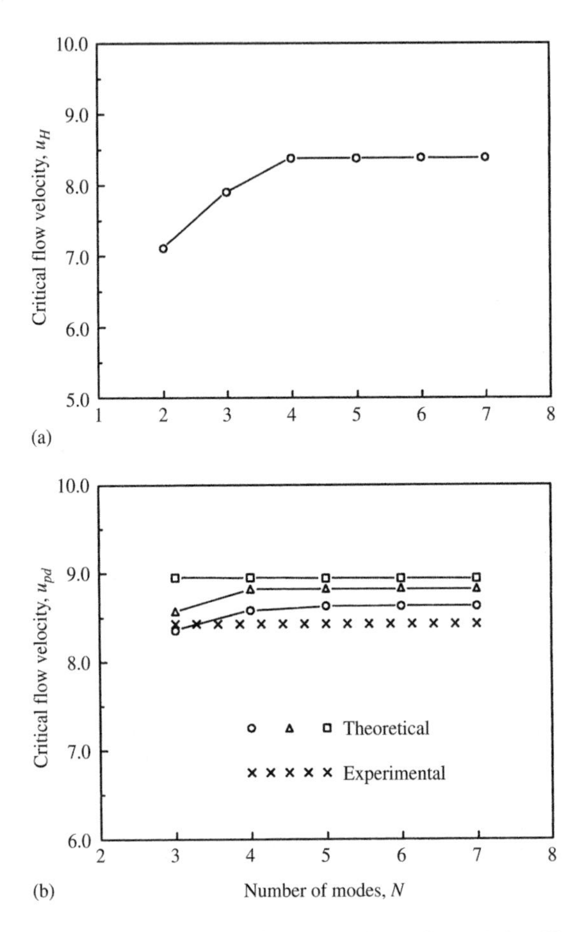

Figure 5.39 Convergence of the critical flow velocities for (a) the Hopf and (b) the first period-doubling bifurcations with the number of modes, N, in the Galerkin discretization of the system of Figure 5.30. In (b): ○, 'quadratic' (i.e. quadratically smoothed; $n = 2$) model; Δ, 'cubic' ($n = 3$); □, quintic ($n = 5$) (Païdoussis, Li & Rand 1991a).

Table 5.4 Theoretical values from (a) Païdoussis *et al*. (1991a) for $N = 5$ and various n (hence the range) and (b) from Païdoussis & Semler (1993a) for $N = 4$ and $n = 3$ compared with experiment, in terms of the main bifurcations of the fluttering cantilevered pipe impacting on motion-limiting restraints.

Bifurcation	u	Theory (a)	Experiment	Theory (b)
Hopf	u_H	8.40	8.04	8.40
1st period doubling	u_{pd}	8.63–8.94	8.43	9.05
Chaos	u_{ch}	8.68–8.97	8.72	9.20
'Restabilization'	u_{rs}	—	~9.0	9.65

is better than expected for a system such as this. However, the amplitude of motion is still much larger than in the experiments.

This final weakness of the model was overcome using the full nonlinear equation of motion, equation (5.42), by Païdoussis & Semler (1993a). Typical results are shown in

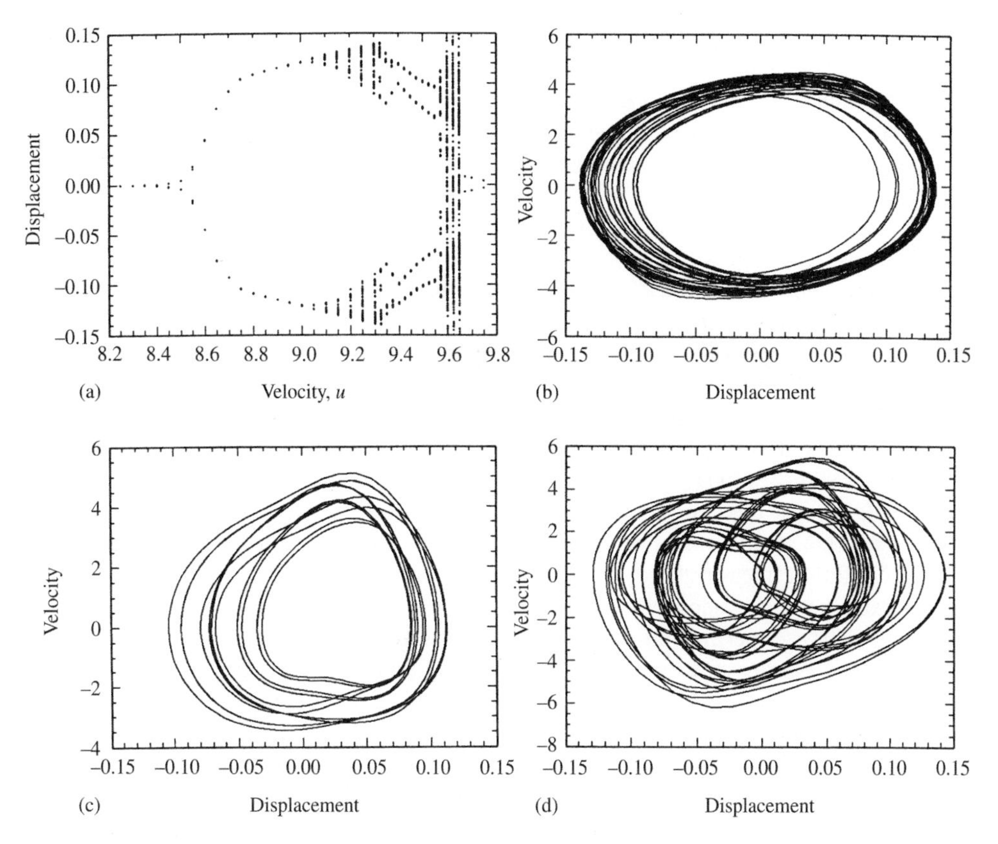

Figure 5.40 (a) Bifurcation diagram and (b–d) some corresponding phase portraits for the $N = 4$ model and the trilinear, smoothed 'cubic' representation of the constraints, with all parameters corresponding to the experimental system of Figure 5.30; (b) $u = 9.3$, (c) $u = 9.57$, (d) $u = 9.6$ (Païdoussis & Semler 1993a).

Figure 5.40; the pipe-tip displacements are now $\eta(1, \tau) \sim \mathbb{O}(0.10)$, i.e. they are of similar magnitude as those in the experiments (contrast with those in Figure 5.35). Again excellent agreement is achieved for some of the key bifurcations, as seen in Table 5.4 (right side); moreover, the final 'restabilization' or 'sticking', where the pipe adheres to the support without further oscillation, is also predicted by theory. Up to this point it was supposed that the inability of the early, $N = 2$ model to converge, when the correct system parameters are used, was due to its low-dimensionality — since the problem disappeared for $N > 2$. However, the real reason is the neglect of the nonlinear terms in the equation of motion, which is also responsible for over-predicting the oscillation amplitudes (Païdoussis & Semler 1993a).

The series of studies into this problem, starting with Païdoussis & Moon's (1988) and ending with Païdoussis & Semler (1993a), serve also as a case study into some of the pitfalls of analytical modelling of nonlinear systems when trying to match experimentally observed behaviour. By 'straining' two physical parameters (κ and ξ_b) so as to circumvent numerical difficulties, the $N = 2$ model with a cubic-spring representation of the constraints could give qualitatively similar behaviour to that observed, as well as reasonable quantitative agreement for the key bifurcations — though with unrealistically large pipe displacements. As a result of fractal dimension calculations, higher-N models were then employed, and convergent solutions were obtained with the correct κ and ξ_b and a better impacting model. Inevitably, this was interpreted to mean that the use of $N = 2$ was the main factor responsible for the original numerical difficulties. However, this was disproved, when it was found that, if the *nonlinear* equation of motion is used, then $N = 2$ with the proper κ, ξ_b and impact model gives perfectly reasonable results (albeit not as accurate as $N = 4$); and, moreover it gives the correct order of magnitude of displacements and retrieves forgotten features of physical behaviour, such as the 'sticking' to the constraints.

These studies, therefore, may be looked upon as providing particular views of the dynamics through different 'sections' of the multidimensional parameter space of this system. Thus, the $N = 2$ model of Païdoussis & Moon (1988) must be judged as so fragile (nonrobust), as to make one wonder if the agreement with experiment were not fortuitous. The Païdoussis *et al*. (1991a) model with $N > 2$ was decidedly more successful and more robust; yet, it too failed for $N = 2$. Finally, the most successful model (Païdoussis & Semler 1993a) is also the most robust: small excursions in the parameter subspace of this model have little effect on the dynamics. It is at this stage only that it can be concluded that the excellent agreement with experiment *cannot* be fortuitous.

Another of the experimental cases of Païdoussis & Moon — pipe 9 (Table 5.3), but with softer, leaf-spring-supported bars, closer to the pipe ($w_b/L = 0.025$) — was studied numerically by Makrides & Edelstein (1992), with the Lundgren *et al*. (1979) nonlinear equations of motion and a finite element and penalty function solution approach — see Section 5.7.1. It is found that the onset of chaos is dependent on the stiffness of the motion constraints (modelled as trilinear springs), in agreement with observations. Furthermore, in this case, the route to chaos is via *quasiperiodicity* (see Section 5.8.3), although in the experiments a period-doubling route appears to be followed. Still, the predicted threshold to chaos, $u_{ch} \simeq 8.05$ for an assumed $\kappa = 10^3$, is not too far from the experimental one, $u_{ch} \simeq 9.1$. One weakness here is that in the theoretical model, as per the original form of the Lundgren *et al*. equations, gravity effects are neglected, whereas in this particular experiment they are not negligible ($\gamma = 26.8$); a nonzero γ would nevertheless raise u_H and hence u_{ch}. One observation that ought to be made here, in view of the foregoing discussion, is that through the use of the nonlinear equations of motion, similarly to Païdoussis & Semler (1993a), no difficulties are reported in obtaining convergent solutions with amplitudes of the correct order of magnitude. In conclusion, Makrides & Edelstein's work shows that, with different motion constraints (and perhaps other parameters), it is possible that a different route to chaos may be followed — cf. Païdoussis & Botez (1995) for a system discussed in Volume 2 — which adds to the interest in this system.

The interested reader is also referred to Miles *et al*. (1992), who demonstrate the power of bispectral analysis techniques to isolate nonlinear phase coupling and energy exchange of the various Fourier components of motion, using this particular system as an example.

5.8.2 Magnetically buckled pipes

Chaotic oscillations of a conservative system with passive damping added, such as a buckled beam subjected to deterministic forced excitations, have been studied by Moon (1980), Holmes & Moon (1983) and Dowell & Pezeshki (1986), among others; rarer are studies of inherently nonconservative such systems, exemplified by Dowell's (1982) work on flutter of a buckled plate, and that described here on a magnetically buckled cantilevered pipe conveying fluid (Tang & Dowell 1988).†

The experimental system, shown in Figure 5.41, consists of a Tygon pipe ($L = 545$ mm, $D_o = 12.7$ mm) conveying water, fitted with a ferromagnetic metal strip, similar to that in Section 5.8.1, and fitted with an end-nozzle. Two permanent magnets on either side of the straight equilibrium position provide two potential wells, into one of which the system buckles statically. The system would ordinarily stay buckled unless excited, either by flow-induced flutter in the case of the autonomous version of the system, or mechanically by a force $F_0\,\delta(\xi - \xi_F)\sin\omega t$. In the experiments this force is provided by a shaker at $\xi_F = 0.11$, while pipe motion is sensed at $\xi_s = 0.92$. In the absence of flow, this system,

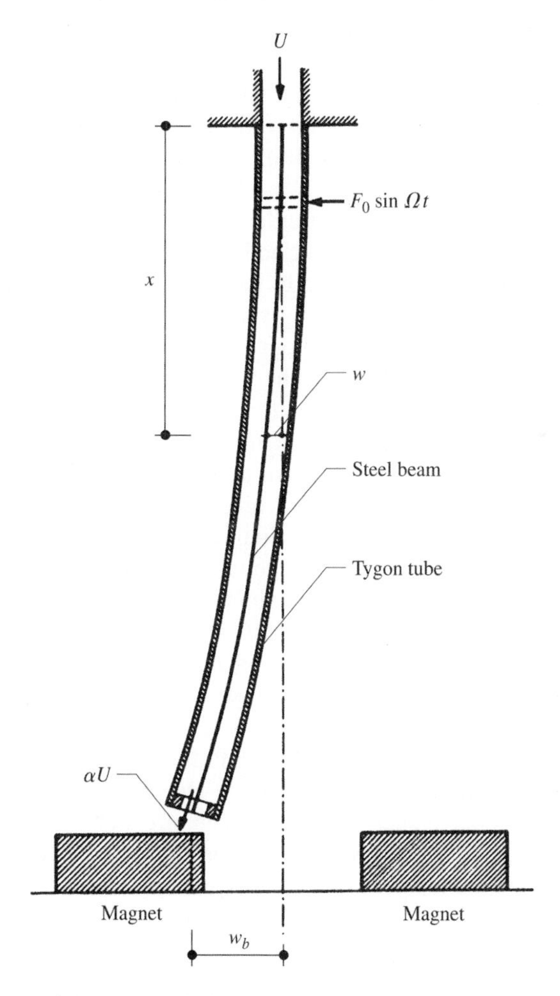

Figure 5.41 Diagram of the magnetically buckled, mechanically excited pipe conveying fluid (Tang & Dowell 1988).

†Apart from the work on the motion-constrained pipe described in Section 5.8.1; Païdoussis & Moon's (1988) paper was published six months after Tang & Dowell's.

if excited sufficiently so that it crosses the vertical (unstable) equilibrium position, may develop chaotic oscillation, as a result of 'hesitation' as to which of the two potential wells it will next gravitate towards. Conceptually, the same may be achieved if the excitation is provided via flow-induced flutter.

The equation of motion is given by the linear equation (3.70), with $k = \Gamma = \Pi = \dot{u} = 0$ and $\alpha, \sigma \neq 0$, as modified by (3.74) to account for the end-nozzle, and the presence of the magnetic forces represented by $\kappa_1 \eta(1, \tau) + \kappa_2 \eta^3(1, \tau)$. In the experiments, κ_1 and κ_2 are determined by measuring the first buckled natural frequency in the post-buckling state of the system, and the location of the statically buckled pipe end. The equations are first discretized by Galerkin's method and then studied by simulation via a fourth-order Runge–Kutta integration scheme, both for the autonomous and the forced system. Useful experimental data, however, could only be obtained for the forced system; flutter of the buckled autonomous system could not be achieved because of experimental limitations (Tang 1997).

Numerical phase-plane plots for an autonomous system similar to the experimental one but with zero damping ($\alpha = \sigma = 0$), discretized to fourth order ($N = 4$), show (a) a stable limit cycle for $u = 3.30$, just beyond the flutter threshold, (b) period-4 motion for $u = 3.89$, and (c) chaotic oscillation for $u = 3.96$ and 4.29. Interestingly, for a model with $N = 2$, periodic rather than chaotic oscillation is displayed for $u = 4.29$. For $N = 4$ it is shown that the dynamics evolves about both potential wells: a small orbit about one of them, followed by a larger orbit leading to the other one.

Additional work on the effect of damping shows that, with increased damping in the pipe material (α), the chaotic attractor is progressively weakened, so that eventually, for $\alpha \sim \mathbb{O}(10^{-2})$, the oscillation becomes periodic.

Typical numerical results for the case of forced oscillation of the system are shown in Figure 5.42 for $N = 2$ (cf. the early work in Section 5.8.1) and a low value of u. It is shown that with increasing F_0 and constant ω, there is an alternation of periodic and chaotic regions, which is mapped in (e) for varying ω. A similar map for $N = 1$ (not shown) is completely different from that of Figure 5.42(e), but another for $N = 3$ is not too radically different from that for $N = 2$, showing the beginnings of convergence.

The phase-plane plots of Figure 5.42(a–d) show clearly that for low $f = F_0 L^2/EI$ the vibration is in the vicinity of the potential well in which the system is buckled. For higher f, it is about both wells, and the motion is essentially as follows: one or more orbits around one of the potential wells, followed by a trajectory over to the other potential well, and so on.

The theoretical ($N = 3$) thresholds for chaos with forced vibration are reasonably close to the experimental ones for a case with $u = 0.35$. However, this does not represent the best test for the theory since, at such low u, the main effect of flow is to contribute some additional damping *vis-à-vis* $u = 0$. Of more interest would have been an experiment at u close to the flutter boundary; however, this was precluded by the apparatus used (Tang 1997).

5.8.3 Pipe with added mass at the free end

As discussed in Section 3.6.3, the linear dynamics of a cantilevered pipe conveying fluid is modified in interesting ways by the addition of a point mass, notably at the free end. The nonlinear dynamics is equally interesting, as will be seen in what follows.

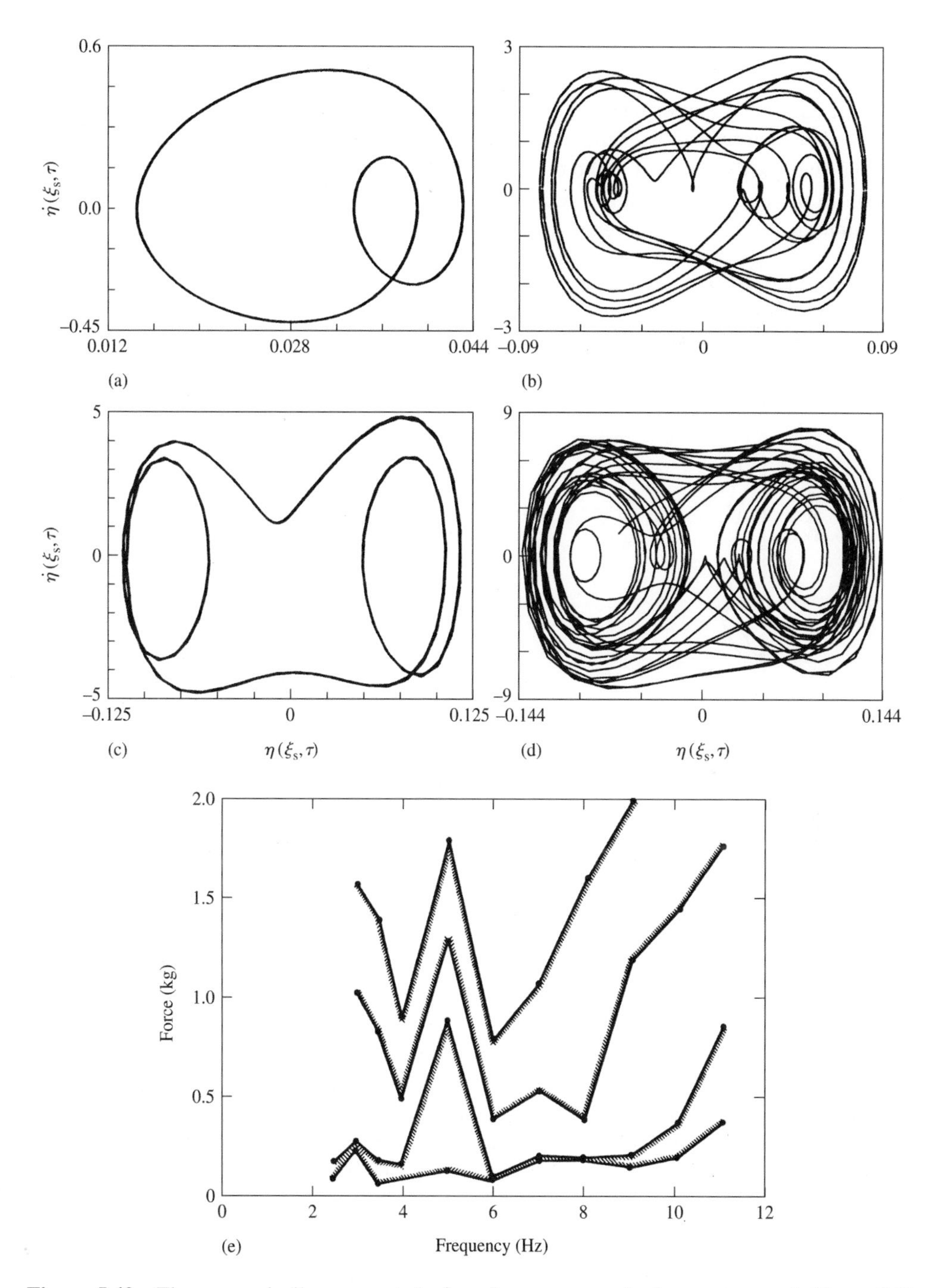

Figure 5.42 The numerically computed forced response of the system of Figure 5.41 for $u = 0.35, N = 2, \omega/2\pi = 4\,\text{Hz}$, pipe/nozzle area-ratio $\alpha_j = 1$, and variable $f = F_0L^2/EI$: (a) $f = 3.0$; (b) $f = 9.0$; (c) $f = 18.1$; (d) $f = 33.1$. (e) The map of chaotic regions (the hatching points *into* the regions) in terms of the excitation force F_0 in kg, where $f = 60.2F_0$, and the frequency of forcing, $\omega/2\pi$ (Tang & Dowell 1988).

Copeland (1992) and Copeland & Moon (1992) studied the 3-D dynamics of the system shown in Figure 5.43(a) and showed that chaotic motions routinely arise, whereas they do not if the additional mass is absent. The 2-D version of the same system was subsequently examined by Païdoussis & Semler (1998), partly to see if chaotic planar motions are possible and partly to shed light on the dynamics of the system in general; the 2-D version in the case of a negative addition of mass (mass deficit) was studied by Semler & Païdoussis (1995). All this work is discussed here, starting with 2-D motions.

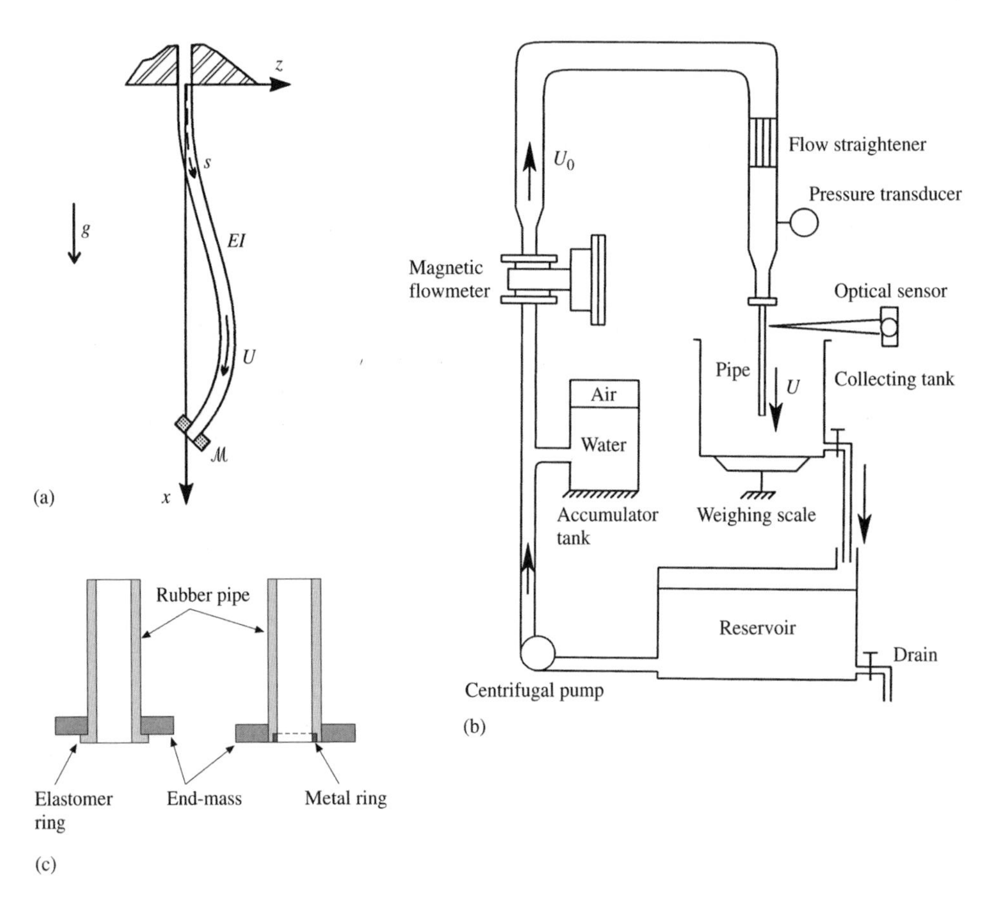

Figure 5.43 (a) Schematic of the system with added end-mass. (b) Schematic of the experimental apparatus, and (c) the two methods used for mounting the end-mass; the 'elastomer ring' in the method on the left is moulded integrally to the pipe (Païdoussis & Semler 1998).

(a) 2-D motions of a pipe with an added end-mass

The work presented in this subsection, unless otherwise attributed, is based on Païdoussis & Semler's (1998). The system studied is shown in Figure 5.43(a), and the experimental set-up and manner of mounting the end-mass in Figure 5.43(b,c). The apparatus, including the noncontacting optical sensor, is similar to that used in Section 5.8.1. The added mass is in two halves, screwed together tightly at the end of the pipe. The rather elaborate alternative schemes for mounting it are necessary (i) to avoid it becoming loose on the pipe in the course of oscillation and (ii) to ensure that it is not overtightened onto the pipe, thus deforming its free end.

Several elastomer pipes were used (typically, $D_i = 6.3$ mm, $D_o = 15.7$ mm, $L = 470$ mm), giving $\beta = 0.125$–0.150 and $\gamma \simeq 20$, and eight different end-masses, made of plastic or metal, $\mathcal{M} = 2.4$–37.8 g; this corresponds to values of the dimensionless added mass,

$$\mu = \frac{\mathcal{M}}{(M+m)L}, \tag{5.139}$$

ranging from $\mu = 0.023$ to 0.380.

Typical results related to the flutter threshold, which exhibits the characteristics of a Hopf bifurcation, are given in Figure 5.44(a). It is remarked that (i) the additional mass destabilizes the system, in agreement with Hill & Swanson's findings (Figure 3.68), and (ii) there is considerable hysteresis in the critical u obtained with increasing and decreasing flow, suggesting that the Hopf bifurcation is subcritical.

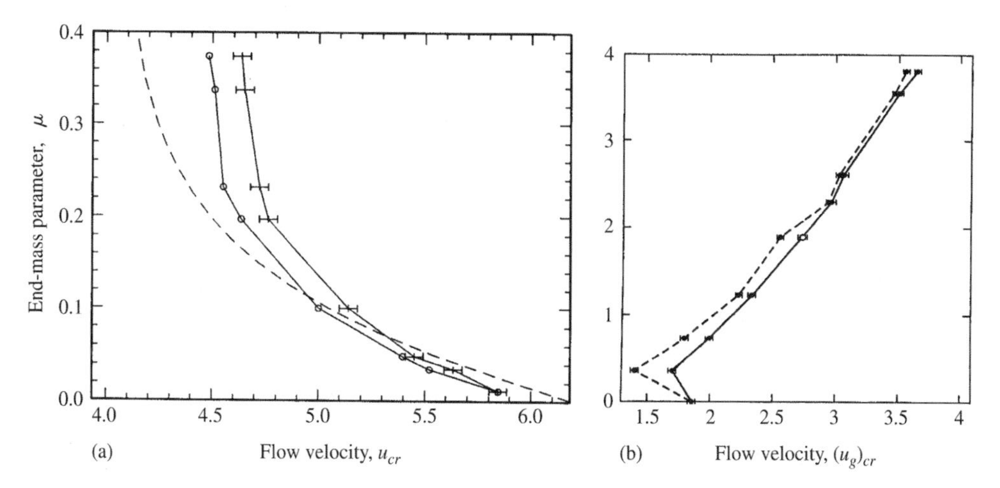

Figure 5.44 (a) Experimental critical flow velocity for the onset of flutter as a function of μ [equation (5.139)] for the system of Figure 5.43(a): – – –, linear theory; +, experiment for increasing u, the error bar indicating maximum repeatability variations; ○, experiment for decreasing u, (mean value) (Païdoussis & Semler 1998). (b) Similar observations over a larger range of μ, with $(u_g)_{cr} = U_{cr}/(gL)^{1/2}$, by Copeland & Moon (1992).

Similar results from Copeland & Moon (1992) are shown in Figure 5.44(b) for a much wider range of μ. Motions in this case too were mostly planar at the onset of flutter [see Section 5.8.3(b)]. It is seen that, contrary to the effect of smaller μ, the presence of end-masses with $\mu > 0.5$ approximately *stabilizes* the system *vis-à-vis* $\mu = 0$. These results also display hysteresis and suggest a subcritical Hopf bifurcation. The pipes in this case were very long and slender ($L \sim 1$ m, $L/D_i \simeq 125$); according to theory (for $\mu = 0$) such slender pipes should lose stability by a supercritical Hopf bifurcation [Figure 5.20 and Bajaj *et al.* (1980)].

Returning to Païdoussis & Semler's experiments, as the flow is increased the system undergoes a secondary bifurcation, shown at the right side of Figure 5.45(a).† For the higher values of μ, approximately $\mu > 0.1$, this secondary bifurcation involves a sudden increase in the frequency of oscillation, as seen in Figure 5.45(b). If u is increased further, the motion becomes chaotic, as confirmed by phase-plane and PSD plots constructed from the experimental signal. At this point the oscillation is three-dimensional and violent, and the pipe impacts on the collecting tank if not restrained.

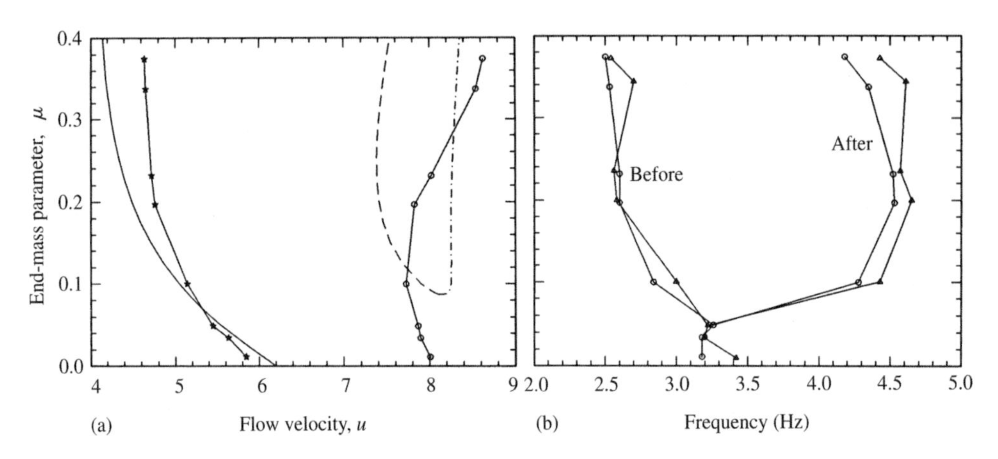

Figure 5.45 (a) Experimental critical flow velocities for the Hopf (—*—), on the left of the figure, and secondary bifurcation (—o—), on the right, for a pipe with $\beta = 0.142$, $\gamma = 18.9$ [Figure 5.43(a)]; ——, first theoretical Hopf bifurcation; – – –, – · · –, onset and cessation of a higher theoretical Hopf bifurcation. (b) o, Dominant experimental frequencies 'before' and 'after' the second bifurcation for the pipe system in (a); ▲, for another pipe system ($\beta = 0.150$, $\gamma = 20.5$); (Païdoussis & Semler 1998).

For $\mu < 0.1$ approximately, the dynamics is rather different. The secondary bifurcation in this case is associated with a change in the *character* of the oscillation rather than a jump in frequency. The oscillating mode becomes distinctly nonlinear: a point along the pipe, at $x \simeq \frac{1}{4}L$, becomes a node,‡ and the upper part oscillates with a smaller amplitude and about half the frequency of the lower part. Two distinct peaks appear in the PSD with a frequency ratio of 2:1; but, with increasing u, the main, higher frequency increases while the lower one decreases slightly, so that the ratio becomes incommensurate. Eventually, in this case also, the motion becomes three-dimensional and chaotic.

Because one of the motives behind this work was to discover whether chaotic oscillation can arise in purely 2-D motions, attempts were made to confine the motion to a plane, even after the onset of chaos. To this end, experiments were done with pipes fitted with a metal strip (as in Section 5.8.1), which in this case were unsuccessful. In the presence of the end-mass, the violence of the chaotic oscillations was such as to quickly destroy

†The dashed and chain-dotted lines are associated with another Hopf bifurcation, in a higher mode, predicted by linear theory. This *may* have something to do with the secondary bifurcation, but it is unlikely (see also Figure 5.48).

‡It is recalled (Section 3.5.6) that there are normally no nodes in the motion, because of travelling-wave components in the oscillation.

the pipe–strip bond and even the pipe as a whole. More successful was the use of plate guides confining the motion to a plane. Under these circumstances the motion did become chaotic in any case.

The route to chaos is not clear from the experimental data. It could be via quasiperiodicity — see Section 5.8.3(b); however, in at least one case, a sequence of two period doublings was observed to precede chaos.

Finally, the case of $\mu = 0$ does appear to be singular, in that no secondary bifurcation arises. The motion remains periodic to the maximum flow available — a conclusion also reached by Copeland & Moon (1992).

In the theoretical study of the system, equation (5.28) or (5.39) is used, rather than (5.42) and (5.43), so that inertial nonlinearities are left intact and no restriction on their magnitude needs be imposed. Of course, equation (5.28) needs be modified to include the effect of the end-mass, which is twofold: (i) the inertial terms now involve $m + M + \mathcal{M}\delta(x - L)$, where $\mathcal{M}$ is the end-mass; (ii) the gravity-induced tension terms are similarly modified. Hence, with $\dot{u} = 0$, the dimensionless equations are

$$\ddot{\eta}[1 + \mu\delta(\xi - 1)] + \eta'' \left[u^2 - \gamma \int_\xi^1 (1 + \mu\delta(\xi - 1))\, \mathrm{d}\xi\right] + \gamma[1 + \mu\delta(\xi - 1)]\eta' + 2u\sqrt{\beta}\dot{\eta}' + \eta'''' + N(\eta) = 0 \tag{5.140}$$

where

$$\begin{aligned} N(\eta) = {} & 2u\sqrt{\beta}\dot{\eta}'\eta'^2 + \eta'' \left[u^2 - \tfrac{3}{2} \int_\xi^1 \gamma[1 + \mu\delta(\xi - 1)\, \mathrm{d}\xi]\right] \eta'^2 + \tfrac{1}{2}\gamma[1 + \mu\delta(\xi - 1)]\eta'^3 \\ & + \eta''''\eta'^2 + 4\eta'\eta''\eta''' + \eta''^3 + \eta'[1 + \mu\delta(\xi - 1)] \int_0^\xi (\dot{\eta}'^2 + \eta'\ddot{\eta}')\, \mathrm{d}\xi \\ & - \eta'' \left[\int_\xi^1 [1 + \mu\delta(\xi - 1)] \int_0^\xi (\dot{\eta}'^2 + \eta'\ddot{\eta}')\, \mathrm{d}\xi\, \mathrm{d}\xi \right. \\ & \left. + \int_\xi^1 \left(2u\sqrt{\beta}\eta'\dot{\eta}' + u^2\eta'\eta''\right) \mathrm{d}\xi \right]; \end{aligned} \tag{5.141}$$

μ is defined by (5.139). Thus, via the use of the Dirac delta function, the effect of μ is incorporated in the equation of motion, while the boundary conditions remain the same as for $\mu = 0$.[†] This facilitates the discretization of the system, via Galerkin's method. As discussed in Section 5.4, care has to be exercised in selecting appropriate numerical methods for the solution of the resultant N second-order ordinary differential equations, because of the presence of the nonlinear inertial terms — methods which should give accurate, convergent solutions. The finite difference (FDM) and the incremental harmonic balance (IHB) methods have been found to be particularly efficient and complementary (Semler *et al*. 1996).

Typical results are shown in Figure 5.46; they are seen to be sensibly the same whether computed with $N = 3$ or 4 in the Galerkin discretization. In (a) it is seen that, for $\mu = 0$, the theory also finds no bifurcation beyond the Hopf bifurcation at $u \simeq 6.2$. The Hopf bifurcation is clearly supercritical, in contrast to the experimental results.

[†]The appropriateness of this formulation and method of solution is demonstrated at the end of Section 4.6.2.

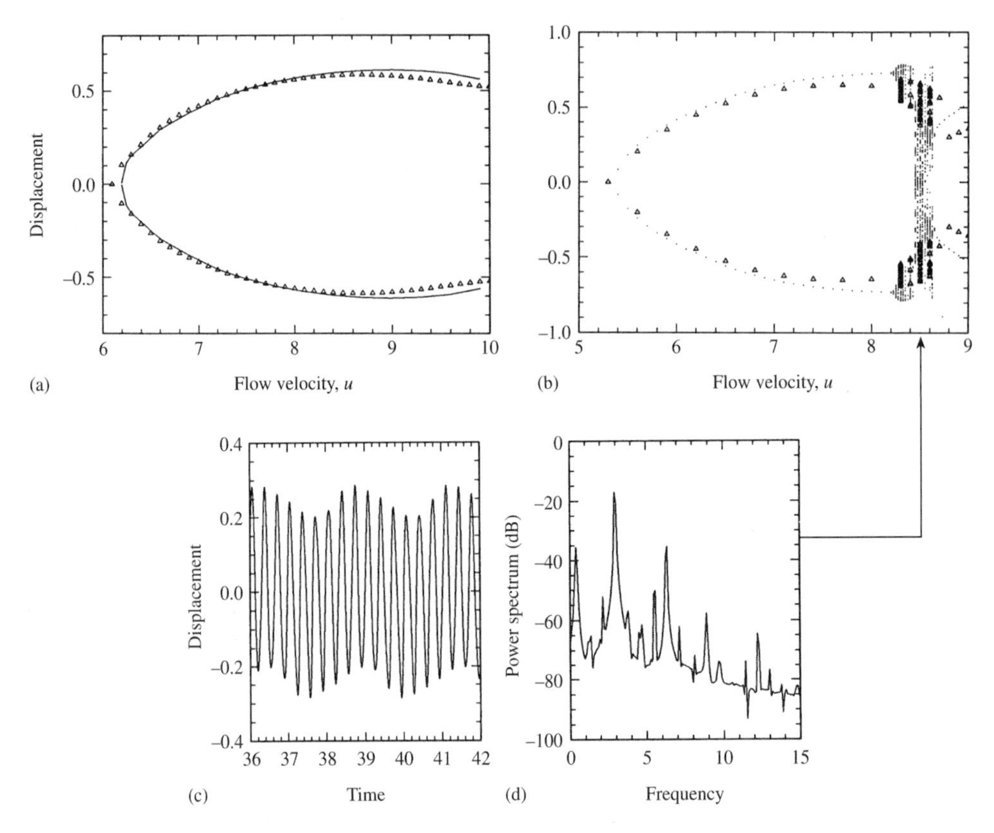

Figure 5.46 Bifurcation diagrams of the dimensionless pipe-end displacement versus u for the system of Figure 5.43(a) by the FDM method for (a) $\mu = 0$ and (b) $\mu = 0.06$; ——, $N = 3$ in the Galerkin discretization; $\Delta, N = 4$. (c) Time trace and (d) PSD in terms of $f = \omega/2\pi$ for $\mu = 0.06, N = 3$ and $u = 8.5$. In all calculations the parameters are as in the experimental system: $\beta = 0.142$, $\gamma = 18.9$, and the measured logarithmic decrements of damping: $\delta_1 = 0.037$, $\delta_2 = 0.108$, $\delta_3 = 0.161$ and extrapolated $\delta_4 = 0.220$ (Païdoussis & Semler 1998).

For $\mu = 0.06$, the dynamical behaviour is initially similar, as shown in Figure 5.46(b). For $8.2 < u < 8.625$, however, there is a band of quasiperiodicity, involving two frequencies, as made clear in the time trace and PSD in (c) and (d) of the figure. The two dimensionless frequencies are $f_1 \simeq 0.5$ and $f_2 \simeq 3$, while the third peak in the PSD is at $2f_2 + f_1$, and so on; $f = \omega/2\pi$, while ω is as defined in equation (3.73). For $u > 8.625$, periodic oscillations resume, but at a smaller amplitude. This theoretical evolution bears some resemblance to the experimental behaviour, but there are some obvious discrepancies as well. Comparing the results of Figure 5.45 for $\mu = 0.06$ with those of Figure 5.46(b), it is seen that theory and experiment agree in the following aspects: (a) the values of u_H for the first Hopf bifurcation are similar ($u_H \simeq 5.3$ in the experiments, compared to $u_H = 5.35$ in theory); (b) the nonlinear model and the experiments both predict a qualitative change in the behaviour of the pipe at a higher flow; and (c) the values of u for the second bifurcation are relatively close ($u_{\text{theory}} \simeq 8.2$ versus $u_{\text{exp}} \simeq 7.8$). On the other hand, only periodic solutions are predicted in the experiment (before the onset of chaos), while the motion is also found to be quasiperiodic in theory, prior to becoming periodic again. Furthermore, there is a frequency jump in the periodic oscillations in the theoretical results before and after the quasiperiodic band: $f = \omega/2\pi = 3.1$ for $u = 8.0$ versus $f = 6.5$ for $u = 8.8$. This corresponds to the jump observed across the second bifurcation in the experiments, Figure 5.45(b), *but only at higher values of* μ.

The maximum tip displacement and frequency of oscillation are shown in Figure 5.47(a,b) for $\mu = 0$–0.10, computed by the IHB method which 'follows' periodic solutions and determines their stability along the way. The following observations may be made.

(i) For $\mu > 0$, the original stable limit cycle loses stability at the points marked with a bullet (•), a pair of complex conjugate Floquet multipliers crossing the unit circle (the modulus becomes greater than 1), which means that quasiperiodic solutions are possible after the bifurcation point (Bergé *et al.* 1984), in agreement with FDM results.

(ii) Following the unstable solutions, two additional saddle-node bifurcations are detected: the first corresponding to a limit or turning point, and the second, represented by

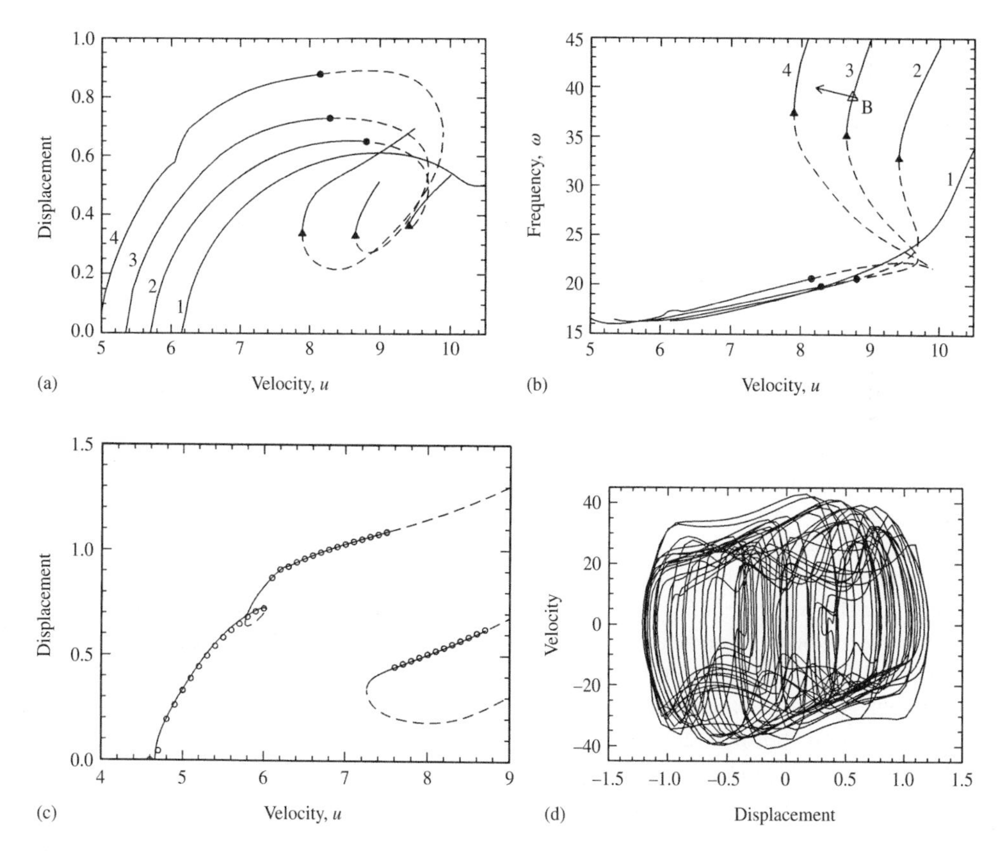

Figure 5.47 Bifurcation diagrams for the system of Figure 5.43(a) via the IHB method: (a) the dimensionless tip displacement and (b) the dimensionless frequency, ω, versus u, for $\mu = 0$–0.10 (curves 1, 2, 3, 4: $\mu = 0, 0.03, 0.06$ and 0.10, respectively). Point B is an arbitrary point on the stable high-frequency solution. (c) Bifurcation diagram for $\mu = 0.15$: ——, stable periodic solution (IHB); – – –, unstable periodic solution (IHB); ∘ , stable periodic solution (FDM). (d) Phase-plane plot for $\mu = 0.15$, $u = 8.8$. All parameters as in the experimental system and $N = 3$ (Païdoussis & Semler 1998).

the triangles (▲) at lower values of u, corresponding at the same time to a turning point *and* a restabilization, for which stable periodic oscillations of small amplitude and high frequency appear. The appearance of the second stable periodic solution explains clearly how the smaller amplitude oscillations at $u > 8.625$ detected by FDM (Figure 5.46) come into being.

(iii) The value of u of the bifurcation point leading to the appearance of stable periodic solutions decreases dramatically with increasing μ, which means that the range where quasiperiodic oscillations can be detected decreases with μ, to a point where they no longer exist. This is confirmed by FDM calculations; for $\mu = 0.1$ there exists only a very narrow range of quasiperiodic solutions.

(iv) The case of $\mu = 0.1$ may be considered as a limiting case for two distinct reasons: firstly because of the previous remark, and secondly because of the small amplitude 'jump' observed in Figure 5.47(a) for $u \simeq 6.1$ or the corresponding small frequency 'hump' in Figure 5.47(b); the evolution of the stable periodic solutions emerging from the Hopf bifurcations is no longer smooth for $\mu > 0.1$, and new phenomena start to occur.

A bifurcation diagram for $\mu = 0.15$ is shown in Figure 5.47(c). It is recalled that for such 'high' μ, there is a very clear frequency jump across the secondary bifurcation in the experiments, succeeded by chaos. The results shown indicate excellent agreement between FDM and IHB — although the former cannot compute unstable solutions, while the latter cannot compute nonperiodic (chaotic) ones. Nevertheless, the two in synergism are a potent tool. Thus, the IHB results reveal that the amplitude jump at $u \simeq 6$ for $\mu = 0.10$ in Figure 5.47(a) is associated with a loop, as shown for $\mu = 0.15$ in Figure 5.47(c).

Unstable solutions emerging at $u = 7.54$ in Figure 5.47(c) exist up to $u = 11.69$, which is beyond the scale of the figure. Furthermore, although the system is always unstable in this big loop, the number of Floquet multipliers inside the unit circle varies several times ($4 \rightarrow 5 \rightarrow 3 \rightarrow 5 \rightarrow 6$). These bifurcations are of no great importance because the system is unstable in any case. Of more interest is the bifurcation occurring in the small-amplitude stable periodic solution at $u = 8.76$: increasing u further, the stable solution becomes unstable, again because two complex conjugate multipliers cross the unit circle, but the solution thereafter is not quasiperiodic but chaotic, as shown in Figure 5.47(d).

Consequently, from a physical viewpoint, four distinct types of solution may be observed for $\mu = 0.15$: (i) solutions converging to the stable equilibrium for $u \leq 4.66$; (ii) periodic solutions whose frequency increases with u for $4.66 < u < 7.54$; (iii) periodic solutions of higher frequency and smaller amplitude for $7.26 < u < 8.76$ (implying a jump in the response); and (iv) chaotic oscillations for $u > 8.76$. This is exactly what is observed in the experiment. As shown in Table 5.5, agreement between theory and experiment is relatively good in terms of the critical flow velocities and the frequency before the 'jump' but not after.

If the value of μ is increased further, the results obtained numerically are qualitatively similar to those for $\mu = 0.15$, except that the number of 'humps' increases, which means that the number of bifurcations in the system increases as well. Alas, the quantitative agreement between theory and experiment deteriorates, since the values of u for the second bifurcation (followed almost immediately by chaotic oscillations) increase in the experiment, while they decrease in theory (Table 5.6). Before giving reasons for this, the effects of the nonlinear inertial terms on the dynamics are investigated; these terms have been included in the analysis so far. The idea here is to compare the results with and without the nonlinear inertial terms. If it is found that these terms do not significantly affect the

Table 5.5 Comparison between theory and experiment of the flow velocity and the frequency of the pipe corresponding to the three bifurcations, for $\mu = 0.15$; the arrow represents the jump in frequency.

	Values of u		Values of f	
	Expt	Theory	Expt	Theory
Hopf bifurcation	4.8	4.66	2.3	2.6
Second bifurcation	7.6	7.26	$2.8 \rightarrow 4.5$	$3.0 \rightarrow 6.3$
Chaos	$\simeq 8$	8.76	—	—

Table 5.6 Flow velocity corresponding to the appearance of chaotic oscillation: comparison between theory and experiment.

	u_{exp}	u_{theory}	
		$N = 3$	$N = 4$
$\mu = 0.2$	8.0	6.9	6.5
$\mu = 0.3$	8.2	6.0	5.9
$\mu = 0.4$	8.6	6.0	5.9

results, at least qualitatively, this would free the way to using AUTO to compute bifurcation diagrams; AUTO cannot handle second-order equations directly, but it is extremely versatile in 'following up' stable and unstable solution branches and their offshoots if these equations can be transformed into first-order form.

Results in the absence of inertial nonlinear terms are given in Figure 5.48 in (a) by AUTO and in (b) both by AUTO and FDM. From (a) it is seen that the original periodic solution loses stability through a subcritical pitchfork bifurcation at $u = 8.6$ (marked by •) prior to the saddle-node bifurcation occurring at $u = 8.7$ (▲). This means that the solution after $u = 8.6$ becomes unstable and that two unstable periodic solutions emerge at the bifurcation point. Following the original solution after the saddle-node bifurcation, three additional limit points are encountered (represented again by filled triangles), the first one at $u = 7.96$, the second at $u = 10.11$ and the third at $u = 8.3$. This last bifurcation point, as in previous cases, corresponds to a restabilization of the periodic solution and the appearance of stable limit cycles of small amplitude and high frequency. This means that the same qualitative results are obtained, whether the nonlinear inertial terms are accounted for or ignored.

On the other hand, the results obtained by FDM in Figure 5.48(b) indicate that not only periodic solutions exist but also chaotic ones, for $8.6 \leq u \leq 9.3$. Consequently, although stable periodic solutions exist for all flow velocities, as demonstrated in Figure 5.48(a), there is a large range of velocity for which these stable periodic solutions are not able to attract the trajectory. This is due to the fact that in the same range, many *unstable* or *repelling* periodic solutions are present, on which the trajectory may 'bounce'. Some of those unstable attractors emerging from the subcritical pitchfork bifurcation have been computed with AUTO [dash-dotted line in Figure 5.48(a)], but there may in fact exist an infinite number of them (indeed, more branch points and period-doubling bifurcation points were detected in this range, but no attempt was made to 'switch' to other solution

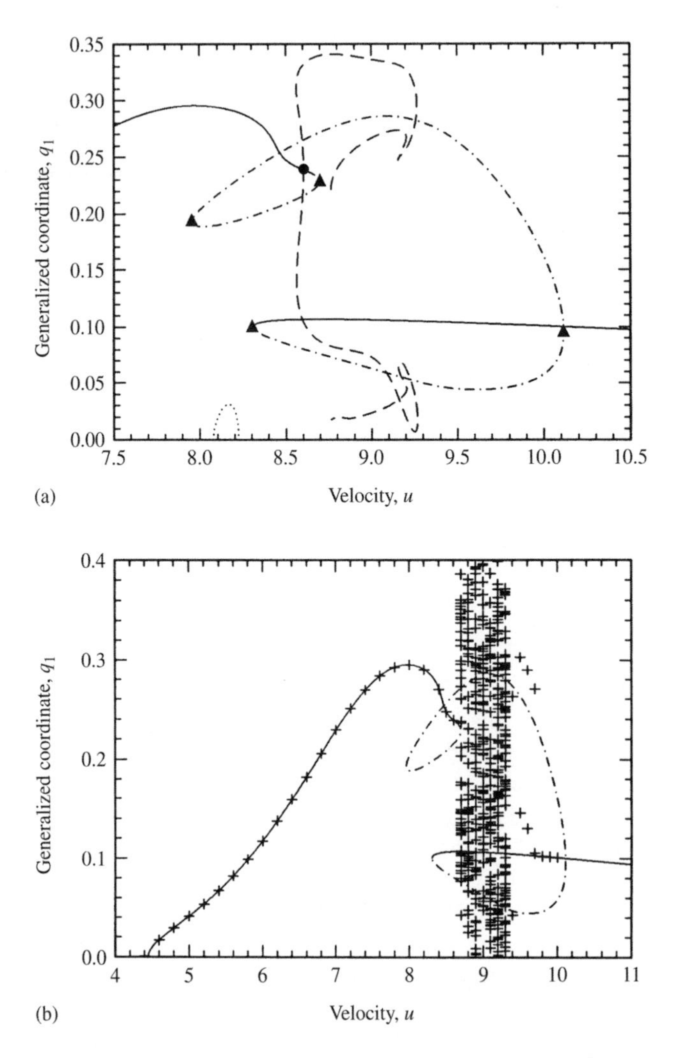

Figure 5.48 (a) Bifurcation diagram for the system of Figure 5.43(a), for $\mu = 0.2$ and $N = 3$, obtained while ignoring the nonlinear inertial terms, showing the maximum generalized coordinate, q_1, versus u obtained with AUTO: ——, stable periodic solution; - · -, main branch of the unstable solution; – – –, unstable branch emerging from the bifurcation point marked by •; · · ·, unstable branch connecting the two Hopf bifurcation points, amplified 20 times; ▲, limit points on the main branch. (b) The main branch of (a), together with results computed by FDM (+) (Païdoussis & Semler 1998).

branches). To give additional proof that the presumed chaotic solutions computed by FDM are really chaotic, the numerical scheme developed by Hairer *et al*. (1993) was used, since it is known to be particularly accurate in chaotic regimes, in the sense that it does not induce artificial chaos numerically. The results obtained confirm that the motion is indeed chaotic.

In conclusion, it may be said that the addition of a small mass at the end of the fluid-conveying pipe enriches the dynamics considerably, in fact revealing the existence of a completely new dynamical system. Not only are different types of periodic solutions detected, but also jump phenomena, quasiperiodic and chaotic oscillations. This conclusion is reinforced by the findings of Sections 5.8.3(b) and (c).

(b) 3-D motions of a pipe with an added end-mass

As discussed in Section 5.8.3(a), there is a natural tendency for motions to be three-dimensional once they become chaotic, even if at the onset of flutter they are planar. Copeland (1992) and Copeland & Moon (1992), whose work is discussed here, observed that for larger end-masses this tendency to three-dimensionality exists even before the onset of chaos. Hence, they studied 3-D motions of the system from the outset, both experimentally and theoretically.

In the experiments, the apparatus is very similar to that of Figure 5.43. The pipes used are also similar to those in Section 5.8.3(a), but very long and slender ($L \sim 1\text{m}$, $L/D_i \simeq 125$, $\beta = 0.219$, $\gamma = 292$), while the end-masses are much larger ($\mathcal{M} = 83.8$–$816.9\,\text{g}$); for the largest, $\mu = 3.81$.

The experimental critical flow velocities for the onset of flutter have already been discussed [Figure 5.44(b)]. For higher flows, there exist a series of increasingly complicated periodic and quasiperiodic motions, eventually leading to chaos; their sequence and range are shown in Figure 5.49 (top), with the motions sketched below — definitely among the most captivating of experimental results with pipes conveying fluid.

As seen in Figure 5.49, rotational motions do not arise for $\mu = 0$ and 0.367, the smallest experimental value of μ. However, they are increasingly evident for higher μ. For $\mu = 3.55$, the response is predominantly circular. It is seen that, in addition to planar and rotational motions, there are three periodic states of greater complexity: *rotating planar, planar and pendular*, and *nutating oscillations*. As evidence of circular symmetry, clockwise and counterclockwise motions may both occur; likewise, the planar oscillations are not biased towards particular vertical planes.

There are three kinds of rotating planar motion. The rotation is either backwards and forwards through a finite angle [PL(R)], as shown in Figure 5.49(c), or more commonly continuous rotation in either the clockwise (PL,CW) or counterclockwise (PL,CCW) sense. Generally, the period of rotation is ten or more times the period of planar oscillation.

For $\mu = 1.24$, there exists a state of motion that appears to be coupled planar oscillation with the pendular mode (PL,P) [Figure 5.49(d)]. The period of pendular oscillation is approximately four times the period of planar oscillation. Finally, the motion described as nutating [Figure 5.49(e)], for its resemblance to the nutation of a spinning rigid body with axial symmetry, is perhaps the most interesting; it occurs for $\mu = 3.81, 3.55$ and 2.30. The motion can be characterized in terms of how many small nutations (the small loops) are made in a single precession (the motion about the vertical axis) and in terms of the relative amplitude of the nutation, R_1/R_2. The number of nutations per precession is generally an irrational number between 4 and 12. The loops are not stationary, but occur at different points for each cycle of precession, suggesting a nonresonant response.

With decreasing flow, the sequence and type of oscillatory states are generally different; e.g. for $\mu = 0.746$, chaos is succeeded by PL, P(R) and PL oscillations, before the pipe regains static equilibrium.

In at least some of the cases, a clear quasiperiodic route to chaos is followed, as put forward by Ruelle, Takens and Newhouse (Newhouse *et al*. 1978; Bergé *et al*. 1984; Moon 1992), observed for example in Taylor–Couette flow. In this scenario, a secondary Hopf bifurcation transforms periodic motions into quasiperiodic ones, involving two

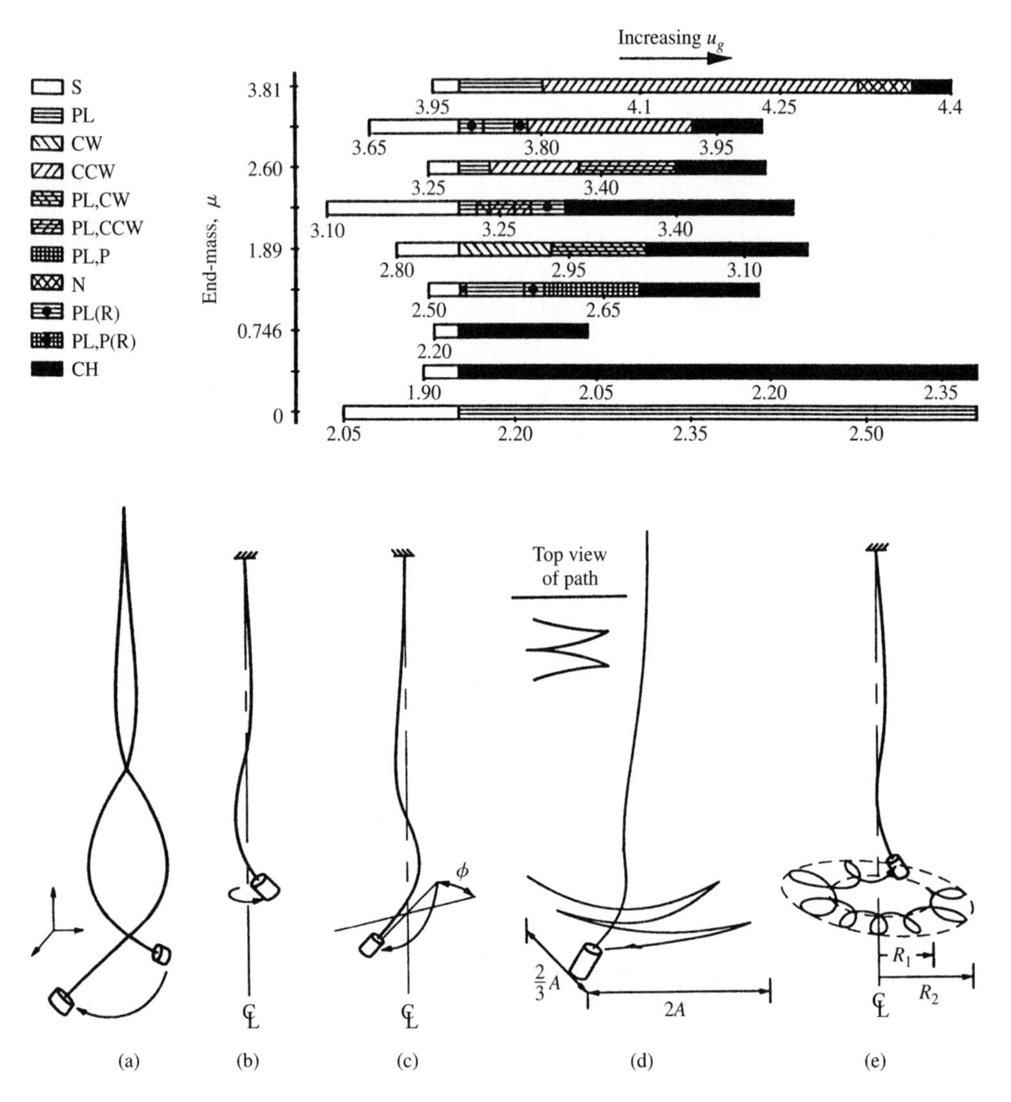

Figure 5.49 Transition from equilibrium to chaos for 3-D motions of the system of Figure 5.43(a), for various end-masses. Top: the ranges of various oscillatory states in terms of increasing u_g. S, stationary pipe; PL, planar oscillation; CW, clockwise rotating motion; CCW, counterclockwise rotation: PL,CW and PL,CCW, clockwise and counterclockwise rotating planar oscillation; PL,P, coupled planar and pendular oscillation; N, nutation; PL(R), planar oscillation rotating through a finite angle; PL, P(R), coupled planar and pendular oscillation rotating through a finite angle; CH, chaos. Bottom: sketches of various periodic motions. (a) PL; (b) CCW; (c) PL(R); (d) PL,P; (e) N (Copeland & Moon 1992).

incommensurate frequencies, so that motions evolve on a 'two-torus, T^2'. Then, a third Hopf bifurcation gives rise to quasiperiodicity involving three frequencies and a 'three-torus, T^3'. This last torus, however, is nonrobust and it can be destroyed by a certain type of perturbation, transforming it into a strange attractor. Thus, the appearance of a third frequency, if it can be captured at all, signals the possible onset of chaos. Therefore, unlike the period-doubling route, this route to chaos involves a finite *and small* number of bifurcations.

For a system with $\mu = 2.60^{\dagger}$ at $u_g = U/(gL)^{1/2} = 3.048$ and in the CCW regime, the motion is periodic and thus involves but one frequency — as found by delay reconstructions of the experimental attractor and associated Poincaré maps and PSD plots, similar to those in Appendix I. For $u_g = 3.080$, however, the motion is in the PL,CW regime, with the pipe oscillating in a plane that is continuously rotating counterclockwise, and hence is associated with two incommensurate frequencies. As seen in Figure 5.50, (a) the pseudo-phase-plane plot, (b) the 'closed curve' in the Poincaré section and (c) the power spectrum with a multitude of combination frequency peaks are all indicative of quasiperiodicity involving two frequencies.

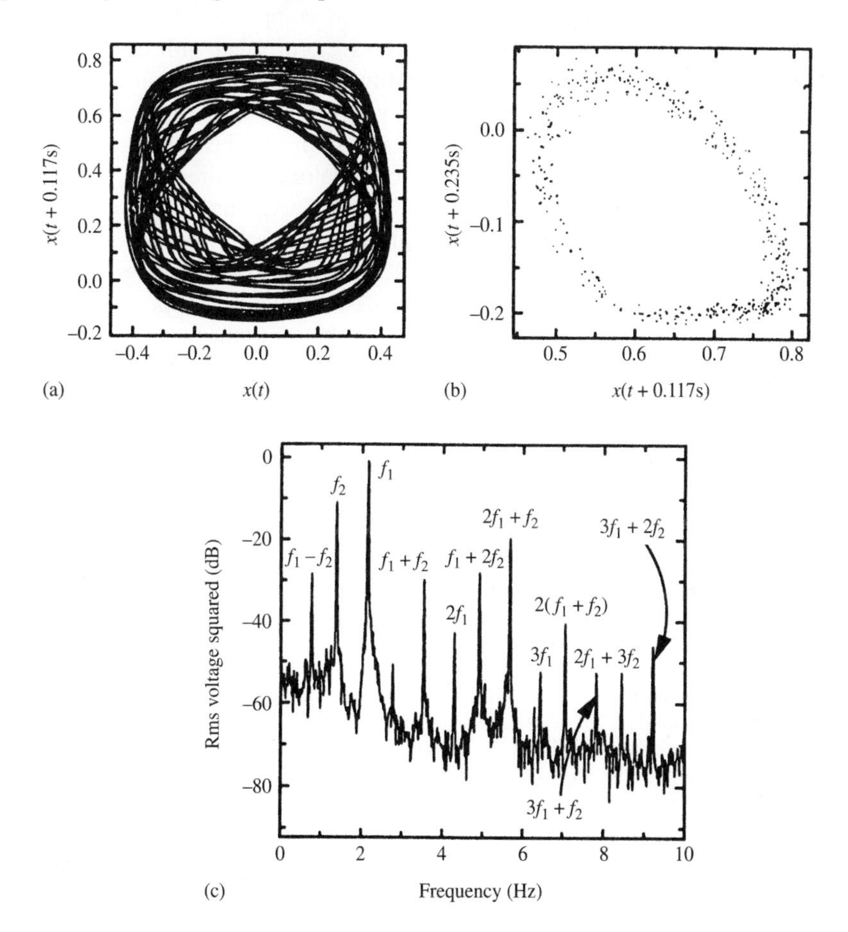

Figure 5.50 Pipe motion in the PL,CCW regime for 3-D motions of the system of Figure 5.43(a), for $\mu = 2.60$, $u_g = 3.080$; scale is arbitrary. (a) Delay reconstruction of attractor, delay = 0.117 s (30 sampling periods), correlation dimension $d_c = 2.105$ for 33 000 data points (b) Poincaré section, at $x(t) = 0.2$; (c) power spectrum (Copeland & Moon 1992).

†End-masses of different aspect ratios and moments of inertia were used, so that μ alone is insufficient to differentiate two experiments. This distinction is not made in this book for simplicity, but it explains why the values of u_g in Figures 5.49 and 5.50 do not correspond.

For $u_g = 3.164$ (Copeland & Moon 1992; Figure 9), the phase-plane plot becomes completely irregular, the Poincaré section shows an unstructured cloud of points, while in the power spectrum the low-frequency background level has risen to almost drown the subharmonic combination peaks, all indicating that the motion is chaotic.

In the analysis (Copeland 1992), motions in both the (x, y) and (x, z) planes are considered — cf. Figure 5.2(b). To simplify things, because the pipes used in the experiments are so slender and flexible, the flexural restoring forces are much smaller than the gravity-induced tensile forces and they are neglected. Thus, the equation of motion is reduced to that of a *pipe-string* with an end-mass. This is the reason for defining $u_g = U/(gL)^{1/2}$ such that it does not involve EI. Furthermore, the effect of the end-mass is not incorporated in the equation of motion but is left in the boundary conditions. Thus, the system is discretized using specially determined comparison functions for a heavy string with an end-mass, involving Bessel functions, to proceed with the analysis.

The linearized system is found to lose stability in its third and fourth modes successively by Hopf bifurcations — of multiplicity two, for each of the two lateral directions. Two reduced forms of the discretized nonlinear system are then analysed: (i) an eight-dimensional invariant manifold, consisting of four centre eigendirections (associated with the two symmetric modes, the third and fourth, first undergoing a Hopf bifurcation) and four stable eigendirections, is obtained and solved numerically; (ii) the further reduced, four-dimensional centre manifold involving but the centre space of the eight-dimensional one, which is analysed further. Then, proceeding essentially as in Appendices F and H via the method of averaging, and assessing stability in the same manner as in Section 5.7.2, the nature of the Hopf bifurcation may be determined (whether sub- or supercritical) and whether the motion is planar or circular. The results are compared in Table 5.7 with those obtained numerically and experimentally. In all cases, the limit cycle is supercritical, in agreement with Bajaj & Sethna's results (Figure 5.20) for $\mu = 0$ and those of Section 5.8.3(a) but in apparent disagreement with the experiments. There is fair agreement between the three sets of results, but some inexplicable differences also. In general, for small μ the motion is planar [which agrees with the results of Section 5.8.3(a)], and interspersed rotational and planar for higher μ. However, the bistable behaviour in some of the numerical results does not exist in the analytical ones.

Table 5.7 Stable limit cycles following the initial instability. C, rotating orbit (CW or CCW); PL, planar orbit; C/PL, bistable orbit, depending on initial conditions; PL→C and C→PL indicate a change in the motion as the flow is increased; PL(R) is defined in Figure 5.49. The numerical results have been computed for $\epsilon = [u_g - (u_g)_H]/(u_g)_H = 0.01$, $(u_g)_H$ being the value for the Hopf bifurcation (Copeland 1992).

μ	Analytical	Numerical	Experimental
0.367	PL	C/PL	PL
0.746	PL	PL(R)	PL
1.24	PL	PL	PL
1.89	PL	PL	C → PL
2.30	C	C/PL	PL → C
2.67	PL	PL	PL → C
3.55	C	C/PL	PL → C

The numerical simulations were pursued to higher u_g by means of the eight-dimensional reduced subsystem. The results for $\mu = 2.30$ indicate that the motion remains basically period-1 for $\epsilon = [u_g - (u_g)_H]/(u_g)_H = 0.11$, which is not in agreement with experiment. Those for $\mu = 3.55$, however, show a clear path to chaos via the quasiperiodic route, which agrees with experiment, as illustrated in Figure 5.51. For $\epsilon = 0.04$ the motion is periodic; for $\epsilon = 0.0485$ it is quasiperiodic with two fundamental frequencies, and for $\epsilon = 0.05$, shown in Figure 5.51(b), with three frequencies; finally, for $\epsilon = 0.065$ the motion is predominantly quasiperiodic, but with a chaotic component. Again, therefore, in view of the results for $\mu = 2.30$, only partial agreement with experiment is obtained.

However, more perplexing is the less than good agreement between the numerical and the analytical results, which must exist at least in some neighbourhood of $\epsilon = 0$: in addition to the bistable behaviour (both rotational and planar oscillations) which occurs only in the former, the variation of frequency with ϵ does not agree. Of course, this has also perplexed Copeland and this author, but no error has been found, though this remains a possibility.

Further experiments on the same system were conducted by Muntean & Moon (1995), in which the system is additionally excited at the support via a shaker, and the 'end-mass' may be a little higher up than the end of the pipe. The objective of this work is to investigate the transition from quasiperiodicity to chaos and this is done by means of multifractal dimensions, or spectra of fractal dimensions, in a similar manner as in Jensen *et al*. (1985) for the forced Rayleigh–Bénard convection experiment. It is shown that the dynamics of the system can be captured by simple maps, and hence the transition to chaos displays remarkable universality irrespective of the physical system.

(c) 2-D motions of a pipe with an end-mass defect

Partly to further explore just how singular the case of $\mathcal{M} = 0$ is, the situation when the additional mass at the free end is *negative* ($\mu < 0$), i.e. when there exists an end-mass defect, was investigated by Semler & Païdoussis (1995) and some interesting results were obtained.

Utilizing equation (5.39) and applying a Galerkin discretization scheme as in equations (5.115) and (5.116a,b), the inertial term is found to have the form $[\delta_{ij} + \mu\phi_i(1)\phi_j(1) + \gamma_{ilkj}q_k q_l]\ddot{q}_j$. Then, making the assumption that μ and the inertial nonlinearities are small, one can write

$$[\delta_{ij} + \mu\phi_i(1)\phi_j(1) + \gamma_{ilkj}q_k q_l]^{-1} \simeq \delta_{ij} - \mu\phi_i(1)\phi_j(1) - \gamma_{ilkj}q_k q_l,$$

and so the nonlinear equation of motion can be recast into first-order form, which may be integrated via a standard Runge–Kutta scheme.

Typical results are shown in Figure 5.52 for a case with $N = 4$ in the discretized system with parameters as for one of the pipes used by Païdoussis & Moon [Section 5.8.1], namely $\beta = 0.216, \gamma = 26.75, \mu = -0.3$,[†] and experimentally determined damping values. The Hopf bifurcation occurs at $u = u_H = 8.7$, so that it is immediately obvious from the figure that subsequent bifurcations begin to occur at much higher values of u. At $u = 19.82$, one of the Floquet multipliers crosses the unit circle at $\lambda = +1$, indicating a

[†]This is a relatively large value of $|\mu|$, but qualitatively similar results are obtained for smaller, more realistic values, e.g. $\mu = -0.085$, although the bifurcations shown in Figure 5.52 then occur at even higher values of u.

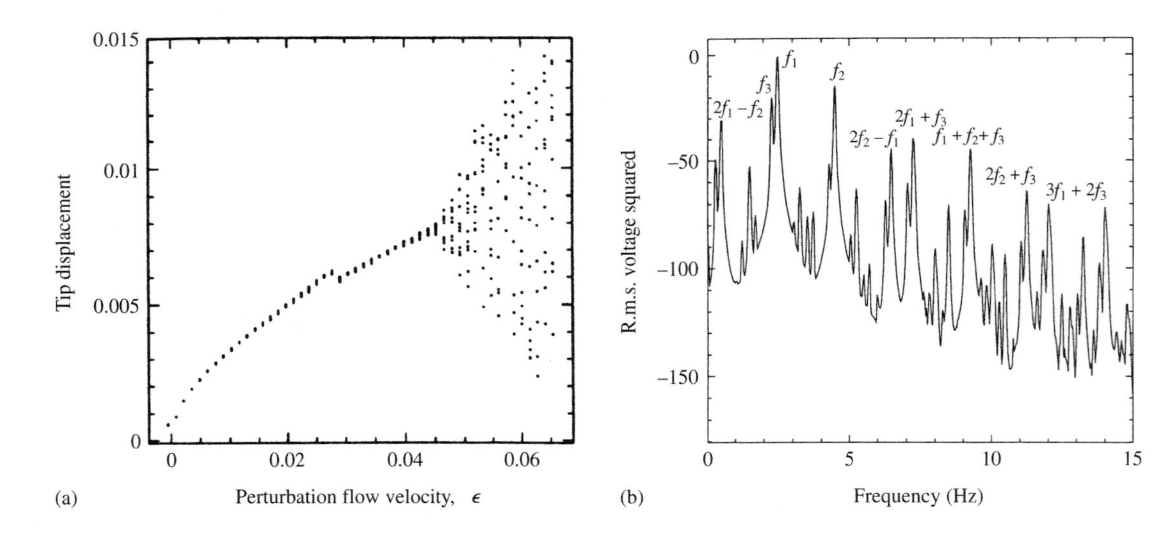

Figure 5.51 Numerical simulations for 3-D motions of the system of Figure 5.43(a): (a) the bifurcation diagram for the system with $\mu = 3.55$ in terms of $\epsilon = [u_g - (u_g)_H]/(u_g)_H$; (b) the power spectrum for $\epsilon = 0.05$, showing quasiperiodic motion with three fundamental frequencies (Copeland 1992).

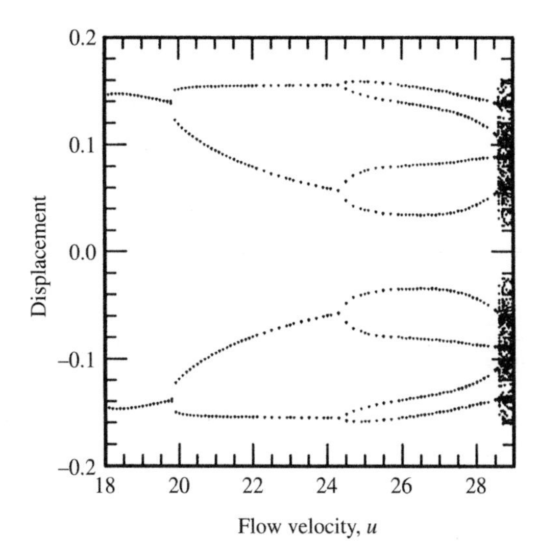

Figure 5.52 Bifurcation diagram of the dimensionless free-end displacement, $\eta(1, \tau)$, when $\dot{\eta}(1, \tau) = 0$, versus u for the system with an end-mass defect, $\mu = -0.3$; the Hopf bifurcation occurs at $u_H = 8.7$ (Semler & Païdoussis 1995).

pitchfork bifurcation, which destroys the symmetry of the limit cycle. At $u = 24.37$, this is followed by another Floquet multiplier crossing the unit circle at $\lambda = -1$, signifying a period-doubling bifurcation. However, this is not followed by another period-doubling. Instead, at $u = 28.56$, a Floquet multiplier crosses the unit circle at $\lambda = +1$ through a saddle-node bifurcation, at which point the oscillation becomes chaotic — as confirmed by a bifurcation diagram of the period of oscillation versus u obtained with AUTO, phase-plane plots, Poincaré maps and Lyapunov exponent calculations. This sequence is characteristic of yet another of the classical routes to chaos, namely that of *intermittency*, in this case of 'type I intermittency' (Bergé *et al.* 1984). A famous system that follows the same route to chaos is the Lorenz model for Rayleigh–Bénard convection.

The best way of understanding the dynamics in this case is by looking at a Lorenz map, otherwise known as a Poincaré *return map*, consisting of successive maxima of the oscillation. Such a map is shown in Figure 5.53(a) for a simple system, to clarify the behaviour, and in Figure 5.53(b) for the problem at hand. In Figure 5.53(a), we have the solution curve for a fictitious problem, nearly tangent to the 45° 'identity line' [whereon $q(n + 1) = q(n)$], which returns the solution on to the next iteration of the map. If the solution curve intersects the identity line at two points, there exist two fixed points on the map, i.e. two limit cycles of the oscillatory system: one stable (the lower one) and the other unstable. The route to chaos involves the gradual lifting of the solution curve away from the identity line; when no intersection exists, then there is no stable oscillatory state. In the figure, it is clear from the iterations that, while 'in the channel' between the solution curve and the identity line, the system performs almost 'regular' oscillations, the amplitude of which varies gradually; this is the so-called *laminar phase* of the oscillation. Once the system 'bursts out' of the channel, it performs an excursion of high irregularity which is called the *turbulent phase*, before it bounces off another part of the solution and is reinjected in the channel, a process known as *relaminarization*.

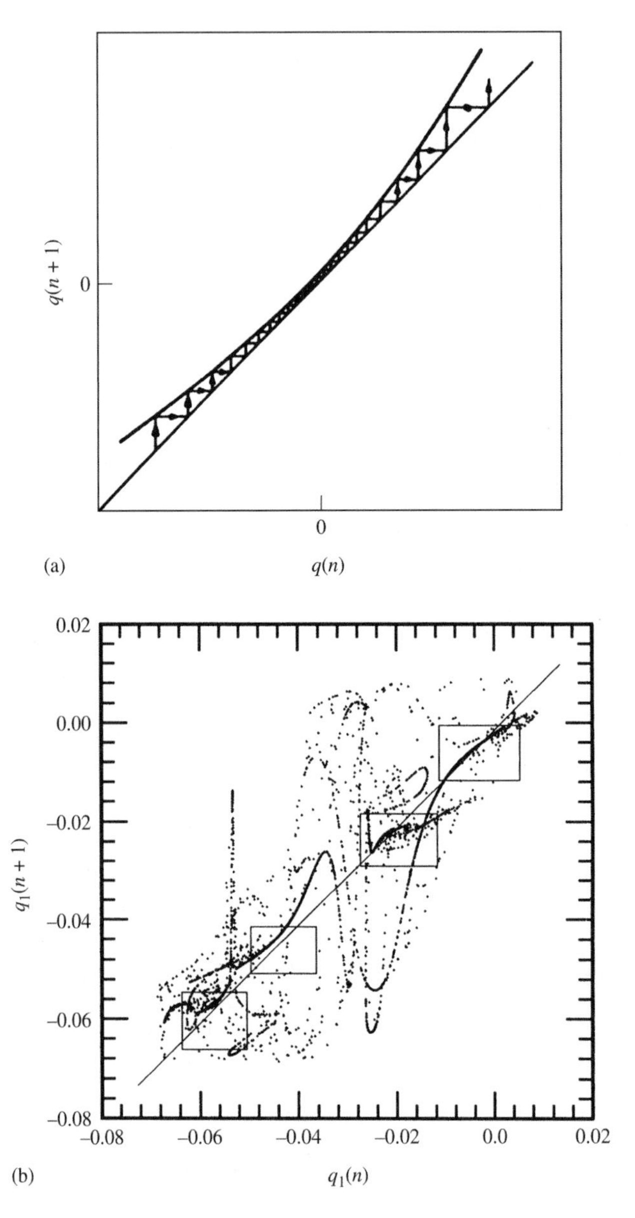

Figure 5.53 (a) One-dimensional Lorenz (return) map of successive maxima of the solution of a system, showing some iterations in 'the channel' wherein the motion is 'laminar'. (b) Lorenz map for the system with an end-mass defect, $\mu = -0.3$, $u = 28.6$, $20 \leq \tau < 1000$ (about 8200 cycles of oscillation); (Semler & Païdoussis 1995).

In the map of Figure 5.53(b), we see four channels. The resulting behaviour is nearly period-2. The system visits two "steady states", but the dynamics is interspersed with bursts of aperiodic motion. According to Manneville & Pomeau (1980), the time between turbulent bursts should scale as $T = [u - u_{\text{int}}]^{-1/2}$ for type I intermittency, where u_{int} is the threshold of intermittency; similarly, the largest Lyapunov exponent, σ, should scale as $\sigma = [u - u_{\text{int}}]^{1/2}$. Plots of T and σ versus u for this problem show excellent agreement with this scaling (Semler & Païdoussis 1995).

One worry concerning this work with $\mu < 0$ is that the robustness of the results for $N > 4$ has not been checked. In view of the high values of u necessary for the post-Hopf bifurcations, especially for very small $|\mu|$, $N = 4$ may well be insufficient.

To conclude, it is clear that the system with $\mu = 0$ is truly singular and relatively dull, capable of limit-cycle oscillations 'only'! In contrast, both for $\mu < 0$ and $\mu > 0$, a succession of interesting dynamical states and chaos generally follow the emergence of limit-cycle oscillations.

5.8.4 Chaos near double degeneracies

As discussed in Section 5.7.3, a number of systems have been studied in the neighbourhood of double degeneracies, in the process determining conditions, e.g. heteroclinic orbits, which when perturbed could lead to chaos; indeed, in several cases, finding chaos was the principal aim.

The first case to be discussed here is that of the so-called up-standing cantilever, in which the double degeneracy involves coincident Hopf and pitchfork bifurcations in the $\{u, \beta, \gamma\}$-space, see Section 5.7.3(d). Keeping β fixed at 0.2, this double degeneracy occurs at $u_c = 2.2458$ and $\gamma_c = -46.0014$ for $N = 2$; the work that follows is for this particular set of parameters. Furthermore, it is recalled that heteroclinic orbits for this system arise on a line in Figure 5.29 defined by equation (5.130), in which the constants are $b = 1.518$ and $c = 3.954$ for $\beta = 0.2$. Thus, the system is studied at

$$u = u_c + \mu \qquad \text{and} \qquad \gamma = \gamma_c + \chi, \tag{5.142}$$

where μ and χ are determined via (5.126) for the system to both be doubly degenerate and to have heteroclinic orbits: $u = 2.2466$, $\gamma = -46.0200$.

The system is perturbed by varying the nonlinear coefficient α_{ijkl} in equation (5.116a), $\alpha_{ijkl} = (u_c + \mu)^2 a_{ijkl} + (\gamma + \chi) b_{ijkl} + c_{ijkl}$, and then varying μ or χ. It is stressed that to keep the characteristics of heteroclinic cycles in the unperturbed system, u and γ in the linear part of the system are kept constant at the values given in the last paragraph.

Simulations have been conducted by using the full nonlinear equations of motion. Variations in u do not lead to chaos, contrary to expectations, but variations in γ do. Results for $\chi \in (13, 14)$ are summarized in the bifurcation diagram of Figure 5.54(a). Note that, although γ is significantly far from γ_c for $\chi = 14$, still $\gamma/\gamma_c \simeq 0.3$ only. It is clear from Figure 5.54 and other calculations given in Li & Païdoussis (1994) that a period-doubling bifurcation occurs for $\chi \simeq 13.4$, and then chaos develops for $\chi \simeq 13.55$. The phase-plane diagram of Figure 5.54(c) is reminiscent of some depicting chaotic oscillation of a two-well-potential oscillator (Moon 1992), which is associated with a homoclinic orbit (two loops connected by a saddle), whereas the analytical subsystem in this case exhibits a heteroclinic orbit. Nevertheless, physically, the existence of homoclinic orbits of the doubly degenerate up-standing cantilever does make sense. Thus, decreasing γ means that two attractors (buckled states) on either side of the straight equilibrium are created, and the oscillator jumps back and forth between the two attracting domains in a stochastic manner.

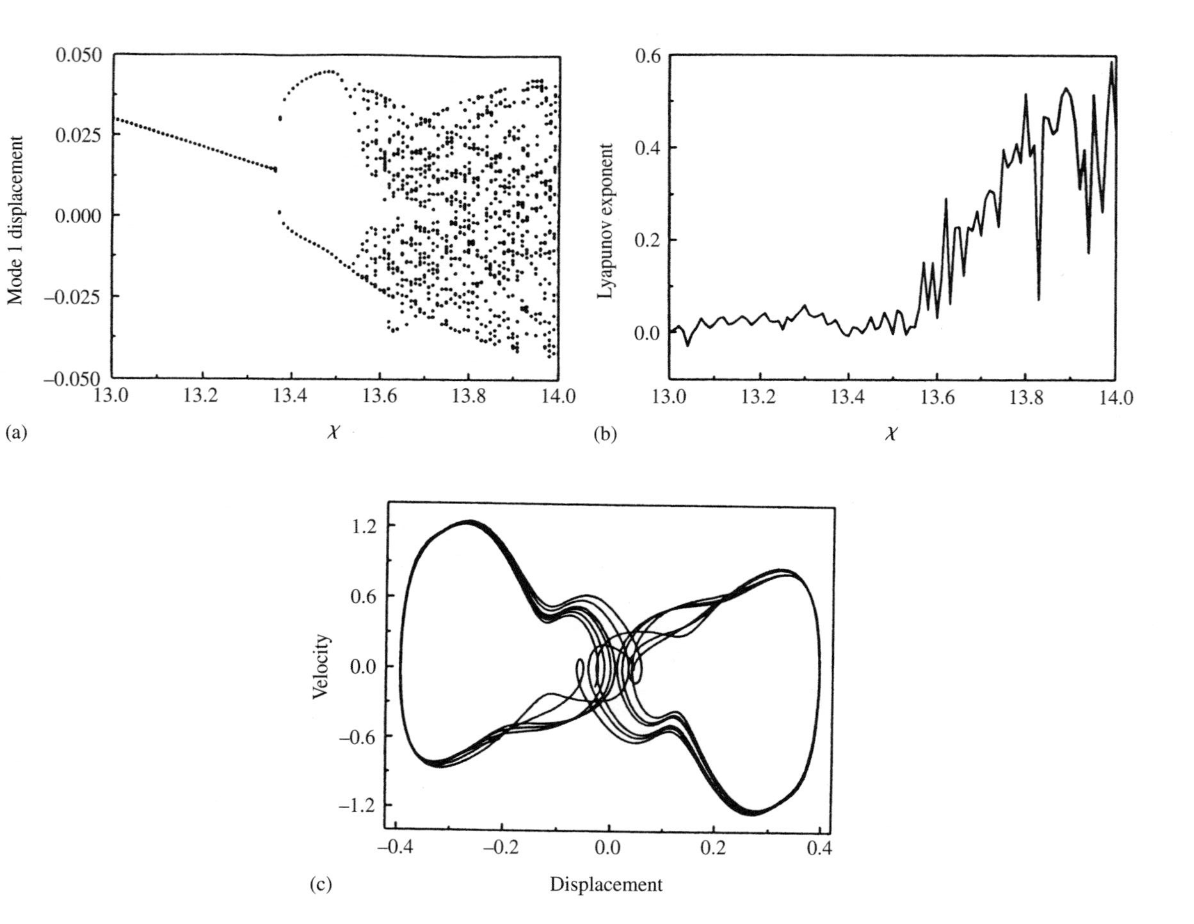

Figure 5.54 (a) Bifurcation diagram for the doubly degenerate up-standing cantilevered system with heteroclinic orbits ($\beta = 0.2$, $u = 2.2466$, $\gamma = -46.020$), perturbed in γ by an amount χ; (b) the corresponding largest Lyapunov exponents; (c) a phase-plane plot in the chaotic regime (Li & Païdoussis 1994).

The system can also be perturbed by adding a small oscillatory component to the flow, so that $u = u_o + \nu \sin \omega t$ — cf. Sections 4.5 and 5.9. Prior to doing this, Li & Païdoussis (1994) conducted a Melnikov analysis on a reduced subsystem, the details of which are not given here; for nonautonomous systems, this can give an indication of the parameters near which chaos can arise. The central idea is to construct a scalar function which gives a measure of the distance between the stable and unstable manifolds when the unperturbed heteroclinic (or homoclinic) orbit is broken by perturbations. If and when this distance vanishes as a system parameter varies, the two manifolds intersect transversally; one such intersection implies infinitely many, yielding horsehoes and chaos (Guckenheimer & Holmes 1983). The analysis in this case gives $\nu > \sigma R(\omega)$ for this to occur, where $\sigma = 0.1$ is the viscous damping coefficient [equation (3.71)], viscous damping being the only dissipation included, and $R(\omega)$ is a function of the forcing frequency of the system, which has a minimum at $\omega = \omega_0 = 2.45$, ω_0 being the frequency of oscillation at the Hopf point. Simulations for a forcing frequency of $\omega = 2.50$ while varying ν show that chaotic oscillations arise for $\nu > 1.22$; this value of ν is very much higher than the minimum required according to the Melnikov calculation. The sequence of oscillatory states for increasing ν is (a) quasiperiodic, (b) periodic and (c) chaotic motions. The results are similar to those discussed next, and hence no further details are given here.

The next case to be considered is that of the pipe–spring system, discussed in relatively great detail in Section 5.7.3(a). In this case, heteroclinic orbits arise along one of the lines shown in Figure 5.27(a). Here, simulations are conducted exclusively with a periodic flow-velocity perturbation of the system, with parameters as given in Figure 5.55, where a bifurcation diagram and a few phase-plane plots are shown; a fuller set, consisting of several power spectra, time traces, phase-plane plots and Lyapunov exponent calculations are given in Païdoussis & Semler (1993b). The sequence of oscillatory states for increasing ν is (i) periodic oscillations around one or the other of the two buckled states (both are shown, obtained via different initial conditions), (ii) quasiperiodic oscillations around both buckled states, (iii) periodic motions with sub-, combination, and super-harmonic content ($3 < \nu < 8$ approximately), and (iv) chaotic oscillations. It is of interest that quasiperiodic and chaotic oscillations in Figure 5.55(a) look not too dissimilar, but the difference in the Lyapunov exponents is quite clear: zero in the former case and positive in the latter.

Three-dimensional motions of the same system are considered by Steindl & Troger (1996), who determine in a map of β versus the location of the spring support, ξ_s, the regions of existence and stability of heteroclinic cycles. Physically, the heteroclinic cycle involves the following set of transitions, as shown in Figure 5.56(a): (i) the system is buckled in one of the two mutually perpendicular planes; (ii) oscillations develop in that plane about the buckled state; (iii) the amplitude of these oscillations increases, while the static deformation due to buckling diminishes, eventually leading to oscillations about the straight equilibrium state (about the origin); (iv) oscillations develop in the perpendicular plane, with decreasing amplitude as the amplitude of buckling increases in that plane; (v) eventually buckling in that plane results, with no oscillation. By symmetry, this state is fundamentally identical to the initial one, and so the sequence just described begins anew.

Steindl (1996) considers another type of heteroclinic cycles for the same system, this time associated with Hopf–Hopf bifurcations, rather than Hopf–pitchfork ones as in the foregoing, again obtaining a map of stable heteroclinic cycles in the $\{\beta, \xi_s\}$ plane, as shown in Figure 5.56(b). In this case, the oscillations in one plane develop a secondary bifurcation at the TB boundary (corresponding to a simultaneous occurrence of a Hopf bifurcation and

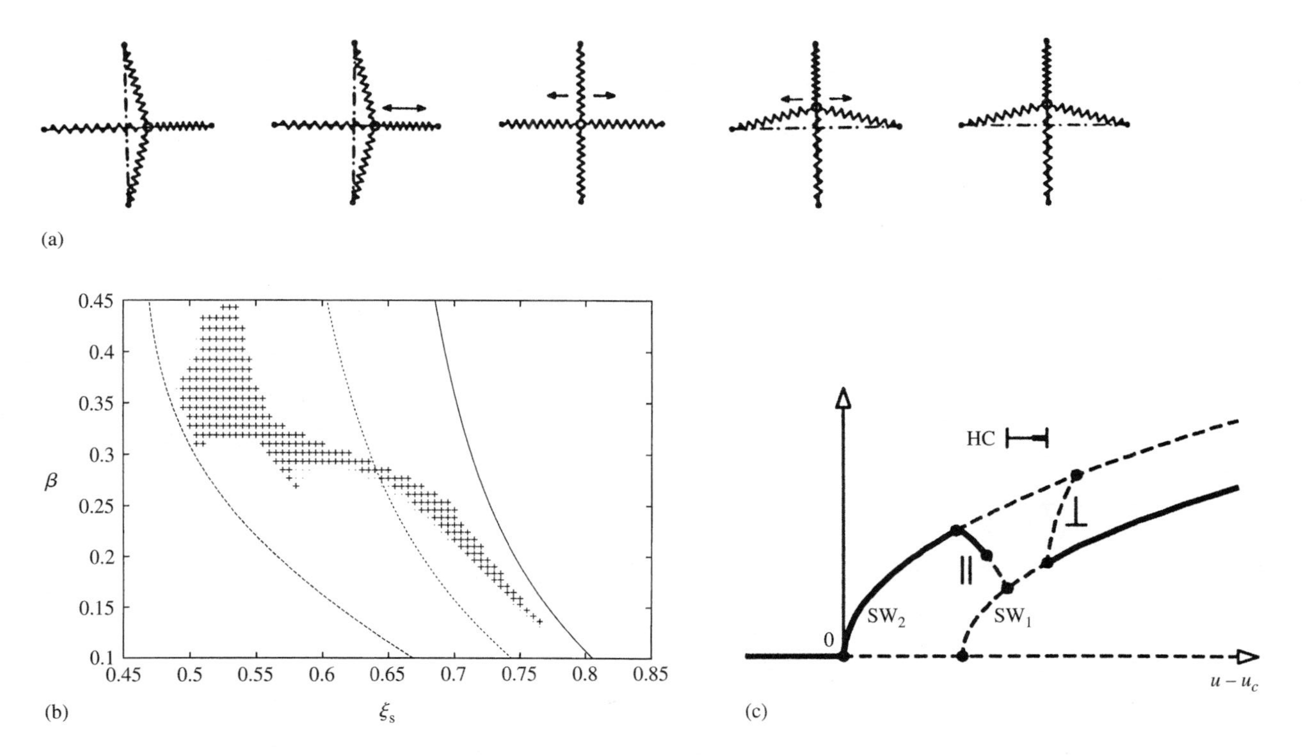

Figure 5.56 (a) The sequence of states associated with Hopf–pitchfork heteroclinic cycles in 3-D motions of the pipe–spring system [from Steindl & Troger (1996)]. (b) Map of stable Hopf–Hopf heteroclinic cycles (+) of the same system; – – –, marks the boundary for the onset of the secondary bifurcation. (c) Qualitative bifurcation diagram, showing the two principal limit-cycle branches for oscillations in one plane (SW_1) or the other perpendicular to it (SW_2): ——, stable solutions; – – –, unstable solutions; HC marks the range of existence of heteroclinic cycles (Steindl 1996).

a 'Takens–Bogdanov' point, where one of the frequencies in the Hopf–Hopf interaction becomes zero); the other lines shown in the figure correspond to features of the dynamics not discussed here. The solution then generally involves two frequencies and the secondary oscillations are in-plane or out-of-plane. As shown in Figure 5.56(c), heteroclinic cycles can occur in the narrow range bounded by the solutions for oscillation in one or the other plane (SW_1 and SW_2) and the secondary branches marked by and $\perp$ in the figure. The same kind of heteroclinic cycles may exist if flutter is rotary rather than planar. These results offer a reasonable qualitative explanation of Copeland & Moon's (1992) experimental observations (Figure 5.49).

5.8.5 Chaos in the articulated system

An extremely complicated bifurcation picture is involved in the nonlinear dynamics of a horizontal two-segment articulated pipe with an asymmetric spring, so that the two pipe segments at equilibrium are at an angle ψ to each other, as shown by Champneys (1991, 1993) and as discussed briefly in Section 5.6.2(a). Because the dynamical theory required to understand the dynamics is beyond the scope of this book, only some selected results are presented and the reader is encouraged to refer to the primary sources.

A typical sequence of dynamical behaviour with increasing u is given in Table 5.8. The 'primary orbit' in the table is a limit cycle due to a supercritical Hopf bifurcation. It is followed by a period-doubling bifurcation (at point 4) and gains amplitude by *a tower*, as shown in Figure 5.57(a); a tower consists of a number of saddle-node bifurcations (the first of which occurs at point 5). (Similar towers show how the period can increase with u.) Eventually, that branch of the curve regains stability at $u = 5.5667$ via a reverse period-doubling bifurcation (at point 6) and remains stable thereafter. It is noted in Figure 5.57(a) that neither the primary nor the period-doubling branches are stable for $u > 5.0865$, and it is for that range of u that interesting dynamics occurs. In fact, the bifurcations at points 4 and 8 are the beginnings of a period-doubling cascade leading for $u \in (5.0865–5.1321)$ to 'small-scale' chaos with periodic windows. This is succeeded for $u \in (5.1321–5.5667)$ by 'large-scale' chaos involving mixed-mode period-(m, n) orbits; these have m large-scale oscillations and n small-scale ones per period, as exemplified in Figure 5.57(b) — see Glendinning (1994). Eventually, the map shown in Figure 5.57(c) is obtained, with the regions I–VI as defined in Table 5.8.

The vertical articulated system when $\psi = 0$ has also been studied by Champneys (1993) close to the pitchfork–Hopf double-degeneracy point. Once more, a very complex and interesting bifurcation structure is revealed. An example of a period-(1,8) figure-of-eight

Table 5.8 The dynamics of the articulated system with $\psi = 0.6$ (Champneys 1991).

Region	u	Behaviour
I	0–4.7072	Stable stationary point
II	4.7072–5.0411	Small-scale primary orbit
III	5.0411–5.0865	Period-doubling cascade
IV	5.0865–5.1321	Small-scale chaos with periodic windows
V	5.1321–5.5667	Large-scale chaos and period-$(1, n)$ orbits
VI	5.5667–∞	Large-scale (1,0)-orbit

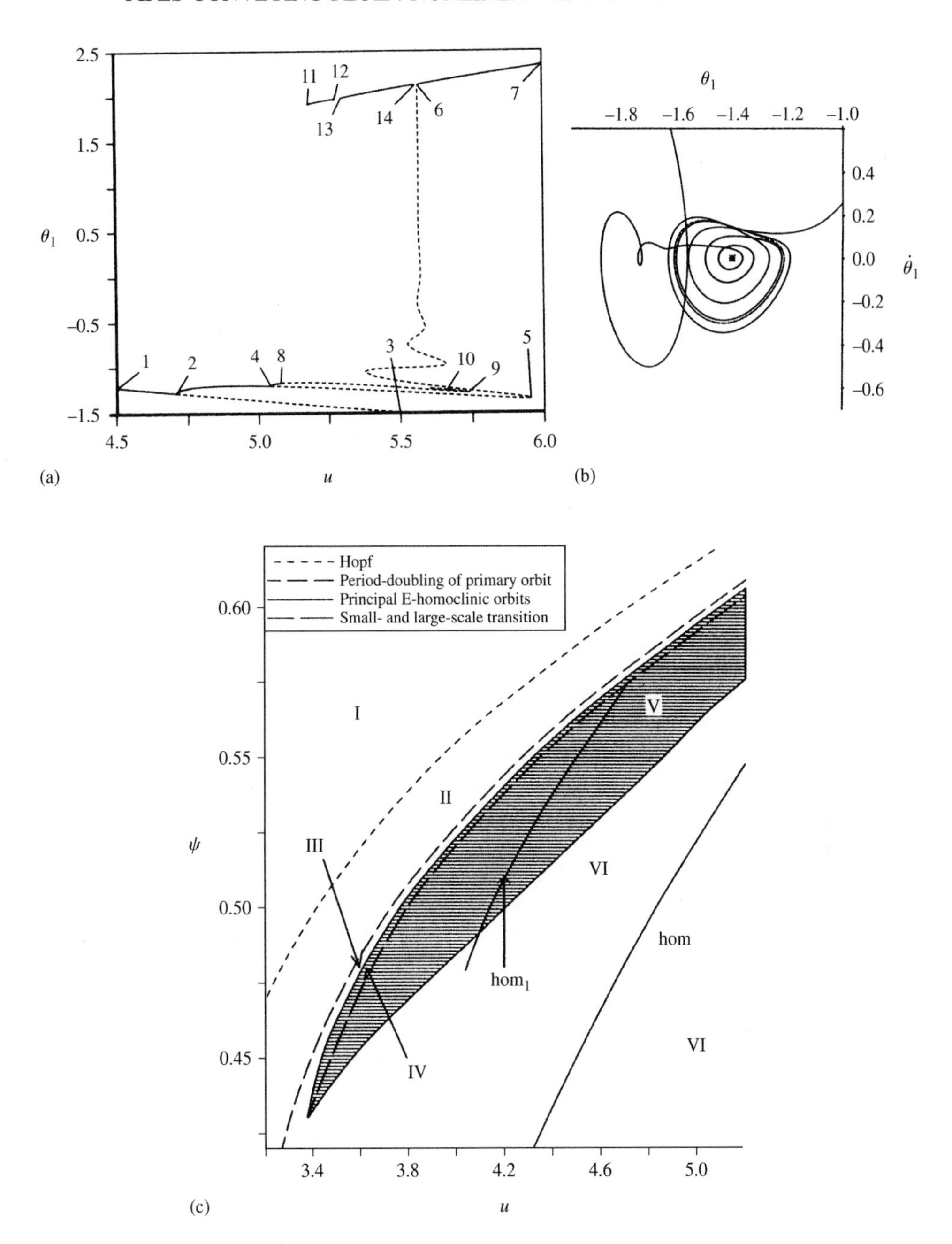

Figure 5.57 (a) A bifurcation diagram for the first-segment angle θ_1 versus u for an initially 'deformed' articulated cantilever with $\psi = 0.6$ (defined in Section 5.8.5): ——, stable orbits; – – –, unstable orbits. (b) Portion of a period-(1, 5) orbit at $u = 5.15$; for the large orbit, $\theta_1, \dot{\theta}_1$ lie within $(-3, 2)$. (c) Map showing the areas of dynamical behaviour defined in Table 5.8; hom_1 stands for the principal E-homoclinic orbits (Champneys 1991, 1993).

homoclinic orbit, along with the corresponding time trace is shown in Figure 5.58. For low enough γ (e.g. for $\gamma = 1.0$), chaos is observed between the stable mixed-mode orbits.

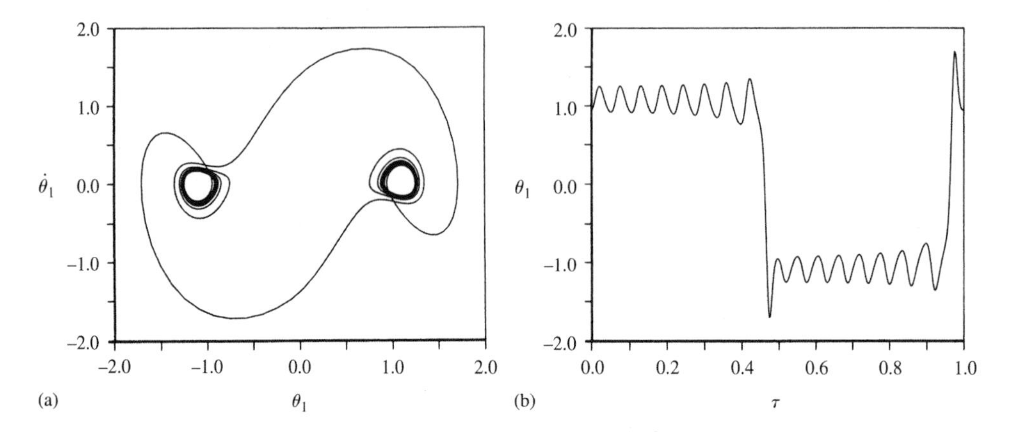

Figure 5.58 (a) A phase portrait of a period-(1, 8) orbit and (b) the corresponding time trace for a vertical articulated system near the point of double degeneracy (Champneys 1993).

5.9 NONLINEAR PARAMETRIC RESONANCES

As shown in Section 4.5 both by linear theory and by experiments conducted in the 1970s, the dynamics of pipes conveying harmonically perturbed flow is quite interesting, especially in the case of cantilevered pipes. It was reasonable to expect, therefore, that nonlinear study of the same system would soon follow — especially in view of the work in Sections 5.5–5.8 conducted in the late 1970s and 1980s, showing that nonlinear dynamics analysis can (i) provide results closer to reality, (ii) elucidate the dynamical behaviour *beyond* the onset of instability, (iii) give new insight into the dynamics even *before* the instability threshold, (iv) reveal more interesting fine structure in the dynamics, and (v) yield entirely new results (e.g. the amplitude of the motion). The nonlinear studies of parametric resonances of pipes conveying fluid, which began appearing in the second half of the 1980s, demonstrate to the full one of the tenets justifying the space allocated to dynamics of pipes conveying fluid in this two-volume book: that this system serves as a crucible for the development, illustration and testing of new dynamical theory. Thus, more or less at the same time and by the same authors, in parallel to the work on the pipe problem to be discussed in what follows, a number of papers have appeared on the general theory of parametrically perturbed generic nonlinear systems subject to Hopf bifurcations, e.g. by Bajaj (1986, 1987a) and Namachchivaya & Ariaratnam (1987).

Most of the work done in this area is analytical, and most of that makes use of the modern methods of nonlinear dynamics theory, but some numerical calculations have also been done, as well as some new experiments.

5.9.1 Pipes with supported ends

Parametric resonances in this inherently conservative system have been studied by Yoshizawa *et al*. (1986), Namachchivaya (1989) and Namachchivaya & Tien (1989a,b), Chang & Chen (1994), and Jayaraman & Narayanan (1996).

In all cases but the first, the pipe is considered to have absolutely fixed ends, i.e. no axial sliding is permitted, and hence the nonlinear equation of motion used is similar to Holmes' and relatively simple, as discussed in what follows. In the case of Yoshizawa *et al*. (1986), however, the downstream end of the clamped–pinned pipe considered is free to slide axially. The equations of motion, similar to Rousselet & Herrmann's (1981), are much more complex: one 'flow equation', similar to those discussed in Section 5.2.8(b,c), in which the pressure itself is pulsatile, $p = p_0(1 + \mu \sin \omega\tau)$,[†] and an equation for the pipe coupled to the first, in which nonlinearities are associated with curvature rather than induced-tension effects, similar to equation (5.43) for cantilevered pipes. This work, being the first to be published and the simplest in terms of methods used, is discussed first. The eigenfunctions $\phi_j(\xi)$ of the subsystem $\ddot{\eta} + \eta'''' - \gamma[(1-\xi)\eta'' - \eta'] + u_0^2\eta'' = 0$ are obtained first, and then the system is discretized via a one-mode Galerkin scheme, so that $\eta(\xi, \tau) = \phi_1(\xi)q(\tau)$, leading to two fairly simple nonlinear coupled ODEs in $u(\tau)$ and $q(\tau)$, involving u_0, μ, p_0, β and γ. Solution of these equations is obtained by the method of multiple scales (Nayfeh & Mook 1979), and the deflection of the pipe is finally expressed as $\eta(\xi, \tau) = \mu^{1/2}\hat{h} \cos[\frac{1}{2}(\omega\tau + \psi)]\phi_1(s) + \mathcal{O}(\mu^{3/2})$, in which it has been assumed that ω is close to $2\omega_1$, ω_1 being the first-mode eigenfrequency associated with $\phi_1(\xi)$. Hence, the first-mode principal parametric resonance is considered (Section 4.5), involving the 'detuning parameter' $\hat{\sigma}$, such that $\omega/\omega_1 = 2(1 + \mu\hat{\sigma})$.[‡]

A number of interesting findings are reported, as follows. (i) Considering no pulsation, the mean-flow nonlinear first-mode eigenfrequency plotted in a u_0^2 versus γ plot shows both softening and hardening spring characteristics, the former for low u_0^2 when inertial nonlinearities are predominant, the latter for larger u_0^2 when nonlinear centrifugal effects are dominant. (ii) The steady-state amplitude, h_s, associated with $\mu^{1/2}\hat{h}$ in the foregoing, is determined and its stability examined, eventually producing the classical plot of h_s versus $\tilde{\sigma} = \mu\hat{\sigma}$ shown in Figure 5.59(a). It is clear that as μ becomes larger, since $\mu < 1$, both h_s and $\tilde{\sigma}$ increase — i.e. the frequency range and amplitude increase with μ. (iii) The extent of the parametric resonance region is larger in terms of $\tilde{\sigma}$ than the linear range, because of the subcritical onset of the oscillation with decreasing $\tilde{\sigma}$, leading to hysteresis (hardening behaviour) as seen in Figure 5.59(a). (iv) The maximum amplitude increases with u_0.

Experiments have also been done by Yoshizawa *et al*. (1985, 1986) using silicone rubber pipes ($D_o = 5$ mm, $L = 600$ mm) stiffened in one plane by wires, to confine the oscillation in the other plane — see also Section 5.5.3. The pulsation was introduced by periodic opening and shutting of a pressure-control valve at the exit of a by-pass line connected to the constant-head tank feeding water into the pipe. The experimental results are in good qualitative agreement with theory. Figure 5.59(b–d) shows the pipe in parametric resonance as $\tilde{\sigma}$ is increased, i.e. as the pulsation frequency is increased, showing that the amplitude increases. For $\tilde{\sigma} = 0.183$ (not shown), the amplitude begins to decrease and soon thereafter the oscillation ceases. The oscillation if excited at or

[†]Throughout Section 5.9, μ denotes the amplitude of harmonic perturbations, usually of the flow velocity, as in Section 4.5 — see equation (5.144). It should not be confused with the dimensionless end-mass parameter in equation (5.139) used in Section 5.8.3.

[‡]It is recalled that, in order to achieve a modicum of uniformity in the book, the notation is sometimes quite different from that in the original papers — though this may be bewildering to their authors! Especially in the case of the detuning parameter, since different definitions are given in virtually every study, the following convention has been used: (i) if the detuning parameter is multiplied by a small parameter, it is denoted by $\hat{\sigma}$, as in Yoshizawa *et al*. (1986) where $\omega/\omega_1 = 2(1 + \mu\hat{\sigma})$ and Bajaj (1987b) where $\omega = \omega_0 - \epsilon\hat{\sigma}$; (ii) otherwise, it is denoted by $\tilde{\sigma}$, as in equation (5.148) where $\omega/\omega_0 = 1 - \tilde{\sigma}$, and in Bajaj (1984) where $\omega = \omega_0 - \tilde{\sigma}$.

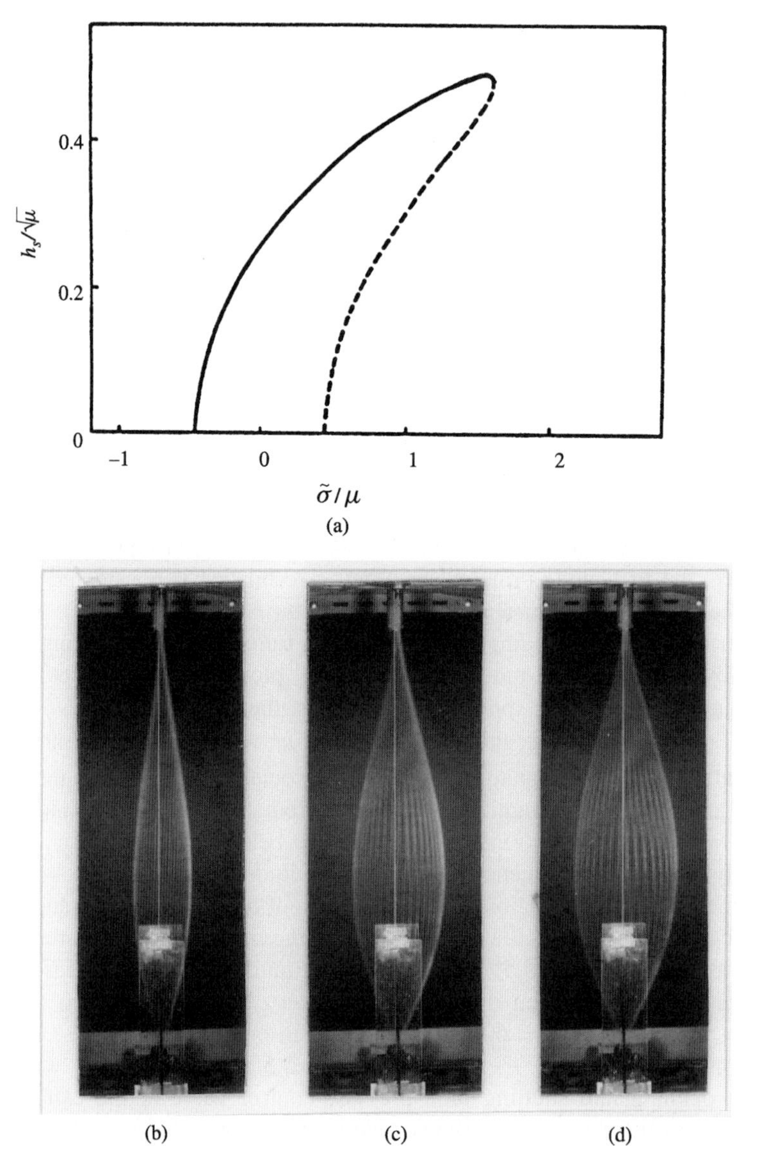

Figure 5.59 (a) The frequency-response curve for the principal parametric resonance of a clamped–pinned system, in terms of the steady-state amplitude h_s and the detuning parameter $\tilde{\sigma}$, for $u_0 = 4.54$, $\beta = 0.273$, $\gamma = 34.4$, friction parameter $\alpha = 4.68$: ——, stable; - - -, unstable. Experimental results for the same parameters and (b) $\tilde{\sigma} = -0.09$, $\eta_{\max} = 0.06$, (c) $\tilde{\sigma} = 0$, $\eta_{\max} = 0.10$, (d) $\tilde{\sigma} = 0.179$, $\eta_{\max} = 0.11$ (Yoshizawa *et al*. 1986).

below $\tilde{\sigma} = 0.17$ persists to $\tilde{\sigma} = 0.18$, but with decreasing ω cannot be excited except at $\tilde{\sigma} = 0.17$ or less — thus displaying the hardening behaviour found in theory.

A more complete and sophisticated analysis of the problem was conducted by Namachchivaya (1989) and Namachchivaya & Tien (1989a,b), utilizing the same basic tools as in Ariaratnam & Namachchivaya (1986a). Chang & Chen (1994) presented an analysis of more limited scope, essentially redoing some of the same work as by Namachchivaya and co-workers, but via more standard and easily accessible terms, in fact following Namachchivaya & Ariaratnam (1987). What is presented below is a melange of all of this, but differentiating according to authors as appropriate; the results are discussed separately.

In all these studies the nonlinear equation used is similar to Holmes' (Sections 5.2.9(b) and 5.5.2), the only nonlinearity taken into account being due to the deformation-induced tension between laterally *and* axially fixed supports; the rest of the equation, including the terms associated with the flow pulsation (i.e. the $\dot{u}$ terms), is linear and as in Section 4.5 and Païdoussis & Issid (1974). Thus, the nonlinear dimensionless equation of motion is given by variants of equation (5.80):

$$\alpha\dot{\eta}'''' + \eta'''' + \left\{u^2 - \Gamma - \gamma(1-\xi) - \tfrac{1}{2}\mathscr{A}\int_0^1 (\eta')^2\,\mathrm{d}\xi\right\}\eta'' \\ - \alpha\mathscr{A}\left\{\int_0^1 (\eta'\dot{\eta}')\,\mathrm{d}\xi\right\}\eta'' + 2\beta^{1/2}u\dot{\eta}' + \gamma\eta' + \sigma\dot{\eta} + \ddot{\eta} = 0, \tag{5.143}$$

in which the parameters are defined in (3.71) and (5.81), subject to

$$u = u_0(1 + \mu\cos\omega\tau). \tag{5.144}$$

Assuming $\gamma = \alpha\mathscr{A} = \sigma = 0$ and taking u as in (5.144), Namachchivaya (1989) discretizes equation (5.143) into one of two degrees of freedom,

$$\ddot{\mathbf{q}} + 2\beta^{1/2}u_0\mathbf{B}\dot{\mathbf{q}} + [\mathbf{\Lambda} + (u_0^2 - \Gamma)\mathbf{C}]\mathbf{q} + \left[\frac{\partial\Phi}{\partial\mathbf{q}}\right] \\ = [\mu\beta^{1/2}u_0\omega(\mathbf{C} - \mathbf{D})\sin\omega\tau - \mu(2u_0^2\mathbf{C} + 2\beta^{1/2}u_0\mathbf{B})\cos\omega\tau]\mathbf{q} - \alpha\mathbf{\Lambda}\dot{\mathbf{q}}, \tag{5.145}$$

in which $\Phi = \frac{1}{2}\mathscr{A}c_{ij}c_{kl}q_iq_jq_kq_l$, the c_{kl} being terms of the type making up $\mathbf{C}$, $\mathbf{\Lambda}$ is a diagonal matrix with elements λ_j^4, where λ_j are the dimensionless beam eigenvalues — cf. equation (4.70); μ and α are assumed to be small ($\ll 1$).† To accentuate its structure, this equation may be written in simplified notation as

$$\ddot{\mathbf{q}} + \mathbf{G}\dot{\mathbf{q}} + \mathbf{K}\mathbf{q} = \mu(\mathbf{E}_1\cos\omega\tau + \mathbf{E}_2\sin\omega\tau)\mathbf{q} - \alpha\mathbf{\Lambda}\dot{\mathbf{q}} + \mathbf{f}(\mathbf{q}). \tag{5.146}$$

This equation is transformed into standard form by an elegant Hamiltonian symplectic transformation (Namachchivaya 1989), a more standard technique via the solutions of the unperturbed system (Namachchivaya & Tien 1986b), and a standard method by Chang & Chen (1994), all leading to variants of the following equation:

$$\dot{\mathbf{u}} = (\mathbf{B}_0 + \alpha\mathbf{B}_1)\mathbf{u} + \mu(\overline{\mathbf{A}}_1\cos\omega\tau + \overline{\mathbf{A}}_2\sin\omega\tau)\mathbf{u} + \bar{\mathbf{f}}(\mathbf{u}), \tag{5.147}$$

†The notation here differs from that in many of this group of papers where, to put in evidence the smallness of μ and α, they are scaled by ϵ, $\epsilon \ll 1$; thus, $\mu = \epsilon\mu^*$ and $\alpha = \epsilon\alpha^*$.

in which

$$\mathbf{B}_0 = \begin{bmatrix} 0 & \omega_1 & 0 & 0 \\ -\omega_1 & 0 & 0 & 0 \\ 0 & 0 & 0 & \omega_2 \\ 0 & 0 & -\omega_2 & 0 \end{bmatrix}, \qquad \mathbf{B}_1 = \begin{bmatrix} \delta_1' & \omega_1' & 0 & 0 \\ -\omega_1' & \delta_1' & 0 & 0 \\ 0 & 0 & \delta_2' & \omega_2' \\ 0 & 0 & -\omega_2' & \delta_2' \end{bmatrix}.$$

The quantities $\omega_i, \omega_i', \delta_i', i = 1, 2$, are related to the eigenvalues of the unperturbed system, $\rho_{1,2} = \alpha\delta_1' \pm \mathrm{i}(\omega_1 + \alpha\omega_1')$ and $\rho_{3,4} = \alpha\delta_2' \pm \mathrm{i}(\omega_2 + \alpha\omega_2')$.

The next step is to introduce the new time $t = \omega\tau$ and the detuning parameter $\tilde{\sigma}$ through $\omega = \omega_0(1 - \tilde{\sigma})$, and to further transform the problem into polar coordinates: $u_1 = a_1 \sin \Phi_1$, $u_2 = a_1 \cos \Phi_1$, $\Phi_1 = (\omega_1/\omega_0)t + \phi_1$; $u_3 = a_2 \sin \Phi_2$, $u_4 = a_2 \cos \Phi_2$, $\Phi_2 = (\omega_2/\omega_0)t + \phi_2$, leading to first-order ODEs in a_1, a_2, ϕ_1 and ϕ_2. These are then reduced by the method of averaging to a set of simpler equations in the averaged $\hat{a}_i$ and $\hat{\phi}_i$ (Appendix F) of the type

$$\begin{aligned} \frac{\mathrm{d}\hat{a}_i}{\mathrm{d}t} &= \frac{1}{\omega_0}[\hat{a}_i\alpha\delta' + \mu(U_i \sin 2\phi + V_i \cos 2\phi)\hat{a}_i + \hat{a}_i^3 R_i], \\ \hat{a}_i \frac{\mathrm{d}\hat{\phi}_i}{\mathrm{d}t} &= \tfrac{1}{2}\hat{a}_i\tilde{\sigma} + \frac{1}{\omega_0}[\hat{a}_i\omega_i'\alpha + \mu(U_i \cos 2\phi - V_i \sin 2\phi)\hat{a}_i + \hat{a}_i^3 S_i], \end{aligned} \tag{5.148}$$

in which U_i, V_i, R_i and S_i are simply constants. This equation is very important, since it represents a parametrically perturbed one-degree-of-freedom oscillator. Most of the nonlinear studies of parametrically excited pipes conveying fluid, whether supported at both ends or cantilevered (but near the Hopf bifurcation point), end up with equation (5.148) or a variant thereof. Therefore, the methodology followed thereafter in most studies is the same: (i) the stability of the origin (trivial solution) is investigated using the linearized version of (5.148) around the origin; (ii) nontrivial solutions or fixed points are sought, determined via (5.148), and their stability is examined. From a physical point of view, a stable (or unstable) nontrivial fixed point in the reduced system (5.148) represents a stable (or, respectively, unstable) periodic solution of the original system; the loss of stability of a fixed point via a Hopf bifurcation signifies the possibility of quasiperiodic motions. The complete study of a periodically perturbed Hopf bifurcation may be found in Bajaj (1986) and Namachchivaya & Ariaratnam (1987).

Namachchivaya (1989) and Namachchivaya & Tien (1989a,b) analyse the averaged equations in the case of the principal primary resonance, $\omega_0 = 2\omega_r$, $r = 1, 2$, and combination resonance $\omega_0 = \omega_1 + \omega_2$.

Typical results are shown in Figure 5.60(a,b) for the principal first-mode resonance of a clamped–clamped pipe. As ω is increased from the left at a fixed μ, the stable trivial equilibrium point of the averaged system becomes unstable through one eigenvalue crossing the origin in the complex plane at point S if dissipation is not zero, or through a double crossing of the origin at $\mathrm{S_D}$ if dissipation is zero (see Figures 2.10 and 3.4). At this point the trivial solution bifurcates into a stable nonzero fixed point, the bifurcation diagram for which is traced in Figure 5.60(b); therefore, the solutions of the original system (5.145) are periodic, of period $2\pi/\omega_r$. It is recalled that averaging provides a solution valid only in the vicinity of $\omega/\omega_r = 2$ and, since the whole bifurcation diagram cannot be traced with high accuracy, its upper part is not given. On the other hand, if ω is reduced from the right, the trivial solution loses stability subcritically, the bifurcating solutions in this case

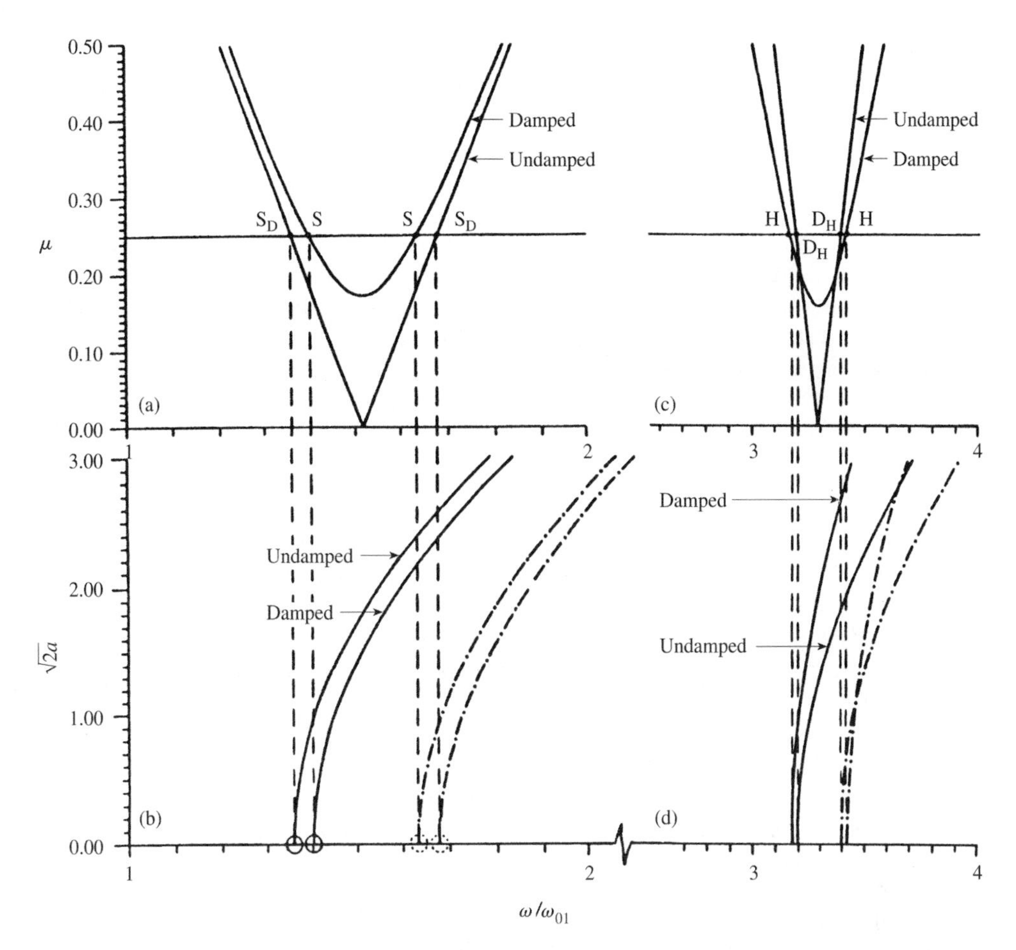

Figure 5.60 (a) Stability boundaries and (b) amplitude–frequency relationships for the principal (subharmonic) resonance of a clamped–clamped pipe ($\omega \simeq 2\omega_1$), for $u_0 = 4$, $\beta = 0.2$, $\alpha = 5 \times 10^{-3}$; $\omega_1 = 16.98$, $\omega_{01} = 22.37$, and $\sqrt{2a}$ is a measure of the dimensionless amplitude. (c,d) Similar diagrams for the combination resonance $\omega \simeq \omega_1 + \omega_2$, where $\omega_2 = 46.77$ (Namachchivaya 1989; Namachchivaya & Tien 1989a,b).

being unstable. Thus, by implication, these results suggest the same type of behaviour as in Figure 5.59(a), but the top of the diagram is missing. Similar results are obtained for the second mode of the system and for pinned–pinned boundary conditions.

Typical results for the combination resonance are shown in Figure 5.60(c,d). In this case, as ω is increased at a constant μ, the averaged system loses stability by a Hopf bifurcation (at H) or a Hamiltonian Hopf bifurcation (at D_H), for dissipation present and absent, respectively. Hence, the motion of the original system becomes amplitude-modulated periodic (quasiperiodic), and the associated bifurcation paths are shown in Figure 5.60(d). On the other hand, as ω is reduced from the right, the system becomes unstable by a subcritical Hopf bifurcation. In the case of subharmonic resonance, the results are supplemented by numerical Floquet analysis of the *averaged* equations, showing excellent agreement.

In a very interesting and well-presented study of the system, Jayaraman & Narayanan (1996) reveal new facets of the nonlinear behaviour of the system and also find chaotic regimes. Equation (5.143) with $u = u_0(1 + \mu \sin \omega\tau)$ is discretized into one- and two-mode Galerkin approximations for a pinned–pinned pipe, the former of which is simply

$$\ddot{q}_1 + (\alpha\pi^4 + \sigma)\dot{q}_1 + \{\pi^4 + \pi^2[\Gamma - u_0^2 + \tfrac{1}{2}\gamma - 2u_0^2\mu \sin \omega\tau - \tfrac{1}{2}\beta^{1/2}u_0\mu\omega \cos \omega\tau]\}q_1 + \tfrac{1}{4}\mathscr{A}\pi^4 q_1^3 + \tfrac{1}{4}\alpha\mathscr{A}\pi^4 q_1^2\dot{q}_1 = 0, \tag{5.149}$$

which, because no sliding of the ends is permitted and hence (5.143) used, is much simpler than that analysed by Yoshizawa *et al.* (1986).† This equation and its two-mode counterpart are studied numerically, while (5.149) is also studied analytically by the multiple time-scale method, in both cases for the fundamental resonance ($\omega \simeq \omega_1$) only. The solution of (5.149) is approximated by

$$q_1(\tau, \mu) = q_{10}(T_0, T_1, T_2) + \mu q_{11}(T_0, T_1, T_2) + \mu^2 q_{12}(T_0, T_1, T_2) + \dots, \tag{5.150}$$

where $T_0 = \tau$ is the fast time-scale associated with frequencies ω and ω_0, and $T_1 = \mu\tau$, $T_2 = \mu^2\tau$, etc. are slow time-scales associated with modulations in amplitude and phase resulting from the nonlinearities and parametric excitation. Assuming $q_{10} = A(T_1, T_2)\exp(\mathrm{i}\omega_0 T_0) + \overline{A}(T_1, T_2)\exp(-\mathrm{i}\omega_0 T_0)$, where $\overline{A}$ is the complex conjugate of A, equations of like powers of μ are solved sequentially while eliminating secular terms, so that equations similar to (5.148) are obtained.

Typical numerical results with the one-mode approximation for $\omega_0 = 8.875$ and $u_0 < u_{cd}$, i.e. before the loss of stability with steady flow, show the following. (i) For $0 \leq \mu \leq 0.259$, the trivial solution converges to zero, while for $\mu > 0.295$ it loses stability. (ii) For $0.229 < \mu < 0.610$, a nontrivial stable periodic solution exists. This implies that for $0.229 < \mu < 0.259$, *two* solutions coexist: the trivial one and a finite-amplitude periodic one, as confirmed by analytical results by the multiple time-scale method for $\omega_0 = 8.8$ shown in Figure 5.61(a). (iii) For $\mu > 0.61$, a period-doubling sequence ensues, a period-8 phase-plane diagram being shown in Figure 5.61(c), leading to chaos at $\mu = 0.7123$. The associated bifurcation diagram is shown in Figure 5.61(b). The extent of the chaotic region is very limited, $0.7123 \leq \mu \leq 0.7162$. For higher μ, transient chaos is observed (Moon 1992): initially, the motion is chaotic with two separate patches in the Poincaré map of the motion; but, as time progresses, these two patches grow and eventually come into contact, whereupon chaotic motion is destroyed and is succeeded by period-1 motion. (iv) As μ is decreased from 0.7164, period-1 motion is found to coexist with the period-8, period-4, and so on, motions found in the foregoing, down to $\mu = 0.66$, as seen in Figure 5.61(b). It should be remarked that values of μ larger than 0.5 ought to be judged as being too large, from both the physical and mathematical viewpoints.

The dynamics obtained with the two-mode approximation is qualitatively similar to that just described. Here it ought to be said that the one-mode approximation is rather hazardous since, as seen in equation (5.149), no Coriolis terms are present because the gyroscopic matrix is skew-symmetric.

For $u_0 > u_{cd} = 4.196$, the system in steady flow becomes a 'buckled beam'. It is not too surprising, therefore, that its dynamical behaviour with pulsating flow is qualitatively similar to a harmonically excited buckled beam, represented by Duffing's equation (Dowell & Pezeshki 1986). For $u_0 = 4.7077$ and $\mu = 0$, the trivial equilibrium is a saddle,

†The unusual $\frac{1}{2}$ factor in some of the terms, e.g. $\frac{1}{2}\gamma$, is due to suppressing the $\sqrt{2}$ in the beam eigenfunctions $\phi_r = \sqrt{2}\sin(r\pi\xi)$ used in the Galerkin scheme.

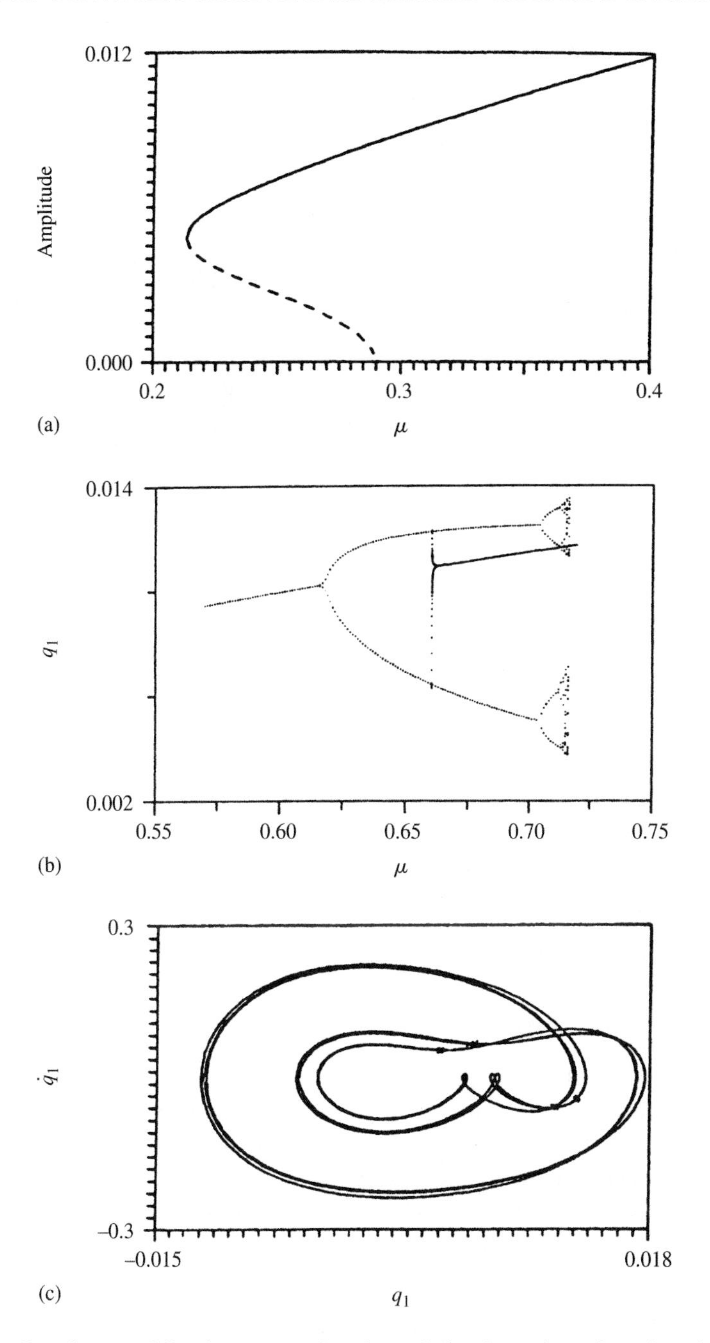

Figure 5.61 One-degree-of-freedom approximation of the dynamics of a pinned–pinned pipe ($\beta = 0.78$, $\gamma = 15.48$, $\alpha = 0$, $\sigma = 0.883$, $\mathscr{A} = 1.81 \times 10^4$, $u_0 = 3.1385$) subjected to pulsating flow: (a) the amplitude versus μ diagram for $\omega_0 = 8.8$ via the multiple scales method; (b,c) numerical results for $\omega_0 = 8.875$ showing the bifurcation diagram and period-8 motion in the phase plane for $\mu = 0.7123$ (Jayaraman & Narayanan 1996).

while two potential wells now exist, centred at $\{\pm 0.01, 0\}$ in the phase plane. With $\mu \neq 0$, period-1 motions exist only for $0 < \mu < 0.0255$, as shown in Figure 5.62(a). For $\mu > 0.0256$, a period-doubling sequence leads to transient chaos (near $\mu = 0.0283$); then to stable period-2 motion, followed by another period-doubling cascade and chaos for $\mu \simeq 0.04$, as seen in Figure 5.62(b). The extremely low values of μ in this case are noted.

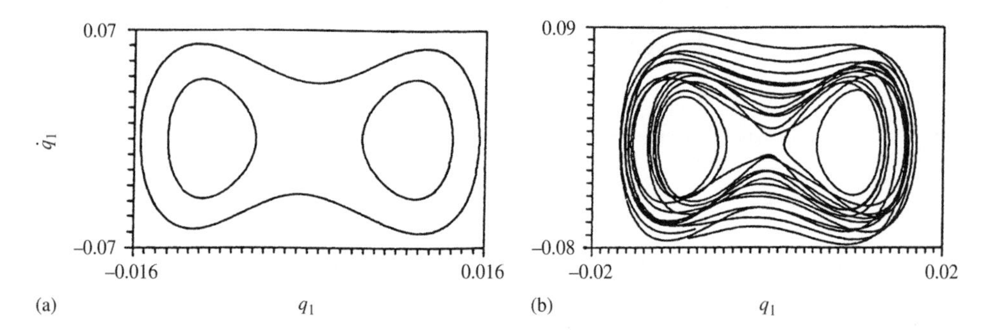

Figure 5.62 Post-divergence phase-plane diagrams for the system of Figure 5.61 for $u_0 = 4.7077$; (a) global view for $\mu = 0.02$, (b) $\mu = 0.04$ (Jayaraman & Narayanan 1996).

Although the results presented in this work are very interesting, it should be remarked that, at least for the parameters chosen, the ranges of u_0 for chaos are extremely narrow, in fact so narrow that their experimental realization could well be problematical.

To conclude, the nonlinear analysis of parametric resonances in pipes with supported ends has brought to light a number of interesting features, e.g. the possible 'subcritical' onset of resonance and the subcritical (hardening) behaviour for its cessation as μ is increased. Furthermore, the amplitude of oscillations can be computed, at least close to the resonance boundaries. Finally, chaotic oscillations have been found to exist in narrow u_0-ranges.

5.9.2 Cantilevered pipes

Two main studies have been conducted in this case: a complete bifurcation analysis of the principal primary resonance in the vicinity of the flutter boundary by Bajaj (1987b) and a combined analytical, numerical and experimental study by Semler & Païdoussis (1996), both for planar motions. The problem in this case is more complex than that of pipes with supported ends, since several modes are required for accuracy in the Galerkin expansion (a one-mode approximation being totally meaningless), and nonlinear inertial terms create additional difficulties in numerical solutions.

The Bajaj (1987b) analysis is at once very powerful, nonstandard and difficult to condense; hence, only an outline of the methods used will be given here. The Lundgren *et al*. (1979) form of the equation of motion is used for motions in a horizontal plane; therefore, apart from harmonic perturbations associated with flow pulsations, the mean flow velocity is steady. Proceeding as in Bajaj *et al.* (1980), the system is re-written in the vector form

$$\frac{\partial \mathbf{u}}{\partial \tau} = \mathbf{L}\mathbf{u} + \epsilon\mu\{\mathbf{L}_1 \cos 2\omega\tau + \mathbf{L}_2 \sin 2\omega\tau\}\mathbf{u} + \epsilon\mathbf{N}(\mathbf{u}, u_0) + \mathcal{O}(\epsilon^2), \tag{5.151}$$

in which $\mathbf{u} = \{(\dot{\eta} + 2u\beta^{1/2}\eta'), \eta\}^{\mathrm{T}}$, $\mathbf{L}$, $\mathbf{L}_1$ and $\mathbf{L}_2$ are linear differential operators involving u_0, β and ω, and $\mathbf{N}$ contains all nonlinear terms; $u = u_0 + \epsilon\mu \cos 2\omega\tau$, reflecting that the main interest is in the *principal* resonance, and it is clear that, in contrast with the foregoing, μ here need not be small — cf. equation (5.144). It has already been shown in Section 4.5.1 that, for the cantilevered system, resonances arise only when u_0 is reasonably close to u_{cf}, where the system loses stability by flutter in steady flow. Hence, solutions are sought close to this flutter boundary, namely for $u_0 = u_{cf} + \epsilon\eta$, $\omega = \omega_0 - \epsilon\hat{\sigma}$, where η denotes the mean-flow velocity variations, and $\hat{\sigma}$ is a detuning parameter. Then, defining $\omega\tau = t$, equation (5.151) is re-written as

$$\omega_0 \frac{\partial \mathbf{u}}{\partial t} = \mathbf{L}_0\mathbf{u} + \frac{\epsilon\hat{\sigma}}{\omega_0}\mathbf{L}_0\mathbf{u} + \epsilon\eta\frac{\partial \mathbf{L}_0}{\partial u_0} + \epsilon\mu(\mathbf{L}_{10}\cos 2t + \mathbf{L}_{20}\sin 2t)\,\mathbf{u} + \epsilon\mathbf{N}(\mathbf{u}, u_{cf}) + \mathcal{O}(\epsilon^2), \tag{5.152}$$

in which $\mathbf{L} = \mathbf{L}_0 + \epsilon\eta(\partial\mathbf{L}_0/\partial u_0) + \mathcal{O}(\epsilon^2)$, $\mathbf{L}_1 = \mathbf{L}_{10} + \mathcal{O}(\epsilon)$, $\mathbf{L}_2 = \mathbf{L}_{20} + \mathcal{O}(\epsilon)$, $\mathbf{N}(\mathbf{u}, u_0) = \mathbf{N}(\mathbf{u}, u_{cf}) + \mathcal{O}(\epsilon)$. For $\epsilon = 0$, equation (5.152) reduces to $\omega_0(\partial\mathbf{u}/\partial t) = \mathbf{L}_0\mathbf{u}$. The operator $\mathbf{L}_0$ (at $u = u_{cf}$) has two pure imaginary eigenvalues, while the others are in the left-hand plane. Hence, the steady-state solutions corresponding to these two eigenvalues may be written as $\mathbf{u}_0 = A\{\mathbf{w}^{(1)} \exp[\mathrm{i}(t + \phi)] + \overline{\mathbf{w}}^{(1)} \exp[-\mathrm{i}(t + \phi)]\}$, where A and ϕ correspond to amplitude and phase, relative to the parametric excitation of the periodic solutions $\mathbf{w}^{(1)}$ and $\overline{\mathbf{w}}^{(1)}$ associated with the two critical eigenmodes. For $\epsilon \neq 0$, the solution is expanded in powers of ϵ as

$$\mathbf{u} = \mathbf{u}_0(A, \phi, t) + \epsilon\mathbf{u}_1(A, \phi, t, \eta) + \epsilon^2\mathbf{u}_2(A, \phi, t, \eta) + \mathcal{O}(\epsilon^2), \tag{5.153}$$

where A and ϕ satisfy

$$\omega_0 \begin{Bmatrix} A' \\ \phi' \end{Bmatrix} = \epsilon \begin{Bmatrix} A_1(A, \phi, \eta) \\ B_1(A, \phi, \eta) \end{Bmatrix} + \epsilon^2 \begin{Bmatrix} A_2(A, \phi, \eta) \\ B_2(A, \phi, \eta) \end{Bmatrix} + \mathcal{O}(\epsilon^3). \tag{5.154}$$

Then, substituting (5.153) into (5.152) and collecting coefficients of equal power of ϵ, a set of new equations with terms which are functions of t and the spatial variable ξ is obtained. They are expanded by Fourier series in time and pertinent comparison functions in ξ. These equations are then averaged, leading to equations of similar form to (5.148).

The results are discussed in terms of *modified* flow-variation, detuning and harmonic flow-perturbation parameters: $\overline{\eta}$, $\overline{\sigma}$ and $\overline{\mu}$. Eventually, the master bifurcation diagram of Figure 5.63(a) is obtained showing, in the $\{\overline{\sigma}, \overline{\eta}\}$-plane, curves across which the trivial equilibrium of the averaged system undergoes a pitchfork or Hopf bifurcation, or the nontrivial fixed points undergo a saddle-node or Hopf bifurcation.

The dynamics is illustrated in three cases by the amplitude(A)–flow($\overline{\eta}$) bifurcation diagrams in Figure 5.63(b–d), each for a constant value of $\overline{\mu}$. In (b), as the mean flow, i.e. $\overline{\eta}$, is increased, at some flow less than u_{cf} ($\overline{\eta} < 0$) the trivial position of the pipe becomes unstable and the pipe performs periodic oscillation at half the excitation frequency. As $\overline{\eta}$ is increased, the amplitude of the oscillation increases, reaches a maximum and then begins to decrease. For $\overline{\eta} = \overline{\eta}_2^*$ the periodic solution becomes unstable, and for $\overline{\eta} > \overline{\eta}_2^*$ there is no stable limit cycle; it is shown that the motion thereafter is amplitude-modulated. The Hopf solution of the unperturbed system is also shown, starting at $\overline{\eta} = 0$, but in this case, since the trivial solution has already become unstable prior to $\overline{\eta} = 0$, this is not a realizable solution. In (c), there is a small region on the left of the figure where both the trivial

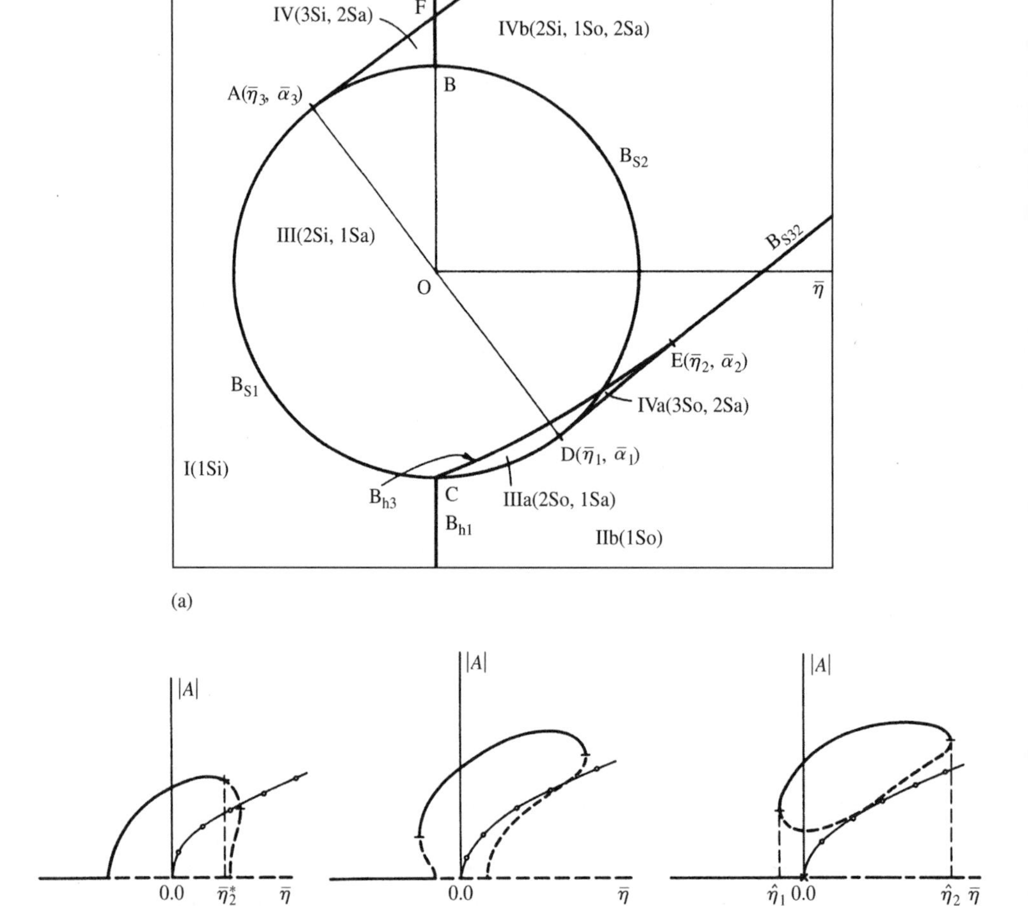

Figure 5.63 (a) Local bifurcation curves in the $\{\bar{\eta}, \bar{\sigma}\}$ plane for a parametrically excited cantilevered pipe. The averaged system undergoes: a pitchfork bifurcation of the trivial solution across B_{s1}, B_{s2}; a saddle-node bifurcation of the nontrivial solution across B_{s31}, B_{s32}; a Hopf bifurcation of the trivial solution across B_{h1}, B_{h2}, and of the nontrivial one across B_{h3}; Si, So and Sa denote 'sink', 'source', and 'saddle', respectively. (b–d) three possible amplitude–flow diagrams for $\beta^{1/2} = 0.65$ (Bajaj 1987b).

and the periodic solutions coexist and are stable, thus implying a subcritical onset of the principal parametric (subharmonic) resonance similar to that found for pipes with supported ends; the dynamics thereafter is similar to that in (b), except that the hysteresis zone is larger. Finally, in (d) we see that for $\bar{\eta} < 0$ the origin is stable, and at $\bar{\eta} = 0$ it becomes unstable by a Hopf bifurcation of the averaged system; hence, for $\bar{\eta} > 0$, we expect the pipe to perform an almost-periodic or amplitude-modulated motion with small amplitude. For $\hat{\eta}_1 < \overline{\eta} < \hat{\eta}_2$, however, these motions coexist with larger-amplitude periodic motions associated with subharmonic resonance which appear as an isolated solution branch. Which one materializes, in that range, depends on the initial conditions.

Points A–F in Figure 5.63(a) are special in that they are intersections of two bifurcation curves. Bajaj undertakes the unfolding of these bifurcations by local analysis. An example is shown in Figure 5.64 for point F. Across B_{h2} a supercritical Hopf bifurcation takes place in the averaged system, leading to the limit cycle in region IIa; the physical system then performs a subharmonic amplitude-modulated motion. However, beyond this limit cycle, other fixed points may exist, and hence periodic solutions of the physical system, as shown in the lower part of this figure.

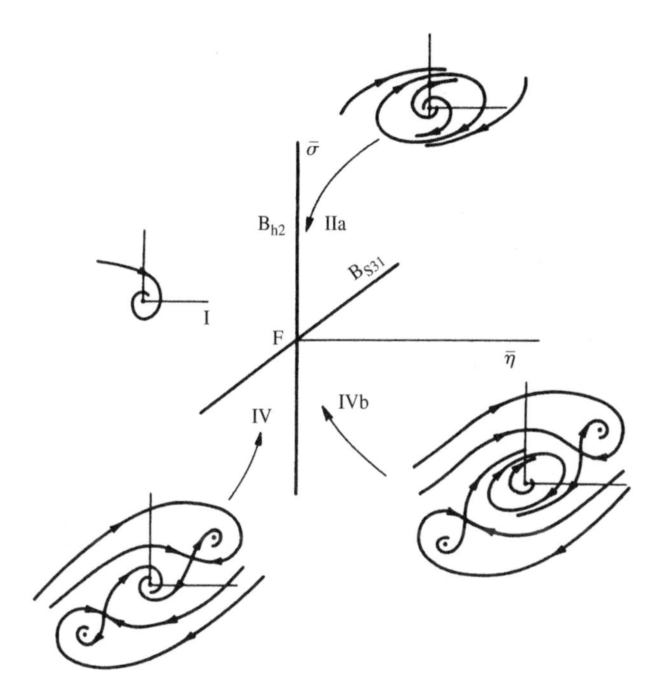

Figure 5.64 The qualitative types of phase portraits of the averaged system around point F of Figure 5.63 (Bajaj 1987b).

To summarize, it is shown that, when the mean flow is below u_{cf}, the pipe can only have periodic solutions which are at half the excitation frequency. Even when the straight position is stable, there are flow fluctuations for which nonzero stable solutions also exist. Thus, a large enough disturbance can force the pipe to perform large steady-state periodic motions.

For mean flow above the critical value, the zero solution is always unstable and the pipe can perform small modulated motions, large periodic motions or large-amplitude modulated motions, depending on the values of flow rate and excitation frequency. Some of these motions coexist for the same values of parameters and then the initial conditions and disturbances determine the motions performed.

A similar study of the articulated system was conducted earlier by Bajaj (1984). Planar motions of a two-segment cantilevered pipe are considered, and hence the 2-D versions of equations (5.74) and (5.76) are used. The periodic solutions for the principal parametric resonance, in the vicinity of the Hopf bifurcation in steady flow (at u_{cf}), are determined by the method of alternate problems (Appendix F.6.3) and their stability assessed from the Floquet exponents of the associated variational equations. The results, in this case also, are discussed in terms of (i) the mean-flow velocity perturbation $\eta = u_0 - u_{cf}$, (ii) the detuning parameter $\tilde{\sigma}$, and (iii) the harmonic flow-perturbation amplitude μ, where $u = u_0 + \mu \cos 2\omega\tau$.

The results are presented in diagrams of (i) μ versus $\tilde{\sigma}$ for a given η, (ii) amplitude A versus η, and (iii) A versus $\tilde{\sigma}$, for $a = \kappa = 1$, $\gamma = 0.25$, and $\beta = \frac{2}{3}$ and $\frac{5}{3} \times 10^{-2}$, some of which are given here in Figure 5.65.

In the A versus η diagrams of Figure 5.65(a–c) the Hopf bifurcation in steady flow (for $\beta = \frac{2}{3}$) is supercritical. For a large negative $\tilde{\sigma}$, e.g. $\tilde{\sigma} = -0.31$ as in (a), there is

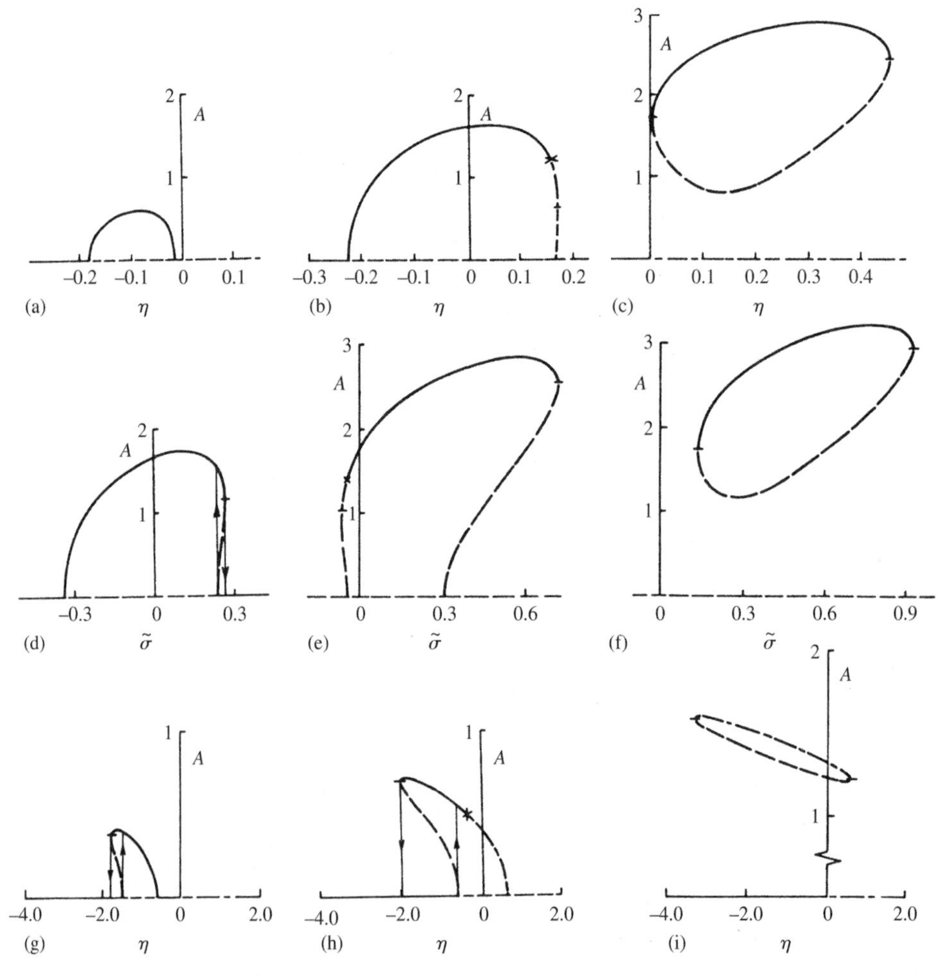

Figure 5.65 Response diagrams for the parametrically excited articulated system. Top: A versus η diagrams for $\beta = \frac{2}{3}$ and (a) $\tilde{\sigma} = -0.31$, (b) $\tilde{\sigma} = -0.1$, (c) $\tilde{\sigma} = 0.4$. Middle: A versus $\tilde{\sigma}$ diagrams for $\beta = \frac{2}{3}$ and (d) $\eta = -0.075$, (e) $\eta = 0.188$, (f) $\eta = 0.30$. Bottom: A versus η diagrams for $\beta = \frac{5}{3} \times 10^{-2}$ and (g) $\tilde{\sigma} = -0.3$, (h) $\tilde{\sigma} = 0$, (i) $\tilde{\sigma} = 1.7$ (Bajaj 1984).

only one nontrivial solution and it is stable. Beyond $\eta = 0$, the trivial solution is unstable and there are no nontrivial periodic solutions. On physical grounds, however, the response should remain bounded; although the methods used preclude finding the solutions existing in this area, by analogy to the results obtained for the continuous cantilevered system, they are expected to be amplitude-modulated motions. As η is increased to $\tilde{\sigma} = -0.1$ in (b), a portion of the nontrivial response for $\eta > 0$ becomes unstable, similarly to the behaviour displayed in Figure 5.63(b). For $\tilde{\sigma} = 0.4$, we see in (c) that the nontrivial periodic solutions 'pinch-off' the trivial one, giving rise to an *isolated* solution.

Similar results are shown in Figure 5.65(d–f) in the A versus $\tilde{\sigma}$ plane. For $\eta = -0.075$ we see in (d) the usual jump phenomenon and the associated hysteresis for high enough $\tilde{\sigma}$. In (e) and (f) we see that the trivial solution is unstable and there are segments of $\tilde{\sigma}$ over which the nontrivial solution also is unstable. In (f) we again see an isolated solution.

Figure 5.65(g–i) shows plots for $\beta = \frac{5}{3} \times 10^{-2}$ where the Hopf bifurcation is subcritical. In (g) and (h) it is seen that for large enough negative detuning one or two nontrivial solutions generally exist, depending on $\tilde{\sigma}$ and η; the lower branch is unstable, while for high enough η a portion of the upper branch becomes unstable also. In (i) the solution is isolated, and both branches are unstable. The remarks already made regarding amplitude-modulated motions apply here too.

It is therefore seen that the dynamics of the system, articulated or continuous, for parameters such that self- and parametrically excited oscillations are close, is very interesting. In most cases the Hopf bifurcation is suppressed and the dynamics is dominated by the parametric resonance. Jump phenomena, subcritical onset of resonance, isolated resonance branches, and amplitude-modulated motions are all possible for given combinations of β, η, $\tilde{\sigma}$ and μ.

Semler & Païdoussis' (1996) contribution is an extension of Bajaj's (1987b) analytical work, but numerical solutions of the full nonlinear equations are also presented, as well as some experiments. Equation (5.39) is utilized, thus retaining the inertial nonlinearities intact, and is discretized by Galerkin's method into

$$\ddot{q}_i + C_{ij}\dot{q}_j + K_{ij}q_j + \alpha_{ijkl}q_jq_kq_l + \beta_{ijkl}q_jq_k\dot{q}_l + \gamma_{ijkl}(q_j\dot{q}_k\dot{q}_l + q_jq_k\ddot{q}_l) = 0, \quad (5.155)$$

where $u = u_0(1 + \mu \sin \omega\tau)$ and

$$C_{ij} = \alpha\lambda_j^4\delta_{ij} + 2\beta^{1/2}u_0(1 + \mu \sin \omega\tau)b_{ij},$$
$$K_{ij} = \lambda_j^4\delta_{ij} + u_0^2(1 + \mu \sin \omega\tau)^2 c_{ij} + \beta^{1/2}\mu u_0\omega \cos \omega\tau(d_{ij} - c_{ij}) + \gamma(b_{ij} - c_{ij} + d_{ij}), \quad (5.156)$$

the λ_j being the dimensionless cantilevered beam eigenvalues and b_{ij}, c_{ij} and d_{ij} are as given in Table 3.1, while α_{ijkl}, β_{ijkl} and γ_{ijkl} are similar to the a_{ijkl}–d_{ijkl} given in Section 5.7.3(a), but different since the inertial nonlinearities here are intact.

For the analytical solution of the problem, we confine ourselves to the vicinity of the Hopf bifurcation in steady flow, i.e. to $u \simeq u_{cf}$, as in Bajaj (1987b), but proceed in a more standard manner, as follows. The system is transformed to first order, such that $\mathbf{y} = \{\mathbf{q}, \dot{\mathbf{q}}\}^{\mathrm{T}}$, and then into standard form via $\mathbf{y} = \mathbf{P}\mathbf{x}$, where $\mathbf{P}$ is a modified modal matrix evaluated at u_{cf}, thus yielding an equation of the form

$$\dot{\mathbf{x}} = \mathbf{A}\mathbf{x} + \mu(\omega \cos \omega\tau\, \mathbf{B}_1 + \sin \omega\tau\, \mathbf{B}_2)\mathbf{x} + \mu^2 \sin^2 \omega\tau\, \mathbf{B}_3\mathbf{x} + F(\mathbf{x}, \dot{\mathbf{x}}), \quad (5.157)$$

where **A** is a matrix, with **J** and **M** nonzero submatrices on its diagonal and zero elsewhere; **J** corresponds to the purely imaginary pair of eigenvalues, and **M** to the $2N - 2$ eigenvalues with negative real parts, so having sub-elements $\mathbf{R}_p$, given as follows:

$$\mathbf{J} = \begin{bmatrix} 0 & -\omega_0 \\ \omega_0 & 0 \end{bmatrix}, \qquad \mathbf{R}_p = \begin{bmatrix} \sigma_p & -\omega_p \\ \omega_p & \sigma_p \end{bmatrix}. \tag{5.158}$$

The system is then projected onto a centre manifold, where special care must be exercised because the system is nonautonomous. To transform the system to an equivalent autonomous one, $\mu \sin \omega\tau$ and $\mu \cos \omega\tau$ are replaced by two new variables,

$$\begin{Bmatrix} \dot{v}_1 \\ \dot{v}_2 \end{Bmatrix} = \begin{bmatrix} \mu^2 & -\omega \\ \omega & \mu^2 \end{bmatrix} \begin{Bmatrix} v_1 \\ v_2 \end{Bmatrix} - \begin{Bmatrix} v_1(v_1^2 + v_2^2) \\ v_2(v_1^2 + v_2^2) \end{Bmatrix}, \tag{5.159}$$

and μ is included in the system of equations as a trivial dependent variable, $\dot{\mu} = 0$. Thus, **J** in (5.158) is replaced by

$$\mathbf{A}_0 = \begin{bmatrix} 0 & 0 & 0 & 0 & 0 \\ 0 & \mu^2 & -\omega & 0 & 0 \\ 0 & \omega & \mu^2 & 0 & 0 \\ 0 & 0 & 0 & 0 & -\omega_0 \\ 0 & 0 & 0 & \omega_0 & 0 \end{bmatrix}, \tag{5.160}$$

the associated vector being $\mathbf{y} = \{\mu, v_1, v_2, x_1, x_2\}^{\mathrm{T}}$. The system is then transformed by defining $x' = \epsilon x$, $\mu' = \epsilon\mu$, $u_0 - u_{cf} = \epsilon\eta$, and the method of normal forms (Appendix F.3) is applied: (a) to find all possible parametric resonances to $\mathcal{O}(\epsilon)$ and $\mathcal{O}(\epsilon^2)$, and (b) to determine the simplest set of equations defining these resonances. Three separate sets of normal forms are determined: (i) for ω away from both $2\omega_0$ and ω_0, (ii) for ω near ω_0, and (iii) for ω near $2\omega_0$. In case (ii) one obtains the fundamental secondary resonance, where the harmonic perturbation in u appears only at the second order, μ^2; in the last case, the principal resonance is obtained, where these terms appear to first order, μ. The results for the principal resonance are identical to those obtained by Bajaj (1986b).

Before presenting any results, the second major component of this study is briefly discussed, namely the solution of the *full equations* by numerical techniques. In this case no restrictions apply as to u_0 being close to u_{cf}. Three such methods are used — see Section 5.4. (a) The nonlinear inertial terms are transformed into equivalent stiffness and velocity-dependent terms [Section 5.2.7(b)], and the resulting equation can then be integrated by a *Runge–Kutta method*; AUTO may also be used in this scheme, once the equations are transformed into those of an equivalent autonomous system, as in equation (5.160). Solutions are also obtained by (b) the *finite difference method* (FDM) based on Houbolt's fourth-order scheme and (c) the *incremental harmonic balance* (IHB) *method*, in both cases with nonlinear inertial terms intact; a complete exposé of the application of the IHB method to the problem at hand may be found in Semler *et al*. (1996).

Typical results for the principal resonance are shown in Figures 5.66 and 5.67. Several observations may be made, as follows: (i) it is seen that there is good agreement between the normal-form and numerical solutions for the resonance boundary in Figure 5.66, but less so for the amplitude in Figure 5.67(b), especially away from the resonance boundaries;

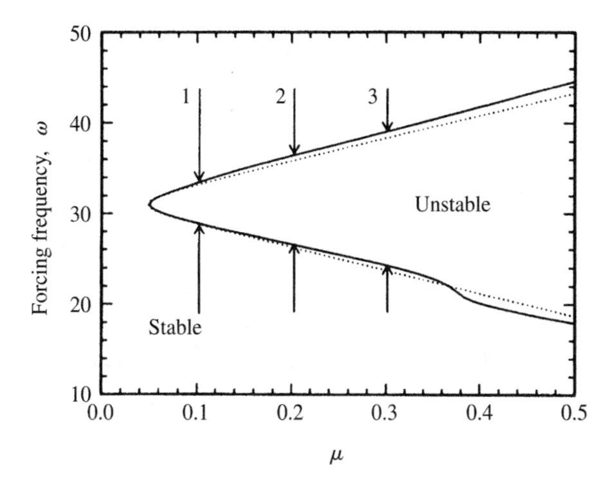

Figure 5.66 Boundaries of the principal parametric resonance of a cantilevered pipe for $u_0 = 6$, $\beta = 0.2$, $\gamma = 10$ and $\alpha = 0$; ——, IHB and AUTO; $\cdots$, normal form theory; $u_{cf} = 6.34$ (Semler & Païdoussis 1996).

(ii) when the nonlinear inertial terms are either transformed or eliminated, the amplitudes are underestimated; furthermore, in the former case, a spurious bulge on the right-hand side of the diagram in Figure 5.67(a) is generated; (iii) the agreement between FDM and the IHB results in Figure 5.67(b) is excellent throughout, although it is noted that FDM can give stable solutions only.

The primary [principal ($\omega/\omega_2 \simeq 2$) and $\omega/\omega_2 \simeq \frac{2}{3}$] and secondary (fundamental, $\omega/\omega_2 \simeq 1$) parametric resonance regions for another system — corresponding to that in Figure 4.33(a) — are shown in Figure 5.68(a), calculated by the same numerical methods as in the foregoing. Agreement with the results obtained by Bolotin's method in Figure 4.33(a) is generally good, although the lower secondary region in that case ($\omega/\omega_2 \simeq \frac{1}{2}$) could not be reproduced unless $\alpha = 0$ is taken, for unknown reasons. Figure 5.68(b) shows clearly that the largest amplitudes are associated with the principal resonance, as observed in the experiments (Section 4.5.3), followed by those of the fundamental resonance.

Some results for $u > u_{cf}$ are given in Figure 5.69 for the same parameters as in Figure 4.29(b). Linear and nonlinear analyses agree for the principal and fundamental resonance boundaries, but of course the nonlinear analysis also gives amplitudes. Of more interest is to compare the regions of combination resonance (quasiperiodic motions), which in the linear results of Figure 4.29(b) almost entirely fill the plane. For $\mu = 0.3$, there are two ranges where the system should execute quasiperiodic motions according to linear theory: for $\omega < 6$ and for $\omega > 38$; there are also two ranges of ω ($6 < \omega < 14.5$ and $18 < \omega < 24.5$) where the system should be stable. These latter are also seen in Figure 5.69. However, the quasiperiodic solutions for $\omega < 6$ are found to be only transient; so are those for $\omega > 38$ if the simulation is allowed to run long enough, although in Figure 5.69 the response is still quasiperiodic. One reason for this might be that $u_0 = 6.50$ is too close to the critical, $u_{cf} = 6.34$. A simulation run for $u_0 = 6.80$ shows stable quasiperiodic oscillations for $\omega = 2$ and $\omega = 40$, so that this at least agrees with the previous linear results. Normal-form theory has been applied in this case also and good

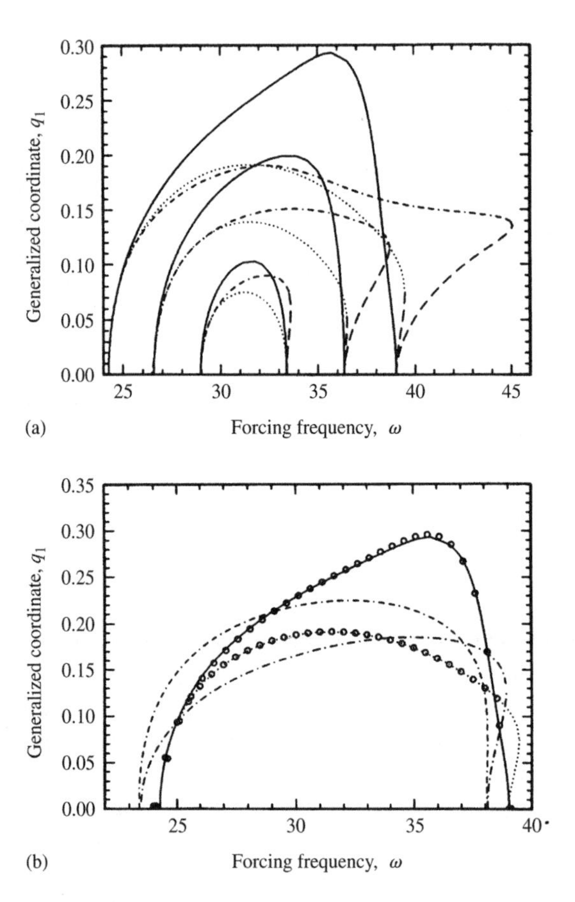

Figure 5.67 (a) Amplitude of periodic solutions for $\mu = 0.1$, 0.2 and 0.3 (corresponding to points 1, 2 and 3 in Figure 5.66): ——, by IHB method with nonlinear inertial terms intact; --·--, by AUTO with these terms transformed; ···, by AUTO with these terms wholly eliminated; in all cases dashed lines (– – –) correspond to unstable solutions: (b) Results for $\mu = 0.3$: ——, by IHB method with nonlinear inertial terms intact; --·--, by normal form method with nonlinear terms intact; ···, by AUTO and IHB with these terms eliminated; -·-, by normal form method, with all such terms eliminated; ∘, solutions by FDM (Semler & Païdoussis 1996).

agreement with the numerical results obtained, for both the parametric resonances and the quasiperiodic regions ('combination resonances').

Experiments were conducted with elastomer pipes similar to those used in Section 5.8.3(a), but without the added end-mass, and a modified form of the apparatus shown in Figure 5.43(b) to allow the addition of a pulsating component to the mean flow; this was provided by the plunger pump shown in Figure 4.30, via a T-junction in the piping. Similar results to those in Section 4.5.3 were obtained, but the main observations are reiterated here with a different emphasis, as follows. (a) The main, most easily excited and pin-pointed resonance region was the principal one associated with the second mode. (b) In contrast, the fundamental resonance, although observed, was difficult to pin-point because of the small but omnipresent forcing component in the response at the pulsation frequency. (c) Quasiperiodic motions were mainly observed for $u > u_{cf}$, with two frequencies in the power spectrum of the response. (d) For sufficiently large

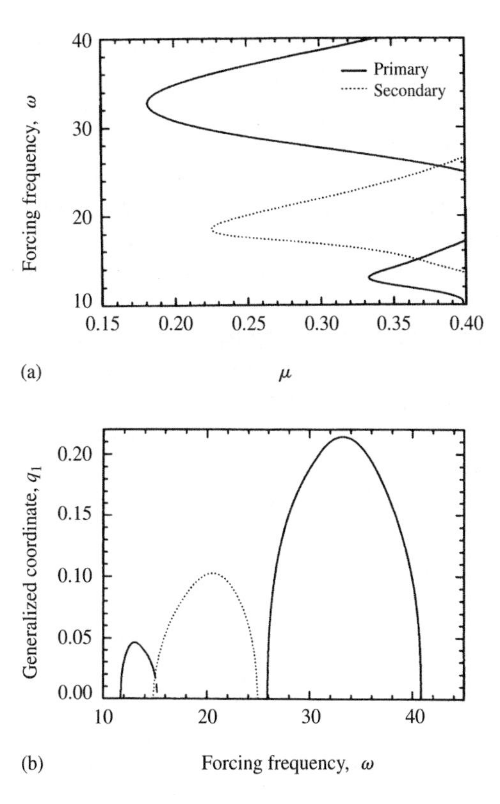

Figure 5.68 (a) Linear stability boundaries for a cantilevered system with $u_0 = 7.68$, $\beta = 0.307$, $\gamma = 16.1$ and $\alpha = 3.65 \times 10^{-3}$. (b) Amplitude versus ω diagram for $\mu = 0.37$: ——, primary resonances ($\omega/\omega_2 \simeq 2$ and $\frac{2}{3}$) associated with the second mode; $\cdots$, secondary fundamental resonance for the second mode (Semler & Païdoussis 1996).

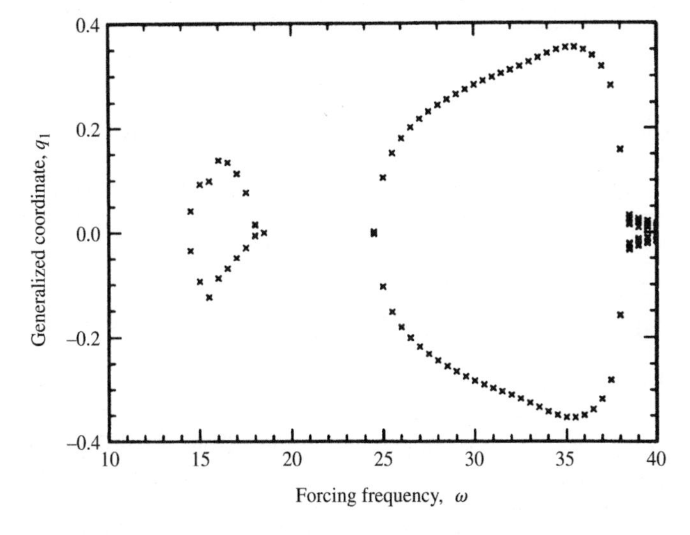

Figure 5.69 Bifurcation diagram for the principal and fundamental resonances for the system of Figure 5.66 but with $u_0 = 6.5$, obtained with the FDM (Semler & Païdoussis 1996).

μ ($\mu > 0.6$) the pipe oscillates about a quasi-stationary deflected shape. (e) Again for $\mu > 0.6$, the oscillation ceases being planar and becomes chaotic.

Quantitative comparisons with theory are undertaken for the principal resonance boundaries for a system at $u = 0.90u_{cf}$ and $0.95u_{cf}$ (Semler & Païdoussis 1996; Figure 12). The experimental boundaries are larger than the theoretical ones, similarly to the results in Figure 4.32, and agreement with theory is similar. In the theoretical results no 'subcritical onset' of the resonance (a behaviour shown in Figure 5.63(c), for instance, but with varying η) has been found; hence, the early appearance of resonance with increasing ω remains unexplainable.

Some experiments were done with varying u_0 around u_{cf} (i.e. varying η) and μ, while keeping the forcing frequency constant, such that $\tilde{\sigma} = \omega/\omega_0 - 1 = -0.14$. The results are shown in Figure 5.70, showing the system to be stable for $u_0 < u_{cf}$, unless μ is large enough to give rise to parametric resonance. If $u > u_{cf}(\eta > 0)$, however, the system executes quasiperiodic motions for low μ, and periodic parametric oscillations for higher μ. Agreement between theory and experiment is reasonably good.

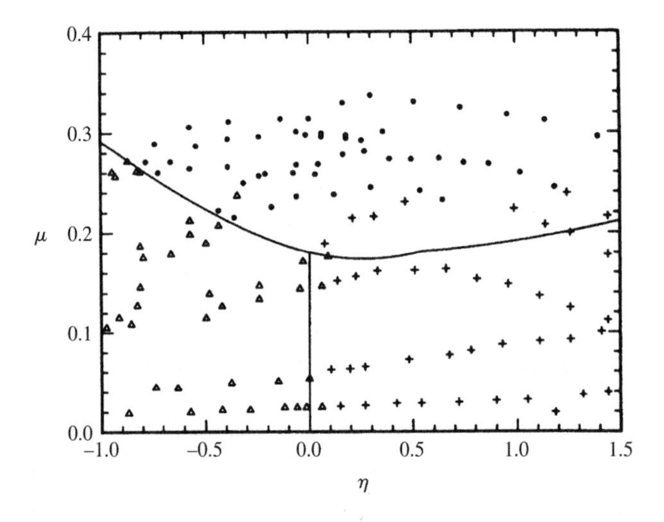

Figure 5.70 Comparison between theory and experiment for the principal parametric resonance of the cantilevered pipe system in the (η, μ) parameter space for $\tilde{\sigma} = -0.14$, $\beta = 0.131$ and $\gamma = 26$. Experimental data points: Δ, the system is stable; $\bullet$, the response is periodic; $+$, the response is quasiperiodic. ——, Theoretical boundaries, via normal form theory and $N = 3$, separating these three dynamical states (Semler & Païdoussis 1996).

5.10 OSCILLATION-INDUCED FLOW

Jensen (1997) discovered yet another new phenomenon, namely that lateral oscillations of the fixed end of a pipe filled with quiescent fluid may induce a mean flow in the pipe. This effectively represents the inverse problem to all of the foregoing: not flow-induced ocillations, but *oscillation-induced flow*!

The equations of motion used are: (i) an equation very similar to (5.39), but with a term $p\omega^2 \cos \omega\tau$ added on the right-hand side (in which $p = w_f(L)/L$ and $w_f(L)$ is the amplitude of motion of the fixed end), as well as $\eta - p \cos \omega\tau$ replacing η, and $\gamma = 0$;

(ii) a ‘flow equation’, similar to those given in Section 5.2.8(b,c), namely

$$\beta^{1/2}\dot{u} + \left(\alpha_j |u^{j-1}| + \tfrac{1}{2}u\right) u = -\beta \int_0^1 \left[\ddot{\eta}\eta' - \left(1 - \tfrac{1}{2}\eta'^2\right) \int_0^\xi \left(\ddot{\eta}'\eta' + \dot{\eta}'^2\right) \mathrm{d}\xi\right] \mathrm{d}\xi$$

$$+ \beta p \omega^2 \sin \omega\tau \int_0^1 \eta' \,\mathrm{d}\xi, \qquad (5.161)$$

where $j = 1$ for laminar and $j = 2$ for turbulent flow. The left-hand side of this equation contains the inertial and frictional terms, while the right-hand side gives the vibratory forcing due to lateral pipe motion; however, unlike in Bajaj *et al*. (1980) and Rousselet & Herrmann (1981), there is no upstream pressurization, so that fluid motion can be induced only because of the mechanical vibration of the pipe.

The system is discretized and then integrated numerically. Appropriate analytical solutions are obtained by the multiple-scales method for near-resonance conditions, i.e. when the forcing frequency is close to one of the natural frequencies of the system.

It is found that, by means of nonlinear interaction, energy is transferred from the vibration exciter to the fluid, resulting in nonzero mean flow velocity from the fixed towards the free end, as well as a small oscillatory component. Typical results are shown in Figure 5.71 for resonant excitation in the first and second modes of two different pipes, showing that a substantial flow may be generated. These results are compared with experiment in the figure, and agreement is exceptionally good.

Fluid flow damps the oscillation of the pipe. This effect is largest for the fundamental mode, due to larger energy transfer to the fluid. Thus, the efficiency, measured as the ratio of the kinetic energy imparted to the fluid compared to the energy supplied by the shaker, decreases as the mode number increases.

Obvious uses of this discovery are for fluid transport or pumping, as well as transport of granular materials — especially in cases where the fact that there are no internal moving parts is important, e.g. in medicine or for corrosive or highly toxic substances.

5.11 CONCLUDING REMARKS

Further work on various aspects of the nonlinear dynamics of the system has been and continues to be done, an attribute of this being *a model system*, of interest not only for its own sake but also for developing theory or exemplifying dynamical behaviour in the broad classes of nonconservative gyroscopic systems and fluidelastic systems.

Thus, in addition to linear studies on control of oscillations in pipes cited in Section 4.8, Yau *et al*. (1995) devise a successful and sophisticated control system for suppressing chaotic oscillation of the constrained system of Figure 5.30, by means of so-called quantitative feedback theory (QFT).

Yoshizawa *et al*. (1997) study theoretically and experimentally the effect of lateral harmonic excitation of a cantilevered pipe with an end-mass performing circular motion. The state-of-the-art experimental set-up involves a laser displacement meter coupled to an FFT analyser, two CCD video cameras and an on-line computer. Both theory and experiments demonstrate a form of *quenching*. This phenomenon occurs when a self-excited system performing limit-cycle oscillations is simultaneously subjected to forced excitation; for sufficiently high amplitude of forcing, the character of the damping changes completely (from ‘normally’ negative to positive) and the self-excited (free) oscillation is

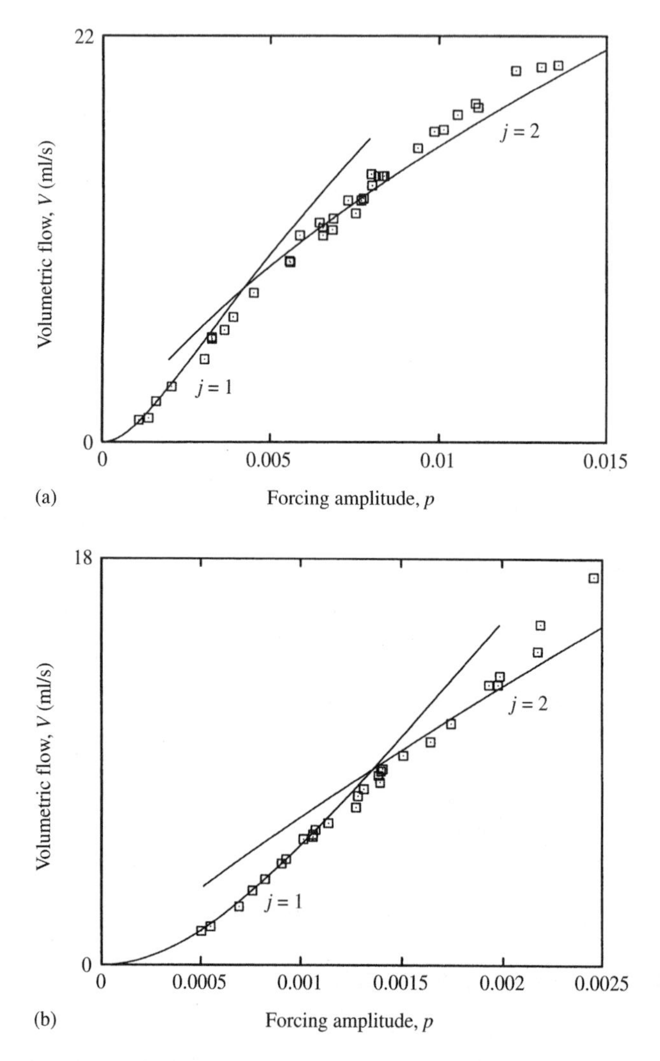

Figure 5.71 Experimental and theoretical results for oscillation-induced volumetric flow-rate, V, versus the dimensionless vibration amplitude of the fixed end for two different pipes: ——, perturbation solution: □, experimental data; (a) first-mode resonance, $\omega = \omega_1$, $\beta = 0.39$, $\alpha_1 = 0.1$, $\alpha_2 = 2.6$, modal viscous damping ratio $\zeta_1 = 0.04$; (b) $\omega = \omega_2$, $\beta = 0.39$, $\alpha_1 = 0.22$, $\alpha_2 = 3.9$, $\zeta_2 = 0.03$ (Jensen 1997).

damped out, while only the forced oscillation remains (Nayfeh & Mook 1979). In this case, with increasing near-resonant forcing, the circular oscillation becomes elliptical and then finally planar, in the direction of forcing. For off-resonant forcing, however, whereas the oscillation in that direction displays only the forced vibration, both forced and self-excited frequencies exist in the other direction, and the self-excited oscillation is not quenched.

The work and the list of contributions go on, but we must stop somewhere; in fact, here! This is the last of the dynamics of straight pipes conveying fluid in this book, which lays the groundwork for the dynamics of shells and slender bodies in axial flow, which are the subjects of Volume 2.

6
Curved Pipes Conveying Fluid

6.1 INTRODUCTION

As may be appreciated from the contents of Chapters 3–5, a great deal of work has been done on the dynamics of *straight* pipes conveying fluid over the past 45 years or so. Relatively less effort has been directed towards the investigation of the dynamics and stability of fluid-conveying *curved* pipes. In general, piping may be curved and twisted into complex spatial forms. In this book, however, reflecting the state of the art, mostly curved pipes which initially lie within a given plane are considered. In this case, one may distinguish motions in the plane of curvature and perpendicular to it, which will be referred to as *in-plane* and *out-of-plane* motions for short; as will be seen, depending on the assumptions made, these two sets of motions are sometimes uncoupled from each other.

Among the first to study the hydroelastic vibration of curved pipes was Svetlitskii (1966). He investigated the out-of-plane motion of a fluid-conveying perfectly flexible hose, treating it as a string, and therefore neglecting the bending rigidity. The ends of the hose were fixed and its initial shape was a catenary. Unny *et al*. (1970) considered the in-plane divergence of initially circular tubular beams with fixed ends. The equations of motion were derived using Hamilton's principle, and critical flow velocities for instability were obtained for pinned and clamped ends; the equations of motion, however, were subsequently shown to be incorrect (Chen 1972b).

The dynamics and stability of curved pipes in the form of circular arcs were studied extensively by Chen (1972b,c, 1973). He derived the equations governing in-plane motions using both the Newtonian (Chen 1972b) and Hamiltonian (Chen 1972c) formulations, and equations governing out-of-plane motions from the Hamiltonian viewpoint (Chen 1972c, 1973). In all cases, it was assumed that the centreline of the pipe is inextensible. It was found that in the case of *clamped–clamped* and *pinned–pinned* boundary conditions, the pipe loses stability by divergence when the flow velocity or the fluid pressure exceeds a certain critical value. This behaviour is qualitatively similar to that of a straight pipe. Chen also studied the stability of *cantilevered* curved pipes. He found that for in-plane motions such pipes are generally subject to both divergence and flutter instabilities, with divergence occurring first, except in cases where the subtended angle is very small (so that the system comes closer to a straight pipe), when only flutter was found to arise (Chen 1972c). In the case of out-of-plane motions, only flutter was predicted, with stability characteristics similar to those of a straight pipe (Chen 1973).

Hill & Davis (1974) studied the dynamics and stability of clamped–clamped pipes conveying fluid, shaped as circular arcs, as well as S-shaped, L-shaped and spiral configurations. Their equations of motion have a significant difference from those of most of the previous studies: they include the effect of the *initial forces* arising from the centrifugal effect and the pressure of the fluid. They obtained the unexpected result that, if the initial forces are taken into account, then pipes with both ends supported do *not* lose stability, no matter how high the flow velocity may be! Similar observations had been made by Svetlitsky (Svetlitskii) in an earlier paper published in Russian (Svetlitskii 1969) and subsequently in English (Svetlitsky 1977); see also Svetlitskii (1982). Svetlitsky considered cantilevered pipes as well and noted that this system loses stability by flutter at sufficiently high flow velocities, even when initial forces are taken into account.

Doll & Mote (1974, 1976) studied a more general case, where the fluid-conveying pipe is both curved and twisted. The equations of motion were derived via Hamilton's principle and solutions were obtained using the finite element method. They considered two cases: (i) the 'constant curvature' case, in which the curvature does not change with flow velocity, and (ii) the 'variable curvature' case in which variations in curvature with changes in flow velocity are accounted for. The first case corresponds to the analyses of Unny *et al*. and Chen, while the second is similar to Hill & Davis. Both Doll & Mote and Hill & Davis take into account the extensibility of the centreline of the curved pipe. An important difference between the two studies is that in the latter the equilibrium configuration and forces are calculated via a linearized set of equations, on the assumption that the initial and the flow-deformed equilibrium configurations are close; in the Doll & Mote study, on the other hand, a cumulative application of linearization is utilized for small flow velocity increments, which is more general. The main conclusions of both studies, however, are the same: the eigenfrequencies of pipes supported at both ends are not sensitive to flow velocity, and hence no instabilities should arise. Doll & Mote also compared their curved-pipe theory to Liu & Mote's (1974) experimental data for nominally straight but actually slightly curved pipes, supported at both ends (see Section 3.4.4). This comparison is discussed in Sections 6.4.5 and 6.6.2.

More recently, Dupuis & Rousselet (1985) have carried out a study on the dynamics of fluid-conveying planar curved pipes modelled as Timoshenko beams. The extension of the centreline was taken into account. This study used the transfer matrix method (Pestel & Leckie 1963), in preference to either analytical or finite element methods. Once more, flutter instabilities are predicted for cantilevered curved pipes, but results for pipes supported at both ends are not presented.

Misra *et al*. (1988a,b,c) re-examined the dynamics and stability of fluid-conveying curved pipes *ab initio*. The main objective was to shed light onto the underlying reasons for the fundamentally different dynamical behaviour for pipes with supported ends as predicted by the extensible theories of Hill & Davis, Doll & Mote and Svetlitsky, on the one hand, and the inextensible theories of Chen and Unny *et al*., on the other: the former predicting no loss of stability, while the latter predict divergence at high enough flow. For this purpose, three theories were formulated and their results were compared:

(i) *the conventional inextensible theory*, in which the centreline of the pipe is assumed to be unstretched, and the steady (initial) flow-induced forces introduced by the pressure and centrifugal forces are entirely neglected;
(ii) *the extensible theory*, in which the shape of the pipe changes with flow velocity under the action of the steady flow-induced forces;
(iii) *the modified inextensible theory*, in which the assumption of inextensibility of the centreline is retained, but the steady flow-related forces are taken into account.

In (ii) and (iii), the steady forces depend on fluid friction. Accordingly, variants of the theory considering the fluid to be inviscid or viscous are formulated.

Of course, theories (i) and (ii) are variants of then already available theories. The strength of the Misra *et al*. work lies in deriving both inextensible and extensible theories from the same basic trunk, and thus having control over the assumptions and parameter differences between them; and hence being able to make meaningful comparisons between their predictions. It should also be mentioned that little cross-comparison between the various theories was done theretofore, and even less systematic analysis of the reasons for the differences between their predictions.

Therefore, since theoretical models (i) and (ii) substantially incorporate the salient features of all the aforementioned inextensible and extensible theories, and since, as will be shown, model (iii) succeeds in isolating the important physical differences between them, this work (Misra *et al*. 1988a,b,c) provides the backbone of the material to be presented in this chapter.

Other work on this topic was undertaken by Fan & Chen (1987), who studied three-dimensionally curved helical pipes (the only such study), and Aithal & Gipson (1990), who studied the effect of dissipative forces on stability, both making the inextensibility assumption. Ko & Bert (1984, 1986) undertook a nonlinear study of the system, and Steindl & Troger (1994) looked into the possibility of chaotic motions of cantilevered curved pipes. More will be said about these studies in the following sections.

Finally, a thorough study of the equations of motion was undertaken by Dupuis & Rousselet (1992), who concluded that 'the stressed-by-flow configuration ... is the only equilibrium state adequate for the study of the linear stability of such pipe-fluid systems' — as Misra *et al*. had concluded and as we, in due course, shall do here.

6.2 FORMULATION OF THE PROBLEM

6.2.1 Kinematics of the system

The system under consideration is shown in Figure 6.1. It consists of a curved pipe of length L, with a uniform cross-sectional area A_p, mass per unit length m, flexural rigidity EI and shear modulus G. The pipe is initially in a plane, having an arbitrary centreline shape, i.e. the radius of curvature is not necessarily constant along its length. It conveys a stream of fluid, of mass M per unit length. The flow is assumed to be a plug flow with constant velocity, U. Furthermore, the pipe is assumed to be fully submerged in a quiescent fluid.

The kinematics of the pipe is developed by the same approach as that used by Love (1927) for a curved rod. This implies an assumption that the pipe diameter is small compared to both the radius of curvature of the centreline and the overall length of the pipe.

To describe the kinematics of the system, it is convenient to use two reference frames (Figure 6.1) — cf. Section 3.3.1. The Lagrangian reference frame (x_0, y_0, z_0) with its origin P_0 is located on the initial, undeformed centreline in such a way that the z_0-axis is tangential to the undeformed centreline, while the axes x_0 and y_0 are directed along the principal normal and binormal directions.[†] At any instant t during the motion

[†]Unusually, *vis-à-vis* most of the foregoing, the long axis along which the fluid flows is here the z_0-axis. This being an intricate analysis, conversion was not attempted, as it might have introduced unwanted errors.

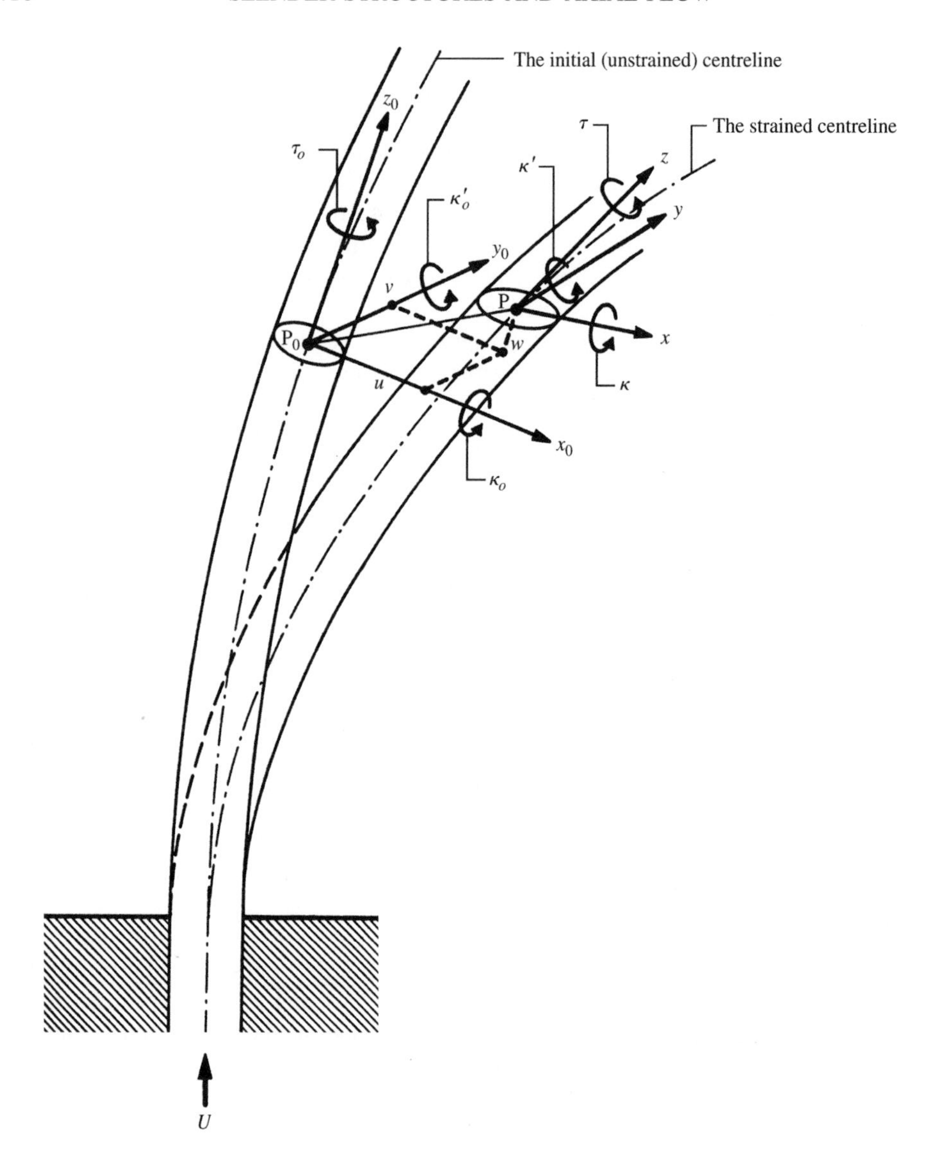

Figure 6.1 Kinematics of the curved pipe system; definition of coordinate systems and of coordinates used.

of the pipe, a particle on the pipe that was initially at P_0, occupies a new position P. The second, moving reference frame (x, y, z) has its origin at P and is attached to the deformed centreline in a manner analogous to the (x_0, y_0, z_0) system. The orientation of the (x, y, z) system is completely determined by the three Eulerian angles $\overline{\psi}, \overline{\theta}$ and $\overline{\phi}$. The orientation of the (x, y, z) system is completely defined by the sequence of small Eulerian rotations: $\overline{\psi}$ about the z_0-axis, $\overline{\theta}$ about the new x-axis, and finally $\overline{\phi}$ about the new z-axis. The components of $\overrightarrow{P_0P}$, referred to the (x_0, y_0, z_0) system, are denoted by u, v and w. Thus u and v are the transverse displacements, while w is the longitudinal displacement. Furthermore, during the motion of the pipe, a plane section at P_0 rotates around the centreline; the angle of rotation is denoted here by ψ and, since $\overline{\theta}$ is here assumed to be small (small deformation assumption), $\psi = \overline{\psi} + \overline{\phi}$. Hence, the deformed state of the pipe is determined by the generalized coordinates u, v, w and ψ.

The relative orientation of the two systems (x, y, z) and (x_0, y_0, z_0) is given by the following transformation (see Appendix J.1):

$$\begin{Bmatrix} x \\ y \\ z \end{Bmatrix} = \begin{bmatrix} 1 & \psi & -\left(\dfrac{\partial u}{\partial s} + \dfrac{w}{R_o}\right) \\ -\psi & 1 & -\dfrac{\partial v}{\partial s} \\ \left(\dfrac{\partial u}{\partial s} + \dfrac{w}{R_o}\right) & \dfrac{\partial v}{\partial s} & 1 \end{bmatrix} \begin{Bmatrix} x_0 \\ y_0 \\ z_0 \end{Bmatrix}, \tag{6.1}$$

where s is the curvilinear coordinate along the deformed centreline.

The components of curvature (around the x_0- and y_0-axes) and the twist (around the z_0-axis) in the undeformed state are given by

$$\kappa_o = 0, \qquad \kappa_o' = \frac{1}{R_o}, \qquad \tau_o = 0, \tag{6.2}$$

while those in the deformed state, assuming small deformations of the pipe, may be expressed in terms of the generalized coordinates as follows (see Appendix J.2):

$$\begin{aligned} \kappa &= \left[\frac{\psi}{R_o} - \frac{\partial^2 v}{\partial s^2}\right], \qquad \kappa' = \left[\frac{1}{R_o} + \frac{\partial^2 u}{\partial s^2} + \frac{1}{R_o}\frac{\partial w}{\partial s}\right], \\ \tau &= \left[\frac{\partial \psi}{\partial s} + \frac{1}{R_o}\frac{\partial v}{\partial s}\right]. \end{aligned} \tag{6.3}$$

To complete the kinematic development, the velocity and acceleration of the internal fluid is derived next. The fluid flow is assumed to be a plug flow (Chapter 3), the fluid being essentially an infinitely flexible rod travelling through the pipe. The radius of curvature is assumed to be very large compared with the pipe radius, and hence the effects of secondary flow are neglected.

The displacement vector of the deformed centreline, expressed in the inertial reference system that coincides with the system (x_0, y_0, z_0), is given by

$$\mathbf{r} = u\,\mathbf{e}_{x_0} + v\,\mathbf{e}_{y_0} + w\,\mathbf{e}_{z_0}. \tag{6.4}$$

By differentiating equation (6.4), the velocity and acceleration of the pipe may be written as

$$\mathbf{V}_p = \frac{\partial u}{\partial t}\,\mathbf{e}_{x_0} + \frac{\partial v}{\partial t}\,\mathbf{e}_{y_0} + \frac{\partial w}{\partial t}\,\mathbf{e}_{z_0}, \tag{6.5}$$

$$\mathbf{a}_p = \frac{\partial^2 u}{\partial t^2}\,\mathbf{e}_{x_0} + \frac{\partial^2 v}{\partial t^2}\,\mathbf{e}_{y_0} + \frac{\partial^2 w}{\partial t^2}\,\mathbf{e}_{z_0}. \tag{6.6}$$

The absolute velocity of the internal fluid is

$$\mathbf{V}_f = \mathbf{V}_p + U\mathbf{e}_z, \tag{6.7}$$

where $\mathbf{e}_z$ is the unit vector along the z-axis, which is tangential to the strained centreline. From equation (6.1), one can obtain

$$\mathbf{e}_z = \left(\frac{\partial u}{\partial s} + \frac{w}{R_o}\right)\mathbf{e}_{x0} + \frac{\partial v}{\partial s}\mathbf{e}_{y0} + \mathbf{e}_{z0}. \tag{6.8}$$

Combining equations (6.5), (6.7) and (6.8) yields

$$\mathbf{V}_f = \left[\frac{\partial u}{\partial t} + U\left(\frac{\partial u}{\partial s} + \frac{w}{R_o}\right)\right]\mathbf{e}_{x0} + \left[\frac{\partial v}{\partial t} + U\frac{\partial v}{\partial s}\right]\mathbf{e}_{y0} + \left[\frac{\partial w}{\partial t} + U\right]\mathbf{e}_{z0}. \tag{6.9}$$

To obtain the acceleration of the fluid, we differentiate $\mathbf{V}_f$, yielding

$$\mathbf{a}_f = \frac{\partial \mathbf{V}_f}{\partial t} + \left(\mathbf{V}_f \cdot \nabla\right)\mathbf{V}_f, \tag{6.10}$$

cf. equations (2.63) and (3.30). Substituting equation (6.7) into (6.10), the acceleration of the fluid may be rewritten as follows:

$$\mathbf{a}_f = \frac{\partial \mathbf{V}_f}{\partial t} + U\left(\mathbf{e}_z \cdot \nabla\right)\mathbf{V}_f + \left(\mathbf{V}_p \cdot \nabla\right)\mathbf{V}_f. \tag{6.11}$$

Now the last term on the right-hand side of this equation may be written as

$$\left(\mathbf{V}_p \cdot \nabla\right)\mathbf{V}_f = \left(\frac{\partial u}{\partial t}\frac{\partial}{\partial x_0} + \frac{\partial v}{\partial t}\frac{\partial}{\partial y_0} + \frac{\partial w}{\partial t}\frac{\partial}{\partial z_0}\right)\mathbf{V}_f; \tag{6.12}$$

by combining equations (6.9) and (6.12) we can see that this term is of higher order and can be neglected. In addition, it is noted that

$$\mathbf{e}_z \cdot \nabla = \frac{\partial}{\partial z} \qquad \text{and} \qquad \frac{\partial}{\partial z} = \frac{\partial}{\partial s}, \tag{6.13}$$

the latter because of the assumption of small motions. Hence, we obtain

$$\mathbf{a}_f = \frac{\partial \mathbf{V}_f}{\partial t} + U\frac{\partial \mathbf{V}_f}{\partial s}. \tag{6.14}$$

Substituting equation (6.9) into (6.14), we can write the fluid acceleration in the x_0-, y_0-, and z_0-directions (see Appendix J.3), as follows:

$$\begin{aligned}
a_{fx0} &= \frac{\partial^2 u}{\partial t^2} + 2U\left(\frac{\partial^2 u}{\partial t\,\partial s} + \frac{1}{R_o}\frac{\partial w}{\partial t}\right) + U^2\left(\frac{\partial^2 u}{\partial s^2} + \frac{1}{R_o}\frac{\partial w}{\partial s} + \frac{1}{R_o}\right),\\
a_{fy0} &= \frac{\partial^2 v}{\partial t^2} + 2U\frac{\partial^2 v}{\partial t\,\partial s} + U^2\frac{\partial^2 v}{\partial x^2},\\
a_{fz0} &= \frac{\partial^2 w}{\partial t^2} + U\left(\frac{\partial^2 w}{\partial t\,\partial s} - \frac{1}{R_o}\frac{\partial u}{\partial t}\right) - \frac{U^2}{R_o}\left(\frac{\partial u}{\partial s} + \frac{w}{R_o}\right).
\end{aligned} \tag{6.15}$$

6.2.2 The equations of motion

Consider an infinitesimal element of the pipe, contained between two cross-sections normal to the deformed centreline, and the forces and moments acting on it, as shown in Figure 6.2. As shown in Appendix J.4, balance of forces and moments along the x_0-, y_0- and z_0-directions yields

$$\begin{aligned}
&\frac{\partial}{\partial s}\mathcal{Q}_{x0} - \tau_o\mathcal{Q}_{y0} + \kappa_o'\mathcal{Q}_{z0} - c\,\frac{\partial u}{\partial t} + R_{x0} + G_{x0} = (m+M_a)a_{px0},\\
&\frac{\partial}{\partial s}\mathcal{Q}_{y0} - \kappa_o\mathcal{Q}_{z0} + \tau_o\mathcal{Q}_{x0} - c\,\frac{\partial v}{\partial t} + R_{y0} + G_{y0} = (m+M_a)a_{py0},\\
&\frac{\partial}{\partial s}\mathcal{Q}_{z0} - \kappa_o'\mathcal{Q}_{x0} + \kappa_o\mathcal{Q}_{y0} - c'\,\frac{\partial w}{\partial t} + R_{z0} + G_{z0} = (m+M_a')a_{pz0},\\
&\frac{\partial}{\partial s}\mathcal{M}_{x0} - \tau_o\mathcal{M}_{y0} + \kappa_o'\mathcal{M}_{z0} + \frac{\partial v}{\partial s}\mathcal{Q}_{z0} - \mathcal{Q}_{y0} = 0,\\
&\frac{\partial}{\partial s}\mathcal{M}_{y0} - \kappa_o\mathcal{M}_{z0} + \tau_o\mathcal{M}_{x0} + \mathcal{Q}_{x0} - \left(\frac{\partial u}{\partial s} + \frac{w}{R_o}\right)\mathcal{Q}_{z0} = 0,\\
&\frac{\partial}{\partial s}\mathcal{M}_{z0} - \kappa_o'\mathcal{M}_{x0} + \kappa_o\mathcal{M}_{y0} + \left(\frac{\partial u}{\partial s} + \frac{w}{R_o}\right)\mathcal{Q}_{y0} - \frac{\partial v}{\partial s}\mathcal{Q}_{x0} = I_z\,\frac{\partial^2\psi}{\partial t^2}.
\end{aligned} \tag{6.16}$$

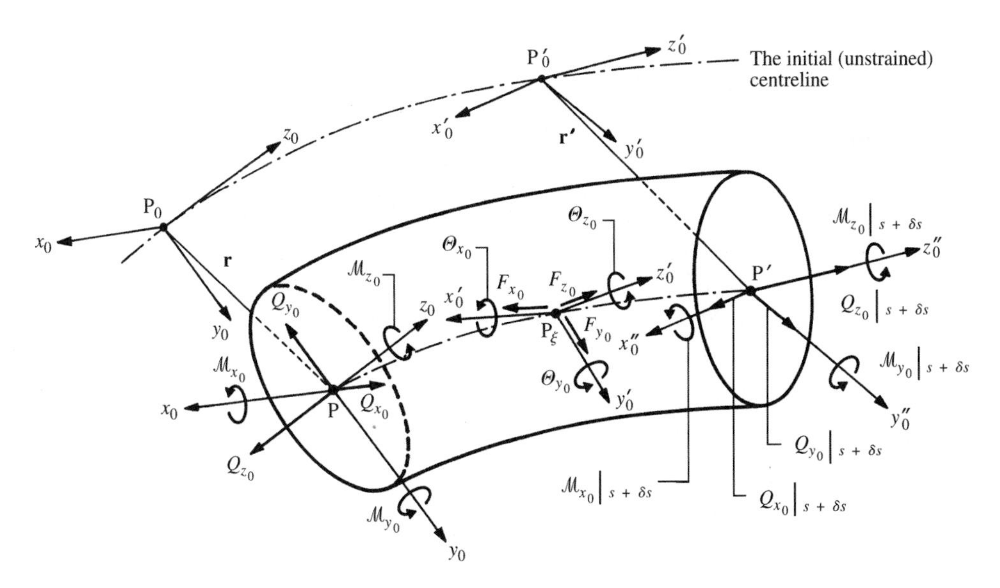

Figure 6.2 Forces and moments acting on a curved pipe element, expressed in the (x_0, y_0, z_0) reference frame.

Here $\mathcal{Q}_{x0}, \mathcal{Q}_{y0}, \mathcal{Q}_{z0}$ are components, along the axes (x_0, y_0, z_0) of the resultant of the transverse shear forces Q_x, Q_y, and of the 'combined force' Q_z^* arising from the axial force Q_z and the external pressure force $A_o p_e$, A_o being the outer cross-sectional area of the pipe and p_e the external pressure;[†] $\mathcal{M}_{x0}, \mathcal{M}_{y0}, \mathcal{M}_{z0}$ are components around the axes (x_0, y_0, z_0) of the resultant of the bending moments $\mathcal{M}_x$ and $\mathcal{M}_y$ and the twisting couple $\mathcal{M}_z$; M_a is the added mass per unit length, and c the coefficient of viscous damping due to the surrounding fluid, associated with transverse motions; M_a' and c' play similar roles for longitudinal motion; R_{x0}, R_{y0}, R_{z0}, are the components of the reaction force per unit length arising from the internal flow, and G_{x0}, G_{y0}, G_{z0} are the components of the effective gravity force, including buoyancy effects.

From the relation between the systems (x_0, y_0, z_0) and (x, y, z) given by equation (6.1), one can obtain

$$\begin{gathered} Q_{x0} = Q_x - \psi Q_y + \left(\frac{\partial u}{\partial s} + \frac{w}{R_o}\right) Q_z^*, \qquad Q_{y0} = Q_y + \psi Q_x + \frac{\partial v}{\partial s} Q_z^*, \\ Q_{z0} = Q_z^* - \left(\frac{\partial u}{\partial s} + \frac{w}{R_o}\right) Q_x - \frac{\partial v}{\partial s} Q_y, \qquad \mathcal{M}_{x0} = \mathcal{M}_x - \psi \mathcal{M}_y + \left(\frac{\partial u}{\partial s} + \frac{w}{R_o}\right) \mathcal{M}_z, \\ \mathcal{M}_{y0} = \mathcal{M}_y + \psi \mathcal{M}_x + \frac{\partial v}{\partial s} \mathcal{M}_z, \qquad \mathcal{M}_{z0} = \mathcal{M}_z - \left(\frac{\partial u}{\partial s} + \frac{w}{R_o}\right) \mathcal{M}_x - \frac{\partial v}{\partial s} \mathcal{M}_y, \end{gathered} \tag{6.17}$$

where

$$Q_z^* = Q_z + A_o p_e. \tag{6.18}$$

Substituting equations (6.17) and the values of κ_o, κ_o' and τ_o from (6.2) into (6.16), the equations of motion for the pipe may be written in the form

$$\begin{aligned} &\frac{\partial}{\partial s}\left[Q_x - \psi Q_y + \left(\frac{\partial u}{\partial s} + \frac{w}{R_o}\right) Q_z^*\right] + \frac{1}{R_o}\left[Q_z^* - \left(\frac{\partial u}{\partial s} + \frac{w}{R_o}\right) Q_x - \frac{\partial v}{\partial s} Q_y\right] - c\,\frac{\partial u}{\partial t} \\ &\quad + R_{x0} + G_{x0} - (m + M_a) a_{px0} = 0, \end{aligned} \tag{6.19}$$

$$\frac{\partial}{\partial s}\left[Q_y + \psi Q_x + \frac{\partial v}{\partial s} Q_z^*\right] - c\,\frac{\partial v}{\partial t} + R_{y0} + G_{y0} - (m + M_a) a_{py0} = 0, \tag{6.20}$$

$$\begin{aligned} &\frac{\partial}{\partial s}\left[Q_z^* - \left(\frac{\partial u}{\partial s} + \frac{w}{R_o}\right) Q_x - \frac{\partial v}{\partial s} Q_y\right] - \frac{1}{R_o}\left[Q_x - \psi Q_y + \left(\frac{\partial u}{\partial s} + \frac{w}{R_o}\right) Q_z^*\right] - c'\,\frac{\partial w}{\partial t} \\ &\quad + R_{z0} + G_{z0} - (m + M_a') a_{pz0} = 0, \end{aligned} \tag{6.21}$$

$$\begin{aligned} &\frac{\partial}{\partial s}\left[\mathcal{M}_x - \psi \mathcal{M}_y + \left(\frac{\partial u}{\partial s} + \frac{w}{R_o}\right) \mathcal{M}_z\right] + \frac{1}{R_o}\left[\mathcal{M}_z - \left(\frac{\partial u}{\partial s} + \frac{w}{R_o}\right) \mathcal{M}_x - \frac{\partial v}{\partial s} \mathcal{M}_y\right] \\ &\quad - Q_y - \psi Q_x = 0, \end{aligned} \tag{6.22}$$

$$\frac{\partial}{\partial s}\left[\mathcal{M}_y + \psi \mathcal{M}_x + \frac{\partial v}{\partial s} \mathcal{M}_z\right] + Q_x - \psi Q_y = 0, \tag{6.23}$$

$$\begin{aligned} &\frac{\partial}{\partial s}\left[\mathcal{M}_z - \left(\frac{\partial u}{\partial s} + \frac{w}{R_o}\right) \mathcal{M}_x - \frac{\partial v}{\partial s} \mathcal{M}_y\right] - \frac{1}{R_o}\left[\mathcal{M}_x - \psi \mathcal{M}_y + \left(\frac{\partial u}{\partial s} + \frac{w}{R_o}\right) \mathcal{M}_z\right] \\ &\quad + \left(\frac{\partial u}{\partial s} + \frac{w}{R_o}\right) Q_y - \frac{\partial v}{\partial s} Q_x - I_z \frac{\partial^2 \psi}{\partial t^2} = 0. \end{aligned} \tag{6.24}$$

[†]The pipe is assumed to be immersed in a fluid of generally nonuniform pressure (e.g. in a hydrostatic environment). Hence, p_e is not absorbed within the internal fluid pressure p_i simply by defining the latter as 'measured above the ambient pressure' as in Chapter 3.

Consider now an infinitesimal element of the fluid and forces acting on it, as shown in Figure 6.3. Let the internal cross-sectional area of the pipe be A_i, and the internal pressure $p_i(s)$. Components of the pressure force $-A_i p_i$ in the directions (x_0, y_0, z_0) can be written in terms of the components of the unit vector $\mathbf{e}_z$, referred to the reference frame (x_0, y_0, z_0), as follows — cf. equation (6.1):

$$P_{x_0} = -A_i p_i \left(\frac{\partial u}{\partial s} + \frac{w}{R_o} \right), \qquad P_{y_0} = -A_i p_i \frac{\partial v}{\partial s}, \qquad P_{z_0} = -A_i p_i. \tag{6.25}$$

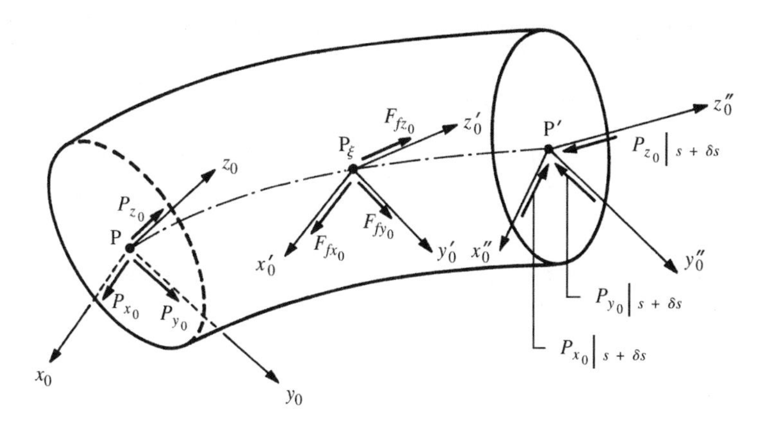

Figure 6.3 Forces acting on an element of the enclosed fluid, expressed in the reference system (x_0, y_0, z_0).

By comparing Figure 6.2 with 6.3, and the forces $(P_{x_0}, P_{y_0}, P_{z_0})$ with $(Q_{x_0}, Q_{y_0}, Q_{z_0})$, and then using the first three of equations (6.16), the equations of motion of the fluid can be written in the form

$$\begin{aligned}
&\frac{\partial P_{x_0}}{\partial s} - \tau_o P_{y_0} + \kappa_o' P_{z_0} - R_{x_0} + G_{fx_0} = M a_{fx_0},\\
&\frac{\partial P_{y_0}}{\partial s} - \kappa_o P_{z_0} + \tau_o P_{x_0} - R_{y_0} + G_{fy_0} = M a_{fy_0},\\
&\frac{\partial P_{z_0}}{\partial s} - \kappa_o' P_{x_0} + \kappa_o P_{y_0} - R_{z_0} + G_{fy_0} = M a_{fz_0},
\end{aligned} \tag{6.26}$$

where $G_{fx_0}, G_{fy_0}, G_{fz_0}$ are components of the gravity force per unit length on the fluid.

Combination of equations (6.25) and (6.26) yields

$$\frac{\partial}{\partial s}\left[-A_i p_i \left(\frac{\partial u}{\partial s} + \frac{w}{R_o} \right) \right] - \frac{1}{R_o} A_i p_i - R_{x_0} + G_{fx_0} - M a_{fx_0} = 0, \tag{6.27}$$

$$\frac{\partial}{\partial s}\left[-A_i p_i \frac{\partial v}{\partial s} \right] - R_{y_0} + G_{fy_0} - M a_{fy_0} = 0, \tag{6.28}$$

$$\frac{\partial}{\partial s}\left[-A_i p_i \right] + \frac{1}{R_o} A_i p_i \left(\frac{\partial u}{\partial s} + \frac{w}{R_o} \right) - R_{z_0} + G_{fz_0} - M a_{fz_0} = 0. \tag{6.29}$$

According to the generalization of the Euler–Bernoulli beam theory, the stress couples $\mathcal{M}_x$, $\mathcal{M}_y$, $\mathcal{M}_z$ in the beam when bent and twisted from the state expressed by $\kappa_o, \kappa'_o, \tau_o$ to that expressed by κ, κ', τ^* are given by

$$\mathcal{M}_x = EI(\kappa - \kappa_o), \qquad \mathcal{M}_y = EI(\kappa' - \kappa'_o), \qquad \mathcal{M}_z = GJ(\tau^* - \tau_o). \tag{6.30}$$

In the discretization to be introduced later, the pipe is divided into a series of constant curvature elements; i.e. within a given element,

$$\frac{\partial R_o}{\partial s} = 0, \tag{6.31}$$

where the number of elements required will depend on the shape of the pipe centreline and the accuracy desired.

Combining equations (6.2), (6.3) and (6.30) yields

$$\mathcal{M}_x = EI\left(\frac{\psi}{R_o} - \frac{\partial^2 v}{\partial s^2}\right), \qquad \mathcal{M}_y = EI\left(\frac{\partial^2 u}{\partial s^2} + \frac{1}{R_o}\frac{\partial w}{\partial s}\right), \qquad \mathcal{M}_z = GJ\left(\frac{\partial \psi}{\partial s} + \frac{1}{R_o}\frac{\partial v}{\partial s}\right). \tag{6.32}$$

Then, substituting equations (6.32) into (6.22) and (6.23), and neglecting the higher-order terms, one obtains

$$Q_x = -EI\left(\frac{\partial^3 u}{\partial s^3} + \frac{1}{R_o}\frac{\partial^2 w}{\partial s^2}\right), \tag{6.33}$$

$$Q_y = EI\left(\frac{1}{R_o}\frac{\partial \psi}{\partial s} - \frac{\partial^3 v}{\partial s^3}\right) + \frac{GJ}{R_o}\left(\frac{\partial \psi}{\partial s} + \frac{1}{R_o}\frac{\partial v}{\partial s}\right). \tag{6.34}$$

By adding equations (6.19), (6.20) and (6.21) to (6.27), (6.28) and (6.29), respectively, one may obtain the equations of motion of the system, which no longer depend on the reaction forces R_{x0}, R_{y0} and R_{z0} between the pipe and the fluid. Then, utilizing equations (6.6), (6.15) and (6.32)–(6.34) and neglecting higher order terms, the governing equations of motion for the dynamical system may be obtained, namely:

$$\begin{aligned} &EI\left(\frac{\partial^4 u}{\partial s^4} + \frac{1}{R_o}\frac{\partial^3 w}{\partial s^3}\right) + \frac{\partial}{\partial s}\left[(A_i p_i - A_o p_e - Q_z)\left(\frac{\partial u}{\partial s} + \frac{w}{R_o}\right)\right] \\ &+ \frac{1}{R_o}(A_i p_i - A_o p_e - Q_z) - (G_{x0} + G_{fx0}) \\ &+ MU^2\left(\frac{\partial^2 u}{\partial s^2} + \frac{1}{R_o}\frac{\partial w}{\partial s} + \frac{1}{R_o}\right) + 2MU\left(\frac{\partial^2 u}{\partial t \partial s} + \frac{1}{R_o}\frac{\partial w}{\partial t}\right) + c\,\frac{\partial u}{\partial t} \\ &+ (m + M + M_a)\frac{\partial^2 u}{\partial t^2} = 0, \end{aligned} \tag{6.35}$$

$$\begin{aligned} &EI\left(\frac{\partial^4 v}{\partial s^4} - \frac{1}{R_o}\frac{\partial^2 \psi}{\partial s^2}\right) - \frac{GJ}{R_o}\left(\frac{\partial^2 \psi}{\partial s^2} + \frac{1}{R_o}\frac{\partial^2 v}{\partial s^2}\right) + \frac{\partial}{\partial s}\left[(A_i p_i - A_o p_e - Q_z)\frac{\partial v}{\partial s}\right] \\ &- (G_{y0} + G_{fy0}) + MU^2\frac{\partial^2 v}{\partial s^2} + 2MU\frac{\partial^2 v}{\partial t \partial s} + c\,\frac{\partial v}{\partial t} + (m + M + M_a)\frac{\partial^2 v}{\partial t^2} = 0, \end{aligned} \tag{6.36}$$

$$\frac{EI}{R_o}\left(\frac{\partial^3 u}{\partial s^3}+\frac{1}{R_o}\frac{\partial^2 w}{\partial s^2}\right)-\frac{\partial}{\partial s}(A_i p_i - A_o p_e - Q_z)$$
$$+\frac{1}{R_o}(A_i p_i - A_o p_e - Q_z)\left(\frac{\partial u}{\partial s}+\frac{w}{R_o}\right)+(G_{z0}+G_{fz0})+\frac{MU^2}{R_o}\left(\frac{\partial u}{\partial s}+\frac{w}{R_o}\right)$$
$$-MU\left(\frac{\partial^2 w}{\partial t\partial s}-\frac{1}{R_o}\frac{\partial u}{\partial t}\right)-c'\frac{\partial w}{\partial t}-(m+M+M_a')\frac{\partial^2 w}{\partial t^2}=0, \tag{6.37}$$
$$-GJ\left(\frac{\partial^2 \psi}{\partial s^2}+\frac{1}{R_o}\frac{\partial^2 v}{\partial s^2}\right)+\frac{EI}{R_o}\left(\frac{\psi}{R_o}-\frac{\partial^2 v}{\partial s^2}\right)+I_z\frac{\partial^2 \psi}{\partial t^2}=0. \tag{6.38}$$

The gravity terms in equations (6.35)–(6.38) are given by

$$G_{x0}+G_{fx0}=(m+M-A_o\rho_e)g\alpha_{x0}, \tag{6.39}$$

and similar ones for the y_0 and z_0 components; α_{x0}, α_{y0} and α_{z0} are the direction cosines of the gravity vector with respect to (x_0, y_0, z_0), and ρ_e is the density of the external fluid.

These equations are, of course, coupled but, similarly to the shell equations (Chapter 7), may each be identified as being principally related to motion in one particular direction; thus, the first is related to in-plane deformations, the second to out-of-plane deformations, the third to deformations along the pipe, and the last to twist of the pipe. Hence, in-plane motions are governed by equations (6.35) and (6.37), and out-of-plane motions by equations (6.36) and (6.38). Note that, if the radius of curvature R_o is made equal to infinity, the curved pipe becomes a straight pipe. Moreover, if the pipe is vertical and the axial motion is ignored, equations (6.35)–(6.38) reduce to

$$EI\frac{\partial^4 u}{\partial s^4}+M\left(\frac{\partial}{\partial t}+U\frac{\partial}{\partial s}\right)^2 u+(m+M_a)\frac{\partial^2 u}{\partial t^2}$$
$$+c\frac{\partial u}{\partial t}+\frac{\partial}{\partial s}\left[(A_i p_i - A_o p_e - Q_z)\frac{\partial u}{\partial s}\right]=0, \tag{6.40}$$
$$EI\frac{\partial^4 v}{\partial s^4}+M\left(\frac{\partial}{\partial t}+U\frac{\partial}{\partial s}\right)^2 v+(m+M_a)\frac{\partial^2 v}{\partial t^2}$$
$$+c\frac{\partial v}{\partial t}+\frac{\partial}{\partial s}\left[(A_i p_i - A_o p_e - Q_z)\frac{\partial v}{\partial s}\right]=0, \tag{6.41}$$
$$\frac{\partial}{\partial s}(Q_z - A_i p_i + A_o p_e)+(m+M)g=0, \tag{6.42}$$

which, as may be verified, are identical to those in Sections 4.2 and 4.3 for the motions of a uniform straight pipe conveying fluid and fully submerged in a quiescent fluid. Equation (6.40) is identical to (6.41) because, for a straight pipe, motions in the x_0- and y_0-directions are uncoupled and identical. Finally, if the surrounding fluid has negligible effect on the dynamics of the system, setting $M_a = 0$, $c = 0$, and $p_e = 0$ *vis-à-vis* the atmospheric pressure, these equations reduce to a version of equation (3.34).

6.2.3 The boundary conditions

The boundary conditions are as follows.

(i) *If an end is clamped,*

$$u = 0, \qquad \frac{\partial u}{\partial s} = 0, \qquad w = 0, \qquad \text{for in-plane motion;} \tag{6.43a}$$

$$v = 0, \qquad \frac{\partial v}{\partial s} = 0, \qquad \psi = 0, \qquad \text{for out-of-plane motion.} \tag{6.43b}$$

(ii) *If an end is pinned,*

$$u = 0, \qquad w = 0, \qquad \mathcal{M}_y = EI\left[\frac{\partial^2 u}{\partial s^2} + \frac{1}{R_o}\frac{\partial w}{\partial s}\right] = 0, \qquad \text{for in-plane motion;} \tag{6.44a}$$

$$v = 0, \qquad \mathcal{M}_x = EI\left[\frac{\psi}{R_o} - \frac{\partial^2 v}{\partial s^2}\right] = 0, \qquad \mathcal{M}_z = GJ\left[\frac{\partial \psi}{\partial s} + \frac{1}{R_o}\frac{\partial v}{\partial s}\right] = 0, \qquad \text{for out-of-plane motion.} \tag{6.44b}$$

(iii) *If an end is free,*

$$Q_x = -EI\left[\frac{\partial^3 u}{\partial s^3} + \frac{1}{R_o}\frac{\partial^2 w}{\partial s^2}\right] = 0, \qquad Q_z = 0,$$

$$\mathcal{M}_y = EI\left[\frac{\partial^2 u}{\partial s^2} + \frac{1}{R_o}\frac{\partial w}{\partial s}\right] = 0, \qquad \text{for in-plane motion;} \tag{6.45a}$$

$$Q_y = EI\left[\frac{1}{R_o}\frac{\partial \psi}{\partial s} - \frac{\partial^3 v}{\partial s^3}\right] + \frac{GJ}{R_o}\left[\frac{\partial \psi}{\partial s} + \frac{1}{R_o}\frac{\partial v}{\partial s}\right] = 0,$$

$$\mathcal{M}_x = EI\left[\frac{\psi}{R_o} - \frac{\partial^2 v}{\partial s^2}\right] = 0,$$

$$\mathcal{M}_z = GJ\left[\frac{\partial \psi}{\partial s} + \frac{1}{R_o}\frac{\partial v}{\partial s}\right] = 0, \qquad \text{for out-of-plane motion.} \tag{6.45b}$$

6.2.4 Nondimensional equations

It is convenient to analyse the equations of motion in dimensionless form by defining the following quantities:

$$\eta_1 = \frac{u}{L}, \qquad \eta_2 = \frac{v}{L}, \qquad \eta_3 = \frac{w}{L}, \qquad \zeta = \frac{s}{L}, \qquad \bar{t} = t\left(\frac{EI}{(M+m)L^4}\right)^{1/2},$$

$$\bar{u} = UL\left(\frac{M}{EI}\right)^{1/2}, \qquad \beta_a = \frac{M_a}{M+m}, \qquad \overline{\beta}_a = \frac{M'_a}{M+m},$$

$$\beta = \frac{M}{M+m}, \qquad \Lambda = \frac{GJ}{EI}, \qquad \sigma = \frac{I_z}{(M+m)L^2}, \qquad \mathscr{A} = \frac{A_p L^2}{I},^{\dagger} \tag{6.46}$$

$$\Theta = \frac{L}{R_o}, \qquad \gamma = (M + m - A_o\rho_e)\frac{gL^3}{EI}, \qquad \Pi_p = (A_i p_i - A_o p_e)\frac{L^2}{EI},$$

$$\Pi = (A_i p_i - A_o p_e - Q_z)\frac{L^2}{EI}, \qquad \Delta = \frac{cL^2}{[EI(M+m)]^{1/2}}, \qquad \overline{\Delta} = \frac{\Delta c'}{c}.$$

Substitution of equations (6.46) into (6.35)–(6.38) yields the following dimensionless equations of motion:

$$\begin{aligned}&\left[(\eta_1^{\text{iv}} + \Theta\eta_3''') + \{\Pi(\eta_1' + \Theta\eta_3)\}' + \Theta\Pi - \gamma\alpha_{x0} + \bar{u}^2(\eta_1'' + \Theta\eta_3' + \Theta)\right]\\ &\qquad + \left[2\beta^{1/2}\bar{u}(\dot{\eta}_1' + \Theta\dot{\eta}_3) + \Delta\dot{\eta}_1 + (1+\beta_a)\ddot{\eta}_1\right] = 0,\end{aligned} \tag{6.47}$$

$$\begin{aligned}&\left[(\eta_2^{\text{iv}} - \Theta\psi'') - \Lambda\Theta(\psi'' + \Theta\eta_2'') + \{\Pi\eta_2'\}' - \gamma\alpha_{y0} + \bar{u}^2\eta_2''\right]\\ &\qquad + \left[2\beta^{1/2}\bar{u}\dot{\eta}_2' + \Delta\dot{\eta}_2 + (1+\beta_a)\ddot{\eta}_2\right] = 0,\end{aligned} \tag{6.48}$$

$$\begin{aligned}&\left[-\Pi' + \Theta(\eta_1''' + \Theta\eta_3'') + \Theta(\eta_1' + \Theta\eta_3)\Pi + \gamma\alpha_{z0} + \bar{u}^2\Theta(\eta_1' + \Theta\eta_3)\right]\\ &\qquad - \left[\beta^{1/2}\bar{u}(\dot{\eta}_3' - \Theta\dot{\eta}_1) + \overline{\Delta}\dot{\eta}_3 + (1+\overline{\beta}_a)\ddot{\eta}_3\right] = 0,\end{aligned} \tag{6.49}$$

$$\left[\Theta(\Theta\psi - \eta_2'') - \Lambda(\psi'' + \Theta\eta_2'')\right] + \left[\sigma\ddot{\psi}\right] = 0, \tag{6.50}$$

where prime and dot denote differentiation with respect to ξ and $\bar{t}$, respectively, $(\ )^{\text{iv}} \equiv (\ )''''$, and α_{x0} etc. have been defined in connection with (6.39).

Each of the displacements and twist η_1, η_2, η_3 and ψ can be imagined to consist of two parts: a steady (static) part, and a perturbation about the steady part, i.e.

$$\eta_1 = \eta_1^o + \eta_1^*, \qquad \eta_2 = \eta_2^o + \eta_2^*, \qquad \eta_3 = \eta_3^o + \eta_3^*, \qquad \psi = \psi^o + \psi^*, \tag{6.51}$$

the superscripts o and $*$ denoting the steady and perturbed parts, respectively. The equations governing the static equilibrium are obtained by deleting the time-dependent terms, i.e. terms contained within the second set of square brackets in each of equations (6.47)–(6.50). If η_1^o and η_3^o are eliminated from the first and third of the resulting set of equations, one obtains

$$\Pi^{o''} + \Theta^2\Pi^o = \Theta\gamma\alpha_{x0} + \gamma\alpha_{z0}' - \Theta^2\bar{u}^2, \tag{6.52}$$

which governs the steady (static) value, Π^o, of the so-called 'combined force' Π. The nomenclature of 'combined force' conveys that it involves both axial tension and pressure forces. It is noted that Π^o depends on the dimensionless fluid flow velocity $\bar{u}$, in addition to the gravity loading and the orientation of the pipe.

Similarly, the dimensionless boundary conditions are given as follows:

---

† A_p arising via J, the torsional area-moment of inertia, is the ring-shaped pipe material cross-sectional area, equal to $A_o - A_i$.

(i) *at a clamped end,*

$$\eta_1 = \eta_1' = \eta_3 = 0, \qquad \text{for in-plane motion;} \tag{6.53}$$

$$\eta_2 = \eta_2' = \psi = 0, \qquad \text{for out-of-plane motion;} \tag{6.54}$$

(ii) *at a pinned end,*

$$\eta_1 = \eta_3 = \eta_1'' + \Theta\eta_3' = 0, \qquad \text{for in-plane motion;} \tag{6.55}$$

$$\eta_2 = \Theta\psi - \eta_2' = \psi' + \Theta\eta_2' = 0, \qquad \text{for out-of-plane motion;} \tag{6.56}$$

(iii) *at a free end,*

$$\eta_1''' + \Theta\eta_3'' = \Pi - \Pi_p = \eta_1'' + \Theta\eta_3' = 0, \quad \text{for in-plane motion;} \tag{6.57}$$

$$\Theta\psi' - \eta_2''' + \Lambda\Theta(\psi' + \Theta\eta_2') = \Theta\psi - \eta_2'' = 0,$$
$$\psi' + \Theta\eta_2' = 0, \qquad \text{for out-of-plane motion.} \tag{6.58}$$

6.2.5 Equations of motion of an inextensible pipe

The formulation carried out so far is valid for both extensible and inextensible pipes. Here, the equations will be simplified by considering the centreline of the pipe to be inextensible — leaving aside for now the question of whether this is justified. The centreline strain is given by equation (J.2), in which $\kappa_o' = 1/R_o$ as in equations (6.2), or in dimensionless terms,

$$\varepsilon = \eta_3' - \Theta\eta_1 = 0, \tag{6.59a}$$

and hence, in accordance with (6.51),

$$\eta_3^{o\prime} - \Theta\eta_1^o = 0, \qquad \eta_3^{*\prime} - \Theta\eta_1^* = 0. \tag{6.59b}$$

Subtracting the steady-state part from equations (6.47) and (6.49) and utilizing (6.59b), one obtains a single sixth-order partial differential equation for η_3^* that governs in-plane motion, namely,

$$\begin{aligned}\left(\eta_3^{*\text{vi}} + 2\Theta^2\eta_3^{*\text{vi}} + \Theta^4\eta_3^{*\prime\prime}\right) &+ \bar{u}^2\left(\eta_3^{*\text{iv}} + 2\Theta^2\eta_3^{*\prime\prime} + \Theta^4\eta_3^*\right) + 2\beta^{1/2}\bar{u}\left(\dot{\eta}_3^{*\prime\prime\prime} + \Theta^2\dot{\eta}_3^{*\prime}\right)\\ &+ (1+\beta_a)\ddot{\eta}_3^{*\prime\prime} - \Theta^2(1+\overline{\beta}_a)\ddot{\eta}_3^* + \Delta\dot{\eta}_3^{*\prime\prime} - \Theta^2\overline{\Delta}\dot{\eta}_3^*\\ &+ \left[\Pi^o(\eta_3^{*\prime\prime} + \Theta^2\eta_3^*)\right]'' + \Theta^2\Pi^o(\eta_3^{*\prime\prime} + \Theta^2\eta_3^*) = 0,\end{aligned} \tag{6.60}$$

where the superscript vi denotes six primes, and so on. It is of interest to note that when the combined force is ignored and external fluid effects are absent ($\beta_a = \overline{\beta}_a = \Delta = \overline{\Delta} = 0$), equation (6.60) reduces to that obtained by Chen (1972c).

Similarly, the equations governing the out-of-plane motion are obtained from equations (6.48) and (6.50) as

$$\begin{aligned}\left(\eta_2^{*\text{iv}} - \Theta\psi^{*\prime\prime}\right) - \Lambda\Theta\left(\psi^{*\prime\prime} + \Theta\eta_2^{*\prime\prime}\right) + \left[\Pi^o\eta_2^{*\prime}\right]' + \bar{u}^2\eta_2^{*\prime\prime} + (1+\beta_a)\ddot{\eta}_2^*\\ + 2\beta^{1/2}\bar{u}\dot{\eta}_2^{*\prime} + \Delta\dot{\eta}_2^* = 0,\end{aligned} \tag{6.61}$$

$$\Theta\left(\Theta\psi^* - \eta_2^{*\prime\prime}\right) - \Lambda\left(\psi^{*\prime\prime} + \Theta\eta_2^{*\prime\prime}\right) + \sigma\ddot{\psi}^* = 0. \tag{6.62}$$

Once again, if the combined force Π^o is suppressed and external fluid effects are absent, the equations of motion are identical to Chen's (1973). It may also be noted that, once the steady combined force Π^o is obtained from equation (6.52), the in-plane and out-of-plane motions can be studied separately by analysing equation (6.60) and equations (6.61)–(6.62), respectively.

The equations obtained here apply to both the conventional and the modified inextensible theories, items (i) and (iii) of the classification in the latter part of Section 6.1. The difference is that, in the conventional inextensible theory, the steady force Π^o is taken to be zero, while in the modified theory, Π^o is fully taken into account.

6.2.6 Equations of motion of an extensible pipe

Unlike the inextensible case, here the longitudinal strain $\varepsilon \neq 0$. Thus, u and w, or equivalently η_1 and η_3, are not directly related to each other in the present case. Furthermore, the axial force Q_z is given by

$$Q_z = EA_p\varepsilon = EA_p\left(\frac{\partial w}{\partial s} - \frac{u}{R_o}\right); \tag{6.63}$$

hence, using equations (6.18) and (6.46), the dimensionless pressure-tension combined force Π may be expressed as

$$\Pi = \Pi_p - \mathscr{A}(\eta_3' - \Theta\eta_1). \tag{6.64}$$

Once η_1 and η_3 are known, Π can be determined from equation (6.64). It may be noted that this procedure can be used only for the extensible case, since the last term vanishes if the pipe is inextensible.

As in the inextensible case, each of the displacements and twist η_1, η_2, η_3, and ψ consists of a steady (or static) part and a perturbation about the static part, as per equations (6.51). The equations governing the steady part are obtained from equations (6.47)–(6.50) by deleting the time-dependent terms and are given by

$$(\eta_1^{o\mathrm{iv}} + \Theta\eta_3^{o'''}) + \left[\Pi^o(\eta_1^{o'} + \Theta\eta_3^o)\right]' + \Theta\Pi^o - \gamma\alpha_{x_0} + \bar{u}^2(\eta_1^{o''} + \Theta\eta_3^{o'} + \Theta) = 0, \tag{6.65}$$

$$(\eta_2^{o\mathrm{iv}} - \Theta\psi^{o''}) - \Lambda\Theta(\psi^{o''} + \Theta\eta_2^{o''}) + (\Pi^o\eta_2^{o'})' - \gamma\alpha_{y_0} + \bar{u}^2\eta_2^{o''} = 0, \tag{6.66}$$

$$-\Pi^{o'} + \Theta(\eta_1^{o'''} + \Theta\eta_3^{o''}) + \Theta\Pi^o(\eta_1^{o'} + \Theta\eta_3^o) + \gamma\alpha_{z_0} + \bar{u}^2\Theta(\eta_1^o + \Theta\eta_3^o) = 0, \tag{6.67}$$

$$\Theta(\Theta\psi^o - \eta_2^{o''}) - \Lambda(\psi^{o''} + \Theta\eta_2^{o''}) = 0. \tag{6.68}$$

It is noted that equations (6.65) and (6.67) governing the static in-plane displacements are decoupled from the equations of out-of-plane static equilibrium (6.66) and (6.68). Furthermore, if the gravity effect is negligible (i.e. $\gamma = 0$), or if the pipe initially lies in a vertical plane (i.e. $\alpha_{y_0} = 0$), then the static out-of-plane deformations η_2^o and ψ^o vanish. All the cases considered here satisfy this requirement and η_2^o and ψ^o are therefore always zero.

Using equation (6.64), the in-plane static equilibrium equations (6.65) and (6.67) may be rewritten after linearization as

$$\begin{aligned}(\eta_1^{o\mathrm{iv}} + \Theta\eta_3^{o'''}) + \left[\Pi_p(\eta_1^{o'} + \Theta\eta_3^o)\right]' - \mathscr{A}\Theta(\eta_3^{o'} - \Theta\eta_1^o) + \bar{u}^2(\eta_1^{o''} + \Theta\eta_3^{o'})\\ + \Theta(\Pi_p + \bar{u}^2) - \gamma\alpha_{x_0} = 0,\end{aligned} \tag{6.69}$$

$$- \mathscr{A}(\eta_3^{o''} - \Theta\eta_1^{o'}) - \Theta(\eta_1^{o'''} + \Theta\eta_3^{o''}) - \Theta\Pi_p(\eta_1^{o'} + \Theta\eta_3^{o}) - \Theta\overline{u}^2(\eta_1^{o'} + \Theta\eta_3^{o})$$
$$+ \Pi_p' - \gamma\alpha_{z0} = 0. \tag{6.70}$$

Once the in-plane static displacements have been evaluated from equations (6.69) and (6.70), the steady combined force Π^o can be determined from equation (6.64). In the derivation of equations (6.69) and (6.70), it has been assumed that the static displacements are small. This assumption is validated later by examining the results.

Substituting equations (6.51) into the equations of motion (6.47)–(6.50), subtracting the static equilibrium equations (6.65)–(6.68), using equation (6.64) and neglecting second- and higher-order perturbation terms, one obtains

$$\left(\eta_1^{*\text{iv}} + \Theta\eta_3^{*'''}\right) + \left[\Pi^o\left(\eta_1^{*'} + \Theta\eta_3^{*}\right)\right]' - \mathscr{A}\left(\eta_1^{o'} + \Theta\eta_3^{o}\right)\left(\eta_3^{*'} - \Theta\eta_1^{*}\right)'$$
$$- \Theta\mathscr{A}\left(\eta_3^{*'} - \Theta\eta_1^{*}\right) + \overline{u}^2\left(\eta_1^{*''} + \Theta\eta_3^{*'}\right) + 2\beta^{1/2}\overline{u}\left(\dot{\eta}_1^{*'} + \Theta\dot{\eta}_3^{*}\right)$$
$$+ \Delta\dot{\eta}_1^{*} + (1 + \beta_a)\ddot{\eta}_1^{*} = 0, \tag{6.71}$$

$$\left(\eta_2^{*\text{iv}} - \Theta\psi^{*''}\right) - \Theta\Lambda(\psi^{*''} + \Theta\eta_2^{*''}) + \left[\Pi^o\eta_2^{*'}\right]' - \mathscr{A}\left[\eta_2^{o'}(\eta_3^{*'} - \Theta\eta_1^{*})\right]'$$
$$+ \overline{u}^2\eta_2^{*''} + 2\beta^{1/2}\overline{u}\dot{\eta}_2^{*'} + \Delta\dot{\eta}_2^{*} + (1 + \beta_a)\ddot{\eta}_2^{*} = 0, \tag{6.72}$$

$$\Theta(\eta_1^{*'''} + \Theta\eta_3^{*''}) + \Theta\Pi^o(\eta_1^{*'} + \Theta\eta_3^{*}) - \Theta\mathscr{A}(\eta_1^{o'} + \Theta\eta_3^{o})(\eta_3^{*'} - \Theta\eta_1^{*})$$
$$+ \mathscr{A}(\eta_3^{*''} - \Theta\eta_1^{*'}) + \Theta\overline{u}^2(\eta_1^{*} + \Theta\eta_3^{*}) - \beta^{1/2}\overline{u}(\dot{\eta}_3^{*'} - \Theta\dot{\eta}_1^{*})$$
$$- \overline{\Delta}\dot{\eta}_3^{*} - (1 + \overline{\beta}_a)\ddot{\eta}_3^{*} = 0, \tag{6.73}$$

$$\Theta\left(\Theta\psi^{*} - \eta_2^{*''}\right) - \Lambda\left(\psi^{*''} + \Theta\eta_2^{*''}\right) + \sigma\ddot{\psi}^{*} = 0, \tag{6.74}$$

where equations (6.71) and (6.73) govern in-plane motion, while equations (6.72) and (6.74) govern out-of-plane motion. Similarly to the case of the static equilibrium equations, the in-plane and out-of-plane perturbation equations may be solved separately; however, the out-of-plane perturbations depend on the in-plane static displacements through Π^o.

It may be noticed that the out-of-plane perturbation equations (6.72) and (6.74) in the extensible case are identical to those for an inextensible pipe, equations (6.61) and (6.62), provided that $\eta_2^o = 0$ and that the steady pressure-tension effects are taken into account. As established earlier in this section, the former condition is satisfied if the pipe initially lies in a vertical plane or if the gravity effects are negligible. Since all the cases considered here satisfy these conditions, out-of-plane motion need not be analysed anew,† and from here on only the analysis of in-plane motion of an extensible pipe needs be considered further.

6.3 FINITE ELEMENT ANALYSIS

Solutions of the equations of motion are obtained by the finite element method. For this problem, this method is preferable to Galerkin's which has been used in most of the foregoing, because of its versatility: once formulated, it can just as easily be applied to

†Although the perturbation equations are identical for inextensible and extensible pipes, the steady combined force Π^o is slightly different in the two cases. This, however, does not change the dynamical behaviour significantly.

pipes of uniform curvature as to S- or Ω-shaped pipes — without having to determine the equivalent of new comparison functions for each case. According to this method, the continuum is subdivided into a number of elements, at the edges of which there are one or more nodes (one for a structure such as this, in which spatial variations involve only one coordinate, ζ). The equations of motion are satisfied at the nodes rigorously and elsewhere approximately via interpolation functions. A variational formulation ensures a systematic minimization of the error for the number of elements utilized. The use of the finite element method has now become routine, and hence the uninitiated reader is referred to one of several texts on the subject, e.g. Zienkiewicz & Cheung (1968), Desai & Abel (1972), Becker *et al*. (1984) or Zienkiewicz & Taylor (1989).

The details of the particular form of finite element analysis employed here may be found in Van (1986). However, different forms may be employed, and this is partly the reason for not presenting the minutiae of the analysis.

6.3.1 Analysis for inextensible pipes

(a) In-plane motion

The pipe is discretized into n elements. The initial curvature of a particular element is constant, although it can vary from element to element. The variational statement used for the finite element discretization is

$$\sum_{j=1}^{n} \int_0^{\zeta_j} \delta\eta_3^* A_{\mathrm{i}}(\eta_3^*)\, \mathrm{d}\zeta = 0, \tag{6.75}$$

where $\delta\eta_3^*$ is an arbitrary variational displacement, $A_{\mathrm{i}}(\eta_3^*)$ represents the left-hand side of equation (6.60) and ζ_j is the length of the jth element. The subscript i in A_{i} stands for 'in-plane'.

The longitudinal displacement η_3^* may be expressed in the form

$$\eta_3^* = [N_3]\{q_{\mathrm{i}}\}^{\mathrm{e}}, \tag{6.76}$$

where $[N_3]$ is a matrix of interpolation functions at the space coordinate ζ, and $\{q_{\mathrm{i}}\}^{\mathrm{e}}$ is the element displacement vector, of appropriate dimension and dependent only on time; the superscript e stands for 'element'. Hence, by using (6.75) and (6.76), integrating by parts, and applying the boundary conditions — thereby eliminating the integrated-out components — one obtains the discretized equation

$$[M_{\mathrm{i}}]^{\mathrm{e}}\{\ddot{q}_{\mathrm{i}}\}^{\mathrm{e}} + [D_{\mathrm{i}}]^{\mathrm{e}}\{\dot{q}_{\mathrm{i}}\}^{\mathrm{e}} + [K_{\mathrm{i}}]^{\mathrm{e}}\{q_{\mathrm{i}}\}^{\mathrm{e}} = \{0\}, \tag{6.77}$$

where

$$[M_{\mathrm{i}}]^{\mathrm{e}} = \int_0^{\zeta_e} \{(1+\beta_a)[N_3]'^{\mathrm{T}}[N_3]' + \Theta^2(1+\overline{\beta}_a)[N_3]^{\mathrm{T}}[N_3]\}\, \mathrm{d}\zeta,$$

$$[D_{\mathrm{i}}]^{\mathrm{e}} = 2\beta^{1/2}\overline{u}\int_0^{\zeta_e} \{[N_3]'^{\mathrm{T}}([N_3]'' + \Theta^2[N_3]) + \Delta[N_3]'^{\mathrm{T}}[N_3]' + \Theta^2\overline{\Delta}[N_3]^{\mathrm{T}}[N_3]\}\, \mathrm{d}\zeta,$$

$$[K_{\rm i}]^{\rm e} = \int_0^{\zeta_e} \Big\{([N_3]''' + \Theta^2[N_3]')^{\rm T}([N_3]''' + \Theta^2[N_3]') \qquad (6.78)$$
$$+ \bar{u}^2 \left\{[N_3]'^{\rm T}([N_3]''' + \Theta^2[N_3]') - \Theta^2[N_3]^{\rm T}([N_3]'' + \Theta^2[N_3])\right\}$$
$$+ [N_3]'^{\rm T} \frac{\partial}{\partial\zeta}[\Pi_o([N_3]'' + \Theta^2[N_3])] - \Theta^2\Pi_o[N_3]^{\rm T}([N_3]'' + \Theta^2[N_3])\Big\}\, {\rm d}\zeta;$$

ζ_e is the length of the element under consideration. The highest order derivative of $[N_3]$ appearing in these expressions is the third. Hence, it is necessary to ensure that η_3^*, $\eta_3^{*\prime}$ and $\eta_3^{*\prime\prime}$ be continuous between elements, which is achieved if the nodal displacements at each node are taken as values of η_3^*, $\eta_3^{*\prime}$ and $\eta_3^{*\prime\prime}$. Thus, for an element of the pipe (Figure 6.4) with a node j at one end and a node $j+1$ at the other,

$$\{q_{\rm i}\}_j = \{\eta_{3,j}^*, \eta_{3,j}^{*\prime}, \eta_{3,j}^{*\prime\prime}\}^{\rm T} \qquad (6.79)$$

and

$$\{q_{\rm i}\}^{\rm e} = \{\{q_{\rm i}\}_j, \{q_{\rm i}\}_{j+1}\}^{\rm T} = \{\eta_{3,j}^*, \eta_{3,j}^{*\prime}, \eta_{3,j}^{*\prime\prime}; \eta_{3,j+1}^*, \eta_{3,j+1}^{*\prime}, \eta_{3,j+1}^{*\prime\prime}\}^{\rm T}. \qquad (6.80)$$

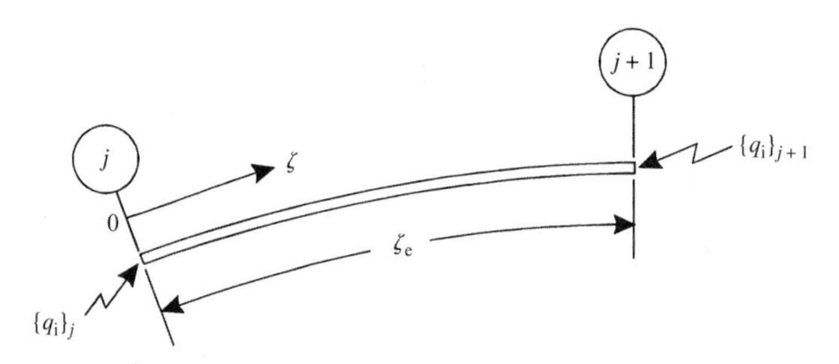

Figure 6.4 Diagram of an element of the pipe with nodes j and $j+1$ at its extremities. For in-plane motion, for example, the nodal displacement vectors are $\{q_{\rm i}\}_j$ and $\{q_{\rm i}\}_{j+1}$, and the element displacement vector is $\{\{q_{\rm i}\}_j, \{q_{\rm i}\}_{j+1}\}^{\rm T}$ — see equations (6.79) and (6.80).

As each element has six degrees of freedom corresponding to the six elements of the vector in (6.80), one can express the deflection by a fifth-order polynomial,

$$\eta_3^* = \alpha_1 + \alpha_2\zeta + \alpha_3\zeta^2 + \alpha_4\zeta^3 + \alpha_5\zeta^4 + \alpha_6\zeta^5, \qquad (6.81)$$

where the α_i are a set of generalized coordinates. Equation (6.81) may be rewritten as

$$\eta_3^* = [\phi_3]\{\alpha\}, \qquad (6.82)$$

where $[\phi_3] = [1, \zeta, \zeta^2, \zeta^3, \zeta^4, \zeta^5]$ is a row-vector, and $\{\alpha\} = \{\alpha_1, \ldots, \alpha_6\}^{\rm T}$. Now, in view of equations (6.80)–(6.82), one can write

$$\{q_{\rm i}\}^{\rm e} = [A_{\rm i}]\{\alpha\}, \qquad (6.83)$$

where

$$[A_i] = \begin{Bmatrix} [\phi_3]_j \\ [\phi_3]'_j \\ [\phi_3]''_j \\ [\phi_3]_{j+1} \\ [\phi_3]'_{j+1} \\ [\phi_3]''_{j+1} \end{Bmatrix} = \begin{bmatrix} 1 & 0 & 0 & 0 & 0 & 0 \\ 0 & 1 & 0 & 0 & 0 & 0 \\ 0 & 0 & 2 & 0 & 0 & 0 \\ 1 & \zeta_e & \zeta_e^2 & \zeta_e^3 & \zeta_e^4 & \zeta_e^5 \\ 0 & 1 & 2\zeta_e & 3\zeta_e^2 & 4\zeta_e^3 & 5\zeta_e^4 \\ 0 & 0 & 2 & 6\zeta_e & 12\zeta_e^2 & 20\zeta_e^3 \end{bmatrix}, \tag{6.84}$$

after substituting $\zeta = 0$ for the jth node and $\zeta = \zeta_e$ for the $j+1$ node.

From equations (6.76), (6.82) and (6.83), one obtains

$$[N_3] = [\phi_3][A_i]^{-1} \qquad \text{and} \qquad \eta_3^* = [\phi_3][A_i]^{-1}\{q_i\}^e. \tag{6.85}$$

Substituting equation (6.85) into (6.78) yields the final form of the element mass, damping and stiffness matrices:

$$\begin{aligned}
[M_i]^e &= ([A_i]^{-1})^T \left[(1+\beta_a)[J_2] + \Theta^2(1+\overline{\beta}_a)[J_1]\right][A_i]^{-1}, \\
[D_i]^e &= ([A_i]^{-1})^T \left[2\beta^{1/2}\overline{u}([J_5] + \Theta^2[J_4]) + \Delta[J_2] + \Theta^2\overline{\Delta}[J_1]\right][A_i]^{-1}, \\
[K_i]^e &= ([A_i]^{-1})^T \left[[J_3] + \Theta^2([J_6] + [J_6]^T) + \Theta^4[J_2]\right. \\
&\quad + \overline{u}^2\{[J_6] + \Theta^2([J_2] - [J_7]) - \Theta^4[J_1]\} \\
&\quad + a_1\{[J_6] + \Theta^2([J_2] - [J_7]) - \Theta^4[J_1]\} + b_1([J_5] + \Theta^2[J_4]) \\
&\quad \left. + a_2\{[J_8] + \Theta^2([J_{10}] - [J_{11}]) - \Theta^4[J_9] + b_2([J_{13}] + \Theta^2[J_{12}])\}\right][A_i]^{-1},
\end{aligned} \tag{6.86}$$

where $[J_1]$ to $[J_{13}]$ are given by

$$\begin{aligned}
[J_1] &= \int_0^{\zeta_e} [\phi_3]^T[\phi_3]\,d\zeta, \quad [J_2] = \int_0^{\zeta_e} [\phi_3]'^T[\phi_3]'\,d\zeta, \quad [J_3] = \int_0^{\zeta_e} [\phi_3]''^T[\phi_3]''\,d\zeta, \\
[J_4] &= \int_0^{\zeta_e} [\phi_3]'^T[\phi_3]\,d\zeta, \quad [J_5] = \int_0^{\zeta_e} [\phi_3]'^T[\phi_3]''\,d\zeta, \quad [J_6] = \int_0^{\zeta_e} [\phi_3]'^T[\phi_3]'''\,d\zeta, \\
[J_7] &= \int_0^{\zeta_e} [\phi_3]^T[\phi_3]''\,d\zeta, \quad [J_8] = \int_0^{\zeta_e} \zeta[\phi_3]'^T[\phi_3]'''\,d\zeta, \quad [J_9] = \int_0^{\zeta_e} \zeta[\phi_3]^T[\phi_3]\,d\zeta, \\
[J_{10}] &= \int_0^{\zeta_e} \zeta[\phi_3]'^T[\phi_3]'\,d\zeta, \quad [J_{11}] = \int_0^{\zeta_e} \zeta[\phi_3]^T[\phi_3]''\,d\zeta, \quad [J_{12}] = \int_0^{\zeta_e} \zeta[\phi_3]'^T[\phi_3]\,d\zeta, \\
[J_{13}] &= \int_0^{\zeta_e} \zeta[\phi_3]'^T[\phi_3]''\,d\zeta.
\end{aligned}$$

The coefficients a_1, b_1, a_2, b_2 are associated with Π^o and its derivative $\Pi^{o\prime}$. These are generally nonlinear functions of ζ, but within each element they are approximated by the linear expressions

$$\Pi^o = a_1 + a_2\zeta, \qquad \Pi^{o\prime} = b_1 + b_2\zeta; \tag{6.87}$$

hence, $a_1 = \Pi^o|_j$, $a_2 = (\Pi^o|_{j+1} - \Pi^o|_j)/\zeta_e$, and similarly for b_1 and b_2.

All the foregoing applies to a single element. The next step is to assemble the global equation of motion, which is similar to (6.77) but the associated vector now covers all the nodes, $j = 1, 2, \ldots, n$; in the corresponding global matrices $[M_{\mathrm{i}}]$, $[C_{\mathrm{i}}]$ and $[K_{\mathrm{i}}]$ there is partial superposition of the element matrices (6.86), since any node, except those at the two ends of the pipe, is shared by two elements. This global equation is then converted to a standard eigenvalue problem, from which the eigenfrequencies may be determined and stability assessed.

(b) Out-of-plane motion

The variational statement used for the finite element model of out-of-plane motion is

$$\sum_{j=1}^{n} \int_0^{\zeta_j} \{\delta\eta_2^* A_{\mathrm{o}1}(\eta_2^*, \psi^*) + \delta\psi^* A_{\mathrm{o}2}(\eta_2^*, \psi^*)\}\, \mathrm{d}\zeta = 0, \tag{6.88}$$

where $\delta\eta_2^*$ and $\delta\psi^*$ are the variations in the out-of-plane transverse displacement and twist, respectively, while $A_{\mathrm{o}1}(\eta_2^*, \psi^*)$ and $A_{\mathrm{o}2}(\eta_2^*, \psi^*)$ represent the left-hand sides of equations (6.61) and (6.62). The subscript o stands for out-of-plane.

The solutions for η_2^* and ψ^* are sought in the form

$$\eta_2^* = [N_2]\{q_{\mathrm{o}}\}^{\mathrm{e}}, \qquad \psi^* = [N_4]\{q_{\mathrm{o}}\}^{\mathrm{e}}, \tag{6.89}$$

where $[N_2]$ and $[N_4]$ are two matrices of interpolation functions of the space coordinates ζ, and $\{q_{\mathrm{o}}\}^{\mathrm{e}}$ is the element-displacement vector for the out-of-plane motion. In this case, a cubic interpolation model for η_2^* and a linear one for ψ^* can guarantee convergence. Hence, proceeding as for in-plane motion, one eventually obtains

$$[M_{\mathrm{o}}]^{\mathrm{e}}\{\ddot{q}_{\mathrm{o}}\}^{\mathrm{e}} + [D_{\mathrm{o}}]^{\mathrm{e}}\{\dot{q}_{\mathrm{o}}\}^{\mathrm{e}} + [K_{\mathrm{o}}]^{\mathrm{e}}\{q_{\mathrm{o}}\}^{\mathrm{e}} = \{0\}, \tag{6.90}$$

where

$$\begin{aligned}
[M_{\mathrm{o}}]^{\mathrm{e}} &= ([A_{\mathrm{o}}]^{-1})^{\mathrm{T}} \left[(1+\beta_a)[I_1] + \sigma[I_4]\right][A_{\mathrm{o}}]^{-1}, \\
[D_{\mathrm{o}}]^{\mathrm{e}} &= ([A_{\mathrm{o}}]^{-1})^{\mathrm{T}} \left[2\beta^{1/2}\bar{u}[I_5] + \Delta[I_1]\right][A_{\mathrm{o}}]^{-1}, \\
[K_{\mathrm{o}}]^{\mathrm{e}} &= ([A_{\mathrm{o}}]^{-1})^{\mathrm{T}} \left[[I_3] - \Theta([I_6] + [I_6]^{\mathrm{T}}) + \Theta^2[I_4] + \bar{u}^2[I_9] + \Lambda[\Theta^2[I_2]\right. \\
&\quad \left. + \Theta([I_7] + [I_7]^{\mathrm{T}}) + [I_8]] + a_1[I_9] + b_1[I_5] + a_2[I_{10}] + b_2[I_{11}]\right][A_{\mathrm{o}}]^{-1},
\end{aligned} \tag{6.91}$$

in which

$$[A_{\mathrm{o}}] = \begin{bmatrix} 1 & 0 & 0 & 0 & 0 & 0 \\ 0 & 1 & 0 & 0 & 0 & 0 \\ 0 & 0 & 0 & 0 & 1 & 0 \\ 1 & \zeta_{\mathrm{e}} & \zeta_{\mathrm{e}}^2 & \zeta_{\mathrm{e}}^3 & 0 & 0 \\ 0 & 1 & 2\zeta_{\mathrm{e}} & 3\zeta_{\mathrm{e}}^2 & 0 & 0 \\ 0 & 0 & 0 & 0 & 1 & \zeta_{\mathrm{e}} \end{bmatrix}, \tag{6.92}$$

and $[I_1]$–$[I_{11}]$ are given by

$$
\begin{aligned}
[I_1] &= \int_0^{\zeta_e} [\phi_2]^{\mathrm{T}}[\phi_2]\,\mathrm{d}\zeta, & [I_2] &= \int_0^{\zeta_e} [\phi_2]'^{\mathrm{T}}[\phi_2]\,\mathrm{d}\zeta, & [I_3] &= \int_0^{\zeta_e} [\phi_2]''^{\mathrm{T}}[\phi_2]''\,\mathrm{d}\zeta,\\
[I_4] &= \int_0^{\zeta_e} [\phi_4]^{\mathrm{T}}[\phi_4]\,\mathrm{d}\zeta, & [I_5] &= \int_0^{\zeta_e} [\phi_2]^{\mathrm{T}}[\phi_2]'\,\mathrm{d}\zeta, & [I_6] &= \int_0^{\zeta_e} [\phi_2]''^{\mathrm{T}}[\phi_4]\,\mathrm{d}\zeta,\\
[I_7] &= \int_0^{\zeta_e} [\phi_2]'^{\mathrm{T}}[\phi_4]'\,\mathrm{d}\zeta\,, & [I_8] &= \int_0^{\zeta_e} [\phi_4]'^{\mathrm{T}}[\phi_4]'\,\mathrm{d}\zeta\,, & [I_9] &= \int_0^{\zeta_e} [\phi_2]^{\mathrm{T}}[\phi_2]''\,\mathrm{d}\zeta,\\
[I_{10}] &= \int_0^{\zeta_e} \zeta[\phi_2]^{\mathrm{T}}[\phi_2]''\,\mathrm{d}\zeta, & [I_{11}] &= \int_0^{\zeta_e} \zeta[\phi_2]^{\mathrm{T}}[\phi_2]'\,\mathrm{d}\zeta,
\end{aligned}
\tag{6.93}
$$

while a_1, b_1, etc. are defined in equations (6.87).

(c) Calculation of Π^o

In the discretization procedure described earlier, the pipe is divided into a series of constant-curvature elements, each of which may be treated as an incomplete circular pipe. Here, the steady combined force Π^o acting on an incomplete circular pipe subtending an angle Θ and conveying fluid at a constant nondimensional velocity $\bar{u}$ is determined.

If the gravity effect is neglected, the solution to equation (6.52) may be written as

$$\Pi^o(\zeta) = C_1 \sin(\zeta\Theta + C_2) - \bar{u}^2, \tag{6.94}$$

where C_1 and C_2 are two constants of integration which can be determined from the boundary conditions. The boundary conditions, in turn, are determined from equilibrium considerations and the application of Castigliano's theorem. They can be shown to be

$$\Pi^o(1) = \Pi_p(1), \qquad \Pi^{o\prime}(1) = -\left(\frac{\Theta - \sin\Theta}{\Theta}\right)[\bar{u}^2 + \Pi_p(1)]^2 \tag{6.95}$$

for a clamped–free incomplete circular pipe, and

$$\Pi^o(1) = -\bar{u}^2, \qquad \Pi^{o\prime}(1) = 0 \tag{6.96}$$

for a clamped–clamped, clamped–pinned or pinned–pinned incomplete circular pipe. In equations (6.95), $\Pi_p = (A_i p_i - A_e p_e)L^2/EI$ and represents the steady-state nondimensional force due to the pressures of the internal and external fluids.

Using the boundary values (6.95) and (6.96) in equation (6.94) one obtains

$$\Pi^o(\zeta) = [\Pi_p(1) + \bar{u}^2]\left(\frac{\sin(\zeta\Theta + C_2)}{\sin(\Theta + C_2)}\right) - \bar{u}^2, \tag{6.97}$$

where

$$C_2 = -\tan^{-1}\left\{\frac{\Theta^2}{(\Theta - \sin\Theta)\,[\Pi_p(1) + \bar{u}^2]}\right\} - \Theta, \tag{6.98}$$

for a clamped–free pipe and

$$\Pi^o = -\bar{u}^2, \tag{6.99}$$

if both ends of the pipe are supported (clamped or pinned).

6.3.2 Analysis for extensible pipes

Because the static displacements appear explicitly in the equations of motion, the static equilibrium is determined first, and then the stability of motions about this position.

(a) Determination of the static equilibrium

In order to discretize the equations governing the static deformations, the following variational statement is utilized

$$\sum_{j=1}^{n}\int_0^{\zeta_j}\{\delta\eta_1^o A_{i1}^o(\eta_1^o,\eta_3^o)+\delta\eta_3^o A_{i3}^o(\eta_1^o,\eta_3^o)\}\,\mathrm{d}\zeta=0, \tag{6.100}$$

where $\delta\eta_1^o$ and $\delta\eta_3^o$ are the variations in the steady-state displacements η_1^o and η_3^o, while $A_{i1}^o(\eta_1^o,\eta_3^o)$ and $A_{i3}^o(\eta_1^o,\eta_3^o)$ represent the left-hand sides of equations (6.69) and (6.70), respectively; n and ζ_j are as in the foregoing.

The solutions for η_1^o and η_3^o are sought in the form

$$\eta_1^o=[N_{1e}]\{q_i^o\}^{\mathrm{e}},\qquad \eta_3^o=[N_{3e}]\{q_i^o\}^{\mathrm{e}}, \tag{6.101}$$

where $[N_{1e}]$ and $[N_{3e}]$ are two matrices of interpolation functions of the space coordinate ζ, and $\{q_i^o\}^{\mathrm{e}}$ is the element in-plane displacement vector. It may be shown that a cubic interpolation model for η_1^o and linear interpolation for η_3^o can guarantee convergence of the finite element scheme. Thus, one can proceed in the same manner as for the *out-of-plane motion in the inextensible case* to obtain a matrix equation governing the static equilibrium of an element as follows:[†]

$$[K_i^o]^{\mathrm{e}}\{q_i^o\}^{\mathrm{e}}=\{F_i^o\}^{\mathrm{e}}, \tag{6.102}$$

where

$$\begin{aligned}
[K_i^o]^{\mathrm{e}}=([A_o]^{-1})^{\mathrm{T}}\big[&\{[I_3]+\Theta([I_{15}]+[I_{15}]^{\mathrm{T}})+\Theta^2[I_8]\}\\
&+\mathscr{A}\{[I_8]-\Theta([I_{12}]+[I_{12}]^{\mathrm{T}})+\Theta^2[I_1]\}\\
&+(\Pi_{p0}+\bar{u}^2)\{[I_9]+\Theta([I_{22}]-[I_{13}])-\Theta^2[I_4]\}\\
&-\bar{\lambda}\bar{u}^2\{[I_5]+[I_{10}]+\Theta([I_{16}]+[I_{14}]-[I_{17}])-\Theta^2[I_{18}]\}\big]\,[A_o]^{-1},\\
\{F_i^o\}^{\mathrm{e}}=([A_o]^{-1})^{\mathrm{T}}\big[&\Theta(\Pi_{p0}+\bar{u}^2)\{F_1\}-\bar{\lambda}\bar{u}^2(\Theta\{F_2\}+\{F_3\})+\{F_4\}\big].
\end{aligned} \tag{6.103}$$

For the integrals $[I_1]$, $\{F_1\}$, etc. and $[A_o]$, see Appendix K.

In deriving equation (6.102) it has been assumed that the pressure in the external fluid is constant, while the internal pressure varies linearly along the centreline. Thus

$$\Pi_p=\Pi_p\big|_0-\bar{\lambda}\bar{u}^2\zeta, \tag{6.104}$$

[†]It is recalled that, for an inextensible pipe, in-plane motion involves sixth-order derivatives in ζ, and hence the corresponding shape functions are not useful here. However, the shape functions for *out-of-plane motions* of the inextensible pipe may be used for this analysis, since the maximum orders of partial derivatives match; hence the appearance of $[A_o]$ in equations (6.103).

where

$$\bar{\lambda} = \lambda(L/2D_i), \tag{6.105}$$

L and D_i being the length and internal diameter of the curved pipe, while λ is the frictional resistance coefficient for turbulent flow in a curved pipe. The resistance coefficient λ for a curved pipe is somewhat larger than that for a straight pipe (λ_o), and according to Schlichting (1960) is given by

$$\lambda = \lambda_o[1 + 0.075\,\mathrm{Re}^{0.25}(D_i/2R_o)^{0.5}], \tag{6.106}$$

where $\mathrm{Re} = UD_i/\nu$ is the Reynolds number, and R_o is the radius of curvature of the pipe segment.

Equation (6.103) corresponds to a single finite element. Such equations for all the elements are assembled to form the global equation of static equilibrium, which is then solved numerically.

(b) Analysis of motion around the static equilibrium

Similarly to the analysis of the static equilibrium equations, the variational statement used for the finite element model of the in-plane perturbations is

$$\sum_{j=1}^{n} \int_0^{\zeta_j} \{\delta\eta_1^* A_{i1}^*(\eta_1^*, \eta_3^*) + \delta\eta_3^* A_{i3}^*(\eta_1^*, \eta_3^*)\}\, \mathrm{d}\zeta = 0, \tag{6.107}$$

where $\delta\eta_1^*$ and $\delta\eta_3^*$ are the variations in the dimensionless in-plane displacement perturbations, while $A_{i1}^*(\eta_1^*, \eta_3^*)$ and $A_{i3}^*(\eta_1^*, \eta_3^*)$ represent the left-hand sides of equations (6.71) and (6.73), respectively; n and ζ_j have the same meaning as in the foregoing.

Proceeding as before, one obtains the matrix differential equation governing the motion of a typical element; the associated matrices are given in Appendix K. Again, the equations for all the finite elements are assembled to form the global equation of motion, which is then converted into an eigenvalue problem that is solved numerically.

6.4 CURVED PIPES WITH SUPPORTED ENDS

Solutions of the global equation of motion yields the system eigenfrequencies, on the basis of which stability also is decided. For convenience of comparison with other results, two forms of nondimensionalization are used for the circular frequency Ω:

$$\omega = \left(\frac{m+M}{EI}\right)^{1/2} \Omega L^2 \qquad \text{and} \qquad \omega^* = \omega\left(\frac{R_o}{L}\right)^2 = \frac{\omega}{\Theta^2}. \tag{6.108a}$$

Similarly, either $\bar{u}$, defined in equations (6.46), or

$$\bar{u}^* = \frac{\bar{u}}{\Theta} \tag{6.108b}$$

is used for the dimensionless flow velocity. In the results to be presented, unless otherwise specified, the values of γ, β_a, $\overline{\beta}_a$, Δ and $\overline{\Delta}$ are zero. Dissipation in the material of the pipe is ignored for simplicity; hence, by the same reasoning as for straight pipes with supported ends, the system is conservative (Section 3.2). Therefore, the eigenfrequencies are wholly real so long as the system remains stable, unless dissipation is taken into account (or the fluid conveyed is viscous).

The dynamics of the system as predicted by inextensible, extensible and modified inextensible theory will be discussed in this section; the same is done in Sections 6.5 and 6.6 for other boundary conditions. It is recalled that in the *conventional inextensible theory* not only is extensibility of the centreline neglected, but also the effect of the steady forces arising from the centrifugal and pressure forces generated by the internal fluid as it moves along the pipe; i.e. Π^o which comprises the axial force Q_z and the pressure force Π_p is neglected. In the *modified inextensible theory* these forces are taken into account. In the *extensible theory*, the extension of the centreline is also taken into account.

6.4.1 Conventional inextensible theory

Typical results for *in-plane motion* of a clamped–clamped semi-circular pipe conveying fluid are presented in Figure 6.5, showing the evolution of the lowest eigenfrequencies of

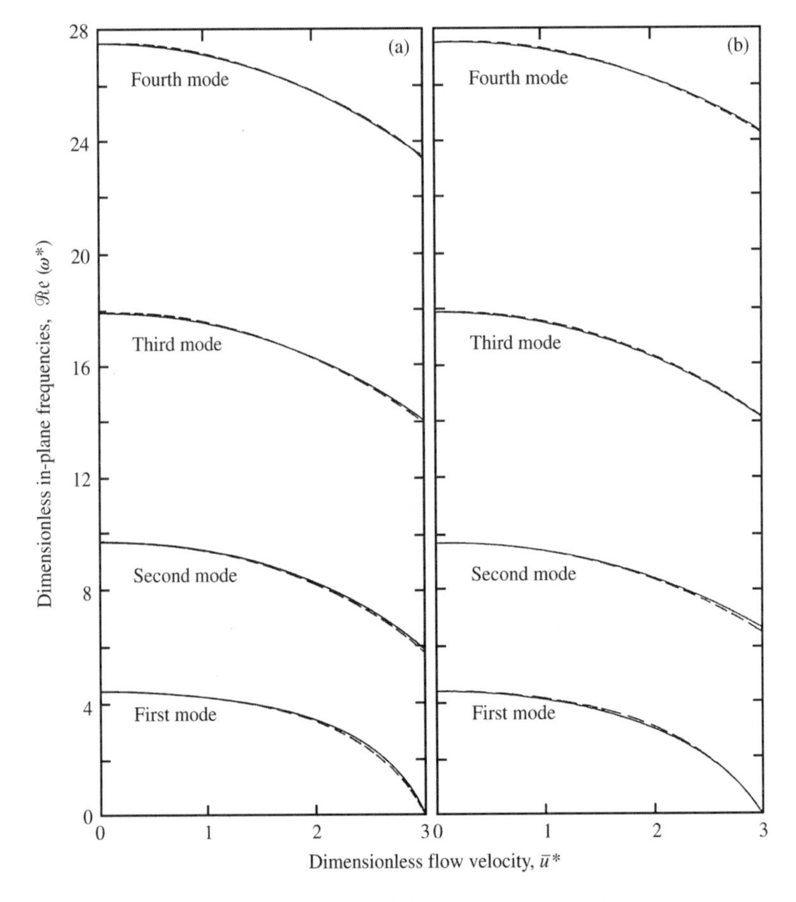

Figure 6.5 Dimensionless eigenfrequencies ω^* by conventional inextensible theory for *in-plane motion* of a clamped–clamped semicircular pipe conveying fluid as functions of the dimensionless flow velocity $\bar{u}^*$, for $\Pi = 0$ and (a) $\beta = 0$; (b) $\beta = 0.5$. - - - Chen (1972a); ——, Misra *et al*. (1988a).

the system with increasing $\bar{u}$. Eight elements are adequate to obtain convergence in the finite element scheme. At $\bar{u} = 0$ the pipe behaves as a semi-circular ring (cf. Archer 1960; Ojalvo 1962; Ojalvo & Newman 1965; Blevins 1979). As the flow velocity increases, the eigenfrequencies become smaller according to this theory, and if the flow velocity exceeds a certain value, the pipe becomes unstable by divergence in the first mode. With further increase in the flow velocity, instability may occur in the higher modes, as well as coupled-mode flutter (not shown). The results are qualitatively similar to those for a straight pipe. It is noted that the finite element results obtained with the present analysis agree very well with those obtained analytically by Chen (1972b). The same is also true for clamped–pinned and pinned–pinned semi-circular pipes (Van 1986).

Similarly to the case of in-plane motion, Figure 6.6 shows the eigenfrequencies for *out-of-plane motion* of a clamped–clamped semi-circular pipe conveying fluid. To obtain convergence, 11 or more finite elements are required, as opposed to eight in the in-plane case; this is because the displacement model is cubic for the out-of-plane motion, whereas it is quintic for the in-plane motion. According to this theory, as the flow velocity increases, the frequencies become smaller for out-of-plane motions as well, and the pipe becomes unstable by divergence in the first mode when a critical flow velocity is exceeded. One may note that the out-of-plane eigenfrequencies are lower than the in-plane ones and

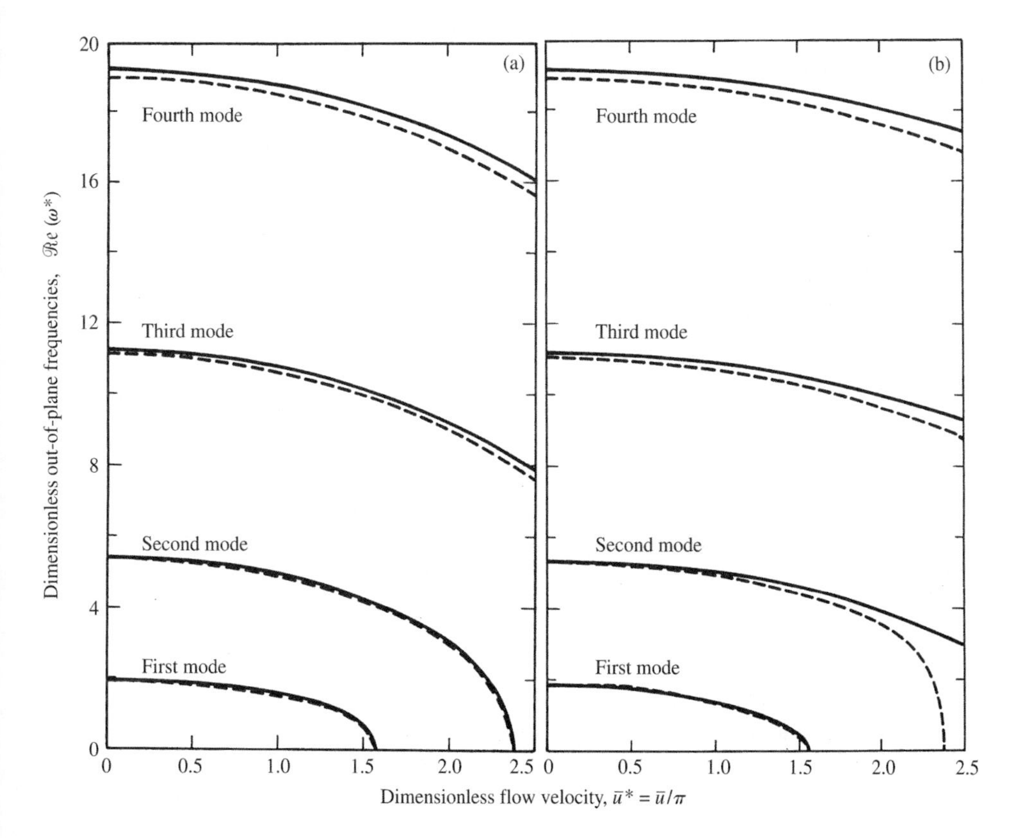

Figure 6.6 Dimensionless eigenfrequencies ω^* versus $\bar{u}^*$ by conventional inextensible theory for *out-of-plane motion* of a clamped–clamped semi-circular pipe conveying fluid, for $\Pi = 0$, $\Lambda = 0.769$ and (a) $\beta = 0$; (b) $\beta = 0.5$. – – –, Chen (1973); ——, Misra *et al*. (1988a).

that the critical flow velocity is also lower, reflecting the relative stiffness in the two directions. These results are also very close to Chen's (1973), except for the second mode in one case†.

Before closing this discussion, it is remarked that, throughout this chapter, the modes are numbered sequentially, strictly in ascending order of frequency, irrespective of whether they are asymmetric or symmetric. For in-plane motions of a semi-circular pipe, modes 1–4 in Figure 6.5 correspond respectively to the modes in Figure 6.7(a–d), i.e. the modes are numbered in ascending order of the number of nodes. Similarly, for out-of-plane motions: the first mode would have no nodes, the second mode a node at mid-point, and so on.

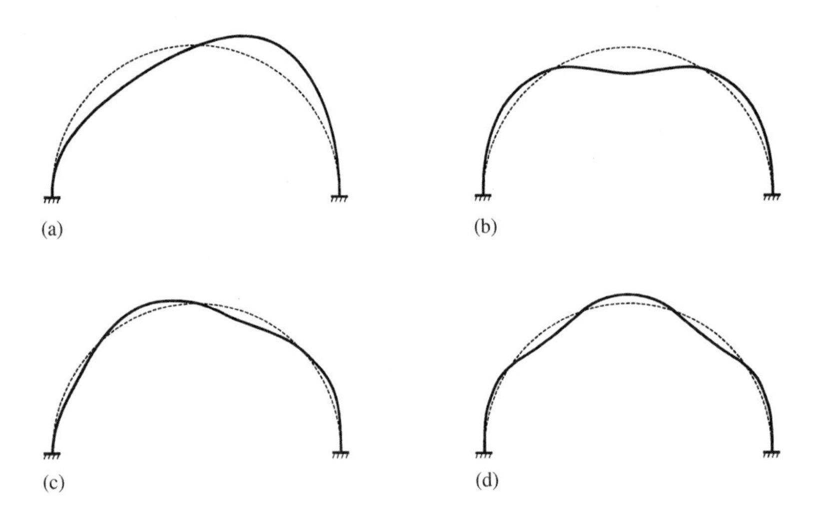

Figure 6.7 Schematics of (a,c) the asymmetric and (b,d) symmetric modes for in-plane motions of an inextensible semi-circular pipe at $\overline{u} = 0$, and approximately for an extensible one.

6.4.2 Extensible theory

As in the previous case, a study of convergence was conducted, to determine what a reasonable number of finite elements would be for accurate computation of the eigenfrequencies. Some results are presented in Figure 6.8 for $\overline{u} = 0$ and various values of $\mathscr{A}$. It may be seen that convergence is very slow, and that it is affected by the slenderness parameter $\mathscr{A}$ (i.e. $A_p L^2/I$); convergence for the third mode is even slower (Misra *et al.* 1988b). For a small number of elements (10 or so), the results for different values of $\mathscr{A}$ are very different. For a larger number of elements (40 or so), the results are comparable. In the curved beam theory used in this work, it has been assumed that the length of the pipe is large in comparison with its radius. This implies that $\mathscr{A}$ must be large; however, calculations with large $\mathscr{A}$ result in high computational cost. Therefore, a value of $\mathscr{A}$ that provides a reasonable trade-off between cost and accuracy has been used in the calculations to be presented, namely $\mathscr{A} = 10^4$.

†In this regard, it is noted that the value of $\overline{u}^*$ at which $\omega^* = 0$ should be independent of β, as is the case in Chen's results but not in those of Misra *et al.* — either due to a plotting error in the latter or, more likely, because of the use of an insufficient number of finite elements to ensure adequate accuracy.

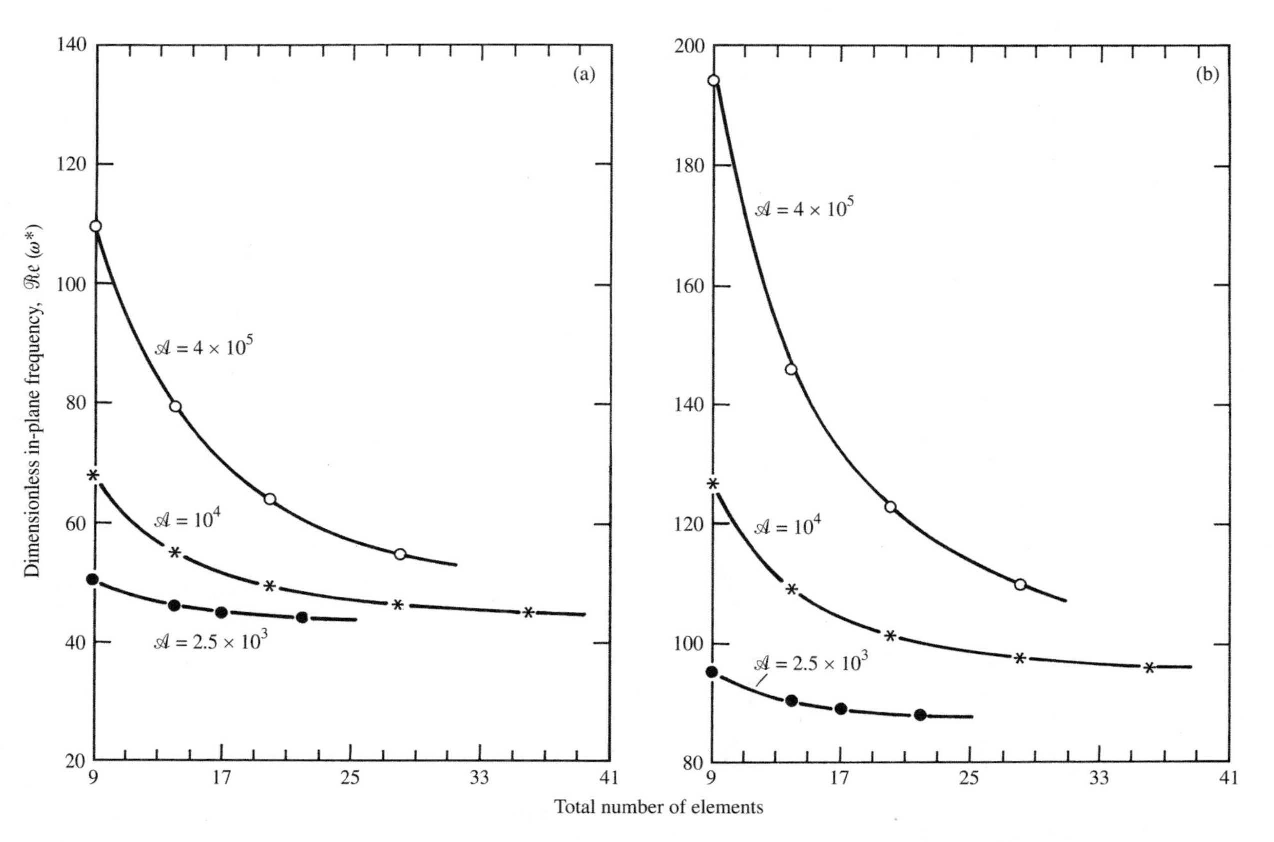

Figure 6.8 Convergence of the in-plane eigenfrequencies of a clamped–clamped semi-circular pipe for different values of $\mathcal{A}$ and $\bar{u} = \Pi = 0$: (a) first (lowest) eigenfrequency; (b) second eigenfrequency (Misra *et al*. 1988b).

It is of interest to recall that for the *inextensible* case, 10 or so elements lead to convergence of the results (Section 6.4.1), as opposed to more than 30 elements required for the extensible case. Thus, the extensible analysis is computationally more demanding.

The steady-state configurations of the system are considered next. Typical results are shown in Figure 6.9, both for inviscid and viscous flow. As indicated in the figure caption, the deformations are exaggerated for clarity. The forms in Figure 6.9(a,b) are for inviscid flow ($\overline{\lambda} = 0$). It may be noted that the stressed shape is symmetric; this is because the steady (static) fluid force acting on the pipe is only the centrifugal force, which is symmetric. When the flow velocity increases, this symmetric deformation away from the initial unstressed shape increases gradually. In the case of viscous flow of Figure 6.9(c), on the other hand, the stressed shape is not symmetric, since the frictional pressure loss causes the pressure to vary along the pipe.

It may be noted that the deformations for both inviscid and viscous flows are fairly small (less than 5%), even for very large flow velocities (up to $\overline{u} = 6\pi$). It is also interesting to note that, beyond a certain $\overline{u}$, the stressed configuration changes to another zero-flow mode shape[†] (see the case of $\overline{u} = 4\pi$ for inviscid flow and $\overline{u} = 3\pi$ for viscous flow).

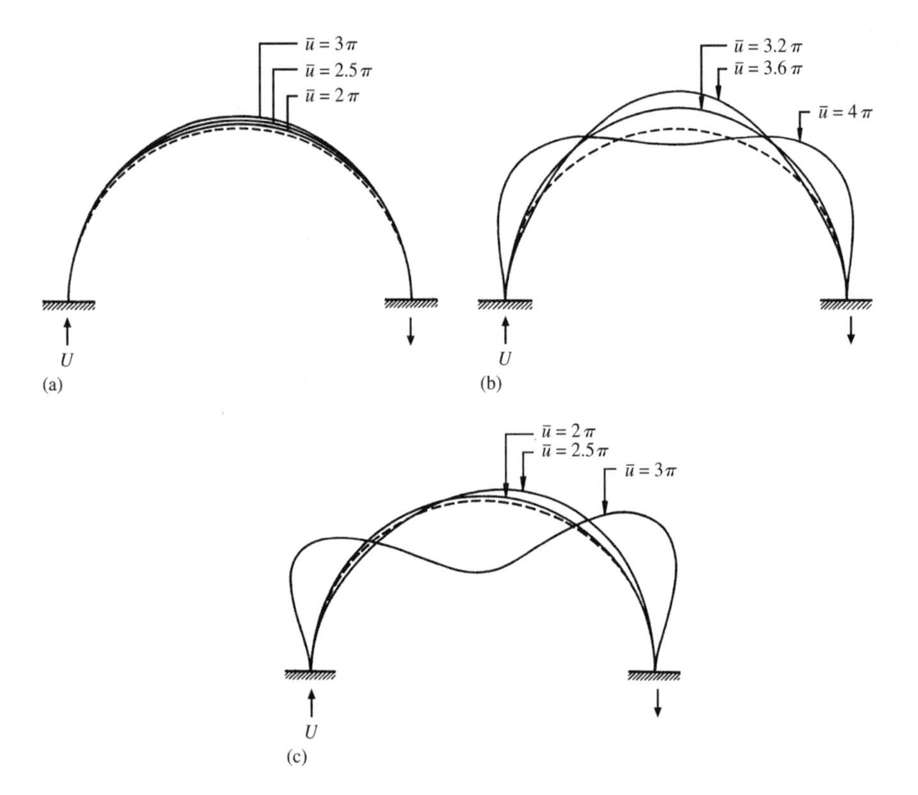

Figure 6.9 Static in-plane equilibrium configurations of a clamped–clamped semi-circular pipe conveying fluid, for $\mathscr{A} = 10^4$ and (a) inviscid fluid, $\overline{u} = 2\pi, 2.5\pi, 3\pi$; (b) inviscid fluid, $\overline{u} = 3.2\pi, 3.6\pi, 4\pi$; (c) viscous fluid, $\overline{u} = 2\pi, 2.5\pi, 3\pi$. In (a), (b) and (c) the deformation is magnified by a factor of 28, 30 and 25, respectively (Misra *et al*. 1988b).

[†]Note that for an extensible pipe, in addition to the shapes in Figure 6.7, there is a zero-node modal shape for in-plane motions: the shape associated with the deflection in Figure 6.9 at small $\overline{u}$.

Nevertheless, divergence according to this theory does *not* occur (as will be shown in what follows); the global stiffness matrix remains positive definite.

For an *inextensible* pipe, there is no difference between the values of Π^o for inviscid and viscous flows; in both cases, it is equal to $-\bar{u}^2$. For an *extensible* pipe, however, there is a difference between viscous and inviscid results: small for low flow velocities, but more significant at higher flows.

The dynamics of *in-plane motion* according to extensible theory is presented next. Several variants of the theory are considered: in one, the steady-state combined force Π^o is neglected; in the second variant, Π^o is taken into account, but the initial (steady) deformations are assumed to be negligible, i.e. the terms involving $\mathscr{A}(\eta_1^{o\prime} + \Theta\eta_3^o)$ in equations (6.69) and (6.70) and (6.71) and (6.73) are set to zero;[†] in the third variant both Π^o and $\mathscr{A}(\eta_1^{o\prime} + \Theta\eta_3^o)$ are nonzero and it is considered to be the complete theory. The first variant is recognized as physically not realizable, but is considered for comparison. The calculations are conducted for a system with $\beta = 0.5$, $\mathscr{A} = 10^4$ [see equations (6.46)].

Figure 6.10 shows the results obtained when the internal fluid is inviscid. It is noted that, generally, the effect of the $\mathscr{A}(\eta_1^{o\prime} + \Theta\eta_3^o)$ term is not very important. This is so

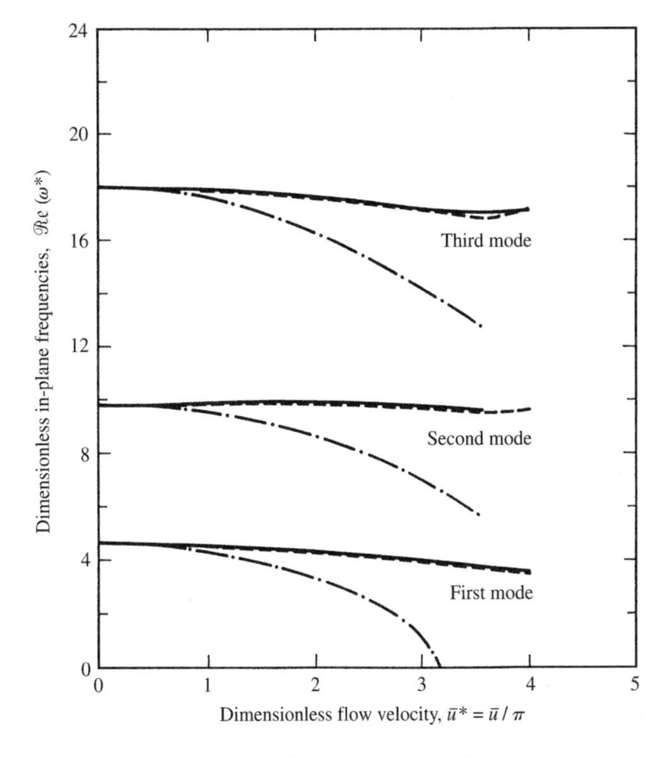

Figure 6.10 Dimensionless eigenfrequencies ω^* versus $\bar{u}^*$ for *in-plane motion* of a clamped–clamped semi-circular pipe conveying *inviscid fluid*, for $\beta = 0.5$ and $\mathscr{A} = 10^4$: — · — , $\Pi^o = 0$, $\mathscr{A}(\eta_1^{o\prime} + \Theta\eta_3^o) = 0$; – – –, $\Pi^o \neq 0$, $\mathscr{A}(\eta_1^{o\prime} + \Theta\eta_3^o) = 0$; ——, $\Pi^o \neq 0$, $\mathscr{A}(\eta_1^{o\prime} + \Theta\eta_3^o) \neq 0$ (Misra *et al*. 1988b).

[†]It is recognized that the first variant corresponds to the *conventional inextensible theory*, whereas the second corresponds to the *modified inextensible theory*, but the calculations were conducted with the equations for extensible theory.

because the static deformations are not very large, as was observed earlier (Figure 6.9). However, for $\bar{u} > 3\pi$, static deformation effects become slightly more pronounced, in the second and third modes particularly, reflecting relatively greater departures from the unstressed state of the pipe.

The most important feature of Figure 6.10 is the fact that extensible theory, properly taking into account the steady-state combined force Π^o, predicts that no instability occurs for a clamped–clamped curved pipe. The frequencies of the system change very slightly with flow, unlike the case of $\Pi^o = 0$ when the system is predicted to lose stability by divergence. This leads to the conclusion that it is the steady flow-related forces, rather than the steady deformations, which are primarily responsible for the inherent stability of fluid-conveying clamped–clamped curved pipes, and this supports the basic tenet for the modified inextensible theory, results for which are presented in Section 6.4.3.

Hill & Davis (1974) and Doll & Mote (1974, 1976) have also presented extensible theories and reached the same general conclusion, namely that curved pipes with clamped or otherwise supported ends do not lose stability when subjected to internal flow. In Figure 6.11 the results obtained by these two sets of investigators are compared with those obtained by the present theory [including Π^o and $\mathscr{A}(\eta_1^{o\prime} + \Theta\eta_3^o)$ terms] for in-plane motion, with the assumption that the fluid is inviscid. It is seen that the general character of the solutions is similar in all three cases, although the results are not identical.

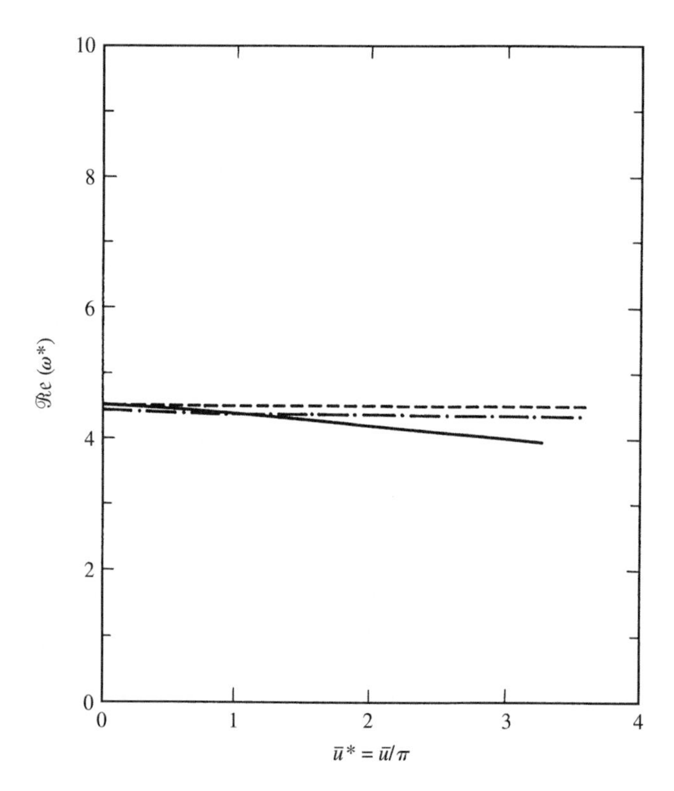

Figure 6.11 Comparison of the fundamental eigenfrequency for *in-plane motion* of a clamped–clamped semi-circular pipe conveying fluid as a function of $\bar{u}^*$ according to extensible theory: – – –, Doll & Mote (1974, 1976) for $\beta = 0.5$, $\mathscr{A} = 1.58 \times 10^4$; — · — , Hill & Davis (1974) for $\beta = 0.43$, $\mathscr{A} = 1.4 \times 10^5$; ——, Misra *et al.* (1988b) for $\beta = 0.5$, $\mathscr{A} = 10^4$ (Misra *et al.* 1988b).

Hill & Davis' equations of motion are perhaps the closest to those utilized here, and the results from these two theories are close, despite some parameters being different: $\beta = 0.43$ and $\mathscr{A} = 1.4 \times 10^5$ in Hill & Davis, as compared to 0.5 and 10^4, respectively, in the present case. Hill & Davis, similarly to the present theory, considered motions about the deformed initial state calculated in a linearized fashion. On the other hand, Doll & Mote calculated the deformed state by a more sophisticated approach, involving a cumulative application of the linearized equations; their β is the same as in the present calculations [note that this is so, despite what appears in their published work ($\beta = 1$), due to a typographical error (Païdoussis 1986b)] and $\mathscr{A}$ was 1.579×10^4.

It should be noted that Doll & Mote and Hill & Davis effectively consider inviscid flow. However, since the steady-state initial forces depend on real flow effects and these forces do work in this case (unlike for straight pipes), this is not necessarily justified. Some calculations with viscous flow are shown in Figure 6.12. It may be seen that frictional effects are not very pronounced for the first mode, but they are more important for the higher modes. The important point is that even for viscous flow, clamped–clamped curved pipes do not lose stability according to the more realistic extensible theory.

We now turn our attention to *out-of-plane motions*. As mentioned in Section 6.2.6, the equations of motion of the extensible theory and the modified inextensible theory (in

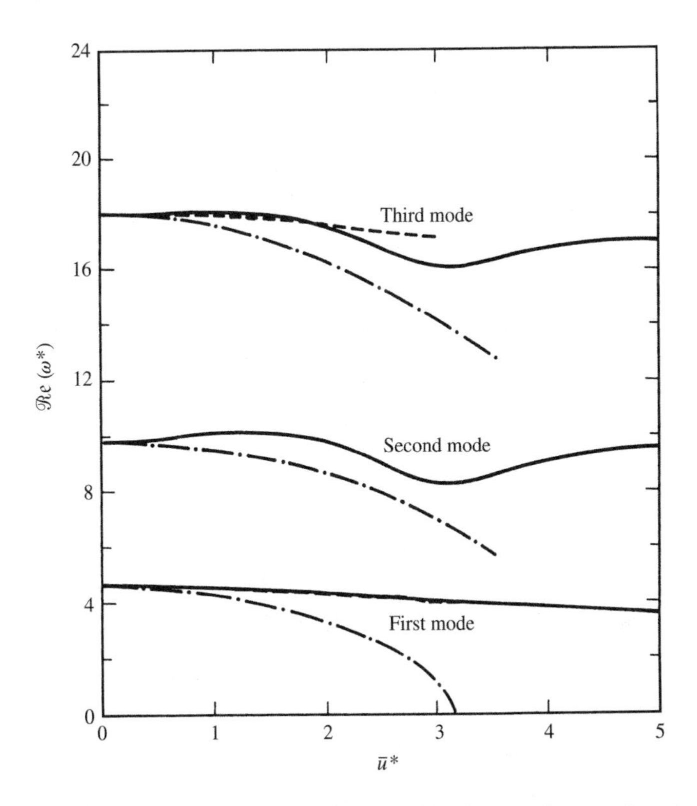

Figure 6.12 The real part of the dimensionless eigenfrequencies as functions of $\bar{u}^*$ for *in-plane motion* of a clamped–clamped semi-circular pipe conveying *viscous fluid*, for $\beta = 0.5$, $\mathscr{A} = 10^4$; — · —, $\Pi^o = 0$, $\mathscr{A}(\eta_1^{o\prime} + \Theta\eta_3^o) = 0$; – – –, $\Pi^o \neq 0$, $\mathscr{A}(\eta_1^{o\prime} + \Theta\eta_3^o) = 0$; ——, $\Pi^o \neq 0$, $\mathscr{A}(\eta_1^{o\prime} + \Theta\eta_3^o) \neq 0$ (Misra *et al*. 1988b).

which the steady fluid loads are *not* neglected) are identical. Hence, the main body of the results will be presented in Section 6.4.3. However, a comparison with Hill & Davis' and Doll & Mote's extensible theories for out-of-plane motions is presented in Figure 6.13. It is clear that the results from the three theories are even closer in this case than for in-plane motion. It is also clear that no divergence occurs for out-of-plane motions either.

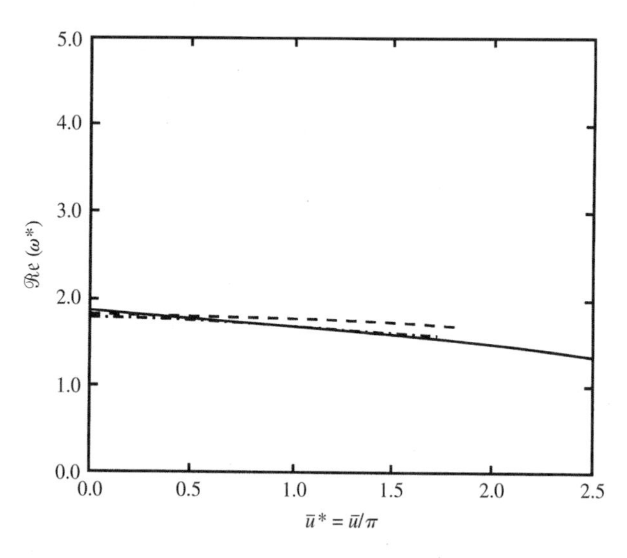

Figure 6.13 Dimensionless eigenfrequencies ω^* versus $\bar{u}^*$ for *out-of-plane motion* of a clamped–clamped semi-circular pipe conveying fluid according to extensible theory: – – –, Hill & Davis (1974); — · —, Doll & Mote (1974, 1976); ——, Misra *et al*. (1988b), for the parameters as in Figure 6.11 (Van 1986).

6.4.3 Modified inextensible theory

Figures 6.14 and 6.15 show the *in-plane* eigenfrequencies of clamped–clamped, pinned–pinned and clamped–pinned semi-circular pipes conveying fluid, as functions of the flow velocity, obtained by both the modified and the conventional inextensible theories. It is obvious that, according to the *modified inextensible theory*, the effect of fluid flow on the eigenfrequencies is not very pronounced. Flow tends to reduce the first-mode eigenfrequency, but does not cause divergence in the flow range investigated (as high as $\bar{u} = 6\pi$). It is also interesting to observe that the eigenfrequencies of some of the higher modes actually increase with flow velocity. Thus, whether the axial force Q_z (or combined force Π) is taken into account or not is *very* important. In the conventional inextensible theory, where Π is neglected, the effect of internal flow on the eigenfrequencies manifests itself via the centrifugal and Coriolis forces, whereas in the modified inextensible theory, where Π is taken into account, the internal flow exerts only a Coriolis force. This is because when both ends are supported, Π^o is a constant equal to $-\bar{u}^2$ [equation (6.99)], and thus in equations (6.47) and (6.49) governing in-plane motion the terms associated with the initial forces cancel out those arising from the centrifugal force. It is recalled that it is the centrifugal forces that are responsible for the divergence instability obtained in the case of pipes with both ends supported (Section 3.2.1).

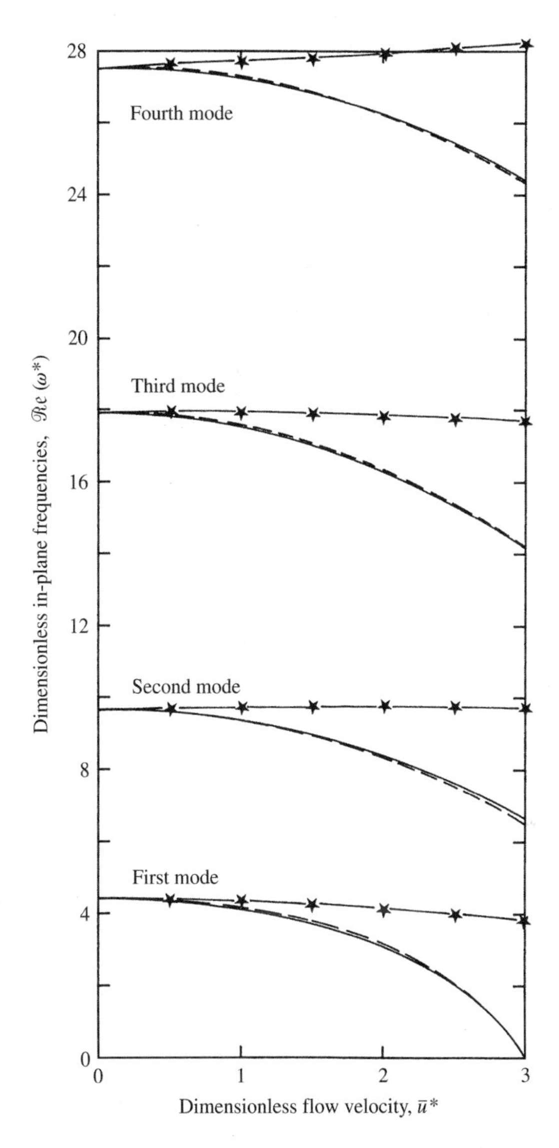

Figure 6.14 Dimensionless eigenfrequencies ω^* for *in-plane motion* of a clamped–clamped semi-circular pipe conveying fluid as functions of the dimensionless flow velocity $\bar{u}^*$, for $\beta = 0.5$: – – –, Chen (1972a); ——, conventional inextensible theory (Misra *et al*. 1988a); — ∗ — , modified inextensible theory (Misra *et al*. 1988a).

Now, as shown in Figures 6.10 and 6.12, the differences between the full extensible theory and the version in which the extension of the centreline $[\mathscr{A}(\eta_1^{o\prime} + \Theta\eta_3^o)]$ is ignored are small, especially in the first mode; and this is the mode in which the system, according to the conventional inextensible theory, would lose stability. Hence, insofar as in-plane motions are concerned, the effect of neglecting extensibility of the centreline is small, and the modified inextensible theory provides a reasonable approximation to the results obtained by the full extensible theory, at considerably smaller computational cost.

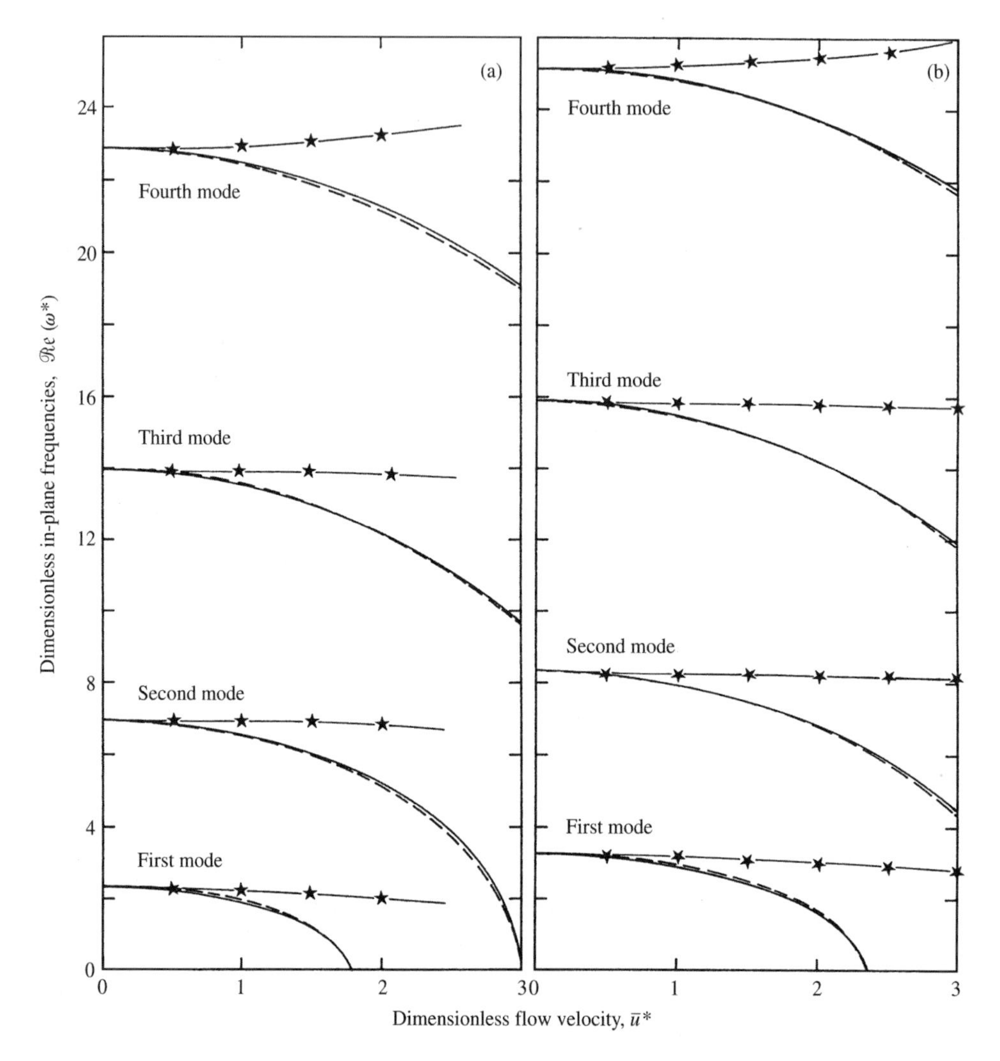

Figure 6.15 Dimensionless eigenfrequencies ω^* for *in-plane motion* of (a) a pinned–pinned and (b) a clamped–pinned semi-circular pipe conveying fluid as functions of the dimensionless flow velocity $\bar{u}^*$, for $\beta = 0.5$: ---, Chen (1972a); ——, conventional inextensible theory (Misra *et al*. 1988a); — ∗ — , modified inextensible theory (Misra *et al*. 1988a).

In the case of *out-of-plane motion* these two theories become identical. A sample result is shown in Figure 6.16. Similarly to in-plane motion, the out-of-plane eigenfrequencies change very little if the combined steady force Π is properly taken into account, and the system does not lose stability by divergence, in contrast to predictions of the conventional inextensible theory.

6.4.4 More intricate pipe shapes and other work

All of the foregoing calculations were for semi-circular pipes, although the theory, as developed, could be applied to any initially planar pipe form. Some calculations for curved

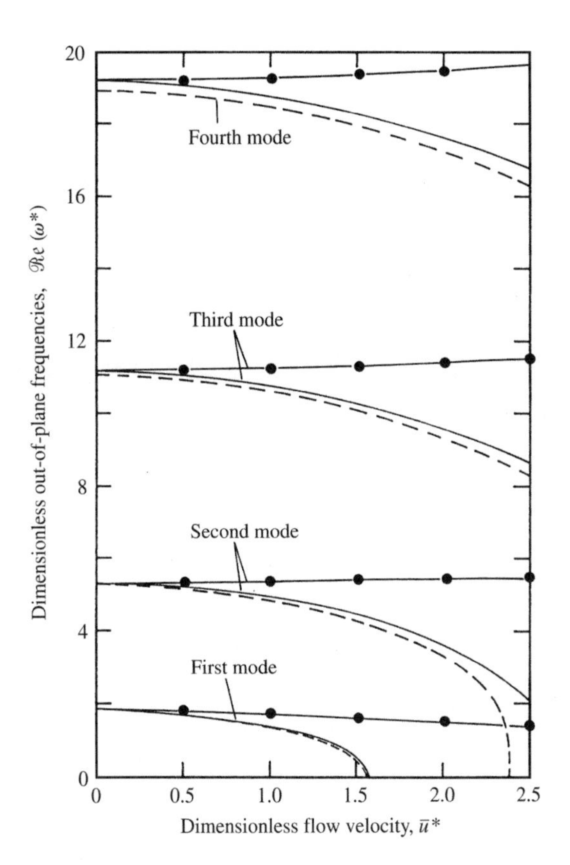

Figure 6.16 Dimensionless eigenfrequencies versus $\bar{u}^*$ for *out-of-plane motion* of a clamped–clamped semi-circular pipe conveying fluid for $\beta = 0.5$, $\Lambda = 0.769$: – – –, Chen (1973); ——, conventional inextensible theory Misra *et al*. (1988a); —•— modified inextensible theory (Misra *et al*. 1988a).

pipes with other arc-angles were conducted by Van (1986) and Misra *et al*. (1988b). However, here a more interesting set of results is presented, obtained by Hill & Davis (1974), who, since they obtained solutions via a finite element method also, could analyse pipes of any shape.

Figure 6.17 shows the evolution of first-mode eigenfrequencies of S-, L- and spiral-shaped pipes with increasing $\bar{u}$, comparing the results of their full extensible theory and those with the initial stresses (the equivalent of Π here) neglected. It is clear that the dynamical behaviour of curved pipes with more complex initial shape is essentially the same as that of semi-circular pipes. The most important result, in view of the results already reported in the foregoing sections, is that if the initial stresses are properly accounted for, (i) there is only small variation of ω with $\bar{u}$, and (ii) there is no loss of stability, even for very large $\bar{u}$.

Fan & Chen (1987) undertook an ambitious study of the dynamics and stability of helical pipes, which may be found in some newer heat-exchanger designs. They obtain the equations of motion in a helical coordinate system via Hamilton's principle and solutions via the finite element method. Unfortunately, however, they make the inextensibility

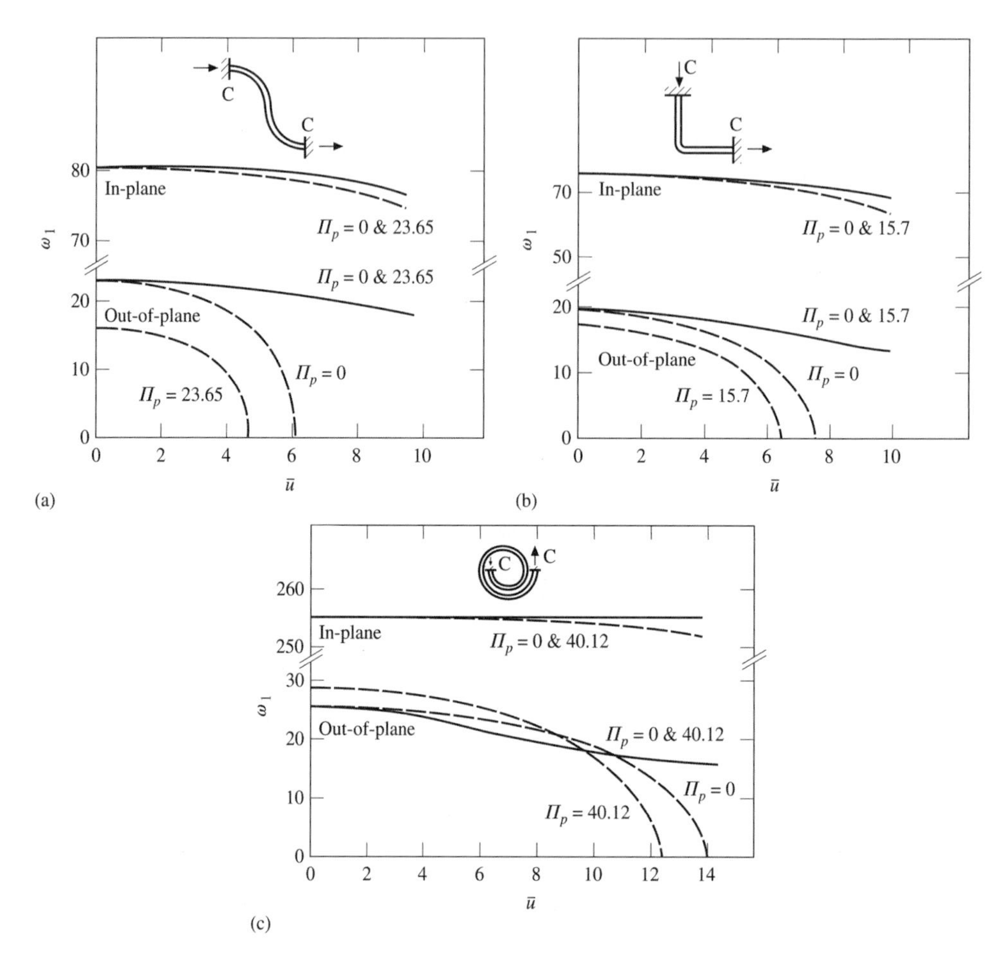

Figure 6.17 First-mode eigenfrequencies ω_1, as functions of $\bar{u}$ for an (a) S-shaped, (b) L-shaped and (c) spiral pipe conveying fluid with $\beta = 0.231$ and $\mathscr{A} = 1.4 \times 10^5$, for various values of Π_p: ——, extensible theory (steady-state forces accounted for); – – –, conventional inextensible theory (steady-state forces other than Π_p neglected); (Hill & Davis 1974).

assumption and neglect the steady fluid loading, so that their results are of limited practical interest.

Aithal & Gipson's (1990) main aim was to examine the effect of dissipation on the in-plane dynamics of planar curved pipes with various boundary conditions. Unfortunately, they too neglect the steady fluid forces and obtain equations similar to Chen's. However, their results are additionally questionable since it is predicted that dissipation (modelled as a Kelvin–Voigt viscoelastic and a viscous model) causes the system to lose stability by flutter rather than divergence at critical flow velocities 35–90% higher than that for divergence of the conservative system. Yet, both physically and mathematically, the effect of dissipation should vanish as $\omega \to 0$.

Al-Jumaily & Al-Saffar (1990) studied an interesting practical problem of a hook-shaped pipe, modelling part of an aircraft fuel line which was prone to failure — but, alas, this too was done while ignoring the effect of steady fluid forces.

Finally, Ko & Bert (1984, 1986) derived a nonlinear equation, under a set of reasonable assumptions, for in-plane motion of a circular-arc pipe conveying fluid,† which they solved for the case of clamped ends by the method of multiple scales. They use the inextensibility assumption but take into account the steady fluid forces — similarly to the Misra *et al.* modified inextensible theory. In a sample calculation, Ko & Bert (1986) find that the frequency of the first asymmetric mode [Figure 6.7(a)] *increases* with the flow velocity. Furthermore, the frequency displays a strong softening behaviour (i.e. it decreases with increasing amplitude).

6.4.5 Concluding remarks

As shown by the results of Figures 6.11 and 6.12 for in-plane motions and Figures 6.13 and 6.16 for out-of-plane motions, differences in the dynamical behaviour as predicted by the modified inextensible and extensible theories are either small or virtually zero, whereas this behaviour is dramatically different from that predicted by the conventional inextensible theory.

It is clear that the main difference between the extensible theories and the 'traditional inextensible' theory is not the extensibility of the centreline at all, but rather whether the combined steady axial force Π^o is taken into account or not. This resolves the apparent paradox that, although it is physically obvious that the actual extension of the centreline cannot be very large, the differences in predicted behaviour between (conventional) inextensible and extensible theory are so profound: the first predicts loss of stability by divergence and pronounced eigenfrequency-flow effects, whereas the second predicts no loss of stability and weak frequency-flow effects. It has now been clarified that the use of the 'inextensible' and 'extensible' labels is rather misleading, as are those of 'constant' and 'variable curvature' utilized by Doll & Mote; the real source of the discrepancy lies in the fact that conventional inextensible theory also neglects all steady stress effects (i.e. all steady flow-induced forces).

Unfortunately, there are no experimental data for curved pipes, apart from those of Liu & Mote (1974) already discussed in Section 3.5.6. In these experiments, however, the curvature was relatively small and inadvertent. The variation of the fundamental eigenfrequency with flow was nevertheless compared with various versions of their theory by Doll & Mote (1976); it was found that, if anything, the experimental results up to a certain maximum $\bar{u}^*$ agreed better with those of Doll & Mote's 'constant curvature' analysis (which corresponds to inextensible theory) than with extensible theory. As seen in Figure 3.26, the frequency varies with $\bar{u}$ essentially as predicted by the conventional inextensible theory! This paradox, which has ever since cast doubt on the validity of the extensible, and hence also the modified inextensible, theory is resolved at the end of Section 6.6. However, proper experiments with curved pipes remain to be done — recognizing, nevertheless, that this is not a simple task.

Until then, since there is no reason why the effect of steady fluid forces on the dynamics of the system should be neglected, and as convincingly argued by Dupuis & Rousselet

†According to Dupuis & Rousselet (1992), their equations of motion 'are free of the Coriolis force and with some linear terms that are neither accounted for in their analysis, nor found in any other analysis'. The Coriolis terms were in fact omitted intentionally (Bert 1996), presumably because the theory was to be applied exclusively to conservative pipe systems (pipes with clamped ends).

(1992), it is concluded that the dynamics of pipes with supported ends is as predicted by extensible or *modified* inextensible theory (Doll & Mote's, Hill & Davis', Misra *et al.*'s and Dupuis & Rousselet's).

6.5 CURVED CANTILEVERED PIPES

It would be tempting to assume in this case that the pipe is inextensible, as for straight pipes conveying fluid, yet to take into account the steady-state initial loads; by this thinking, the use of the modified inextensible theory would at first sight appear to be ideal. It should be realized, however, that under the action of the flow, the shape of the curved pipe varies continuously and substantially (not as shown in Figure 6.9, because here one end of the pipe is unrestrained); thus, an initially semi-circular pipe will become considerably shallower as the critical flow velocity is approached. Hence, properly, the shape and the loads for any given $\bar{u}$ should be determined first, and then the stability of the deformed pipe assessed. Furthermore, since deformed and initial shapes are likely not close, a nonlinear analysis is called for in determining the deformed shape and the steady-state stresses in that state,† which is not a trivial task; as a result, this type of analysis has virtually never been done in its entirety.

In this light, the analysis of stability of a semi-circular pipe by means of inextensible theory amounts to saying that it is the study of stability of a family of pipes, each of a *different and unspecified initial shape*, which, under the action of flow, all become semi-circular at the appropriate set of values of $\bar{u}$. With this artifice, one *could* consider the dynamical behaviour as predicted by the *modified* inextensible theory developed in the foregoing. The weakness in this, however, is that the steady-state loads would be determined on the assumption of small deformations away from a semi-circular shape, initially unstressed at $\bar{u} = 0$, which is at variance with the assumption made regarding shape, increasingly as $\bar{u}$ is augmented.

On the other hand, the use of the conventional inextensible theory is wholly inappropriate because, in addition to the question of shape of the pipe, one would have to imagine that the system is magically annealed or otherwise massaged at each $\bar{u}$ concerned so as to eliminate the steady stresses in the deformed pipe. For this reason, no results obtained by the conventional inextensible theory are presented, except by way of comparison with those of the modified theory. Otherwise, suffice it to say that Argand diagrams obtained by Misra *et al.* and Chen (1973) via the conventional inextensible theory of a semi-circular pipe are in qualitative but not quantitative agreement [e.g. Misra *et al.* (1988a; Figure 5)].

6.5.1 Modified inextensible and extensible theories

Some results are presented, obtained via the modified inextensible theory and the extensible theory. Before doing so, however, it is of interest to show one typical Argand diagram for a straight pipe, obtained by the methods developed by Misra *et al.* for $R_o \to \infty$, shown in Figure 6.18. These results (i) lend further support to the validation of the finite element scheme, by showing near-perfect agreement with analytical results for a cantilevered pipe, and (ii) demonstrate the power of this finite element scheme, in the following sense: with only 6 finite elements, the eigenfrequencies of the lowest three modes could be predicted

†These comments agree with the careful analysis of the problem by Dupuis & Rousselet (1992).

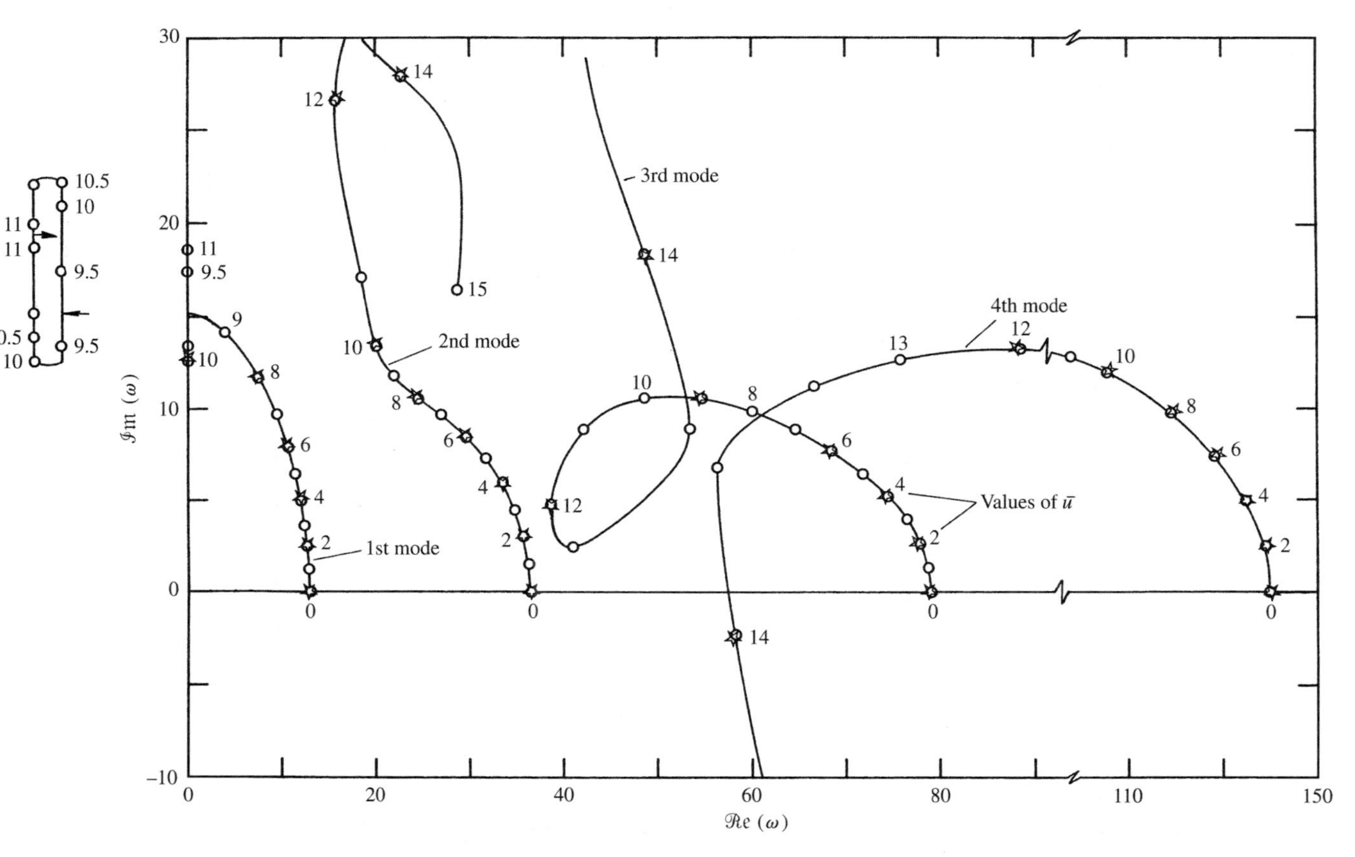

Figure 6.18 Argand diagram of the lowest four modes of a vertical *straight* tubular cantilever conveying fluid as functions of $\bar{u}$, for $\beta = 0.4$ and $\gamma = 100$: ∘, Galerkin (Païdoussis 1970); ⋆, finite element discretization (Misra *et al*. 1988a).

to within 2.5%. Calculations were done with both in-plane and out-of-plane versions of the theory, which for straight pipes should give identical results. The in-plane version was nevertheless found to give superior agreement with analytical results for the same number of elements, presumably because of the use of quintic as opposed to cubic interpolation functions.

Sample Argand diagrams for in-plane and out-of-plane motions of a semi-circular pipe obtained by the modified inextensible theory are shown in Figures 6.19 and 6.20, where they are compared with those obtained by the conventional inextensible theory. Both theories predict divergence followed by flutter at higher $\bar{u}^*$ for in-plane motions, and only flutter for out-of-plane motions (although divergence in the first mode almost occurs). The critical flow velocity for in-plane divergence is approximately the same (Figure 6.19) according to the two theories, $\bar{u}^*_{cd} \simeq 0.7$, in contrast to the results for clamped–clamped pipes. However, the critical flow velocities for flutter are much lower according to the modified inextensible theory: $\bar{u}^*_{cf} \simeq 1.3$ for in-plane motions and $\bar{u}^*_{cf} \simeq 0.8$ for out-of-plane motions, versus 4.2 and 3.5, respectively. The differences are large but not surprising, in view of the dramatic effect that accounting for the steady fluid forces has been found to have on the dynamics of pipes with both ends supported (Section 6.4).

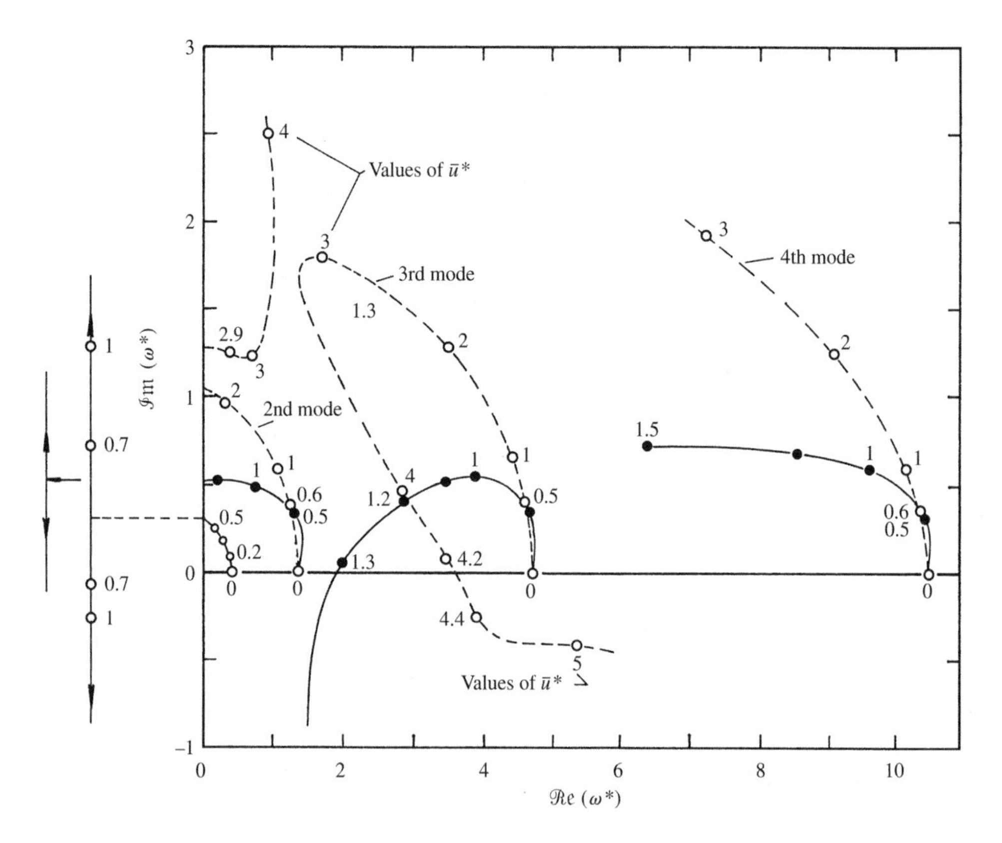

Figure 6.19 Argand diagram for the lowest four eigenfrequencies for *in-plane motion* of a cantilevered semi-circular pipe conveying fluid for $\beta = 0.75$: -○-, conventional inextensible theory ($\Pi = 0$); — • — , *modified* inextensible theory ($\Pi \neq 0$). The two sets of results sensibly coincide for the first mode, so only one is shown (Misra *et al*. 1988a).

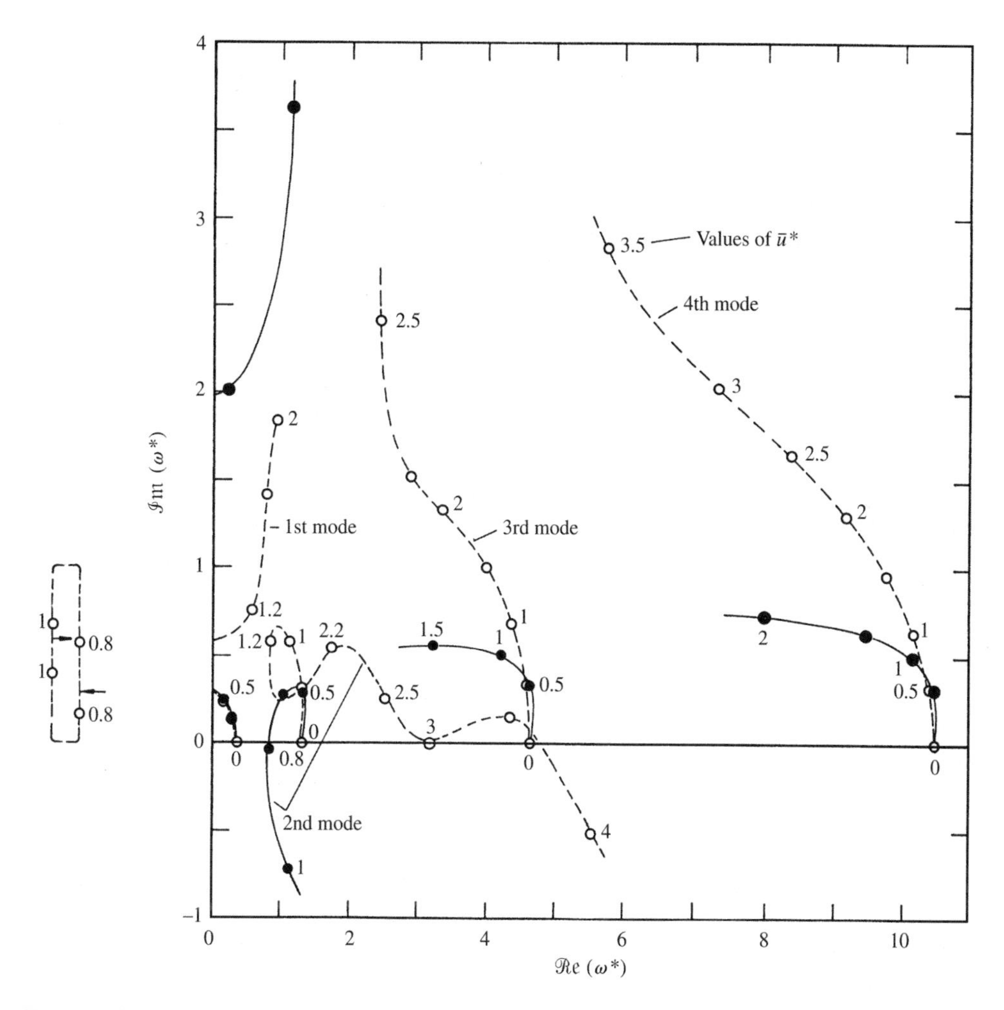

Figure 6.20 Argand diagram for the lowest four eigenfrequencies for *out-of-plane motion* of a cantilevered semi-circular pipe conveying fluid for $\beta = 0.75$ and $\Lambda = 0.769$: -○-, conventional inextensible theory; — • — , *modified* inextensible theory (Misra *et al*. 1988a).

These results, together with others obtained by the modified inextensible theory, are summarized in Table 6.1. It is seen that a smaller subtended angle Θ has a strong stabilizing influence. On the other hand, $\bar{u}^*_{cf}$ appears to be a weak function of β, in contrast to straight pipes. The results obtained are likely qualitatively sound (cf. those obtained by extensible theory, to be discussed next), except for the prediction of loss of stability by divergence for in-plane motions. This may well be a by-product of the limitations of the theoretical model; it appears more physically reasonable that, if the pipe can deform freely under the action of the steady flow, the predicted divergence will devolve into a gradual and continuous change of shape with increasing flow.

One case of in-plane motions of a semi-circular pipe was analysed by Doll & Mote (1974) by means of their extensible theory, for $\beta = 0.5$ and $R/a = 40$, where a is the radius of gyration of the pipe about its centreline. It is predicted that the system loses stability by flutter in its second mode at $\bar{u}^*_{cf} = 0.6$, as well as by divergence in its first

Table 6.1 Critical flow velocities for divergence, $\overline{u}^*_{cd}$, and flutter, $\overline{u}^*_{cf}$, of a cantilevered pipe according to the modified inextensible theory for $\Lambda = 0.769$ and varying β and the angle Θ subtended by the curved pipe (Van 1986; Barbeau 1987; Misra *et al*. 1988a); the asterisk denotes that the result is unavailable.

Θ	β	In-plane motion		Out-of-plane motion
		$\overline{u}^*_{cd}$	$\overline{u}^*_{cf}$	$\overline{u}^*_{cf}$
$\frac{1}{2}\pi$	0.25	1.5	2.2	1.7
π	0.25	0.7	1.2	*
π	0.50	0.7	1.3	*
π	0.75	0.7	1.3	0.8
$\frac{3}{2}\pi$	0.50	0.4	1.5	*
1.9π	0.50	0.3	1.9	*

mode at $\overline{u}^*_{cf} \simeq 0.9$; these values are of the same order of magnitude as those in Table 6.1, although the sequence of the instabilities is reversed. However, these results are questionable, as pointed out by Dupuis & Rousselet (1985, 1986): (i) there appears to be an error in the nondimensionalization, so that the values of ω^* (even for $\overline{u}^* = 0$) are quite different from those of Dupuis & Rousselet (1985) and Misra *et al*. (1988a,b,c), which agree; (ii) more seriously, even the ratio ω^*_2/ω^*_1 at $\overline{u}^* = 0$, which should be $\simeq 3$, is $\simeq 5.7$ in Doll & Mote's results. This is why the figure in question is not presented here.

Dupuis & Rousselet (1985) attempted to reproduce Doll & Mote's results, using their own extensible theory, without success. This was partly because of the aforementioned discrepancy in the values of ω, but also because they were unaware of (a) a typographical error in Doll & Mote (1974) which made it appear that $\beta = 1$ instead of $\beta = 0.5$, and (b) the fact that, despite using $\overline{v} = (M/EI)^{1/2}Ua$ as the dimensionless flow velocity in their analysis, a being the radius of gyration of the pipe about its centreline, Doll & Mote used $\overline{u}^*$ as defined in equation (6.108b) in the presentation of their results. The latter can easily be fixed, since $\overline{u}^* = (R/a)\overline{v}$. The former, however, meant that Dupuis & Rousselet's Argand diagram was for $\beta = 1$.

Once the typographical error was pointed out in discussion by Païdoussis (1986b), a new *eigenvalue* Argand diagram (cf. Figure 2.10) was generated in Dupuis & Rousselet's response, given here as Figure 6.21. Once converted, the critical flow velocities are $\overline{u}^*_{cf} = 0.44$ and, possibly, $\overline{u}^*_{cd} \simeq 0.64$; thus the dynamical behaviour is qualitatively similar to Doll & Mote's, but the critical values of $\overline{u}^*$ are considerably lower. In this regard it should be mentioned that in Dupuis & Rousselet's original calculation for $\beta = 1$ a 10^{-2} factor was forgotten in the Argand diagram presented (Dupuis 1997).[†]

Finally, Aithal & Gipson (1990) looked into the effect of dissipation on the dynamics of cantilevered systems. Although they use conventional inextensible theory, their results are nevertheless discussed here because they are so bizarre; so much so, that the authors themselves characterize them as 'highly intuitive' and 'anomalous'. For instance, for $\Theta = \frac{1}{4}\pi$ and $\frac{1}{2}\pi$, they find that some modes 'fail', so that in these modes 'it is not possible to sustain flow' and 'the pipe will experience a flutter type oscillation under arbitrarily small values of fluid velocity'; the authors, however, insist that these results are correct (Dupuis & Rousselet 1991b).

[†]As presented, the critical flutter flow velocity of $\overline{v}_{cf} \simeq 6.5$, when converted, results in the enormous value of $\overline{u}^* \simeq 280$!

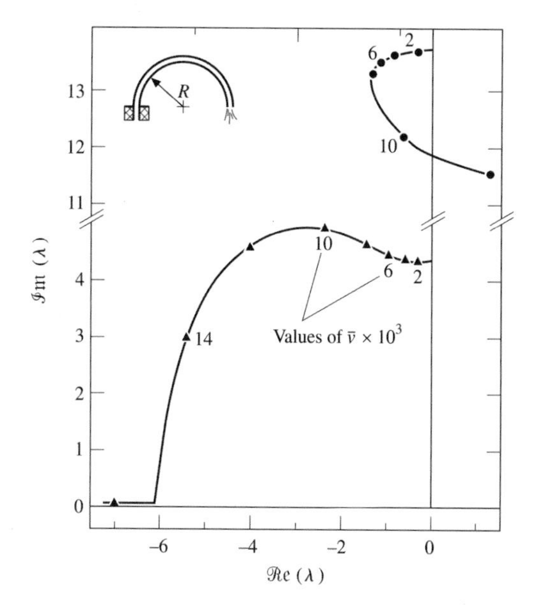

Figure 6.21 Argand diagram of the lowest two eigenvalues $\lambda_j = \omega_j \mathrm{i}$, $j = 1, 2$, for in-plane motion of a semi-circular cantilevered pipe conveying fluid, as functions of $\overline{v}$ defined in the text, for $\beta = 0.5$, $R/a = 40$, according to extensible theory (Dupuis & Rousselet 1986).

The purpose of this rather tedious discussion is to show that most of the results for cantilevered pipes conveying fluid are tinged with uncertainty: those obtained by the modified inextensible theory (Misra *et al*. 1988a) because of the limitations of that theory, those by the extensible theory (Doll & Mote 1974; Dupuis & Rousselet 1985, 1986) by other worrisome features, and those on the effect of dissipation (Aithal & Gipson 1990) for several reasons.

However, taking all the results together, a number of common features emerge which lead to the following consensual, reasonably well-founded conclusions: (i) unlike for curved pipes with supported ends, the eigenfrequencies of cantilevered pipes are strongly dependent on $\overline{u}^*$, just as they are for straight cantilevered pipes conveying fluid; (ii) for sufficiently high $\overline{u}^*$, the system loses stability by divergence or flutter depending on the theory used for in-plane motions, and by flutter for out-of-plane motions; (iii) for reasonable values of β, the critical flow velocities for loss of stability are in the range of $\overline{u}_c^* \simeq 0.4$–$0.8$.

6.5.2 Nonlinear and chaotic dynamics

Steindl & Troger (1994) studied the nonlinear in-plane dynamics of curved pipes as an extension of Champneys' (1991) work discussed in Sections 5.6.2 and 5.8.5. Instead of an articulated system they use a continuously flexible one, and instead of the initial angle between the two articulations as the secondary bifurcation parameter (the primary being the flow) they use the pipe initial curvature, $\kappa_o' = L/R_o \equiv \Theta_o$. The equations of motion are derived by means of director rod theory (Buzano *et al*. 1985; Simo 1985). The pipe centreline is assumed to be inextensible, but changes in shape with increasing flow (i.e. the effects of steady fluid forces on the dynamics) are taken into account, as shown in Figure 6.22.

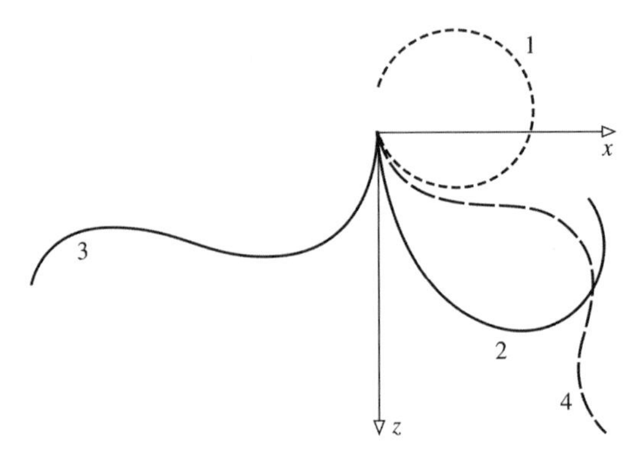

Figure 6.22 Planar states of the initially curved pipe, showing (1) the initial, unstrained shape ($\Theta_o = 6$) under zero gravity and $\overline{u} = 0$; (2) the shape under gravity and $\overline{u} = 0$; (3) the shape at the Hopf bifurcation, $\overline{u} = \overline{u}_H = 5.9$; (4) the shape where the homoclinic orbit occurs, $\overline{u} = \overline{u}_h \simeq 8.5$ (Steindl & Troger 1994).

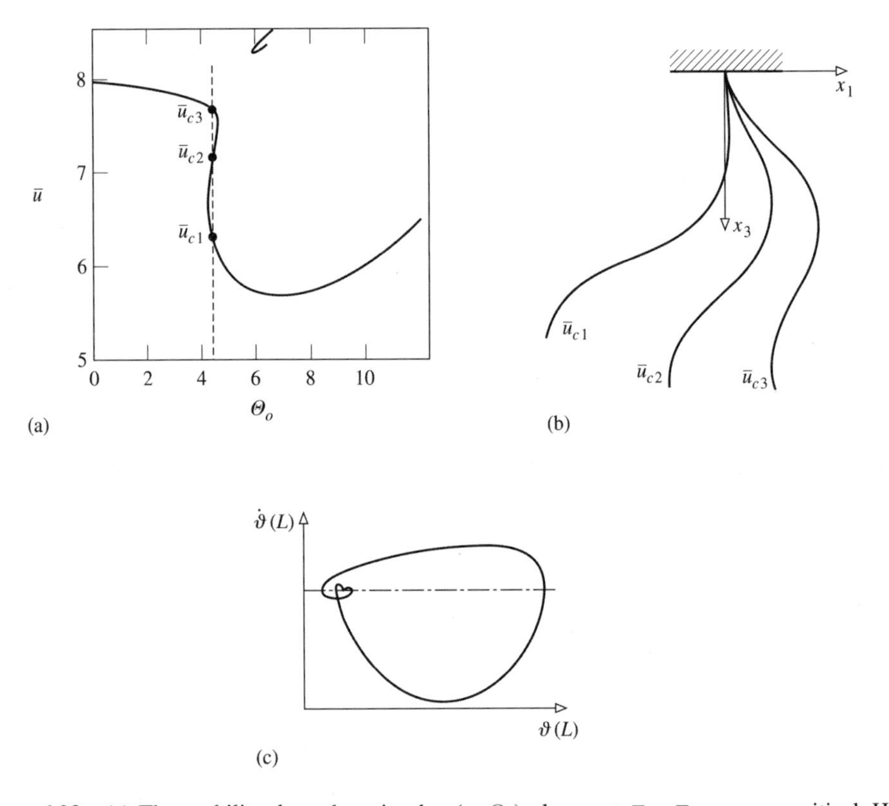

Figure 6.23 (a) The stability boundary in the (u, Θ_o)-plane; at $\overline{u} = \overline{u}_{c1}$ a supercritical Hopf bifurcation occurs; the system is stable for $[\overline{u}_{c1}, \overline{u}_{c2}]$, and then loses stability again, in the same way, at $\overline{u} = \overline{u}_{c3}$; (b) the corresponding equilibrium pipe shapes. (c) The phase-plane diagram of the tangent to the end of the pipe $\vartheta(L)$ at $\overline{u} = \overline{u}_h \simeq 8.5$, corresponding to the small region $\overline{u} = 8.5$, $\Theta_o \simeq 6$ in (a), where a homoclinic orbit occurs (Steindl & Troger 1994).

The derivation of the equations of motion is very compact, in six short steps, and so are the calculations of the equilibrium state leading to Figure 6.22 and of the stability boundary for motions about the equilibrium. It is shown that stability is lost by a supercritical Hopf bifurcation, which in the $(\bar{u}, \Theta_o)$-plane of Figure 6.23(a) displays interesting behaviour for $\Theta_o \simeq 4.5$. The initial Hopf bifurcation is at $\bar{u}_{c1}$; the system regains stability between $\bar{u}_{c2}$ and $\bar{u}_{c3}$ and then loses it again at $\bar{u}_{c3}$ via another supercritical Hopf bifurcation.

The infinite dimensional system is then discretized into a 10-degree-of-freedom one by a finite difference scheme and reduced to a four-dimensional inertial manifold (Foias *et al.* (1988); Brown *et al.* 1990; Dubussche & Marion 1992; Foale *et al.* 1998). Then, making use of the similarity in shape between curve 4 in Figure 6.22 and that at $\bar{u}_{c3}$ in Figure 6.23(b), it is shown that a homoclinic orbit exists in the small isolated curve on the upper part of Figure 6.23(a) near $\bar{u} = 8.5$, $\Theta_o = 6$, signalling the possibility of chaotic motions in that neighbourhood. In Figure 6.23(c) is shown a phase-plane diagram characteristic of homoclinic behaviour: the pipe oscillates about the focus with increasing amplitude at one frequency, then makes a large amplitude excursion and returns back to the focus, oscillating now with decreasing amplitude at another frequency.

Steindl & Troger's is an important contribution, for not only does it demonstrate the possibility of interesting nonlinear dynamical behaviour, but it also reinforces the view expressed elsewhere in Section 6.5: the shape of the pipe is a strong function of the flow velocity and, hence, linear analysis on its own cannot hope to capture the essential dynamics of cantilevered curved pipes conveying fluid.

6.6 CURVED PIPES WITH AN AXIALLY SLIDING END

Since fully clamped pipes are always stable if steady forces are properly accounted for, whereas cantilevered ones are not, the question arises as to the dynamical behaviour of the intermediate case of a pipe with a transversely or axially sliding end. This question is also of some practical interest; for example, U- or Ω-shaped thermal expansion joints are by design not fully clamped. Some such cases were considered by Barbeau (1987) and Misra *et al.* (1988b).

Four different types of sliding ends were studied, shown in Figure 6.24: a transversely sliding end, and three slightly different types of axial sliding; they were analysed either by the modified inextensible theory or by the fully extensible form of the theory, essentially as in the foregoing. The equations are the same as in Section 6.2 and only the boundary conditions for in-plane motion differ. For example, the boundary conditions for the system of Figure 6.24(a) are $\partial\eta_1/\partial\zeta = \partial\eta_2/\partial\zeta = 0$, while those for (b) are zero rotation $(\partial\eta_1/\partial\zeta + \Theta\eta_3 = 0)$ and moment $(\mathcal{M}_y = 0)$; after physical interpretation and use of the inextensibility condition, these lead to

$$\begin{aligned}
&\text{(a)}\quad \eta_3 = \frac{\partial^2\eta_3}{\partial\zeta^2} = \frac{\partial^4\eta_3}{\partial\zeta^4} = 0,\\
&\text{(b)}\quad \frac{\partial\eta_3}{\partial\zeta} = \frac{\partial^2\eta_3}{\partial\zeta^2} + \Theta^2\eta_3 = \Pi^o - \Pi_p = 0,
\end{aligned} \tag{6.109}$$

at $\zeta = 1$; the corresponding values of the combined force Π are also generally different. Out-of-plane sliding has not been considered: the boundary conditions for out-of-plane

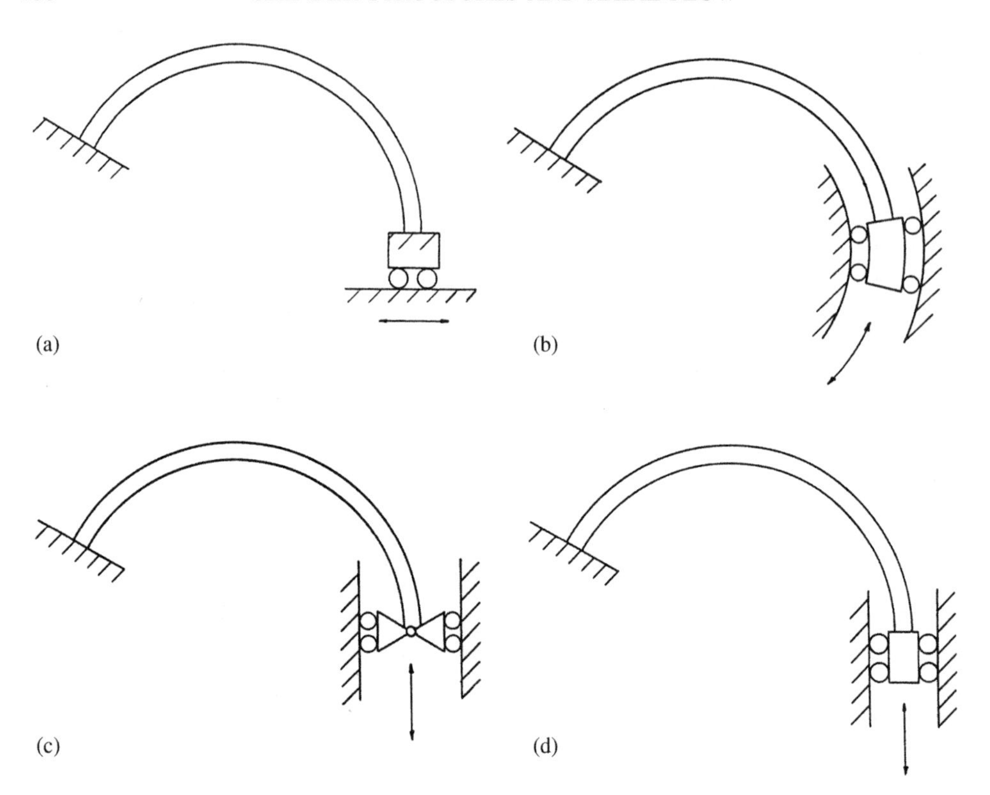

Figure 6.24 Curved pipes with one end clamped and the other sliding: (a) transverse sliding; (b)–(d) three variants of axial sliding.

motion are the same as for clamped–clamped pipes; hence, out-of-plane motions will not be discussed here further.

6.6.1 Transversely sliding downstream end

The dynamics of this system is very similar to that of clamped–clamped pipes, showing: (a) very slight variation of the eigenfrequencies with flow, and (b) no loss of stability as $\overline{u}$ is increased. This is surprising at first sight. However, on reflection this is not quite so, since (i) the steady forces prevent loss of stability by divergence in a similar way as for clamped–clamped pipes, and (ii) the slope at the sliding end remains zero and hence, by similarity to straight cantilevered pipes, the system cannot develop flutter.

6.6.2 Axially sliding downstream end

A typical Argand diagram obtained by the modified inextensible theory for in-plane motions of a quarter-circular pipe conveying fluid and supported as in Figure 6.24(b) is shown in Figure 6.25. It is seen that the system loses stability by divergence at $\overline{u}^*_{cd} \simeq 3.7$ and by flutter at $\overline{u}^*_{cf} \simeq 6.0$. The dynamics of a semi-circular pipe is qualitatively similar,

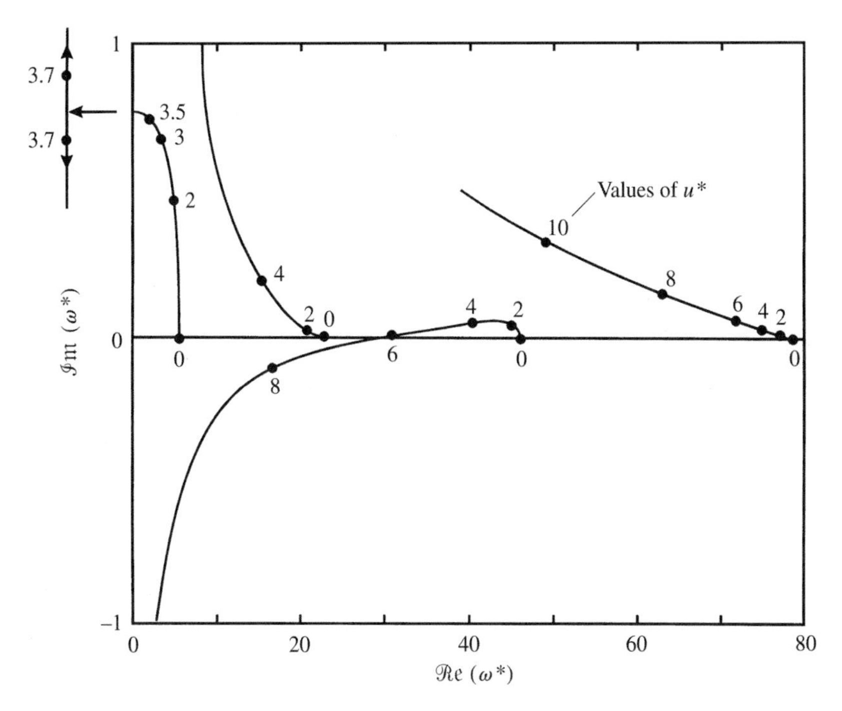

Figure 6.25 Argand diagram of the four lowest eigenfrequencies for in-plane motion of a clamped–axially-sliding quarter-circular pipe of the type of the type of Figure 6.24(b) conveying fluid, as functions of $\overline{u}^*$, for $\beta = 0.5$ (Barbeau 1987).

but in this case $\overline{u}^*_{cd} \simeq 1.5$ and $\overline{u}_{cf} \simeq 5.5$. The behaviour of pipes with a sliding downstream support of the type shown in Figure 6.24(c,d) is similar, but quantitatively a little different. However, a disturbing aspect of these results is that they have been found to depend (quantitatively only) on the method of calculation of Π^o — two methods having been considered, apparently both correct (Misra *et al*. 1988b); this casts some doubt as to the quantitative aspects of the results.

Calculations with the full extensible theory $[\Pi^o \neq 0, \mathcal{A}(\eta_1^{o\prime} + \Theta\eta_3^o) \neq 0]$ show only flutter: for the quarter-circular pipe at $\overline{u}^*_{cf} \simeq 2.9$; for the semi-circular pipe at $\overline{u}^*_{cf} \simeq 0.99$, as shown in Figure 6.26 (Misra *et al*. 1988b). Thus, the predicted dynamical behaviour is quite different.

In conclusion, it may be said that, despite several questions remaining unresolved, it is clear that, if axial sliding is permitted, the system behaves in a manner reminiscent of a curved cantilevered pipe: its eigenfrequencies are strongly dependent on the flow velocity and the system eventually loses stability at high enough flow.

Incidentally, it is also observed that the variation of the first-mode frequency with flow in Figure 6.25 (and similar ones for other Θ) as predicted by the modified inextensible theory and for $\overline{u}^* \leq 2.8$ in Figure 6.26 is qualitatively similar to that of the *conventional inextensible theory* for clamped–clamped pipes. This offers a plausible explanation as to why the dynamics of slightly curved pipes in the experiments by Liu & Mote (1974) paradoxically appears to be in better agreement with conventional inextensible than with extensible theory [see Figure 3.26 and Doll & Mote (1976)]: in both theories axial sliding

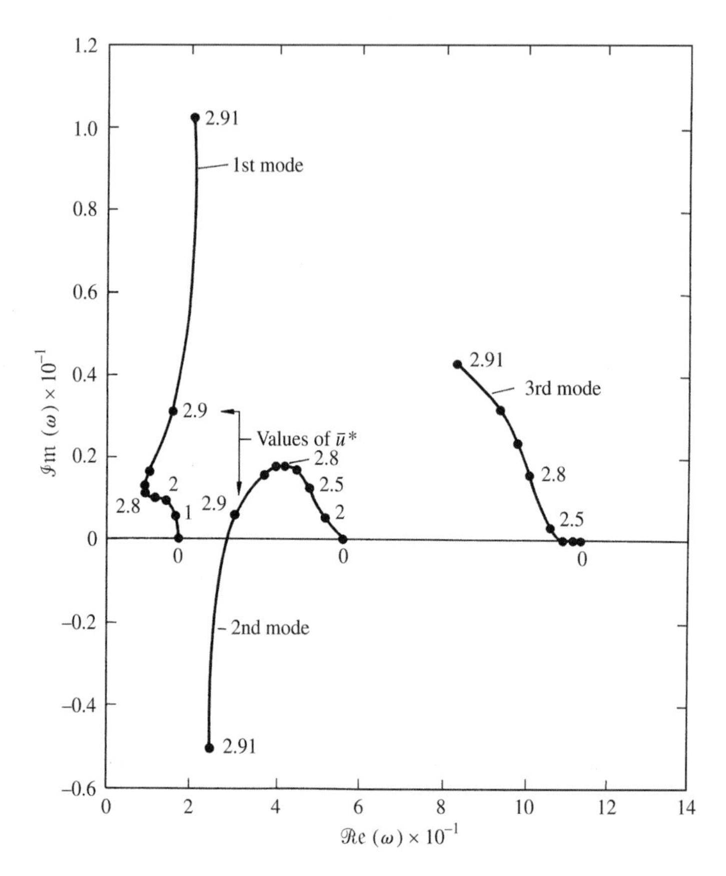

Figure 6.26 Argand diagram for the lowest three eigenfrequencies ω for in-plane motion of a quarter-circular clamped–axially-sliding pipe conveying viscous fluid as functions of $\bar{u}^*$ for $\beta = 0.5$, $\mathscr{A} = 10^4$, according to extensible theory $[\Pi^o \neq 0,\ \mathscr{A}(\eta_1^{o\prime} + \Theta\eta_3^o) \neq 0]$, (Misra *et al*. 1988b).

was prevented, while the experiments were designed to permit it (Section 3.4.4). In view of the foregoing, it is now clear that, once axial sliding is permitted in the theory also (while still accounting for steady forces), the theoretical variation of frequency with flow will be much more like the experimental one:[†] experimental observations no longer disagree with extensible and modified inextensible theory.

[†]Note that $\bar{u}^*$ in Figures 6.25 and 6.26 corresponds to u/π in Figure 3.26.

Appendix A
First-principles Derivation of the Equation of Motion of a Pipe Conveying Fluid

Consider the system of Figure A.1(a), free to oscillate in the horizontal $\{X, Z\}$-plane, so that gravity is inoperative. Externally imposed tension and pressurization effects are not present and, for simplicity, dissipative effects are neglected. Elements of the fluid and the pipe of length δx are shown in Figure A.1(c,d), with the forces and moments at the ends apportioned slightly differently from Figure 3.6.

The acceleration of the fluid element (still making the plug-flow approximation) is derived by the standard dynamics approach, following Ginsberg (1973). An inertial reference frame $\{X, Y, Z\}$ with Y into the plane of the paper and unit vectors **I, J, K**, and an $\{x, y, z\}$ frame embedded in the pipe element with unit axes **i, j, k** are utilized [Figure A.1(b)], together with the expression

$$\mathbf{a}_f = \mathbf{a}_0 + \dot{\boldsymbol{\omega}} \times \mathbf{r} + 2\boldsymbol{\omega} \times \mathbf{v}_{\mathrm{rel}} + \boldsymbol{\omega} \times (\boldsymbol{\omega} \times \mathbf{r}) + \mathbf{a}_{\mathrm{rel}}, \tag{A.1}$$

which may be found in any book on dynamics [e.g. Meriam (1980)]; $\boldsymbol{\omega}$ is the angular velocity of the pipe (and of the $\{x, y, z\}$ frame) with respect to the inertial frame, and the subscript 'rel' denotes quantities relative to the $\{x, y, z\}$ frame. The various components of (A.1) may be expressed and then approximated according to the assumptions made in Section 3.3.1 as follows:

$$\mathbf{a}_0 = \frac{\partial^2 u}{\partial t^2}\mathbf{I} + \frac{\partial^2 w}{\partial t^2}\mathbf{K} \simeq \frac{\partial^2 w}{\partial t^2}\mathbf{K}, \qquad \mathbf{v}_{\mathrm{rel}} = U\mathbf{i} = U\cos\phi\,\mathbf{I} + U\sin\phi\,\mathbf{K} \simeq U\mathbf{I} + U\frac{\partial w}{\partial s}\mathbf{K},$$

$$\boldsymbol{\omega} = -\frac{\partial \phi}{\partial s}\mathbf{J} \simeq -\frac{\partial}{\partial t}\left(\frac{\partial w}{\partial s}\right)\mathbf{J}, \qquad \mathbf{a}_{\mathrm{rel}} = \frac{\mathrm{d}U}{\mathrm{d}t}\mathbf{i} + \frac{U^2}{\mathcal{R}}\mathbf{k} \simeq \frac{\mathrm{d}U}{\mathrm{d}t}\mathbf{I} + \frac{\mathrm{d}U}{\mathrm{d}t}\frac{\partial w}{\partial s}\mathbf{K} + U^2\frac{\partial^2 w}{\partial s^2}\mathbf{K}, \tag{A.2}$$

assuming a positive (counterclockwise) rotation, $\partial\phi/\partial s$; it is also noted that **r**, the distance from the origin of $\{x, y, z\}$ to other points within the element is of second order smallness, so that the second and fourth terms of (A.1) are negligible. Hence, (A.1) may be written as

$$\mathbf{a}_f = \left(\frac{\mathrm{d}U}{\mathrm{d}t}\right)\mathbf{I} + \left(\frac{\partial^2 w}{\partial t^2} + 2U\frac{\partial^2 w}{\partial s \partial t} + U^2\frac{\partial^2 w}{\partial s^2} + \frac{\mathrm{d}U}{\mathrm{d}t}\frac{\partial w}{\partial s}\right)\mathbf{K}, \tag{A.3}$$

correct to $\mathcal{O}(\epsilon)$ — which is the same as equation (3.28).

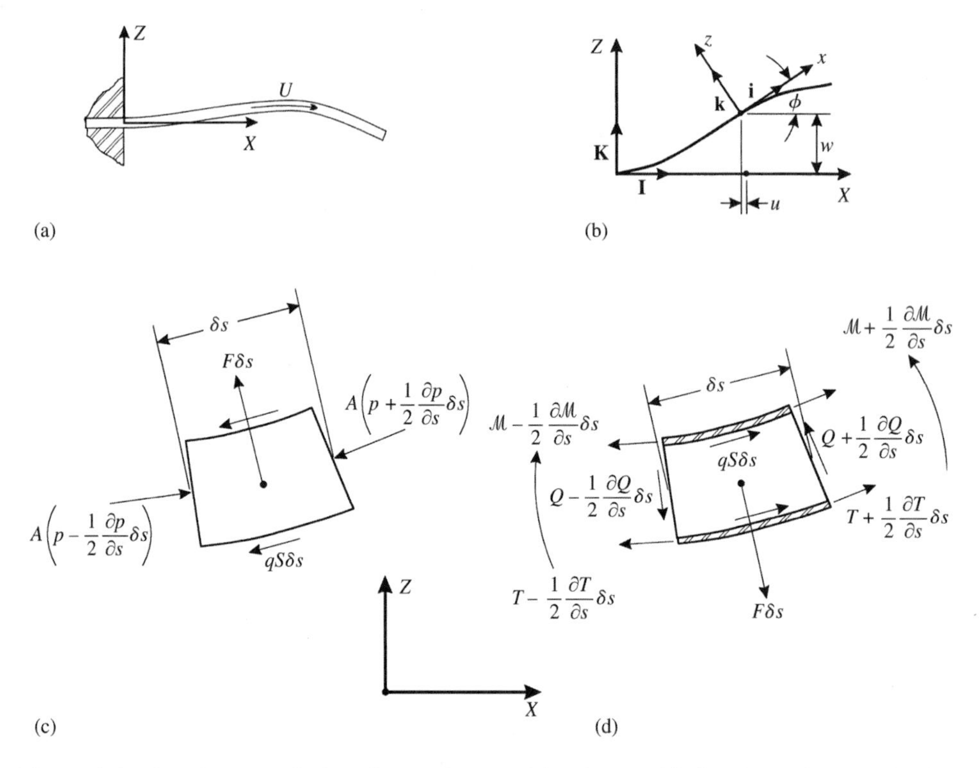

Figure A.1 (a) Diagram of the pipe under consideration oscillating in a horizontal plane; (b) definition of coordinate systems and displacements; (c,d) an element of the fluid and of the pipe, respectively, with the forces and moments acting thereon.

Force and moment balances in the z- and x-directions, as in Section 3.3.2, after simplification and substitution of $s \simeq x$, give

$$EI\,\frac{\partial^4 w}{\partial x^4} + (pA - T)\,\frac{\partial^2 w}{\partial x^2} + \left[\frac{\partial}{\partial x}(pA - T)\right]\frac{\partial w}{\partial x}$$
$$+M\left(\frac{\partial^2 w}{\partial t^2} + 2U\,\frac{\partial^2 w}{\partial x \partial t} + U^2\,\frac{\partial^2 w}{\partial x^2} + \frac{\mathrm{d}U}{\mathrm{d}t}\,\frac{\partial w}{\partial x}\right) + m\,\frac{\partial^2 w}{\partial t^2} = 0, \tag{A.4}$$

$$\frac{\partial}{\partial x}(pA - T) + M\,\frac{\mathrm{d}U}{\mathrm{d}t} = 0. \tag{A.5}$$

Integration of (A.5) from x to L and substitution in (A.4) gives

$$EI\,\frac{\partial^4 w}{\partial x^4} + \left[M\,U^2 + M\,\frac{\mathrm{d}U}{\mathrm{d}t}(L - x)\right]\frac{\partial^2 w}{\partial x^2} + 2MU\,\frac{\partial^2 w}{\partial x \partial t} + (M + m)\,\frac{\partial^2 w}{\partial t^2} = 0, \tag{A.6}$$

which is the same as equation (3.38) once terms involving $\overline{T}, \overline{p}, g, c$ and E^* have been deleted.

The same equation was obtained by Ginsberg (1973). His derivation, however, is flawed in two ways. (i) Having approximated $\mathbf{i} \simeq \mathbf{I}$, irrespective of order-of-magnitude considerations, the second terms of $\mathbf{v}_{\text{rel}}$ and $\mathbf{a}_{\text{rel}}$ in (A.2) are absent, and hence so is the last term in (A.3); this missing term leads to one equal to $M(\mathrm{d}U/\mathrm{d}t)(\partial w/\partial x)$ in the equation of motion. (ii) The third term in the equivalent to (A.4) in Ginsberg's derivation is also missing, which, after substitution of (A.5) in it, leads to a term $-M(\mathrm{d}U/\mathrm{d}t)(\partial w/\partial x)$ in the equation of motion. These two missing terms cancel each other out and hence, fortuitously but fundamentally erroneously, the correct equation of motion was obtained!

Appendix B
Analytical Evaluation of b_{sr}, c_{sr} and d_{sr}

The method — or at least a method — for the analytical evaluation of the constants defined by equation (3.87) is illustrated here, first for b_{sr}.

Let us re-write

$$b_{sr} = \int_0^1 \phi_s \phi_r' \, \mathrm{d}\xi = \frac{1}{\lambda_r^4} \int_0^1 \phi_s \phi_r''''' \, \mathrm{d}\xi, \tag{B.1}$$

which, after successive integration by parts, yields

$$\frac{1}{\lambda_r^4} \left\{ [\phi_r'''' \phi_s - \phi_r''' \phi_s' + \phi_r'' \phi_s'' - \phi_r' \phi_s'''] \Big|_0^1 + \int_0^1 \phi_r' \phi_s'''' \, \mathrm{d}\xi \right\};$$

the last integrand may be written as $\phi_r' \lambda_s^4 \phi_s$, which leads to

$$(\lambda_r^4 - \lambda_s^4) b_{sr} = [\lambda_r^4 \phi_r \phi_s - \phi_r''' \phi_s' + \phi_r'' \phi_s'' - \phi_r' \phi_s'''] \big|_0^1 . \tag{B.2}$$

This can be evaluated for any particular set of the standard boundary conditions. Thus, for a cantilevered pipe, $\phi_r(1) = 2(-1)^r$, $\phi_r''(1) = \phi_r'''(1) = 0$, and $\phi_r(0) = \phi_r'(0) = 0$, $\phi_r''(0) = 2\lambda_r^2$, and similarly for ϕ_s (Bishop & Johnson 1960; Blevins 1979). Hence, after some manipulation, equation (B.2) gives

$$b_{sr} = \frac{4}{(-1)^{r+s} + (\lambda_s/\lambda_r)^2}. \tag{B.3}$$

For $r = s$, this clearly gives

$$b_{rr} = 2. \tag{B.4}$$

Working in a similar manner, the other entries of Table 3.1 may be determined — at least for $r \neq s$. For $r = s$, however, some of the expressions obtained with $r \neq s$ become singular and have to be determined in another way. An example is c_{sr} which is zero for pinned (simply-supported) ends and

$$c_{sr} = \frac{4\lambda_r^2 \lambda_s^2}{\lambda_r^4 - \lambda_s^4} (\lambda_r \sigma_r - \lambda_s \sigma_s) \{(-1)^{r+s} + 1\}$$

for clamped ends — clearly indeterminate for $r = s$. Hence, here a method will be presented for the evaluation of c_{rr}, which may be written as

$$c_{rr} = \int_0^1 \phi_r \phi_r'' \, d\xi \tag{B.5a}$$

$$= \phi_r' \phi_r \Big|_0^1 - \int_0^1 \phi_r' \phi_r' \, d\xi \tag{B.5b}$$

$$= \frac{1}{\lambda_r^4} \int_0^1 \phi_r'' \phi_r'''' \, d\xi \tag{B.5c}$$

$$= \frac{1}{\lambda_r^4} \left[\phi_r'' \phi_r''' \Big|_0^1 - \int_0^1 (\phi_r''')^2 \, d\xi \right]. \tag{B.5d}$$

Multiplying each of these by λ_r^4 and adding them together gives

$$4\lambda_r^4 c_{rr} = [\lambda_r^4 \phi_r' \phi_r + \phi_r'' \phi_r''']\Big|_0^1 + \int_0^1 [2\lambda_r^4 \phi_r'' \phi_r - \lambda_r^4 (\phi_r')^2 - (\phi_r''')^2] \, d\xi. \tag{B.6}$$

Now, for any $\phi_r = A \cos(\lambda_r \xi) + B \sin(\lambda_r \xi) + C \cosh(\lambda_r \xi) + D \sinh(\lambda_r \xi)$, it is easy to verify that the integrand in (B.6) is equal to $2\lambda_r^6[-A^2 - B^2 + C^2 - D^2]$. Hence,

$$4\lambda_r^4 c_{rr} = [\lambda_r^4 \phi_r' \phi_r + \phi_r'' \phi_r''']\Big|_0^1 + 2\lambda_r^6[-A^2 - B^2 + C^2 - D^2]. \tag{B.7}$$

For a clamped–clamped pipe, ϕ_r and ϕ_r' are zero at both limits, while $\phi_r''(1) = 2\lambda_r^2(-1)^{r+1}$, $\phi_r'''(1) = 2\lambda_r^3 \sigma_r (-1)^{r+1}$, $\phi_r''(0) = 2\lambda_r^2$, $\phi_r'''(0) = -2\lambda_r^3 \sigma_r$, $A = -1$, $B = \sigma_r$, $C = 1$, $D = -\sigma_r$, leading to

$$c_{rr} = \lambda_r \sigma_r (2 - \lambda_r \sigma_r). \tag{B.8}$$

For a pipe with pinned ends, $\phi_r = \sqrt{2} \sin \lambda_r \xi$, with $\lambda_r = r\pi$, the $\sqrt{2}$ factor ensuring orthonormality. In this case, it follows easily from (B.7) that

$$c_{rr} = -\lambda_r^2. \tag{B.9}$$

Appendix C
Destabilization by Damping: T. Brooke Benjamin's Work

An attempt to explain the phenomenon in simple terms was made by Benjamin (1963). A one-degree-of-freedom mechanical system subject to fluid flow is considered,

$$m\ddot{q} + c\dot{q} + kq = Q, \qquad Q = M\ddot{q} + C\dot{q} + Kq, \tag{C.1}$$

where the generalized force Q is associated with fluid forces. Consider then an impulsive disturbance applied to the solid at $t = 0$; the work done on the solid by the fluid forces is

$$W = \int_0^t Q\dot{q}\,\mathrm{d}t = \tfrac{1}{2}M\dot{q}^2 + \tfrac{1}{2}Kq^2 + C\int_0^t \dot{q}^2\,\mathrm{d}t. \tag{C.2}$$

This is also the energy lost by the fluid, from the unbounded store of kinetic energy possessed by the flow, so that

$$\mathscr{E} = T + V - W = \tfrac{1}{2}(m - M)\dot{q}^2 + \tfrac{1}{2}(k - K)q^2 - C\int_0^t \dot{q}^2\,\mathrm{d}t \tag{C.3}$$

is the total energy of the whole system relative to the original quiescent state.

Assuming that the fluid is inviscid, energy can only be dissipated by the solid, and so $\mathrm{d}\mathscr{E}/\mathrm{d}t \leq 0$ or

$$\mathscr{E} = \mathscr{E}_0 - c\int_0^t \dot{q}^2\,\mathrm{d}t, \tag{C.4}$$

where $\mathscr{E}_0$ is the energy level immediately after the initial disturbance. As compared to the total energy $\mathscr{E}$, which is not directly changed by the irreversible energy transfer proportional to C, a more useful measure of the degree of excitation is what may suitably be termed as 'the activation energy' E, which is the sum of $\mathscr{E}$ and the energy transferred to the solid by the nonconservative hydrodynamic forces, i.e.

$$E = \mathscr{E} + C\int_0^t \dot{q}^2\,\mathrm{d}t = \tfrac{1}{2}(m - M)\dot{q}^2 + \tfrac{1}{2}(k - K)q^2. \tag{C.5}$$

This is also the energy, relative to the quiescent state, involved in *conservative* energy exchanges between the kinetic and potential energies during oscillation. Combining (C.4) and (C.5),

$$E - \mathscr{E}_0 = (C - c)\int_0^t \dot{q}^2\,\mathrm{d}t, \tag{C.6}$$

which is the difference between the nonconservative energy transfer to the solid and dissipation within it. Hence, this represents the balance of energy converted *irreversibly* by the disturbance (not the actual gain in energy by the solid, since the conservative forces may also contribute to this).

Benjamin (1963) then considered three cases, corresponding to his three classes, A, B and C, of instability of compliant surfaces subjected to fluid flow.

(i) *Case of* $m > M$, $k > K$. If $c = C = 0$, a simple harmonic solution with $\omega = [(k - K)/(m - M)]^{1/2}$ is obtained, and $\mathscr{E} = E = \frac{1}{2}(k - K)\hat{q}^2$, where $\hat{q}$ is the amplitude. The total energy level is positive. For finite but small c and C, on the other hand, the frequency is little changed, but the oscillation is amplified for $c < C$, which means that the rate of irreversible energy transfer from the fluid to the solid exceeds the mean rate of dissipation — by reference to (C.6). The activation energy, $E \simeq \frac{1}{2}(k - K)\hat{q}^2$, must be positive to begin with (i.e. a positive $\mathscr{E}_0$ must be added in generating the disturbance) and if $C > c$ it steadily increases, even though $\mathscr{E}$ steadily decreases in view of (C.4). The energy of the initial excitation $\mathscr{E}_0$ is eventually lost and $\mathscr{E}$ becomes negative, but the disturbance continues to grow, because this is more than compensated by the transfer to the disturbance of energy by the infinite store in the fluid. This mechanism exemplifies Benjamin's (1960, 1963) class B instability, in which dissipation is stabilizing.

The case of class C, or Kelvin–Helmholtz, instability will not be considered here and we go directly to a situation exemplifying class A instability.

(ii) *Case of* $m < M$ *and* $k < K$. For $c = C = 0$ we once more have simple harmonic motion with frequency ω, but now the energy level of the disturbance is $\mathscr{E} = -\frac{1}{2}(K - k)\hat{q}^2$ and so is negative. This means that the absolute energy level of the whole system must be reduced in the process of creating a free oscillation: i.e. the system must be allowed *to do work* against the external forces providing the excitation. For small and finite c and C, oscillations are now amplified if $c > C$ and damped if $c < C$. Thus dissipation and energy transfer in this case have opposite effects as compared to (i). In particular, the effect of dissipation is always destabilizing. A physical interpretation is again provided by (C.6). The activation energy $E \simeq -\frac{1}{2}(K - k)\hat{q}^2$ is negative when the disturbance is first created (i.e. $\mathscr{E}_0 < 0$) and the amplitude of oscillation grows progressively by increases in the negative magnitude of E for $c > C$. The significance of E is perhaps made clearest as follows. Suppose that the irreversible processes were suddenly stopped, so that the oscillation continued at constant amplitude $\hat{q}$. Then E is the absolute energy level of the system if the same oscillation had been excited by external forces, and we know from the discussion above that E is essentially negative, increasing in magnitude with q. Hence it is readily appreciated that dissipation is destabilizing since it lowers the absolute energy level.

The preceding theoretical model provides the simplest possible demonstration of how the removal of energy by dissipation may destabilize a system. However, the system of equations (C.1) for case (ii) is more mathematical than physical since, for destabilization, $m < M$ is required; but M is the negative of the added mass [Section 2.2.1(a)], and so for a physical system $M < 0$ always, while $m > 0$, rendering $m < M$ impossible. Hence, an 'ordinary' one-degree-of-freedom system cannot be destabilized by dissipation; two modes and a travelling wave component in the motion are necessary (cf. Sections 3.2.2 and 3.5.6).

Benjamin (1963) recognized this and so considered next a system which is unbounded in the flow direction, x, and which is disturbed by a sinusoidal wave travelling in that direction — see also Yeo & Dowling (1987). The motion within an interval of x may be considered to comprise two modes $q = q_1(t) \sin \alpha x$ and $q = q_2(t) \cos \alpha x$, in which q_1 and q_2 are oscillations in quadrature. Through the action of the flow there may be coupling between these two modes, and so q is generally taken to be complex, on the understanding that $\mathcal{R}e\{q \exp(\mathrm{i}\alpha x)\}$ describes the physical disturbance. The equation of motion is still of the form of (C.1), but now we insist that $M < 0$, so that $m - M > 0$ always; k and K are real, but C can now be complex, $C = C_r + \mathrm{i}C_i$ (cf. the Coriolis term in the pipe problem).

Corresponding to (C.2), the energy transfer W averaged over x is given by the real part of the integral of $\frac{1}{2}Q^*\dot{q}$, where Q^* is the complex conjugate of Q. The term $\mathrm{i}C_i\dot{q}$ in Q makes no contribution to W, and so the expressions for $\mathscr{E}$ and E in (C.4)–(C.6) are as before, except that C is now replaced by C_r. Thus, $\mathrm{d}E/\mathrm{d}t$ takes the sign of $C_r - c$.

Representing the solutions of (C.1) by $q \exp(-\mathrm{i}\nu t)$, where ν is complex, we get

$$\nu = \frac{C_i}{2(m-M)}\left[-1 - \frac{\mathrm{i}(c - C_r)}{C_i} \pm \sqrt{1 + R + \frac{2\mathrm{i}(c - C_r)}{C_i}}\right], \tag{C.7}$$

where $R = [4(m-M)(k-K) + (c - C_r)^2]/C_i^2$. It is recognized that instability is indicated by $\mathcal{I}m(\nu) > 0$, where

$$\mathcal{I}m(\nu) = \frac{C_i}{2(m-M)}\left[-\frac{c - C_r}{C_i} \pm \left\{\frac{1}{2}\left[\sqrt{(1+R)^2 + \frac{4(c-C_r)^2}{C_i^2}} - (1+R)\right]\right\}^{1/2}\right],$$

$$\mathcal{R}e(\nu) = \frac{C_i}{2(m-M)}\left[-1 \pm \left\{\frac{1}{2}\left[\sqrt{(1+R)^2 + \frac{4(c-C_r)^2}{C_i^2}} + (1+R)\right]\right\}^{1/2}\right]. \tag{C.8}$$

Since $m - M > 0$, R may be positive or negative, depending on whether $k > K$ or otherwise. The following three cases may be distinguished. (a) When $R > 0$, $\mathcal{I}m(\nu) > 0$ for both solutions if $C_r > c$, which from (C.6) corresponds to $\mathrm{d}E/\mathrm{d}t > 0$ and hence to class B instability, i.e. to case (i) in the foregoing. (b) When $-1 < R < 0$, one solution is again of class B, but the other one is of class A, being unstable for $c > C_r$. (c) The case of $R < -1$ corresponds to class C instability, not considered here. Therefore, it is clear that for $-1 < R < 0$ and $c > C_r$ the physical system obeys the arguments given in (ii) in the foregoing and is thus destabilized by damping.

Appendix D
Experimental Methods for Elastomer Pipes

The purpose of this appendix is to present some of the techniques, developed over the past 30 years in the author's laboratories, for manufacturing elastomer pipes, shells or cylinders, as well as for determining some of their key physical properties. It is recalled that experimentation with elastomer flexible structures translates into low-pressure test-rigs, and hence easier experiments than with stiffer bodies, e.g. made of metal. In the case of pipes, the home-made ones are far superior to those commercially available, as explained for example in Section 3.5.6.

D.1 MATERIALS, EQUIPMENT AND PROCEDURES

In most cases, the material used for making the flexible bodies is a room-temperature vulcanizing (RTV) silicone rubber (e.g. 'Silastic E RTV' made by Dow Corning). It is supplied in a two-component kit, and the two fluids, one of which is the catalyst, are mixed in the prescribed ratio (typically 10:1) just before manufacture. The mixture is poured into a mould, cured, and then extracted, as described in what follows. Another essential item is a liquid agent supplied by the manufacturer for coating surfaces to which the silicone rubber should adhere; e.g. the edges of the metal strip sometimes embedded in pipes (Figure D.1), which are thus constrained to oscillate in 2-D. A 'releasing agent' is also available, for coating surfaces on which the silicone rubber should *not* adhere at all, e.g. the inner surface of the mould and the middle length of the metal strip. In what follows, we shall continue using the moulding of a pipe as an example; a few words on other structures are given in Section D.2.

Five basic pieces of equipment are required: (i) the mould; (ii) a large injector syringe; (iii) a vacuum pump; (iv) a supply of compressed air; (v) a temperature-controlled oven (optional).

A schematic of a mould for a pipe with an embedded metal strip is shown in Figure D.2(a). It is basically composed of (a) a split outer mould and (b) a split cylindrical core (or, when no metal strip exists, a whole cylinder). The outer mould is made of two solid-block halves; after the interfaces are ground flat, semicircular grooves are carefully milled in each with a ball end-mill, so as to produce a fine finish. Similar care should be taken to make sure that the split core when sandwiching the metal strip is cylindrical and of the required diameter. The mould can be made of Plexiglas to allow viewing while casting or, for better dimensional tolerances and robustness, of brass. The alignment of

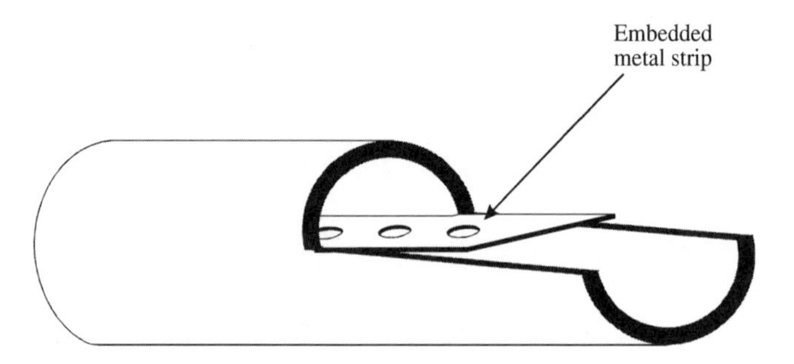

Figure D.1 Cut-away view of a pipe with an embedded metal strip — typically of 0.005 in (0.127 mm) feeler gauge. The holes are for equalizing the pressure in the two channels during flow testing, in case of small asymmetries.

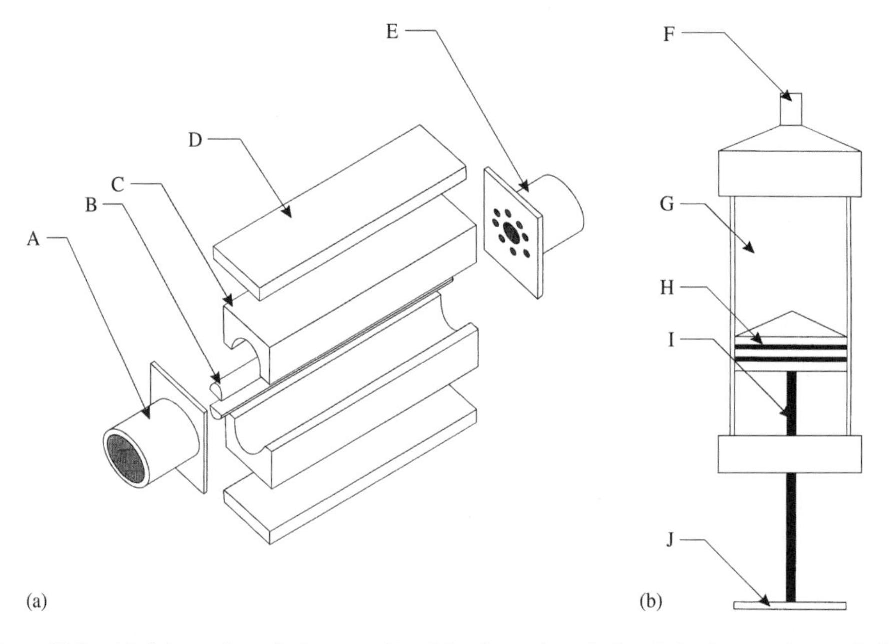

Figure D.2 (a) Schematic of the mould; (b) schematic of the injection syringe. A, lower end-support for connection with injector outlet; B, split cylinder core; C, split outer mould; D, reinforcing plate; E, upper end-support for holding overflow; F, injector outlet; G, transparent-wall injector; H, injector piston with O-rings; I, threaded rod; J, injector handle.

the two halves of the mould and of the components of the core is crucial, since it controls the quality of the final product: axial symmetry, straightness, central positioning of the metal strip, and so on. Hence, tight tolerances should be imposed, and dowel pins used to ensure correct assemblage every time. Long Plexiglas moulds should be reinforced with metal reinforcing plates. The end-supports serve (a) to support the central core and (b) to connect to the injector or collect some overflow (since the silicone rubber contracts a little during curing). All surfaces must be thoroughly cleaned and then treated with a *thin* film of either adhering or releasing agent, just before manufacture of the pipe.

The injector, Figure D.2(b), is an elephant-size syringe — typically 10 cm in diameter and 30 cm long. The two components of the silicone rubber are mixed in a beaker with the aid of an electric drill, and then poured into the syringe, typically filling $\frac{1}{10}$–$\frac{1}{8}$ of its volume. Then the top of the injector is connected to a vacuum pump, capable of generating a pressure of 0.1 atm approximately, 'to boil off' trapped air in the viscous mixture (of the consistency of bread dough), but not low enough to reach the boiling point of the silicone rubber itself; hence the piston in the injector must be leak-proof. Air is trapped not only by the folding of the mixture during pouring, but also in the form of small bubbles trapped during mixing, which cannot rise to the surface fast enough. The vacuum is applied and held long enough for the mixture to expand, filling half or two-thirds of the injector volume, allowing the larger bubbles to burst and the mixture to collapse. This cycle of (a) application of the vacuum, (b) holding it, and (c) releasing it gradually has to be repeated — perhaps up to 50 times — until application of the vacuum results in no noticeable change in volume.

The 'working time' available before the mixture begins to set varies from one silicone rubber to another, but it is typically 1–2 hours. Room-temperature curing takes about 72 hours, but in a temperature-controlled oven at 160°C this can be accelerated to 1 hour.

Once the mixture is de-aerated, it is injected into the lower end of the mould slowly, so as to rise in it at no more than 0.5 mm/s. The mould and injector are arranged in a vertical configuration and, usually, remain so connected during curing.

Extracting the casting from the mould is perhaps the most challenging aspect of the manufacturing process. Even with the mould-release agent, the casting does not simply slide off the central core, because of the vacuum that needs to be broken between the surfaces. An effective way is to put the pipe, with the core in it, on a long V-block and then apply compressed air (at no more than $\sim$ 140 psi or 1 MPa) at one end, to slightly expand and lift the pipe off the core; a little water lubrication helps to then draw the core out from the other end. In the case of a split core, the first half is removed in this way, but the second one has to be painstakingly eased out mechanically, by tapping it with a smaller rod carefully, so as not to damage the bond between the metal strip and the pipe.

Lower-quality, but easier to manufacture pipes and cylinders may be cast in glass tubes which, after curing, are broken and the core removed in the manner just described. The weakness here is the imperfect uniformity and straightness of the glass tubes.

For cantilevered pipe experiments, it is best to make the free end 'square' to the long pipe-axis at manufacture. If cutting a piece of the free end becomes necessary, however, it should be done with great care. A good way is to sandwich the pipe between a close-fitting rod inside and a shorter pipe outside with a square-cut end, then to slice the elastomer pipe with a sharp razor, slowly and with minimum local deformation.

D.2 SHORT PIPES, SHELLS AND CYLINDERS

For short pipes and shells it is more important than for other pipes that, in the experiments with flow, the transition from the metal supporting structure upstream to the flexible pipe be smooth and as disturbance-free as possible. Hence, in such cases an upstream adapter is actually cast integral to the elastomer pipe; in the experiments, the adapter is then screwed directly into the fluid-supply piping.

For obvious reasons, cylinders are the easiest to cast, unless they are instrumented; instrumented cylinders will be discussed separately in Volume 2. Finally, conical cylinders and pipes (such as those in Section 4.1) present no special difficulties.

D.3 FLEXURAL RIGIDITY AND DAMPING CONSTANTS

A very important thing to know is that boxes or drums of a given type of silicone rubber (e.g. Silastic E) have virtually the same physical properties, *only so long as they have the same 'lot number'*; otherwise, the properties of items cast from different boxes vary a great deal.

For pipes and cylinders, the two essential quantities to know are (i) the flexural rigidity, EI, and (b) the damping constants: α for viscoelastic Kelvin–Voigt damping and/or μ for hysteretic damping (Sections 3.3.2 and 3.3.5). In some cases, and for shells in all cases, the Poisson ratio, ν, is also needed.

The most convenient method for determining EI and the damping constants is from planar free-vibration tests on empty cantilevered vertical pipes — vertical because of inevitable sagging otherwise[†] — in which, typically, the first-mode natural frequency,

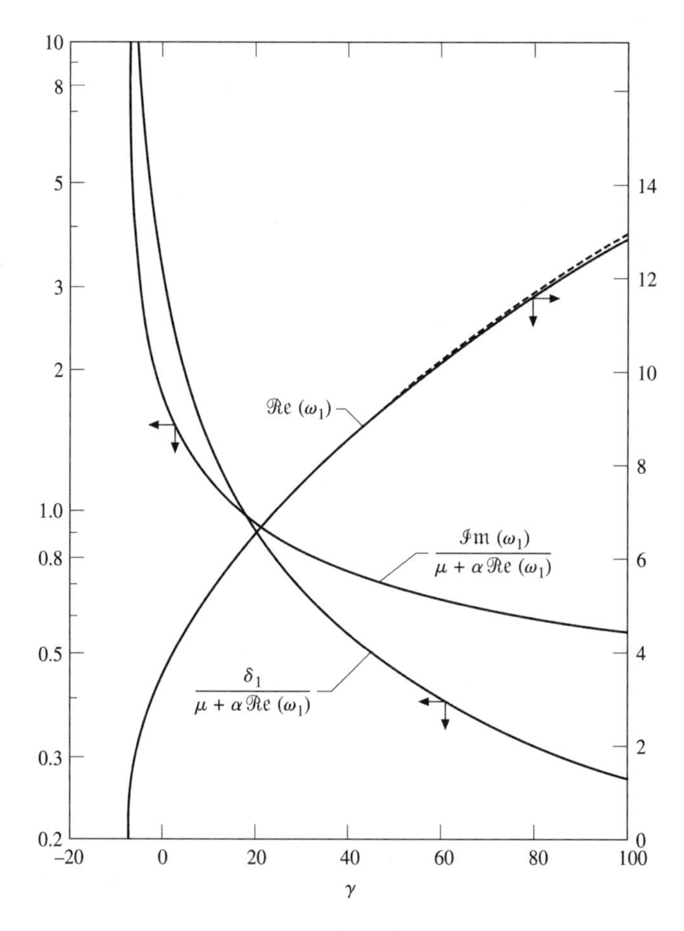

Figure D.3 Real and imaginary components of the first-mode frequency, ω_1, and the corresponding logarithmic decrement, δ_1: ——, 'exact' Galerkin solution; – – –, approximate Rayleigh method solution (Païdoussis & Des Trois Maisons 1971).

[†]If using very short horizontal pipes to avoid sagging, they may not fulfil the slenderness requirements for Euler–Bernoulli theory to apply.

Ω_1, and the logarithmic decrement, δ_1, are measured. These are then compared with the theoretical values, to determine EI and the damping constants; this is done in an indirect manner, as described in what follows, since Ω_1 and δ_1 are functions of the gravity parameter, γ. In the experiments the decaying pipe vibration can be sensed by a fibre-optic sensor or an optical tracking system (Section 5.8.1), both noncontacting, the signal from which can be processed electronically; see also Section D.4.

The equation of motion of the vertical empty pipe is a simplified form of (3.70), namely

$$(\alpha + \mu/\omega)\dot{\eta}'''' + \eta'''' - \gamma(1 - \xi)\eta'' + \gamma\eta' + \ddot{\eta} = 0, \tag{D.1}$$

the complex eigenfrequencies of which, $\omega_i = \mathcal{R}e(\omega_i) + i\mathcal{I}m(\omega_i)$, and hence the logarithmic decrement $\delta_i = 2\pi\mathcal{I}m(\omega_i)/\mathcal{R}e(\omega_i)$, may be found for any γ by the method of Section 3.3.6(b). In this way, Figure D.3 is constructed, for the first mode, $i = 1$. The dashed line in this figure is from a Rayleigh method approximation, yielding $[\mathcal{R}e(\omega_1)]^2/\gamma = (81/52) + (162/13\gamma)$.

However, Figure D.3 is not convenient for determining EI, since both the abscissa and ordinate, i.e. both ω_1 and γ, are functions of EI — cf. equations (3.71) and (3.73). Figure D.4 is therefore needed, where it is noted that

$$\gamma/[\mathcal{R}e(\omega_1)]^2 = g/[\mathcal{R}e(\Omega_1)]^2 L, \tag{D.2}$$

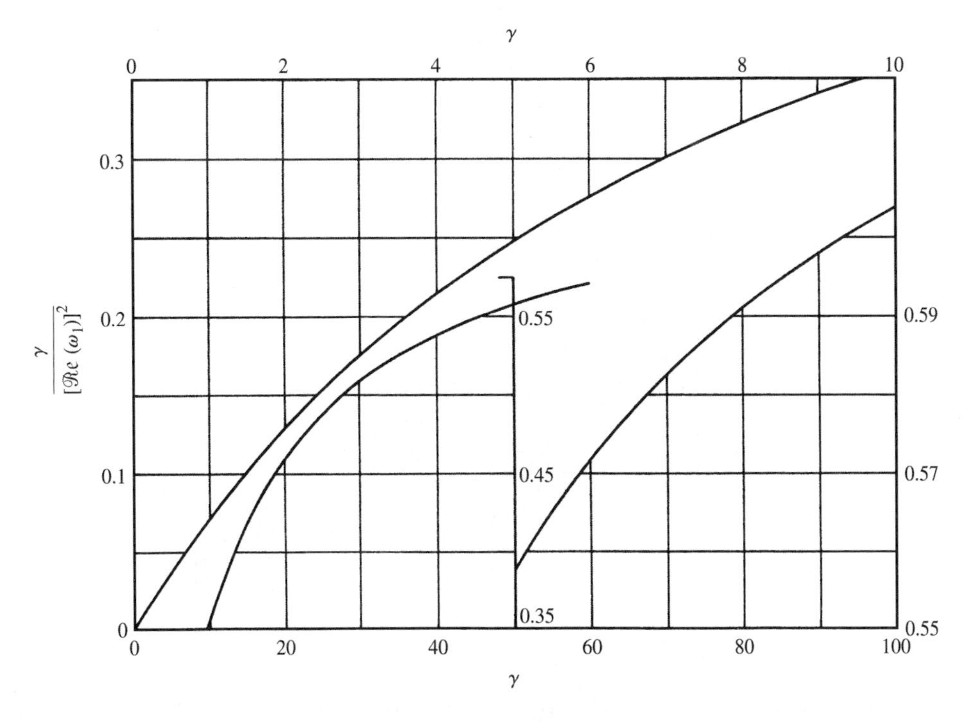

Figure D.4 Special diagram for determining the flexural rigidity of heavy, lightly damped cantilevers; note split scale for three different ranges of γ (Païdoussis & Des Trois Maisons 1971).

Ω_1 being the dimensional first-mode radian frequency. Hence, from the measured Ω_1 and equation (D.2), the ordinate in Figure D.4 is known, and hence γ may be determined; then, from the definition of γ, here mgL^3/EI, so can EI. Similarly, from the measured δ_1 and Figure D.3, $\mu + \alpha\mathcal{Re}(\omega_1)$ may be found. Thus, μ may be determined if the damping is supposed to be purely hysteretic ($\alpha = 0$); and so can α, if $\mu = 0$ is taken. If both are required for a more realistic representation of the damping, then two experiments are necessary with different pipes, e.g. pipes of different length.

The robustness of a particular damping model may be assessed by determining μ and/or α for second- and third-mode vibration also, and then utilizing another figure for these higher modes, similar to Figure D.4 and given in Païdoussis & Des Trois Maisons (1971).

In some cases, rather than commit oneself to a particular damping model, δ_1, δ_2, δ_3 and so on are determined separately and used directly when comparing with theory — see Section D.4.

Finally, to determine the Poisson ratio, ν, a sufficiently large cube of silicone rubber is cast, and is then weighed down with progressively heavier blocks, while its vertical compression and lateral expansion are measured with dial gauges and fibre-optic sensors.

D.4 MEASUREMENT OF FREQUENCIES AND DAMPING

Many different ways for measuring the damping are possible, but the methods described here are both simple and efficient. The following pieces of equipment are required: (i) a sensor, (ii) a signal recording device; also, optional but very useful are (iii) a small shaker, (iv) a band-pass filter, and (v) a digital signal analyser.

Exciting the pipe by flexing it and then releasing it generally works well for the first mode only. Trying to excite the second and third modes in this way is difficult if not

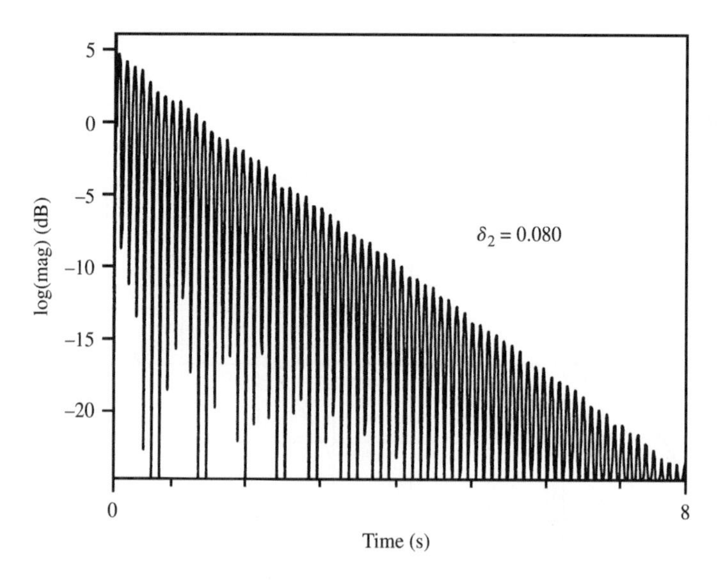

Figure D.5 The recorded signal for vibration of a pipe in its second mode (in dB) versus time, after filtering, from which $\delta_2 = [(\ln\ 10)/20](\text{slope})/f_2$ may be found, where 'slope' is the linear slope of the decaying peaks.

impossible. The optional shaker (which can simply be a small DC motor with a cam and a slider) is used to excite the pipe at precisely the frequency of the mode of interest. This way, when the shaker is abruptly removed, the pipe oscillates in only the mode concerned.

The sensor is calibrated to provide a linear response in the range of pipe motion, but calibration to real units is not necessary. As mentioned before, a noncontacting fibre-optic ('Fotonic') or optical-tracking ('Optron') system is ideal. The band-pass filter is used to remove from the recorded vibration signal components from modes other than that being measured.

A digital signal analyser with FFT capabilities is useful for (i) determining the modal frequency from PSD plots and (ii) providing log-amplitude versus time plots, such as that shown in Figure D.5, for the determination of δ_i. In cases where only a few cycles of free oscillation are possible, a Hilbert-transform of the signal can be helpful in the determination of the decay-envelope slope. In cases where adulteration from other modes is strong, the FFT of the signal can be edited to remove the unwanted components, and then an inverse FFT used to rebuild a clean wave form.[†]

Free-vibration tests are usually sufficiently accurate, but more sophisticated transfer-function Nyquist-type analysis can be done to find the logarithmic decrement in the lowest few modes with forced vibration tests, using a small shaker and a force transducer (Ewins 1975, 1985).

[†]The operation is given mathematically by 'envelope' $= |a + \mathrm{i}\{\mathrm{FFT}^{-1}[\mathrm{FFT}(a) * (-\mathrm{i})]\}|$, where a is the real amplitude.

Appendix E

The Timoshenko Equations of Motion and Associated Analysis

E.1 THE EQUATIONS OF MOTION

It is of interest to compare the equations of motion obtained by means of the Newtonian approach, equations (4.35) [see also Païdoussis & Laithier (1976) and Païdoussis *et al.* (1986)] with those developed from Hamilton's principle by Laithier & Païdoussis (1981). These two sets of equations are not identical.

The derivation of the equations of motion by Hamilton's principle is not a trivial task; indeed it is much more complex and laborious than the derivation relying on Newtonian mechanics. From Laithier & Païdoussis (1981), these equations are as follows:

$$
\begin{aligned}
&F_A - m\frac{\partial^2 w}{\partial t^2} - (M+m)g\psi \\
&\quad +(M+m)(L-x)g\,\frac{\partial \psi}{\partial x} + \delta\, T(L)\frac{\partial^2 w}{\partial x^2} + k'GA_p\left(\frac{\partial^2 w}{\partial x^2} - \frac{\partial \psi}{\partial x}\right) = 0, \qquad \text{(E.1)}\\
&\left[E + \frac{T(L)}{A_p}\right] I_p\,\frac{\partial^2 \psi}{\partial x^2} + [k'GA_p - (M+m)g(L-x)]\left(\frac{\partial w}{\partial x} - \psi\right) - (\bar{I}_f + \bar{I}_p)\frac{\partial^2 \psi}{\partial t^2} = 0,
\end{aligned}
$$

where $T(L)$ may also be expressed as $T(L) = \sigma_o A_p$, σ_o being the stress induced by externally imposed tension at $x = L$.

Comparing equations (4.35) and (E.1), one can see that (i) all the terms associated with fluid flow and gravity are identical, and (ii) the principal differences are associated with the tension term, $T(L)$. The differing terms may be summarized as in Table E.1 (for $\delta = 1$). Here it should be noted that in the dimensionless form of equations (E.1), the term $T_o \equiv \sigma_o/E$ is introduced — and it appears in the last line of terms of Table E.1; of course, since both σ_o and $\mathcal{T}_L$ are functions of $T(L)$, T_o and $\mathcal{T}_L$ are not independent, but are related through

$$\mathcal{T}_L = T_o(E/k'G)\Lambda. \qquad \text{(E.2)}$$

The differences in the tension terms appear to be inherently associated with the method of derivation of the equations: whether by the Newtonian or by the Hamiltonian approach, as discussed by Laithier & Païdoussis (1981). Since it has been impossible to reconcile these differences, a sensitivity analysis was undertaken to quantify their importance, insofar as the dynamical behaviour of a pipe conveying fluid is concerned, as discussed

Table E.1 Terms with differences in equations (4.35) and (E.1).

Equations	First equation	Second equation	
Equation (4.35)	$T(L)(\partial\psi/\partial x)$	$EI_p(\partial^2\psi/\partial x^2)$	$-T(L)(\partial w/\partial x - \psi)$
Dimensionless form	$\mathcal{T}_L(\partial\psi/\partial\xi)$	$\partial^2\psi/\partial\xi^2$	$-\mathcal{T}_L(\partial\eta/\partial\xi - \psi)$
Equations (E.1)	$T(L)(\partial^2 w/\partial x^2)$	$(E+\sigma_o)I_p\partial^2\psi/\partial x^2$	0
Dimensionless form	$\mathcal{T}_L(\partial^2\eta/\partial\xi^2)$	$(1+T_o)\partial^2\psi/\partial\xi^2$	0

below. Katsikadelis & Kounadis (1983) have conducted a similar exercise in the case of a Timoshenko column subjected to a follower force and reached similar conclusions.

In the calculations (conducted with the TRF theory) a tensile force is imposed on a clamped–clamped short pipe and then the eigenfrequencies of the lowest few modes are calculated, first with equations (4.35) and then with equations (E.1), to assess the importance of the differences in the two sets of equations, as shown in Table E.2. The calculations have been conducted for $\Lambda = 10$, $\beta = 0.5$, $\gamma = 10$, $\mu = \sigma = 0$ and $T_o = \sigma_o/E = 10^{-3}$; this corresponds to $\mathcal{T}_L \simeq 4.88 \times 10^{-2}$. It is noted that $T_o = 10^{-3}$ is an extremely high value; for ordinary steel, for instance, this tensile load is of the order of the yield strength of the material.

The first-mode eigenfrequencies obtained by the two sets of equations are compared in Table E.2. It is noted that the absolute values of the discrepancies remain of the same order as u is increased (they do not exceed 0.122 for $u \leq 3.4$); however, because the frequencies themselves tend to zero, the percentage discrepancies increase with flow, reaching 22% just prior to divergence. However, in terms of the critical flow velocity, the two sets of equations give virtually the same answer: $u_{cd} = 3.42$ by the Newtonian equations and $u_{cd} = 3.43$ by the Hamiltonian ones. Bearing in mind the extremely high value of tension utilized in these calculations, it may be said that the differences in the results for clamped–clamped pipes — at least from a practical viewpoint — are negligible.

Similar calculations have been conducted for cantilevered pipes, for the same set of parameters, except $\beta = 0.3$. For the third (critical) mode of the system, the absolute difference in the eigenfrequencies is less than 0.20 for $u \leq 4$; however, because the absolute values of the frequencies in this case do not tend to zero, the percentage differences do not increase dramatically with flow, and they remain less than 1%. The differences in the critical conditions are also quite small: $u_{cf} = 3.96$ by the Newtonian equations and $u_{cf} = 3.97$ by the Hamiltonian ones.

Table E.2 Comparison of the first-mode eigenfrequencies of a short clamped–clamped pipe under an initial tension obtained by Newtonian and Hamiltonian approaches.

Flow velocity u	Newtonian approach	Hamiltonian approach	Absolute difference (Relative difference, %)
0.01	$\omega = 9.8414$	$\omega = 9.8629$	0.0215 (0.22)
2.5	$\omega = 5.7465$	$\omega = 5.7788$	0.0323 (0.56)
3.4	$\omega = 0.5565$	$\omega = 0.6778$	0.1214 (21.8)

Therefore, the overall conclusion is that, except for extreme conditions, either set of equations may be used — leaving aside the thorny question as to which set is the correct one.

E.2 THE EIGENFUNCTIONS OF A TIMOSHENKO BEAM

Neglecting rotatory inertia and flow effects, equations (4.38) reduce to

$$\Lambda \frac{\partial^2 \eta}{\partial \xi^2} - \Lambda \frac{\partial \psi}{\partial \xi} - \frac{\partial^2 \eta}{\partial \tau^2} = 0, \qquad \frac{\partial^2 \psi}{\partial \xi^2} + \Lambda\left(\frac{\partial \eta}{\partial \xi} - \psi\right) = 0. \tag{E.3}$$

Eliminating η or ψ from one of these equations, one obtains

$$\frac{\partial^4 \eta}{\partial \xi^4} - \frac{1}{\Lambda}\frac{\partial^4 \eta}{\partial \xi^2\,\partial \tau^2} + \frac{\partial^2 \eta}{\partial \tau^2} = 0, \qquad \frac{\partial^4 \psi}{\partial \xi^4} - \frac{1}{\Lambda}\frac{\partial^4 \psi}{\partial \xi^2\,\partial \tau^2} + \frac{\partial^2 \psi}{\partial \tau^2} = 0, \tag{E.4}$$

and hence the eigenvalue problem associated with just one of them needs to be considered. Letting $\eta = Y(\xi)\exp(\mathrm{i}\omega\tau)$, $\psi = \Psi(\xi)\exp(\mathrm{i}\omega\tau)$, this is associated with

$$\frac{\mathrm{d}^4 Y}{\mathrm{d}\xi^4} + \frac{\omega^2}{\Lambda}\frac{\mathrm{d}^2 Y}{\mathrm{d}\xi^2} - \omega^2 Y = 0, \tag{E.5}$$

and the same for Ψ. Proceeding as for an Euler–Bernoulli beam (cf. Section 2.1.3), the solution of (E.5) is of the form

$$Y = \cosh q\xi + B \sinh q\xi + C \cos p\xi + D \sin p\xi, \tag{E.6}$$

where

$$2p^2 = \frac{\omega^2}{\Lambda} + \left(\frac{\omega^4}{\Lambda^2} + 4\omega^2\right)^{1/2}, \qquad 2q^2 = -\frac{\omega^2}{\Lambda} + \left(\frac{\omega^4}{\Lambda^2} + 4\omega^2\right)^{1/2}; \tag{E.7}$$

and similarly for Ψ. After some manipulation, making use of (E.3), it is easy to obtain† 'separated forms' of boundary conditions (4.39a,b) as follows:

(i) displacement zero:

$$Y = 0 \quad \text{and} \quad \Psi''' = 0; \tag{E.8a}$$

(ii) slope zero:

$$\Psi = 0 \quad \text{and} \quad \left(\frac{1}{\Lambda}\right) Y''' + \left[1 + \left(\frac{\omega}{\Lambda}\right)^2\right] Y' = 0; \tag{E.8b}$$

(iii) bending moment zero:

$$\Psi' = 0 \quad \text{and} \quad Y'' + \left(\frac{\omega^2}{\Lambda}\right) Y = 0; \tag{E.8c}$$

†For example, if $\Psi = 0$, differentiating the first of (E.3) with respect to ξ and then substituting the second one (with $\Psi = 0$) into it leads to the second of (E.8a).

(iv) shear zero:

$$\Psi'' = 0 \qquad \text{and} \qquad Y''' + \left(\frac{\omega^2}{\Lambda}\right) Y' = 0; \tag{E.8d}$$

where $(\ )' = \mathrm{d}/\mathrm{d}\xi$. It is evident that the boundary conditions in Ψ are simpler and, hence, for convenience, we proceed to determine the eigenfunctions associated with Ψ first.

For a *clamped–clamped beam*, after application of boundary conditions (E.8a) and (E.8b) one obtains the characteristic function

$$-2 - \frac{p^6 - q^6}{\omega^2}\frac{\sinh q}{q}\frac{\sin p}{p} + 2\cosh q \cos p = 0, \tag{E.9}$$

from which the eigenfrequencies ω_j, $j = 1, 2, \ldots$, may be obtained. The corresponding eigenfunctions are

$$\Psi_j(\xi) = -q_j^3(\cosh q_j - \cos p_j)\cosh(q_j\xi) + (q_j^3 \sinh q_j - p_j^3 \sin p_j)\sinh(q_j\xi)$$
$$+ q_j^3(\cosh q_j - \cos p_j)\cos(p_j\xi) + \frac{q_j^3}{p_j^3}(q_j^3 \sinh q_j - p_j^3 \sin p_j)\sin(p_j\xi), \tag{E.10}$$

where p_j and q_j are as in (E.7), but with ω_j replacing ω; these eigenfunctions are not normalized. Ψ_j and Y_j are related via

$$\omega_j^2 Y_j = \Psi_j'''. \tag{E.11}$$

Similarly, for a *cantilevered beam* one obtains

$$-2 + \frac{\omega^2}{\Lambda}\frac{\sinh q}{q}\frac{\sin p}{p} - \left(\frac{\omega^2}{\Lambda^2} + 2\right)\cosh q \cos p = 0, \tag{E.12}$$

and

$$\Psi_j(\xi) = \left(q_j \cosh q_j + \frac{q_j^3}{p_j^2}\cos p_j\right)\cosh q_j\xi - (q_j \sinh q_j + p_j \sin p_j)\sinh q_j\xi$$
$$- \left(q_j \cosh q_j + \frac{q_j^3}{p_j^2}\cos p_j\right)\cos p_j\xi - \frac{q_j^3}{p_j^3}(q_j \sinh q_j + p_j \sin p_j)\sin p_j\xi. \tag{E.13}$$

E.3 THE INTEGRALS I_{kn}

These integrals, appearing in equation (4.45), have been evaluated analytically by Luu (1983). A sample is given here

$$I_{kn}^{(4)} = \int_0^1 \Psi_n' Y_k \,\mathrm{d}\xi = \frac{1}{\omega_k^2}\int_0^1 \Psi_n' \Psi_k''' \,\mathrm{d}\xi = \frac{1}{\omega_k^2}(\Psi_n \Psi_k''')\Big|_0^1 - \mathcal{F}_{0,4},$$

$$\mathscr{F}_{0,4} = \int_0^1 \Psi_n \Psi_k'''' \, \mathrm{d}\xi = \frac{\omega_k^2 \omega_n^2}{\omega_n^2 - \omega_k^2} \left[-\frac{1}{\Lambda} (\Psi_n \Psi_k' - \Psi_n' \Psi_k) \Big|_0^1 \right.$$

$$\left. - \frac{1}{\omega_n^2} (\Psi_n''' \Psi_k - \Psi_n'' \Psi_k' + \Psi_n' \Psi_k'' + \Psi_n \Psi_k''') \Big|_0^1 \right]. \qquad \text{(E.14)}$$

However, some are much more complex and this is why they are not all given here. In any case, they may all be determined numerically.

Appendix F
Some of the Basic Methods of Nonlinear Dynamics

The purpose of this appendix is to outline some of the methods utilized in modern nonlinear dynamics. It is intended to help those not already familiar with them.

A common feature of analytical methods in nonlinear dynamics is the transformation of complicated dynamical systems into simpler ones. Another aspect is that nonlinear analysis often emphasizes qualitative features of system dynamics, frequently in the neighbourhood of critical parameter values. After introducing the concept of stability, we briefly go over several of the most commonly used methods.

F.1 LYAPUNOV METHOD

F.1.1 The concept of Lyapunov stability

Consider a system of differential equations of the form

$$\dot{\mathbf{x}} = \mathbf{f}(\mathbf{x}, t), \qquad \mathbf{x} \in \mathbb{R}^n. \tag{F.1}$$

It is assumed that there exists a unique solution $\overline{\mathbf{x}}(t)$ of (F.1) that is determined by the initial condition $\mathbf{x}_0$ at t_0. This solution is said to be stable if, starting close to $\overline{\mathbf{x}}(t)$ at a given time, it remains close to $\overline{\mathbf{x}}(t)$ for all later times. More precisely, $\overline{\mathbf{x}}(t)$ is stable if for any other solution $\mathbf{y}(t)$ of (F.1) and for every (arbitrarily small) $\epsilon > 0$ there exists a $\delta(\epsilon) > 0$, such that

$$|\mathbf{x}_0 - \mathbf{y}(t_0)| < \delta(\epsilon) \Rightarrow |\overline{\mathbf{x}}(t) - \mathbf{y}(t)| < \epsilon, \qquad \forall t \geq t_0. \tag{F.2}$$

The norm here may refer to the Euclidian or any other norm. Within this definition it makes no sense to use terms such as 'stable system' or 'stable differential equation', since one and the same differential equation may have stable as well as unstable solutions.

Unfortunately, with this definition of stability, periodic solutions of equation (F.1) are not stable! This is because a small change in the initial conditions may produce a slight change in the period of oscillations and for a reasonably large time, two solutions starting from nearby points will not remain nearby. It is therefore necessary to enlarge the concept of stability to cover also the case in which the phase trajectories remain close to each other. This is the purpose of the concept of a *stable trajectory* or of *orbital stability*.

The solution $\overline{\mathbf{x}}(t)$ has a stable trajectory, or is *orbitally stable*, if for every (arbitrarily small) $\epsilon > 0$ there exists a $\delta(\epsilon) > 0$ and a function $t_1(t)$ such that

$$|\mathbf{x}_0 - \mathbf{y}(t_0)| < \delta(\epsilon) \Rightarrow |\overline{\mathbf{x}}(t) - \mathbf{y}(t_1(t))| < \epsilon, \qquad \forall t \geq t_0. \tag{F.3}$$

In other words, if for every $\epsilon > 0$ there exists a δ-sphere about $\mathbf{x}_0$ such that all solutions which begin in this sphere at $t = t_0$ never leave the ϵ-tube about $\overline{\mathbf{x}}(t)$, then $\overline{\mathbf{x}}(t)$ is orbitally stable.

A solution $\overline{\mathbf{x}}(t)$ is attractive if there exists a $\delta > 0$ such that

$$|\mathbf{x}_0 - \mathbf{y}(t_0)| < \delta(\epsilon) \Rightarrow \lim_{t\to\infty} |\overline{\mathbf{x}}(t) - \mathbf{y}(t)| = 0. \tag{F.4}$$

A solution which is both stable and attractive is called asymptotically stable. It may very well be that a solution is attractive without being stable.

The stability of any given solution of (F.1) may be determined, without difficulty, if the general solution is known. However, for nonlinear systems this is almost never the case. One generally knows only certain particular solutions, usually stationary or periodic, whose stability is of interest. It has thus become necessary to search for means of determining stability without actually solving the differential equation.

Before proceeding further, it is noted that, by a simple coordinate transformation $\mathbf{y} = \mathbf{x} - \overline{\mathbf{x}}(t)$, it is easy to transform the original equation (F.1) into

$$\dot{\mathbf{y}} = \mathbf{g}(\mathbf{y}, t), \tag{F.5}$$

so that the solution $\overline{\mathbf{x}}(t)$ of (F.1) now corresponds to the trivial solution $\mathbf{y} = \mathbf{0}$ of (F.5); the stability of this solution corresponds to that of $\overline{\mathbf{x}}(t)$.

There are at least two different methods for determining the stability of a solution without actually solving the differential equations, both developed by Lyapunov.

F.1.2 Linearization

In Lyapunov's first method the right-hand side of equation (F.5) may be developed in a Taylor series with respect to $\mathbf{y}$,

$$\dot{\mathbf{y}} = A(t)\mathbf{y} + \mathbf{h}(\mathbf{y}, t), \tag{F.6}$$

where $\mathbf{h}(\mathbf{y}, t)$ includes all the nonlinear terms in equation (F.5). It is much easier to investigate the stability of the trivial solution of the linearized differential equation

$$\dot{\mathbf{y}} = A(t)\mathbf{y}, \tag{F.7}$$

rather than the solution $\mathbf{y} = \mathbf{0}$ of (F.5).

The method of first approximation is used to obtain results concerning the stability of the trivial solution of (F.6) by making use of the linearized equation (F.7). It can be applied differently in the following three cases: (i) A is not time-dependent (autonomous case); (ii) $A(t)$ is periodic; (iii) $A(t)$ is nonperiodic.

Autonomous case

If A is a *constant real-valued* matrix, then the solution of (F.7) is asymptotically stable if all the eigenvalues of A have negative real parts. On the other hand, if at least one eigenvalue of A has a positive real part, then the solution is unstable.

If there exist real numbers $\beta > 1$, $\alpha \geq 0$, such that the condition

$$|\mathbf{h}(\mathbf{y}, t)| \leq \alpha |\mathbf{y}|^{\beta} \tag{F.8}$$

is satisfied in a neighbourhood of $\mathbf{y} = \mathbf{0}$, then the stability of trivial solutions of the nonlinear system (F.6) can be obtained from the eigenvalues of A in the following form:

(i) if all of the eigenvalues of A have negative real parts, then the equilibrium solution of (F.6) is asymptotically stable;
(ii) if at least one eigenvalue of A has a positive real part, then the trivial solution of (F.6) is unstable.

These statements are valid, independently of the higher order terms; $\mathbf{h}(\mathbf{y}, t)$ need only satisfy the inequality (F.8). In cases where A has at least one eigenvalue with vanishing real part, then the effect of nonlinear terms must be taken into account in the stability analysis.

Periodic case

In the case where $A(t)$ is a periodic function of time, $A(t + T) = A(t)$, the stability of the trivial solution of (F.7) is obtained using Floquet theory: for the system (F.7), it can be shown that a *fundamental solution matrix* can be found, in the form

$$Y(t) = Z(t) \exp(tR), \tag{F.9}$$

where $Z(t)$ is also periodic of period T, $Z(t + T) = Z(t)$, and R is a (nonunique) constant *matrix* (Nayfeh & Mook 1979). Furthermore, if $Z(0)$ is equal to the identity matrix, then $Y(T) = \mathrm{e}^{TR}$. It thus becomes obvious that the stability of the trivial solution is related to the eigenvalues of the matrix e^{TR}, since after n periods the trivial solution will be related to e^{nTR}. These eigenvalues are called the *characteristic* or *Floquet multipliers*. Consequently, the trivial solution of (F.7) is asymptotically stable if and only if all of the eigenvalues of the matrix e^{TR} have absolute values (modulus) less than unity, while it is unstable if one of the eigenvalues has a modulus greater than 1. In the case where one or several eigenvalues have modulus equal to 1, then the trivial solution of (F.7) may be stable or unstable, depending on the structure of the Jordan normal form[†] corresponding to e^{TR}. Furthermore, linearization theorems as in the case of systems with constant coefficients can be proved, which means that it is possible to relate the stability of the nonlinear system to the stability of the linearized one.

In practice, an analytical determination of the fundamental matrix is very difficult, except in some special cases. Nevertheless, it can be found using perturbation methods, or using numerical schemes.

[†] The Jordan form is the simplest form a matrix can take, when transformed in the appropriate vector space (Hirsch & Smale 1974).

It should be mentioned that the Floquet theory presented here may be used equally well to determine the stability of periodic solutions. Indeed, let us consider again the original system of equation (F.1) when a *periodic* solution $\overline{\mathbf{x}}(t) = \overline{\mathbf{x}}(t + T)$ exists. To study the stability of this periodic orbit, we again linearize or perturbe the differential equation about $\overline{\mathbf{x}}$, $\mathbf{x}(t) = \overline{\mathbf{x}}(t) + \mathbf{u}(t)$, to obtain

$$\dot{\mathbf{u}} = \mathrm{D}\mathbf{f}\,\overline{\mathbf{x}}(t)\mathbf{u}, \tag{F.10}$$

where $\mathrm{D}\mathbf{f}\,\overline{\mathbf{x}}(t)$ is the Jacobian matrix function of the vector field $\mathbf{f}$ evaluated along $\overline{\mathbf{x}}(t)$. Since $\overline{\mathbf{x}}(t)$ is periodic, the linear system (F.10) has exactly the same form as before, which means that the perturbation $\mathbf{u}$(t) will grow or decay depending on the Floquet multipliers.

As an example, consider the case discussed in the early part of Section 5.8.1. The system is of fourth-order, so that, to determine the stability of the periodic solution, four independent initial conditions are chosen corresponding to the identity matrix,

$$u_0^1 = \{1, 0, 0, 0\}^{\mathrm{T}}, \quad u_0^2 = \{0, 1, 0, 0\}^{\mathrm{T}}, \quad u_0^3 = \{0, 0, 1, 0\}^{\mathrm{T}}, \quad u_0^4 = \{0, 0, 0, 1\}^{\mathrm{T}}, \tag{F.11}$$

and a numerical solution is obtained for each of them, after one period T: $u^i(T)$, $i = 1, \ldots, 4$. The fundamental matrix $u^i(T)$ is then constructed,

$$[Y] \equiv [u^1(T), u^2(T), u^3(T), u^4(T)], \tag{F.12}$$

and the eigenvalues of $[Y]$ determine the stability of the periodic trajectory. It should be mentioned that in the case of a periodic orbit, one multiplier associated with the periodicity of the orbit $\overline{\mathbf{x}}(t)$ is always unity, so that the stability is determined by the remaining eigenvalues. In the case of a cubic nonlinearity, as in the case of the pipe conveying fluid, if a second multiplier crosses the unit circle in the complex plane at $+1$, either a *transcritical* or a *pitchfork* bifurcation occurs, and the original orbit $\overline{\mathbf{x}}(t)$ becomes unstable while a new periodic orbit is created. If a multiplier crosses the unit circle at -1, then a *period-doubling* bifurcation is indicated, i.e. a new periodic orbit with twice the original period T emerges (Guckenheimer & Holmes 1983; Kubicek & Marek 1983).

Aperiodic case — Lyapunov exponents

In the last two sections, it was shown that the asymptotic stability of the linear autonomous or periodic system implied the asymptotic stability of the trivial solution of the complete nonlinear system, but this is no longer the case when the coefficients of the linear system are arbitrary functions of time (Hagedorn 1981). However, the notions of 'local stability' or sensitivity to initial conditions may still be important. They are both related to the Lyapunov exponents that are discussed in Section 5.8.1.

F.1.3 Lyapunov direct method

This method of Lyapunov can often be used to determine the stability of the trivial solution of equation (F.5) when the information obtained from the linearization is inconclusive. Lyapunov theory covers a large area (Lasalle & Lefschetz 1961; Hagedorn 1981), and we shall examine only a very small part of it. In the following, and in the rest of this appendix, we consider only autonomous systems of the type

$$\dot{\mathbf{x}} = \mathbf{f}(\mathbf{x}), \quad \mathbf{x} \in \mathbb{R}^n. \tag{F.13}$$

The basic idea of the method is as follows. Suppose that we wish to determine the stability of a fixed point $\overline{\mathbf{x}}$ of the vector field (F.13). Roughly speaking, according to the previous definitions of stability it would be sufficient to find a neighbourhood U of $\overline{\mathbf{x}}$, for which orbits starting in U remain in U for all positive time. This condition would be satisfied if it could be shown that the vector field is either tangent to the boundary of U or pointing inwards towards $\overline{\mathbf{x}}$. This situation should remain true even as U is shrunk down into $\overline{\mathbf{x}}$. Lyapunov's method provides a way of making this precise. Let $V(\mathbf{x})$ be a scalar function with $V(\overline{\mathbf{x}}) = 0$, such that $V(\mathbf{x}) = \text{constant}$ is a hypersurface encircling $\overline{\mathbf{x}}$, with $V(\mathbf{x}) > 0$ in a neighbourhood of $\overline{\mathbf{x}}$. Now recall that the gradient of V, ∇V, is a vector perpendicular to the surface in the direction of increasing V. So, if the vector field were always to be either tangent or pointing inwards for each of these surfaces surrounding $\overline{\mathbf{x}}$, one would have

$$\nabla V \cdot \dot{\mathbf{x}} \leq 0. \tag{F.14}$$

The following theorem makes these ideas more precise.

Theorem. Consider the vector field (F.13). Let $\overline{\mathbf{x}}$ be a fixed point and let V: $U \rightarrow \mathbb{R}$ be a C^1 function† defined on some neighbourhood U of $\overline{\mathbf{x}}$. If
(i) $V(\overline{\mathbf{x}}) = 0$, and $V(\mathbf{x}) > 0$ for $\mathbf{x} \neq \overline{\mathbf{x}}$,
(ii) $\dot{V}(\mathbf{x}) \leq 0$ in $U - \{\overline{\mathbf{x}}\}$,
then $\overline{\mathbf{x}}$ is stable. Moreover, if
(iii) $\dot{V}(\mathbf{x}) < 0$ in $U - \{\overline{\mathbf{x}}\}$,
then $\overline{\mathbf{x}}$ is asymptotically stable. We remark that if U can be chosen to be all of $\mathbb{R}^n$, then $\overline{\mathbf{x}}$ is said to be *globally asymptotically* stable if (i) and (iii) hold.

Functions which satisfy the theorem above are called Lyapunov functions. The theorem, however, contains no hint as to how a function $V(\mathbf{x})$ may be found in any given case. For differential equations which describe the behaviour of a physical system, it is often possible to deduce a suitable Lyapunov function by using general physical principles. It can be proved (Krasovskii 1963) that for every differential equation with trivial solution $\mathbf{x} = \mathbf{0}$, there indeed exists a Lyapunov function which may determine the stability or instability of the solution. In many cases, however, it just cannot be found. There are a multitude of procedures which have been proposed for the systematic construction of these functions (Hagedorn 1981), but they are either too complicated or suited only for certain classes of differential equations.

F.2 CENTRE MANIFOLD REDUCTION

Centre manifold reduction is basically a process of reducing the dimension of a system of ordinary differential equations in the neighbourhood of an equilibrium point (Carr 1981; Guckenheimer & Holmes 1983). The method involves restricting attention to an invariant

†That is, let V be a real continuous function defined in an open subset U of $\mathbb{R}^n$ that includes $\overline{\mathbf{x}}$.

subspace, *the centre manifold*, which contains all of the essential behaviour of the system in this neighbourhood as $t \to \infty$.

This method is applicable to systems which, when linearized about an equilibrium point, have some eigenvalues with zero real parts, and others with negative real parts (if an eigenvalue has a positive real part, the centre manifold will not be attractive as $t \to \infty$ and therefore becomes useless); the general framework is the following:

$$\begin{aligned} \dot{\mathbf{x}} &= \mathbf{A}\mathbf{x} + \mathbf{f}(\mathbf{x}, \mathbf{y}), \qquad \mathbf{x} \in \mathbb{R}^{n_c}, \\ \dot{\mathbf{y}} &= \mathbf{B}\mathbf{y} + \mathbf{g}(\mathbf{x}, \mathbf{y}), \qquad \mathbf{y} \in \mathbb{R}^{n_s}, \end{aligned} \tag{F.15}$$

where n_c is the dimension of $\mathbf{A}$ which has eigenvalues with zero real part, and n_s is the dimension of $\mathbf{B}$ which has eigenvalues with negative real part. Furthermore, it is assumed that $\mathbf{f}(\mathbf{0}, \mathbf{0}) = \mathbf{g}(\mathbf{0}, \mathbf{0}) = \mathrm{D}\mathbf{f}(\mathbf{0}, \mathbf{0}) = \mathrm{D}\mathbf{g}(\mathbf{0}, \mathbf{0}) = \mathbf{0}$.

It is obvious that the components of the solution of the *linearized* equations corresponding to $\mathbf{y}$ will decay as $t \to \infty$ and hence the motion of the linearized system will asymptotically approach the space E^c spanned by the eigenvectors of $\mathbf{A}$. Centre manifold theory ensures that this picture (based so far on the linearized equations) extends to the full nonlinear equations, as follows.

There exists a subspace $W^c(\mathbf{0})$, the centre manifold, which is tangent to the subspace E^c at the equilibrium point and which is invariant under the flow generated by the nonlinear equations. All solutions which start sufficiently close to the equilibrium point will tend asymptotically to the centre manifold. Furthermore, the stability of the equilibrium point in the full nonlinear equations is the same as its stability when restricted to the flow on the centre manifold. Also, any additional equilibrium points or limit cycles which emerge in the neighbourhood of the given equilibrium point on the centre manifold are guaranteed to exist in the full nonlinear equations (Carr 1981).

The next question now is: how to find the centre manifold? To answer this question, we introduce the function $\mathbf{h}$:

$$\mathbf{y} = \mathbf{h}(\mathbf{x}), \tag{F.16}$$

with

$$\mathbf{h}(\mathbf{0}) = \mathbf{0}, \qquad \mathrm{D}\mathbf{h}(\mathbf{0}) = \mathbf{0}, \tag{F.17}$$

such that it defines an invariant centre manifold for (F.15). Differentiating (F.16) with respect to time implies that

$$\dot{\mathbf{y}} = \mathrm{D}\mathbf{h}(\mathbf{x})\dot{\mathbf{x}}, \tag{F.18}$$

with

$$\dot{\mathbf{x}} = \mathbf{A}\mathbf{x} + \mathbf{f}(\mathbf{x}, \mathbf{h}(\mathbf{x})), \qquad \dot{\mathbf{y}} = \mathbf{B}\mathbf{h}(\mathbf{x}) + \mathbf{g}(\mathbf{x}, \mathbf{h}(\mathbf{x})). \tag{F.19}$$

Substituting (F.18) and the first of equations (F.19) into the second gives

$$\mathrm{D}\mathbf{h}(\mathbf{x})[\mathbf{A}\mathbf{x} + \mathbf{f}(\mathbf{x}, \mathbf{h}(\mathbf{x}))] = \mathbf{B}\mathbf{h}(\mathbf{x}) + \mathbf{g}(\mathbf{x}, \mathbf{h}(\mathbf{x})). \tag{F.20}$$

The first of equations (F.19) captures the essential dynamics of (F.15), and other techniques are necessary to investigate the flow on the centre manifold; on the other hand, (F.20) represents a partial differential equation that $\mathbf{h}(\mathbf{x})$ must satisfy. Consequently, to find a centre manifold it is necessary to solve (F.20), which can be a more difficult problem than the original one! Fortunately, only an approximate solution of (F.20) needs be computed, usually in terms of power series of $\mathbf{x}$ to any desired degree of accuracy. Examples of how the centre manifold is determined can be found in Rand & Armbruster (1987) and Wiggins (1990), and one specific example is given in Appendix H.

F.3 NORMAL FORMS

The central idea of the method of normal forms is to use a coordinate transformation to simplify or eliminate nonlinear terms in a dynamical system (Arnold 1983; Guckenheimer & Holmes 1983; Wiggins 1990). To demonstrate how the method works, we consider a differential equation of the form of the first of (F.19), with the nonlinear terms representing homogeneous polynomials of order k,

$$\dot{\mathbf{x}} = \mathbf{A}\mathbf{x} + \epsilon \mathbf{f}_k(\mathbf{x}), \tag{F.21}$$

where $\epsilon \ll 1$ is a real number used as a book-keeping device. In other words, $\mathbf{f}_k(\mathbf{x})$ belongs to the space H_k which is spanned by the vector-valued mononomials

$$\mathbf{e}_{i,k} = x_1^{k_1} x_2^{k_2} \cdots x_n^{k_n} \mathbf{e}_i, \tag{F.22}$$

where $k = k_1 + k_2 + \cdots + k_n$ is the order of the polynomial $\mathbf{f}_k(\mathbf{x})$ and $\mathbf{e}_i$ are unit orthogonal vectors in $\mathbb{R}^n$ (for example, H_2 is spanned by the three 'vectors' $\mathbf{e}_{1,2} = x^2\mathbf{e}_1$, $\mathbf{e}_{2,2} = xy\mathbf{e}_2$, $\mathbf{e}_{3,2} = y^2\mathbf{e}_3$).

The aim of the method is to find a coordinate transformation,

$$\mathbf{x} = \mathbf{y} + \epsilon \mathbf{h}_1(\mathbf{y}), \tag{F.23}$$

such that (F.21) takes the 'simplest possible form', the so-called *normal form*,

$$\dot{\mathbf{y}} = \mathbf{A}\mathbf{y} + \epsilon \mathbf{g}_1(\mathbf{y}). \tag{F.24}$$

Substituting (F.23) into (F.21) yields

$$\dot{\mathbf{y}} + \epsilon \mathbf{D}\mathbf{h}_1(\mathbf{y})\dot{\mathbf{y}} = \mathbf{A}\mathbf{y} + \epsilon \mathbf{A}\mathbf{h}_1(\mathbf{y}) + \epsilon \mathbf{f}_k(\mathbf{y} + \epsilon \mathbf{h}_1(\mathbf{y})), \tag{F.25}$$

where $\mathbf{D}\mathbf{h}_1$ is the Jacobian of $\mathbf{h}_1$. Using (F.24) to eliminate $\dot{\mathbf{y}}$ and equating coefficients of like powers in ϵ leads to

$$\mathbf{g}_1(\mathbf{y}) + \mathbf{D}\mathbf{h}_1(\mathbf{y})\mathbf{A}\mathbf{y} - \mathbf{A}\mathbf{h}_1(\mathbf{y}) = \mathbf{f}_k(\mathbf{y}) \tag{F.26}$$

or

$$\mathcal{L}_A[\mathbf{h}_1(\mathbf{y})] = \mathbf{D}\mathbf{h}_1(\mathbf{y})\mathbf{A}\mathbf{y} - \mathbf{A}\mathbf{h}_1(\mathbf{y}) = \mathbf{f}_k(\mathbf{y}) - \mathbf{g}_1(\mathbf{y}); \tag{F.27}$$

$\mathcal{L}_A$ is known as the Lie or Poisson bracket of the vector fields $\mathbf{A}\mathbf{y}$ and $\mathbf{h}_1(\mathbf{y})$ (Arnold 1988).

Ideally, one would like to remove all nonlinear terms using successive transformations, in order to reduce the vector field to its linear part, i.e. to transform (F.21) into $\dot{\mathbf{y}} = \mathbf{A}\mathbf{y}$. This condition means that $\mathbf{g}_1(\mathbf{y})$ in equation (F.27) is zero, i.e.

$$\mathbf{f}_k(\mathbf{y}) = \mathcal{L}_A[\mathbf{h}_1(\mathbf{y})]. \tag{F.28}$$

It can be shown easily that $\mathcal{L}_A$ is a linear map from $H_k \rightarrow H_k$ (i.e. it can be represented by a matrix), and that the eigenvalues $\Lambda_{k,i}$ of this linear map are related to the eigenvalues

λ_i of $\mathbf{A}$:

$$\Lambda_{k,i} = k_1\lambda_1 + k_2\lambda_2 + \cdots + k_k\lambda_k - \lambda_i. \tag{F.29}$$

Consequently, the inverse operator $\mathscr{L}_A^{-1}$ exists if and only if $\Lambda_{k,i} \neq 0$. Introducing the subset X_k of H_k representing the eigenvectors of $\mathscr{L}_A$ with nonzero eigenvalue, it can be said that the components of $\mathbf{f}_k(\mathbf{y})$ lying in X_k can be eliminated by a proper choice of $\mathbf{h}_1(\mathbf{y})$, while the components of $\mathbf{f}_k(\mathbf{y})$ not lying in X_k cannot be eliminated. In other words, the terms of $\mathbf{f}_k(\mathbf{y})$ lying in the range of $\mathscr{L}_A$ can be eliminated, while those lying in the kernel of $\mathscr{L}_A$ have to stay.

From the analysis, three important characteristics become apparent: (i) the normal form method is *local*, since the coordinate transformation is generated in the neighbourhood of a known solution (usually a fixed point for vector fields); (ii) the coordinate transformation is a nonlinear function of the dependent variables, but it is found by solving a *linear* problem; (iii) the structure of the normal form is determined entirely by $\mathbf{A}$, since the transformation depends only on the eigenvalues of $\mathbf{A}$ — see equation (F.29).

Examples of how to find normal forms may be found in many books, e.g. Guckenheimer & Holmes (1983), Wiggins (1990), Arrowsmith & Place (1990). Because of its importance, here we consider the case of a two-dimensional system with purely imaginary eigenvalues:

$$\dot{\mathbf{x}} = \begin{bmatrix} 0 & 1 \\ -1 & 0 \end{bmatrix} \mathbf{x} + \epsilon \begin{bmatrix} \alpha_1 x_1^3 + \alpha_2 x_1^2 x_2 + \alpha_3 x_1 x_2^2 + \alpha_4 x_2^3 \\ \alpha_5 x_1^3 + \alpha_6 x_1^2 x_2 + \alpha_7 x_1 x_2^2 + \alpha_8 x_2^3 \end{bmatrix} \tag{F.30}$$

and, because of its simplicity, we follow the methodology developed by Nayfeh (1993). Equation (F.30) is first transformed into a single complex-valued equation using the transformation

$$x_1 = \zeta - \bar{\zeta}, \qquad x_2 = \mathrm{i}(\zeta - \bar{\zeta}), \tag{F.31}$$

to obtain

$$\begin{aligned} \dot{\zeta} = \mathrm{i}\zeta + \tfrac{1}{2}\epsilon \left[(\alpha_1 - \mathrm{i}\alpha_5)(\zeta + \bar{\zeta})^3 + \mathrm{i}(\alpha_2 - \mathrm{i}\alpha_6)(\zeta + \bar{\zeta})^2(\zeta - \bar{\zeta}) \right. \\ \left. -(\alpha_3 - \mathrm{i}\alpha_7)(\zeta + \bar{\zeta})(\zeta - \bar{\zeta})^2 - \mathrm{i}(\alpha_4 - \mathrm{i}\alpha_8)(\zeta - \bar{\zeta})^3\right]. \end{aligned} \tag{F.32}$$

Using the methodology described previously, we assume

$$\zeta = \eta + \epsilon h(\eta, \bar{\eta}) \qquad \text{and} \qquad \eta = \mathrm{i}\eta + \epsilon g(\eta, \bar{\eta}). \tag{F.33}$$

Substituting (F.33) into (F.32) and equating the coefficients of ϵ on both sides yields

$$\begin{aligned} g + \mathrm{i}\left(\frac{\partial h}{\partial \eta}\eta - \frac{\partial h}{\partial \bar{\eta}}\bar{\eta} - h\right) \\ = \frac{1}{2}[(\alpha_1 - \mathrm{i}\alpha_5)(\eta + \bar{\eta})^3 + \mathrm{i}(\alpha_2 - \mathrm{i}\alpha_6)(\eta + \bar{\eta})^2(\eta - \bar{\eta}) \\ - (\alpha_3 - \mathrm{i}\alpha_7)(\eta + \bar{\eta})(\eta - \bar{\eta})^2 - \mathrm{i}(\alpha_4 - \mathrm{i}\alpha_8)(\eta - \bar{\eta})^3]. \end{aligned} \tag{F.34}$$

Next, the function h must be chosen so as to eliminate the nonresonance terms, i.e. all nonlinear terms that do not produce secular terms (which implies that resonance terms *are* those producing secular terms). The form of equation (F.34) suggests choosing h in the form

$$h = \Gamma_1\eta^3 + \Gamma_2\eta^2\bar{\eta} + \Gamma_3\eta\bar{\eta}^2 + \Gamma_4\bar{\eta}^3. \tag{F.35}$$

Substituting (F.35) into (F.34) leads to

$$g - 4(a + \mathrm{i}b)\eta^2\overline{\eta} + f_1(\Gamma_1)\eta^3 + f_3(\Gamma_3)\eta\overline{\eta}^2 + f_4(\Gamma_4)\overline{\eta}^3 = 0, \tag{F.36}$$

where $f_i(\Gamma_i)$ represent functions of Γ_i, $i = 1, 3, 4$ (not given here for brevity) and of the coefficients α_j, and

$$a = (3\alpha_1 + \alpha_3 + \alpha_6 + 3\alpha_8)/8, \qquad b = (\alpha_2 + 3\alpha_4 - 3\alpha_5 - \alpha_7)/8. \tag{F.37}$$

Equation (F.36) is independent of Γ_2, indicating that $\eta^2\overline{\eta}$ is a resonance term. This can be explained by using a 'multiple scales' approach: carrying out a straighforward expansion to order ϵ in (F.32) by putting $\zeta = A\ \exp(\mathrm{i}t)$, one finds that the term proportional to $\zeta^2\overline{\zeta}$ produces a secular term proportional to $A^2\overline{A}\ t\ \exp(\mathrm{i}t)$, whereas the remaining ones do not.

It is therefore possible to choose Γ_1, Γ_3 and Γ_4 in (F.36) to eliminate the nonresonance terms, thereby reducing g to the form

$$g = 4(a + \mathrm{i}b)\eta^2\overline{\eta}, \tag{F.38}$$

and the normal form, to first order, is

$$\dot{\eta} = \mathrm{i}\eta + 4\epsilon(a + \mathrm{i}b)\eta^2\overline{\eta}. \tag{F.39}$$

The normal form (F.39) can now be expressed in polar coordinates, using

$$\eta = \tfrac{1}{2}r\ \exp(\mathrm{i}\beta). \tag{F.40}$$

Substituting (F.40) into (F.39) and separating real and imaginary parts yields

$$\dot{r} = \epsilon a r^3, \qquad \dot{\beta} = 1 + \epsilon b r^2. \tag{F.41}$$

The analysis in which equation (F.30) is treated with real rather than complex variables can be found in Nayfeh (1993). The final result is of course the same, but the algebra involved is much more complicated.

F.4 THE METHOD OF AVERAGING

The method of averaging, originally due to Krylov & Bogoliubov (1947) is particularly useful for determining periodic solutions of weakly nonlinear problems or small perturbations of a linear oscillator. In contrast to the treatment of normal forms presented previously, the method of averaging can be extended easily to the case of *nonautonomous* systems, i.e. systems in which time appears explicitly [see, e.g. Semler & Païdoussis (1996), for the treatment of a nonautonomous system with normal form theory]. To begin with, let us consider the nonlinear harmonic oscillator

$$\ddot{x} + \omega_0^2 x = \epsilon f(x, \dot{x}). \tag{F.42}$$

When $\epsilon = 0$, the solution of (F.42) can be written as

$$x(t) = r\ \cos(\omega_0 t + \beta), \tag{F.43}$$

where r and β are constants. When $\epsilon \neq 0$, the solution of (F.42) can still be expressed in the form of (F.43) provided that r and β are considered to be functions of t rather than constants. Since (F.42) and (F.43) constitute two equations for the three variables x, r and β, we have to find an additional equation or impose a constraint condition. For example, it is convenient to assume that the velocity has the same form as for the case when $\epsilon = 0$, i.e.

$$\dot{x}(t) = -r(t)\omega_0 \sin(\omega_0 t + \beta(t)). \tag{F.44}$$

Differentiating (F.43) with respect to time t and comparing the result with (F.44) leads to

$$\dot{r}(t) \cos(\omega_0 t + \beta(t)) - r\dot{\beta}(t) \sin(\omega_0 t + \beta(t)) = 0. \tag{F.45}$$

Equation (F.42) is now written in terms of $r(t)$ and $\beta(t)$ by finding $\ddot{x}$ through differentiation of (F.44) with respect to t. By using the resulting equation together with the constraint relation (F.45) and solving for $\dot{r}$ and $\dot{\beta}$, after some algebra we obtain

$$\dot{r} = -(\epsilon/\omega_0) f(r, \beta) \sin\psi, \qquad r\dot{\beta} = -(\epsilon/\omega_0) f(r, \beta) \cos\psi, \tag{F.46}$$

where $\psi = \omega_0 t + \beta$. For small ϵ, $\dot{r}$ and $\dot{\beta}$ are small; this means that r and β vary much more slowly with t than ψ. In other words, r and β hardly change during the period of oscillation $2\pi/\omega_0$ of $\sin\psi$ and $\cos\psi$. This enables us to 'average out' the variations of ψ in (F.46). Averaging these equations over the period $2\pi/\omega_0$ and considering r, β, $\dot{r}$ and $\dot{\beta}$ to be constants while performing the integrations, one obtains

$$\begin{aligned} \dot{r}_{\text{av}} &= -\frac{\epsilon}{2\pi\omega_0} \int_0^{2\pi} f(r, \psi) \sin\psi \, \mathrm{d}\psi, \\ \dot{\psi}_{\text{av}} &= \omega_0 - \frac{\epsilon}{2\pi r\omega_0} \int_0^{2\pi} f(r, \psi) \cos\psi \, \mathrm{d}\psi. \end{aligned} \tag{F.47}$$

The full description of the method may be found in, e.g. Nayfeh & Mook (1979), Guckenheimer & Holmes (1983) or Sanders & Verhulst (1985). One advantage of the averaging method over the normal form method is that it is based on several basic comparison theorems which compare solutions of the original equation (F.42) to those of the averaged equations (F.47). For solutions valid for time of $\mathbb{O}(\epsilon^{-1})$, any solution of (F.47) can be shown to be close to those of (F.42) for sufficiently small ϵ. Also, all the qualitative local behaviour of the dynamics of the averaged equations (F.47) corresponds to the same qualitative and local behaviour of periodic orbits of (F.42). In particular, a stable (unstable) fixed point of (F.47) corresponds to a stable (unstable) limit cycle in (F.42), and a Hopf bifurcation giving rise to an attracting (repelling) limit cycle in (F.47) corresponds to a bifurcation to a stable (unstable) invariant torus in (F.42), and so on.

To see in practice how the method works, let us consider again equation (F.30). The solution when $\epsilon = 0$ is simply $x_1 = r\cos(t + \beta) = rC$, $x_2 = r\sin(t + \beta) = rS$, where C stands for $\cos(t + \beta)$ and S for $\sin(t + \beta)$. Following the methodology described in the foregoing leads to

$$\dot{r}_{\text{av}} = \frac{\epsilon}{2\pi} \int_0^{2\pi} (f_1 C + f_2 S) \, \mathrm{d}\psi, \qquad (r\dot{\beta})_{\text{av}} = \frac{\epsilon}{2\pi} \int_0^{2\pi} (f_2 C - f_1 S) \, \mathrm{d}\psi, \tag{F.48}$$

in which f_1 and f_2 represent the cubic nonlinearities in (F.30). The algebra involved may be carried out easily using, for example, MATHEMATICA. This leads finally to the same results as obtained by the normal form, equation (F.41).

A derivation of the averaging method, more adapted to the case of PDEs is given in Section F.6.1; but, before this, bifurcation theory is discussed briefly, together with the calculation of the unfolding parameters.

F.5 BIFURCATION THEORY AND UNFOLDING PARAMETERS

As mentioned already, systems of physical interest typically have parameters which appear in their defining equations; in fluid–structure interaction systems, for example, one of the major concerns is the study of the effect of increasing flow velocity on the dynamics. The original problem (F.13) may therefore be rewritten as

$$\dot{\mathbf{x}} = \mathbf{f}(\mathbf{x}, \mu), \qquad \mathbf{x} \in \mathbb{R}^n, \qquad \mu \in \mathbb{R}^k, \tag{F.49}$$

where μ represents all the parameters. As these parameters are varied, changes may occur in the *qualitative* structure of the solutions for certain parameter values μ_0 (Chow & Hale 1982). These changes are called *bifurcations* and the parameter values are called *bifurcation values*.

In light of the stability theory that was introduced in Section F.1, it is clear that in the case of a fixed point $\overline{\mathbf{x}}$ of (F.49), a necessary condition for a bifurcation to occur is that the Jacobian $D_x\mathbf{f}(\overline{\mathbf{x}}, \mu_0)$ have at least one eigenvalue with a zero real part, in which case $\overline{\mathbf{x}}$ is referred to as being *nonhyperbolic*. This of course is the interesting case from the nonlinear dynamics point of view. In the simplest cases, these bifurcations have been classified and are now well known. For example, if the linearized system contains a pair of purely imaginary eigenvalues at the critical parameter, u_c, it is called a Hopf bifurcation; if it contains a single zero eigenvalue, it may be a saddle-node, a transcritical or a pitchfork bifurcation, depending on the nonlinear terms, or there might not even exist a bifurcation (e.g. for the system defined by $\dot{x} = \mu - x^3$, $x \in \mathbb{R}$, $\mu \in \mathbb{R}$). In the case of higher degeneracy (e.g. when zero and purely imaginary eigenvalues occur simultaneously, as discussed in Section 5.7.3), the situation is even more complicated.

Here, we show how to take into account the variation of system parameters in the neighbourhood of critical values for cases where the linearized system has a single pair of imaginary eigenvalues or a single zero eigenvalue. To this end, the ordinary differential equation (F.49) is replaced by

$$\dot{\mathbf{x}} = \mathbf{L}(u)\mathbf{x} + \mathbf{f}(\mathbf{x}, u), \qquad \mathbf{x} \in \mathbb{R}^n, \tag{F.50}$$

where $u \in \mathbb{R}$ represents the system parameter in question (the dimensionless flow velocity), $\mathbf{L}$ is an $n \times n$ matrix, and $\mathbf{f}$ contains all the nonlinear terms. At $u = u_c$, suppose that $\mathbf{L}(u_c)$ contains a pair of purely imaginary eigenvalues, $\lambda = \pm i\omega_0$. Then, when u is varied by a small amount, $u = u_c + \mu$, $\mu \ll 1$, and the new eigenvalues corresponding to $\pm i\omega_0$ can be expressed as $\lambda = \sigma \pm i\omega$, with $\sigma = \mu_1$ and $\omega = \omega_0 + \mu_2$. Here, μ_1 and μ_2 are called the *unfolding parameters* and can be determined from the following characteristic equation:

$$\det[\mathbf{L}(u) - \lambda\mathbf{I}] = 0 \equiv \mathcal{R}(\sigma, \omega, u) + i\mathcal{I}(\sigma, \omega, u) = 0. \tag{F.51}$$

Performing variational calculations in the neighbourhood of u_c, a relationship between μ_1, μ_2 and μ can be found as

$$\frac{\partial\mathcal{R}}{\partial\sigma}\mu_1 + \frac{\partial\mathcal{R}}{\partial\omega}\mu_2 + \frac{\partial\mathcal{R}}{\partial u}\mu = 0, \qquad \frac{\partial\mathcal{I}}{\partial\sigma}\mu_1 + \frac{\partial\mathcal{I}}{\partial\omega}\mu_2 + \frac{\partial\mathcal{I}}{\partial u}\mu = 0, \tag{F.52}$$

where all the partial derivatives are evaluated at the critical point: $u = u_c$, $\sigma = 0$, $\omega = \omega_0$. If one constructs a modal matrix by using eigenvectors determined from (F.51), then at $u = u_c + \mu$, the original dynamical system of (F.50) may be transformed into the form

$$\dot{\mathbf{y}} = \mathbf{A}\mathbf{y} + \mathbf{g}(\mathbf{y}) + \text{h.o.t.}, \tag{F.53}$$

where

$$\mathbf{A} = \begin{bmatrix} \mu_1 & -(\omega_0 + \mu_2) & \mathbf{0} \\ \omega_0 + \mu_2 & \mu_1 & \mathbf{0} \\ \mathbf{0} & \mathbf{0} & \mathbf{B} \end{bmatrix},$$

with $\mathbf{B}$, the remaining matrix, having eigenvalues with negative real part; h.o.t. stands for ‘higher-order terms’, and $\mathbf{g}$ represents nonlinear terms. With proper order analysis, the final reduced form of (F.53), in the case of cubic nonlinearities, may be shown to be

$$\dot{r} = (\mu_1 + ar^2)r + \text{h.o.t.}, \qquad \dot{\theta} = \omega_0 + \mu_2 + br^2 + \text{h.o.t.}, \tag{F.54}$$

in polar coordinates, which may be compared to equation (F.41).

For the case with a single zero eigenvalue at $u = u_c + \mu$, $\mu \ll 1$, the unfolding parameter can be calculated in a similar fashion. Letting

$$\mathcal{R}(\sigma, u) = \det[\mathbf{L}(u) - \lambda\mathbf{I}] = 0, \tag{F.55}$$

one obtains

$$\mu = -\mu_1 \frac{\partial\mathcal{R}}{\partial\sigma} \Big/ \frac{\partial\mathcal{R}}{\partial u}, \tag{F.56}$$

where $\sigma = \mu_1$ was used. The final reduced form in the case of cubic nonlinearities becomes

$$\dot{x} = x(\mu_1 + ax^2) + \text{h.o.t.} \tag{F.57}$$

Finally, for the doubly degenerate system, two unfolding parameters need be introduced, since two parameters need be varied to ‘unfold’ the *double* degeneracy. Here, let us take these two parameters to be the velocity u and the gravity parameter γ. The critical values are denoted by u_c and γ_c, so that the bifurcation parameters μ and χ are defined by

$$u = u_c + \mu, \qquad \gamma = \gamma_c + \chi. \tag{F.58}$$

At the critical point (u_c, γ_c), the three critical eigenvalues may be expressed as

$$\lambda_{1,2} = (0 + \mu_1) \pm \mathrm{i}(\omega_0 + \mu_3) + \mathcal{O}(\mu^2), \qquad \lambda_3 = (0 + \mu_2) + \mathrm{i}(0) + \mathcal{O}(\mu^2); \tag{F.59}$$

hence, the matrix $\mathbf{A}$ in equation (F.53) becomes

$$\begin{bmatrix} \mu_1 & -(\omega_0+\mu_3) & 0 & \mathbf{0} \\ \omega_0+\mu_3 & \mu_1 & 0 & \mathbf{0} \\ 0 & 0 & \mu_2 & \mathbf{0} \\ \mathbf{0} & \mathbf{0} & \mathbf{0} & \mathbf{B} \end{bmatrix}.$$

Following the methodology developed previously, the bifurcation parameters can be related to the unfolding parameters μ_1, μ_2 and μ_3, by requiring that

$$\begin{aligned} \det[\mathbf{A}-\lambda_{1,2}\mathbf{I}] &= \mathcal{R}_1(\sigma_1,\omega_1,u)+\mathrm{i}\mathcal{I}_1(\sigma_1,\omega_1,u)=0, \\ \det[\mathbf{A}-\lambda_3\mathbf{I}] &= \mathcal{R}_2(\sigma_3,u)=0, \end{aligned} \tag{F.60}$$

which lead to

$$\begin{aligned} &\mu_1\frac{\partial\mathcal{R}_1}{\partial\sigma_1}+\mu_3\frac{\partial\mathcal{R}_1}{\partial\omega_1}+\mu\frac{\partial\mathcal{R}_1}{\partial u}+\chi\frac{\partial\mathcal{R}_1}{\partial\gamma}=0, \\ &\mu_1\frac{\partial\mathcal{I}_1}{\partial\sigma_1}+\mu_3\frac{\partial\mathcal{I}_1}{\partial\omega_1}+\mu\frac{\partial\mathcal{I}_1}{\partial u}+\chi\frac{\partial\mathcal{I}_1}{\partial\gamma}=0, \\ &\mu_2\frac{\partial\mathcal{R}_2}{\partial\sigma_3}+\mu\frac{\partial\mathcal{R}_2}{\partial u}+\chi\frac{\partial\mathcal{R}_2}{\partial\gamma}=0. \end{aligned} \tag{F.61}$$

Computing the derivatives in equation (F.61) numerically and eliminating μ_3 leads to a linear relationship of the form

$$\mu=\alpha_1\mu_1+\alpha_2\mu_2, \qquad \chi=\alpha_3\mu_1+\alpha_4\mu_2, \tag{F.62}$$

where α_i, $i=1,\ldots,4$, are real constants.

F.6 PARTIAL DIFFERENTIAL EQUATIONS

F.6.1 The method of averaging revisited

In order to adapt the averaging method to PDEs, an alternative form of averaging for a system of ODEs is first given. Thus, consider the ODE of the form

$$\dot{\mathbf{x}}=\mathbf{A}_0\mathbf{x}+\epsilon\mathbf{A}_1\mathbf{x}+\epsilon\mathbf{f}(\mathbf{x}), \qquad \mathbf{x}\in\mathbb{R}^2, \tag{F.63}$$

where

$$\mathbf{A}_0=\begin{bmatrix} 0 & -\omega_0 \\ \omega_0 & 0 \end{bmatrix}, \qquad \mathbf{A}_1=\begin{bmatrix} \mu_1 & -\mu_2 \\ \mu_2 & \mu_1 \end{bmatrix}.$$

When $\epsilon=0$, the solution to the above system has the form

$$\mathbf{x}_0=r\begin{Bmatrix} \cos(\omega_0 t+\beta) \\ \sin(\omega_0 t+\beta) \end{Bmatrix}\equiv r\begin{Bmatrix} \cos\psi \\ \sin\psi \end{Bmatrix}\equiv\tfrac{1}{2}r\begin{Bmatrix} 1 \\ -\mathrm{i} \end{Bmatrix}\mathrm{e}^{\mathrm{i}\psi}+\tfrac{1}{2}r\begin{Bmatrix} 1 \\ \mathrm{i} \end{Bmatrix}\mathrm{e}^{-\mathrm{i}\psi}, \tag{F.64}$$

where r and β are integration constants; $\mathbf{x}_0$ is 2π-periodic in ψ. When $\epsilon \neq 0$, but $\epsilon \ll 1$, it is expected that both r and β are slowly varying with time, such that

$$\dot{r} = \epsilon A \qquad \text{and} \qquad \dot{\psi} = \omega_0 + \epsilon B. \tag{F.65}$$

The purpose here is to find A and B in terms of other system parameters.

Recall that system (F.63) is autonomous. Therefore, the time variable may be eliminated with the use of the following chain rule:

$$\frac{\mathrm{d}\mathbf{x}}{\mathrm{d}t} = \epsilon A \frac{\partial \mathbf{x}}{\partial r} + (\omega_0 + \epsilon B)\frac{\partial \mathbf{x}}{\partial \psi}. \tag{F.66}$$

Let $\mathbf{x} = \mathbf{x}_0 + \epsilon \mathbf{x}_1$; then, expanding and collecting coefficients of equal powers of ϵ,

$$\omega_0 \frac{\partial \mathbf{x}_0}{\partial \psi} - \mathbf{A}_0 \mathbf{x}_0 = \mathbf{0}, \tag{F.67a}$$

$$\omega_0 \frac{\partial \mathbf{x}_1}{\partial \psi} - \mathbf{A}_0 \mathbf{x}_1 = -A \frac{\partial \mathbf{x}_0}{\partial r} - B \frac{\partial \mathbf{x}_0}{\partial \psi} + \mathbf{A}_1 \mathbf{x}_0 + \mathbf{f}(\mathbf{x}_0) \tag{F.67b}$$

are obtained. Obviously, expression (F.64) is the solution of (F.67a). In order to get a periodic solution of $\mathbf{x}_1$, secular terms on the right-hand side of (F.67b) must be eliminated. This requirement is guaranteed if

$$\frac{1}{2\pi}\int_0^{2\pi} \left[-A \frac{\partial \mathbf{x}_0}{\partial r} - B \frac{\partial \mathbf{x}_0}{\partial \psi} + \mathbf{A}_1 \mathbf{x}_0 + \mathbf{f}(\mathbf{x}_0)\right] \begin{Bmatrix} 1 \\ \mathrm{i} \end{Bmatrix} \mathrm{e}^{-\mathrm{i}\psi}\, \mathrm{d}\psi = 0. \tag{F.68}$$

Substituting $\mathbf{x}_0$ and $\mathbf{A}_1$ into the above and integrating, yields

$$A + Br\mathrm{i} = r(\mu_1 + \mathrm{i}\mu_2) + \frac{1}{2\pi}\int_0^{2\pi} \mathbf{f}(r, \psi) \begin{Bmatrix} 1 \\ \mathrm{i} \end{Bmatrix} \mathrm{e}^{-\mathrm{i}\psi}\, \mathrm{d}\psi. \tag{F.69}$$

A and B are obtained by separating the real and imaginary parts of (F.69). Thus, (F.65) becomes

$$\begin{aligned} \dot{r} &= \epsilon \left[r\mu_1 + \frac{1}{2\pi}\int_0^{2\pi} (f_1 \cos\psi + f_2 \sin\psi)\, \mathrm{d}\psi\right], \\ \dot{\psi} &= \omega_0 + \epsilon \left[\mu_2 + \frac{1}{2\pi r}\int_0^{2\pi} (f_2 \cos\psi - f_1 \sin\psi)\, \mathrm{d}\psi\right]. \end{aligned} \tag{F.70}$$

Revisiting system (F.42), it is noted that it can be transformed into two first-order equations: $\dot{x} = -\omega_0 y$, $\dot{y} = \omega_0 x - \epsilon f/\omega_0$, where it is seen that $f_1 = 0$, $f_2 = -f/\omega_0$ and $\mu_1 = \mu_2 = 0$. Thus, the result in (F.70) is identical to (F.47). Note also the similarity to equation (F.48).

We now consider the method of averaging for PDEs of the form

$$\frac{\partial \mathbf{y}(x, t)}{\partial t} = \mathbf{L}(u)\mathbf{y}(x, t) + \epsilon \mathbf{f}(\mathbf{y}(x, t), u), \tag{F.71}$$

where both $\mathbf{L}$ and $\mathbf{f}$ are differential operators of $x \in (0, 1)$; u once again represents the varied system parameter. It is further assumed that at $u = u_c$ the linear system contains a pair of purely imaginary eigenvalues,

$$\pm \mathrm{i}\omega_0 \mathbf{v} = \mathbf{L}(u_c)\mathbf{v}, \tag{F.72}$$

and the other eigenvalues (infinitely many!) have negative real parts. At u_c, the steady-state solution, which lies in the centre manifold, may be expressed as

$$\mathbf{y}_0 = r(\mathbf{v}\mathrm{e}^{\mathrm{i}\psi} + \overline{\mathbf{v}}\mathrm{e}^{-\mathrm{i}\psi}), \tag{F.73}$$

with r and $\beta = \psi - \omega_0 t$ being arbitrary, and $\overline{\mathbf{v}}$ being the complex conjugate of $\mathbf{v}$. Similarly to the case of ordinary differential equations, both r and β will be slowly varying in time as u is adjusted slightly away from u_c. At $u = u_c + \epsilon\mu$, expressing $\mathbf{y}(x,t) = \mathbf{y}_0(x,t) + \epsilon\mathbf{y}_1(x,t)$, and

$$\frac{\mathrm{d}r}{\mathrm{d}t} = \epsilon A \qquad \text{and} \qquad \frac{\mathrm{d}\psi}{\mathrm{d}t} = \omega_0 + \epsilon B,$$

and equating coefficients of equal powers of ϵ, we obtain

$$\omega_0 \frac{\partial \mathbf{y}_0}{\partial \psi} - \mathbf{L}(u_c)\mathbf{y}_0 = \mathbf{0}, \tag{F.74a}$$

$$\omega_0 \frac{\partial \mathbf{y}_1}{\partial \psi} - \mathbf{L}(u_c)\mathbf{y}_1 = -A\frac{\partial \mathbf{y}_0}{\partial r} - B\frac{\partial \mathbf{y}_0}{\partial \psi} + \mu\frac{\partial \mathbf{L}}{\partial u}(u_c)\mathbf{y}_0 + \mathbf{f}(\mathbf{y}_0, u_c). \tag{F.74b}$$

Expression (F.73) is the solution for (F.74a). Unlike equations (F.67a,b), the left-hand side of the above equations involves both the cyclic variable ψ and the spatial variable x. Introducing a linear operator $\mathbf{E}$ defined by

$$\mathbf{E}\mathbf{y} \equiv \omega_0 \frac{\partial \mathbf{y}}{\partial \psi} - \mathbf{L}(u_c)\mathbf{y} \equiv \mathbf{F}(\mathbf{y}) \tag{F.75}$$

and its adjoint

$$\mathbf{E}^*\mathbf{y}^* = \mathbf{0}, \tag{F.76}$$

then the necessary and sufficient condition for the existence of a solution for (F.75) is

$$\frac{1}{2\pi}\int_0^{2\pi}\int_0^1 \mathbf{F}(\mathbf{y})\mathbf{y}^*\,\mathrm{d}\psi\,\mathrm{d}x = 0. \tag{F.77}$$

Comparing (F.74b) with (F.75), the necessary and sufficient condition for finding a periodic solution of $\mathbf{u}_1$ can be derived from (F.77). By substituting $\mathbf{u}_0$ of (F.73) into (F.74b) and applying condition (F.77), one obtains

$$A + Br\mathrm{i} = \mu br + cr^3, \tag{F.78}$$

where $\mathbf{f}(\mathbf{y}_0, u_c)$ has been assumed to be homogeneous and cubic, and

$$b = \frac{1}{2\pi}\int_0^{2\pi}\int_0^1 \frac{\partial \mathbf{L}}{\partial u}(u_c)(\mathbf{v}\mathrm{e}^{\mathrm{i}\psi} + \overline{\mathbf{v}}\mathrm{e}^{-\mathrm{i}\psi})\cdot\mathbf{v}^*\mathrm{e}^{\mathrm{i}\psi}\,\mathrm{d}\psi\,\mathrm{d}x,$$

$$c = \frac{1}{2\pi}\int_0^{2\pi}\int_0^1 \mathbf{f}(\mathbf{v}\mathrm{e}^{\mathrm{i}\psi} + \overline{\mathbf{v}}\mathrm{e}^{-\mathrm{i}\psi}, u_c)\cdot\mathbf{v}^*\mathrm{e}^{\mathrm{i}\psi}\,\mathrm{d}\psi\,\mathrm{d}x.$$

Thus the slowly varying variables r and ψ are governed by

$$\dot{r} = \epsilon(\mu b_r + c_r r^2)r + \mathcal{O}(\epsilon^2), \quad \dot{\psi} = \omega_0 + \epsilon(\mu b_i + c_i r^2) + \mathcal{O}(\epsilon^2), \tag{F.79}$$

where b_r and c_r are the real parts of b and c, and b_i and c_i the corresponding imaginary parts.

F.6.2 The Lyapunov–Schmidt reduction

Similarly to centre manifold reduction, the Lyapunov–Schmidt reduction is a method which replaces a large and complicated set of equations by a simpler and smaller one which contains all the essential information concerning a bifurcation. The method is applicable to the system of equation (F.49), but here the nonlinear functional $\mathbf{f}(\mathbf{x}, \mu)$ may be either finite dimensional (ordinary differential equations) or infinite dimensional (partial differential or integro-differential equations). However, we shall restrict the analysis to the case of *nondegenerate* bifurcations from a *stationary* solution to another stationary solution, which excludes the important case of Hopf bifurcations; in principle, this case can be treated with similar methods (Golubitsky & Schaeffer 1985) — see also the next section. Thus, we shall be interested in determining steady-state solutions $\bar{\mathbf{y}}$ of the system defined by

$$\mathbf{f}(\mathbf{y}, u) = \mathbf{0}, \tag{F.80}$$

when the linearized system in the neighbourhood of $\bar{\mathbf{y}}$ has a single zero eigenvalue at the critical parameter, u_c, while the other eigenvalues (infinitely many for a PDE) have negative real parts. Without loss of generality, we shall also assume that the equilibrium is zero, $\bar{\mathbf{y}} = \mathbf{0}$, since this can be accomplished by a simple coordinate transformation.

As mentioned in Section F.5 for the bifurcation analysis, we would like to find an equilibrium solution when the parameter u is varied. From the *implicit function theorem* (Sattinger 1980), we know that there is *no unique solution* of (F.80) in the form of $\mathbf{y} = \mathbf{y}(u)$ in a small neighbourhood of $(\mathbf{0}, u_c)$, since we are exactly in the situation where the linear operator $\mathbf{L} = \mathrm{D}_y\mathbf{f}(\mathbf{0}, u_c)$ is singular, because the Jacobian matrix has a zero eigenvalue. The basic idea in the Lyapunov–Schmidt method is to decompose the space in which the solution lies into two subspaces, in order to 'remove' this singularity.

Formally, this may be expressed as follows. First, we define $\mathbf{v}_0$ as the eigenfunction of $\mathbf{L}$ corresponding to the zero eigenvalue, i.e.

$$\mathbf{L}(u_c)\mathbf{v}_0 = \lambda\mathbf{v}_0 = \mathbf{0}, \tag{F.81}$$

and $\mathbf{v}_0^*$ the eigenfunction of the adjoint system

$$\mathbf{L}^*(u_c)\mathbf{v}_0^* = \mathbf{0}, \tag{F.82}$$

where $\mathbf{L}^*$ is the adjoint operator of $\mathbf{L}$. Furthermore, we assume that the original Banach space, $\mathscr{E}$, in which the solution of (F.80) lies, can be decomposed into two subspaces: $\mathscr{E} = \ker\mathbf{L} + \mathscr{M}$, where $\ker\mathbf{L}$ is the kernel of $\mathbf{L}$, and $\mathscr{M}$ is a subspace perpendicular to $\ker\mathbf{L}$; in practice, $\mathscr{M} = \mathrm{range}\,\mathbf{L}^*$ [see Kolmogorov & Fomin (1970) for details]. Then, it is possible to express the solution $\mathbf{y}$ as

$$\mathbf{y} = \mathbf{y}_c + \mathbf{y}_s, \tag{F.83}$$

where $\mathbf{y}_c \in \ker \mathbf{L}$ and $\mathbf{y}_s \in \operatorname{range} \mathbf{L}^*$, and to define two orthogonal operators, $\mathbf{P}$ onto the range of $\mathbf{L}$, and the complementary operator $\mathbf{I} - \mathbf{P}$, where $\mathbf{I}$ is the identity operator, such that (Golubitsky & Schaeffer 1985)

$$\mathbf{Pf}(\mathbf{y}, u) = \mathbf{0}, \tag{F.84a}$$

$$(\mathbf{I} - \mathbf{P})\mathbf{f}(\mathbf{y}, u) = \mathbf{0}, \tag{F.84b}$$

in which $\mathbf{y}$ is replaced by (F.83). We can then apply the implicit function theorem to find $\mathbf{y}_s$ in (F.84a) as a function of $\mathbf{y}_c$ and u, since the mapping $\mathbf{Pf}(\mathbf{y}_c + \mathbf{y}_s, u_c)$ acting onto the range of $\mathbf{L}$ is regular. This means that it is possible to express $\mathbf{y}_s$ as

$$\mathbf{y}_s = \mathbf{h}(\mathbf{y}_c, u_c). \tag{F.85}$$

Equation (F.85) is valid only locally, in a neighbourhood of $(\mathbf{0}, u_c)$ with tangency properties similar to (F.17), i.e.

$$\mathbf{h}(\mathbf{y}_c, u_c) = \mathbf{0}, \qquad \mathbf{h}_u(\mathbf{y}_c, u_c) = \mathbf{0}, \tag{F.86}$$

which means that $\mathbf{h}$ is at least second order in $\mathbf{y}_c$, $\mathbf{h} = \mathbb{O}|\mathbf{y}_c|^2$. To obtain the bifurcation equations, we introduce (F.83) into (F.84b) and use (F.85) to find

$$(\mathbf{I} - \mathbf{P})\mathbf{f}(\mathbf{y}_c + \mathbf{h}(\mathbf{y}_c, u_c), u_c) = \mathbf{0}. \tag{F.87}$$

Usually, $\mathbf{h}(\mathbf{y}_c, u_c)$ cannot be found analytically in closed form, but it can be obtained using Taylor series in terms of $\mathbf{y}_c$ and $\mu = u - u_c$, to any desired order. Note that to determine (F.87) to a certain order n, we need to find $\mathbf{h}(\mathbf{y}_c, u_c)$ only to order $n - 1$, since $\mathbf{h}$ is a nonlinear function of $\mathbf{y}_c$. In fact, in many cases, e.g. under certain symmetry conditions, it is not necessary to solve (F.84a) to find the bifurcation equation to order n.

To show the reader how such an operation can be carried out, the following simple nondimensional system is considered (Troger & Steindl 1991):

$$f(\psi, u) = \frac{\mathrm{d}^2\psi}{\mathrm{d}\xi^2} + u \sin \psi = 0. \tag{F.88}$$

It corresponds to the buckling of a rod under a compressive load $u = P/EI$. We assume that the rod is simply supported at both ends, i.e. $\psi'(0) = \psi'(1) = 0$, and we want to find the bifurcation equation for the solution $\psi = 0$. The linear operator $\mathbf{L}$ is defined by taking the (Fréchet) derivative of (F.88) at $\psi = 0$:

$$\mathbf{L}\chi = f_\chi(0, u)\chi = \left[\frac{\mathrm{d}^2}{\mathrm{d}\xi^2} + u \cos 0\right]\chi = \chi'' + u\chi, \tag{F.89}$$

with boundary conditions $\chi'(0) = \chi'(1) = 0$. Solving the eigenvalue problem $\mathbf{L}\chi = 0$ leads to the well-known critical parameter $u_c = \pi^2$, with the corresponding eigenfunction $B \cos(\pi\xi)$. Consequently, $\ker \mathbf{L} = \operatorname{span}\{\cos \pi\xi\}$ and since $\mathbf{L}^* = \mathbf{L}$, $\dim \ker \mathbf{L} = \dim \ker \mathbf{L}^* = 1$. The range of $\mathbf{L}$ is given by the function $g(x)$ orthogonal to the range of $\mathbf{L}^*$, i.e. $\int_0^1 g(\xi) \cos(\pi\xi)\, \mathrm{d}\xi = 0$.

We can now decompose the image space: to find the projection $\mathbf{P}$ onto the range of $\mathbf{L}$, we recall that for any function $f(\xi)$, the projection $(\mathbf{I} - \mathbf{P})f(\xi)$ must be in $\ker \mathbf{L}^*$, i.e. $(\mathbf{I} - \mathbf{P})f(\xi) = C\cos(\pi\xi)$, where C is a constant to be determined. Therefore, the projection $\mathbf{P}$ is defined by

$$\mathbf{P}f(\xi) = f(\xi) - C\cos(\pi\xi). \tag{F.90}$$

The constant C may be calculated from the condition that each element in the range of $\mathbf{L}$ is orthogonal to $\ker\mathbf{L}^*$,

$$\int_0^1 [f(\xi) - C\cos(\pi\xi)]\cos(\pi\xi)\,\mathrm{d}\xi = 0, \tag{F.91}$$

which leads to $C = 2\int_0^1 f(\xi)\cos(\pi\xi)\,\mathrm{d}\xi$. Thus, the projection of any function $f(\xi)$ onto $\ker\mathbf{L}^*$ is given by

$$(\mathbf{I} - \mathbf{P})f(\xi) = \left[2\int_0^1 f(\xi)\cos(\pi\xi)\,\mathrm{d}\xi\right]\cos(\pi\xi). \tag{F.92}$$

Now we decompose the solution according to (F.83), $\psi = \psi_c + \psi_s$, where

$$\psi_c = q\cos(\pi\xi) \in \ker\mathbf{L} \qquad \text{and} \qquad \psi_s = h(q, u, \xi) \in \operatorname{range}\mathbf{L}^*. \tag{F.93}$$

In (F.93), q is the amplitude of the buckling mode. The bifurcation equation is obtained by projecting the original equation (F.88) onto the kernel of $\mathbf{L}^*$ through (F.92), which leads to

$$\int_0^1 [\psi'' + u(\psi - \psi^3/6 + \cdots)]\cos(\pi\xi)\,\mathrm{d}\xi = 0. \tag{F.94}$$

Equation (F.94) is solved by making use of (F.93), and by introducing the unfolding parameter $\mu = u - u_c$. From the symmetry of the problem, it can be seen that it is not necessary to find h explicitly if one wants to find a bifurcation equation to the third-order only. After some manipulations, this leads to

$$\frac{1}{2}\mu q - \frac{3}{8}\left(\frac{u_c + \mu}{6}\right)q^3 = 0. \tag{F.95}$$

The unfolding parameter μ represents the small deviation from the bifurcation point, and from equation (F.95), it is obvious that $\mu = \mathcal{O}(q^2)$. Therefore the term μq^3 can be neglected in comparison to $u_c q^3$, so that the final bifurcation equation then becomes

$$(8\mu/\pi^2)q - q^3 = 0. \tag{F.96}$$

F.6.3 The method of alternate problems

Although the method of alternate problems is usually applied to finite dimensional problems, it is introduced very briefly here because of its similarity with the Lyapunov–Schmidt method presented in Section F.6.2. A detailed presentation may be found in Hale (1969) and Bajaj (1982). The method is particularly useful if the scaling relationship between the small parameters and the amplitude is *a priori* unknown, this scaling being suggested ultimately by the bifurcation equation, as in the example of the previous section. In spirit, the method of alternate problems is very similar to the

Lyapunov–Schmidt method, since projection operators as defined in (F.84a) and (F.84b) are also introduced in this case. The purpose of the method is to find a *periodic* solution of a nonlinear system in the neighbourhood of a Hopf bifurcation, which is defined by the occurrence of one or several pairs of complex eigenvalues with zero real parts, $\pm i\omega_0$. Introducing a new time scale, $\tau = \omega_0 t$, the original equation (F.49) takes the form

$$\omega_0 \dot{\mathbf{x}} = \mathbf{A}\mathbf{x} + \mathbf{F}(\mathbf{x}, \mu), \tag{F.97}$$

where μ is a small parameter. The idea is to transform the original *differential* equation into two *algebraic* ones, of the form

$$\mathbf{y} = \mathbf{U}\mathbf{y} + \mathcal{K}(\mathbf{I} - \mathbf{V})\mathbf{F}(\mathbf{y}, \mu), \tag{F.98a}$$

and

$$\mathbf{V}\mathbf{F}(\mathbf{y}, \mu) = \mathbf{0}, \tag{F.98b}$$

where $\mathbf{U}$, $\mathbf{V}$ and $\mathcal{K}$ are some projection operators, and $\mathbf{I}$ is the identity operator [see Bajaj (1982) for their definition]. Note the similarity with equations (F.84a) and (F.84b). The solution $\mathbf{y}$ is split into two components, $\mathbf{y} = \mathbf{y}_1 + \mathbf{y}_2$, so that the $\mathbf{y}_1$ component is made up of the solutions of the homogeneous part of equation (F.97), whereas the $\mathbf{y}_2$ component contains the higher harmonics. It is then possible to solve for $\mathbf{y}_2$ explicitly in equation (F.98a) using the implicit function theorem and, assuming $\mathbf{y} = \mathbf{y}_1 + \mathbf{y}_2(\mathbf{y}_1, \mu)$, equation (F.98b) becomes the bifurcation equation.

Appendix G
Newtonian Derivation of the Nonlinear Equations of Motion of a Pipe Conveying Fluid

G.1 CANTILEVERED PIPE

This derivation is based on the work of Lundgren *et al*. (1979) but here it is developed further to lead to a single equation of motion, better suited for later analysis.

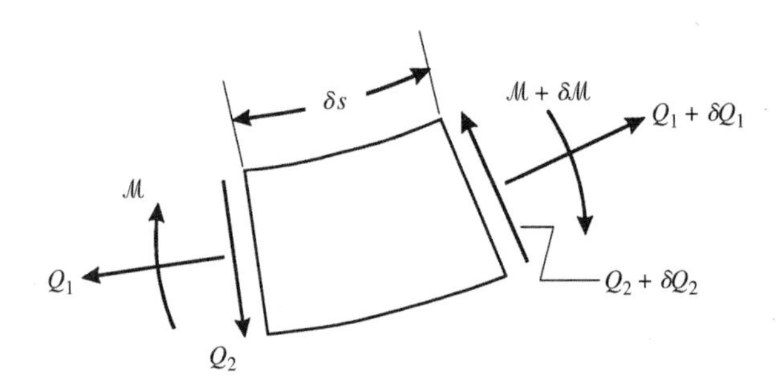

Figure G.1 Free-body diagram of an element of the pipe; gravity and flow-velocity-dependent forces are not included for clarity.

Consider an element of the pipe of length δs (Figure G.1). Let $\mathbf{Q}$ and $\mathbf{M}$ represent the resultant force and bending moment on the left cross-section, and $\mathbf{Q} + \delta\mathbf{Q}$ and $\mathbf{M} + \delta\mathbf{M}$ on the right cross-section. A force balance leads to

$$\frac{\partial \mathbf{Q}}{\partial s} + (m + M)\, g\mathbf{i} = m\,\frac{\partial^2 \mathbf{r}}{\partial t^2} + M\,\frac{\mathrm{D}^2\mathbf{r}}{\mathrm{D}t^2}, \tag{G.1}$$

and a moment balance to

$$\frac{\partial \mathbf{M}}{\partial s} + \boldsymbol{\tau} \times \mathbf{Q} = 0, \tag{G.2}$$

where $\boldsymbol{\tau}$ is the unit vector tangential to the pipe centreline.

As the effect of rotatory motion is neglected, and due to the assumptions associated with Euler–Bernoulli beam theory, the following moment–curvature relation holds:

$$\mathbf{M} = EI\boldsymbol{\tau} \times \frac{\partial \boldsymbol{\tau}}{\partial s} = EI\boldsymbol{\tau} \times \boldsymbol{\kappa}. \tag{G.3}$$

We next decompose $\mathbf{Q}$ along the tangential and normal directions,

$$\mathbf{Q} = (T_0 - P)\boldsymbol{\tau} + \boldsymbol{\tau} \times \frac{\partial \mathbf{M}}{\partial s}, \tag{G.4}$$

where $(T_0 - P)$ is the axial force due to tension and fluid pressure. By combining (G.3) with (G.4) one obtains

$$\begin{aligned} \mathbf{Q} &= (T_0 - P)\boldsymbol{\tau} + EI\boldsymbol{\tau} \times \frac{\partial}{\partial s}\left(\boldsymbol{\tau} \times \frac{\partial \boldsymbol{\tau}}{\partial s}\right) \\ &= (T_0 - P)\boldsymbol{\tau} + EI\left[\left(\boldsymbol{\tau} \cdot \frac{\partial^2 \boldsymbol{\tau}}{\partial s^2}\right)\boldsymbol{\tau} - \frac{\partial^2 \boldsymbol{\tau}}{\partial s^2}\right]. \end{aligned} \tag{G.5}$$

After some further manipulation involving the use of properties of $\boldsymbol{\tau}$ and its derivatives (Semler 1991), and projecting along x and z, one obtains the following equations:

$$(m + M)g - EI\frac{\partial^4 x}{\partial s^4} + \frac{\partial}{\partial s}\left[(T_0 - P - EI\kappa^2)\frac{\partial x}{\partial s}\right] = m\frac{\partial^2 x}{\partial t^2} + M\frac{\mathrm{D}^2 x}{\mathrm{D}t^2}, \tag{G.6a}$$

$$-EI\frac{\partial^4 z}{\partial s^4} + \frac{\partial}{\partial s}\left[(T_0 - P - EI\kappa^2)\frac{\partial z}{\partial s}\right] = m\frac{\partial^2 z}{\partial t^2} + M\frac{\mathrm{D}^2 z}{\mathrm{D}t^2}. \tag{G.6b}$$

These two equations are coupled through the curvature κ and the axial force $(T_0 - P)$. In order to derive a single equation of motion in terms of $z \equiv w$, the first equation is integrated from s to L, divided by $\partial x/\partial s$ to yield $(T_0 - P - EI\kappa^2)$, and x is eliminated through the inextensibility condition. After many straightforward but tedious manipulations, one finally finds the same equation as that obtained by the energy method, equation (5.28). Note that, in this derivation, the terms need to be correct to $\mathcal{O}(\epsilon^3)$ only, and higher order terms have been neglected.

G.2 PIPE FIXED AT BOTH ENDS

Recalling that the forces and moments can also be defined in terms of the original coordinate x_0, equation (G.1) becomes

$$\frac{\partial \mathbf{Q}}{\partial x_0} + (M + m)g\mathbf{i} = m\frac{\partial^2 \mathbf{r}}{\partial t^2} + M\frac{\mathrm{D}^2 \mathbf{r}}{\mathrm{D}t^2}, \tag{G.7}$$

where the material derivative is defined as in (5.30). By taking into account the force due to $(T_0 - P)$ and the extensibility of the pipe, the force $\mathbf{Q}$ may be expressed as

$$\mathbf{Q} = \mathbf{Q}_1 + \mathbf{Q}_2. \tag{G.8}$$

From expression (5.34), the axial force $\mathbf{Q}_1$ is

$$\mathbf{Q}_1 = (T_0 - P + EA\varepsilon)\,\boldsymbol{\tau}, \tag{G.9}$$

while the shear force $\mathbf{Q}_2$, perpendicular to $\mathbf{Q}_1$ (see Figure G.1), is given by

$$\mathbf{Q}_2 = -\frac{\partial \mathcal{M}}{\partial s}\,\mathbf{n} = -\frac{1}{1+\varepsilon}\frac{\partial \mathcal{M}}{\partial x_0}\,\mathbf{n}, \tag{G.10}$$

where $\mathbf{n}$ is the unit vector normal to $\boldsymbol{\tau}$. As the effect of rotatory motion is neglected, the moment due to bending has a contribution only in the $\mathbf{n}$ direction. Moreover, the moment in its scalar form simply becomes

$$\mathcal{M} = EI(1+\varepsilon)\frac{\partial \theta}{\partial s} = EI\,\frac{\partial \theta}{\partial x_0}. \tag{G.11}$$

Therefore, decomposing $\mathbf{Q}$ along $\boldsymbol{\tau}$ and $\mathbf{n}$, one obtains

$$\mathbf{Q} = \mathbf{Q}_1 + \mathbf{Q}_2 = (T_0 - P + EA\varepsilon)\boldsymbol{\tau} - \frac{EI}{1+\varepsilon}\frac{\partial^2 \theta}{\partial x_0^2}\,\mathbf{n}. \tag{G.12}$$

By decomposing these two components along the x- and z-directions, recalling the expressions of the accelerations obtained in (5.18), extending the results of (5.30), and introducing again the angle θ, one obtains

$$(m+M)g + \frac{\partial}{\partial x_0}(Q_1 \cos\theta) - \frac{\partial}{\partial x_0}(Q_2 \sin\theta) = m\,\frac{\partial^2 u}{\partial t^2} + M\,\frac{\mathrm{D}^2(x_0+u)}{\mathrm{D}t^2}, \tag{G.13a}$$

$$\frac{\partial}{\partial x_0}(Q_1 \sin\theta) + \frac{\partial}{\partial x_0}(Q_2 \cos\theta) = m\,\frac{\partial^2 w}{\partial t^2} + M\,\frac{\mathrm{D}^2 w}{\mathrm{D}t^2}, \tag{G.13b}$$

where $\sin\theta$ and $\cos\theta$ are defined by equations (5.4).

Here, an order of magnitude analysis is useful, so as to simplify the algebra as much as possible. The first equation (in the x_0-direction) is of second order, and the second (in the z-direction) of third order. Hence, all the terms have to be exact up to third order. For example,

$$\sin\theta = w'\left(1 - u' - \tfrac{1}{2}w'^2\right) + \mathcal{O}(\epsilon^4),$$
$$\cos\theta = 1 - \tfrac{1}{2}w'^2 + \mathcal{O}(\epsilon^4), \qquad \varepsilon = u' + \tfrac{1}{2}w'^2 + \mathcal{O}(\epsilon^4).$$

After some further manipulations, the governing equations obtained are found to be the same as those derived by the energy method, equations (5.36a,b).

Finally, it can be shown easily that equations (G.13a,b) are equivalent to equations (G.6a,b) simply by letting $\varepsilon = 0$ and replacing x_0 by s. In other words, the equations of the cantilevered pipe can be obtained from those of a pipe fixed at both ends by imposing the inextensibility condition.

Appendix H
Nonlinear Dynamics Theory Applied to a Pipe Conveying Fluid

The purpose of this appendix is to show how the methods utilized in modern nonlinear dynamics can be applied to the problem of a pipe conveying fluid. More particularly, how the centre manifold and the normal form theories can be used to characterize the dynamical behaviour of a physical system in the neighbourhood of two types of instabilities. In the first section, the centre manifold theory is used to show how to reduce the dimension of the original system; in the second part, the flow on the centre manifold is found for both the static and the dynamic case.

H.1 CENTRE MANIFOLD

In this section, we shall show how to find the centre manifold for the 'static' instability, i.e. when the linearized system has a zero eigenvalue. The case of dynamic instability can be treated very similarly. Let us consider equation (F.15) with the nonlinear functions **f** and **g** being cubic,

$$\dot{x} = \mathbf{A}x + f_{i,k}x^k y^{\bar{k}}, \qquad \dot{y} = \mathbf{B}y + g_{j,k}x^k y^{\bar{k}}. \tag{H.1}$$

where $k + \bar{k} = 3$. We want to find the centre manifold in the neighbourhood of the origin, so that we can assume $x = \sqrt{\epsilon}\, u$, $y = \sqrt{\epsilon}\, v$, where ϵ is a small parameter. Furthermore, we assume that we are close to the static instability, so that the main parameter, the dimensionless flow velocity $\mathcal{U}$, is such that $\mathcal{U} - \mathcal{U}_c = \epsilon\mu$.[†] Consequently, as shown in Appendix F, equation (H.1) can be replaced by

$$\dot{\mu} = 0, \qquad \dot{u} = \epsilon\alpha\mu u + \epsilon f_{1,k}u^k v^{\bar{k}}, \qquad \dot{v} = \mathbf{B}v + \epsilon g_{j,k}u^k v^{\bar{k}}, \tag{H.2}$$

where α is a real constant. As can be seen, the parameter μ in equation (H.2) has been converted into a state variable. Hence, the first two equations linearized around the origin represent the 2-D centre eigenspace (zero eigenvalue), while the last one represents the

[†]The dimensionless flow velocity is denoted here by $\mathcal{U}$ to avoid confusion with u, as in $\{u, v\}$. In any case, in what follows in this section, only μ appears explicitly.

j-dimensional stable one. Following (F.16), the centre manifold is found in the form

$$v = h(u, \mu), \tag{H.3}$$

with boundary conditions similar to (F.17): $h(0, 0) = \mathrm{D}h/\mathrm{D}u(0, 0) = \mathrm{D}h/\mathrm{D}\mu(0, 0) = 0$. Consequently, to keep the nonlinear terms cubic in the second of equations (H.2), and to satisfy the boundary conditions, we must have h be a linear function of u, with $\epsilon\mu$ as a coefficient, or $v = \epsilon\mu[C]u$, where $[C]$ here is a $j \times 1$ matrix that has to be determined. To find $[C]$, an equation similar to (F.20) must be sought, so that the flow on the centre manifold can be found:

$$\begin{aligned} \dot{u} &= \epsilon\alpha\mu u + \epsilon f_{1,k} u^k (\epsilon\mu C u)^{\bar{k}} \\ &= \epsilon\alpha\mu u + \epsilon f_{1,3} u^3 + \mathcal{O}(\epsilon^2). \end{aligned} \tag{H.4}$$

This means that, to order ϵ, the centre manifold can be approximated as $h(u) = 0$, and hence, the flow on the centre manifold can be approximated by

$$\dot{\mathbf{x}} = \mathbf{A}\mathbf{x} + \mathbf{f}(\mathbf{x}, \mathbf{0}). \tag{H.5}$$

This, of course, is a straighforward operation since one simply has to ignore the stable component in the equation on the centre, once the original system of equations has been put in standard form.

H.2 NORMAL FORM

H.2.1 Dynamic instability

In this section, the different manipulations leading to the equation of motion on the centre manifold are given for the pipe conveying fluid. As will be seen, most of them are straighforward. The different parameters are the same as in Païdoussis & Semler (1993): the gravity parameter $\gamma = 25$, the mass parameter $\beta = 0.2$, and the viscoelastic damping $\alpha = 0.005$. The number of modes is equal to $N = 2$. It can be shown easily that for these parameters, a dynamic instability occurs for $\mathcal{U}_c = 7.093$; this, in fact, is also shown in the computer program, written in MATHEMATICA, which follows. Once the nonlinear equation of motion is set up, the approximation for the centre manifold is made, $v = 0$, which corresponds in the program to $x_3 = x_4 = 0$. Then, the method of averaging is applied, as outlined in Appendix F. Once the normal form is found,

$$\dot{r} = 2.27\mu r - 0.31 r^3, \tag{H.6}$$

corresponding to equation Out [121] in the listing, the limit-cycle amplitude is computed and, converting to the original coordinates, the phase-plane plot for $\mu = \mathcal{U} - \mathcal{U}_c = 0.3$ (shown at the end of the program) is obtained.

The amplitude of the first generalized coordinate, q_1, and the frequency of the motion are given versus μ in Figure H.1 for another set of parameters: $\beta = 0.2$, $\gamma = 10$. Agreement with numerically computed results, especially for $\mu < 0.2$, is remarkably good.

(Local B) In[44]:=

```
(* NORMAL FORM OF THE HOPF BIFURCATION using Mathematica *)

(* ------------------------------------------------------ *)
(*      Main parameters                                   *)
(* ------------------------------------------------------ *)

nmode = 2;       (* Number of modes *)
ar = 0.005;      (* Damping coefficient alpha *)
uhb = 7.093;     (* Critical flow velocity *)
beta = 0.2;      (* Mass parameter beta *)
gama = 25;       (* Gravity parameter *)

u = uhb + eps mu;

(* ------------------------------------------------------ *)
(* SIGMA COEFFICIENTS AND EIGENFUNCTIONS OF THE BEAM      *)
(* ------------------------------------------------------ *)

rl[1] = 1.875104043341462;
rl[2] = 4.694091054370627;

xb = 1.0;
Do [{
  si[i] = (Sinh[rl[i]]-Sin[rl[i]])/
                          (Cosh[rl[i]]+Cos[rl[i]]),
  ph[i] = Cosh[rl[i]*xb]-Cos[rl[i]*xb]-
             si[i]*(Sinh[rl[i]*xb]-Sin[rl[i]*xb])
     },{i,nmode}]

(* ------------------------------------------------------ *)
(* LINEAR COEFFICIENTS -- see Paidoussis and Issid (1974)*)
(* ------------------------------------------------------ *)

Do [ Do [{
  tau[i,j] = (rl[i]/rl[j])^2,
  one[i,j] = (-1)^(i+j),
  If [j==i,
    bb[i,j] = 2.0,
    bb[i,j] = 4.0/(tau[i,j]+one[i,j])],
  If [j==i,
    cc[i,j] = rl[j]*si[j]*(2.0-rl[j]*si[j]),
    cc[i,j] = 4.0*(rl[j]*si[j]-rl[i]*si[i])/
                 (one[i,j]-tau[i,j])],
  If [j==i,
    ee[i,j] = 2.0 - 0.5*cc[i,j],
    ee[i,j] = (4.0*(rl[j]*si[j]-rl[i]*si[i]+2.0)*
                                          one[i,j]-
     +2.0*(1.0+tau[i,j]^2)*bb[i,j])/
```

```
          (1.0-tau[i,j]^2)-cc[i,j]],
    ff[i,j] = bb[i,j] - ee[i,j]},
{j,nmode}],{i,nmode}];

(* ------------------------------------------------------ *)
(* THE DAMPING AND STIFFNESS MATRICES                      *)
(* ------------------------------------------------------ *)

 Do [ Do [{
    If [i==j,
      rc[i,j] = -(ar*rl[j]^4 + 2.0*Sqrt[beta]*u*bb[i,j]),
      rc[i,j] = -(2.0*Sqrt[beta]*u*bb[i,j])],
    If [i==j,
      rk[i,j] = -(rl[j]^4 +u^2*cc[i,j] + gama*ee[i,j]),
      rk[i,j] = -(u^2*cc[i,j] + gama*ee[i,j])]},
{j,nmode}],{i,nmode}];

(* ------------------------------------------------------ *)
(* LINEAR MATRIX A                                         *)
(* ------------------------------------------------------ *)

Do [ Do [{
  If [i==j, delta[i,j] = 1, delta[i,j] = 0],
  r[i,j] = 0,
  r[i,j+nmode] = delta[i,j],
  r[i+nmode,j] = rk[i,j],
  r[i+nmode,j+nmode] = rc[i,j]},
{j,nmode}],{i,nmode}]
a = Table[r[i,j],{i,2*nmode},{j,2*nmode}];
mu = 0;

(* ------------------------------------------------------ *)
(* EIGENVALUES AND TRANSFORMATION MATRIX P                 *)
(* ------------------------------------------------------ *)

{lam,vecs} = Eigensystem[a];
Clear[mu];
Do [Print[lam[[i]]],{i,2*nmode}]
ptp[1] = Re[vecs[[1]]];
ptp[2] = Im[vecs[[2]]];
ptp[3] = Re[vecs[[3]]];
ptp[4] = Im[vecs[[4]]];
ptem = Table[ptp[i],{i,2*nmode}];
p = Transpose[ptem];
aa = Inverse[p].a.p;

-0.000373452 + 16.1603 I
-0.000373452 - 16.1603 I
-13.9327 + 3.25323 I
-13.9327 - 3.25323 I
```

```
(* ----------------------------------------------------------- *)
(* NONLINEAR COEFFICIENTS                                       *)
(* ----------------------------------------------------------- *)

toto1 = ReadList["/u/chris/math/rm01.dat",Number];
toto2 = ReadList["/u/chris/math/rm02.dat",Number];
toto3 = ReadList["/u/chris/math/rm03.dat",Number];
toto4 = ReadList["/u/chris/math/rm04.dat",Number];
toto5 = ReadList["/u/chris/math/rm05.dat",Number];

Do [rm10[i]=Part[toto1,i],{i,Length[toto1]}]
Do [rm20[i]=Part[toto2,i],{i,Length[toto2]}]
Do [rm30[i]=Part[toto3,i],{i,Length[toto3]}]
Do [rm40[i]=Part[toto4,i],{i,Length[toto4]}]
Do [rm50[i]=Part[toto5,i],{i,Length[toto5]}]

it = 1;
Do [ Do [ Do [ Do [ {
  rm1[i,j,k,l] = rm10[it],
  rm2[i,j,k,l] = rm20[it],
  rm3[i,j,k,l] = rm30[it],
  rm4[i,j,k,l] = rm40[it],
  rm5[i,j,k,l] = rm50[it],
  it = it + 1},
{l,nmode}],{k,nmode}],{j,nmode}],{i,nmode}]

Do [ Do [ Do [ Do [{
  al[i,j,k,l] = -(uhb^2*rm1[i,j,k,l]+gama*rm2[i,j,k,l]+
                                          rm3[i,j,k,l]),
  be[i,j,k,l] = -(2.0*uhb*Sqrt[beta]*rm4[i,j,k,l]),
  ga[i,j,k,l] = -rm5[i,j,k,l]},
{l,nmode}],{k,nmode}],{j,nmode}],{i,nmode}]

xx = Table[x[i],{i,2*nmode}];
z = p.xx;
Do [tp[i] = 0, {i,nmode}]
Do [ Do [ Do [ Do [{
  tp[i] = tp[i]+ al[i,j,k,l]*z[[j]]*z[[k]]*z[[l]]+
                 be[i,j,k,l]*z[[j]]*z[[k]]*z[[l+nmode]]+
                 ga[i,j,k,l]*z[[j]]*z[[k+nmode]]*z[[l+nmode]]
{l,nmode}],{k,nmode}],{j,nmode}],{i,nmode}]
Do [{f[i] = 0,
     f[i+nmode] = Expand[tp[i]]},{i,nmode}]
(* ----------------------------------------------------------- *)
(* STANDART FORM : Xdot = aa.X + Inverse[p].f                  *)
(* where f represents the nonlinear terms                       *)
(* ----------------------------------------------------------- *)
```

```
ff = Array[f,{4}];
xx = Array[x,{4}];

(* linear part *)

equ = aa.xx;

(* ------------------------------------------------------- *)
(* This corresponds to the approximation of the centre     *)
(* manifold                                                *)
(* ------------------------------------------------------- *)

x[3] = 0;
x[4] = 0;

(* ------------------------------------------------------- *)
(* TRANSFORMATION OF COORDINATES (polar)                   *)
(* w0 is the natural frequency                             *)
(* ------------------------------------------------------- *)

c = Cos[kap t + phi];
s = Sin[kap t + phi];
w0 =  Im[lam[[1]]];
w = nu;
nu = w0/kap;

equnl = Expand[Inverse[p].ff];

(* ------------------------------------------------------- *)
(* Set the nonlinear equation of motion                    *)
(* ------------------------------------------------------- *)

toto = Chop[Expand[equnl + equ],0.001];
x1d = toto[[1]];
x2d = toto[[2]];

rd = Expand[x1d c + x2d s];
rtheta = Expand[x2d c - x1d s];
x[1] = r c;
x[2] = r s;
```

(Local B) In[108]:=

```
(* ---------------------------------------------------------- *)
(* Trigonometric rules needed to simplify the algebra     *)
(* ---------------------------------------------------------- *)

TrigSimpRules = {
   Sin[x_+y_] :> Sin[x] Cos[y] + Sin[y] Cos[x],
   Cos[x_+y_] :> Cos[x] Cos[y] - Sin[x] Sin[y]}

TrigSimpSign = {
    (* Sin is an odd function *)
    Sin[n_?Negative x_.]     :> -Sin[-n x],
    Sin[n_?Negative x_ + y_] :> -Sin[-n x - y] /;
        Order[x, y] == 1 && NumberQ[n],

    (* Cos is an even function *)
    Cos[n_?Negative x_.]     :>  Cos[-n x],
    Cos[n_?Negative x_ + y_] :>  Cos[-n x - y] /;
    Order[x, y] == 1 && NumberQ[n]}

kap = 1/2;
```

(Local B) In[114]:=

```
(* ---------------------------------------------------------- *)
(* PERFORM THE INTEGRATION                                     *)
(* ---------------------------------------------------------- *)

rdav[t_] = Integrate[rd,t] /. TrigSimpSign
rdavf = Expand[rdav[2 Pi] - rdav[0]] /. TrigSimpRules
rdavf= Simplify[1/2/Pi/nu * rdavf];
```

(Local B) In[117]:=

```
rth[t_] = Integrate[rtheta,t];
rthf = Expand[rth[2 Pi] - rth[0]] /. TrigSimpRules
rthf = Chop[Simplify[
        1/nu*(1/2/Pi*rthf - r w0 (1-eps sig))],0.0001]
```

(Local B) In[122]:=

```
(* ---------------------------------------------------------- *)
(* This corresponds to the equation (43) of the article
   by Paidoussis and Semler; as can be seen, the nonlinear
   terms are not the same, but the relative magnitude is
   exactly the same                                          *)
(* ---------------------------------------------------------- *)

Simplify[rdavf nu]

Simplify[rthf nu]
```

(Local B) Out[121]=

```
                               2   2                3
2.27738 eps mu r + 0.273127 eps  mu  r - 0.312835 r
```

(Local B) Out[122]=

```
                                  2   2                3
-0.90322 eps mu r - 0.0823938 eps  mu  r + 0.371685 r  +
  16.1603 eps r sig
```

(Local B) In[124]:=

```
eps = 1;
rl2 = Simplify[rdavf/r]
```

(Local B) Out[124]=

```
                             2                2
0.0704619 mu + 0.00845055 mu  - 0.00967909 r
```

(Local B) In[128]:=

```
(* ----------------------------------------------------- *)
(* Compute the limit cycle amplitude                      *)
(* ----------------------------------------------------- *)

rlc = Sqrt[- Coefficient[rl2,mu] mu / Coefficient[rl2,r^2]]
```

(Local B) Out[128]=

```
2.69811 Sqrt[mu]
```

(Local B) In[130]:=

```
x[1] = rlc Cos[w0 t];
x[2] = rlc Sin[w0 t];
```

(Local B) In[134]:=

```
(* ----------------------------------------------------- *)
(* Transform back into original coordinates               *)
(* ----------------------------------------------------- *)

y = p.xx;
```

(Local B) In[139]:=

```
(* ----------------------------------------------------- *)
(* Compute the displacement and velocticy of the tip end *)
(* ----------------------------------------------------- *)

disp = Simplify[y[[1]] ph[1] + y[[2]] ph[2]];
velo = Simplify[y[[3]] ph[1] + y[[4]] ph[2]];
```

```
(* ------------------------------------------------------ *)
(* Plot the phase plane plot for mu = 0.3; corresponds    *)
(* to Figure 10(a) of the article by Paidoussis and       *)
(* Semler (1993)                                          *)
(* ------------------------------------------------------ *)

mu = 0.3;
ParametricPlot[{disp,velo},{t,0,2 Pi/w0},
        Frame->True,
        FrameLabel -> {Displacement, Velocity},
        PlotRange -> {{-0.3,0.3},{-4,4}}]
```

(Local B) Out[146]=
```
-Graphics-
```

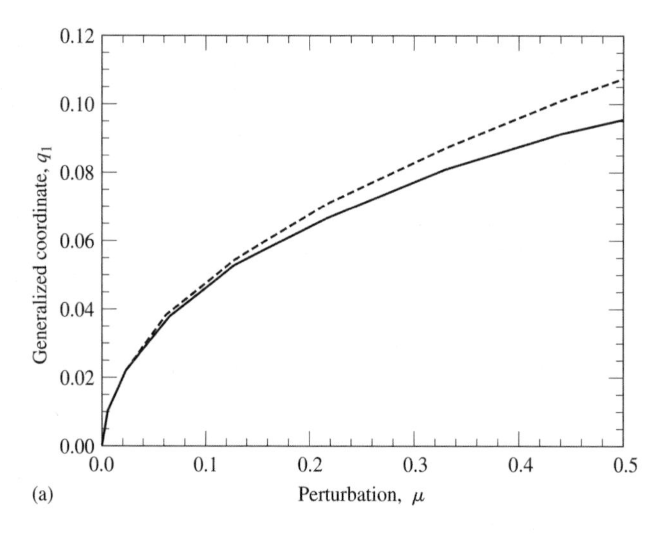

Figure H.1 (a) The amplitude of the first generalized coordinate, q_1, and (b) the frequency, ω, for limit-cycle motion as a function of $\mu = \mathcal{U} - \mathcal{U}_c$, for a system with $\beta = 0.2$ and $\gamma = 10$, for which $\mathcal{U}_c = 6.2708$. - - -, by centre manifold theory; ——, by AUTO.

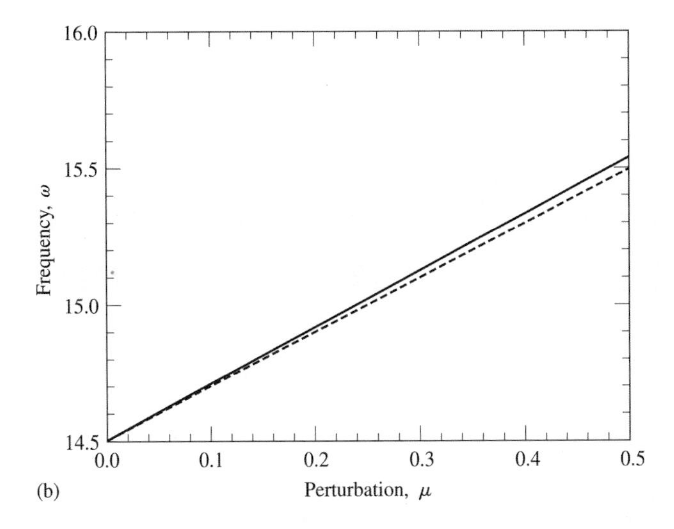

Figure H.1 *(continued)*.

H.2.2 Static instability

The procedure to characterize the static instability is very similar to the one presented for the dynamic instability. It is even simpler, since no integration is needed: once the equation of motion is found and the centre manifold approximation applied (by setting $x_2 = x_3 = x_4 = 0$), the normal form arises 'naturally'. This is applied to the case of a standing pipe conveying fluid which is represented by a negative gravity parameter, $\gamma < 0$. For $\gamma = -25$ and $\beta = 0.2$, for example (in fact, for any β), it can be shown that there is a zero eigenvalue at $\mathcal{U}_c = 3.05$. After some manipulation, the flow on the centre manifold is found to be

$$\dot{x} = (-4.44\mu - 10.85x^2)x, \tag{H.7}$$

which shows clearly that the static instability corresponds to a supercritical pitchfork bifurcation: when $\mu < 0$ ($\mathcal{U} < \mathcal{U}_c$), the pipe diverges to one or the other stable equilibrium, depending on the initial conditions; when $\mu > 0$ ($\mathcal{U} > \mathcal{U}_c$), the origin becomes stable and the two symmetric equilibrium positions disappear, thus the system regains its undeformed equilibrium state.

Appendix I
The Fractal Dimension from the Experimental Pipe-vibration Signal

The delay-embedding method and the computation of what are generically called fractal dimensions are relatively recent developments in dynamical systems theory. They are briefly reviewed here, following Païdoussis, Cusumano & Copeland (1992). A full introduction may be found in Moon (1992), while a more theoretical review is given by Eckmann & Ruelle (1985).

The basic assumption is that the dynamical steady state being analysed is evolving on a low-dimensional manifold in the full phase space (which itself can have many, possibly infinite dimensions). Knowledge of the dimensions of attractors over the operating range of a system yields a firm estimate of the number of degrees of freedom needed to model observed dynamics.

Here we shall use the *correlation dimension* developed by Grassberger & Proccacia (1983a,b), which is the most widely applied dimension measure — largely because of the ease with which it can be computed — see, e.g. Malraison *et al.* (1983), Brandstäter *et al.* (1983) and Cusumano & Moon (1995a,b).

To define the correlation dimension, let $\mathbf{x}(t)$ denote the steady-state solution under consideration. It is assumed that $\mathbf{x}(t)$ is a finite dimensional state vector. We sample the data at a fixed time-step Δt and obtain a data record

$$\{\mathbf{x}_1, \mathbf{x}_2, \mathbf{x}_3, \ldots, \mathbf{x}_N\},$$

where $\mathbf{x}_i \equiv \mathbf{x}(i\Delta t)$ and the time origin is taken to be zero. To measure the dimension of this set, Grassberger & Proccacia define the *correlation integral* $C(r)$ as

$$C(r) = \lim_{N\to\infty} \frac{1}{N^2} \sum_{i=1}^{N} \sum_{\substack{j=1 \\ j\neq i}}^{N} H(r - \|\mathbf{x}_i - \mathbf{x}_j\|) \simeq \frac{1}{N_{\text{pairs}}} \sum_{i=1}^{N} \sum_{j>i} H(r - \|\mathbf{x}_i - \mathbf{x}_j\|), \qquad \text{(I.1)}$$

where H is the Heaviside step function, r is a scalar length scale and $N_{\text{pairs}} = \frac{1}{2}(N^2 - N)$. Note that in the limit, as $N \to \infty$, the two expressions (I.1) become equal. $C(r)$ is the cumulative distribution of length scales on the attractor; this statistical interpretation is important for efficient computation. Grassberger & Proccacia define the correlation dimension d_c by

$$\lim_{r\to 0} \frac{\ln C(r)}{\ln r} = d_c. \qquad \text{(I.2)}$$

The main problem that must be addressed in applying equations (I.1) and (I.2) to experimental data is that the data record (e.g. the sampled output from the optical tracking system in Figure 5.30) is of the form

$$\{x_1, x_2, \ldots, x_N\}, \qquad x_i = x(i\Delta t), \tag{I.3}$$

where the x_i are now simply *scalar* quantities. Using the delay embedding procedure, however, one can reconstruct the phase space of the underlying system. This technique was first used by Packard *et al.* (1980) and put on a sound mathematical foundation by Takens (1980). To construct vectors $\mathbf{x}_i \in \mathbb{R}^m$ from the scalar series $[x_i]_{i=1}^N$ for some fixed m, one simply forms m-tuples from the scalar series by defining

$$\begin{aligned}\mathbf{x}_i &\triangleq \{x(i\Delta t), x((i+d)\Delta t), \ldots, x((i+(m-1)d)\Delta t)\} \\ &= \{x_i, x_{i+d}, \ldots, x_{i+(m-1)d}\},\end{aligned} \tag{I.4}$$

where $d \in \mathbb{N}$ and $(\Delta t)d$ is called the *delay*. The set of all vectors so constructed are called *pseudovectors*, and the dimension m used in their construction is called the *embedding dimension*. For m sufficiently large, this procedure leaves the topological type and dimension of the underlying attractor invariant. Thus, one can use the collection of pseudovectors to obtain an estimate for d_c. Note, however, that one must pick m and d to implement the method.

Selection of a delay is a subtle issue and the reader is referred to papers by Broomhead & King (1986) and Fraser & Swinney (1986) for examples of how the idea of 'optimality' in d might be approached. Here, suitable delays are found by plotting (x_i, x_{i+d}) and choosing values for d that expand the pseudo-orbit as much as possible with respect to the noise amplitude in the system while maintaining a deterministic orbit structure. Nearby values for d are then used to check that consistent results are obtained, following a simple trial-and-error approach for finding delays, as originally used by Malraison *et al.* (1983).

The overall strategy for finding the dimension of the attractor is to pick m, construct the m-dimensional pseudovectors, and compute $d_c = d_c(m)$; m is then incremented and the procedure is repeated. For a deterministic signal, d_c will level out at some critical value of m; whereas for a random signal it will grow indefinitely, and in the limit of an infinite number of data points, $d_c(m) = m$.

The statistical nature of $C(r)$ may be used to efficiently compute d_c. For a given embedding dimension, all pseudovectors are constructed and stored; then, a random subset thereof (with N_{subs} elements) is selected from the total population of approximately N pseudovectors ($N \gg N_{\text{subs}}$). All distances in the subset are computed, sorted, normalized so that the largest distance is equal to 1, and stored in a one-dimensional array with N_{pairs} elements, where $N_{\text{pairs}} = \frac{1}{2}(N_{\text{subs}}^2 - N_{\text{subs}})$. This array is used to obtain an approximate cumulative distribution $C_j(r_i)$ evaluated at 500 values of r_i which are equally spaced on a logarithmic scale. Another subset is chosen and the procedure is repeated N_{avg} times for the same embedding dimension. Then the average $C_j(r_i)$ is obtained:

$$\overline{C}_i \triangleq \overline{C}(r_i) = \frac{1}{N_{\text{avg}}} \sum_{j=1}^{N_{\text{avg}}} C_j(r_i) \simeq C(r_i). \tag{I.5}$$

This algorithm is repeated for each embedding dimension, giving an entire family of $\ln\overline{C}(r)$ versus $\ln r$ curves. The scaling regions in the $\ln\overline{C}(r)$ versus $\ln r$ curves are identified, and a least-squares fit is used to obtain an estimate of $d_c(m)$ for each m. Error estimates for d_c are obtained using standard methods (Bevington 1969). We utilize the expression for the variance of the mean,

$$\sigma^2_{\overline{C}(r_i)} = \frac{1}{N_{\text{avg}}(N_{\text{avg}} - 1)} \sum_{j=1}^{N_{\text{avg}}} \left[C_j(r_i) - \overline{C}(r_i)\right]^2; \tag{I.6}$$

then, from the least-squares fit for the slope,

$$d_c = \frac{\sum u_i \sum v_i - N_{\text{scale}} \sum u_i \sum v_i}{\left(\sum u_i\right)^2 - N_{\text{scale}} \sum u_i^2}, \tag{I.7}$$

where N_{scale} is the number of points in the scaling region, $u_i = \ln r_i$ and $v_i = \ln\overline{C}_i$. Then, the measurement error in d_c is given by

$$\delta d_c \leq \sum_{i=1}^{N_{\text{scale}}} \left|\frac{\partial d_c}{\partial v_i}\right| \delta v_i = \sum_{i=1}^{N_{\text{scale}}} \left|\frac{\partial d_c}{\partial v_i}\right| \frac{\sigma_{\overline{C}_i}}{\overline{C}_i}. \tag{I.8}$$

For all the results presented here, 68% confidence limits are used for all error estimates.

For the work of Section 5.8.1, the data from the noncontacting optical probe were recorded, and 32 000-point records sampled at 50 Hz were used in the analysis in each case. In all cases, the data were low-pass filtered by a Butterworth filter with a knee frequency of 25 Hz. The results for another run with the same system as in Figures 5.31 and 5.32 (pipe #9 of Table 5.3, water flow) are shown in Figures I.1–I.3. In each case, (a) and (b) are the power spectrum and autocorrelation, respectively; (c) shows a pseudo-phase portrait of the reconstructed orbit and a Poincaré map; (d) is a plot of the correlation integral $C(r)$ versus the length scale r, for various embedding dimensions m. It is clear that in Figure I.1 the system executes periodic (period-1) motion. In Figure I.2, the oscillation is of period-2 but, as shown from components (b) and (c) of the figure, there is already a small but nonnegligible chaotic component to the motion; we call this 'fuzzy period-2' oscillation. The oscillation in Figure I.3 is clearly chaotic. It is of interest to note that the 'knee' at $\ln r \simeq -3.8$ in Figure I.1(d), and at less well-defined points in the other figures, corresponds to the point below which random noise is important. For $\ln r > -3.8$, however, the curves for the higher m converge and their slopes may be used to provide an estimate of d_c via equation (I.2). The results are shown in Figure 5.37.

The techniques for analysing observed chaotic data have developed rapidly in the last few years — since the work just described was done. The interested reader is referred to, for instance, Parker & Chua (1989), Ott (1993) and Abarbanel (1994, 1996). A useful classification has been provided by Cusumano (1997), summarized in this appendix as follows.

Improvements in the dimension estimation, or more specifically the delay-reconstruction part of it, relate to more reliable methods for selecting the embedding dimension, d_e (m in the foregoing), and the delay, τ. 'Singular systems analysis', which is essentially

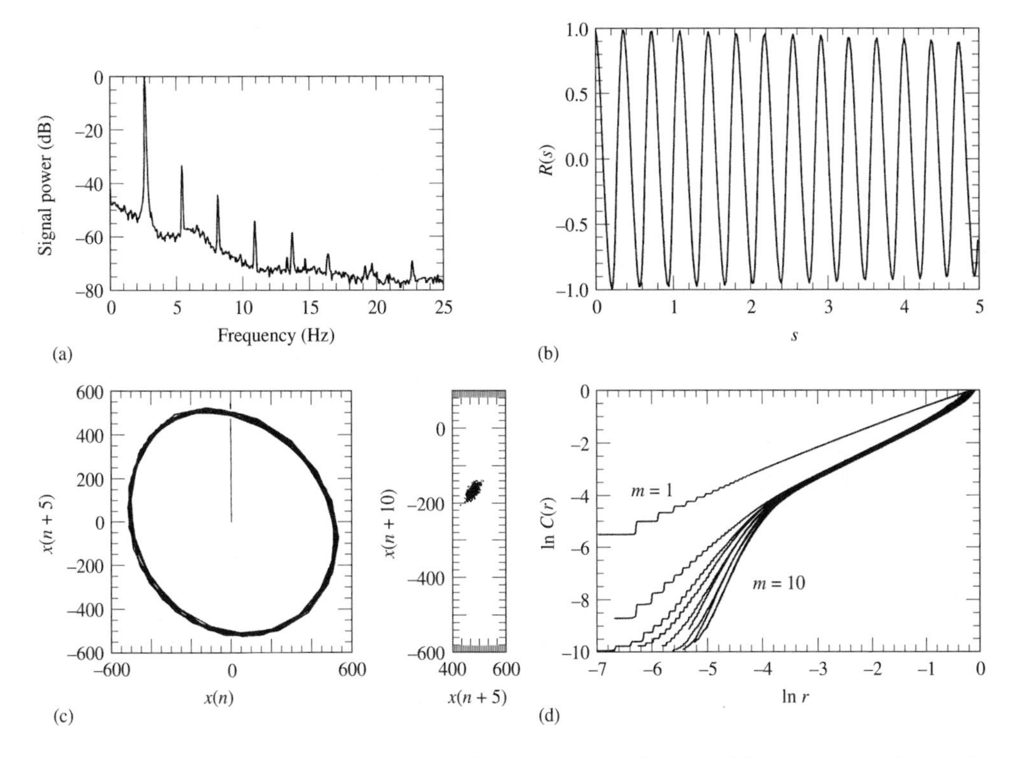

Figure I.1 (a) Power spectrum; (b) normalized autocorrelation; (c) delay reconstruction of the orbit and corresponding Poincaré section; (d) correlation integral $C(r)$ versus length scale r for embedding dimensions $m = 1$–10; for pipe #9 (Table 5.3) and water flow with $U = 6.77$ m/s. The vertical line cutting the orbit in (c), marks $\{x(n) = 0; x(n+5) > 0\}$, used for the construction of the Poincaré section. In (d), $d = 5$, $N_{\text{subs}} = 300$, $N_{\text{avg}} = 50$.

the application of the Karhunen–Loève (KL) decomposition to the delay-reconstructed vectors, fixes both d_e and τ by finding the maximum number of singular values above the noise floor in the covariance matrix of the delay vectors (Broomhead & King 1986; Cusumano & Sharkady 1995).

Another approach centres around a combination of the mutual information (MI) algorithm and the method of 'false nearest neighbours' (FNN) (Fraser & Swinney 1986; Kennel *et al.* 1992). MI is used to select a τ large enough to make the delay coordinates independent (in an information theoretical sense), but not so large that sensitive dependence on initial conditions (positive Lyapunov exponents) hides the deterministic relationship between successive coordinates. FNN finds the minimum global embedding dimension by checking to make sure that parts of the attractor are not folded over on themselves: when the embedding dimension is sufficiently large, the delay reconstructions will generically *not* do this; thus, the method will not create 'false neighbours' in the delay-reconstructed space.

Since one is primarily concerned with using dimensionality for the purpose of constructing low-dimensional models of continua, the fractal dimension estimates are not as important as the embedding dimension estimate. Thus, reliable techniques, such as singular systems analysis or FNN are of more than theoretical interest, since they

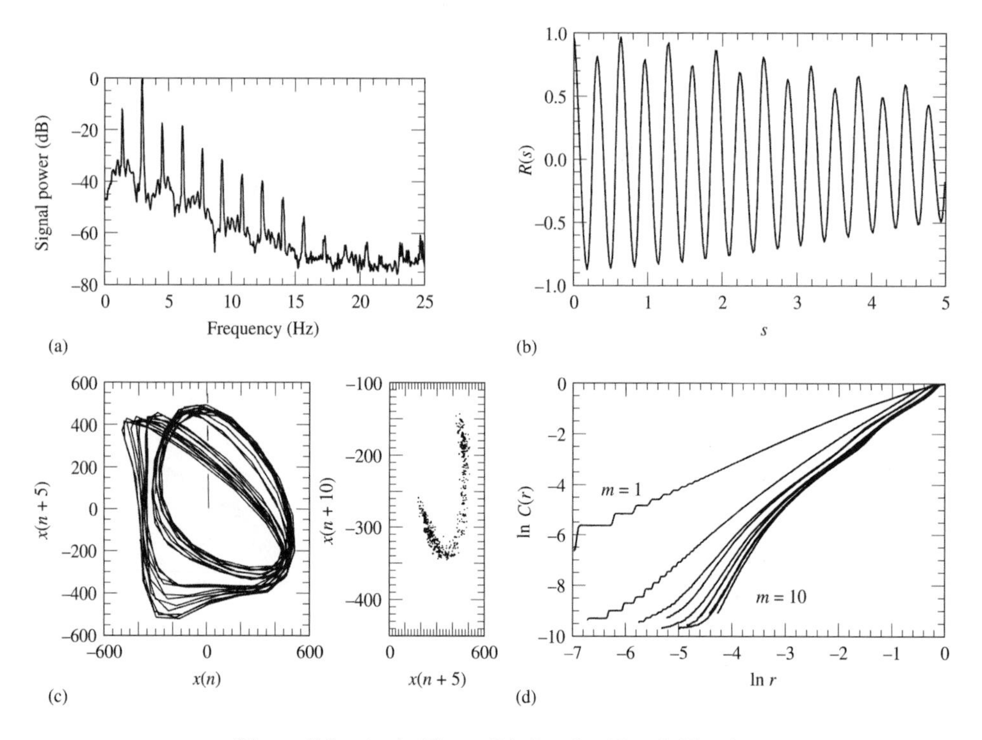

Figure I.2 As in Figure I.1, but for $U = 7.27$ m/s.

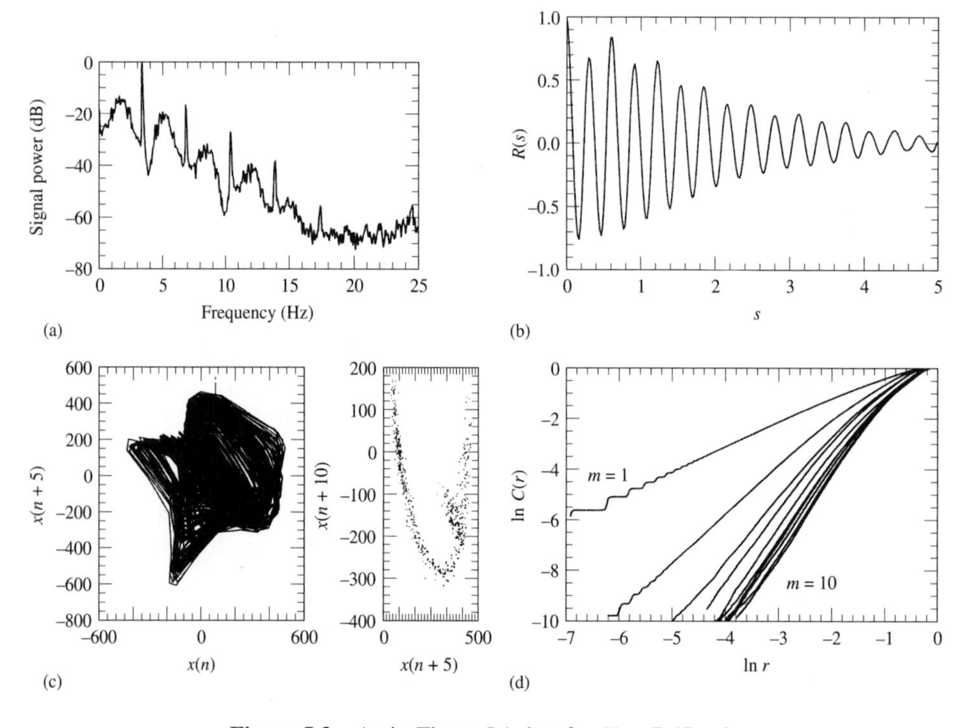

Figure I.3 As in Figure I.1, but for $U = 7.47$ m/s.

estimate the phase space dimension of the required model. The KL decomposition method addresses the problem of dimensionality from the perspective of 'modes' or 'shape functions' needed to describe a given motion (Loève 1963) — specifically by obtaining shape functions, or 'KL modes', from experimental data. This method is more generally applicable than conventional modal analysis, since it does not require that the system be linear. The interested reader is referred to Cusumano *et al*. (1994) and Cusumano (1996).

Appendix J

Detailed Analysis for the Derivation of the Equations of Motion of Chapter 6

J.1 RELATIONSHIP BETWEEN (x_0, y_0, z_0) AND (x, y, z)

The derivation of this relationship is given by Love (1927; Chapter XXI) for the analysis of the 'Small deformation of naturally curved rods'. The detailed derivation, specifically for the curved pipe problem, may be found in Van (1986; Appendix A). Here, only some definitions and the final result are given.

Let us define a so-called Frenet–Serret reference frame (x_0, y_0, z_0) centered at P_0, consisting of the principal axes of the undeformed cross-section of the pipe, z_0 being tangent to the centreline (Figure 6.1); also, a so-called flexure-torsion reference frame (x, y, z) associated with the deformed centreline. Further, let the unit vectors associated with the (x_0, y_0, z_0) and (x, y, z) systems be $(\mathbf{e}_{x_0}, \mathbf{e}_{y_0}, \mathbf{e}_{z_0})$ and $(\mathbf{e}_x, \mathbf{e}_y, \mathbf{e}_z)$, respectively.

The initial curvature is defined by κ_o and κ_o' and the initial twist by τ_o; for the initially planar [in the (x_0, z_0) plane], untwisted pipe, these are

$$\kappa_o = 0, \qquad \kappa_o' = 1/R_o, \qquad \tau_o = 0. \tag{J.1}$$

After deformation, point P_0 moves to P through displacements u, v and w, referred to the (x_0, y_0, z_0) system, as shown in Figure 6.1. The angle between x_0 and x is ψ, which is the angle of rotation about the z-axis of a plane section at P_0 due to deformation.

The centreline strain is given by

$$\varepsilon = \frac{\partial w}{\partial s} - \kappa_o' u + \kappa_o v, \tag{J.2}$$

where s is the curvilinear coordinate along $\mathbf{e}_z$ referred to the (x_0, y_0, z_0) system. Since $\kappa_o = 0$, if the centreline is inextensible, then clearly

$$\frac{\partial w}{\partial s} - \kappa_o' u = 0. \tag{J.3}$$

For small deformations, the relationship between the unit axes is found to be (Love 1927)

$$\begin{Bmatrix} \mathbf{e}_x \\ \mathbf{e}_y \\ \mathbf{e}_z \end{Bmatrix} = \begin{bmatrix} 1 & \psi & -\left(\dfrac{\partial u}{\partial s} + \dfrac{w}{R_o}\right) \\ -\psi & 1 & -\dfrac{\partial v}{\partial s} \\ \left(\dfrac{\partial u}{\partial s} + \dfrac{w}{R_o}\right) & \dfrac{\partial v}{\partial s} & 1 \end{bmatrix} \begin{Bmatrix} \mathbf{e}_{x0} \\ \mathbf{e}_{y0} \\ \mathbf{e}_{z0} \end{Bmatrix}. \tag{J.4}$$

The same transformation matrix relates (x, y, z) to (x_0, y_0, z_0). In (J.4) it is noted that $-\partial v/\partial s$, $\partial u/\partial s + w/R_o$ and ψ are simply the angles of rotation about the x, y and z axes, respectively; hence (J.4) could have been obtained by a sequence of small rotations via the corresponding transformation (rotation) matrices.

Equation (J.4) has been obtained assuming no centreline extension; but, for small deformations, the two frames obey the same relationship, even for the extensible case.

J.2 THE EXPRESSIONS FOR CURVATURE AND TWIST

In this case, three coordinate systems are utilized: the Frenet–Serret and flexure-torsion reference frames used in Section J.1, as well as an inertial system coincident with the former. Then, after introducing the direction cosines relating these reference frames and considering the derivatives of $(\mathbf{e}_x, \mathbf{e}_y, \mathbf{e}_z)$ with respect to s, which are related to curvature and twist, after very lengthy but straightforward manipulation (Love 1927, Chapter XXI; Van 1986, Appendix B) one finds the curvature and twist of the deformed pipe in terms of the deformation:

$$\kappa = \frac{\psi}{R_o} - \frac{\partial^2 v}{\partial s^2}, \qquad \kappa' = \frac{1}{R_o} + \frac{\partial^2 u}{\partial s^2} + \frac{1}{R_o}\frac{\partial w}{\partial s}, \qquad \tau^* = \frac{\partial \psi}{\partial s} + \frac{1}{R_o}\frac{\partial v}{\partial s}. \tag{J.5}$$

The axes associated with κ, κ' and τ^* are defined in Figure 6.1.

J.3 DERIVATION OF THE FLUID-ACCELERATION VECTOR

Recall equation (6.14) in the main text,

$$\mathbf{a}_f = \frac{\partial \mathbf{V}_f}{\partial t} + U\frac{\partial \mathbf{V}_f}{\partial s}, \tag{J.6}$$

where $\mathbf{V}_f$ is given by equation (6.9). By differentiating $\mathbf{V}_f$ with respect to t and s yields

$$\frac{\partial \mathbf{V}_f}{\partial t} = \left[\frac{\partial^2 u}{\partial t^2} + U\left(\frac{\partial^2 u}{\partial t\,\partial s} + \frac{1}{R_o}\frac{\partial w}{\partial t}\right)\right]\mathbf{e}_{x0} + \left[\frac{\partial^2 v}{\partial t^2} + U\frac{\partial^2 v}{\partial t\,\partial s}\right]\mathbf{e}_{y0} + \frac{\partial^2 w}{\partial t^2}\mathbf{e}_{z0}, \tag{J.7}$$

$$\begin{aligned} \frac{\partial \mathbf{V}_f}{\partial s} &= \left[\frac{\partial^2 u}{\partial t\,\partial s} + U\left(\frac{\partial^2 u}{\partial s^2} + \frac{1}{R_o}\frac{\partial w}{\partial s}\right)\right]\mathbf{e}_{x0} + \left[\frac{\partial^2 v}{\partial t\,\partial s} + U\frac{\partial^2 v}{\partial s^2}\right]\mathbf{e}_{y0} + \frac{\partial^2 w}{\partial t\,\partial s}\mathbf{e}_{z0} \\ &+ \left[\frac{\partial u}{\partial t} + U\left(\frac{\partial u}{\partial s} + \frac{w}{R_o}\right)\right]\frac{\partial \mathbf{e}_{x0}}{\partial s} + \left[\frac{\partial v}{\partial t} + U\frac{\partial v}{\partial s}\right]\frac{\partial \mathbf{e}_{y0}}{\partial s} + \left[\frac{\partial w}{\partial t} + U\right]\frac{\partial \mathbf{e}_{z0}}{\partial s}. \end{aligned} \tag{J.8}$$

The derivatives $\partial \mathbf{e}_{x0}/\partial s$, $\partial \mathbf{e}_{y0}/\partial s$, $\partial \mathbf{e}_{z0}/\partial s$ may be found by the same principle as those of a rotating vector. Thus, $\partial \mathbf{e}_{x0}/\partial s = \boldsymbol{\Omega} \times \mathbf{e}_{x0} = -\kappa_o' \mathbf{e}_{z0} + \tau_o \mathbf{e}_{y0}$, where $\boldsymbol{\Omega} = \kappa_o \mathbf{e}_{x0} + \kappa_o' \mathbf{e}_{y0} + \tau_o \mathbf{e}_{z0}$; and similarly for the others. Then, using (J.1), one obtains

$$\frac{\partial \mathbf{e}_{x0}}{\partial s} = -\frac{1}{R_o}\mathbf{e}_{z0}, \qquad \frac{\partial \mathbf{e}_{y0}}{\partial s} = \mathbf{0}, \qquad \frac{\partial \mathbf{e}_{z0}}{\partial s} = \frac{1}{R_o}\mathbf{e}_{x0}. \tag{J.9}$$

Combining equations (J.6)–(J.8) with (J.9), the components of the fluid acceleration vector given in equations (6.15) may be obtained.

J.4 THE EQUATIONS OF MOTION FOR THE PIPE

Consider an infinitesimal element of the pipe contained between the cross-section through P_1 and P_1' on the strained centreline, and the forces and moments acting on it, as shown in Figure J.1. $\mathcal{Q}_{x0}, \mathcal{Q}_{y0}$ and $\mathcal{Q}_{z0}$ are components, referred to the (x_0, y_0, z_0) frame, of the resultant of the transverse shear forces $\mathcal{Q}_x, \mathcal{Q}_y$ and $\mathcal{Q}_z^* = \mathcal{Q}_z + A_o p_e$ [see equation (6.18)]; $\mathcal{M}_{x0}, \mathcal{M}_{y0}$ and $\mathcal{M}_{z0}$ are the components of the resultant of the bending moments $\mathcal{M}_x, \mathcal{M}_y$ and the twist couple $\mathcal{M}_z$ in the x_0, y_0 and z_0 directions; F_{x0}, F_{y0} and F_{z0} are the components, referred to the (x_0', y_0', z_0') frame [defined in Figure J.1(b)] of the force resultant at P_ξ per unit length of the centreline, which includes the inertial and gravity forces, the viscous damping and pressure forces associated with the surrounding fluid and the reaction force associated with the internal flow; Θ_{x0}, Θ_{y0} and Θ_{z0} are components, referred to the (x_0', y_0', z_0') frame of the moment resultant at P_ξ per unit length, which include the moments of rotatory inertia and external moments, if any.

We next consider an inertial coordinate frame (X_0, Y_0, Z_0), relative to which the (x_0, y_0, z_0) and (x_0', y_0', z_0') frames have direction cosines l_{ij}^* and $l_{ij}^{*\prime}$, such that, for instance, $\mathcal{Q}_{y0}$ along Y_0 is given by $l_{12}^*\mathcal{Q}_{x0} + l_{22}^*\mathcal{Q}_{y0} + l_{32}^*\mathcal{Q}_{z0}$. In the limit of $\delta s \to 0$, $l_{ij}^* = l_{ij}^{*\prime}$. After projecting all forces and moments in Figure J.1(b) on the inertial frame (X_0, Y_0, Z_0) with the aid of diagrams such as Figure J.2, and balancing forces and moments along X_0, Y_0 and Z_0, one obtains

$$\begin{aligned}
&\frac{\partial}{\partial s}\left[l_{11}^*\mathcal{Q}_{x0} + l_{21}^*\mathcal{Q}_{y0} + l_{31}^*\mathcal{Q}_{z0}\right] + l_{11}^*F_{x0} + l_{21}^*F_{y0} + l_{31}^*F_{z0} = 0,\\
&\frac{\partial}{\partial s}\left[l_{12}^*\mathcal{Q}_{x0} + l_{22}^*\mathcal{Q}_{y0} + l_{32}^*\mathcal{Q}_{z0}\right] + l_{12}^*F_{x0} + l_{22}^*F_{y0} + l_{32}^*F_{z0} = 0,\\
&\frac{\partial}{\partial s}\left[l_{13}^*\mathcal{Q}_{x0} + l_{23}^*\mathcal{Q}_{y0} + l_{33}^*\mathcal{Q}_{z0}\right] + l_{13}^*F_{x0} + l_{23}^*F_{y0} + l_{33}^*F_{z0} = 0,
\end{aligned} \tag{J.10}$$

$$\begin{aligned}
&\frac{\partial}{\partial s}\left[l_{11}^*\mathcal{M}_{x0} + l_{21}^*\mathcal{M}_{y0} + l_{31}^*\mathcal{M}_{z0}\right] + l_2\left(l_{13}^*\mathcal{Q}_{x0} + l_{23}^*\mathcal{Q}_{y0} + l_{33}^*\mathcal{Q}_{z0}\right)\\
&\qquad - l_3\left(l_{12}^*\mathcal{Q}_{x0} + l_{22}^*\mathcal{Q}_{y0} + l_{32}^*\mathcal{Q}_{z0}\right) + \left(l_{11}^*\Theta_{x0} + l_{21}^*\Theta_{y0} + l_{31}^*\,\Theta_{z0}\right) = 0,\\
&\frac{\partial}{\partial s}\left[l_{12}^*\mathcal{M}_{x0} + l_{22}^*\mathcal{M}_{y0} + l_{32}^*\mathcal{M}_{z0}\right] + l_3\left(l_{11}^*\mathcal{Q}_{x0} + l_{21}^*\mathcal{Q}_{y0} + l_{31}^*\mathcal{Q}_{z0}\right)\\
&\qquad - l_1\left(l_{13}^*\,\mathcal{Q}_{x0} + l_{23}^*\mathcal{Q}_{y0} + l_{33}^*\mathcal{Q}_{z0}\right) + \left(l_{12}^*\Theta_{x0} + l_{22}^*\Theta_{y0} + l_{32}^*\Theta_{z0}\right) = 0,\\
&\frac{\partial}{\partial s}\left[l_{13}^*\mathcal{M}_{x0} + l_{23}^*\,\mathcal{M}_{y0} + l_{33}^*\mathcal{M}_{z0}\right] + l_1\left(l_{12}^*\mathcal{Q}_{x0} + l_{22}^*\mathcal{Q}_{y0} + l_{32}^*\mathcal{Q}_{z0}\right)\\
&\qquad - l_2\left(l_{11}^*\mathcal{Q}_{x0} + l_{21}^*\mathcal{Q}_{y0} + l_{31}^*\mathcal{Q}_{z0}\right) + \left(l_{13}^*\Theta_{x0} + l_{23}^*\Theta_{y0} + l_{33}^*\Theta_{z0}\right) = 0,
\end{aligned} \tag{J.11}$$

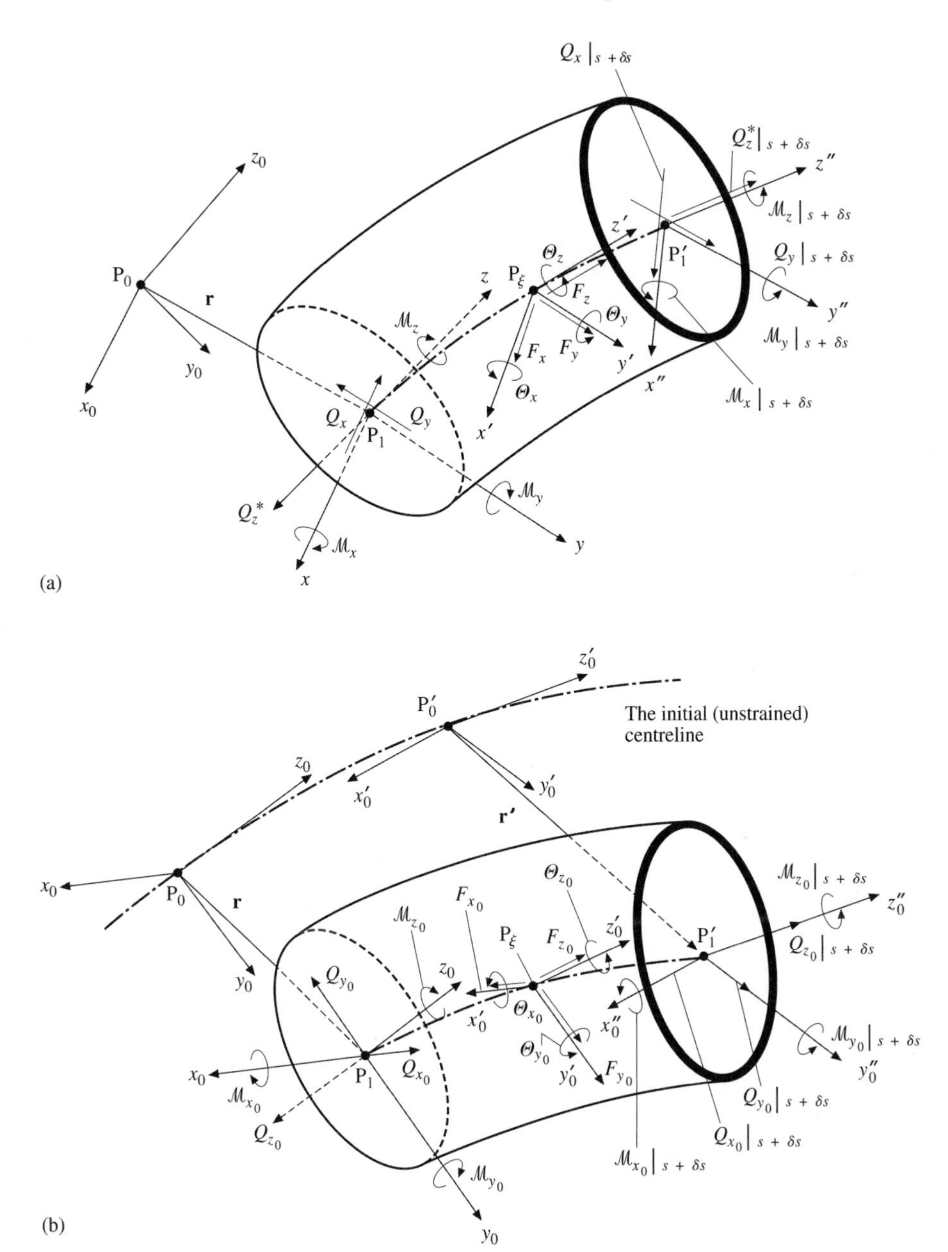

Figure J.1 Forces and moments acting on a pipe element expressed (a) in the reference frame (x, y, z) and (b) in the reference frame (x_0, y_0, z_0).

where

$$l_1 = \frac{\partial X_0}{\partial s}, \qquad l_2 = \frac{\partial Y_0}{\partial s}, \qquad l_3 = \frac{\partial Z_0}{\partial s}. \tag{J.12}$$

Now, in conjunction with the derivation of (J.9), we have obtained (a) $\partial \mathbf{e}_{x0}/\partial s = -\kappa'_o \mathbf{e}_{z0} + \tau_o \mathbf{e}_{y0}$ and similar expressions for $\partial \mathbf{e}_{y0}/\partial s$ and $\partial \mathbf{e}_{z0}/\partial s$; but, writing $\mathbf{e}_{x0} = l^*_{11}\mathbf{i} +$

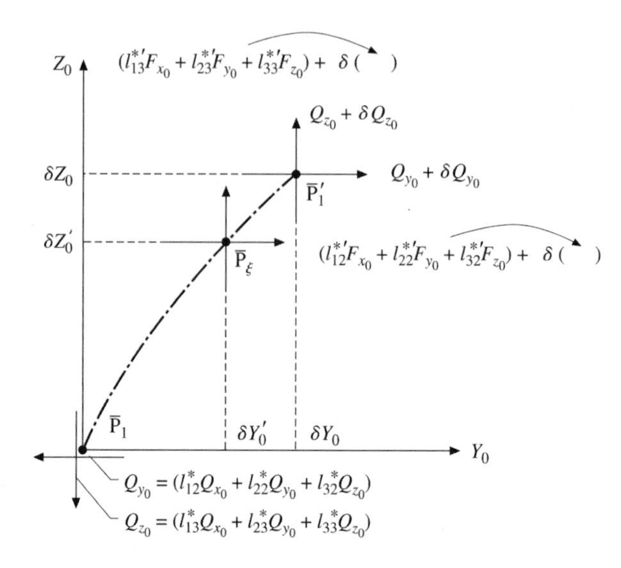

Figure J.2 Forces in Figure J.1(b) projected onto the Z_0 and Y_0 axes.

$l_{12}^*\mathbf{j} + l_{13}^*\mathbf{k}$, $\mathbf{e}_{y0} = l_{21}^*\mathbf{i} + l_{22}^*\mathbf{j} + l_{23}^*\mathbf{k}$, $\mathbf{e}_{z0} = l_{31}^*\mathbf{i} + l_{32}^*\mathbf{j} + l_{33}^*\mathbf{k}$, we also have (b) $\partial\mathbf{e}_{x0}/\partial s = (\partial l_{11}^*/\partial s)\mathbf{i} + (\partial l_{12}^*/\partial s)\mathbf{j} + (\partial l_{13}^*/\partial s)\mathbf{k}$, etc. Then, combining the two different forms (a) and (b) of the expressions for the derivatives in each case, one can obtain the derivatives of l_{ij}^*, as follows:

$$\begin{aligned}
\frac{\partial l_{11}^*}{\partial s} &= l_{21}^*\tau_o - l_{31}^*\kappa_o', & \frac{\partial l_{12}^*}{\partial s} &= l_{22}^*\tau_o - l_{32}^*\kappa_o', & \frac{\partial l_{13}^*}{\partial s} &= l_{23}^*\tau_o - l_{33}^*\kappa_o';\\
\frac{\partial l_{21}^*}{\partial s} &= l_{31}^*\kappa_o - l_{11}^*\tau_o, & \frac{\partial l_{22}^*}{\partial s} &= l_{32}^*\kappa_o - l_{12}^*\tau_o, & \frac{\partial l_{23}^*}{\partial s} &= l_{33}^*\kappa_o - l_{13}^*\tau_o;\\
\frac{\partial l_{31}^*}{\partial s} &= l_{11}^*\kappa_o' - l_{21}^*\kappa_o, & \frac{\partial l_{32}^*}{\partial s} &= l_{12}^*\kappa_o' - l_{22}^*\kappa_o, & \frac{\partial l_{33}^*}{\partial s} &= l_{13}^*\kappa_o' - l_{23}^*\kappa_o.
\end{aligned} \tag{J.13}$$

The (X_0, Y_0, Z_0) frame is now set to coincide with (x_0, y_0, z_0), so that $l_{ij}^* = \delta_{ij}$; also, $l_i = L_{3i}$, $i = 1, 2, 3$, in (J.12) become the direction cosines of the z-axis referred to the (x_0, y_0, z_0) frame as given by

$$L_{31} = \frac{\partial u}{\partial s} - \tau_o v + \kappa_o' w, \qquad L_{32} = \frac{\partial v}{\partial s} - \kappa_o w + \tau_o u, \qquad L_{33} = \frac{\partial w}{\partial s} - \kappa_o' u + \kappa_o v + 1. \tag{J.14}$$

Then, substituting relations (J.13) into equations (J.10) and (J.11), the equations of motion of the pipe along the x_0-, y_0- and z_0-axes may be written as

$$\begin{aligned}
&\frac{\partial Q_{x0}}{\partial s} - \tau_o Q_{y0} + \kappa_o' Q_{z0} + F_{x0} = 0, \qquad \frac{\partial Q_{y0}}{\partial s} - \kappa_o Q_{z0} + \tau_o Q_{x0} + F_{y0} = 0,\\
&\frac{\partial Q_{z0}}{\partial s} - \kappa_o' Q_{x0} + \kappa_o Q_{y0} + F_{z0} = 0,\\
&\frac{\partial \mathcal{M}_{x0}}{\partial s} - \tau_o \mathcal{M}_{y0} + \kappa_o' \mathcal{M}_{z0} + \Theta_{x0} + L_{32} Q_{z0} - L_{33} Q_{y0} = 0,
\end{aligned} \tag{J.15}$$

$$\frac{\partial \mathcal{M}_{y0}}{\partial s} - \kappa_o \mathcal{M}_{z0} + \tau_o \mathcal{M}_{x0} + \Theta_{y0} + L_{33} Q_{x0} - L_{31} Q_{z0} = 0,$$

$$\frac{\partial \mathcal{M}_{z0}}{\partial s} - \kappa_o' \mathcal{M}_{x0} + \kappa_o \mathcal{M}_{y0} + \Theta_{z0} + L_{31} Q_{y0} - L_{32} Q_{x0} = 0.$$

We now consider the force per unit length of the centreline due to gravity and the pressure due to the surrounding fluid. For convenience, the pressure distribution of the surrounding fluid acting on the external lateral surface per unit length of the pipe may be replaced by the buoyancy force **B** (i.e. $\mathbf{B} = A_o \rho_e \mathbf{g}$) and the tensions $A_o p_e$ and $A_o p_e'$ applied on the top and bottom faces, where p_e and p_e' are the pressures at levels P_1 and P_1'. The buoyancy **B** and gravity forces can be combined into a single force, called the effective gravity force **G**, and the pressure force $A_o p_e$ and the tension Q_z can also be combined into a single term Q_z^*. Let (G_{x0}, G_{y0}, G_{z0}) denote components, referred to the system (x_0, y_0, z_0), of the effective gravity force **G**; then, we can write

$$G_{x0} = (m - A_o \rho_e) g \alpha_{x0}, \qquad G_{y0} = (m - A_o \rho_e) g \alpha_{y0},$$
$$G_{z0} = (m - A_o \rho_e) g \alpha_{z0}, \qquad Q_z^* = Q_z + A_o p_e, \tag{J.16}$$

where m is the mass per unit length of the pipe, A_o is the external cross-sectional area of the pipe, ρ_e is the density of the surrounding fluid, g is the acceleration due to gravity and $\alpha_{x0}, \alpha_{y0}, \alpha_{z0}$ are the direction cosines, referred to the system (x_0, y_0, z_0) of the gravitational acceleration.

For the pipe vibrating in a quiescent fluid, fluid damping arises due to viscous effects and due to the energy carried away by acoustic waves. The damping force arising from these effects may be considered to be proportional to the pipe velocity. The components of this force, referred to the system (x_0, y_0, z_0), may be written as

$$f_{x0} = -c \frac{\partial u}{\partial t}, \qquad f_{y0} = -c \frac{\partial v}{\partial t}, \qquad f_{z0} = -c' \frac{\partial w}{\partial t}, \tag{J.17}$$

where c and c' are the coefficients of viscous damping due to the surrounding fluid, associated with the lateral and axial motion of the pipe, respectively, and u, v, w are the displacements of the pipe along the x_0-, y_0-, z_0-axes.

Finally, components of the force resultant per unit length of the pipe centreline can be written as follows:

$$F_{x0} = -(M_a + m) a_{px0} - c \frac{\partial u}{\partial t} + R_{x0} + G_{x0},$$
$$F_{y0} = -(M_a + m) a_{py0} - c \frac{\partial v}{\partial t} + R_{y0} + G_{y0}, \tag{J.18}$$
$$F_{z0} = -(M_a' + m) a_{pz0} - c' \frac{\partial w}{\partial t} + R_{z0} + G_{z0},$$

where $a_{px0}, a_{py0}, a_{pz0}$ are components of the pipe acceleration, M_a and M_a' represent the added mass per unit length, and R_{x0}, R_{y0}, R_{z0} are components of the reaction force arising from the internal flow.

Subject to the limitation that the cross-sectional dimensions of the pipe are small as compared with the overall length of the pipe, the rotatory inertia about axes x and y can be neglected. Therefore, if external moments are absent one obtains

$$\Theta_{x_0} = 0, \qquad \Theta_{y_0} = 0, \qquad \Theta_{z_0} = I_z \frac{\partial^2 \psi}{\partial t^2}, \tag{J.19}$$

where I_z is the moment of inertia of the pipe about the z-axis.

Substituting (J.18), (J.19), (J.14) and (J.1) into equations (J.15), the equations of motion for the pipe, i.e. equations (6.16), may be obtained.

Appendix K
Matrices for the Analysis of an Extensible Curved Pipe Conveying Fluid

The equation governing the in-plane motion of a typical element is

$$[M_{\rm i}]^{\rm e}\{\ddot{q}_{\rm i}\}^{\rm e} + [D_{\rm i}]^{\rm e}\{\dot{q}_{\rm i}\}^{\rm e} + [K_{\rm i}]^{\rm e}\{q_{\rm i}\}^{\rm e} = \{0\}, \tag{K.1}$$

the matrices $[M_{\rm i}]^{\rm e}$, $[D_{\rm i}]^{\rm e}$ and $[K_{\rm i}]^{\rm e}$ in (K.1) being given by

$$\begin{aligned}
[M_{\rm i}]^{\rm e} &= ([A_{\rm o}]^{-1})^{\rm T}\{(1+\beta_a)[I_1] + (1+\overline{\beta}_a)[I_4]\}[A_{\rm o}]^{-1},\\
[D_{\rm i}]^{\rm e} &= ([A_{\rm o}]^{-1})^{\rm T}\left[\beta^{1/2}\overline{u}\left\{2([I_5]+\Theta[I_{14}]) + ([I_{20}]-[I_{14}]^{\rm T})\right\} + \Delta[I_1] + \overline{\Delta}[I_4]\right][A_{\rm o}]^{-1},\\
[K_{\rm i}]^{\rm e} &= ([A_{\rm o}]^{-1})^{\rm T}\left[\left\{[I_3] + \Theta([I_{15}]+[I_{15}]^{\rm T}) + \Theta^2[I_8]\right\} + \mathscr{A}\{[I_8]\right.\\
&\quad - \Theta([I_{12}]+[I_{12}]^{\rm T}) + \Theta^2[I_1]\} + \mathscr{A}\left(c_1\left\{[I_7] - \Theta([I_5]^{\rm T}-[I_{20}]) - \Theta^2[I_{14}]\right\}\right.\\
&\quad \left. + c_2\left\{[I_{22}] - \Theta([I_{11}]^{\rm T}-[I_{21}]) - \Theta^2[I_{23}]\right\}\right)\\
&\quad + \overline{u}^2\left\{[I_9] + \Theta([I_{12}]-[I_{13}]) - \Theta^2[I_4]\right\}\\
&\quad + a_1\left\{[I_9] + \Theta([I_{12}]-[I_{13}]) - \Theta^2[I_4]\right\} + b_1\{[I_5]+\Theta[I_{14}]\}\\
&\quad \left. + a_2\left\{[I_{10}] + \Theta([I_{16}]-[I_{17}]) - \Theta^2[I_{18}]\right\} + b_2\{[I_{11}]+\Theta[I_{23}]\}\right][A_{\rm o}]^{-1},
\end{aligned} \tag{K.2}$$

where $[A_{\rm o}]$ is given by equation (6.92), the same as for out-of-plane motions of an inextensible pipe. The coefficients a_1, a_2, etc. are associated with linear interpolation of the following static parameters:

$$\Pi^o = a_1 + a_2\zeta, \qquad \Pi^{o\prime} = b_1 + b_2\zeta, \qquad \eta^{o\prime} + \Theta\eta_3^o = c_1 + c_2\zeta; \tag{K.3}$$

hence, in terms of the values of Π^o at nodes j and $j+1$,

$$a_1 = \Pi^o|_j, \qquad a_2 = (\Pi^o|_{j+1} - \Pi^o|_j)/\zeta_e, \tag{K.4}$$

where ζ_e is the length of the element in question. Similar expressions hold good for b_1, b_2 and c_1, c_2. The integrals $[I_1]$–$[I_{11}]$ are the same as in equations (6.93), while $[I_{12}]$–$[I_{23}]$

are defined as follows:

$$
\begin{aligned}
&[I_{12}] = \int_0^{\zeta_e} [\phi_2]^{\mathrm{T}}[\phi_4]'\,\mathrm{d}\zeta, \quad [I_{13}] = \int_0^{\zeta_e} [\phi_4]^{\mathrm{T}}[\phi_2]'\,\mathrm{d}\zeta, \quad [I_{14}] = \int_0^{\zeta_e} [\phi_2]'\,[\phi_4]\,\mathrm{d}\zeta,\\
&[I_{15}] = \int_0^{\zeta_e} [\phi_2]''^{\mathrm{T}}[\phi_4]'\,\mathrm{d}\zeta, \quad [I_{16}] = \int_0^{\zeta_e} \zeta[\phi_2]^{\mathrm{T}}[\phi_4]'\,\mathrm{d}\zeta, \quad [I_{17}] = \int_0^{\zeta_e} \zeta[\phi_4]^{\mathrm{T}}[\phi_2]'\,\mathrm{d}\zeta,\\
&[I_{18}] = \int_0^{\zeta_e} \zeta[\phi_4]^{\mathrm{T}}[\phi_4]\,\mathrm{d}\zeta, \quad [I_{20}] = \int_0^{\zeta_e} [\phi_4]^{\mathrm{T}}[\phi_4]'\,\mathrm{d}\zeta, \quad [I_{21}] = \int_0^{\zeta_e} \zeta[\phi_4]^{\mathrm{T}}[\phi_4]'\,\mathrm{d}\zeta,\\
&[I_{22}] = \int_0^{\zeta_e} \zeta[\phi_2]'^{\mathrm{T}}[\phi_4]'\,\mathrm{d}\zeta, \quad [I_{23}] = \int_0^{\zeta_e} \zeta[\phi_2]^{\mathrm{T}}[\phi_4]\,\mathrm{d}\zeta,
\end{aligned}
\tag{K.5}
$$

where

$$
[\phi_2] = [1, \zeta, \zeta^2, \zeta^3, 0, 0] \qquad \text{and} \qquad [\phi_4] = [0, 0, 0, 0, 1, \zeta]. \tag{K.6}
$$

Finally, $[F_1]$–$[F_4]$ appearing in equations (6.103) are defined as follows:

$$
\begin{aligned}
&\{F_1\} = \int_0^{\zeta_e} [\phi_2]^{\mathrm{T}}\,\mathrm{d}\zeta, \qquad \{F_2\} = \int_0^{\zeta_e} \zeta[\phi_2]^{\mathrm{T}}\,\mathrm{d}\zeta,\\
&\{F_3\} = \int_0^{\zeta_e} [\phi_4]^{\mathrm{T}}\,\mathrm{d}\zeta, \qquad \{F_4\} = -\int_0^{\zeta_e} \gamma(\alpha_{x_0}[\phi_2]^{\mathrm{T}} + \alpha_{z_0}[\phi_4]^{\mathrm{T}})\,\mathrm{d}\zeta.
\end{aligned}
\tag{K.7}
$$

References[†]

ABARBANEL, H.D.I. (1994) Tools for analysing observed chaotic data. In *Small Structures, Nonlinear Dynamics, and Control* (eds A. Guran & D.J. Inman), pp. 1–95. Englewood Cliffs, NJ: Prentice-Hall.

ABARBANEL, H.D.I. (1996) *Analysis of Observed Chaotic Data*. New York: Springer-Verlag.

AITHAL, R. & GIPSON, G.S. (1990) Instability of internally damped curved pipes. *ASCE Journal of Engineering Mechanics* **116**, 77–90.

AITKEN, J. (1878) An account of some experiments on rigidity produced by centrifugal force. *Philosophical Magazine*, Series V **5**, 81–105.

AL-JUMAILY, A.M. & AL-SAFFAR, Y.M. (1990) Out-of-plane vibration of an intermediately supported curved-straight tube conveying fluid subjected to a constant thermal force. In *Flow-Induced Vibration — 1990* (eds S.S. Chen, K. Fujita & M.K. Au-Yang), pp. 245–252. New York: ASME.

ANDERSON, G.L. (1972) A comparison of approximate methods for solving non-conservative problems of elastic stability. *Journal of Sound and Vibration* **22**, 159–168.

ANDRONOV, A.A., VITT, A.A. & KHAIKIN, S.E. (1966) *Theory of Oscillators*. New York: Dover.

ANOSOV, D.V., ARNOLD, V.I., SINAI, I.G. & NOVIKOV, S.P. (1988) *Dynamical Systems*. New York: Springer-Verlag.

ARCHER, R.R. (1960) Small vibrations of thin incomplete circular rings. *International Journal of Mechanical Science* **1**, 45–56.

AREF, H. (ed.) (1995) *Chaos Applied to Fluid Mixing*. Oxford: Pergamon.

ARIARATNAM, S.T. & NAMACHCHIVAYA, N.S. (1986a) Dynamic stability of pipes conveying pulsating fluid. *Journal of Sound and Vibration* **107**, 215–230.

ARIARATNAM, S.T. & NAMACHCHIVAYA, N.S. (1986b) Dynamic stability of pipes conveying fluid with stochastic flow velocity. In *Random Vibration — Status and Recent Developments* (eds I. Elishakoff & R.H. Lyon), pp. 1–17. Amsterdam: Elsevier.

ARNOLD, V.I. (1988) *Geometrical Methods in the Theory of Ordinary Differential Equations*, 2nd edition. New York: Springer-Verlag.

[†]The references are generally arranged alphabetically. However, for single-, double- and multiple-author papers *with the same first author*, they are listed as follows:

(i) the single-author papers first: e.g. SMITH, A. 1990 before SMITH, A. 1991;

(ii) the double-author papers next, according to the second author's name: e.g. SMITH, A. & BROWN, G. 1991 before SMITH, A. & GREEN, S. 1979;

(iii) multiple-author papers, which will be cited in the text as, e.g. Smith *et al*. (1979), are listed last, *strictly chronologically*.

ARROWSMITH, D.K. & PLACE, C.M. (1990) *An Introduction to Dynamical Systems*. Cambridge: Cambridge University Press.

ASHLEY, H. & HAVILAND, G. (1950) Bending vibrations of a pipe line containing flowing fluid. *Journal of Applied Mechanics* **17**, 229–232.

ASO, K. & KAN, K. (1986) Behavior of a pipe string in the deep sea. In *Proceedings 5th OMAE Symposium*, Vol. II, pp. 491–498. New York: ASME.

AXISA, F. (1988) Private communication.

BAJAJ, A.K. (1981) Bifurcation to periodic solutions in rotationally symmetric discrete mechanical systems. Ph.D. Thesis, University of Minnesota.

BAJAJ, A.K. (1982) Bifurcating periodic solutions in rotationally symmetric systems. *SIAM Journal of Applied Mathematics* **42**, 1078–1098.

BAJAJ, A.K. (1984) Interactions between self and parametrically excited motions in articulated tubes. *Journal of Applied Mechanics* **51**, 423–429.

BAJAJ, A.K. (1986) Resonant parametric perturbations of the Hopf bifurcation. *Journal of Mathematical Analysis and Applications* **115**, 214–224.

BAJAJ, A.K. (1987a) Bifurcations in a parametrically excited non-linear oscillator. *International Journal of Non-Linear Mechanics* **22**, 47–59.

BAJAJ, A.K. (1987b) Nonlinear dynamics of tubes carrying a pulsatile flow. *Dynamics and Stability of Systems* **2**, 19–41.

BAJAJ, A.K. (1998) Private communication (April 1998).

BAJAJ, A.K. & SETHNA, P.R. (1982a) Bifurcations in three dimensional motions of articulated tubes. Part 1: Linear systems and symmetry. *Journal of Applied Mechanics* **49**, 606–611.

BAJAJ, A.K. & SETHNA, P.R. (1982b) Bifurcations in three dimensional motions of articulated tubes. Part 2: Non-linear analysis. *Journal of Applied Mechanics* **49**, 612–618.

BAJAJ, A.K. & SETHNA, P.R. (1984) Flow induced bifurcations to three-dimensional oscillatory motions in continuous tubes. *SIAM Journal of Applied Mathematics* **44**, 270–286.

BAJAJ, A.K. & SETHNA, P.R. (1991) Effect of symmetry-breaking perturbations on flow-induced oscillations in tubes. *Journal of Fluids and Structures* **5**, 651–679.

BAJAJ, A.K., SETHNA, P.R. & LUNDGREN, T.S. (1980) Hopf bifurcation phenomena in tubes carrying fluid. *SIAM Journal of Applied Mathematics* **39**, 213–230.

BALL, J.E. (1973a) Stability theory for an extensible beam. *Journal of Differential Equations* **14**, 399–418.

BALL, J.E. (1973b) Saddle point analysis for an ordinary differential equation in a Banach space, and an application to dynamic buckling of a beam. In *Nonlinear Elasticity* (ed. R.W. Dickey). London: Academic Press.

BARBEAU, N. (1987) Dynamics of curved pipes conveying fluid. B.Eng. Honours Thesis, Department of Mechanical Engineering, McGill University, Montreal, Québec, Canada.

BARNES, H.A., HUTTON, J.F. & WALTERS, K. (1989) *An Introduction to Rheology*. Amsterdam: Elsevier.

BARNETT, S. & STOREY, C. (1970) *Matrix Methods in Stability Theory*. London: Thomas Nelson & Sons.

BATCHELOR, G.K. (1960) *The Theory of Homogeneous Turbulence*. Cambridge: Cambridge University Press.

BATCHELOR, G.K. (1967) *An Introduction to Fluid Dynamics*. Cambridge: Cambridge University Press.

BECKER, E.B., CAREY, G.F. & ODEN, J.T. (1984) *Finite Elements: An Introduction*. Englewood Cliffs, NJ: Prentice-Hall.

BECKER, M., HAUGER, W. & WINZEN, W. (1978) Exact stability analysis of uniform cantilevered pipes conveying fluid or gas. *Archives of Mechanics (Warsaw)* **30**, 757–768.

BECKER, O. (1979) Zum Stabilitätsverhalten des durchströmten geraden Rohres mit elastischer Querstützung. *Maschinenbautechnik* **28**, 325–327.

BECKER, O. (1981) Das durchströmte Rohr — Literaturbericht. Report IHZ-M-80-212, der Ingenieurhochschule Zittau, Germany.

BEJAN, A. (1987) Buckling flows: a new frontier in Fluid Mechanics. In *Annual Review of Numerical Fluid Mechanics and Heat Transfer* (ed. T.C. Chawla), pp. 262–304. Washington: Hemisphere.

BEJAN, A. (1989) Exploring the origins and structure of turbulence. *Mechanical Engineering (ASME)*, Nov. 1989, pp. 70–74.

BENDIKSEN, O.O. (1987) Mode localization phenomena in large space structures. *AIAA Journal* **25**, 1241–1248.

BENJAMIN, T.B. (1960) Effects of a flexible boundary on hydrodynamic stability. *Journal of Fluid Mechanics* **9**, 513–532.

BENJAMIN, T.B. (1961a) Dynamics of a system of articulated pipes conveying fluid. I. Theory. *Proceedings of the Royal Society (London)* A **261**, 457–486.

BENJAMIN, T.B. (1961b) Dynamics of a system of articulated pipes conveying fluid. II. Experiments. *Proceedings of the Royal Society (London)* A **261**, 487–499.

BENJAMIN, T.B. (1963) The threefold classification of disturbances in flexible surfaces bounding inviscid flows. *Journal of Fluid Mechanics* **16**, 436–450.

BERGÉ, P., POMEAU, Y. & VIDAL, C. (1984) *Order within Chaos*. New York: John Wiley & Sons.

BERT, C.W. (1996) Private communication (23 May 1996).

BEVINGTON, P.R. (1969) *Data Reduction and Error Analysis for the Physical Sciences*. New York: McGraw-Hill.

BISHOP, R.E.D. & FAWZY, I. (1976) Free and forced oscillation of a vertical tube containing a flowing fluid. *Philosophical Transactions of the Royal Society (London)* **284**, 1–47.

BISHOP, R.E.D. & JOHNSON, D.C. (1960) *The Mechanics of Vibration*. Cambridge: Cambridge University Press.

BISHOP, R.E.D., GLADWELL, G.M.L. & MICHAELSON, S. (1965) *The Matrix Analysis of Vibration*. Cambridge: Cambridge University Press.

BLEVINS, R.D. (1979) *Formulas for Natural Frequency and Mode Shape*. New York: Van Nostrand Reinhold.

BLEVINS, R.D. (1990) *Flow-Induced Vibration*, 2nd edition. New York: Van Nostrand Reinhold.

BOHN, M.P. & HERRMANN, G. (1974a) The dynamic behaviour of articulated pipes conveying fluid with periodic flow rate. *Journal of Applied Mechanics* **41**, 55–62.

BOHN, M.P. & HERRMANN, G. (1974b) Instabilities of a spatial system of articulated pipes conveying fluid. *ASME Journal of Fluids Engineering* **96**, 289–296.

BOLOTIN, V.V. (1956) End deformations of flexible pipelines. *Trudy Moskovskogo Energeticheskogo Instituta* **19**, 272–291.

BOLOTIN, V.V. (1963) *Nonconservative Problems of the Theory of Elastic Stability*. London: Pergamon.

BOLOTIN, V.V. (1964) *The Dynamic Stability of Elastic Systems*. San Francisco: Holden Day.

BOLOTIN, V.V. & ZHINZHER, N.I. (1969) Effects of damping on stability of elastic systems subjected to non-conservative forces. *International Journal of Solids and Structures* **5**, 965–989.

BOURRIÈRES, F.-J. (1939) Sur un phénomène d'oscillation auto-entretenue en mécanique des fluides réels. *Publications Scientifiques et Techniques du Ministère de l'Air*, No. 147.

BRANDSTÄTER, A., SWIFT, J., SWINNEY, H., WOLF, A., FARMER, J.D., JEN, E. & CRUTCHFIELD, J.P. (1983) Low-dimensional chaos in a hydrodynamic system. *Physical Review Letters* **51**, 1442–1445.

BROADBENT, E.G. & WILLIAMS, M. (1956) The effect of structural damping on binary flutter. Aeronautical Research Council (U.K.) A.R.C. R & M 3169.

BROOMHEAD, D.S. & KING, G.P. (1986) Extracting qualitative dynamics from experimental data. *Physica* D **20**, 217–236.

BROWN, H.S., JOLLY, M.S., KEVREKIDIS, I.G. & TITI, E.S. (1990) Use of appropriate inertial manifolds in bifurcation calculations. In *Continuation and Bifurcations: Numerical Techniques and Applications* (eds D. Roose *et al*.), pp. 9–23. Dordrecht: Kluwer.

BUZANO, E., GEYMONAT, G. & POSTON, T. (1985) Post-buckling behavior of a non-linearly hyperelastic thin rod with cross-section invariant under the dihedral group D_n. *Archive of Rational Mechanics and Analysis* **89**, 307–388.

CAI, Y. & CHEN, S.S. (1995) Numerical analysis for dynamic instability of electrodynamic Maglev systems. *Shock and Vibration* **2**(4), 339–349.

CAI, Y., CHEN, S.S., ROTE, D.M. & COFFEY, H.T. (1994) Vehicle/guideway interaction for high speed vehicles on a flexible guideway. *Journal of Sound and Vibration* **175**, 625–646.

CAI, Y., CHEN, S.S., ROTE, D.M. & COFFEY, H.T. (1996) Vehicle/guideway interaction in Maglev systems. *ASME Journal of Dynamic Systems, Measurement, and Control* **118**, 526–530.

CARLUCCI, L.N. (1980) Damping and hydrodynamic mass of a cylinder in simulated two-phase flow. *ASME Journal of Mechanical Design* **102**, 597–602.

CARLUCCI, L.N. & BROWN, J.D. (1983) Experimental studies of damping and hydrodynamic mass of a cylinder in confined two-phase flow. *ASME Journal of Vibration, Stress, and Reliability in Design* **105**, 83–89.

CARR, J. (1981) *Applications of Center Manifold Theory*. New York: Springer-Verlag.

CAUGHEY, T.K. & O'KELLEY, M.E.J. (1965) Classical normal modes in damped linear dynamic systems. *Journal of Applied Mechanics* **32**, 583–588.

CHAMPNEYS, A.R. (1991) Homoclinic orbits in the dynamics of articulated pipes conveying fluid. *Nonlinearity* **4**, 747–774.

CHAMPNEYS, A.R. (1993) Homoclinic tangencies in the dynamics of articulated pipes conveying fluid. *Physica* D **62**, 347–359.

CHANG, C.O. & CHEN, K.C. (1994) Dynamics and stability of pipes conveying fluid. *ASME Journal of Pressure Vessel Technology* **116**, 57–66.

CHEN, S.S. (1970) Forced vibration of a cantilevered tube conveying fluid. *Journal of the Acoustical Society of America* **48**, 773–775.

CHEN, S.S. (1971a) Flow-induced instability of an elastic tube. *ASME Paper* No. 71-Vibr.-39.

CHEN, S.S. (1971b) Dynamic stability of a tube conveying fluid. *ASCE Journal of the Engineering Mechanics Division* **97**, 1469–1485.

CHEN, S.S. (1972a) Vibrations of continuous pipes conveying fluid. In *Flow-Induced Structural Vibrations* (ed. E. Naudascher), pp. 663–675. Berlin: Springer-Verlag.

CHEN, S.S. (1972b) Vibration and stability of a uniformly curved tube conveying fluid. *Journal of the Acoustical Society of America* **51**, 223–232.

CHEN, S.S. (1972c) Flow-induced in-plane instabilities of curved pipes. *Nuclear Engineering and Design* **23**, 29–38.

CHEN, S.S. (1973) Out-of-plane vibration and stability of curved tubes conveying fluid. *Journal of Applied Mechanics* **40**, 362–368.

CHEN, S.S. (1981) Fluid damping for circular cylindrical structures. *Nuclear Engineering and Design* **63**, 81–100.

CHEN, S.S. (1987) *Flow-Induced Vibration of Circular Structures*. Washington: Hemisphere.

CHEN, S.S. (1995) Private communication (20 September 1995); also responses by telephone to a number of oral or written queries throughout 1995.

CHEN, S.S. & JENDRZEJCZYK, J.A. (1985) General characteristics, transition, and control of instability of tubes conveying fluid. *Journal of the Acoustical Society of America* **77**, 887–895.

CHEN, S.S. & ROSENBERG, G.S. (1971) Vibrations and stability of a tube conveying fluid. Argonne National Laboratory Report ANL-7762, Argonne, Illinois, U.S.A.

CHEN, S.S., WAMBSGANSS, M.W. & JENDRZEJCZYK, J.A. (1976) Added mass and damping of a vibrating rod in confined viscous fluids. *Journal of Applied Mechanics* **43**, 325–329.

CHEN, W.-H. & FAN, C.-N. (1987) Stability analysis with lumped mass and friction effects in elastically supported pipes conveying fluid. *Journal of Sound and Vibration* **119**, 429–442.

CH'NG, E. (1977) The original version of Ch'ng (1978).

CH'NG, E. (1978) A theoretical analysis of nonlinear effects on the flutter and divergence of a tube conveying fluid. Dept of Mechanical and Aerospace Engineering, Princeton University, AMS Report No. 1343 (revised).

CH'NG, E. & DOWELL, E.H. (1979) A theoretical analysis of nonlinear effects on the flutter and divergence of a tube conveying fluid. In *Flow-Induced Vibrations* (eds S.S. Chen & M.D. Bernstein), pp. 65–81. New York: ASME.

CHOW, S.N. & HALE, J.K. (1982) *Methods of Bifurcation Theory*. New York: Springer-Verlag.

CHUNG, J.S. & WHITNEY, A.K. (1983) Axial stretching oscillation of an 18,000-ft vertical pipe in the ocean. *ASME Journal of Energy Resources Technology* **105**, 195–200.

CHUNG, J.S., WHITNEY, A.K. & LODEN, W.A. (1981) Nonlinear transient motion of deep ocean mining pipe. *ASME Journal of Energy Resources Technology* **103**, 2–10.

COLLAR, A.R. & SIMPSON, A. (1987) *Matrices and Engineering Dynamics*. Chichester, U.K.: Ellis Horwood Ltd; New York: John Wiley & Sons.

COPELAND, G.S. (1992) Flow-induced vibration and chaotic motion of a slender tube conveying fluid. Ph.D. dissertation, Cornell University, Ithaca, NY, U.S.A.

COPELAND, G.S. & MOON, F.C. (1992) Chaotic flow-induced vibration of a flexible tube with end mass. *Journal of Fluids and Structures* **6**, 705–718.

COWPER, G.R. (1966) The shear coefficient in Timoshenko's beam theory. *Journal of Applied Mechanics* **33**, 335–340.

CRAIK, A.D.D. (1985) *Wave Interactions and Fluid Flows*. Cambridge: Cambridge University Press.

CRANDALL, S.H. (1995a) The effect of damping on the stability of gyroscopic pendulums. *Zeitschrift für angewandte Mathematik und Physik* **46**, S761–S780.

CRANDALL, S.H. (1995b) Canonical physical models of dynamic instability. In *Proceedings of CANCAM 95 (Canadian Congress of Applied Mechanics)*. Victoria, B.C., Canada, pp. 1–12.

CRUICKSHANK, J.O. & MUNSON, B.R. (1981) Viscous fluid buckling of viscous and axisymmetric jets. *Journal of Fluid Mechanics* **113**, 221–239.

CUI, H.-W., TANI, J. & QIU, J.H. (1994) Flutter robust control of a pipe conveying fluid. *Proceedings 1st World Conference on Structural Control*, Los Angeles, CA, U.S.A.; TP4, pp. 83–91.

CUI, H.-W., TANI, J. & OHTOMO, K. (1995) Robust flutter control of vertical pipe conveying fluid using gyroscopic mechanism. *Transactions of JSME*, Series C **61**(585), 1822–1826.

CURTAIN, R.F. & PRITCHARD, A.J. (1977) *Functional Analysis in Modern Applied Mathematics*. London: Academic Press.

CUSUMANO, J.P. (1996) Experimental application of the Karhunen–Loève decomposition to the study of modal interactions in a mechanical oscillator. In *Chaotic, Fractal, and Nonlinear Signal Processing* (ed. R. Katz). American Institute of Physics.

CUSUMANO, J.P. (1997) Private communication (2 April 1997).

CUSUMANO, J.P. & MOON, F.C. (1995a) Chaotic non-planar vibrations of the thin elastica. Part I: Experimental observation of planar instability. *Journal of Sound and Vibration* **179**, 185–208.

CUSUMANO, J.P. & MOON, F.C. (1995b) Chaotic non-planar vibrations of the thin elastica. Part II: Derivation and analysis of a low dimensional model. *Journal of Sound and Vibration* **179**, 209–226.

CUSUMANO, J.P. & SHARKADY, M.T. (1995) An experimental study of bifurcation, chaos, and dimensionality in a system forced through a bifurcation parameter. *Nonlinear Dynamics* **8**, 467–489.

CUSUMANO, J.P., SHARKADY, M.T. & KIMBLE, B.W. (1994) Experimental measurements of dimensionality and spatial coherence in the dynamics of a flexible-beam impact oscillator. *Philosophical Transactions of the Royal Society (London)* A **347**, 421–438.

DANG, X.Q., LIU, W.M. & ZHENG, T.S. (1989) Efficient numerical analysis for dynamic stability of pipes conveying fluids. *ASME Journal of Pressure Vessel Technology* **111**, 300–303.

DEN HARTOG, J.P. (1956) *Mechanical Vibrations*, 4th edition. New York: McGraw-Hill.

DEN HARTOG, J.P. (1969) John Orr Memorial Lecture: Recent cases of mechanical vibration. *The South African Mechanical Engineer* **19**, 53–68.

DESAI, C.S. & ABEL, J.F. (1972) *Introduction to the Finite Element Method.* New York: Van Nostrand Reinhold.

DESSLER, A.J. (1967) Solar wind and interplanetary magnetic field. *Reviews of Geophysics* **5**, 1–41.

DEVANEY, R.L. (1989) *An Introduction to Chaotic Dynamical Systems*, 2nd edition. Redwood City, CA: Addison-Wesley.

DODDS, H.L. Jr. & RUNYAN, H.L. (1965) Effect of high velocity fluid flow on the bending vibrations and static divergence of a simply supported pipe. *NASA Technical Note* D-2870.

DOEDEL, E. (1981) AUTO, a program for automatic bifurcation analysis of autonomous systems. *Congressus Numerantium* **30**, 265–284.

DOEDEL, E.J. & KERNÉVES, J.P. (1986) AUTO: software for continuation and bifurcation problems in ordinary differential equations. Applied Mathematics Report, California Institute of Technology, Pasadena, CA, U.S.A. (procurable from doedel@cs.concordia.ca).

DOKI, H., HIRAMOTO, K. & SKELTON, R.E. (1998) Active control of cantilevered pipes conveying fluid with constraints on import energy. *Journal of Fluids and Structures* **12**(5) (in press).

DOLL, R.W. & MOTE, C.D. Jr. (1974) The dynamic formulation and the finite element analysis of curved and twisted tubes transporting fluids. Report to the National Science Foundation, Dept of Mechanical Engineering, University of California, Berkeley.

DOLL, R.W. & MOTE, C.D. Jr. (1976) On the dynamic analysis of curved and twisted cylinders transporting fluids. *ASME Journal of Pressure Vessel Technology* **98**, 143–150.

DONE, G.T.S. (1963) The effect of linear damping on flutter speed. Aeronautical Research Council (U.K.) A.R.C. R & M 3396.

DONE, G.T.S. & SIMPSON, A. (1977) Dynamic stability of certain conservative and non-conservative systems. *I.Mech.E. Journal of Mechanical Engineering Science* **19**, 251–263.

DOWELL, E.H. (1975) *Aeroelasticity of Plates and Shells*. Leyden: Noordhoff.

DOWELL, E.H. (1982) Flutter of a buckled plate as an example of chaotic motion of a deterministic autonomous system. *Journal of Sound Vibration* **85**, 333–344.

DOWELL, E.H. & PEZESHKI, C. (1986) On the understanding of chaos in Duffing's equation including a comparison with experiment. *Journal of Applied Mechanics* **53**, 5–9.

DOWELL, E.H. & WIDNALL, S.E. (1966) Generalized aerodynamic forces on oscillating cylindrical shell: subsonic and supersonic flow. *AIAA Journal* **4**, 607–610.

DOWELL, E.H., CRAWLEY, E.F., CURTISS, H.C. Jr., PETERS, D.A., SCANLAN, R.H. & SISTO, F. (1995) *A Modern Course in Aeroelasticity*, 3rd edition. Dordrecht: Kluwer Academic.

DRAZIN, P.G. & REID, W.H. (1981) *Hydrodynamic Stability*. Cambridge: Cambridge University Press.

DUBUSSCHE, A. & MARION, M. (1992) On the construction of families of approximate inertial manifolds. *Journal of Differential Equations* **100**, 173–201.

DUNCAN, W.J., THOM, A.S. & YOUNG, A.D. (1970) *Mechanics of Fluids*, 2nd edition. London: Edward Arnold.

DUPUIS, C. (1997) Private communication (17 July 1997).

DUPUIS, C. & ROUSSELET, J. (1985) Application of the transfer matrix method to non-conservative systems involving fluid flow in curved pipes. *Journal of Sound and Vibration* **98**, 415–429.

DUPUIS, C. & ROUSSELET, J. (1986) Response to Païdoussis' discussion of the paper by Dupuis & Rousselet (1985). *Journal of Sound and Vibration* **111**, 168–170.

DUPUIS, C. & ROUSSELET, J. (1991a) Discussion to the papers by Sällström & Åkesson (1990) and Sällström (1990), with a note by M.P. Païdoussis. *Journal of Fluids and Structures* **5**, 597–600.

DUPUIS, C. & ROUSSELET, J. (1991b) Discussion to the paper by Aithal & Gipson (1990) and authors' closure. *ASCE Journal of Engineering Mechanics* **117**, 2456–2457.

DUPUIS, C. & ROUSSELET, J. (1992) The equations of motion of curved pipes conveying fluid. *Journal of Sound and Vibration* **153**, 473–489.

DWIGHT, H.B. (1961) *Tables of Integrals and Other Mathematical Data*, 4th edition. New York: Macmillan Publishing.

ECKMANN, J.-P. & RUELLE, D. (1985) Ergodic theory of chaos and strange attractors. *Reviews of Modern Physics* **57**, 617–656.

EDELSTEIN, W.S. & CHEN, S.S. (1985) Flow-induced instability of an elastic tube with a variable support. *Nuclear Engineering and Design* **84**, 1–11.

EDELSTEIN, W.S., CHEN, S.S. & JENDRZEJCZYK, J.A. (1986) A finite element computation of the flow-induced oscillations in a cantilevered tube. *Journal of Sound and Vibration* **107**, 121–129.

ENGELSTAD, R.L. (1988) Vibration and stability of vertical tubes conveying fluid subjected to planar excitation. Ph.D. Thesis, Department of Engineering Mechanics, University of Wisconsin-Madison, U.S.A.

ENGELSTAD, R.L. & LOVELL, E.G. (1985) Vibration analysis of LIBRA INPORTs. *Fusion Technology* **8**, 1884–1889.

ENGELSTAD, R.L. & LOVELL, E.G. (1995) Dynamic response of flexible tubes conveying fluid subjected to planar sequential impulses. In *Applied Mechanics in the Americas* (eds L.A. Godoy, S.R. Idelsohn, P.A.A. Laura & D.T. Mook), Vol. II, pp. 334–339. Santa Fe, Argentina: American Academy of Mechanics & AMCA.

ERINGEN, A.C. (1987) *Mechanics of Continua*. New York: John Wiley.

EVAN-IWANOWSKI, R.M. (1976) *Resonance Oscillations in Mechanical Systems*. Amsterdam: Elsevier.

EWINS, D.J. (1975) Measurement and application of mechanical impedance data. *Journal of the Society of Environmental Engineers* **14**, 3–12.

EWINS, D.J. (1985) *Modal Testing: Theory and Practice*. New York: John Wiley; Letchworth, U.K.: Research Studies Press.

FAN, C.-N. & CHEN, W.-H. (1987) Vibration and stability of helical pipes conveying fluid. *ASME Journal of Pressure Vessel Technology* **109**, 402–410.

FEIGENBAUM, M.J. (1978) Qualitative universality for a class of nonlinear transformations. *Journal of Statistical Physics* **19**, 25–52.

FELIPPA, C.A. & CHUNG, J.S. (1981) Nonlinear static analysis of deep ocean mining pipe. Part I: modeling and formulation. *ASME Journal of Energy Resources Technology* **103**, 11–15.

FEODOS'EV, V.P. (1951) Vibrations and stability of a pipe when liquid flows through it. *Inzhenernyi Sbornik* **10**, 169–170.

FERRI, A.A. (1986) On the equivalence of the incremental harmonic balance method and the harmonic balance Newton-Raphson method. *Journal of Applied Mechanics* **53**, 455–457.

FINLAYSON, B.A. & SCRIVEN, L.E. (1966) The method of weighted residuals — a review. *Applied Mechanics Reviews* **19**, 735–748.

FLÜGGE, W. (1960) *Stresses in Shells*. Berlin: Springer-Verlag.

FOALE, S., McROBIE, F.A. & THOMPSON, J.M.T. (1998) Numerical dimension-reduction methods for nonlinear shell vibrations. *Journal of Sound and Vibration* (to be published).

FOIAS, C., JOLLY, M.S., KEVREKIDES, I.G., SELL, G.R. & TITI, E.S. (1988) On the computation of inertial manifolds. *Physics Review Letters* A **131**, 433–436.

FRASER, A.M. & SWINNEY, H.L. (1986) Independent coordinates for strange attractors from mutual information. *Physical Review* **33**A, 1134–1140.

FRIEDMAN, B. (1956) *Principles and Techniques of Applied Mathematics*. New York: John Wiley & Sons.

FUNG, Y.C. (1969) *A First Course in Continuum Mechanics*. Englewood Cliffs, NJ: Prentice-Hall.

GEAR, C.W. (1971) *Numerical Initial Value Problems in Ordinary Differential Equations*. Englewood Cliffs, NJ: Prentice-Hall.

GIBERT, R.J. (1988) *Vibrations des structures*. Paris: Eyrolles.

GINSBERG, J.H. (1973) The dynamic stability of a pipe conveying a pulsatile flow. *International Journal of Engineering Science* **11**, 1013–1024.

GLEICK, J. (1987) *Chaos*. New York: Viking.

GLEICK, J. (1992) *Genius*. New York: Pantheon Books.

GLENDINNING, P. (1994) *Stability, Instability and Chaos: An Introduction to the Theory of Nonlinear Differential Equations*. Cambridge: Cambridge University Press.

GOLDSTEIN, H. (1950) *Classical Mechanics*. Reading, MA: Addison-Wesley.

GOLUB, G.H. & VAN LOAN, C.F., (1989) *Matrix Computations*. Baltimore: John Hopkins University Press.

GOLUBITSKY, M. & SCHAEFFER, D. (1985) *Singularities and Groups in Bifurcation Theory*. New York: Springer-Verlag.

GOLUBITSKY, M. & STEWART, I. (1986) Symmetry and stability in Taylor–Couette flow. *SIAM Journal of Applied Mathematical Analysis* **17**, 249–288.

GRASSBERGER, P. & PROCCACIA, I. (1983a) Measuring the strangeness of strange attractors. *Physica* D **9**, 189–208.

GRASSBERGER, P. & PROCCACIA, I. (1983b) Characterization of strange attractors. *Physical Review Letters* **50**, 346–349.

GRAY, J. (1968) *Animal Locomotion*. London: Weidenfeld & Nicolson.

GREENWALD, A.S. & DUGUNDJI, J. (1967) Static and dynamic instabilities of a propellant line. MIT Aeroelastic and Structures Research Lab, AFOSR Sci. Report: AFOSR 67–1395.

GREGORY, R.W. & PAÏDOUSSIS, M.P. (1966a) Unstable oscillation of tubular cantilevers conveying fluid. I. Theory. *Proceedings of the Royal Society (London)* A **293**, 512–527.

GREGORY, R.W. & PAÏDOUSSIS, M.P. (1966b) Unstable oscillation of tubular cantilevers conveying fluid. II. Experiments. *Proceedings of the Royal Society (London)* A **293**, 528–542.

GROH, G.G. (1992) Computation of hydrodynamic mass for general configurations: by integral equation method. *Proceedings International Symposium on Flow-Induced Vibration and Noise*, Vol. 7 (eds M.P. Païdoussis, T. Akylas & P.B. Abraham), pp. 159–172. New York: ASME.

GUCKENHEIMER, J. & HOLMES, P. (1983) *Nonlinear Oscillations, Dynamical Systems, and Bifurcations of Vector Fields*. New York: Springer-Verlag.

GURAN, A. & PLAUT, R.H. (1994) Stability boundaries for fluid conveying pipes with flexible support under axial load. *Archive of Applied Mechanics* **64**, 417–422.

HAGEDORN, P. (1981) *Non-linear Oscillations*. Oxford: Clarendon.

HAHN, W. (1963) *Theory and Application of Liapunov's Direct Method*. Englewood Cliffs, NJ: Prentice-Hall.

HAIRER, E., NORSETT, S.P. & WANNER, G. (1993) *Solving Ordinary Differential Equations*. Berlin: Springer-Verlag.

HALE, J.K. (1969) *Ordinary Differential Equations*. New York: Wiley-Interscience.

HANDELMAN, G.H. (1955) A note on the transverse vibration of a tube containing flowing fluid. *Quarterly of Applied Mathematics* **13**, 326–330.

HANNOYER, M.J. (1972) A solution to linear differential equations in the field of dynamics of continuous systems. Mechanical Engineering Research Laboratories Report MERL 72-5, McGill University, Montréal, Québec, Canada.

HANNOYER, M.J. (1977) Instabilities of slender, tapered tubular beams induced by internal and external axial flow. Ph.D. Thesis, Department of Mechanical Engineering, McGill University, Montreal, Québec, Canada.

HANNOYER, M.J. & PAÏDOUSSIS, M.P. (1978) Instabilities of tubular beams simultaneously subjected to internal and external axial flows. *ASME Journal of Mechanical Design* **100**, 328–336.

HANNOYER, M.J. & PAÏDOUSSIS, M.P. (1979a) Dynamics of slender tapered beams with internal or external axial flow. Part 1: Theory. *Journal of Applied Mechanics* **46**, 45–51.

HANNOYER M.J. & PAÏDOUSSIS, M.P. (1979b) Dynamics of slender tapered beams with internal or external axial flow. Part 2: Experiments. *Journal of Applied Mechanics* **46**, 52–57.

HAO, B.-L. (1990) *Chaos II*. Teaneck, NJ: World Scientific.

HARA, F. (1977) Two-phase-flow-induced vibrations in a horizontal piping system. *Bulletin of the JSME* **20**, 419–427.

HARA, F. (1980) Two-phase flow induced parametric vibration in structural systems — pipes and nuclear pins. The Institute of Industrial Sciences, The University of Tokyo, **20**, No. 4 (Serial No. 183).

HARINGX, J.A. (1952) Instability of thin-walled cylinders subjected to internal pressure. *Philips Research Reports* **7**, 112–118. [Also, in *De Ingenieur* **63**, O39–O41 in 1951 (in Dutch).]

HAUGER, W. & VETTER, K. (1976) Influence of an elastic foundation on the stability of a tangentially loaded column. *Journal of Sound and Vibration* **47**, 296–299.

HAYASHI, C. (1964) *Nonlinear Oscillations in Physical Systems*. New York: McGraw-Hill.

HEINRICH, G. (1956) Vibrations of tubes with flow. *Zeitschrift für Angewandte Mathematik und Mechanik* **36**, 417–427.

HERRMANN, G. (1967) Stability of equilibrium of elastic systems subjected to nonconservative forces. *Applied Mechanics Reviews* **20**, 103–108.

HERRMANN, G. & BUNGAY, R.W. (1964) On the stability of elastic systems subjected to nonconservative forces. *Journal of Applied Mechanics* **31**, 435–440.

HERRMANN, G. & JONG, I.C. (1965) On the destabilizing effect of damping in nonconservative elastic systems. *Journal of Applied Mechanics* **32**, 592–597.

HERRMANN, G. & JONG, I.C. (1966) On nonconservative stability problems of elastic systems with slight-damping. *Journal of Applied Mechanics* **33**, 125–133.

HERRMANN, G. & NEMAT-NASSER, S. (1967) Instability modes of cantilevered bars induced by fluid flow through attached pipes. *International Journal of Solids and Structures* **3**, 39–52.

HERRMANN, G., NEMAT-NASSER, S. & PRASAD, S.N. (1966) Models demonstrating instability in nonconservative mechanical systems. The Technological Institute, Dept of Civil Engineering, Northwestern University, Technical Report No. 66–4.

HILL, J.L. & DAVIS, C.G. (1974) The effect of initial forces on the hydroelastic vibration and stability of planar curved tubes. *Journal of Applied Mechanics* **41**, 355–359.

HILL, J.L. & SWANSON, C.P. (1970) Effects of lumped masses on the stability of fluid conveying tubes. *Journal of Applied Mechanics* **37**, 494–497.

HINZE, J.O. (1975) *Turbulence*, 2nd edition. New York: McGraw-Hill.

HIRSCH, M.W. & SMALE, S. (1974) *Differential Equations, Dynamical Systems, and Linear Algebra*. Orlando: Academic Press.

HODGE, P.G. (1970) *Continuum Mechanics*. New York: McGraw-Hill.

HOLMES, P.J. (1977) Bifurcations to divergence and flutter in flow-induced oscillations: a finite dimensional analysis. *Journal of Sound and Vibration* **53**, 471–503.

HOLMES, P.J. (1978) Pipes supported at both ends cannot flutter. *Journal of Applied Mechanics* **45**, 619–622.

HOLMES, P.J. & MARSDEN, J.E. (1978) Bifurcation to divergence and flutter in flow-induced oscillations: an infinite dimensional analysis. *Automatica* **14**, 367–384.

HOLMES, P.J. & MOON, F.C. (1983) Strange attractors and chaos in nonlinear mechanics. *Journal of Applied Mechanics* **50**, 1021–1032.

HOUSNER, G.W. (1952) Bending vibrations of a pipe line containing flowing fluid. *Journal of Applied Mechanics* **19**, 205–208.

HU, H.-H. & TSOON, W.-S. (1957) On the flexible vibrations of a pipe containing flowing fluid. In *Proceedings of Theoretical and Applied Mechanics (India)*, pp. 203–216.

HUGHES, P.C. (1986) *Spacecraft Attitude Dynamics*. New York: John Wiley & Sons.

HUNT, K. & CROSSLEY, F. (1975) Coefficient of restitution interpreted as damping in vibroimpacting. *Journal of Applied Mechanics* **42**, 440–445.

HUSEYIN, K. & PLAUT, R.H. (1974/5) Transverse vibrations and stability of systems with gyroscopic forces. *Journal of Structural Mechanics* **3**, 163–177.

ILGAMOV, M.A., TANG, D.M. & DOWELL, E.H. (1994) Flutter and forced response of a cantilevered pipe: the influence of internal pressure and nozzle discharge. *Journal of Fluids and Structures* **8**, 139–156.

INAGAKI, T., UMEOKA, T., FUJITA, K., NAKAMURA, T., SHIRAISHI, T., KIYOKAWA, T. & SUGIYAMA, Y. (1987) Flow induced vibration of inverted U-shaped piping containing flowing fluid of top entry system for LMFBR. In *Proceedings SMiRT-9*, Vol. E, pp. 295–302.

IOOSS, G. & JOSEPH, D.D. (1980) *Elementary Stability and Bifurcation Theory*. New York: Springer-Verlag.

IWATSUBO, T., SUGIYAMA, Y. & OGINO, S. (1974) Simple and combination resonances of columns under periodic axial loads. *Journal of Sound and Vibration* **33**, 211–221.

JAHNKE, E. & EMDE, F. (1945) *Tables of Functions*, 4th edition. New York: Dover Publications.

JAYARAMAN, K. & NARAYANAN, S. (1996) Chaotic oscillations in pipes conveying pulsating fluid. *Nonlinear Dynamics* **10**, 333–357.

JENDRZEJCZYK, J.A. & CHEN, S.S. (1985) Experiments on tubes conveying fluid. *Thin-Walled Structures* **3**, 109–134.

JENSEN, J.S. (1997) Fluid transport due to nonlinear fluid–structure interaction. *Journal of Fluids and Structures* **11**, 327–344.

JENSEN, M.H., KADANOFF, L.P., LIBCHABER, A., PROCACCIA, I. & STAVANS, J. (1985) Global universality at the onset of chaos: results of a Rayleigh–Bénard experiment. *Physical Review Letters* **55**, 2798–2801.

JOHNSON, R.O., STONEKING, J.E. & CARLEY, T.G. (1987) The stability of simply-supported tubes conveying a compressible fluid. *Journal of Sound and Vibration* **117**, 335–350.

JONES, L.H. & GOODWIN, B.E. (1971) The transverse vibrations of a pipe containing flowing fluid: methods of integral equations. *Quarterly of Applied Mathematics* **29**, 363–374.

JONES, W.P. & LAUNDER, B.E. (1972) The prediction of laminarization with a two-equation model for turbulence. *International Journal of Heat and Mass Transfer* **15**, 301–314.

KANGASPUOSKARI, M., LAUKKANEN, J. & PRAMILA, A. (1993) The effect of feedback control on critical velocity of cantilevered pipes aspirating fluid. *Journal of Fluids and Structures* **7**, 707–715.

KARAMCHETI, K. (1966) *Principles of Ideal Fluid Aerodynamics*. New York: John Wiley & Sons.

KATSIKADELIS, J.T. & KOUNADIS, A.N. (1983) Flutter loads of a Timoshenko beam-column under a follower force governed by two variants of equations of motion. *Acta Mechanica* **48**, 209–217.

KENNEL, M.B., BROWN, R. & ABARBANEL, H.D.I. (1992) Determining embedding dimension using a geometrical construction. *Physical Review* A **45**, 3403–3411.

KLEIN, C. (1981) The effect of randomly varying added mass on the dynamics of a flexible cylinder in two-phase axially flowing fluid. M.Eng. Thesis, Department of Mechanical Engineering, McGill University, Montreal, Québec, Canada.

KO, C.L. & BERT, C.W. (1984) Nonlinear vibration of uniformly curved, fluid-conveying pipes. In *Proceedings 5th ICPVT, San Francisco*, pp. 469–477.

KO, C.L. & BERT, C.W. (1986) A perturbation solution for non-linear vibration of uniformly curved pipes conveying fluid. *International Journal of Non-Linear Mechanics* **21**, 315–325.

KOEHNE, M. (1978) The control of vibrating elastic systems. In *Distributed Parameter Systems* (eds W.H. Ray & D.G. Lainiotis), Chapter 7. New York: Marcel Dekker.

KOEHNE, M. (1982) Modelling and simulation of distributed-parameter mechanical systems. In *Proceedings of 10th IMACS World Congress on System Simulation and Scientific Computation*, Vol. 3, pp. 357–361.

KOLMOGOROV, A.N. & FOMIN, S.V. (1970) *Introductory Real Analysis*. New York: Dover.

KRASSOVSKII, N.N. (1963) *Stability of Motion*. Stanford, CA: Stanford University Press.

KRYLOV, N.M. & BOGOLIUBOV, N.N. (1947) *Introduction to Nonlinear Mechanics*. Princeton: Princeton University Press (Russian original: Moscow, 1937).

KUBICEK, M. & MAREK, M. (1983) *Computational Methods in Bifurcation Theory and Dissipative Structures*. New York: Springer-Verlag.

LAITHIER, B.E. (1979) Dynamics of Timoshenko tubular beams conveying fluid. Ph.D. Thesis, Department of Mechanical Engineering, McGill University, Montreal, Québec, Canada.

LAITHIER, B.E. & PAÏDOUSSIS, M.P. (1981) The equations of motion of initially stressed Timoshenko tubular beams conveying fluid. *Journal of Sound and Vibration* **79**, 175–195.

LAMB, Sir Horace (1957) *Hydrodynamics*, 6th edition. Cambridge: Cambridge University Press.

LAMBERT, J.D. (1973) *Computational Methods in Ordinary Differential Equations*. New York: Academic Press.

LANDAHL, M.T. (1962) On the stability of a laminar incompressible boundary layer over a flexible surface. *Journal of Fluid Mechanics* **13**, 609–632.

LANDAU, L.D. & LIFSHITZ, E.M. (1959) *Fluid Mechanics*. Oxford: Pergamon Press.

LANGTHJEM, M.A. (1995) On dynamic stability of an immersed fluid-conveying tube. Danish Center for Applied Mathematics and Mechanics, Report No. 512. An up-dated version to be published in the *Journal of Fluids and Structures*.

LASALLE, J.P. & LEFSCHETZ, S. (1961) *Stability by Liapounov's Direct Method*. New York: Academic Press.

LAU, S.L. & YUEN, S.W. (1993) Solution diagram of non-linear dynamic systems by the IHB method. *Journal of Sound and Vibration* **167**, 303–316.

LAU, S.L., CHEUNG, Y.K. & WU, S.Y. (1982). A variable parameter incrementation method for dynamic instability of linear and nonlinear vibration of elastic systems. *Journal of Applied Mechanics* **49**, 849–853.

LAU, S.L., CHEUNG, Y.K. & WU, S.Y. (1983) Incremental harmonic balance method with multiple time scales for nonlinear dynamics systems. *Journal of Applied Mechanics* **50**, 871–876.

LAUNDER, B.E. & SHARMA, B.I. (1974) Application of the energy dissipation model of turbulence to the calculation of flow near a spinning disc. *Letters in Heat and Mass Transfer* **1**, 131–138.

LAUNDER, B.E. & SPALDING, D.B. (1972) *Mathematical Models of Turbulence*. London: Academic Press.

LEIPHOLZ, H. (1970) *Stability Theory*. New York: Academic Press.

LESIEUR, M. (1990) *Turbulence in Fluids: Stochastic and Numerical Modelling*, 2nd edition. Dordrecht: Kluwer Academic Publishers.

LEVY, S. & WILKINSON, J.P.D. (1975) Calculation of added water mass effects for reactor system components. *Transactions 3rd International Conference on Structural Mechanics in Reactor Technology (SMiRT)*, Vol. 2, Paper F 2/5, London, U.K.

LI, G.X. & PAÏDOUSSIS, M.P. (1994) Stability, double degeneracy and chaos in cantilevered pipes conveying fluid. *International Journal of Non-Linear Mechanics* **29**, 83–107.

LI, T. & DIMAGGIO, O.D. (1964) Vibration of a propellant line containing flowing fluid. In *Proceedings AIAA 5th Annual Structures and Materials Conference*, CP-8, pp. 194–199.

LIGHTHILL, M.J. (1960) Note on the swimming of slender fish. *Journal of Fluid Mechanics* **9**, 305–317.

LIGHTHILL, M.J. (1969) Hydromechanics of aquatic animal propulsion. *Annual Review of Fluid Mechanics* **1**, 413–446.

LIN, H.C. & CHEN, S.S. (1976) Vibration and stability of fluid-conveying pipes. *Shock & Vibration Bulletin* **46**(2), 267–283.

LIN, Y.-H. & CHU, C.-L. (1996) Active flutter control of a cantilever tube conveying fluid using piezoelectric actuators. *Journal of Sound and Vibration* **196**, 97–105.

LIU, H.-S. & MOTE, C.D. Jr. (1974) Dynamic response of pipes transporting fluids. *ASME Journal of Engineering for Industry* **96**, 591–596.

LOÈVE, M. (1963) *Probability Theory*. Princeton, NJ: Van Nostrand.

LONG, R.H. Jr. (1955) Experimental and theoretical study of transverse vibration of a tube containing flowing fluid. *Journal of Applied Mechanics* **22**, 65–68.

LONG, Y.G., NAGAYA, K. & NIWA, H. (1993) Vibration conveyance of a continuous long beam in a spatially curved tube. *Transactions of JSME*, Series C **59**(568), 3658–3667.

LONG, Y.G., NAGAYA, K. & NIWA, H. (1994) Vibration conveyance in spatial-curved tubes. *ASME Journal of Vibration and Acoustics* **116**, 38–46.

LOTTATI, I. & KORNECKI, A. (1985) The effect of an elastic foundation and of dissipative forces on the stability of fluid conveying pipes. Technion Report TAE No. 563, Haifa, Israel.

LOTTATI, I. & KORNECKI, A. (1986) The effect of an elastic foundation and of dissipative forces on the stability of fluid conveying pipes. *Journal of Sound and Vibration* **109**, 327–338.

LOVE, A.E.H. (1927) *A Treatise on the Mathematical Theory of Elasticity*, 4th edition. Cambridge: Cambridge University Press; New York: Dover (1944).

LU, S., SEMERCIGIL, S.E. & TURAN, Ö.F. (1993) Employing fluid flow in a cantilever pipe for vibration control. *Proceedings 14th Canadian Congress of Applied Mechanics*, Vol. 1, pp. 219–220.

LUNDGREN, T.S., SETHNA, P.R. & BAJAJ, A.K. (1979) Stability boundaries for flow induced motions of tubes with an inclined terminal nozzle. *Journal of Sound and Vibration* **64**, 553–571.

LUNN, T.S. (1982) Flow-induced instabilities of fluid-conveying pipes. Ph.D. Thesis, University College London, London, U.K.

LUU, T.P. (1983) On the dynamics of three systems involving tubular beams conveying fluid. M.Eng. Thesis, Department of Mechanical Engineering, McGill University, Montreal, Québec, Canada.

LYNN, J. & JAY, A. (1989) *The Complete Yes Minister*. London: BBC Books.

MAKRIDES, G.A. & EDELSTEIN, W.S. (1992) Some numerical studies of chaotic motions in tubes conveying fluid. *Journal of Sound and Vibration* **152**, 517–530.

MALRAISON, G., ATTEN, P., BERGÉ, P. & DUBOIS, M. (1983) Dimension of strange attractors: An experimental determination of the chaotic regime of two convective systems. *Journal of Physics Letters* **44**, 897–902.

MAÑÉ, R. (1981) On the dimension of the compact invariant sets of certain non-linear maps. In *Dynamical Systems and Turbulence*, Springer Lecture Notes in Mathematics Vol. 898 (eds D.A. Rand & L.-S. Young), pp. 230–242. New York: Springer-Verlag.

MANNEVILLE, P. & POMEAU, Y. (1980) Different ways to turbulence in dissipative dynamical systems. *Physica* D **1**, 219–226.

MASSEY, B.S. (1979) *Mechanics of Fluids*, 4th edition. New York: Van Nostrand Reinhold.

MATEESCU, D., PAÏDOUSSIS, M.P. & BÉLANGER, F. (1994a) Unsteady annular viscous flows between oscillating cylinders. Part I: Computational solutions based on a time-integration method. *Journal of Fluids and Structures* **8**, 489–507.

MATEESCU, D., PAÏDOUSSIS, M.P. & BÉLANGER, F. (1994b) Unsteady annular viscous flows between oscillating cylinders. Part II: A hybrid time-integration solution based on azimuthal Fourier expansions for configurations with annular backsteps. *Journal of Fluids and Structures* **8**, 509–528.

MATEESCU, D., PAÏDOUSSIS, M.P. & SIM, W.-G. (1994a) A spectral collocation method for confined unsteady flows with oscillating boundaries. *Journal of Fluids and Structures* **8**, 157–181.

MATEESCU, D., PAÏDOUSSIS, M.P. & SIM, W.-G. (1994b) Spectral solutions for unsteady annular flows between eccentric cylinders induced by transverse oscillations. *Journal of Sound and Vibration* **177**, 635–649.

McIVER, D.B. (1973) Hamilton's principle for systems of changing mass. *Journal of Engineering Mathematics* **7**, 249–261.

MEAD, D.J. (1970) Free waves in periodically supported, infinite beams. *Journal of Sound and Vibration* **11**, 181–197.

MEAD, D.J. (1973) A general theory of harmonic wave propagation in linear periodic systems with multiple coupling. *Journal of Sound and Vibration* **27**, 235–260.

MEIROVITCH, L. (1967) *Analytical Methods in Vibrations*. New York: The Macmillan Co.

MEIROVITCH, L. (1970) *Methods of Analytical Dynamics*. New York: McGraw-Hill.

MERIAM, J.L. (1980) *Engineering Mechanics*. Vol. 2: *Dynamics*. New York: John Wiley & Sons.

MILES, W.H., PEZESHKI, C. & ELGAR, S. (1992) Bispectral analysis of a fluid elastic system: the cantilevered pipe. *Journal of Fluids and Structures* **6**, 633–640.

MILNE-THOMSON, L.M. (1949) *Theoretical Hydrodynamics*, 2nd edition; 1968, 5th edition. London: Macmillan & Co.

MILNE-THOMSON, L.M. (1958) *Theoretical Aerodynamics*, 3rd edition. London: Macmillan & Co.

MINORSKY, N. (1962) *Nonlinear Oscillations*. Princeton, NJ: Van Nostrand.

MISRA, A.K., PAÏDOUSSIS, M.P. & VAN, K.S. (1988a) On the dynamics of curved pipes transporting fluid. Part I: inextensible theory. *Journal of Fluids and Structures* **2**, 211–244.

MISRA, A.K., PAÏDOUSSIS, M.P. & VAN, K.S. (1988b) On the dynamics of curved pipes transporting fluid. Part II: extensible theory. *Journal of Fluids and Structures* **2**, 245–261.

MISRA, A.K., PAÏDOUSSIS, M.P. & VAN, K.S. (1988c) Dynamics and stability of fluid conveying curved pipes. In *Proceedings International Symposium on Flow-Induced Vibration and Noise*, Vol. 4 (eds M.P. Païdoussis, M.K. Au-Yang & S.S. Chen), pp. 1–24. ASME: New York.

MOE, G. & CHUCHEEPSAKUL, S. (1988) The effect of internal flow on marine risers. *Proceedings 7th International Offshore Mechanics and Arctic Engineering Symposium*, Vol. 1, pp. 375–382. New York: ASME.

MOE, G., STROMSEM, K.C. & FYLLING, I. (1994) Behaviour of risers with internal flow under various boundary conditions. *Proceedings of 4th International Offshore and Polar Engineering Conference*, Osaka, Japan, pp. 258–262. ISOPE.

MONTGOMERY, D.C. & TIDMAN, D.A. (1964) *Plasma Kinetic Theory*. New York: McGraw-Hill.

MOODY, F.J. (1990) *Introduction to Unsteady Thermofluid Mechanics*. New York: John Wiley & Sons.

MOON, F.C. (1980) Experiments on chaotic motion of a forced nonlinear oscillator: strange attractors. *Journal of Applied Mechanics* **47**, 638–644.

MOON, F.C. (1992) *Chaotic and Fractal Dynamics*. New York: John Wiley & Sons.

MORSE, P.M. (1948) *Vibration and Sound*, 2nd edition. New York: McGraw Hill; American Institute of Physics (1976).

MOTE, C.D. Jr. (1968) Dynamic stability of an axially moving band. *Journal of the Franklin Institute* **285**, 329–346.

MOTE, C.D. Jr. (1972) Dynamic stability of axially moving material. *Shock and Vibration Digest* **14**(4), 2–11.

MOVCHAN, A.A. (1959) The direct method of Liapunov in stability problems of elastic systems. *Prikladnaia Matematika i Mekhanika* **23**, 483–493.

MOVCHAN, A.A. (1965) On the problem of stability of a pipe with fluid flowing through it. *Prikladnaia Matematika i Mekhanika* **29**, 760–762.

MUKHIN, O.N. (1965) Stability of a pipeline and some methods in nonconservative problems. *Vestnik Moskovskovo Universiteta*, Series I, Mathematics, No. 2, pp. 76–87.

MUNTEAN, G.G. & MOON, F.C. (1995) Multifractals in elastic tube vibrations due to internal flow. *Journal of Fluids and Structures* **9**, 787–799.

NAGULESWARAN, S. (1996) Private communication (22 October 1996).

NAGULESWARAN, S. & WILLIAMS, C.J.H. (1968) Lateral vibrations of a pipe conveying fluid. *Journal of Mechanical Engineering Science* **10**, 228–238.

NAKRA, B.C. & KOHLI, A.K. (1984) Vibration analysis of straight and curved tubes conveying fluid by means of straight beam finite elements. *Journal of Sound and Vibration* **93**, 307–311.

NAMACHCHIVAYA, N.S. (1989) Non-linear dynamics of supported pipe conveying pulsating fluid. 1. Subharmonic resonance. *International Journal of Non-Linear Mechanics* **24**, 185–196.

NAMACHCHIVAYA, N.S. & ARIARATNAM, S.T. (1987) Periodically perturbed Hopf bifurcation. *SIAM Journal of Applied Mathematics* **47**, 15–39.

NAMACHCHIVAYA, N.S. & TIEN, W.M. (1989a) Non-linear dynamics of supported pipe conveying pulsating fluid. 2. Combination resonance. *International Journal of Non-Linear Mechanics* **24**, 197–208.

NAMACHCHIVAYA, N.S. & TIEN, W.M. (1989b) Bifurcation behavior of nonlinear pipes conveying pulsating flow. *Journal of Fluids and Structures* **3**, 609–629.

NARAYANAN, S. (1983) Stochastic stability of fluid conveying tubes. In *Random Vibrations and Reliability* (ed. K. Hennig), pp. 273–283. Berlin: Akademie-Verlag.

NAUDASCHER, E. & ROCKWELL, D. (1980) Oscillator-model approach to the identification and assessment of flow-induced vibrations in a system. *Journal of Hydraulic Research* **18**, 59–82.

NAUDASCHER, E. & ROCKWELL, D. (1994) *Flow-Induced Vibrations: An Engineering Guide*. Rotterdam: A.A. Balkema.

NAYFEH, A.H. (1973) *Perturbation Methods*. New York: John Wiley & Sons.

NAYFEH, A.H. (1981) *Introduction to Perturbation Techniques*. New York: Wiley-Interscience.

NAYFEH, A.H. (1985) *Problems in Perturbation*. New York: John Wiley & Sons.

NAYFEH, A.H. (1993) *Method of Normal Forms*. New York: Wiley-Interscience.

NAYFEH, A.H. & MOOK, D.T. (1979) *Nonlinear Oscillations*. New York: Wiley-Interscience.

NEĬMARK, JU.I. & FUFAEV, N.A. (1972) *Dynamics of Nonholonomic Systems*, English Translation. Providence, RI: American Mathematical Society. (Original in Russian, 1967.)

NEMAT-NASSER, S. & HERRMANN, G. (1966) Torsional instability of cantilevered bars subjected to nonconservative loading. *Journal of Applied Mechanics* **33**, 102–104.

NEMAT-NASSER, S., PRASAD, S.N. & HERRMANN, G. (1966) Destabilizing effect of velocity-dependent forces in non-conservative continuous systems. *AIAA Journal* **4**, 1276–1280.

NEWHOUSE, S.E., RUELLE, D. & TAKENS, F. (1978) Occurrence of strange axiom A attractors near quasiperiodic flows on T^m, $m \geq 3$. *Communications in Mathematical Physics* **64**, 35–40.

NGUYEN, V.B., PAÏDOUSSIS, M.P. & MISRA, A.K. (1993) A new outflow model for cylindrical shells conveying fluid. *Journal of Fluids and Structures* **7**, 417–419.

NIORDSON, F.I. (1953) Vibrations of a cylindrical tube containing flowing fluid. *Kungliga Tekniska Hogskolans Handlingar (Stockholm)* No. 73.

NISSIM, E. (1965) Effect of linear damping on flutter speed. Part I: binary systems. *The Aeronautical Quarterly* **16**, 159–178.

NOAH, S.T. & HOPKINS, G.R. (1980) Dynamic stability of elastically supported pipes conveying pulsating fluid. *Journal of Sound and Vibration* **71**, 103–116.

ODEN, J.T. (1979) *Applied Functional Analysis*. Englewood Cliffs, NJ: Prentice-Hall.

OJALVO, I.U. (1962) Coupled twist-bending vibrations of incomplete elastic rings. *International Journal of Mechanical Science* **4**, 53–72.

OJALVO, I.U. & NEWMAN, N. (1965) Natural frequencies of cantilevered ring segments. *Machine Design* **37**, 191–195.

OTT, E. (1993) *Chaos in Dynamical Systems*. Cambridge: Cambridge University Press.

PACKARD, N.H., CRUTCHFIELD, J.P., FARMER, J.D. & SHAW, R.S. (1980) Geometry from a time series. *Physical Review Letters* **45**, 712–716.

PAÏDOUSSIS, M.P. (1963) Oscillations of liquid-filled flexible tubes. Ph.D. Thesis, University of Cambridge.

PAÏDOUSSIS, M.P. (1966) Dynamics of flexible slender cylinders in axial flow. Part 1: theory. *Journal of Fluid Mechanics* **26**, 717–736.

PAÏDOUSSIS, M.P. (1969) Dynamics of vertical tubular cantilevers conveying fluid. Mechanical Engineering Research Laboratories Report MERL 69-3, Department of Mechanical Engineering, McGill University, Montreal, Québec, Canada.

PAÏDOUSSIS, M.P. (1970) Dynamics of tubular cantilevers conveying fluid. *Journal of Mechanical Engineering Science* **12**, 85–103.

PAÏDOUSSIS, M.P. (1973a) Vibration of tubes containing fluid flow. In *Aéro-Hydro-Elasticité*, pp. 442–500. Paris: Eyrolles.

PAÏDOUSSIS, M.P. (1973b) Dynamics of cylindrical structures subjected to axial flow. *Journal of Sound and Vibration* **29**, 365–385.

PAÏDOUSSIS, M.P. (1975) Flutter of conservative systems of pipes conveying incompressible fluid. *Journal of Mechanical Engineering Science* **17**, 19–25.

PAÏDOUSSIS, M.P. (1976) Hydroelastic ichthyoid propulsion. *AIAA Journal of Hydronautics* **10**, 30–32.

PAÏDOUSSIS, M.P. (1979) The dynamics of clusters of flexible cylinders in axial flow: theory and experiments. *Journal of Sound and Vibration* **65**, 391–417.

PAÏDOUSSIS, M.P. (1980) Flow-induced vibrations in nuclear reactors and heat exchangers: Practical experiences and state of the knowledge. In *Practical Experiences with Flow-Induced Vibrations* (eds E. Naudascher & D. Rockwell), pp. 1–81. Berlin: Springer-Verlag.

PAÏDOUSSIS, M.P. (1986a) Flow-induced instabilities of cylindrical structures. In *Proceedings 10th U.S. National Congress of Applied Mechanics* (ed. J.P. Lamb), pp. 155–170.

PAÏDOUSSIS, M.P. (1986b) Discussion to the paper by Dupuis & Rousselet (1985). *Journal of Sound and Vibration* **111**, 167–168.

PAÏDOUSSIS, M.P. (1987) Flow-induced instabilities of cylindrical structures. *Applied Mechanics Reviews* **40**, 163–175.

PAÏDOUSSIS, M.P. (1991) Pipes conveying fluid: a model dynamical problem. In *Proceedings of CANCAM'91 (Canadian Congress of Applied Mechanics)*, Winnipeg, Manitoba, Canada, pp. 1–33.

PAÏDOUSSIS, M.P. (1993) *The 1992 Calvin Rice Lecture*: Some curiosity-driven research in fluid–structure interactions and its current applications. *ASME Journal of Pressure Vessel Technology* **115**, 2–14.

PAÏDOUSSIS, M.P. (1997) Fluid–structure interactions between axial flows and slender structures, *Proceedings of XIX ICTAM*, Kyoto, Japan, 1996. In *Theoretical and Applied Mechanics 1996* (eds T. Tatsumi, E. Watanabe & T. Kambe), pp. 427–442. Amsterdam: Elsevier.

PAÏDOUSSIS, M.P. & BOTEZ, R.M. (1995) Three routes to chaos for a three-degree-of-freedom articulated cylinder system subjected to annular flow and impacting on the outer pipe. *Nonlinear Dynamics* **7**, 429–450.

PAÏDOUSSIS, M.P. & DEKSNIS, E.B. (1969) Stability of articulated cantilevers conveying fluid. MERL Report 69-11, Dept of Mechanical Engineering, McGill University, Montreal, Québec, Canada.

PAÏDOUSSIS, M.P. & DEKSNIS, E.B. (1970) Articulated models of cantilevers conveying fluid: the study of a paradox. *Journal of Mechanical Engineering Science* **12**, 288–300.

PAÏDOUSSIS, M.P. & DES TROIS MAISONS, P.E. (1971) Free vibration of a heavy, damped, vertical cantilever. *Journal of Applied Mechanics* **38**, 524–526.

PAÏDOUSSIS, M.P. & ISSID, N.T. (1974) Dynamic stability of pipes conveying fluid. *Journal of Sound and Vibration* **33**, 267–294.

PAÏDOUSSIS, M.P. & ISSID, N.T. (1976) Experiments on parametric resonance of pipes containing pulsatile flow. *Journal of Applied Mechanics* **43**, 198–202.

PAÏDOUSSIS, M.P. & LAITHIER, B.E. (1976) Dynamics of Timoshenko beams conveying fluid. *Journal of Mechanical Engineering Science* **18**, 210–220.

PAÏDOUSSIS, M.P. & LI, G.X. (1993) Pipes conveying fluid: a model dynamical problem. *Journal of Fluids and Structures* **7**, 137–204.

PAÏDOUSSIS, M.P. & LUU, T.P. (1985) Dynamics of a pipe aspirating fluid, such as might be used in ocean mining. *ASME Journal of Energy Resources Technology* **107**, 250–255.

PAÏDOUSSIS, M.P. & MOON, F.C. (1988) Nonlinear and chaotic fluidelastic vibrations of a flexible pipe conveying fluid. *Journal of Fluids and Structures* **2**, 567–591.

PAÏDOUSSIS, M.P. & OSTOJA-STARZEWSKI, M. (1981) Dynamics of a flexible cylinder in subsonic axial flow. *AIAA Journal* **19**, 1467–1475.

PAÏDOUSSIS, M.P. & SEMLER, C. (1993a) Nonlinear and chaotic oscillations of a constrained cantilevered pipe conveying fluid: a full nonlinear analysis. *Nonlinear Dynamics* **4**, 655–670.

PAÏDOUSSIS, M.P. & SEMLER, C. (1993b) Nonlinear dynamics of a fluid-conveying cantilevered pipe with an intermediate spring support. *Journal of Fluids and Structures* **7**, 269–298; addendum in **7**, 565–566.

PAÏDOUSSIS, M.P. & SEMLER, C. (1998) Nonlinear dynamics of a fluid-conveying cantilevered pipe with a small mass attached at the free end. *International Journal of Non-Linear Mechanics* **33**, 15–32.

PAÏDOUSSIS, M.P. & SUNDARARAJAN, C. (1975) Parametric and combination resonances of a pipe conveying pulsating fluid. *Journal of Applied Mechanics* **42**, 780–784.

PAÏDOUSSIS, M.P., SUSS, S. & PUSTEJOVSKY, M. (1977) Free vibration of clusters of cylinders in liquid-filled channels. *Journal of Sound and Vibration* **55**, 443–459.

PAÏDOUSSIS, M.P., CHAN, S.P. and MISRA, A.K. (1984) Dynamics and stability of coaxial cylindrical shells containing flowing fluid. *Journal of Sound and Vibration* **97**, 201–235.

PAÏDOUSSIS, M.P., LUU, T.P. & LAITHIER, B.E. (1986) Dynamics of finite-length tubular beams conveying fluid. *Journal of Sound and Vibration* **106**, 311–331.

PAÏDOUSSIS, M.P., LI, G.X. & MOON, F.C. (1989) Chaotic oscillations of the autonomous system of a constrained pipe conveying fluid. *Journal of Sound and Vibration* **135**, 1–19.

PAÏDOUSSIS, M.P., LI, G.X. & RAND, R.H. (1991a) Chaotic motions of a constrained pipe conveying fluid: comparison between simulation, analysis and experiment. *Journal of Applied Mechanics* **58**, 559–565.

PAÏDOUSSIS, M.P., NGUYEN, V.B. & MISRA, A.K. (1991b) A theoretical study of the stability of cantilevered coaxial cylindrical shells conveying fluid. *Journal of Fluids and Structures* **5**, 127–164.

PAÏDOUSSIS, M.P., CUSUMANO, J.P. & COPELAND, G.S. (1992) Low-dimensional chaos in a flexible tube conveying fluid. *Journal of Applied Mechanics* **59**, 196–205.

PARKER, E.N. (1963) *Interplanetary Dynamical Processes*. New York: Interscience.

PARKER, T.S. & CHUA, L. (1989) *Practical Numerical Algorithms for Chaotic Systems*. New York: Springer-Verlag.

PARKS, P.C. (1967) A stability criterion for a pannel flutter problem via the second method of Liapunov. In *Differential Equations and Dynamical Systems* (eds J.K. Hale & J.P. LaSalle). New York: Academic Press.

PAYNE, A.R. & SCOTT, J.R. (1960) *Engineering Design with Rubber*. London: MacLaren and Sons.

PESTEL, E.C. & LECKIE, F.A. (1963) *Matrix Methods in Elastomechanics*. New York: McGraw-Hill.

PIERRE, C. & DOWELL, E.H. (1987) Localization of vibrations by structural irregularity. *Journal of Sound and Vibration* **114**, 549–564.

PIET-LAHANIER, N. & OHAYON, R. (1990) Finite element analysis of a slender fluid structure system. *Journal of Fluids and Structures* **4**, 631–645.

PIPES, L.A. (1963) *Matrix Methods for Engineering*. Englewood Cliffs, N.J.: Prentice-Hall.

PLACHE, K.O. (1979) Coriolis/gyroscopic flow meter. *Mechanical Engineering* **101**, 36–41.

PLAUT, R.H. (1995) Private e-mail communication (23 September 1995).

PLAUT, R.H. & HUSEYIN, K. (1975) Instability of fluid conveying pipes under axial load. *Journal of Applied Mechanics* **42**, 889–890.

PRAMILA, A., LAUKKANEN, J. & LIUKKONEN, S. (1991) Dynamics and stability of short fluid-conveying Timoshenko element pipes. *Journal of Sound and Vibration* **144**, 421–425.

PRANDTL, L. (1952) *Essentials of Fluid Dynamics*. London: Blackie.

PRESS, W.H., TEUKOLSKY, S.A., VETTERING, W.T. & FLANNERY, B.P. (1992) *Numerical Recipes in FORTRAN*, 2nd Edition. Cambridge: Cambridge University Press.

RAND, R.H. (1984) *Computer Algebra in Applied Mathematics: An Introduction to MACSYMA*. Boston: Pitman.

RAND, R.H. & ARMBRUSTER, D. (1987) *Perturbation Methods, Bifurcation Theory, and Computer Algebra*. New York: Springer-Verlag.

RASZILLIER, H. & DURST, F. (1991) Coriolis-effect in mass flow metering. *Archive of Applied Mechanics* **61**, 192–214.

RASZILLIER, H., ALLEBORN, N. & DURST, F. (1993) Mode mixing in Coriolis flowmeters. *Archive of Applied Mechanics* **63**, 219–227.

RODI, W. (1980) Turbulence models and their applications in hydraulics — a state-of-the-art review. IAHR Section on Fundamentals of Division II: Experimental and Mathematical Fluid Dynamics. Delft: IAHR.

ROGERS, L. (ed.) (1984) *Proceedings Vibration Damping 1984 Workshop*. (U.S.) Air Force Wright Aeronautical Laboratories Report AFWAL-TR-84-3064.

ROTH, W. (1964) Instabilität durchströmter Rohre. *Ingenieur-Archiv* **33**, 236–263.

ROTH, W. (1965a) Transversalschwingungen durchströmten Saiten. *Zeitschrift für angewandte Mathematik und Physik* **16**, 201–214.

ROTH, W. (1965b) Einfluß der Rotationsträgheit auf die Stabilität des durchströmten Schlauches. *Zeitschrift für angewandte Mathematik und Mechanik* **45**, T133–T135.

ROTH, W. (1966) Instabilität des durchströmten, einseitig eingespannten Rohres. *Ölhydraulik und Pneumatik* **10**, 58–64.

ROTH, W. & CHRIST, H. (1962) Ausknicken eines abgesetzten Teleskoprohres durch Innendruck. *Konstruktion* **14**, 109–111.

ROTTA, J.C. (1962) Turbulent boundary layers in incompressible flow. In *Progress in Aeronautical Sciences*, Vol. 2 (ed. A. Ferri), pp. 1–20. New York: The Macmillan Co. and Pergamon Press.

ROUSSELET, J. (1975) Dynamic behavior of pipes conveying fluid near critical velocities. Ph.D. dissertation, Stanford University, Stanford, CA, U.S.A.

ROUSSELET, J. & HERRMANN, G. (1977) Flutter of articulated pipes at finite amplitude. *Journal of Applied Mechanics* **44**, 154–158.

ROUSSELET, J. & HERRMANN, G. (1981) Dynamic behaviour of continuous cantilevered pipes conveying fluid near critical velocities. *Journal of Applied Mechanics* **48**, 943–947.

SÄLLSTRÖM, J.H. (1990) Fluid-conveying damped Rayleigh–Timoshenko beams in transverse vibration analyzed by use of an exact finite element. Part II: applications. *Journal of Fluids and Structures* **4**, 573–582.

SÄLLSTRÖM, J.H. (1993) Fluid-conveying damped Rayleigh–Timoshenko beams in transient transverse vibration studied by use of complex modal synthesis. *Journal of Fluids and Structures* **7**, 551–563.

SÄLLSTRÖM, J.H. & ÅKESSON, B.Å. (1990) Fluid-conveying damped Rayleigh–Timoshenko beams in transverse vibration analyzed by use of an exact finite element. Part I: theory. *Journal of Fluids and Structures* **4**, 561–572.

SANDERS, J.A. & VERHULST, F. (1985) *Averaging Methods in Nonlinear Dynamical Systems*. New York: Springer-Verlag.

SARPKAYA, T. & ISAACSON, M. (1981) *Mechanics of Wave Forces on Offshore Structures*. New York: Van Nostrand Reinhold.

SATTINGER, D.H. (1980) Bifurcation and symmetry breaking in applied mathematics. *Bulletin of the American Mathematical Society* **3**, 779–819.

SCHETZ, J.A. (1993) *Boundary Layer Analysis*. Englewood Cliffs, NJ: Prentice-Hall.

SCHLICHTING, H. (1960) *Boundary Layer Theory*, 4th edition. New York: McGraw-Hill.

SCHMIDT, G. & TONDL, A. (1986) *Non-linear Vibrations*. Berlin: Academie-Verlag.

SECHLER, E.E. (1952) *Elasticity in Engineering*. New York: Dover.

SEMLER, C. (1991) Nonlinear dynamics and chaos of pipes conveying fluid. M.Eng. Thesis, Faculty of Engineering, McGill University, Montreal, Québec, Canada.

SEMLER, C. & PAÏDOUSSIS, M.P. (1995) Intermittency route to chaos of a cantilevered pipe conveying fluid with a mass defect at the free end. *Journal of Applied Mechanics* **62**, 903–907.

SEMLER, C. & PAÏDOUSSIS, M.P. (1996) Nonlinear analysis of the parametric resonances of a planar fluid-conveying cantilevered pipe. *Journal of Fluids and Structures* **10**, 787–825.

SEMLER, C., LI, G.X. & PAÏDOUSSIS, M.P. (1994) The nonlinear equations of motion of pipes conveying fluid. *Journal of Sound and Vibration* **169**, 577–599.

SEMLER, C., GENTLEMAN, W.C. & PAÏDOUSSIS, M.P. (1996) Numerical solutions of second order implicit non-linear ordinary differential equations. *Journal of Sound and Vibration* **195**, 553–586.

SEMLER, C., ALIGHANBARI, H. & PAÏDOUSSIS, M.P. (1998) A physical explanation of the destabilizing effect of damping. *Journal of Applied Mechanics* (in press).

SETHNA, P.R. & GU, X.M. (1985) On global motions of articulated tubes carrying fluid. *International Journal of Non-Linear Mechanics* **20**, 453–469.

SETHNA, P.R. & SHAW, S.W. (1987) On codimension-three bifurcations in the motion of articulated tubes conveying a fluid. *Physica* D **24**, 305–327.

SEYRANIAN, A.P. (1994) Collision of eigenvalues in linear oscillatory systems. *Journal of Applied Mathematics and Mechanics* **58**, 805–813.

SHAMES, I.H. (1964) *Mechanics of Deformable Solids*. Englewood Cliffs, NJ.: Prentice-Hall.

SHAMES, I.H. (1992) *Mechanics of Fluids*, 3rd edition. New York: McGraw-Hill.

SHAYO, L.K. & ELLEN, C.H. (1974) The stability of finite length circular cross-section pipes conveying inviscid fluid. *Journal of Sound and Vibration* **37**, 535–545.

SHAYO, L.K. & ELLEN, C.H. (1978) Theoretical studies of internal flow-induced instabilities of cantilevered pipes. *Journal of Sound and Vibration* **56**, 463–474.

SHIEH, R.C. (1971) Energy and vibrational principles for generalized (gyroscopic) conservative systems. *International Journal of Non-Linear Mechanics* **5**, 495–509.

SHILLING, R. III & LOU, Y.K. (1980) An experimental study on the dynamic response of a vertical cantilever pipe conveying fluid. *ASME Journal of Energy Resources Technology* **102**, 129–135.

SILVA, M.A.G. (1979) Flow induced vibrations of pipes with attached valves. *Transactions 5th International Conference on Structural Mechanics in Reactor Technology (SMiRT)*, Berlin, Paper B6/4.

SILVA, M.A.G. (1981) Influence of eccentric valves on the vibration of fluid conveying pipes. *Nuclear Engineering and Design* **64**, 129–134.

SIMO, J.C. (1985) A finite strain beam formulation. The three-dimensional dynamic problem. Part I. *Computer Methods in Applied Mechanics and Engineering* **49**, 55–70.

SINGH, K. & MALLIK, A.K. (1977) Wave propagation and vibration response of a periodically supported pipe conveying fluid. *Journal of Sound and Vibration* **54**, 55–66.

SINGH, K. & MALLIK, A.K. (1979) Parametric instabilities of a periodically supported pipe conveying fluid. *Journal of Sound and Vibration* **62**, 379–397.

SINYAVSKII, V.F., FEDOTOVSKII, V.S. & KUKHTIN, A.B. (1980) Oscillation of a cylinder in a viscous liquid. *Prikladnaya Mekhanika* **16**, 62–67.

SMITH, T.E. & HERRMANN, G. (1972) Stability of a beam on an elastic foundation subjected to a follower force. *Journal of Applied Mechanics* **39**, 628–629.

SMITH, L. & RUESCH, J.R. (1991) Mass flow meters. In *Flow Measurement* (ed. D.W. Spitzer), pp. 221–247. Research Triangle Park, NC: Instrument Society of America.

SNOWDON, J.C. (1968) *Vibration and Shock in Damped Mechanical Systems*. New York: John Wiley & Sons.

SNOWDON, J.C. (ed.) (1975) *Proceedings Seminar on the Vibration of Damped Structures*. Applied Research Laboratory, The Pennsylvania State University, 22–26 September 1975.

SO, R.M.C., LAI, Y.G., ZHANG, H.S. & HWANG, B.C. (1991) Second-order near-wall turbulence closures: a review. *AIAA Journal* **29**, 1819–1835.

SPARKS, C.P. (1983) The influence of tension, pressure and weight on pipe riser deformations and stresses. *Proceedings 2nd International Offshore Mechanics and Arctic Engineering Symposium*, pp. 46–62. New York: ASME.

STEIN, R.A. & TOBRINER, W.M. (1970) Vibrations of pipes containing flowing fluids. *Journal of Applied Mechanics* **37**, 906–916.

STEINDL, A. (1996) Heteroclinic cycles in the dynamics of a fluid conveying tube. In *Proceedings ICIAM Conference* (eds K. Kirchgässner, O. Mahrenholtz & R. Mennicken), pp. 529–532. Berlin: Akademie-Verlag.

STEINDL, A. & TROGER, H. (1988) Flow induced bifurcations to 3-dimensional motions of tubes with an elastic support. In *Trends in Applications of Mathematics to Mechanics* (eds J.F. Besserling & W. Eckhaus), pp. 128–138. Berlin: Springer-Verlag.

STEINDL, A. & TROGER, H. (1994) Chaotic oscilations of a fluid-conveying viscoelastic tube. In *Nonlinearity and Chaos in Engineering Dynamics* (eds J.M.T. Thompson & S.R. Bishop), pp. 231–240. Chichester: John Wiley & Sons.

STEINDL, A. & TROGER, H. (1995) Nonlinear three-dimensional oscillations of elastically constrained fluid conveying viscoelastic tubes with perfect and broken $\mathbb{O}(2)$-symmetry. *Nonlinear Dynamics* **7**, 165–193.

STEINDL, A. & TROGER, H. (1996) Heteroclinic cycles in the three-dimensional post-bifurcation motion of $\mathbb{O}(2)$-symmetrical fluid conveying tubes. *Applied Mathematics and Computation* **78**, 269–277.

STOKER, J.J. (1968) *Nonlinear Elasticity*. New York: Gordon and Breach.

STREETER, V.L. (1948) *Fluid Dynamics*. New York: McGraw-Hill.

SUGIYAMA, Y. (1984) Studies on stability of two-degree-of-freedom articulated pipes conveying fluid (the effect of a spring support and a lumped mass). *Bulletin of JSME* **27**, 2658–2663.

SUGIYAMA, Y. & NODA, T. (1981) Studies on stability of two-degree-of-freedom articulated pipes conveying fluid (effect of attached mass and damping). *Bulletin of JSME* **24**(194), 1354–1362.

SUGIYAMA, Y. & PAÏDOUSSIS, M.P. (1982) Studies on the stability of two-degree-of-freedom articulated pipes conveying fluid: the effect of characteristic parameter ratios. In *Theoretical and Applied Mechanics*, Vol. 31, pp. 333–342. Tokyo: University of Tokyo Press.

SUGIYAMA, Y., TANAKA, Y., KISHI, T. & KAWAGOE, H. (1985a) Effect of a spring support on the stability of pipes conveying fluid. *Journal of Sound and Vibration* **100**, 257–270.

SUGIYAMA, Y., KUMAGAI, Y., KISHI, T. & KAWAGAOE, H. (1985b) Studies on the stability of pipes conveying fluid (the combined effect of a lumped mass and damping). *Transactions of JSME*, Series C **51**(467), 1506–1514.

SUGIYAMA, Y., MATSUMOTO, S. & IWATSUBO, T. (1986a) Studies on stability of two-degree-of-freedom articulated pipes conveying fluid. *Transactions of JSME*, Series C **52**(473), 264–270.

SUGIYAMA, Y., MATSUMOTO, S. & IWATSUBO, T. (1986b) A theoretical and experimental study of the effect of damping in nonconservative stability problems. *Transactions of JSME*, Series C **52**(476), 1058–1065.

SUGIYAMA, Y., KAWAGOE, H., KISHI, T. & NISHIYAMA, S. (1988a) Studies on the stability of pipes conveying fluid (the combined effect of a spring support and a lumped mass). *JSME International Journal*, Series 1 **31**, 20–26.

SUGIYAMA, Y., CHIBA, M., KATAYAMA, T., SHIRAKI, K. & FUJITA, K. (1988b) Studies on the stability of pipes conveying fluid (the effect of a damper). *Transactions of JSME*, Series C **54**(498), 353–356.

SUGIYAMA, Y., KATAYAMA, K. & KINOI, S. (1990) Experiment on flutter of cantilevered columns subjected to rocket thrust. *In Proceedings 31st AIAA et al. Structures, Structural Dynamics & Materials Conference*, pp. 1893–1898. Paper AIAA-90-0948-CP.

SUGIYAMA, Y., KATAYAMA, T., ÅKESSON, B. & SÄLLSTRÖM, J.H. (1991) Stability of cantilevered pipes conveying fluid and having intermediate spring support. *Transactions 11th International Conference on Structural Mechanics in Reactor Technology (SMiRT)*, Tokyo, Paper J10/1.

SUGIYAMA, Y., KATAYAMA, T., KANKI, E., NISHINO, K. & ÅKESSON, B. (1992) Active flutter suppression of a vertical pipe conveying fluid. *Proceedings 3rd International Symposium on Flow-Induced Vibration and Noise*, Vol. 8: Stability and Control of Pipes Conveying Fluid (eds M.P. Païdoussis & N.S. Namachchivaya), pp. 76–86. New York: ASME.

SUGIYAMA, Y., KATAYAMA, K. & KINOI, S. (1995) Flutter of cantilevered column under rocket thrust. *ASCE Journal of Aerospace Engineering* **8**, 9–15.

SUGIYAMA, Y., KATAYAMA, T., KANKI, E., CHIBA, M., SHIRAKI, K. & FUJITA, K. (1996a) Stability of vertical fluid-conveying pipes having the lower end immersed in fluid. *JSME International Journal*, Series B **39**, 57–65.

SUGIYAMA, Y., KATAYAMA, T., KANKI, E., NISHINO, K. & ÅKESSON, B. (1996b) Stabilization of cantilevered flexible structures by means of an internal flowing fluid. *Journal of Fluids and Structures* **11**, 653–661.

SULEIMAN, S.M. & MUNSON, B.R. (1981) Buckling of a thin sheet of a viscous fluid. *Physics of Fluids* **24**, 1–5.

SULTAN, G. & HEMP, J. (1989) Modelling of the Coriolis mass flowmeter. *Journal of Sound and Vibration* **132**, 473–489.

SVETLISKII, V.A. (1966) Vibrations of flexible hoses filled with a moving fluid (fuel). *Izvestiya Vysshikh Vchebrykh Zavedenii, Mashinostroyenye* **3**, 22–30.

SVETLISKII, V.A. (1969) Statics, stability and small vibrations of the flexible tubes conveying ideal incompressible fluid. *Raschetv na Prochnost* **14**, 332–351.

SVETLISKY (SVETLISKII), V.A. (1977) Vibration of tubes conveying fluids. *Journal of the Acoustical Society of America* **62**, 595–600.

SVETLISKII, V.A. (1982) *Mechanika Truboprovodov i Shlangov* (in Russian). Moscow: Machinostronye.

TABARROK, B., LEECH, C.M. & KIM, Y.I. (1974) On the dynamics of an axially moving beam. *Journal of the Franklin Institute* **297**, 201–220.

TAI, C.-T. (1992) *Generalized Vector and Dyadic Analysis*. New York: IEEE.

TAKAHASHI, F., NAKAMURA, S. & TANI, J. (1990) Dynamic stability and active control of a cantilevered pipe conveying fluid by piezoelectric actuators. *Transactions of JSME*, Series C **56**(526), 1481–1487.

TAKENS, F. (1974) Singularities of vector fields. *Publications Mathématiques* (Institut des Hautes Etudes Scientifiques) **43**, 47–100.

TAKENS, F. (1980) Detecting strange attractors in turbulence. In *Dynamical Systems and Turbulence*, Springer Lecture Notes in Mathematics Vol. 898 (eds D.A. Rand & L.-S. Young), pp. 366–381. New York: Springer-Verlag.

TANAKA, M., TANAKA, S. & SEGUCHI, Y. (1993) Optimal and robust shapes of a pipe conveying fluid. *Proceedings Asia-Pacific Vibration Conference*, Kitakyushu, Japan, Vol. 4, pp. 1757–1762. Tokyo: JSME.

TANG, D. (1997) Private communication (1 April 1997).

TANG, D.M. & DOWELL, E.H. (1988) Chaotic oscillations of a cantilevered pipe conveying fluid. *Journal of Fluids and Structures* **2**, 263–283.

TANI, J. & SUDANI, Y. (1995) Active flutter suppression of a vertical pipe conveying fluid. *JSME International Journal*, Series C **38**, 55–58.

TAYLOR, G.I. (1952) Analysis of the swimming of long and narrow animals. *Proceedings of the Royal Society (London) A* **214**, 158–183.

TAYLOR, G.I. (1969) Instability of jets, threads, and sheets of viscous fluids. In *Proceedings 12th International Congress of Applied Mechanics*, pp. 382–388. Berlin: Springer-Verlag.

TELIONIS, D.P. (1981) *Unsteady Viscous Flows*. New York: Springer-Verlag.

THOM, R. (1972) *Stabilité Structurelle et Morphogénèse*. Reading, MA: Benjamin.

THOMPSON, J.M.T. (1979) Stability predictions through a succession of folds. *Philosophical Transactions of the Royal Society (London)* A **292**, 1–23.

THOMPSON, J.M.T. (1982a) *Instabilities and Catastrophies in Science and Engineering*. Chichester, U.K.: John Wiley & Sons.

THOMPSON, J.M.T. (1982b) 'Paradoxial' mechanics under fluid flow. *Nature* **296**, 135–137.

THOMPSON, J.M.T. & HUNT, G.W. (1973) *A General Theory of Elastic Stability*. London: John Wiley & Sons.

THOMPSON, J.M.T. & LUNN, T.S. (1981) Static elastica formulations of a pipe conveying fluid. *Journal of Sound and Vibration* **77**, 127–132.

THOMPSON, J.M.T. & STEWART, H.B. (1986) *Nonlinear Dynamics and Chaos*. Chichester: John Wiley & Sons.

THOMSON, W. & TAIT, G. (1879) *A Treatise on Natural Philosophy*, Vol. 1, Part I, New Edition, pp. 387–391. Cambridge: Cambridge University Press.

THURMAN, A.L. & MOTE, C.D. Jr. (1969a) Free, periodic, nonlinear oscillation of an axially moving strip. *Journal of Applied Mechanics* **36**, 83–91.

THURMAN, A.L. & MOTE, C.D. Jr. (1969b) Non-linear oscillation of a cylinder containing flowing fluid. *ASME Journal of Engineering for Industry* **91**, 1147–1155.

TIJSSELING, A.G. (1996) Fluid–structure interaction in liquid-filled pipe systems: a review. *Journal of Fluids and Structures* **10**, 109–146.

TIMOSHENKO, S. (1955) *Strength of Materials*, Part I, 3rd edition. Princeton, NJ: Van Nostrand.

TIMOSHENKO, S. & GERE, J. (1961) *Theory of Elastic Stability*, 2nd edition. New York: McGraw-Hill.

TING, E.C. & HOSSEINIPOUR, A. (1983) A numerical approach for flow-induced vibration of pipe structures. *Journal of Sound and Vibration* **88**, 289–298.

TOWNSEND, A.A. (1961) Turbulence. In *Handbook of Fluid Dynamics* (ed. V.L. Streeter), Section 10, pp. 10.1–10.33. New York: McGraw-Hill.

TOWNSEND, A.A. (1976) *The Structure of Turbulent Shear Flow*, 2nd edition. Cambridge: Cambridge University Press.

TRIANTAFYLLOU, G.S. (1992) Physical condition for absolute instability in inviscid hydroelastic coupling. *Physics of Fluids* **4**, 544–552.

TRIANTAFYLLOU, G.S., TRIANTAFYLLOU, M.S. & GROSENBOUGH, M.A. (1993) Optimal thrust development in oscillating foils with application to fish propulsion. *Journal of Fluids and Structures* **7**, 205–224.

TRITTON, D.J. (1988) *Physical Fluid Dynamics*, 2nd edition. Oxford: Clarendon Press.

TROGER, H. & STEINDL, A. (1991) *Nonlinear Stability and Bifurcation Theory*. Wien: Springer-Verlag.

TSAI, Y.-K. & LIN, Y.-H. (1997) Adaptive modal vibration control of a fluid-conveying cantilever pipe. *Journal of Fluids and Structures* **11**, 535–547.

TSONIS, A.A. (1992) *Chaos: From Theory to Applications*. New York: Plenum Press.

TSUTSUI, H. & TOMIKAWA, Y. (1993) Coriolis force mass-flow meter composed of a straight pipe and an additional resonance-vibrator. *Japanese Journal of Applied Physics* **32**, 2369–2371.

UNNY, T.E., MARTIN, E.L. & DUBEY, R.N. (1970) Hydroelastic instability of uniformly curved pipe-fluid system. *Journal of Applied Mechanics* **37**, 617–622.

VAKAKIS, A.F. (1994) Passive spatial confinement of impulsive excitations in coupled nonlinear beams. *AIAA Journal* **32**, 1902–1910.

VAN, K.S. (1986) Dynamics and stability of curved pipes conveying fluid. M.Eng. Thesis, Department of Mechanical Engineering, McGill University, Montreal, Québec, Canada.

VOGEL, H. & NATVIG, B.J. (1987) Dynamics of flexible hose riser systems. *ASME Journal of Offshore Mechanics and Arctic Engineering* **109**, 244–248.

WAMBSGANSS, M.W., CHEN, S.S. & JENDRZEJCZYK, J.A. (1974) Added mass and damping of a vibrating rod in confined viscous fluids. Argonne National Laboratory report ANL-CT-75-08, Argonne, Illinois, U.S.A.

WARD, G.N. (1955) *Linearized Theory of Steady High Speed Flow*. Cambridge: Cambridge University Press.

WASHIZU, K. (1966) Note on the principle of stationary complementary energy applied to free vibration of an elastic body. *International Journal of Solids and Structures* **2**, 27–35.

WASHIZU, K. (1968) *Variational Methods in Elasticity and Plasticity*. New York: Pergamon Press.

WEAVER, D.S. (1976) On flow induced vibrations in hydraulic structures and their alleviation. *Canadian Journal of Civil Engineering* **3**, 126–137.

WHITE, F.M. (1974) *Viscous Fluid Flow*. New York: McGraw-Hill.

WHITNEY, A.K., CHUNG, J.S. & YU, B.K. (1981) Vibrations of long marine pipes due to vortex shedding. *ASME Journal of Energy Resources Technology* **103**, 231–236.

WICKERT, J.A. & MOTE, C.D. Jr (1990) Classical vibration analysis of axially moving continua. *Journal of Applied Mechanics* **57**, 738–744.

WIDNALL, S.E. & DOWELL, E.H. (1967) Aerodynamic forces on an oscillating cylindrical duct with an internal flow. *Journal of Sound and Vibration* **6**, 71–85.

WIGGERT, D.C. (1986) Coupled transient flow and structural motion in liquid-filled piping systems. ASME Paper 86-PVP-4.

WIGGERT, D.C. (1996) Fluid transients in flexible piping systems. In *Proceedings VIII IAHR Symposium on Hydraulic Machinery and Cavitation*, Valencia, Spain, pp. 58–67. Delft: IAHR.

WIGGINS, S. (1988) *Global Bifurcations and Chaos: Analytical Methods*. New York: Springer-Verlag.

WIGGINS, S. (1990) *Introduction to Applied Nonlinear Dynamical Systems and Chaos*. New York: Springer-Verlag.

WILCOX, D.C. (1993) *Turbulence Modelling for CFD*. La Cañada, CA: DCW Industries.

WILEY, J.C. & FURKERT, R.E. (1972) Stability of beams subjected to a follower force within the span. *ASCE Journal of the Engineering Mechanics Division* **98**, 1353–1364.

WILLS, A.P. (1958) *Vector Analysis with an Introduction to Tensor Analysis*. New York: Dover Publications.

WYLIE, E.B. & STREETER, V.L. (1978) *Fluid Transients*. New York: McGraw-Hill; also 1983, Ann Arbor: FEB Press.

YANG, C.I. & MORAN, T.J. (1979) Finite element solution of added mass and damping of oscillating rods in viscous fluids. *Journal of Applied Mechanics* **46**, 519–523.

YAU, C.-H., BAJAJ, A.K. & NWOKAH, O.D.I. (1995) Active control of chaotic vibration in a constrained flexible pipe conveying fluid. *Journal of Fluids and Structures* **9**, 99–122.

YEH, T.T. & CHEN, S.S. (1978) The effect of fluid viscosity on coupled tube/fluid vibration. *Journal of Sound and Vibration* **59**, 453–467.

YEO, K.S. & DOWLING, A.P. (1987) The stability of inviscid flows over passive compliant walls. *Journal of Fluid Mechanics* **183**, 265–292.

YOSHIZAWA, M., NAO, H., HASEGAWA, E. & TSUJIOKA, Y. (1985) Buckling and postbuckling behavior of a flexible pipe conveying fluid. *Bulletin of JSME* **28**(240), 1218–1225.

YOSHIZAWA, M., NAO, H., HASEGAWA, E. & TSUJIOKA, Y. (1986) Lateral vibration of a flexible pipe conveying fluid with pulsating flow. *Bulletin of JSME* **29**, 2243–2250.

YOSHIZAWA, M., WATANABE, M., HASHIMOTO, K. & TAKAYANAGI, M. (1997) Nonlinear lateral vibration of a vertical fluid-conveying pipe with end mass (effects of horizontal excitation at the upper end). *Proceedings 4th International Symposium on FSI, AE & FIV+N*, Dallas, TX, U.S.A., Vol. I, pp. 281–288. New York: ASME.

ZIEGLER, H. (1952) Die Stabilitätskriterien der Elasto-mechanik. *Ingenieur-Archiv* **20**, 49–56.

ZIEGLER, H. (1968) *Principles of Structural Stability*. Waltham, MA: Blaisdell.

ZIENKIEWICZ, O.C. & CHEUNG, Y.K. (1968) *The Finite Element Method in Structural and Continuum Mechanics*. New York: McGraw-Hill.

ZIENKIEWICZ, O.C. & TAYLOR, R.L. (1989) *The Finite Element Method*, 4th edition, Vols. 1 and 2. New York: McGraw-Hill.

Index

RELATED JOURNAL

Journal of Fluids and Structures

Editor
Michael Païdoussis
McGill University, Canada

JFS is a high quality outlet for original full-length papers, brief communications, review articles and special issues in any aspect of fluid–structure interaction.

With excellent web coverage through IDEAL, the Academic Press Online Journal Library, publication of supplementary data and images through this media is also considered.

Research areas

- Flow-induced excitation mechanisms
- Solid–fluid interactions
- Response of mechanical, civil, marine and physiological structures to flow and flow-acoustic excitation
- Unsteady fluid dynamics
- Aeroelasticity and other relevant aeronautical applications

www.academicpress.com/jfs